THE NATIONAL BROADBAND PLAN: ANALYSIS AND STRATEGY FOR CONNECTING AMERICA

MEDIA AND COMMUNICATIONS – TECHNOLOGIES, POLICIES AND CHALLENGES

Additional books in this series can be found on Nova's website under the Series tab.

Additional E-books in this series can be found on Nova's website under the E-books tab.

INTERNET POLICIES AND ISSUES

Additional books in this series can be found on Nova's website under the Series tab.

Additional E-books in this series can be found on Nova's website under the E-books tab.

THE NATIONAL BROADBAND PLAN: ANALYSIS AND STRATEGY FOR CONNECTING AMERICA

DANIEL M. MORALES
EDITOR

Nova Science Publishers, Inc.
New York

NOTICE TO THE READER

The Publisher has taken reasonable care in the preparation of this book, but makes no expressed or implied warranty of any kind and assumes no responsibility for any errors or omissions. No liability is assumed for incidental or consequential damages in connection with or arising out of information contained in this book. The Publisher shall not be liable for any special, consequential, or exemplary damages resulting, in whole or in part, from the readers' use of, or reliance upon, this material. Any parts of this book based on government reports are so indicated and copyright is claimed for those parts to the extent applicable to compilations of such works.

Independent verification should be sought for any data, advice or recommendations contained in this book. In addition, no responsibility is assumed by the publisher for any injury and/or damage to persons or property arising from any methods, products, instructions, ideas or otherwise contained in this publication.

This publication is designed to provide accurate and authoritative information with regard to the subject matter covered herein. It is sold with the clear understanding that the Publisher is not engaged in rendering legal or any other professional services. If legal or any other expert assistance is required, the services of a competent person should be sought. FROM A DECLARATION OF PARTICIPANTS JOINTLY ADOPTED BY A COMMITTEE OF THE AMERICAN BAR ASSOCIATION AND A COMMITTEE OF PUBLISHERS.

Additional color graphics may be available in the e-book version of this book.

LIBRARY OF CONGRESS CATALOGING-IN-PUBLICATION DATA
The national broadband plan : analysis and strategy for connecting America / editor, Daniel M. Morales.
p. cm.
Includes index.
ISBN 978-1-61122-024-7 (hardcover)
1. Telecommunication policy--United States. 2. Internet service providers--Government policy--United States. 3. Broadband communication systems--Government policy--United States. 4. United States. Federal Communications Commission. I. Morales, Daniel M.
HE7781.N393 2010
384.5'30973--dc22
2010035913

Published by Nova Science Publishers, Inc. † New York

CONTENTS

PREFACE

Broadband is the great infrastructure challenge of the early 21st century. Like electricity a century ago, broadband is a foundation for economic growth, job creation, global competitiveness and a better way of life. It is enabling entire new industries and unlocking vast new possibilities for existing ones. It is changing how we educate children, deliver health care, manage energy, ensure public safety, engage government, and access, organize and disseminate knowledge. But broadband in America is not all it needs to be. Approximately 100 million Americans do not have broadband at home. This book explores and analyzes The National Broadband Plan.

Chapter 1- On March 16, 2010, the Federal Communications Commission (FCC) released *Connecting America: The National Broadband Plan*. Mandated by the American Recovery and Reinvestment Act of 2009 (ARRA, P.L. 111-5), the FCC's National Broadband Plan (NBP) is a 360-page document composed of 17 chapters containing 208 specific recommendations directed to the FCC, to the Executive Branch (both to individual agencies and to Administration as a whole), to Congress, and to nonfederal and nongovernmental entities. The ARRA specified that the NBP should "seek to ensure that all people of the United States have access to broadband capability."

The NBP identified significant gaps in broadband availability and adoption in the United States. In order to address these gaps and other challenges, the NBP set six specific goals to be achieved by the year 2020. These six goals are discussed further in this chapter, and an outline of the NBP is provided at the end of this chapter.

Chapter 2- Broadband is the great infrastructure challenge of the early 21st century.

Like electricity a century ago, broadband is a foundation for economic growth, job creation, global competitiveness and a better way of life. It is enabling entire new industries and unlocking vast new possibilities for existing ones. It is changing how we educate children, deliver health care, manage energy, ensure public safety, engage government, and access, organize and disseminate knowledge.

Fueled primarily by private sector investment and innovation, the American broadband ecosystem has evolved rapidly. The number of Americans who have broadband at home has grown from eight million in 2000 to nearly 200 million last year. Increasingly capable fixed and mobile networks allow Americans to access a growing number of valuable applications through innovative devices.

But broadband in America is not all it needs to be. Approximately 100 million Americans do not have broadband at home. Broadband-enabled health information technology (IT) can improve care and lower costs by hundreds of billions of dollars in the coming decades, yet the United States is behind many advanced countries in the adoption of such technology. Broadband can provide teachers with tools that allow students to learn the same course material in half the time, but there is a dearth of easily accessible digital educational content required for such opportunities. A broadband-enabled Smart Grid could increase energy independence and efficiency, but much of the data required to capture these benefits are inaccessible to consumers, businesses and entrepreneurs. And nearly a decade after 9/11, our first responders still lack a nationwide public safety mobile broadband communications network, even though such a network could improve emergency response and homeland security.

Chapter 3- The convergence of wireless telecommunications technology with the Internet Protocol (IP) is fostering new generations of mobile technologies. This transformation has created new demands for advanced communications infrastructure and radio frequency spectrum capacity that can support high-speed, content-rich uses. Furthermore, a number of services, in addition to consumer and business communications, rely at least in part on wireless links to broadband backbones. Wireless technologies support public safety communications, sensors, smart grids, medicine and public health, intelligent transportation systems, and many other vital communications.

Existing policies for allocating and assigning spectrum rights may not be sufficient to meet the future needs of wireless broadband. A challenge for Congress is to provide decisive policies in an environment where there are many choices but little consensus. In formulating spectrum policy, mainstream viewpoints generally diverge on whether to give priority to market economics or social goals. Regarding access to spectrum, economic policy looks to harness market forces to allocate spectrum efficiently, with spectrum license auctions as the driver. Social policy favors ensuring wireless access to support a variety of social objectives where economic return is not easily quantified, such as improving education, health services, and public safety. Both approaches can stimulate economic growth and job creation.

Deciding what weight to give to specific goals and setting priorities to meet those goals pose difficult tasks for federal administrators and regulators and for Congress. Meaningful oversight or legislation may require making choices about what goals will best serve the public interest. Relying on market forces to make those decisions may be the most efficient and effective way to serve the public but, to achieve this, policy makers may need to broaden the concept of what constitutes competition in wireless markets.

Chapter 4- Over the past decade, the telecommunications sector has undergone a vast transformation fueled by rapid technological growth and subsequent evolution of the marketplace. Much of the U.S. policy debate over the evolving telecommunications infrastructure is framed within the context of a "national broadband policy." The way a national broadband policy is defined, and the particular elements that might constitute that policy, determine how and whether various stakeholders might support or oppose a national broadband initiative. The issue for policymakers is how to craft a comprehensive broadband strategy that not only addresses broadband availability and adoption problems, but also addresses the long term implications of next-generation networks on consumer use of the Internet and the implications for a regulatory framework that must keep pace with evolving telecommunications technology.

In: The National Broadband Plan: Analysis and Strategy for...   ISBN: 978-1-61122-024-7
Editor: Daniel M. Morales   © 2011 Nova Science Publishers, Inc.

Chapter 1

THE NATIONAL BROADBAND PLAN

*Lennard G. Kruger, Angele A. Gilroy, Charles B. Goldfarb,
Linda K. Moore and Kathleen Ann Ruane*

SUMMARY

On March 16, 2010, the Federal Communications Commission (FCC) released
Connecting America: The National Broadband Plan. Mandated by the American Recovery
and Reinvestment Act of 2009 (ARRA, P.L. 111-5), the FCC's National Broadband Plan
(NBP) is a 360-page document composed of 17 chapters containing 208 specific
recommendations directed to the FCC, to the Executive Branch (both to individual agencies
and to Administration as a whole), to Congress, and to nonfederal and nongovernmental
entities. The ARRA specified that the NBP should "seek to ensure that all people of the
United States have access to broadband capability."

The NBP identified significant gaps in broadband availability and adoption in the United
States. In order to address these gaps and other challenges, the NBP set six specific goals to
be achieved by the year 2020. These six goals are discussed further in this chapter, and an
outline of the NBP is provided at the end of this chapter.

It is important to note that many aspects of telecommunications policies, regulations, and
legal issues would be affected by the NBP. For example:

- The Universal Service Fund (USF) is a fund that was created to provide universal
  availability and affordability of communications throughout the United States; the
  issue is whether or how the universal service concept should embrace access to
  broadband as one of its policy objectives.
- Because wireless broadband can play a key role in the deployment of broadband
  services, the NBP extensively addresses spectrum policy and the issue of how to
  make more spectrum available and usable for mobile broadband applications.
- Issues such as intercarrier compensation and set-top boxes are identified by the NBP
  as having potential significant impact on broadband availability and adoption.

- Broadband will likely play a role in addressing critical national challenges in areas such as health care, education, energy, environment, and public safety; the issue is how, for each national purpose, the existing legislative and regulatory framework and trends in the field might best benefit from better broadband access and services.
- Finally, one potential issue the FCC may face in its attempts to achieve NBP goals is the scope of the agency's authority to regulate broadband Internet access and management.

A major issue for Congress will be how to shape the Plan's various initiatives when and if they go forward, either through oversight, through consideration of specific legislation, or in the context of comprehensive telecommunications reform. A key challenge for Congressional policymakers will be to assess whether an appropriate balance is maintained between the public and private sectors, and the extent to which government intervention in the broadband marketplace would help or hinder private sector investment and competition.

BACKGROUND

Signed into law on February 17, 2009, Section 6001(k) of the American Recovery and Reinvestment Act of 2009 (ARRA, P.L. 111-5) mandated the Federal Communications Commission (FCC) to prepare a report containing a national broadband plan. The impetus behind mandating a national broadband plan was derived from the widely accepted view in Congress of broadband as a critical public infrastructure, increasingly important to the nation's economic development. Broadband was also viewed as playing an increasingly critical role in addressing specific challenges facing the nation in areas such as health care, energy, education, public safety, and others.

In the United States, broadband infrastructure is constructed, operated, and maintained primarily by the private sector, including telephone, cable, satellite, wireless, and other information technology companies. Although broadband is primarily deployed by private sector providers, federal and state regulation of the telecommunications industry, as well as government financial assistance programs, can have a significant impact on private sector decisions to invest in and deploy broadband infrastructure, particularly in underserved and unserved areas of the nation. When considering broadband policy, the ongoing challenge for Congressional policymakers is how to strike a balance between providing federal assistance for unserved and underserved areas where the private sector may not be providing acceptable levels of broadband service, while at the same time minimizing any deleterious effects that government intervention in the marketplace may have on competition and private sector investment.

The ARRA specified that the national broadband plan "shall seek to ensure that all people of the United States have access to broadband capability and shall establish benchmarks for meeting that goal," and that the plan should include:

- an analysis of the most effective and efficient mechanisms for ensuring broadband access by all people of the United States;

- a detailed strategy for achieving affordability of such service and maximum utilization of broadband infrastructure and service by the public;
- an evaluation of the status of deployment of broadband service, including progress of projects supported by the grants made pursuant to this section; and
- a plan for use of broadband infrastructure and services in advancing consumer welfare, civic participation, public safety and homeland security, community development, health care delivery, energy independence and efficiency, education, worker training, private sector investment, entrepreneurial activity, job creation and economic growth, and other national purposes.

Starting in the summer of 2009, an FCC task force embarked on a massive information gathering effort consisting of 36 public workshops, 9 field hearings, 31 public notices producing 75,000 pages of public comments, and 131 online blog postings triggering almost 1,500 comments.[1]

On March 16, 2010, the FCC publically released its report, *Connecting America: The National Broadband Plan.*[2] As mandated by the ARRA, the report was formally submitted to the House Committee on Energy and Commerce and the Senate Committee on Commerce, Science, and Transportation. At the March 16, 2010 Open Commission Meeting, the FCC Commissioners voted to approve a Broadband Mission Statement containing goals for a U.S. broadband policy. However, the FCC Commissioners did not vote on whether to approve the plan itself.

OVERVIEW OF PLAN[3]

The FCC's National Broadband Plan (NBP) is a 360-page document composed of 17 chapters containing 208 specific recommendations. Table 1, at the end of this chapter, is an outline of the National Broadband Plan. *Connecting America: The National Broadband Plan* begins with an introduction, a statement of goals, and a discussion of the current state of the broadband "ecosystem." This is followed by three parts containing the bulk of the plan's recommendations: Part I, "Innovation and Investment," Part II, "Inclusion," and Part III, "National Purposes." The NBP concludes with a chapter on implementation and benchmarks.

Goals to Create a "High-Performance America"

The NBP seeks to "create a high-performance America" which the FCC defines as "a more productive, creative, efficient America in which affordable broadband is available everywhere and everyone has the means and skills to use valuable broadband applications."[4] In order to achieve this mission, the NBP recommends that the country set six goals for 2020:

- **Goal No. 1: At least 100 million U.S. homes should have affordable access to actual download speeds of at least 100 megabits per second and actual upload speeds of at least 50 megabits per second.** Speeds of 100 Mbps in 100 million homes (popularly referred to as "100 squared") would constitute next-generation broadband[5] in most U.S. households. As a milestone, the FCC has set an interim goal of 100 million homes with actual download speeds of 50 Mbps and actual upload speeds of 20 Mbps by 2015. The FCC notes that existing providers are in the process of upgrading their networks, and it is likely that 90% of the country will have access to advertised peak download speeds of more than 50 Mbps by 2013.[6]

- **Goal No. 2: The United States should lead the world in mobile innovation, with the fastest and most extensive wireless networks of any nation.** According to November 2009 data from American Roamer, 3G wireless service covers roughly 60% of U.S. land mass.[7] Approximately 77% of the U.S. population live in an area served by three or more 3G service providers, 12% live in an area served by two, 9% live in an area served by one, and about 2% live in an area with no provider.[8] The FCC currently has 50 MHz of spectrum that it can assign for broadband use. The NBP recommends making 500 MHz of spectrum available for broadband by 2020, with an interim benchmark of 300 MHz by 2015.

- **Goal No. 3: Every American should have affordable access to robust broadband service, and the means and skills to subscribe if they so choose.** There are two aspects to the goal of universal broadband: availability and adoption. Regarding *broadband availability*, 290 million Americans—95% of the U.S. population—currently live in housing units with access to terrestrial, fixed broadband infrastructure capable of supporting actual download speeds of at least 4 Mbps. This leaves a "gap" of 14 million people in the United States living in 7 million housing units that do not have access to terrestrial broadband infrastructure capable of this speed.[9] The FCC has estimated that $24 billion in additional funding would be necessary to fill what it refers to as the "broadband availability gap."[10] Regarding *broadband adoption*, the NBP sets an adoption goal of "higher than 90%" by 2020. Currently, broadband adoption stands at 67%, about two-thirds of the adult population. Certain demographic groups exhibit significantly lower rates of broadband adoption, for example: 40% of adults making less than $20,000 per year have adopted terrestrial broadband at home, 50% of adults in rural areas, 24% of those with less than a high school degree, 35% of those older than 65, 59% of African Americans, 49% of Hispanics, 42% of people with disabilities, and fewer than 10% of residents on Tribal lands.[11]

- **Goal No. 4: Every American community should have affordable access to at least 1 gigabit per second broadband service to anchor institutions such as schools, hospitals and government buildings.** The NBP notes that while 99% of all health care locations with physicians have access to an actual download speed of at least 4 Mbps, and while 97% of schools are connected to the Internet (many supported by the federal E-rate program), more than 50% of teachers say slow or unreliable Internet access presents obstacles to their use of technology in classrooms, and only 71% of rural health clinics have access to mass-market broadband solutions. Further, many business locations, schools and hospitals often have connectivity requirements that cannot be met by mass-market DSL, cable modems, satellite or wireless providers, and must buy dedicated high-capacity circuits such as T-1 or Gigabit Ethernet service. The availability and price of such circuits vary greatly across different geographies.[12]

- **Goal No. 5: To ensure the safety of the American people, every first responder should have access to a nationwide, wireless, interoperable broadband public safety network.** Nearly nine years after 9/11, first responders from different jurisdictions and agencies still often cannot communicate with each other during emergencies and continue to operate outdated communications systems, most of which do not have broadband capability.[13]
- **Goal No. 6: To ensure that America leads in the clean energy economy, every American should be able to use broadband to track and manage their realtime energy consumption.** According to the FCC, "broadband and advanced communications infrastructure will play an important role in achieving national goals of energy independence and efficiency."[14]

Recommendations

Chapters 4 through 17 constitute the heart of the National Broadband Plan and contain 208 specific recommendations intended to help achieve the Plan's goals. The NBP's recommendations are directed to the FCC, to the Executive Branch (both to individual agencies and to Administration as a whole), to Congress, and to nonfederal and nongovernmental entities. Table 2 (at the end of this chapter) provides a listing of recommendations specifically directed to Congress.

The NBP is categorized into three parts:

- **Part I (Innovation and Investment)** which "discusses recommendations to maximize innovation, investment and consumer welfare, primarily through competition. It then recommends more efficient allocation and management of assets government controls or influences."[15] The recommendations address a number of issues, including: spectrum policy, improved broadband data collection, broadband performance standards and disclosure, special access rates, interconnection, privacy and cybersecurity, child online safety, poles and rights-of-way, research and experimentation (R&E) tax credits, R&D funding.
- **Part II (Inclusion)** which "makes recommendations to promote inclusion—to ensure that all Americans have access to the opportunities broadband can provide."[16] Issues include: reforming the Universal Service Fund, intercarrier compensation, federal assistance for broadband in Tribal lands, expanding existing broadband grant and loan programs at the Rural Utilities Service, enable greater broadband connectivity in anchor institutions, and improved broadband adoption and utilization especially among disadvantaged and vulnerable populations.
- **Part III (National Purposes)** which "makes recommendations to maximize the use of broadband to address national priorities. This includes reforming laws, policies and incentives to maximize the benefits of broadband in areas where government plays a significant role."[17] National purposes include: health care, education, energy and the environment, government performance, civic engagement, and public safety. Issues include: telehealth and health information technology, online learning and modernizing educational broadband infrastructure, digital literacy and job training,

smart grid and smart buildings, federal support for broadband in small businesses, telework within the federal government, cybersecurity and protection of critical broadband infrastructure, copyright of public digital media, interoperable public safety communications, next generation 911 networks and emergency alert systems.

Implementation

The NBP discusses an implementation strategy intended to carry out the recommendations. First, because many of the recommendations are directed towards the Executive Branch, the NBP recommends the creation of an interagency Broadband Strategy Council to coordinate implementation of the NBP. Second, given that approximately half the recommendations are directed to the FCC, the NBP calls on the FCC to quickly publish a timetable or proceedings to implement those NBP recommendations that fall under FCC authority. On April 8, 2010, the FCC released a Broadband Action Agenda explaining the purpose and timing of more than 60 rulemakings and other notice-and-comment proceedings.[18]

Additionally, Congress is seen as playing a major role in implementing the National Broadband Plan, both by considering legislation to implement NBP recommendations, and by overseeing (and possibly funding) broadband activities conducted by the FCC and Executive Branch agencies.

As telecommunications technologies increasingly converge onto a broadband platform, many of the issues traditionally regarded as part of "telecommunications policy" are becoming viewed as part of "broadband policy." Accordingly, the NBP addresses many of the ongoing major telecommunications policy issues, such as the reform and reorientation of the Universal Service Fund, reform of intercarrier compensation, the possible mandating of "gateway" set-top boxes, spectrum and wireless policy, and the appropriate regulatory framework for an evolving information infrastructure. Some of these issues will likely be addressed in subsequent FCC proceedings, and all may be debated and considered by Congress.

BROADBAND ADOPTION AND AVAILABILITY AND THE FEDERAL UNIVERSAL SERVICE FUND[19]

The NBP states that "Everyone in the United States today should have access to broadband services supporting a basic set of applications that include sending and receiving e-mail, downloading web pages, photos and video, and using simple video conferencing."[20] A universalization target of 4 Mbps of actual download speed and 1 Mbps of actual upload speed has been set as the initial target rate for public investment to ensure that these expectations will be met. The NBP calls upon the Federal Universal Service Fund (USF) to undertake a major role to ensure that this goal is achieved.

The Evolution of the Universal Service Concept

Since its creation in 1934 the Federal Communications Commission has been tasked with "... mak[ing] available, so far as possible, to all the people of the United States,... a rapid, efficient, Nation-wide, and world-wide wire and radio communications service with adequate facilities at reasonable charges...."[21] This mandate led to the development of what has become known as the universal service concept. The universal service concept, as originally designed, called for the establishment of policies to ensure that telecommunications services are available to all Americans, including those in rural, insular, and high cost areas, by ensuring that rates remain affordable.

The term universal service, when applied to telecommunications, refers to the ability to make available a basket of telecommunications services to the public, across the nation, at a reasonable price. Over time, access to the public switched network through a single wireline connection, enabling voice service, became the standard of communications. Currently the basic universal service package, which was established in 1997, is comprised of:

- voice grade access to and some usage of the public switched network;
- single line service;
- dual tone signaling;
- access to directory assistance;
- emergency service such as 911;
- operator services;
- access and interexchange (long distance) service.

Since the U.S. household telephone connection rate is 95.6% of homes, an all-time high, some might say that the program has been a success and nothing more needs to be done.[22] The universal service concept, however, is an evolving one, and consequently so are universal service policies and goals. The initial focus of universal service support targeted eligible telecommunications carriers usually serving rural, insular, or other high cost areas by providing funds to help offset higher than average costs of providing telephone service (e.g., the High-Cost Program). Changes in expectations by policymakers and consumers have led to an expansion of universal service programs as well as to the establishment of a Federal USF to administer them.[23] For example, the passage of the Telecommunications Act of 1996 (P.L. 104-104) codified the universal service concept and expanded the concept to include, among other principles, that elementary and secondary schools and classrooms, libraries, and rural health care providers have access to telecommunications services for specific purposes at discounted rates as well as access to advanced telecommunications and information services.[24] This led to the establishment by the FCC of the Schools and Libraries and the Rural Health Care Programs. Earlier policy decisions by the FCC led to the development, in the mid-1980s, of a needs based Low-Income Program to assist economically needy individuals to join and remain on the telecommunications network.[25]

Universal Service and Broadband

Over the past decade the telecommunications sector has undergone a vast transformation fueled in particular by the deployment of and access to broadband infrastructure and applications. One of the challenges facing this transition is the desire to ensure that all consumers have access to an affordable and advanced broadband infrastructure so that all members of society may derive its social and economic benefits. Broadband adoption rates are estimated at 67%, representing about two-thirds of the adult population, but these rates are uneven and significant gaps exist.[26] For example, those who: live in rural areas, have low education and income levels, have disabilities, are elderly, are African Americans, are Hispanics, and are living on Tribal lands all have significantly lower broadband adoption rates than the national average. Furthermore, approximately 5% of the U.S. population, equivalent to 14 million people living in 7 million housing units, do not have access to terrestrial broadband infrastructure capable of supporting the NBP's recommended 4 Mbps actual download speed.[27]

One of the major policy debates surrounding universal service policy is whether the universal service concept should embrace access to broadband as one of its policy objectives. The 1934 Communications Act, as amended, does take into consideration the changing nature of the telecommunications sector and allows for the mix of services eligible for universal service support to be modified. In particular, provisions in the universal service section state that "universal service is an evolving level of telecommunications services" and the FCC is tasked with "periodically" reevaluating this definition "taking into account advances in telecommunications and information technologies and services."[28]

There is a growing consensus among policymakers that the FCC should change the mix of services eligible for universal service support to include the universal availability of broadband services and use federal universal service funds to help eliminate broadband adoption and availability gaps.[29] The FCC's NBP recommends that access to and adoption of broadband be a national goal and has proposed that the USF be restructured to become a vehicle to help reach this goal.

The National Broadband Plan and the USF

The USF will be a key component in this transition as the NBP would reorient its programs to address the deployment, affordability, and connectivity of broadband. To enable the USF to take on this role the NBP calls for the USF to be transformed, in three stages over a ten-year period, from a mechanism that largely supports voice telephone service to one that supports the deployment, adoption, and utilization of broadband.[30]

The Connect America and Mobility Funds

The NBP calls on the existing High Cost program to transition from one that supports voice communications to one that supports a broadband platform that enables multiple applications, including voice. The NBP recommends that the High Cost program be phased out and replaced in stages, over the next ten years, to directly support high-capacity broadband networks through newly created Connect America and Mobility Funds.[31]

The Connect America Fund (CAF) would be the major vehicle to ensure the universal availability of affordable broadband by addressing the gaps in broadband deployment and adoption. The NBP adopts a new expanded USF definition embracing affordable broadband with at least 4 Mbps actual download speed and 1 Mbps of actual upload speed. Examples of the applications that could be supported by such a download speed include: advanced Web-browsing; e-mail; Voice over the Internet Protocol (VOIP); multimedia; streamed audio; streamed video lectures; and lower definition telemedicine.[32]

The NBP also calls for the USF to move from a largely fixed model to incorporate a mobile model. A Mobility Fund (MF) would be created to target funding to ensure that all states achieve the national average for 3G wireless coverage for both voice and data. The MF would provide one-time support for deployment of 3G networks.

Low Income Program

According to an FCC conducted broadband consumer survey, 36% of non-adopters of broadband cite a financial reason as the main reason they do not have broadband at home.[33] To address this barrier the NBP recommends that the existing Low Income Program (Lifeline and Link Up) be expanded to address low broadband access levels in low-income households.[34] The NBP also calls for the integration of Federal Low Income programs with state and local efforts as well as the establishment of pilot programs to gain information to help develop a future full-scale low income program for broadband.

Schools and Libraries and Rural Health Programs

Included in the national purposes stated in the NBP are those that address the role of broadband in the delivery of education and health care. The USF has two programs: the Schools and Libraries Program (also known as the E-rate program); and the Rural Health Care Program, which address the telecommunications needs of eligible schools, libraries, and rural health care providers respectively. The NBP contains almost a dozen recommendations to modernize and improve the Schools and Libraries program.[35] These recommendations focus on three goals: improve flexibility, deployment, and use of infrastructure; improve program efficiency; and foster innovation. Included among the recommendations are those that: raise the yearly funding cap to account for inflation; set minimum broadband connectivity rates; and expand support for internal connections.

Citing the importance of health care to the lives of consumers and its importance to the national economy the NBP also calls for the reform of the USF's Rural Health Care program. The major focus of the reform calls for the restructuring and expansion of its program components.[36] Included among the recommendations to modify the program are those that: expand eligibility to include urban as well as rural providers based on need; increase subsidy support beyond the current 25%; expand the definition for eligibility to include certain for-profit entities; replace the existing Internet Access Fund with a Health Care Broadband Access Fund; simplify the application process; and establish a Health Care Broadband Infrastructure Fund.

Funding

While the NBP calls for a major restructuring of the USF, it recommends that the funding level be maintained close to its current size (in 2010 dollars). The NBP recommends that

$15.5 billion be shifted, through selected reforms, over the next decade from the existing USF High Cost program to support the transition to broadband.[37] However, the NBP also recommends that if Congress wishes to accelerate this transition it could allocate to the CAF additional general funds of "... a few billion dollars a year over a two or three year period."[38] Additional comments regarding funding include a recommendation that the USF contribution base be broadened, that the contribution methodology rules be revised to ensure sustainability, and acknowledgement of the potential negative impact that increasing USF funding rates may have on consumers.[39]

REFORM OF INTERCARRIER COMPENSATION[40]

Most telephone calls and other electronic communications travel over more than one carrier's network to get from the originating (or calling) party to the terminating (or receiving) party, thus requiring the facilities of an originating network, a terminating network, and perhaps one or more intermediate networks. Intercarrier compensation (ICC) is the system of rates that service providers are charged for the use of these networks to provide service to their subscribers.

There is a monopoly element to terminating a communication. Once the receiving party has chosen her local carrier (say Verizon or Comcast or, in a rural area, the local rural telephone company), the originating party's carrier has no choice but to pay the rate charged by the terminating carrier to complete the communication. Therefore these rates are subject to price regulation—the FCC regulates interstate rates and state public utility commissions regulate intrastate rates.

Intercarrier compensation rates have developed over time in an *ad hoc* fashion, and often were set to help foster a particular policy objective. For example, both the FCC and state regulatory commissions purposely set the rates for terminating long distance calls to subscribers of rural telephone companies significantly above cost in order to provide those rural carriers with a large revenue source that would allow them to keep local rates low. As a result, the revenues generated from intercarrier compensation charges imposed on long distance carriers represent approximately 25% of total rural telephone carrier revenues, but only approximately 10% of the revenues of other local telephone companies.[41] As another example of regulators setting intercarrier compensation rates to help meet a public policy objective, the FCC purposely has treated Internet Service Provider (ISP)-bound traffic differently from other traffic, not imposing termination charges on ISP-bound traffic or setting lower rates for terminating ISP-bound traffic than other traffic, in order to foster the development of Internet services.

Although the use of a terminating network's facilities is similar for each type of communication, the rate charged for terminating a communication ranges from zero to 35.9 cents a minute[42] depending on the jurisdiction of the communication,[43] the type of traffic carried,[44] and the regulatory status of the terminating carrier.[45]

These regulatory-mandated distinctions create inefficient market signals that, in addition to imposing artificial advantages or disadvantages on certain categories of services or providers,[46] are skewing investment decisions, in some cases perhaps retarding the migration from legacy circuit-switched voice networks to Internet protocol (IP) broadband networks.

At the same time, these artificially high termination charges generate revenues that may be needed for rural companies to be able to offer basic telephone service at affordable rates comparable to those for urban subscribers. These high charges represent an implicit universal service subsidy imposed on long distance users. Intercarrier compensation reform that eliminates this implicit subsidy by moving terminating access rates toward cost may have to be accompanied by the creation of new sources of explicit universal service funding.

The Broadband Plan proposes that "the FCC should adopt a framework for long-term intercarrier compensation reform that creates a glide path to eliminate per-minute charges while providing carriers an opportunity for adequate cost recovery."[47] Changes would be transitioned in over ten years, starting with reductions in the highest intercarrier compensation rates, which generally are intrastate rates.

Since the federal courts have ruled that the FCC does not have authority over intrastate rates, however, legislation may be needed to give the FCC *explicit* authority to reform intrastate intercarrier rates as well as interstate rates.

FOSTERING A MARKET FOR SET-TOP BOXES

Universal access to broadband networks is not an end in itself; it is a means to give consumers access to the applications that are provided over those networks. Consumers need devices—computers, smart phones, set-top boxes—to reach both their broadband network and the applications riding over the network. Those devices help consumers navigate to the many applications. Today, consumers can turn to many different manufacturers and retailers of computers and smart phones, but they generally have few options for set-top boxes (or, more broadly, for smart video devices). Virtually all such devices are provided by the consumer's multichannel video programming distributor (MVPD—cable or satellite video service provider).

Section 629(a) of the Communications Act[48] directs the Federal Communications Commission (FCC),

> in consultation with appropriate industry standard-setting organizations, [to] adopt regulations to assure the commercial availability ... of converter boxes, interactive communications equipment, and other equipment used by consumers to access multichannel video programming [cable or satellite video service].

There is a consensus that FCC and industry efforts to date have not achieved this goal. There are few devices available in the retail market; these devices cannot work with all MVPDs and, even where compatible, cannot be used to identify many video signals, particularly those of high-definition cable offerings.

In its NBP, the FCC concluded that access to video services drives broadband usage and thus actions that would foster a market for smart video devices that make it easier for consumers to access broadband-enabled video would increase consumer adoption of broadband.[49] The FCC therefore adopted a Notice of Inquiry on April 21, 2010,[50] to explore the feasibility of:

- developing a nationally supported standard interface that is common across all MVPDs, thus allowing independent equipment manufacturers to produce smart video devices that could be used by end users without regard to their choice of MVPD, *and*
- requiring each MVPD to develop a complying adapter or gateway that would allow end users to purchase in a retail market smart video devices capable of searching for all available video options—from their MVPD, from the Internet, etc.—in one place.

As envisioned by the FCC:

- The smart video devices could be used with the services of any MVPD and without the need to coordinate or negotiate with MVPDs.
- The adapter or gateway would communicate with the MVPD service, performing the tuning and security decryption functions specific to a particular MVPD.
- The smart video devices would perform navigation functions, including presentation of programming guides and search functionality.

The envisioned "AllVid" solution would employ a nationwide interoperability standard analogous to how Ethernet and the IEEE industry standards have led to nationwide interoperability for customer data networks despite broadband service providers deploying differing proprietary network technologies.

Critics of this gateway concept argue that the cable industry already has spent almost $1 billion attempting to implement the FCC's earlier, unsuccessful efforts to create a market for video devices and that there is no demonstrated consumer demand for these set-top boxes. They claim that most consumers would prefer to lease set-top boxes that might become obsolete in an environment characterized by rapid product and service innovation. The additional expenditures to construct the adapter, they claim, would be passed through to consumers in higher MVPD rates. The critics further contend that technological change inherently outpaces any government rule or one-size-fits-all requirement. They also question whether households that do not currently subscribe to broadband service would be motivated to do so simply because they had access to smart video devices that provide them with greater, but perhaps more complicated, access to video services. The critics suggest that the AllVid mandate would primarily serve tech-savvy individuals who already subscribe to broadband service.

Proponents of the AllVid solution respond that, in the absence of standards and an interoperability requirement, there have been no incentives for MVPDs, manufacturers, and consumers to cooperate. They claim that consumer demand for smart video devices and manufacturer incentives to produce such devices have been constrained by the lack of a universal standard that would allow a single device to serve all MVPDs and by MVPD resistance to a device that would make it easier for consumers to access non-MVPD sources of video. They claim simple to use smart video devices would appeal particularly to non-tech savvy households. They dispute that the AllVid solution would constrain MVPD innovation or consumer choice by prescribing a single technical solution. Rather, they claim, it would foster innovation and choice by developing an industry-wide interoperability standard open to independent equipment manufacturers and applications providers without placing restrictions on MVPD networks.

The FCC recognizes that it may be especially challenging to develop an adapter for satellite video providers because, unlike in cable networks where the intelligence resides deep in the network at the head-end, in satellite networks the intelligence resides in the set-top box at the customer premise. It therefore might be more difficult for the satellite company to troubleshoot whether the source of a customer complaint lies in equipment under the control of the satellite operator or in the smart video device.

One NBP recommendation is that the FCC initiate a proceeding to ensure that all MVPDs install a gateway device or equivalent functionality in all new subscriber homes and in all homes requiring replacement set-top boxes by December 31, 2012.[51] Many observers question whether the technical and market challenges to accomplish this can be performed in that period of time.

An AllVid solution will not be available for several years. Currently, CableCARD technology—which only works for cable, not for satellite—is available to separate the system that customers use to gain access to video programming (called the conditional element) from the device customers use to navigate the programming. This allows independent smart video device manufacturers, such as TiVo, to serve end users, but there are problems with the technology, notably, it does not allow customers to receive certain high-definition cable channels. The FCC therefore has adopted a Notice of Proposed Rulemaking to expeditiously address some of the current problems with cableCARDs.[52]

SPECTRUM POLICIES FOR WIRELESS BROADBAND[53]

Wireless broadband[54] plays a key role in the deployment of broadband services. Because of the importance of wireless connectivity, radio frequency spectrum policy is deemed by the NBP to be a critical factor in successful planning for a national policy. Mobile broadband provides high-speed Internet connectivity on the move. Other wireless technologies complement needed infrastructure for a host of national broadband goals for education, health, energy efficiency, public safety, and other social benefits. Mobile and fixed wireless broadband communications, with their rich array of services and content, require new spectrum capacity to accommodate growth. Although radio frequency spectrum is abundant, usable spectrum is limited by the constraints of technology and the cost of investment.[55]

The NBP proposes to increase spectrum capacity by:

- Making more spectrum licenses available for mobile broadband.
- Increasing the amount of spectrum available for shared use.
- Encouraging and supporting the development of spectrum-efficient technologies, particularly those that facilitate sharing spectrum bands.
- Instituting new policies for spectrum management, such as assessing fees on some spectrum licenses.

To facilitate the deployment of broadband in rural areas, the NBP also proposes:

- Improving the environment for providing fixed wireless services.

Many of the NBP proposals for wireless broadband may be achieved through changes in FCC regulations governing spectrum allocation and assignment. Other actions may require changes by federal agencies, state authorities, and commercial owners of spectrum licenses. To assist the implementation of the NBP there are also a number of areas where congressional action might be required to change existing statutes or to give the FCC new powers. The NBP includes the announcement of plans for the FCC to create what it refers to as a Spectrum Dashboard.[56] The initial release of the FCC's Spectrum Dashboard provided an interactive tool to search for information about how some non-federal frequency assignments are being used.[57] The dashboard could be used to meet requirements set by Congress for a spectrum inventory. In addition to the dashboard, the NBP proposes that the FCC and the National Telecommunications and Information Administration (NTIA) should create methods for recovering spectrum[58] and that the FCC maintain an ongoing spectrum strategy plan.[59] The NTIA manages federal use of spectrum, among other responsibilities. All of these steps will facilitate decisions about spectrum management by providing detailed information about the current and potential use of spectrum resources.

From a policy perspective, the NBP recommendations that would speed the arrival of new, spectrally efficient technologies might have the most impact over the long term. In particular, support for exploring ways to use technologies that enable sharing could pave the way for dramatically different ways of managing the nation's spectrum resources.

The need for a robust plan to accelerate the adoption of new technologies has, however, been eclipsed by public debate over the plan's proposed steps to add 300 MHz[60] of licensed spectrum for broadband within five years. All of the spectrum assignment proposals put forth in the NBP are contentious in that the various parties affected by the decisions have diverging views on how technology should be used to provide access to these frequencies. The disagreements may be in part over the cost of implementing different technological solutions, or about a shift in who controls access, but these are associated with the technical fixes the FCC has proposed to facilitate the spectrum assignment.

The NBP has laid out several opportunities for the FCC, the NTIA, and other government agencies to contribute to and encourage the development of new technologies for more efficient spectrum access.[61] For example, Congress might choose to require performance goals for improved spectrum efficiency, not unlike the way federal goals have been set for energy conservation or transportation safety. Congress might also evaluate how a detailed plan to encourage new technologies might assist in resolving current disagreements about spectrum assignment and use. The impact of evolving technologies on spectrum management is discussed in the section "Technology and Spectrum Management."

Spectrum Assignment

One of the management tools available to the FCC is its power to assign spectrum licenses through auctions. Auctions are regarded as a market-based mechanism for assigning spectrum. Before auctions became the primary method for distributing spectrum licenses the FCC used a number of different approaches, primarily based on perceived merit, to select license-holders. The FCC was authorized to organize auctions to award spectrum licenses for certain wireless communications services in the Omnibus Budget Reconciliation Act of 1993

(P.L. 103-66). Following passage of the act, subsequent laws that dealt with spectrum policy and auctions included the Balanced Budget Act of 1997 (P.L. 105-33), the Auction Reform Act of 2002 (P.L. 107-195), the Commercial Spectrum Enhancement Act of 2004 (P.L. 108-494, Title II), and the Deficit Reduction Act of 2005 (P.L. 109-171). The Balanced Budget Act of 1997 (1997 Act) gave the FCC auction authority until September 30, 2007. This authority was extended to September 30, 2011, by the Deficit Reduction Act of 2005 and to 2012 by the DTV Delay Act (P.L. 111-4).

In the NBP, the FCC has proposed taking steps to add 300 MHz of licensed spectrum for broadband within five years and a total of 500 MHz of new frequencies in ten years.[62] Approximately 50 MHz would be released in the immediate future by the completion of existing auction plans. An additional 40 MHz would be made available for auction, of which 20 MHz would be reassigned from federal to commercial use. Reallocating some spectrum from over-the-air broadcasting to commercial spectrum might provide an additional 120 MHz of spectrum. Final rulings on existing proceedings would release 110 MHz, of which 90 MHz would be for Mobile Satellite Services (MSS). Resolution of interference issues between Wireless Communications Services (WCS) and satellite radio would free up 20 MHz of new capacity.

Although Congress has shown interest in all of these debates, two proposals that are the most likely to generate interest in congressional action are repurposing and auctioning an estimated 120 MHz of airwaves assigned to over-the-air digital television broadcasting and the plans for auctioning the D Block (10 MHz in the 700 MHz band). These proposals are discussed below.

Television Broadcast Spectrum

The Balanced Budget Act of 1997 represented the legislative culmination of over a decade of policy debates and negotiations between the FCC and the television broadcast industry on how to move the industry from analog to digital broadcasting technologies. To facilitate the transition, the FCC provided each qualified broadcaster with 6 MHz of spectrum for digital broadcasting to replace licenses of 6 MHz that were needed for analog broadcasting. The analog licenses would be yielded back when the transition to digital television was concluded. The completed transition freed up the 700 MHz band for mobile communications and public safety in 2009.

The FCC has revisited the assumptions reflected in the 1997 Act and has made new proposals, and decisions based on, among other factors, changes in technology and consumer habits. The NBP announced that a new proceeding would be initiated to recapture up to 120 MHz of spectrum from broadcast TV allocations for re-assignment to broadband communications. This proceeding would propose four sets of actions to achieve the goal; a fifth set of actions to increase efficiency would be pursued separately.[63] The FCC stipulated in the NBP that its recommendations "seek to preserve [over-the-air television] as a healthy, viable medium going forward, in a way that would not harm consumers overall, while establishing mechanisms to make available additional spectrum for flexible broadband uses."[64]

Many of the proposals for redirecting TV broadcast capacity are based on refinements in the way frequencies are managed and are procedural in nature. Because over-the-air digital broadcasting does not necessarily require 6 MHz of spectrum, the NBP has proposed that some stations could share a single 6 MHz band without significantly reducing service to over-the-air TV viewers. The NBP also has proposed that broadcasters might form partnerships to provide other communications services using licenses assigned to TV. Among the proposals for how broadcasters might make better use of their TV licenses, the NBP has raised the possibility of auctioning unneeded spectrum and sharing the proceeds between the TV license-holder and the U.S. Treasury. The FCC has called on Congress to provide new legislation that would allow these "incentive auctions." Although most spectrum license auction revenues are deposited as general funds, Congress has passed laws that permit the proceeds to be used for other purposes. The plan suggests the Commercial Spectrum Enhancement Act could provide a model for sharing auction proceeds. The act created the Spectrum Relocation Fund to provide a mechanism whereby federal agencies could recover the costs of moving from one spectrum band to another.

D Block

The D Block refers to a set of frequencies within the 700 MHz band that were among the frequencies made available after the transition from analog to digital television in 2009. In compliance with instructions from Congress to auction all unallocated spectrum in this band, the FCC conducted an auction, which concluded on March 18, 2008. As part of its preparation for the auction (Auction 73), the FCC sought to increase the amount of spectrum available to public safety users in the 700 MHz band. Congress had previously designated 24 MHz of radio frequencies in the 700 MHz band for public safety channels. In 2007, the FCC proposed to allocate 10 MHz of the public safety frequencies specifically for broadband communications and to match the allocation with 10 MHz of commercial spectrum. This commercial license, known as the D Block, was to be auctioned under rules that would require the creation of a public-private partnership to develop the two 10-MHz assignments as a single broadband network, available to both public safety users and commercial customers. The D Block license was offered for sale in Auction 73 but did not find a buyer. The FCC then set about the task of writing new service rules for a re-auction of the D Block.[65]

In the NBP, the FCC announced its decision to auction the D Block under rules that would not require a partnership with public safety but would establish a framework for priority access to the D Block network by public safety users.[66] Generally, public safety officials had anticipated that the D Block would be an integral part of a public safety broadband network. Since the failed D Block auction of 2008, there has also been growing pressure on the FCC and on Congress to take the steps necessary to reallocate the D Block from commercial to public safety use. The NBP announcement regarding the D Block has increased that pressure. Although funding and control are critical elements of the debate, the controversy is rooted in contradictory assumptions about the level of service and reliability that new, largely untried, and in some cases undeveloped technology will be able to deliver for public safety broadband communications.

The FCC would address public safety needs such as developing standards and establishing procedures through the newly established Emergency Response Interoperability Center (ERIC).[67] ERIC would work closely with the Public Safety Communications Research program, jointly managed by the National Institute of Standards and Technology (NIST) and the NTIA, to develop and test the technological solutions needed for public safety broadband communications.[68]

Wireless Backhaul

Most mobile communications depend on fixed infrastructure to relay calls to and from wireless networks. The infrastructure that links wireless communications to the wired world is commonly referred to as backhaul. In situations where installing communications cables is impractical, fixed wireless infrastructure may be used to provide the needed backhaul. Microwave technologies, for example, are used in a number of applications to extend coverage to areas not served by fiberoptic or other wire links.

The NBP has predicted that the importance of backhaul will increase with the implementation of 4G technologies, as mobile access to the Internet and other wired networks becomes increasingly prevalent.[69] The FCC therefore has proposed to take a number of procedural steps to increase the capacity of point-to-point wireless technologies.[70]

Technology and Spectrum Management

The NBP spectrum assignment proposals are based on managing radio channels as the way to maximize spectral efficiency while meeting common goals such as minimizing interference among devices operating on the same or nearby frequencies. Today, channel management is a significant part of spectrum management; many of the FCC dockets deal with assigning channels and resolving the issues raised by these decisions. In the future, channel management is likely to be replaced by technologies that operate without the need for designated channels. The primary benefit from these new technologies will be the significant increase in available spectrum but new efficiencies in operational and regulatory costs will also be realized. The question for policy makers might be: has the time come to take actions so that future technologies—many of which are viewed as being within reach—become an integral part of planning for mobile broadband?

The concept of channel management dates to the development of the radio telegraph by Guglielmo Marconi and his contemporaries. In the age of the Internet, however, channel management is an inefficient way to provide spectrum capacity for mobile broadband. Innovation points to network-centric spectrum management as an effective way to provide spectrum capacity to meet the bandwidth needs of fourth-generation wireless devices.[71] Network-centric technologies organize the transmission of radio signals along the same principle as the Internet. A transmission moves from origination to destination not along a fixed path but by passing from one available node to the next. Pooling resources, one of the concepts that powers the Internet now, is likely to become the dominant principle for spectrum management in the future.

New Technologies

The iPhone 3G and 3GS provide early examples of how the Internet is likely to change wireless communications as more and more of the underlying network infrastructure is converted to IP-based standards. The iPhone uses the Internet Protocol to perform many of its functions; these require time and space—spectrum capacity—to operate. The next generation of wireless networks, 4G, for Fourth Generation, will be supported by technologies structured and managed to emulate the Internet. The wireless devices that operate on these new, IP-powered networks will be able to share spectrum capacity in ways not currently used on commercial networks, greatly increasing network availability on licensed bandwidths. Another technological boost will come from improved ways to use unlicensed spectrum. Unlicensed spectrum refers to bands of spectrum designated for multiple providers, multiple uses, and multiple types of devices that have met operational requirements set by the FCC. Wi-Fi is an example of a current use of unlicensed spectrum.

The FCC refers to the new technologies for licensed and unlicensed spectrum as "opportunistic." Identifying an opportunity to move to an open radio frequency is more flexible—and therefore more productive—than operating on a set of pre-determined frequencies.

New technologies that can use unlicensed spectrum without causing interference are being developed for vacant spectrum designated to provide space between the broadcasting signals of digital television, known as white spaces. On September 11, 2006, the FCC announced a timetable for allowing access to the spectrum so that devices could be developed.[72] One of the recommendations of the NBP is that the FCC complete the proceeding that would allow use of the white spaces for unlicensed devices.

More efficient spectrum use can be realized by integrating adaptive networking technologies, such as dynamic spectrum access (DSA),[73] with IP-based, 4G commercial network technologies such as Long Term Evolution (LTE). Adaptive networking has the potential to organize wireless communications to achieve the same kinds of benefits that have been seen to accrue with the transition from proprietary data networks to the Internet. These enabling technologies allow communications to switch instantly among network frequencies that are not in use and therefore available to any wireless device equipped with cognitive technology. Adaptive technologies are designed to use pooled spectrum resources. Pooling spectrum licenses goes beyond sharing. Licenses are aggregated and specific ownership of channels becomes secondary to the common goal of maximizing network performance.

New Policies

The NBP has laid out several opportunities for the FCC, the NTIA, and other government agencies to contribute to and encourage the development of new technologies for more efficient spectrum access.[74] Among the technologies that facilitate spectrum sharing are cognitive radio and Dynamic Spectrum Access (DSA).[75] Enabling technologies such as these allow communications to switch instantly among network frequencies that are not in use and therefore available to any radio device equipped with cognitive technology. Among the steps that might be taken to encourage spectrum-efficient technologies, the NBP has recommended that the FCC identify and free up a "new, contiguous nationwide band for unlicensed use" by 2020[76] and provide spectrum and take other steps to "further development and deployment"of new technologies that facilitate sharing.[77] Unlike its recommendations for auctioning spectrum licenses in the near future, the FCC's plans for bringing new technologies into play provide few details. The NBP provides a glimpse through the keyhole of the horizons beyond, but not the key that might open the door.

The NTIA has recommended exploring "ways to create incentives for more efficient use of limited spectrum resources, such as dynamic or opportunistic frequency sharing arrangements in both licensed and unlicensed uses."[78] This suggestion was incorporated into the FY2011 Federal Budget prepared by the Office of Management and Budget. The budget document directed the NTIA to collaborate with the FCC "to develop a plan to make available significant spectrum suitable for both mobile and fixed wireless broadband use over the next ten years. The plan is to focus on making spectrum available for exclusive use by commercial broadband providers or technologies, or for dynamic, shared access by commercial and government users."[79]

A Presidential Memorandum[80] has directed the NTIA to take a number of actions in support of NBP goals, including taking the lead in creating and implementing a plan that will facilitate the exploration of "innovative spectrum-sharing technologies."

The NTIA's Commercial Spectrum Management Advisory Committee is actively looking at policy and technology issues in a series of subcommittee reports. The reports are addressing spectrum inventory, transparency, dynamic spectrum access, incentives, unlicensed spectrum, and sharing.[81]

The widespread adoption of opportunistic technologies would likely require a re-thinking of spectrum management policies and tools. Policies for channel management to control interference could give way to standards for preventing interference by managing networks and devices. The assignment and supervision of licenses would be replaced by policies and procedures for managing pooled resources. If opportunistic technologies are adopted, auctioning licenses might be replaced by auctioning access; the static event of selling a license replaced by the dynamic auctioning of spectrum access on a moment-by-moment basis.

The testing of new technologies that increase spectrum capacity, and the policy changes they are likely to bring, has been designated by the NBP as a future event. Its immediate plans for spectrum policy are to fine-tune existing spectrum assignments to increase the availability of licensed capacity. The level of opposition to most of these spectrum assignment plans might suggest that current spectrum management practices have reached the point of diminishing returns.

NATIONAL PURPOSES[82]

Among the requirements for the NBP, Congress specified that it should include

> a plan for use of broadband infrastructure and services in advancing consumer welfare, civic participation, public safety and homeland security, community development, health care delivery, energy independence and efficiency, education, worker training, private sector investment, entrepreneurial activity, job creation and economic growth, and other national purposes.[83]

In the plan, the FCC has made recommendations that might fulfill both social and economic goals. In the section of the plan titled "National Purposes," it has focused on social goals with an agenda of actions for federal, state, and local agencies. The areas covered in this section are

- Health care. The NBP identifies stated goals of the Department of Health and Human Services that might be effectively supported with technologies that are enhanced by access to broadband communications.
- Education. The NBP proposes that broadband can provide an effective tool for meeting the educational needs and ambitions of educators, students, and parents of young children as well as support the Department of Education's strategies to improve educational achievement.
- Energy and the Environment. According to the NBP, broadband has multiple applications in the field of energy, conservation, and environmental protection. For example, SmartGrid goals set by Congress[84] might not be achievable without broadband communications.
- Economic Opportunity. Actions proposed in the NBP to further economic opportunity are centered on increasing access to Information Technology for small and medium-sized businesses. The role of broadband in providing job training and employment services and supporting telework are also addressed in recommendations.
- Government Performance. The NBP recommendations for federal government actions encompass both ways that broadband might improve the effectiveness of government and also steps the federal government might take to increase the availability of broadband networks. The latter included federal actions to improve cybersecurity and ways that federal agencies might assist communities and state and local governments in building broadband infrastructure.
- Civic Engagement. The NBP describes concepts such as government transparency that can lead to greater participation by all in the democratic process. Broadband access has been described in the plan as a useful tool for encouraging civic engagement because of the part it plays in interactive communication and providing information.
- Public Safety. The NBP recommendations primarily address delivering wireless broadband to the radios of first responders. It also considered the role of broadband in upgrading the nation's 911 services and emergency alert systems.

Meeting Policy Goals

Each of the sections on national purposes has mentioned the existing legislative and regulatory framework and trends in the field that might benefit from better broadband access and services. Although each sector serves different needs and goals, the NBP recommendations are fairly similar for each. In general, stakeholders have been encouraged to

- Create incentives to achieve broadband goals.
- Leverage broadband technology, including wireless broadband.
- Encourage innovation and improved productivity.
- Provide or increase funding for programs that support broadband policy goals.
- Modify regulations.

Each policy slice addresses aspects of the needs and services for the national purposes identified in the NBP. Considering all these slices as a single policy pie may be difficult. However, there are some common ingredients that each slice shares that could be addressed as a single policy. Connectivity through broadband networks represents an area of convergence that might benefit from a shared technology policy. The NBP observes that "... in many cases today's connectivity levels are insufficient for current use, let alone the needs of potential future applications."[85]

The NBP discusses some of the ways that federal investment in broadband infrastructure might be leveraged for community and state broadband services.[86] The plan has recognized many of the common elements of broadband use in the federal government but it has not explicitly addressed the possibility of unifying them as a common infrastructure project with many applications riding on a shared grid.

Development of the grid concept reflects recent trends in the expansion of the capabilities of the Internet and its feeder networks. The grid supports applications of any type, at any endpoint. Its strength derives in good measure from the imposition of the Internet Protocol. Not only can the grid accommodate any IP-based plug-in device but also it can route communications along any link within the grid operations. Some technologies can operate within the grid network without necessarily depending on terminals or switches. Software can reside anywhere and move around inside the grid as needed.

The Internet is typically described as comprised of three main parts: the Internet backbone, backbone access networks, and retail access networks—the services that link consumers and small businesses to the Internet. In business theory, the components of Internet service might be described as a distribution channel; the product—Internet access—is delivered to the end user through wholesalers and retailers. Increasingly, the backbone access networks—the wholesalers—are diversifying to accommodate new services that may never travel over the Internet backbone. The more technologically sophisticated wholesalers are expanding through internetworking to create powerful grids that run many applications to meet specific markets. These advanced communications grids might provide the technology needed to coordinate federal efforts to meet the goals laid out in "National Purposes." An IP-enabled communications grid could, for example, support next-generation 911 call centers and public safety radios, enable parts of utility company smart grids, and deliver telehealth services.

The NBP recommends that the Executive Branch create a Broadband Strategy Council.[87] This council would coordinate efforts by the many agencies that the FCC has identified as having a role in the plan's implementation. The NBP has suggested that the President could require that federal departments and agencies submit broadband implementation plans to the council. The council could also act as an intermediary between the agencies and Congress regarding legislation that might facilitate meeting the NBP's goals. Another recommendation of the NBP would require the FCC to track progress in meeting the plan's goals.[88]

THE FCC'S AUTHORITY TO IMPLEMENT THE NATIONAL BROADBAND PLAN[89]

One potential issue the FCC may face in its attempts to implement the NBP is the scope of the agency's authority to regulate broadband Internet access and management. The decision of the D.C. Circuit earlier this year in *Comcast v. FCC*[90] has thrown the agency's current authority to regulate these practices into doubt. Broadband Internet services are currently classified as information services, to which Title I of the Communications Act applies.[91] The FCC does not possess direct authority to regulate services classified under Title I.[92] The FCC has announced the possibility of reclassifying the transmission component of broadband Internet services as a telecommunications service under Title II of the Communications Act.[93] The FCC hopes this potential reclassification would ground the FCC's authority to regulate broadband Internet services more firmly in the governing law.

In order to understand the current uncertainty surrounding the FCC's authority over broadband Internet services, some background is needed. After the passage of the Telecommunications Act of 1996, the FCC found it necessary to determine what kind of service broadband Internet service was.[94] The agency's choices were to classify broadband Internet access as an information service,[95] over which it would have no direct authority to regulate under Title I, or as a telecommunications service,[96] over which it would have extensive authority to regulate under Title II. There was also an intermediate option. The FCC contemplated classifying the transmission component of a broadband Internet service as a telecommunications service, while classifying the processing component as an information service.[97] The FCC ultimately chose to classify broadband Internet services as information services only.[98]

At the time (2002), the provision of broadband Internet services arguably was still a nascent industry, and the FCC expressed a desire to avoid introducing into the developing market what it thought at the time could be too many regulations.[99] However, this was a contentious question. The Supreme Court, in *NCTA v. Brand X*, made the final decision.[100] The question before the court was whether the FCC could define cable-modem services (i.e., cable broadband services) as information services. Opponents of that classification argued that the FCC did not have discretion to define cable modem services as an information service. The Court, however, sided with the FCC. What is important for the purposes of this discussion is that the Court did not say that cable modem services are clearly and unambiguously information services. Instead, the court said that the definitions of telecommunications services and of information services were ambiguous as they related to cable modem services, and that the FCC, as the agency with jurisdiction under the Communications Act, had the authority to interpret those definitions.[101] The Court gave deference to the FCC's determination that cable modem services should be defined as information services and determined that the FCC's classification of cable modem services in this way was reasonable.[102]

However, three Justices dissented. Justice Scalia authored the dissent, concluding that cable modem services were actually two separate services: the computing service which was an information service, and the transmission service, which was a telecommunications service.[103] The classification that these Justices believe the Communications Act clearly mandates is the classification that the FCC now proposes to apply to broadband Internet services.[104]

Chairman Genachowski has announced his intention to pursue what he has termed "light touch" Title II regulation of broadband services.[105] As explained in the statement of the FCC's General Counsel, it is the intention of the FCC to commence a rulemaking to reclassify only the transmission component of broadband access services ("Internet connectivity") as a telecommunications service, while the data processing portion of the service would remain an information service.[106] The Chairman argues that, in choosing only to reclassify the transmission component of broadband access services, the reach of the FCC's jurisdiction will be sufficiently narrowed so as to avoid giving the agency the authority to regulate Internet content. This plan would also avoid the imposition of regulation so pervasive as to become burdensome.[107]

In keeping with this announcement, on June 17, 2010, the FCC released a notice of inquiry (NOI) into the framework of broadband Internet services.[108] In the NOI, the agency asked for comment on a number of questions. The FCC made clear that its ultimate goal in issuing the NOI was to determine the best avenue for restoring the agency's previous understanding of its authority to regulate broadband Internet services.[109] In other words, the FCC is seeking firmer ground for its authority to continue rulemakings along the lines of the broadband network management rulemakings[110] and the order it issued in 2007 finding Comcast to be in violation of the FCC's network management policies.[111] In doing so, the FCC recognizes that the D.C. Circuit's decision in *Comcast v. FCC* has thrown the agency's assertions of ancillary authority over broadband network management into considerable doubt.[112]

The NOI lists three main potential paths forward and seeks comment on the feasibility of each. The first question the NOI asks is whether the FCC may find a better way to assert ancillary authority over broadband Internet services.[113] The D.C. Circuit did not foreclose on the possibility of the FCC asserting ancillary authority in other ways. It merely rejected the FCC's argument in that particular case.[114] Therefore, the FCC asks whether broadband Internet services may continue to be classified as information services while the agency asserts a different statutory basis for exercising ancillary jurisdiction. There are a number of potential theories for ancillary jurisdiction for which the FCC seeks comment.[115]

The other two potential paths towards firmer authority to regulate would involve direct regulation under Title II of the Communications Act. Therefore, it would be necessary to reclassify at least the Internet connectivity portion of broadband Internet services as a telecommunications service, because only telecommunications services are governed by Title II. The FCC asks for comment on how to define Internet connectivity for reclassification.[116] Assuming the FCC chooses one of these two paths, this reclassification would likely be reviewed by the courts, in light of the fact that the Supreme Court upheld the agency's previous classification of broadband Internet services as a unified information service. However, as discussed earlier, *Brand X* gave deference to the FCC's interpretation of the Communications Act in this area.[117] Furthermore, in the recent case *FCC v. Fox Television*, the Supreme Court held that when an agency issues a new (and different from its previous) interpretation of a statute it has the authority to implement, the agency "need not demonstrate to a court's satisfaction that the reasons for the new policy are better than the reasons for the old one."[118] The agency must show only that its current interpretation is reasonable, though in some circumstances a more detailed justification for the change must be made than would otherwise be necessary if the agency was rulemaking on a blank slate.[119]

Assuming that such a reclassification is upheld by the courts, the second potential path forward would be to apply the full force of Title II regulation to broadband Internet connectivity (as the FCC would define it). The FCC seeks comment on the potential effects of such a decision.[120] However, the Chairman and General Counsel have expressed that this is not the approach the agency is likely to take.[121] Rather, they have announced that their intention is to forbear from applying the portions of Title II to broadband access services that the FCC deems contrary to the public interest. Section 401 of the Telecommunications Act of 1996 requires the FCC to forbear from applying any regulation or provision under Title II to a provider of telecommunications services if the Commission determines that:

> (1) enforcement of such regulation or provision is not necessary to ensure that the charges, practices, classifications, or regulations by, for, or in connection with that telecommunications carrier or telecommunications service are just and reasonable and are not unjustly or unreasonably discriminatory; (2) enforcement of such regulation or provision is not necessary for the protection of consumers; and (3) forbearance from applying such provision or regulation is consistent with the public interest.[122]

The Chairman and General Counsel argued, in their statements, that this provision would require forbearance from many of Title II's more onerous provisions, such as the rate regulation and tariff provisions, because applying those provisions would not be consistent with the public interest.[123]

The NOI asks for comment on this potential action.[124] It further asks for comment on the provisions on Title II from which the agency should not forbear. In particular, the NOI asks for comment on applying the provisions of Title II that the FCC had identified as likely to be needed to have adequate enforcement authority in its earlier press releases on this issue.[125] These provisions are Sections 201 (requiring service upon request and reasonable rates),[126] 202 (prohibiting unreasonable discrimination),[127] 208 (granting the FCC authority to act upon complaints),[128] 222 (protecting privacy),[129] 254 (universal service),[130] and 255[131] (access for disabled persons).[132] In the FCC's announcements, the General Counsel identified these provisions as potentially sufficient to "do the job" of providing enough authority to accomplish the FCC's goals.[133] However, the NOI asks for comment on other provisions that may be necessary to assert jurisdiction.[134]

The NOI also asks for comment on a number of other issues, including the method of forbearing. Currently, companies seeking forbearance from a provision of Title II (which had heretofore presumably applied to such companies) apply to the FCC seeking such forbearance. Under the FCC's proposal, the FCC would forbear under its own motion to maintain what is currently the status quo.[135] The FCC seeks comment on the process the agency should adopt for accomplishing this plan. The agency also seeks comment on how to treat wireless broadband services (terrestrial and satellite). The agency notes that "there are technological, structural, consumer usage, and historical differences between mobile wireless and wireline/cable networks" that may require different statutory and regulatory treatment.[136] Furthermore, the agency seeks comment on other open questions, such as the implications changes may have for state and local regulators,[137] and the effect any action taken to reclassify might have on the Communications Assistance for Law Enforcement Act.[138]

Comments are due July 15, 2010, and reply comments are due by August 12, 2010. Any decisions that the FCC may make as a result of this proceeding will likely face legal challenge.

TOWARDS A NATIONAL BROADBAND POLICY?

Policy issues discussed in the previous sections—universal service reform, intercarrier compensation, mandating of gateway set-top devices, spectrum policy for wireless broadband, and national purposes—all seek to address the NBP's availability and adoption goals, each in their own way. At the same time, the debate over the FCC's authority to regulate broadband services will likely impact the FCC's ability to achieve many of the goals of the NBP.

The cumulative effect of these and other discrete policies and initiatives proposed by and related to the NBP can be viewed as pieces of an overall strategy towards achieving NBP goals. The release of the NBP is seen by many as a precursor towards the development of a national broadband policy—whether comprehensive or piecemeal—that will likely be shaped and developed by Congress, the FCC, and the Administration.

Viewed holistically, several themes emerge from the NBP, with each theme having implications for policymakers with respect to a national broadband policy as it goes forward:

- *Government-private sector balance*—the NBP acknowledges that the growth of broadband in the U.S. has been "fueled primarily by private sector investment and innovation,"[139] and that "the role of government is and should remain limited."[140] However, given the identified gaps in broadband availability and adoption, the NBP envisions an active role for government, saying that "we must strike the right balance between the public and private sectors," and that "done right, government policy can drive and has driven progress."[141] Specifically, the NBP states that "instead of choosing a specific path for broadband in America, this plan describes actions government should take to encourage more private innovation and investment."[142] The challenge for broadband policymakers will be to assess whether an appropriate balance is maintained between the public and private sectors, and the extent to which government intervention in the broadband marketplace will help or hinder private sector investment and competition.

- *Interconnectedness*—the NBP views broadband as an "ecosystem" and suggests that many of the diverse topics and issues covered in the NBP, though seemingly distinct and separate, are in fact interconnected. For example, the NBP contains recommendations intended to allay consumers' concerns over Internet privacy, which in turn could lead to higher adoption rates and greater broadband and Internet utilization, which in turn could help provide more market incentive for private sector providers to deploy broadband infrastructure. The NBP identifies applications, devices, and networks as the key forces shaping the broadband ecosystem, and states that these three forces "drive each other in a virtuous cycle."[143] The NBP's focus on the quality of interconnectedness—and the central metaphor of broadband as an ecosystem—implies that policymakers should consider the various issues not in a vacuum, but as part of an integrated whole.

- *National purposes*—as directed by the ARRA, the NBP emphasizes that broadband infrastructure and services should be utilized to advance important national purposes including health care, education, energy and the environment, economic opportunity, government performance, civic engagement, and public safety. Broadband availability and adoption could both drive and be driven by the growth of these national purpose applications. Recommendations addressing national purpose applications impact different sectors of society (e.g., health care, education, energy), and in turn call for action by different agencies of the federal government (e.g., Department of Health and Human Services, Department of Education, Department of Energy). A challenge for policymakers will be to ensure adequate coordination among the disparate agencies and entities implementing various broadband-related policies.

To achieve the goals it has set for the year 2020, the NBP has called for moving forward on a number of specific initiatives, many of which address some of the ongoing major telecommunications policy issues likely to be debated and considered by Congress, the Administration, and the FCC. A major issue for Congress will be how to shape the Plan's various initiatives when and if they go forward, either through oversight, through consideration of specific legislation, or in the context of comprehensive telecommunications reform.

While most agree with the general goals of the NBP—for example, that robust and affordable broadband should be available and utilized throughout the United States—disagreement persists on the best ways to reach those goals. A key challenge for Congressional policymakers will be to assess whether an appropriate balance is maintained between the public and private sectors, and the extent to which government intervention in the broadband marketplace would help or hinder private sector investment and competition.

Table 1. Outline of National Broadband Plan

1. Establishing competition policies	Collect, analyze, benchmark and publish detailed, market-by-market information on broadband pricing and competition	
	Develop disclosure requirements for broadband service providers	
	Undertake a comprehensive review of wholesale competition rules	
	Free up and allocate additional spectrum for unlicensed use	
	Update rules for wireless backhaul spectrum	
	Expedite action on data roaming	
	Change rules to ensure a competitive and innovative video set-top box market	

Table 1. (Continued)

	Clarify the Congressional mandate allowing state and local entities to provide broadband in their communities	
	Clarify the relationship between users and their online profiles to enable continued innovation and competition in applications and ensure consumer privacy	
2. Ensuring efficient allocation and use of government-owned and government-influenced assets	Spectrum	Make 500 megahertz of spectrum newly available
		Enable incentives and mechanisms to repurpose spectrum
		Ensure greater transparency
		Expand opportunities for innovative spectrum access models
	Infrastructure	Establish low and more uniform rental rates for access to poles
		Improve rights-of-way management for cost and time savings
		Facilitate efficient new infrastructure construction
		Provide ultra-high-speed broadband connectivity to select U.S. Department of Defense installations
3. Creating incentives for universal availability and adoption of broadband	Ensure universal access to broadband network services	Create the Connect America Fund (CAF)
		Create a Mobility Fund to provide targeted funding
		Transition the "legacy" High-Cost component of the USF
		Reform intercarrier compensation
		Design the new Connect America Fund and Mobility Fund in a tax-efficient manner
		Broaden the USF contribution base
	Create mechanisms to ensure affordability to low-income Americans	
	Expand the Lifeline and Link-Up programs by allowing subsidies provided to low-income Americans to be used for broadband	Consider licensing a block of spectrum with a condition to offer free or low-cost service
	Ensure every American has the opportunity to become digitally literate	Launch a National Digital Literacy Corps

Table 1. (Continued)

4. Updating policies, setting standards and aligning incentives to maximize use for national priorities	Health care	
	Education	
	Energy and the environment	
	Economic opportunity	
	Government performance and civic engagement	
	Public safety and homeland security	

Source: Compiled by CRS from the National Broadband Plan.

Table 2. Recommendations of the National Broadband Plan to Congress

Chapter	Recommendation
Broadband Competition and Innovation Policy	4.14: Congress, the Federal Trade Commission (FTC) and the FCC should consider clarifying the relationship between users and their online profiles.
	4.15: Congress should consider helping spur development of trusted "identity providers" to assist consumers in managing their data in a manner that maximizes the privacy and security of the information.
Spectrum	5.4: Congress should consider expressly expanding the FCC's authority to enable it to conduct incentive auctions in which incumbent licensees may relinquish rights in spectrum assignments to other parties or to the FCC.
	5.5: Congress should consider building upon the success of the Commercial Spectrum Enhancement Act (CSEA) to fund additional approaches to facilitate incumbent relocation.
	5.6: Congress should consider granting authority to the FCC to impose spectrum fees on license holders and to NTIA to impose spectrum fees on users of government spectrum.
Infrastructure	6.5: Congress should consider amending Section 224 of the act to establish a harmonized access policy for all poles, ducts, conduits and rights-of-way.
	6.8: Congress should consider enacting "dig once" legislation applying to all future federally funded projects along rights-of-way (including sewers, power transmission facilities, rail, pipelines, bridges, tunnels and roads).
	6.9: Congress should consider expressly authorizing federal agencies to set the fees for access to federal rights-of-way on a management and cost recovery basis.
Research and Development	7.2: Congress should consider making the Research and Experimentation (R&E) tax credit a long-term tax credit to stimulate broadband R&D.
Availability	8.15: To accelerate broadband deployment, Congress should consider providing optional public funding to the Connect America Fund, such as a few billion dollars per year over a two to three year period.
	8.16: Congress should consider expanding combination grant-loan programs.
	8.17: Congress should consider expanding the Community Connect program.
	8.18: Congress should consider establishing a Tribal Broadband Fund to support sustainable broadband deployment and adoption in Tribal lands, and all federal agencies that upgrade connectivity on Tribal lands should

Table 2. (Continued)

Chapter	Recommendation
	coordinate such upgrades with Tribal governments and the Tribal Broadband Fund grant-making process.
	8.19: Congress should make clear that state, regional and local governments can build broadband networks.
	8:21: Congress should consider amending the Communications Act to provide discretion to the FCC to allow anchor institutions on Tribal lands to share broadband network capacity that is funded by the E-rate or the Rural Health Care program with other community institutions designated by Tribal governments.
Adoption and Utilization	9.10: Congress, the FCC and the U.S. Department of Justice (DOJ) should modernize accessibility laws, rules and related subsidy programs.
	9.12: Congress and federal agencies should promote third-party evaluation of future broadband adoption programs.
	9.14: The Executive Branch, the FCC and Congress should consider making changes to ensure effective coordination and consultation with Tribes on broadband-related issues.
Health Care	10.1: Congress and the Secretary of Health and Human Services (HHS) should consider developing a strategy that documents the proven value of e-care technologies, proposes reimbursement reforms that incent their meaningful use and charts a path for their widespread adoption.
	10.2: Congress, states and the Centers for Medicare & Medicaid Services (CMS) should consider reducing regulatory barriers that inhibit adoption of health IT solutions.
	10.5: Congress should consider providing consumers access to—and control over—all their digital health care data in machine-readable formats in a timely manner & at a reasonable cost.
	10.10: Congress should consider providing an incremental sum (up to $29 million a year) for the Indian Health Service for the purpose of upgrading its broadband service to meet connectivity requirements.
Education	11.4: Congress should consider taking legislative action to encourage copyright holders to grant educational digital rights of use, without prejudicing their other rights.
	11.22: Congress should consider amending the Communications Act to help Tribal libraries overcome barriers to E-rate eligibility arising from state laws.
	11.25: Congress should consider providing additional public funds to connect all public community colleges with high-speed broadband and maintain that connectivity.
Energy and the Environment	12.4: Congress should consider amending the Communications Act to enable utilities to use the proposed public safety 700MHz wireless broadband network.
	12.7:States should require electric utilities to provide consumers access to, and control of, their own digital energy information, including real-time information from smart meters and historical consumption, price and bill data over the Internet. If states fail to develop reasonable policies over the next 18 months, Congress should consider national legislation to cover consumer privacy and the accessibility of energy data.

Table 2. (Continued)

Chapter	Recommendation
Economic Opportunity	13.4: Congress should consider additional funds for the Economic Development Administration (EDA) to bolster entrepreneurial development programs with broadband tools and training.
	13.6: Congress should consider eliminating tax and regulatory barriers to telework.
Government Performance	14.2: When feasible, Congress should consider allowing state and local governments to get lower service prices by participating in federal contracts for advanced communications services.
	14.4: The Executive Branch and Congress should consider using federal funding to encourage cities and counties to gather information on initiatives enabled by broadband in ways that allow for rigorous evaluation and lead to an understanding of best practices.
	14.17: Congress should consider re-examining the Privacy Act to facilitate the delivery of online government services and to account for changes in technology.
Civic Engagement	15.6: Congress should consider increasing funding to public media for broadband-based distribution and content.
	15.7: Congress should consider amending the Copyright Act to provide for copyright exemptions to public broadcasting organizations for online broadcast and distribution of public media.
	15.9: Congress should consider amending the Copyright Act to enable public and broadcast media to more easily contribute their archival content to a digital national archive and grant reasonable noncommercial downstream usage rights for this content to the American people.
Public Safety	16.4: Preserve broadband communications during emergencies.
	16.14: Congress should consider enacting of federal regulatory framework.

Source: Compiled by CRS from the National Broadband Plan.

End Notes

[1] Federal Communications Commission, *News Release*, "FCC Sends National Broadband Plan to Congress," March 16, 2010, p. 2, available at http://hraunfoss.fcc.gov/edocs_public/attachmatch/DOC-296880A1.pdf.

[2] Available at http://www.broadband.gov/plan/.

[3] Prepared by Lennard G. Kruger, Specialist in Science and Technology Policy.

[4] Federal Communications Commission, *Connecting America: The National Broadband Plan*, March 17, 2010, p. 9.

[5] A distinction is often made between "current generation" and "next generation" broadband (commonly referred to as next generation networks or NGN). "Current generation" typically refers to currently deployed cable, DSL, and many wireless systems, while "next generation" refers to dramatically faster download and upload speeds offered by fiber technologies and also potentially by future generations of cable, DSL, and wireless technologies.

[6] According to the FCC, "the affordability and actual performance of these networks will depend on many factors such as usage patterns, investment in infrastructure, and service take-up rates." See *Connecting America*, p. 21.

[7] *Connecting America: The National Broadband Plan*, p. 22.

[8] According to the FCC, "these measures likely overstate the coverage actually experienced by consumers, since American Roamer reports advertised coverage as reported by many carriers who all use different definitions of coverage. In addition, these measures do not take into account other factors such as signal strength, bitrate or in-building coverage, and may convey a false sense of consistency across geographic areas and service providers." See *Connecting America: The National Broadband Plan*, p. 22.

[9] *Connecting America: The National Broadband Plan*, p. 20.

[10] Ibid., p. 136.

[11] Ibid., p. 23.

[12] Ibid., p. 20.

[13] Ibid., p. 313.

[14] Ibid., p. 265.

[15] Ibid., p. 11.

[16] Ibid.

[17] Ibid.

[18] FCC, *Broadband Action Agenda*, available at http://www.broadband.gov/plan/broadband-action-agenda.html.

[19] Prepared by Angele A. Gilroy, Specialist in Telecommunications Policy.

[20] *Connecting America: The National Broadband Plan,* Chapter 8, p.135.

[21] Communications Act of 1934, as amended, title I sec.1 [47 U.S.C. 151].

[22] *Telephone Subscribership in the United States*, Federal Communications Commission. Released February 2010. Table 1, p. 7. Data as of November 2009.

[23] The Federal USF provides support and discounts for providers and subscribers through four programs: the High-Cost Program, the Low-Income Program, the Schools and Libraries Program; and the Rural Health Care Program.

[24] See sections 254(b)(6) and 254(h) of the 1996 Telecommunications Act, incorporated in to the Communications Act of 1934, 47 U.S.C. 254.

[25] For a further discussion of the programs, funding, and policy issues relating to universal service see CRS Report RL33979, *Universal Service Fund: Background and Options for Reform*, by Angele A. Gilroy.

[26] *Connecting America: The National Broadband Plan*, p. 23.

[27] Ibid., p. 20.

[28] Section 254(c) of the 1996 Telecommunications Act, incorporated into the Communications Act of 1934, 47 U.S.C. 254.

[29] Some, however, have cautioned that a more modest approach is appropriate given the "universal mandate" associated with this definition, the uncertainty and costs associated with mandating nationwide deployment, and the stress currently facing the USF.

[30] For a more detailed analysis of the transition of the USF to accommodate the inclusion of broadband and the implementation and funding of the USF provisions contained in the NBP see CRS Report RL33979, *Universal Service Fund: Background and Options for Reform*, by Angele A. Gilroy.

[31] Much of this transition is detailed in Chapter 8, *Availability*, of the NBP.

[32] It would not support high definition video or high definition video conferencing.

[33] *Connecting America: The National Broadband Plan*, Chapter 9, p. 165.

[34] See ibid., Chapter 9, *Adoption and Utilization*, for details on this expansion.

[35] See ibid., Chapter 11, *Education*, for a detailed discussion of these recommendations.

[36] See ibid., Chapter 10, *Health Care*, for a detailed discussion of these recommendations.

[37] Ibid., Chapter 8, Recommendation 8.6.

[38] Ibid., Chapter 8, Recommendation 8.15.

[39] Ibid., Chapter 8, Recommendations 8.10 and 8.12.

[40] Prepared by Charles B. Goldfarb, Specialist in Telecommunications Policy. For a detailed discussion of issues relating to intercarrier compensation, including how it has developed historically and the market incentives created under alternative intercarrier compensation schemes, see CRS Report RL32889, *Intercarrier Compensation: One Component of Telecom Reform*, by Charles B. Goldfarb.

[41] These figures were cited by the FCC in *In the Matter of Developing a Unified Intercarrier Compensation Regime*, Further Notice of Proposed Rulemaking, adopted February 10, 2005, and released March 3, 2005, at para. 107. These percentages probably have fallen since then as the minutes of long distance traffic have fallen substantially in the past six years, but access charges still represent a far larger portion of rural telephone company revenues than urban telephone company revenues.

[42] *Connecting America: The National Broadband Plan,* p. 142 and footnote 42, citing a letter submitted by AT&T.

[43] For example, whether a wireline call is interstate or intrastate, or whether a wireless call crosses Metropolitan Trading Area (MTA) boundaries.

[44] For example, whether it is interexchange (long distance) traffic, local exchange (local) traffic, or ISP-bound traffic.

[45] For example, whether the terminating carrier is an incumbent wireline carrier subject to rate of return regulation, an incumbent wireline carrier subject to price cap regulation, a competitive wireline carrier, or a wireless carrier.

[46] For example, some service providers have created "free" teleconference services by having end users call in to a telephone number in the service area of a rural telephone company that has very high terminating access charges. The caller's long distance carrier must pay the high per minute terminating access charges to the rural telephone company and that rural telephone company in turn shares those revenues with the service provider. The terminating access charges are so high that both the rural telephone company and the teleconference

service provider can profit. But the end users' long distance carriers end up bearing the costs for the "free" teleconference service.

[47] *Connecting America: The National Broadband Plan*, p. 148.

[48] 47 U.S.C. § 549(a).

[49] *Connecting America: The National Broadband Plan*, pp. 49-52.

[50] Federal Communications Commission, *In the Matter of Video Device Competition; Implementation of Section 304 of the Telecommunications Act of 1996; Commercial Availability of Navigation Devices; Compatibility Between Cable Systems and Consumer Electronics Equipment*, MB Docket No. 10-91, CS Docket No. 97-80, and PP Docket No. 00-67, Notice of Inquiry, adopted and released on April 21, 2010.

[51] *Connecting America: The National Broadband Plan*, Recommendation 4.12.

[52] Federal Communications Commission, *In the Matter of Implementation of Section 304 of the Telecommunications Act of 1996; Commercial Availability of Navigation Devices; Compatibility Between Cable Systems and Consumer Electronics Equipment*, CS Docket No. 97-80 and PP Docket No. 00-67, Fourth Further Notice of Proposed Rulemaking, adopted and released on April 21, 2010.

[53] Prepared by Linda K. Moore, Specialist in Telecommunications Policy.

[54] Broadband refers here to the capacity of the radio frequency channel. A broadband channel can quickly transmit live video, complex graphics, and other data-rich information as well as voice and text messages, whereas a narrowband wireless channel might be limited to handling voice, text, and some graphics.

[55] Many of the spectrum policies and proposals discussed in this section are covered in CRS Report R40674, *Spectrum Policy in the Age of Broadband: Issues for Congress* , by Linda K. Moore

[56] *Connecting America: The National Broadband Plan*, Recommendation 5.1.

[57] For more information on the Spectrum Dashboard, go to http://reboot.fcc.gov/reform/systems/spectrum-dashboard/ about.

[58] *Connecting America: The National Broadband Plan*, Recommendation 5.2.

[59] Ibid., Recommendation 5.3.

[60] Spectrum is segmented into bands of radio frequencies and typically measured in cycles per second, or hertz. Standard abbreviations for measuring frequencies include kHz—kilohertz or thousands of hertz; MHz—megahertz, or millions of hertz; and GHz—gigahertz, or billions of hertz.

[61] *Connecting America: The National Broadband Plan*, Recommendations 5.13 and 5.14. The NBP proposed that the National Science Foundation "should fund wireless research and development that will advance the science of spectrum access." p. 96.

[62] Ibid., Recommendation 5.8.

[63] Ibid., Recommendation 5.8.5.

[64] Ibid., p 89.

[65] Background information regarding the D Block is provided in CRS Report R40859, *Public Safety Communications and Spectrum Resources: Policy Issues for Congress* , by Linda K. Moore.

[66] *Connecting America: The National Broadband Plan*, Recommendation 5.8.2.

[67] FCC News, "The Federal Communications Commission Establishes New Emergency Response Interoperability Center," April 23, 2010, at http://hraunfoss.fcc.gov/edocs_public/attachmatch/DOC-297707A1.pdf.

[68] NIST, "Demonstration Network Planned for Public Safety 700 MHz Broadband," December 15, 2009 at http://www.nist.gov/eeel/oles/network_121509.cfm.

[69] *Connecting America: The National Broadband Plan*, p. 93.

[70] Ibid., Recommendations 5.9 and 5.10.

[71] A leading advocate for replacing channel management of radio frequency with network-centric management is Preston Marshall, the source for much of the information about network-centric technologies in this chapter. Mr. Marshall is Director, Information Sciences Institute, University of Southern California, Viterbi School of Engineering, Arlington, Virginia.

[72] FCC, *First Report and Order and Further Notice of Proposed Rule Making*, ET Docket No. 04-186, released October 18, 2006.

[73] Dynamic Spectrum Access, Content-Based Networking, and Delay and Disruption Technology Networking, along with cognitive radio, and decision-making software, are examples of technologies that can enable Internet-like management of spectrum resources.

[74] *Connecting America*, Recommendations 5.13 and 5.14. The NBP proposed that the National Science Foundation "should fund wireless research and development that will advance the science of spectrum access." p. 96.

[75] Dynamic Spectrum Access, Content-Based Networking, and Delay and Disruption Technology Networking, along with cognitive radio, and decision-making software, are examples of technologies that can enable Internet-like management of spectrum resources. DSA is part of the neXt Generation program, or XG, a technology development project sponsored by the Strategic Technology Office of the Defense Advanced Research Projects Agency (DARPA). The main goals of the program include developing both the enabling technologies and system concepts that dynamically redistribute allocated spectrum.

[76] *Connecting America*, Recommendation 5.11.

[77] *Connecting America*, Recommendation 5.13.

[78] Letter to the FCC, Re: National Broadband Plan, GN Doc. No. 09-51, January 4, 2010 at http://www.ntia.doc.gov/ filings/2009/FCCLetter_Docket09-51_20100104.pdf.

[79] Office of Management and Budget, *Budget of the U.S. Government, Fiscal Year 2011, Appendix*, "Other Independent Agencies," p. 1263. See also, FCC, *Fiscal Year 2011 Budget Estimates Submitted to Congress*, February 2010 at http://hraunfoss.fcc.gov/edocs_public/attachmatch/DOC-296111A1.pdf.

[80] The White House, *Presidential Memorandum: Unleashing the Wireless Broadband Revolution*, June 28, 2010 at http://www.whitehouse.gov/the-press-office/presidential-memorandum-unleashing-wireless-broadband-revolution.

[81] See Spectrum Management Advisory Committee website at http://www.ntia.doc.gov/advisory/spectrum/.

[82] Prepared by Linda K. Moore, Specialist in Telecommunications Policy.

[83] P.L. 111-5, § 6001 (k) (2) (D); 123 STAT. 516.

[84] P.L. 110-140, Sec. 1301; 123 STAT. 1783.

[85] *Connecting America: The National Broadband Plan*, p. 193.

[86] Ibid., Recommendation 14.1.

[87] Ibid., Recommendation 17.1.

[88] Ibid., Recommendation 17.2.

[89] Prepared by Kathleen Ann Ruane, Legislative Attorney.

[90] Comcast v. Federal Communications Commission, 600 F.3d 642 (D.C. Cir. 2010). (*Comcast*) CRS Report R40234, *The FCC's Authority to Regulate Net Neutrality after Comcast v. FCC* , by Kathleen Ann Ruane.

[91] *See*, Inquiry Concerning High-Speed Access to the Internet Over Cable & Other Facilities; Internet Over Cable Declaratory Ruling; Appropriate Regulatory Treatment for Broadband Access to the Internet Over Cable Facilities, 17 FCC Rcd 4798 (2002) (Cable Modem Declaratory Ruling).

[92] *Id.*

[93] Press Release, Chairman Julius Genachowski, FCC, The Third Way: A Narrowly Tailored Broadband Framework (May 6, 2010). ["Genachowski Statement"]. Press Release, Austin Schlick, FCC, A Third-Way Legal Framework for Addressing the *Comcast* Dilemma (May 6, 2010). ["Schlick Statement"].

[94] It is worth noting that the Ninth Circuit Court of Appeals had issued a ruling declaring that cable modem Internet service was a telecommunications service, prior to the FCC's decision to implement a rulemaking on this issue. AT&T Corp. v. City of Portland, 216 F.3d 871, 877-79 (9th Cir. 2002). However, as discussed *infra*, despite the FCC reaching the opposite conclusion, the Supreme Court upheld the FCC's interpretation of the Communications Act.

[95] Information services are defined as:
the offering of a capability for generating, acquiring, storing, transforming, processing, retrieving, utilizing, or making available information via telecommunications, and includes electronic publishing, but does not include any use of any such capability for the management, control or operation of a telecommunications system or the management of a telecommunications service.
47 U.S.C. § 153(20).

[96] Telecommunications services are defined as:
the offering of telecommunications for a fee directly to the public, or to such classes of users as to be effectively available directly to the public, regardless of the facilities used.
47 U.S.C. § 153(46).

[97] The agency identified a portion of cable modem Internet services as "Internet connectivity," which is the portion the agency would seek to redefine as a telecommunications service today. See *Cable Modem Declaratory Ruling*, 17 FCC Rcd at 4809-11.

[98] *Cable Modem Declaratory Ruling*, 17 FCC Rcd at 4819.

[99] Ibid. at 14856.

[100] Nat'l Cable & Telecomms. Ass'n v. Brand X Internet Servs., 545 U.S. 967 (2005) (*Brand X*).

[101] Ibid. at 987.

[102] Ibid. at 991, 1002-03.

[103] Ibid. at 1005 (Scalia, J., dissenting).

[104] See Genachowski Statement, *supra* note 4; Schlick Statement, *supra* note 4.

[105] Genachowski Statement, *supra* note 4.

[106] Schlick Statement, *supra* note 4.

[107] Genachowski Statement, *supra* note 4.

[108] In the Matter of Framework for Broadband Internet Service, Notice of Inquiry, GN Docket No. 10-127 (2010) available at http://hraunfoss.fcc.gov/edocs_public/attachmatch/FCC-10-114A1.pdf. ["NOI"]

[109] Ibid. at ¶¶ 1-2.

[110] See Preserving the Open Internet: Broadband Industry Practices, GN Docket no. 09-191, WC Docket No. 07-52, Notice of Proposed Rulemaking, 24 FCC Rcd 13064 (2009).

[111] See Formal Complaint of Free Press and Public Knowledge Against Comcast Corporation for Secretly Degrading Peer-to-Peer Applications; Broadband Industry Practices et al., WC Docket No. 07-52, Memorandum Opinion and Order, 23 FCC Rcd 13028 (2008).

[112] NOI, at ¶ 1.

[113] Ibid. at ¶ 30.

[114] See CRS Report R40234, *The FCC's Authority to Regulate Net Neutrality after Comcast v. FCC* , by Kathleen Ann Ruane.

[115] NOI, at ¶¶ 32-51.

[116] Ibid. at ¶¶ 52-66.

[117] *Brand X*, 545 U.S. at 991.

[118] FCC v. Fox Television Stations, Inc. 129 S. Ct. 1800, 1811 (2009).

[119] Ibid.

[120] NOI, at ¶ 52.

[121] See Genachowski Statement, *supra* note 4; Schlick Statement, *supra* note 4.

[122] Codified at 47 U.S.C. § 160.

[123] See Genachowski Statement, *supra* note 4; Schlick Statement, *supra* note 4.

[124] NOI, at ¶ 74. The Chairman and General Counsel analogized this approach to its regulation of wireless voice communications. In 1993, Congress specified that Title II applies to wireless communications, such as cellular phone service. 47 U.S.C. § 332(c). Section 332(c) gave the FCC the discretion to determine which regulations under Title II should be inapplicable to wireless voice services; however, the FCC could not forbear from applying Sections 201, 202, or 208 to wireless voice services. *Id.* Similarly, the statement of the Chairman has pledged to apply Sections 201, 202, and 208 to broadband access services

[125] NOI, at ¶¶ 74-85.

[126] 47 U.S.C. § 201

[127] 47 U.S.C. § 202.

[128] 47 U.S.C. § 208.

[129] 47 U.S.C. § 222.

[130] 47 U.S.C. § 254.

[131] 47 U.S.C. § 255.

[132] Genachowski Statement, *supra* note 4.

[133] Schlick Statement, *supra* note 4.

[134] NOI, at ¶¶ 86-7.

[135] Ibid. at ¶¶ 69-70.

[136] Ibid. at ¶¶ 101-05.

[137] Ibid. at ¶¶ 109-10.

[138] Ibid. at ¶¶ 88-9.

[139] *Connecting America: The National Broadband Plan*, p. xi.

[140] Ibid., p. 5.

[141] Ibid.

[142] Ibid.

[143] Ibid., p. 15.

In: The National Broadband Plan: Analysis and Strategy for...   ISBN: 978-1-61122-024-7
Editor: Daniel M. Morales   © 2010 Nova Science Publishers, Inc.

Chapter 2

CONNECTING AMERICA: THE NATIONAL BROADBAND PLAN

Federal Communications Commission

EXECUTIVE SUMMARY

Broadband is the great infrastructure challenge of the early 21st century.

Like electricity a century ago, broadband is a foundation for economic growth, job creation, global competitiveness and a better way of life. It is enabling entire new industries and unlocking vast new possibilities for existing ones. It is changing how we educate children, deliver health care, manage energy, ensure public safety, engage government, and access, organize and disseminate knowledge.

Fueled primarily by private sector investment and innovation, the American broadband ecosystem has evolved rapidly. The number of Americans who have broadband at home has grown from eight million in 2000 to nearly 200 million last year. Increasingly capable fixed and mobile networks allow Americans to access a growing number of valuable applications through innovative devices.

But broadband in America is not all it needs to be. Approximately 100 million Americans do not have broadband at home. Broadband-enabled health information technology (IT) can improve care and lower costs by hundreds of billions of dollars in the coming decades, yet the United States is behind many advanced countries in the adoption of such technology. Broadband can provide teachers with tools that allow students to learn the same course material in half the time, but there is a dearth of easily accessible digital educational content required for such opportunities. A broadband-enabled Smart Grid could increase energy independence and efficiency, but much of the data required to capture these benefits are inaccessible to consumers, businesses and entrepreneurs. And nearly a decade after 9/11, our first responders still lack a nationwide public safety mobile broadband communications network, even though such a network could improve emergency response and homeland security.

Fulfilling the Congressional Mandate

In early 2009, Congress directed the Federal Communications Commission (FCC) to develop a National Broadband Plan to ensure every American has "access to broadband capability." Congress also required that this plan include a detailed strategy for achieving affordability and maximizing use of broadband to advance "consumer welfare, civic participation, public safety and homeland security, community development, health care delivery, energy independence and efficiency, education, employee training, private sector investment, entrepreneurial activity, job creation and economic growth, and other national purposes."

Broadband networks only create value to consumers and businesses when they are used in conjunction with broadband- capable devices to deliver useful applications and content. To fulfill Congress's mandate, the plan seeks to ensure that the entire broadband ecosystem—networks, devices, content and applications—is healthy. It makes recommendations to the FCC, the Executive Branch, Congress and state and local governments.

The Plan

Government can influence the broadband ecosystem in four ways:

1. Design policies to ensure robust competition and, as a result maximize consumer welfare, innovation and investment.
2. Ensure efficient allocation and management of assets government controls or influences, such as spectrum, poles, and rights-of-way, to encourage network upgrades and competitive entry.
3. Reform current universal service mechanisms to support deployment of broadband and voice in high-cost areas; and ensure that low-income Americans can afford broadband; and in addition, support efforts to boost adoption and utilization.
4. Reform laws, policies, standards and incentives to maximize the benefits of broadband in sectors government influences significantly, such as public education, health care and government operations.

1. Establishing competition policies

Policymakers, including the FCC, have a broad set of tools to protect and encourage competition in the markets that make up the broadband ecosystem: network services, devices, applications and content. The plan contains multiple recommendations that will foster competition across the ecosystem. They include the following:

- **Collect, analyze, benchmark and publish detailed, market-by-market information on broadband pricing and competition**, which will likely have direct impact on competitive behavior (e.g., through benchmarking of pricing across geographic markets). This will also enable the FCC and other agencies to apply appropriate remedies when competition is lacking in specific geographies or market segments.

- **Develop disclosure requirements for broadband service providers** to ensure consumers have the pricing and performance information they need to choose the best broadband offers in the market. Increased transparency will incent service providers to compete for customers on the basis of actual performance.
- **Undertake a comprehensive review of wholesale competition rules** to help ensure competition in fixed and mobile broadband services.
- **Free up and allocate additional spectrum for unlicenseduse,** fostering ongoing innovation and competitive entry.
- **Update rules for wireless backhaul spectrum** to increase capacity in urban areas and range in rural areas.
- **Expedite action on data roaming** to determine how best to achieve wide, seamless and competitive coverage, encourage mobile broadband providers to construct and build networks, and promote entry and competition.
- **Change rules to ensure a competitive and innovative video set-top box market,** to be consistent with Section 629 of the Telecommunications Act. The Act says that the FCC should ensure that its rules achieve a competitive market in video "navigation devices," or set-top boxes—the devices consumers use to access much of the video they watch today.
- **Clarify the congressional mandate allowing state and local entities to provide broadband in their communities** and do so in ways that use public resources more effectively.
- **Clarify the relationship between users and their online profiles to enable continued innovation and competition in applications and ensure consumer privacy,** including the obligations of firms collecting personal information to allow consumers to know what information is being collected, consent to such collection, correct it if necessary, and control disclosure of such personal information to third parties.

2. Ensuring efficient allocation and use of government- owned and government-influenced assets

Government establishes policies for the use of spectrum and oversees access to poles, conduits, rooftops and rights-of-way, which are used in the deployment of broadband networks. Government also finances a large number of infrastructure projects. Ensuring these assets and resources are allocated and managed efficiently can encourage deployment of broadband infrastructure and lower barriers to competitive entry. The plan contains a number of recommendations to accomplish these goals. They include the following:

- **spectrum** is a major input for providers of broadband service. Currently, the FCC has only 50 megahertz in inventory, just a fraction of the amount that will be necessary to match growing demand. More efficient allocation and assignment of spectrum will reduce deployment costs, drive investment and benefit consumers through better performance and lower prices. The recommendations on spectrum policy include the following:

- **Make 500 megahertz of spectrum newly available** for broadband within 10 years, of which 300 megahertz should be made available for mobile use within five years.
- **Enable incentives and mechanisms to repurpose spectrum** to more flexible uses. Mechanisms include incentive auctions, which allow auction proceeds to be shared in an equitable manner with current licensees as market demands change. These would benefit both spectrum holders and the American public. The public could benefit from additional spectrum for high-demand uses and from new auction revenues. Incumbents, meanwhile, could recognize a portion of the value of enabling new uses of spectrum. For example, this would allow the FCC to share auction proceeds with broadcasters who voluntarily agree to use technology to continue traditional broadcast services with less spectrum.
- **Ensure greater transparency** of spectrum allocation, assignment and use through an FCC-created spectrum dashboard to foster an efficient secondary market.
- **Expand opportunities for innovative spectrum access models** by creating new avenues for opportunistic and unlicensed use of spectrum and increasing research into new spectrum technologies.

- **infrastructure** such as poles, conduits, rooftops and rights-of-way play an important role in the economics of broadband networks. Ensuring service providers can access these resources efficiently and at fair prices can drive upgrades and facilitate competitive entry. In addition, testbeds can drive innovation of next-generation applications and, ultimately, may promote infrastructure deployment. Recommendations to optimize infrastructure use include:
  - **Establish low and more uniform rental rates for access to poles**, and simplify and expedite the process for service providers to attach facilities to poles.
  - **Improve rights-of-way management for cost and time savings,** promote use of federal facilities for broadband, expedite resolution of disputes and identify and establish "best practices" guidelines for rights-ofway policies and fee practices that are consistent with broadband deployment.
  - **Facilitate efficient new infrastructure construction**, including through "dig-once" policies that would make federal financing of highway, road and bridge projects contingent on states and localities allowing joint deployment of broadband infrastructure.
  - **Provide ultra-high-speed broadband connectivity to select u.s. Department of Defense installations** to enable the development of next-generation broadband applications for military personnel and their families living on base.

3. Creating incentives for universal availability and adoption of broadband

Three elements must be in place to ensure all Americans have the opportunity to reap the benefits of broadband. All Americans should have access to broadband service with sufficient capabilities, all should be able to afford broadband and all should have the opportunity to develop digital literacy skills to take advantage of broadband. Recommendations to promote universal broadband deployment and adoption include the following:

- **Ensure universal access to broadband network services.**
  - **Create the connect America Fund (CAF)** to support the provision of affordable broadband and voice with at least 4 Mbps *actual* download speeds and shift up to $15.5 billion over the next decade from the existing Universal Service Fund (USF) program to support broadband. If Congress wishes to accelerate the deployment of broadband to unserved areas and otherwise smooth the transition of the Fund, it could make available public funds of a few billion dollars per year over two to three years.
  - **Create a Mobility Fund to provide targeted funding** to ensure no states are lagging significantly behind the national average for 3G wireless coverage. Such 3G coverage is widely expected to be the basis for the future footprint of 4G mobile broadband networks.
  - **Transition the "legacy" High-Cost component of the USF** over the next 10 years and shift all resources to the new funds. The $4.6 billion per year High Cost component of the USF was designed to support primarily voice services. It will be replaced over time by the CAF.
  - **Reform intercarrier compensation,** which provides implicit subsidies to telephone companies by eliminating per-minute charges over the next 10 years and enabling adequate cost recovery through the CAF.
  - **Design the new connect America Fund and Mobility Fund in a tax-efficient manner** to minimize the size of the broadband availability gap and thereby reduce contributions borne by consumers.
  - **Broaden the USF contribution base** to ensure USF remains sustainable over time.
- **Create mechanisms to ensure affordability to low-income Americans.**
- **Expand the Lifeline and Link-up programs by allowing subsidies provided to low-income Americans to be used for broadband.**
  - **consider licensing a block of spectrum with a condition to offer free or low-cost service** that would create affordable alternatives for consumers, reducing the burden on USF.
- **Ensure every American has the opportunity to become digitally literate.**
  - **Launch a National Digital Literacy corps** to organize and train youth and adults to teach digital literacy skills and enable private sector programs addressed at breaking adoption barriers.

4. Updating policies, setting standards and aligning incentives to maximize use for national priorities

Federal, Tribal, state and local governments play an important role in many sectors of our economy. Government is the largest health care payor in the country, operates the public education system, regulates many aspects of the energy industry, provides multiple services to its citizens and has primary responsibility for homeland security. The plan includes recommendations designed to unleash increased use, private sector investment and innovation in these areas. They include the following:

- **Health care.** Broadband can help improve the quality and lower the cost of health care through health IT and improved data capture and use, which will enable clearer understanding of the most effective treatments and processes. To achieve these objectives, the plan has recommendations that will:
  - Help ensure health care providers have access to affordable broadband by transforming the FCC's Rural Health Care Program.
  - Create incentives for adoption by expanding reimbursement for e-care.
  - Remove barriers to e-care by modernizing regulations like device approval, credentialing, privileging and licensing.
  - Drive innovative applications and advanced analytics by ensuring patients have control over their health data and ensuring interoperability of data.
- **Education.** Broadband can enable improvements in public education through e-learning and online content, which can provide more personalized learning opportunities for students. Broadband can also facilitate the flow of information, helping teachers, parents, schools and other organizations to make better decisions tied to each student's needs and abilities. To those ends, the plan includes recommendations to:
- Improve the connectivity to schools and libraries by upgrading the FCC's E-Rate program to increase flexibility, improve program efficiency and foster innovation by promoting the most promising solutions and funding wireless connectivity to learning devices that go home with students.
  - Accelerate online learning by enabling the creation of digital content and learning systems, removing regulatory barriers and promoting digital literacy.
  - Personalize learning and improve decision–making by fostering adoption of electronic educational records and improving financial data transparency in education.
- **Energy and the environment**. Broadband can play a majorrole in the transition to a clean energy economy. America can use these innovations to reduce carbon pollution, improve our energy efficiency and lessen our dependence on foreign oil. To achieve these objectives, the plan has recommendations that will:
  - Modernize the electric grid with broadband, making it more reliable and efficient.
  - Unleash energy innovation in homes and buildings by making energy data readily accessible to consumers.
  - Improve the energy efficiency and environmental impact of the ICT sector.
- **Economic opportunity.** Broadband can expand access to jobs and training, support entrepreneurship and small business growth and strengthen community development efforts. The plan includes recommendations to:
  - Support broadband choice and small businesses' use of broadband services and applications to drive job creation, growth and productivity gains.
  - Expand opportunities for job training and placement through an online platform.
  - Integrate broadband assessment and planning into economic development efforts.
- **Government performance and civic engagement.** Within government, broadband can drive greater efficiency and effectiveness in service delivery and internal

operations. It can also improve the quantity and quality of civic engagement by providing a platform for meaningful engagement with representatives and agencies. Through its own use of broadband, government can support local efforts to deploy broadband, particularly in unserved communities. To achieve these goals, the plan includes recommendations to:

- Allow state and local governments to purchase broad-band from federal contracts such as Networx.
- Improve government performance and operations through cloud computing, cybersecurity, secure authentication and online service delivery.
- Increase civic engagement by making government moreopen and transparent, creating a robust public media ecosystem and modernizing the democratic process.

- **Public safety and homeland security.** Broadband can bolster efforts to improve public safety and homeland security by allowing first responders to send and receive video and data, by ensuring all Americans can access emergency services and improving the way Americans are notified about emergencies. To achieve these objectives, the plan makes recommendations to:

- Support deployment of a nationwide, interoperable public safety mobile broadband network, with funding of up to $6.5 billion in capital expenditures over 10 years, which could be reduced through cost efficiency measures and other programs. Additional funding will be required for operating expenses.
- Promote innovation in the development and deployment of next-generation 911 and emergency alert systems.
- Promote cybersecurity and critical infrastructure survivability to increase user confidence, trust and adoption of broadband communications.

Long-Term Goals

In addition to the recommendations above, the plan recommends that the country adopt and track the following six goals to serve as a compass over the next decade.

Goal No. 1: At least 100 million u.s. homes should have affordable access to actual download speeds of at least 100 megabits per second and actual upload speeds of at least 50 megabits per second.

Goal No. 2: The united states should lead the world in mobile innovation, with the fastest and most extensive wireless networks of any nation.

Goal No. 3: Every American should have affordable access to robust broadband service, and the means and skills to subscribe if they so choose.

Goal No. 4: Every American community should have affordable access to at least 1 gigabit per second broadband service to anchor institutions such as schools, hospitals and government buildings.

Goal No. 5: To ensure the safety of the American people, every first responder should have access to a nationwide, wireless, interoperable broadband public safety network.

Goal No. 6: To ensure that America leads in the clean energy economy, every American should be able to use broadband to track and manage their real-time energy consumption.

Meeting these six goals will help achieve the Congressional mandate of using broadband to achieve national purposes, while improving the economics of deployment and adoption. In particular, the first two goals will create the world's most attractive market for broadband applications, devices and infrastructure and ensure America has the infrastructure to attract the leading communications and IT applications, devices and technologies. The third goal, meanwhile, will ensure every American has the opportunity to take advantage of the benefits broadband offers, including improved health care, better education, access to a greater number of economic opportunities and greater civic participation.

Budget Impact of Plan

Given the plan's goal of freeing 500 megahertz of spectrum, future wireless auctions mean the overall plan will be revenue neutral, if not revenue positive. The vast majority of recommendations do not require new government funding; rather, they seek to drive improvements in government efficiency, streamline processes and encourage private activity to promote consumer welfare and national priorities. The funding requests relate to public safety, deployment to unserved areas and adoption efforts. If the spectrum auction recommendations are implemented, the plan is likely to offset the potential costs.

Implementation

The plan is in beta, and always will be. Like the Internet itself, the plan will always be changing—adjusting to new developments in technologies and markets, reflecting new realities, and evolving to realize the unforeseen opportunities of a particular time.

As such, implementation requires a long-term commitment to measuring progress and adjusting programs and policies to improve performance.

Half of the recommendations in this plan are offered to the FCC. To begin implementation, the FCC will:

- Quickly publish a timetable of proceedings to implement plan recommendations within its authority.
- Publish an evaluation of plan progress and effectiveness as part of its annual 706 Advanced Services Inquiry.
- Create a Broadband Data Depository as a public resource for broadband information.

The remaining half of the recommendations are offered to the Executive Branch, Congress and state and local governments. Policymakers alone, though, cannot ensure success. Industry, non-profits, and government together with the American people, must now act and rise to our era's infrastructure challenge.

1. INTRODUCTION

IN EVERY ERA, AMERICA MUST CONFRONT THE CHALLENGE OF CONNECTING OUR NATION ANEW.

In the 1860s, we connected Americans to a transcontinental railroad that brought cattle from Cheyenne to the stockyards of Chicago. In the 1930s, we connected Americans to an electric grid that improved agriculture and brought industry to the Smoky Mountains of Tennessee and the Great Plains of Nebraska. In the 1950s, we connected Americans to an interstate highway system that fueled jobs on the line in Detroit and in the warehouse in L.A.

Infrastructure networks unite us as a country, bringing together parents and children, buyers and sellers, and citizens and government in ways once unimaginable. Ubiquitous access to infrastructure networks has continually driven American innovation, progress, prosperity and global leadership.

Communications infrastructure plays an integral role in this American story. In the 1920s, '30s, '40s and '50s, telephony, radio and television transformed America, unleashing new opportunities for American innovators to create products and industries, new ways for citizens to engage their elected officials and a new foundation for job growth and international competitiveness.

Private investment was pivotal in building most of these networks, but government actions also played an important role. Treasury bonds and land grants underwrote the railroad,[1] the Rural Electrification Act brought electricity to farms and the federal government funded 90% of the cost of the interstate highways.[2]

In communications, the government stimulated the construction of radio and television facilities across the country by offering huge tracts of the public's airwaves free of charge. It did the same with telephony through a Universal Service Fund, fulfilling the vision of the Communications Act of 1934 "to make available, so far as possible, to all the people of the United States, a rapid, efficient, Nation-wide, and world-wide wire and radio communication service with adequate facilities at reasonable charges."[3]

Today, high-speed Internet is transforming the landscape of America more rapidly and more pervasively than earlier infrastructure networks. Like railroads and highways, broadband accelerates the velocity of commerce, reducing the costs of distance. Like electricity, it creates a platform for America's creativity to lead in developing better ways to solve old problems. Like telephony and broadcasting, it expands our ability to communicate, inform and entertain.

Broadband is *the* great infrastructure challenge of the early 21st century.

But as with electricity and telephony, ubiquitous connections are means, not ends. It is what those connections enable that matters. Broadband is a platform to create today's high-performance America—an America of universal opportunity and unceasing innovation, an America that can continue to lead the global economy, an America with world-leading, broadband-enabled health care, education, energy, job training, civic engagement, government performance and public safety.

Due in large part to private investment and market-driven innovation, broadband in America has improved considerably in the last decade. More Americans are online at faster speeds than ever before. Yet there are still critical problems that slow theprogress of availability, adoption and utilization of broadband.

Recognizing this, one year ago Congress echoed the Communications Act of 1934 and directed the FCC to develop a National Broadband Plan ensuring that every American has "access to broadband capability." Specifically, the statute dictates:

"The national broadband plan required by this section shall seek to ensure that all people of the United States have access to broadband capability and shall establish benchmarks for meeting that goal. The plan shall also include:

- *an analysis of the most effective and efficient mechanisms for ensuring broadband access by all people of the United States,*
- *a detailed strategy for achieving affordability of such service and maximum utilization of broadband infrastructure and service by the public,*
- *an evaluation of the status of deployment of broadband service, including progress of projects supported by the grants made pursuant to this section, and*
- *a plan for use of broadband infrastructure and services in advancing consumer welfare, civic participation, public safety and homeland security, community development, health care delivery, energy independence and efficiency, education, worker training, private sector investment, entrepreneurial activity, job creation and economic growth, and other national purposes."[4]*

This is a broad mandate. It calls for broadband networks that reach higher and farther, filling the troubling gaps we face in the deployment of broadband networks, in the adoption of broadband by people and businesses and in the use of broadband to further our national priorities.

Nearly 100 million Americans do not have broadband today.[5] Fourteen million Americans do not have access to broadband infrastructure that can support today's and tomorrow's applications.[6] More than 10 million school-age children[7] do not have home access to this primary research tool used by most students for homework.[8] Jobs increasingly require Internet skills; the share of Americans using high-speed Internet at work grew by 50% between 2003 and 2007,[9] and the number of jobs in information and communications technology is growing 50% faster than in other sectors.[10] Yet millions of Americans lack the skills necessary to use the Internet.[11]

What's more, there are significant gaps in the utilization of broadband for other national priorities. In nearly every metric used to measure the adoption of health information technology (IT), the United States ranks in the bottom half among comparable countries,[12] yet electronic health records could alone save more than $500 billion over 15 years.[13] Much of the electric grid is not connected to broadband, even though a Smart Grid could prevent 360 million metric tons of carbon emissions per year by 2030, equivalent to taking 65 million of today's cars off the road.[14] Online courses can dramatically reduce the time required to learn a subject while greatly increasing course completion rates,[15] yet only 16% of public community colleges—which have seen a surge in enrollment[16]—have high-speed connections comparable to our research universities.[17] Nearly a decade after 9/11, our first responders still require access to better communications.

Unless we reform our approach to these gaps, we will fail to seize the opportunity to improve our nation, and we will fall behind those countries that do. In fact, other countries already have adopted plans to address these gaps.

The ways that other countries have confronted this challenge help inform how we might approach the problem. But each country's experiences and challenges have critical differences. Our solutions must reflect the unique economic, institutional and demographic conditions of our country.

The United States is distinct in many ways. For example, many countries have a single, dominant nationwide fixed telecommunications provider; the United States has numerous providers. Cable companies play a more prominent role in our broadband system than in other countries. The U.S. is less densely populated than other countries. Unlike most other countries, we regulate at both the state and federal levels. Our plan should learn from international experiences, but must also take into account the distinguishing realities of broadband in the United States.

Our plan must be candid about where current government policies hinder innovation and investment in broadband. Government or influences critical inputs needed to build broadband networks— such as spectrum, universal service funds and rights-of-way—yet all are structured to serve the priorities of the past, not the opportunities of the future. In addition, current government policies maintain incentives for our schools, hospitals and other public interest institutions to use outdated technologies and practices, disadvantaging our people and hindering our economy. Just as this plan should build on the distinctive attributes of the American market, it should also correct the problematic policies found here.

Above all, an American plan should build on American strengths.

The first of these strengths is innovation. The United States maintains the greatest tradition of innovation and entrepreneurship in the world—one that combines creativity with engineering to produce world-leading applications, devices and content, as well as the businesses that bring them to market.

Our national plan must build on this strength to ensure that the next great companies, technologies and applications are developed in the United States. U.S. leadership in these spheres will advance our most important public purposes. A healthy environment for innovation will enable advances in health care, energy, education, job training, public safety and all of our national priorities. Creativity is a national virtue that has catalyzed American leadership in many sectors. America's plan should unlock that creativity to transform the public sector, too.

We have just begun to benefit from the ways broadband unleashes innovations to improve American lives: a job seeker in South Bend telecommuting for a company in the Deep South; a medical specialist in Chapel Hill providing medical consultations to a patient in the Hill Country; grandparents in Cleveland video-chatting with their grandchildren in Colorado Springs; firefighters downloading blueprints of a burning building. The applications that broadband enables provide innovative, efficient solutions to challenges Americans confront every day.

Many international broadband plans emphasize speeds and networks, focusing only on technical capacity as a measure of a successful broadband system. Our plan must go beyond that. While striving for ubiquitous and fast networks, we must also strive to use those networks more efficiently and effectively than any other country. We should lead the world where it counts—in the use of the Internet and in the development of new applications that provide the tools that each person needs to make the most of his or her own life.

The United States is well positioned to lead in creating those applications. We have leading health research centers; we should also lead the world in effective health care applications. We have leading educational institutions; we should also lead the world in effective educational applications. We should seize this opportunity to lead the world in applications that serve public purposes.

The second great American strength is inclusion. As a country, we believe that to march ahead we don't need to leave anyone behind. We believe that all deserve the opportunity to improve their lives. We believe that where you start shouldn't dictate where you finish, that demography isn't destiny, that privilege isn't a necessary prologue to success.

This ideal doesn't just compel us to rebuke discrimination; it compels us to be proactive. It inspires us to live up to an obligation we have to each other—to ensure that everyone has an opportunity to succeed.

This desire for equal opportunity has long guided our efforts to make access to technologies universal, from electricity to telephony, from television to radio. Today, as technology continues to change the way the world interacts, to be on the outside is to live in a separate, analog world, disconnected from the vast opportunities broadband enables.

While broadband adoption has grown steadily, it is still far from universal. It lags considerably among certain demo-graphic groups, including the poor, the elderly, some racial and ethnic minorities, those who live in rural areas and those with disabilities. Many of these Americans already struggle to succeed. Unemployment rates are high, services like job training are difficult to obtain and schools are substandard.

Broadband can help bridge these gaps. Today, millions of students are unprepared for college because they lack access to the best books, the best teachers and the best courses. Broadband- enabled online learning has the power to provide high-quality educational opportunities to these students—opportunities to which their peers at the best public and private schools have long had access. Similarly, with broadband, people with disabilities can live more independently, wherever they choose.[18] They can telecommute and run businesses from their homes or receive rehabilitation therapy in remote and rural areas.

Of course, access to broadband is not enough. People still need to work hard to benefit from these opportunities. But universal broadband, and the skills to use it, can lower barriers of means and distance to help achieve more equal opportunity.

Absent action, the individual and societal costs of digital exclusion will grow. With so many Americans lacking broadband access or the skills to make it matter, the Internet has the potential to exacerbate inequality. If learning online accelerates your education, if working online earns you extra money, if searching for jobs online connects you to more opportunities, then for those offline, the gap only widens. If political dialogue moves to online forums, if the Internet becomes the comprehensive source of real-time news and information, if the easiest way to contact your political representatives is through e-mail or a website, then those offline become increasingly disenfranchised.

Until recently, not having broadband was an inconvenience. Now, broadband is essential to opportunity and citizenship.

While we must build on our strengths in innovation and inclusion, we need to recognize that government cannot predict the future. Many uncertainties will shape the evolution of broadband, including the behavior of private companies and consumers, the economic environment and technological advances.

As a result, the role of government is and should remain limited. We must strike the right balance between the public and private sectors. Done right, government policy can drive, and has driven, progress. In the 1960s and '70s, government research funding supported the development of the technology on which the Internet is based.[19] In the 199 0s, the Federal Communications Commission acted to ensure that telephone providers would not stall use of the Internet.[20] An act of Congress stimulated competition that caused cable companies to upgrade their networks and, for the first time, offer broadband to many Americans.[21] Auctions for public spectrum promoted competitive wireless markets, prompting continual upgrades that first delivered mobile phones and, now, mobile broadband.[22]

Instead of choosing a specific path for broadband in America, this plan describes actions government should take to encourage more private innovation and investment. The policies and actions recommended in this plan fall into three categories: fostering innovation and competition in networks, devices and applications; redirecting assets that government controls or influences in order to spur investment and inclusion; and optimizing the use of broadband to help achieve national priorities.

A thoughtful approach to the development of electricity, telephony, radio and television transformed the United States and, in turn, helped us transform the world. Broadband will be just as transformative.

The consequences of our digital transformation may not be uniformly positive. But the choice is not whether the transformation will continue. It will. The choice is whether we, as a nation, will understand this transformation in a way that allows us to make wise decisions about how broadband can serve the public interest, just as certain decisions decades ago helped communications and media platforms serve public interest goals. This plan is the first attempt to provide that understanding—to clarify the choices and to point to paths by which all Americans can benefit.

2. GOALS FOR A HIGH-PERFORMANCE AMERICA

The mission of this plan is to create a high-performance America—a more productive, creative, efficient America in which affordable broadband is available everywhere and everyone has the means and skills to use valuable broadband applications.

The importance of broadband continues to grow around the world. High-performing companies, countries and citizens are using broadband in new, more effective ways. Some countries have recognized this already and are trying to get ahead of the curve. South Korea, Japan, Australia, Sweden, Finland and Germany, among others, have already developed broadband plans.

A high-performance America cannot stand by as other countries charge into the digital era. In the country where the Internet was born, we cannot watch passively while other nations lead the world in its utilization. We should be the leading exporter of broadband technology—high-value goods and services that drive enduring economic growth and job creation. And we should be the leading user of broadband-enabled technologies that help businesses increase their productivity, help government improve its openness and efficiency, and give consumers new ways to communicate, work and entertain themselves.

To ensure we lead the world, this plan addresses the troubling gaps and unrealized opportunities in broadband in America by recommending ways federal, state and local governments can unleash private investment, innovation, lower prices and better options for consumers. Its recommendations fall into four general categories:

- Design policies to ensure robust competition and, as a result, maximize consumer welfare, innovation and investment.
- Ensure efficient allocation and management of assets government controls or influences, such as spectrum, poles, and rights-of-way, to encourage network upgrades and competitive entry.
- Reform current universal service mechanisms to support deployment of broadband and voice in high-cost areas; and ensure that low-income Americans can afford broadband; and in addition, support efforts to boost adoption and utilization.
- Reform laws, policies, standards and incentives to maximize the benefits of broadband in sectors government influences significantly, such as public education, health care and government operations.

Across these categories, this plan offers recommendations for the Federal Communications Commission (FCC), the Executive Branch, Congress, states and other parties. But to ensure we are on the right path, the country should set longterm goals and benchmarks to chart our progress. The plan recommends that the country set the following six goals for 2020 to serve as a compass over the next decade.

GOAL NO. 1: At least 100 million U.S. homes should have affordable access to actual download speeds of at least 100 megabits per second and actual upload speeds of at least 50 megabits per second.

The United States must lead the world in the number of homes and people with access to affordable, world-class broadband connections. As such, 100 million U.S. homes should have affordable access to actual download speeds of at least 100 Mbps and actual upload speeds of at least 50 Mbps by 2020. This will create the world's most attractive market for broadband applications, devices and infrastructure.

The plan has recommendations to foster competition, drive demand for increased network performance and lower the cost of deploying infrastructure. These recommendations include providing consumers with information about the actual performance of broadband services, reviewing wholesale access policies and conducting more thorough data collection to monitor and benchmark competitive behavior. Reforming access to rights-of-way can lower the cost of upgrades and entry for all firms. Increased spectrum availability and use for backhaul can enable more capable wireless networks that will drive wired providers to improve network performance and ensure service is affordable.

Government can also help create demand for more broadband by enabling new applications across our most important national priorities, including health care, education and energy, and by ensuring consumers have full control of their personal data.

As a milestone, by 2015, 100 million U.S. homes should have affordable access to actual download speeds of 50 Mbps and actual upload speeds of 20 Mbps.

GOAL NO. 2: The united states should lead the world in mobile innovation, with the fastest and most extensive wireless networks of any nation.

Mobile broadband is growing at unprecedented rates. From smartphones to app stores to e-book readers to remote patient monitoring to tracking goods in transit and more, mobile services and technologies are driving innovation and playing an increasingly important role in our lives and our economy. Mobile broadband is the next great challenge and opportunity for the United States. It is a nascent market in which the United States should lead.

Spectrum policy is the most important lever government has to help ensure wireless and mobile broadband thrive. Efficient allocation of spectrum consistent with the public interest will maximize its value to society. It will lower network deployment costs, making it easier for new companies to compete and enabling lower prices, more investment and better performance.

Today, the FCC has only 50 megahertz of spectrum in the pipeline that it can assign for broadband use, just a fraction of the amount that will be necessary to match growing demand. As a result, companies representing 5% of the U.S. economy asked the FCC to make more spectrum available for mobile broadband, saying that "without more spectrum, America's global leadership in innovation and technology is threatened."[23]

To achieve this goal of leading the world in mobile broadband, the plan recommends making 500 megahertz of spectrum newly available for broadband by 2020, with a benchmark of making 300 megahertz available by 2015. In addition, we should ensure greater transparency in spectrum allocation and utilization, reserve spectrum for unlicensed use and make more spectrum available for opportunistic and secondary uses.

GOAL NO. 3: Every American should have affordable access to robust broadband service, and the means and skills to subscribe if they so choose.

Not having access to broadband applications limits an individual's ability to participate in 21st century American life. Health care, education and other important aspects of American life are moving online. What's more, government services and democratic participation are shifting to digital platforms. This plan recommends government use the Internet to increase its own transparency and make more of its data available online. Getting everyone online will improve civic engagement—a topic this plan also addresses by recommending a more robust digital public ecosystem.

Three requirements must be satisfied to ensure every American can take advantage of broadband. First, every American home must have access to network services. Second, every household should be able to afford that service. Third, every American should have the opportunity to develop digital skills.

The plan recommends reforming existing support mechanisms to foster deployment of broadband in high-cost areas: specifically, the Universal Service Fund and intercarrier compensation. The plan outlines a 10-year, three-stage course of action to transform these programs to connect those who do not have access to adequate broadband infrastructure.[24] Rather than add new burdens to the already strained contribution base, we must make the tough choice to shift existing support that is not advancing public policy goals in order to directly focus those resources on communities unserved by broadband.

To promote affordability, this plan also proposes extending the Lifeline and Link-Up programs to support broadband. To promote digital skills, we need to ensure every American has access to relevant, age-appropriate digital literacy education, for free, in whatever language they speak, and we neeed to create a Digital Literacy Corps.

Achieving this goal will likely lead to an adoption rate higher than 90% by 2020 and reduced differences in broadband adoption among demographic groups.

To the end, government can make broadband more accessible to people with disabilities. It can also work with Tribal governments to finally improve broadband deployment and adoption on Tribal lands.[25] And it can ensure small businesses— many of which are owned by women and minorities—have the opportunity to purchase broadband service at reasonable rates.

GOAL NO. 4: Every American community should have affordable access to at least 1 gigabit per second broadband service to anchor institutions such as schools, hospitals and government buildings.

Schools, libraries and health care facilities must all have the connectivity they need to achieve their purposes. This connectivity can unleash innovation that improves the way we learn, stay healthy and interact with government.

If this plan succeeds, every American community will have affordable access to far better broadband performance than they enjoy today. To do so, the plan makes recommendations about reforming the E-rate and the Rural Health Care support programs. Second, non-profit and public institutions should be able to find efficient alternatives for greater connectivity through aggregated efforts.

What's more, unleashing the power of new broadband applications to solve previously intractable problems will drive new connectivity demands. The plan makes numerous recommendations, including reforming incentive structures, licensing and data interoperability, to ensure public priorities take advantage of the benefits broadband networks, applications and devices offer. If they are implemented, demand for connectivity in hospitals, schools, libraries and government buildings will soar.

In some communities, gigabit connectivity may not be limited to anchor institutions. Certain applications could also require ultra-high-speed connectivity at home. And once community anchors are connected to gigabit speeds, it would presumably become less expensive and more practical to get the same speeds to homes.

GOAL NO. 5: To ensure the safety of the American people, every first responder should have access to a nationwide, wireless, interoperable broadband public safety network.

In June 2004, the 9/11 Commission released its final report about events of September 11, 2001. The report found that "the inability to communicate was a critical element" at each of the "crash sites, where multiple agencies and multiple jurisdictions responded." They concluded: "Compatible and adequate communications among public safety organizations at the local, state, and federal levels remains an important problem."[26]

It remains a problem more than five years later. Often, first responders from different jurisdictions cannot communicate at the scene of an emergency. Federal officials can rarely communicate with state and local officials. Officials from different towns and cities have difficulties communicating with each other. What's more, with few exceptions, current networks do not take advantage of broadband capability, limiting their capacity to transmit data and hindering potential innovations in public safety that could save lives.

The country should create a nationwide, wireless, interoperable broadband public safety network by 2020. The network should be robust enough to maintain performance in the aftermath of a disaster, and should allow every first responder, regardless of jurisdiction or agency, to communicate with each other and share real-time data over high-speed connections. Chapter 16 outlines recommendations to make this goal a reality.

GOAL No. 6: To ensure that America leads in the clean energy economy, every American should be able to use broadband to track and manage their real-time energy consumption.

America can no longer rely on fossil fuels and imported oil. To improve national security, reduce pollution and increase national competitiveness, the United States must lead, not follow, in the clean energy economy. Encouraging renewable power, grid storage and vehicle electrification are important steps to improve American energy independence and energy efficiency; to enable these technologies at scale, the country will need to modernize the electric grid with broadband and advanced communications.

Studies have repeatedly demonstrated that when people get feedback on their electricity usage, they make simple behavioral changes that save energy.[27] Real-time data can also inform automated thermostats and appliances, allowing consumers to save energy and money while helping the country reduce the need for expensive new power plants.

Chapter 12 outlines specific recommendations to ensure that consumers can use broadband to gain access to and improve their control of their real-time energy information. With strong cybersecurity and privacy protections, consumers and their authorized third parties should be able to get access to real-time usage information from smart meters and historical billing information over the Internet.

Conclusion

To achieve these goals, it is not enough to simply state where we wish to be.[*] America needs a plan that creates a process to meet these targets and look beyond them. The chapters that follow offer specific recommendations to launch that process.

Part I of this plan makes recommendations to ensure that America has a world-leading broadband ecosystem for both fixed and mobile service. It discusses recommendations to maximize innovation, investment and consumer welfare, primarily through competition. It then recommends more efficient allocation and management of assets government controls or

---

[*] In Shakespeare's *Henry IV*, Welsh rebel Glendower tells his co-conspirator Hotspur: "I can call spirits from the vasty deep." Hotspur responds, "Why, so can I, or so can any man; But will they come when you do call for them?" William Shakespeare, *Henry IV*, pt. I, act 3, sc. 1, 52–58.

influences, such as spectrum, poles and rights-of-way, to maximize private sector investment and facilitate competition.

Part II makes recommendations to promote inclusion—to ensure that all Americans have access to the opportunities broadband can provide. These include reforming the Universal Service Fund and intercarrier compensation. It also makes recommendations to promote broadband affordability, adoption and digital literacy.

Part III makes recommendations to maximize the use of broadband to address national priorities. This includes reforming laws, policies and incentives to maximize the benefits of broadband in areas where government plays a significant role. This part makes recommendations to unleash innovation in health care, energy, education, government performance, civic engagement, job training, economic development and public safety.

Finally, the plan outlines an implementation strategy to ensure the country executes these recommendations, creates a dynamic process and meets each of the goals outlined here.

Before exploring any of these recommendations further, though, it is important to understand the current state of broadband in the United States, which is described in Chapter 3.

3. CURRENT STATE OF THE BROADBAND ECOSYSTEM

To see how broadband is transforming American life, walk down a busy street or pay a visit to any school, business or airport. parents on business trips use their smartphones to check e-mail or watch short videos of their children playing soccer, hundreds, if not thousands, of miles away. Americans work together in real time on complex documents from different desks in the same office, and workers in different offices around the world collaborate via videoconferencing technology. Sales and field maintenance personnel use mobile devices to access inventory information in their businesses, place orders and update records, increasing efficiency and productivity. Students draw on the richness of the Internet to research historical events or watch simulations of challenging math problems. people are using broadband in ways they could not imagine even a few years ago.

To understand how this transformation will evolve, it is important to understand the forces shaping the broadband ecosystem in America today (see Exhibit 3-A).

The broadband ecosystem includes applications and content: e-mail, search, news, maps, sales and marketing applications used by businesses, user-generated video and hundreds of thousands of more specialized uses. Ultimately, the value of broadband is realized when it delivers useful applications and content to end-users.

Applications run on devices that attach to the network and allow users to communicate: computers, smartphones, set-top boxes, e-book readers, sensors, private branch exchanges (PBX), local area network routers, modems and an ever-growing list of other devices. New devices mean new opportunities for applications and content.

Finally, broadband networks can take multiple forms: wired or wireless, fixed or mobile, terrestrial or satellite. Different types of networks have different capabilities, benefits and costs.

The value of being connected to the network increases as more people and businesses choose to adopt broadband and use applications and devices that the network supports.

Several factors contribute to their decisions. These include whether they can afford a connection, whether they are comfortable with digital technology and whether they believe broadband is useful.

Networks, devices and applications drive each other in a virtuous cycle. If networks are fast, reliable and widely available, companies produce more powerful, more capable devices to connect to those networks. These devices, in turn, encourage innovators and entrepreneurs to develop exciting applications and content. These new applications draw interest among end- users, bring new users online and increase use among those who already subscribe to broadband services. This growth in the broadband ecosystem reinforces the cycle, encouraging service providers to boost the speed, functionality and reach of their networks.

While the explosive growth in the use of broadband suggests that many aspects of the American broadband ecosystem are healthy, there are many ways America can do better.

3.1. Applications

Users benefit directly from the applications and content they access through broadband networks. Applications help people purchase products, search for jobs, interact with government agencies and find information related to their health.[28] Users also spend considerable time using broadband for banking, shopping, entertainment, social networking and communication (see Exhibit 3-B).[29]

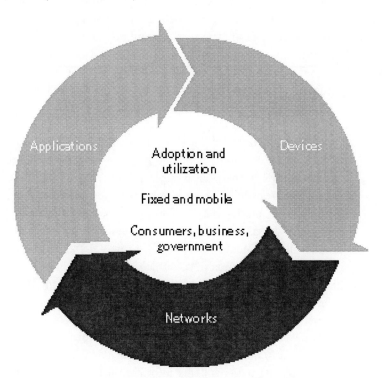

Exhibit 3-A. Forces Shaping the Broadband Ecosystem in the United States

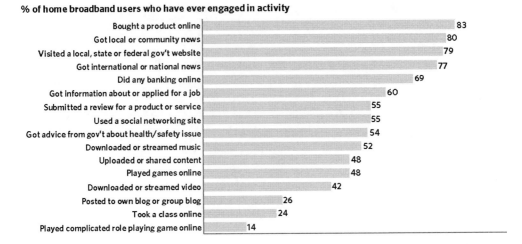

% of home broadband users who have ever engaged in activity

Activity	%
Bought a product online	83
Got local or community news	80
Visited a local, state or federal gov't website	79
Got international or national news	77
Did any banking online	69
Got information about or applied for a job	60
Submitted a review for a product or service	55
Used a social networking site	55
Got advice from gov't about health/safety issue	54
Downloaded or streamed music	52
Uploaded or shared content	48
Played games online	48
Downloaded or streamed video	42
Posted to own blog or group blog	26
Took a class online	24
Played complicated role playing game online	14

Exhibit 3-B. Percentage of Home Broadband Users Who Have Ever Engaged in Selected Online Activities[30]

Home broadband use has increased from roughly 1 hour per month in 1995, to more than 15 hours per month in 2000, to almost 29 hours per month today, as consumers find more valuable applications and content online.[31] Increased hours of use are correlated with increased actual speeds of broadband connections to the home.[32] As connection speeds have grown and more applications have been developed, the amount of data consumers download has increased. Today, the average Internet user with a fixed connection consumes 9 gigabytes of data per month over that connection. But that consumption varies significantly across user types, with some heavy users consuming upwards of 1,000 GB or more each month. Total data use per fixed residential connection is growing quickly, by roughly 30% annually.[33]

Almost two-thirds of the time users spend online is focused on communication, information searching, entertainment or social networking.[34] However, use patterns vary significantly. Except for high-definition video, most applications in use today can be supported by actual download speeds of about 1 Mbps (see Exhibit 3-C).

Broadband applications are helping businesses improve internal productivity and reach customers. Many businesses use at least basic applications: 97% of small businesses use e-mail; 74% have a company website.[35] There is evidence that broadband applications may improve individual companies' productivity.[36] Though gains vary drastically depending on the size and type of firm, as well as breadth of implementation, broadband-based applications may allow faster product development cycles, access to new geographic markets, and more efficient business processes and allocation of resources.

These productivity gains benefit the entire economy. Investment in information and communications technologies accounted for almost two-thirds of all economic growth attributed to capital investment in the United States between 1995 and 2005.[37]

Businesses also find it valuable to collect and aggregate information derived from use of broadband applications. More sophisticated digital profiles of Internet users allow businesses to better understand user buying patterns. This information is also useful for advertising or other purposes. Businesses are creating services tailored to individual consumers that improve their health, help them reduce their carbon footprint, track students' educational progress and target appeals for charitable, social and political causes.

Businesses often use broadband in ways that are fundamentally different from how consumers use it. For example, high-capacity broadband service is often used to connect PBX's for business voice and local area networks. These mission critical uses require broadband service with business-grade performance and customer support levels.

Both consumers and businesses are turning to applications and content that use video. Video is quickly becoming an important element of many applications, including desktop videoconference calls between family members and online training applications for businesses. Cisco forecasts that video consumption on fixed and mobile networks will grow at over 40% and 120% per year, respectively, through 2013.[38]

User-generated video and entertainment—from sites such as YouTube and Hulu—are a large portion of the total video traffic over broadband connections. Increasingly, video is embedded in traditional websites, such as news sites, and in applications such as teleconferencing. Skype reports that video calls account for over one-third of its total calls, and that number is growing rapidly.[39]

Video, television (TV) and broadband are converging in the home and on mobile handsets. The presence of broadband connections and TVs in the home could facilitate the development of a new medium for accessing the Web and watching video content. Traditional, or "linear," television still accounts for more than 90% of all time spent watching video.[40] Video consumed over the Internet still represents a small portion of overall video consumption at less than 2% of all time spent viewing.

Broadband-enabled video could grow as more innovative and user-friendly devices reach the home, allowing access to both traditional linear and Internet content via the TV.

Cloud computing—accessing applications from the Internet instead of on one's own computer—is also growing as more companies migrate to hosted solutions. Software based in the cloud may allow more small businesses and consumers to access applications that were once only available to large corporations with sophisticated information technology departments in the applications and content markets.

There are several issues that are important for the development of applications and content.

Illegal distribution of copyright-protected content over the Internet continues to be an issue. Although there have been promising results from technologies such as content fingerprinting and from industry-led initiatives to develop guidelines for dealing with illegal content, piracy is still present in the broadband ecosystem.[41]

Increased use of personal data raises material privacy and security concerns. Almost half of all consumers have concerns about online privacy and security, which may limit their adoption or use of broadband.[42] Better security and more control over private information may trigger a more robust applications market.

By making more of its information freely available, government can make it easier for companies to develop applications and content. The Global Positioning System (GPS) industry was born after the U.S. Department of Defense opened its fleet of GPS navigational satellites to the public and the National Oceanic and Atmospheric Administration made public its satellite data.[43] More recently, Sunlight Labs sponsored Apps for America, a competition to build useful applications with federal government data available on Data.gov. One application was FlyOnTime.us, which gives average flight delay information by airline and between U.S. cities.[44] Moving forward, government information can unleash additional new applications that help drive the growth of the broadband ecosystem.

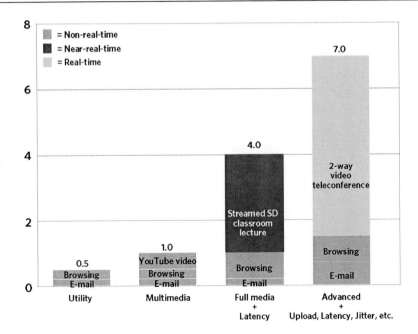

Exhibit 3-C. Actual Download Speeds Necessary to Run Concurrent Applications (Mbps)

3.2. Devices

Devices continue to grow in number and variety as more computers, phones and other machines connect to the Internet. New devices have repeatedly revolutionized the personal computer (PC) market in the past three decades. Today, about 80% of U.S. households have some sort of personal computer.[45] Although desktops initially dominated the market, 74% of all new personal computers sold today are laptops.[46] Many predict that, over the next 5 years, growth in the netbook and tablet markets will far outpace growth in the traditional PC market.[47]

The mobile phone market has also seen robust innovation. There were more than 850 different certified mobile products in the United States in 2009.[48] In that same year, approximately 172 million mobile phones were sold in the United States. Of these, 27% were Internet-capable smartphones manufactured by a wide variety of firms, including Apple, HTC, LG, Motorola, Nokia, Palm, RIM, Samsung and Sony-Ericsson. Analysts expect smart-phone sales to overtake standard mobile phone sales soon.[49]

Countless other Internet-capable devices come to the market each year. Companies are building smart appliances that notify owners of maintenance issues over broadband networks and communicate with the electric grid to run at off-peak hours when prices are lowest. E - book readers deliver books almost instantly to consumers anytime and anywhere, often at lower prices than traditional editions. Devices monitor patients at home and wirelessly transmit data to doctors' offices, so problems can be identified before they become too serious.

Devices already are starting to communicate with each other, keeping humans out of the loop. Increasing machineto-machine (M2M) interaction will occur over the network, particularly for mobile broadband. A pioneering example of machine-to-machine communication for consumer use is General Motors' OnStar, an M2M system for automobiles in which an onboard sensor automatically notifies OnStar's network if there is an accident or system failure.[50] M2M communications are used in many industries, often to collect information from sensors deployed remotely. For example, devices tracking the heart rate or blood-sugar level of patients with chronic conditions can transmit the information to a monitoring station that will trigger an alarm for a nurse or doctor where an abnormal pattern is detected. Networked sensors in a power plant can collect and transmit data on how genera-tors are operating, to allow analysis by sophisticated predictive methods that will diagnose potential faults and schedule preventive maintenance automatically.

The emergence and adoption of new technologies such as radiofrequency identification and networked micro-electromechanical sensors, among others, will give rise to the "Internet of Things." Billions of objects will be able to carry and exchange information with humans and with other objects, becoming more useful and versatile. For example, the Internet of Things will likely create whole new classes of devices that connect to broadband, and has the potential to generate fundamentally different requirements on the fixed and mobile networks: they will require more IP addresses, will create new traffic patterns possibly demanding changes in Internet routing algorithms, and potentially drive demand for more spectrum for wireless communications.

Significant competition and innovation exist for most classes of devices that interact with broadband networks. But one class of devices has not faced substantial competition in recent years: the television set-top box. The Telecommunications Act of 1996 contained provisions designed to stimulate competition and innovation in set-top boxes. Two years later, the FCC, in partnership with industry, developed the CableCARD standard to incent competition in the set-top box market.[51] Yet by 2008, two manufacturers shared 92% of the market, up from 87% in 2006.[52] Only 11 set-top boxes have been certified for retail sale, in contrast to the more than 850 unique handsets that were certified to operate on mobile networks in 2009 alone.[53] In addition, 97% of CableCARD -deployed set-top boxes installed between July 2007 and November 2009 were leased from operators rather than purchased at retail.[54]

Set-top boxes are an important part of the broadband ecosystem. An estimated 39 million set-top boxes were shipped in the United States in 2007 and 2008 combined.[55] The lack of innovation in set-top boxes limits what consumers can do and their choices to consume video, and the emergence of new uses and applications. It may also be inhibiting business models that could serve as a powerful driver of adoption and utilization of broadband, such as, models that integrate traditional television and the Internet.

3.3. Networks

Network service providers are an important part of the American economy. The 10 largest providers have combined annual revenue of more than $350 billion and annual capital investments in excess of $50 billion.[56] These investments have led to the deployment of multiple networks that today bring fixed and mobile broadband to end-users via the telephone, cable television, satellite and third-generation (3G) and fourth- generation (4G) mobile networks.

Terrestrial Fixed Broadband Availability

Today, 290 million Americans—95% of the U.S. population— live in housing units[57] with access to terrestrial, fixed broadband infrastructure capable of supporting actual download speeds of at least 4 Mbps.[58] Of those, more than 80% live in markets with more than one provider capable of offering actual download speeds of at least 4 Mbps.[59] Meanwhile, 14 million people in the United States living in 7 million housing units do not have access to terrestrial broadband infrastructure capable of this speed.[60] Although housing units without access to terrestrial broadband capable of 4 Mbps download speeds exist throughout the country, they are more common in rural areas (see Exhibit 3-D).[61]

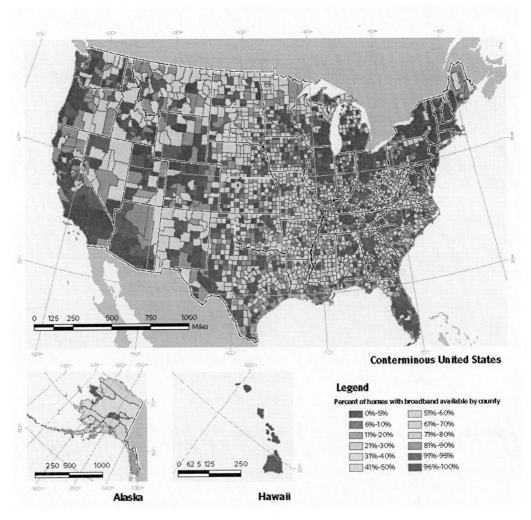

Exhibit 3-D. Availability of 4 Mbps-Capable Broadband Networks in the United States by County[63]

Exhibit 3-E. Announced Upgrades to the U.S. Fixed Broadband Network (Millions of Households Covered)[78]

	Companies	2009	2010	2011
FTTP	• Verizon • Cincinnati Bell • Tier 3 ILECs	• All providers (17.2 million–Sept) • Verizon FiOS (14.5 million–June)	• Verizon FiOS (17 million)	
FTTN	• AT&T • Qwest	• Qwest (3 million)	• Qwest (5 million)	• AT&T U-verse (30 million)
DOC-SIS 3.0	• Comcast • Cablevision • Cox • Knology • Time Warner • Charter • Mediacom • RCN	• Comcast (40 million) • Charter (St. Louis) • Mediacom (50% of footprint) • Knology (50% of footprint) • RCN (begin deployment)	• Comcast (50 million) • Cablevision (entire footprint) • Cox (entire footprint) • Time Warner (New York City) • Knology (entire footprint)	

Businesses and community anchor institutions are often served by broadband. Ninety-six percent of all business locations have access to Digital Subscriber Line (DSL) service, and 92% have access to cable broadband service.[62] In addition, 99% of all health care locations with physicians have access to actual download speed of at least 4 Mbps (see Exhibit 3-D). Finally, 97% of schools are connected to the Internet,[64] many supported by the federal E -rate connectivity programs. But crucial gaps exist: More than 50% of teachers say slow or unreliable Internet access presents obstacles to their use of technology in classrooms,[65] and only 71% of rural health clinics have access to mass-market broadband solutions.[66] Further, many business locations, schools and hospitals often have connectivity requirements that cannot be met by mass-market DSL, cable modems, satellite or wireless offers, and must buy dedicated high-capacity circuits such as T-1 or Gigabit Ethernet service.

The availability and price of such circuits vary greatly across different geographies, and many businesses and anchor institutions face challenges acquiring the connectivity to support their needs.

Typical advertised broadband speeds that consumers purchase have grown approximately 20% each year. This growth has been driven by a shift in consumer preferences to faster, more advanced technologies, improved performance of different technologies and large investments by service providers in network upgrades.[67]

Both telephone and cable companies continue to upgrade their networks to offer higher speeds and greater capacities. Many have announced specific upgrades. For example, Verizon plans to pass over 17 million homes by the end of 2010 with its FiOS fiber-to-the-premises (FTTP) service, three million more than today.[68] AT&T has announced it will build fiber-to-thenode (FTTN) infrastructure to serve 30 million homes by 2011, 11 million more than today. In addition, many smaller companies plan to aggressively build FTTP networks. If the targets in these public announcements are met, at least 50 million homes will be able to receive peak download speeds of 18 Mbps or more from their telephone company within the next 2 years.[69]

Cable companies have also announced that over the next 2–3 years they will upgrade their networks to DOCSIS 3.0 technology, which is capable of maximum download speeds of more than 50 Mbps. One analyst predicts that by 2013, leading cable companies will cover 100% of the homes they pass with DOCSIS 3.0. The top five cable companies currently pass 103 million housing units, or about 80% of the country's homes.[70]

As noted in a recent report from the Columbia Institute for Tele-Information (CITI), history suggests that service providers will meet these announced targets. So it is likely that 90% of the country will have access to advertised peak download speeds of more than 50 Mbps by 2013.[71] The affordability and actual performance of these networks will depend on many factors such as usage patterns, investment in infrastructure, and service take-up rates.

However, these major announced buildouts target areas already served by broadband. It is unlikely there will be a significant change in the number of unserved Americans based on planned upgrades over the next few years, although some small companies may upgrade their networks to support broadband in currently unserved areas.

The performance of fixed broadband connections is often advertised in terms of maximum "up to" download and upload speeds. For example, an end-user with a connection for which download speeds are "up to 8 Mbps" can expect to reach 8 Mbps download speeds, but not necessarily reach and sustain that speed all or even most of the time. Data show that actual speeds experienced by end-users differ considerably from the "up to" speeds advertised by service providers. This distinction is important because it is the actual experience of the consumer (not theoretical technical capabilities) that enables or limits the use of different applications by end-users.

Estimates of the average advertised "up to" download speed that Americans currently purchase range from 6.7 Mbps to 9.6 Mbps,[72] with the most detailed data showing an average of approximately 8 Mbps and a median of approximately 7 Mbps.[73] As noted, the average advertised speed purchased by broadband users has grown approximately 20% each year for the last decade. Upload speeds are significantly lower, as the advertised "up to" upload speed typically is closer to 1.0 Mbps.[74]

However, the actual experienced speeds for both downloads and uploads are materially lower than the advertised speeds. Data indicates the average *actual* download speed in American households for broadband is 4 Mbps (median *actual* is 3.1 Mbps) (see Exhibit 3-G).[75] Therefore, the actual download speed experienced on broadband connections in American households is approximately 40–50% of the advertised "up to" speed to which they subscribe. The same data suggest that for upload speeds, actual performance is approximately 45% of the "up to" advertised speed (closer to 0.5 Mbps).

Actual download speeds vary by technology as well.[77] While median actual download speeds for fiber and cable are 5–6 Mbps, median actual download speeds for DSL are 1.5–2 Mbps, and under 1 Mbps for satellite (see Exhibit 3-F). Despite this variation in performance across technologies, on a percentage basis, the gap between advertised and actual speeds experienced by consumers is consistent and prevalent across all types of connection technologies.[79]

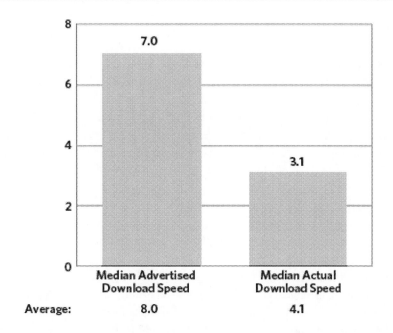

Exhibit 3-G. Advertised Versus Actual U.S. Fixed Broadband Residential Download Speeds (Mbps)

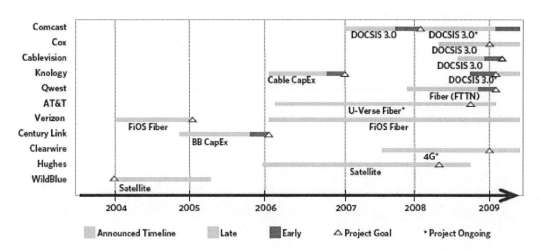

Exhibit 3-F. Timeline of Fixed Broadband Industry Network Upgrades[76]

This performance gap between advertised "up to" speeds and actual performance is consistent with reports published in a number of other countries. A study in the United Kingdom found that average actual speeds were typically about 57% of average advertised speeds.[80] Studies in New Zealand, Australia, Italy and Ireland have shown similar results.[81]

Mobile Broadband Availability

As of November 2009, according to data from American Roamer, 3G service covers roughly 60% of U.S. land mass.[82] In addition, approximately 77% of the U.S. population lived in an area served by three or more 3G service providers, 12% lived in an area served by two, and 9% lived in an area served by one. About 2% lived in an area with no provider.[83]

These measures likely overstate the coverage actually experienced by consumers, since American Roamer reports *advertised* coverage as reported by many carriers who all use different definitions of coverage. In addition, these measures do not take into account other factors such as signal strength, bitrate or in-building coverage, and may convey a false sense of consistency across geographic areas and service providers.[84] As with fixed broadband, most areas without mobile broadband coverage are in rural or remote areas. In fact, 3G build out is significantly lower in several states—in West Virginia, only 71% of the population has 3G coverage and in Alaska only 77% have coverage.[85]

Additionally, American Roamer also suggests that 98% of businesses have 3G coverage today, although the data have similar limitations regarding signal strength, bitrate and in-building coverage.[86] While most businesses have wireless broadband coverage,[87] nearly 9% of rural business sites still do not have access, compared to less than 1% of business sites in urban or suburban areas.[88] Finally, while a business location may have coverage, the value in mobile broadband comes when employees can access applications everywhere, which limits the importance of this particular coverage metric.

Several operators have announced upgrades to 4G broadband networks. CITI notes that by 2013, Verizon Wireless plans to roll out Long Term Evolution (LTE)—a 4G mobile broadband technology—to its entire footprint, which currently covers more than 285 million people.[89] AT&T has announced it will test LTE in 2010 and begin rollout in 2011. Through its partnership with Clearwire, Sprint plans to use WiMAX as its 4G technology. WiMAX has been rolled out in a few markets already, and Clearwire plans to cover 120 million people with WiMAX by the end of 2010.[90]

Mobile broadband network availability will change rapidly because of these deployments. Improved spectral efficiencies and significantly lower network latencies are some of the features of 4G networks that could lead to a better mobile broadband experience. For example, the spectral efficiency of mobile broadband networks could improve by over 50% with a transition from early 3G networks to 4G, while improvements relative to state-of-the-art 3G networks are likely to be a more modest 10–30%.[91] The extent to which the effect of these advances are reflected in users' experiences will depend on a variety of factors, including the total amount of spectrum dedicated to mobile broadband and the availability of high-speed backhaul connections from cellular sites.[92]

Evaluating network availability and performance is much harder for mobile than for fixed broadband. For instance, the quality of the signal depends on how far the user is from the cell tower, and how many users are using the network at the same time. Therefore, the fact that users are in the coverage area of a 3G network does not mean they will get broadband-quality performance. Still, as with fixed broadband, it is clear that the speeds experienced on mobile broadband networks are generally less than advertised. Actual average download speeds have been reported to be as low as 245 kbps, while speeds in excess of 600 kbps are advertised. Actual average upload speeds as low as 106 kbps have been reported, versus advertised rates of 220 kbps or higher.[93]

Both mobile network performance and the availability of mobile broadband rely on the availability of spectrum. Carriers and other broadband-related companies agree that more spectrum will be needed to maintain robust, high-performing wireless broadband networks in the near future.[94]

Exhibit 3-H. Announced Upgrades to the U.S. Mobile Broadband Network (Persons Covered)[95]

Technology	Companies	2009	2010	2011	By 2013
LTE	• Verizon • AT&T • MetroPCS • Cox		• Verizon (100 million) • AT&T (trials)	• AT&T (start deployment) • Cox (start deployment) • MetroPCS (start deployment)	• Verizon (entire network)
WiMAX	• Clearwire • Open Range • Small wireless Internet service providers (WISPs)	• Clearwire (30 million) • WISPs (2 million)	• Clearwire (120 million)		• Open Range (6 million)

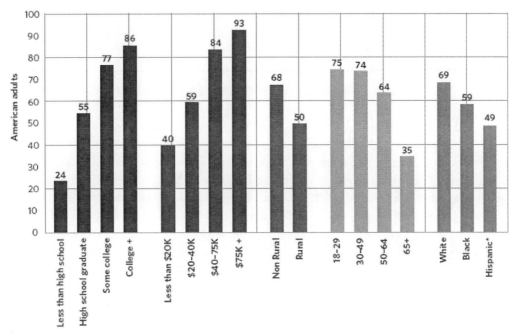

*Hispanics includes both English and Spanish-speaking Hispanics

Exhibit 3-I. Broadband Adoption by American Adults by Socio-Economic and Demographic Factors

3.4. Adoption and Utilization

Nearly two-thirds of American adults have adopted broadband at home. While adoption likely will continue to increase, different demographic groups adopt at significantly different rates (see Exhibit 3-I). For example, only 40% of adults making less than $20,000 per year have adopted terrestrial broadband at home, while 93% of adults earning more than $75,000 per year have adopted broadband at home (see Exhibit 3-H). Only 24% of those with less than a high school degree, 35% of those older than 65, 59% of African Americans and 49% of Hispanics have adopted broadband at home.[96] Among people with disabilities, who face distinctive barriers to using broadband, only 42% have adopted.[97] Those living on Tribal lands have very low adoption rates, mainly due to a lack of available infrastructure.

What little data exist on broadband deployment in Tribal lands suggest that fewer than 10% of residents on Tribal lands have terrestrial broadband available.[98]

While it is important to respect the choices of those who prefer not to be connected, the different levels of adoption across demographic groups suggest that other factors influence the decision not to adopt. Hardware and service are too expensive for some. Others lack the skills to use broadband.

Broadband adoption among businesses, by contrast, is quite strong: Ninety-five percent of America's small and medium- sized businesses have adopted broadband.[99] Only 10% of small businesses are planning to upgrade to a faster Internet connection in the next 12 months.[100]

Subsequent chapters address adoption as well as the other elements of the broadband ecosystem that can help ensure America captures the full promise of broadband.

PART I— INNOVATION AND INVESTMENT

Broadband is changing many aspects of life—increasing business productivity, improving health care and education, enabling a smarter and more efficient power grid and creating more opportunities for citizens to participate in the democratic process. It is also fueling large global markets for high-value-added goods and services and creating high-paying jobs in important sectors such as information and communications technology (ICT).

The U.S. must lead the world in broadband innovation and investment and take all appropriate steps to ensure all Americans have access to modern, high-performance broadband and the benefits it enables. Broadband has been a main driver of growth and innovation in the ICT industry, generating demand for semiconductors, consumer and enterprise software, computers, devices, applications, networking equipment and many different types of services. A world-class broadband ecosystem will help ensure that America's ICT sector continues to lead the world—creating jobs, tapping American ingenuity and allowing American consumers to receive the substantial benefits that flow from the evolution of ICT.

BOX I-1. BROADBAND AS A TRANSFORMATIVE GENERAL PURPOSE TECHNOLOGY

Technological progress drives long-term economic growth.[101] As economists Timothy Bresnahan and Manuel Trajtenberg explained in a 1995 paper, "Whole eras of technical progress and economic growth appear to be driven by a few key technologies, which we call General Purpose Technologies (GPTs). The steam engine and the electric motor may have played such a role in the past, whereas semiconductors and computers may be doing as much in our era. GPTs are characterized by pervasiveness (they are used as inputs by many downstream sectors), inherent potential for technical improvements, and innovational complementarities, meaning that the productivity of R&D in downstream sectors increases as a consequence of innovation in the GPT. Thus, as GPTs improve they spread throughout the economy, bringing about generalized productivity gains."[102] The report continued, "As use of the GPT grows, its effects become significant at the aggregate level, thus affecting overall growth."[103]

The Internet has the characteristics of a GPT.[104] Businesses of all kinds and sizes use it to improve their processes, from procurement to supply chain management, market research to sales and asset management to customer support. It has driven performance improvements; for example, the average U.S. broadband connection speed has grown more than 20% per year for the last several years. These improvements are driving technology and business innovation in several other sectors, including health care,[105] education,[106] energy,[107] online commerce[108] and the government.[109]

Today's broadband ecosystem is vibrant and healthy in many ways. In numerous communities, consumer demand is strong. Service providers are investing in upgrades of fixed and mobile networks. New devices, and even new device categories—such as e-book readers, tablets and netbooks—are being created. New applications keep emerging, and more and more content is available online. However, there are some areas where America can and should do better. Government policies and actions can foster innovation and investment across the ecosystem in four key areas:

- *Enacting policies to foster competition.* Competition is a major driver of innovation and investment, and the Federal Communications Commission (FCC) and other agencies have many tools to influence competition in different areas of the broadband ecosystem. These tools are best applied on a fact-driven, case-by-case basis. Therefore, continuous collection and analysis of detailed data on competitive behavior must be the linchpin of effective competition policy. This plan establishes a process for such collection and, in addition, proposes several specific actions that will foster competition.

- *Freeing up more spectrum.* The federal government controls and influences the availability and cost of spectrum. Spectrum plays an important role in the economics of broadband networks. By ensuring spectrum is allocated and managed as efficiently as possible, the government can help reduce the costs borne by firms deploying network infrastructure, thus encouraging both competitive entry and increased investment by incumbent firms. The plan highlights actions that the FCC, the

National Telecommunications and Information Administration and Congress can take to enable more productive uses of spectrum and make more spectrum available for broadband.

- *Lowering infrastructure costs.* Government also controls and influences the availability and cost of other resources, such as pole attachments and rights-of-way. As with spectrum, ensuring these assets are allocated and managed as efficiently as possible can reduce the costs borne by firms and foster competition and investment. The plan outlines infrastructure policies that lower the cost of network deployment.
- *Investing directly through research and development.* Government should invest directly in areas where the return on investment to society as a whole is greater than the return for individual firms. Research and development (R&D) is one of these areas, as the effects of R&D often extend beyond those anticipated by its funders in unanticipated ways.[110] The plan contains specific recommendations for the creation of a broadband R&D agenda, including development of ultra-high-speed testbeds to drive new innovations in broadband and applications.

Since the Telecommunications Act of 1996, U.S. policy has embraced competition as the best means to bring the fruits of investment and innovation—including lower prices, new services and features, higher service quality and choice—to the American people. This plan follows in that tradition. The four chapters that comprise Part I of the National Broadband Plan contain more than 40 recommendations that directly spur competition. But the plan as a whole helps to promote competition in other areas. A small sampling of the pro-competition, pro-consumer initiatives outside of Part I include:

- Enable competition in digital educational content by setting standards for content created by the federal government and proposing sharing of procurement information among local education agencies (see Chapter 11).
- Ensure greater competition and innovation in broadband-enabled Smart Grid information services and related devices by providing secure access to digital electric information for consumers and authorized third parties (see Chapter 12).
- Ensure first responders reap the benefits of competition in choosing handsets and wireless broadband technology, allowing them to take advantage of advances in the commercial wireless ecosystem (see Chapter 16).

Part I of the plan (Innovation and Investment) begins with Chapter 4, which contains recommendations to drive innovation through competition in networks, devices and applications. Chapters 5 and 6 contain recommendations to lower the cost of inputs such as spectrum and infrastructure and to maximize private sector investment and competitive entry. Chapter 7 proposes a process to create an agenda for government-sponsored R&D to support broadband.

4. BROADBAND COMPETITION AND INNOVATION POLICY

Twenty-five years ago, the world wide web did not exist. Very few Americans had even seen a mobile phone, and broadband networks were available only to a few businesses and research institutions.

Today, innovations such as broadband and others like it drive the creation of a wide variety of products and services. The competitive forces that sparked these breakthroughs need to be nurtured, so that the United States can continue to reap the benefits of its unrivaled culture of innovation.

This chapter examines innovation and competition in the broadband ecosystem. First, it discusses each of the three elements of the broadband ecosystem—networks, devices and applications. Then it addresses competition for value across the ecosystem, the transition from a circuit-switched network to an all-Internet Protocol (IP) network and the leveraging of the benefits of innovation and investment internationally.

Section 4.1 approaches network competition in three ways. First, it addresses the state of competition in residential broadband and makes recommendations to bolster consumer benefits by developing data-driven competition policies for broadband services. Second, it makes recommendations intended to ensure that consumers have the information they need to make decisions that maximize benefits from these services. Increased transparency will likely drive service providers to deliver better value to consumers through better services. Third, it focuses on competition in the wholesale broadband market—including issues associated with high-capacity circuits, copper retirement, interconnection and data roaming. All are crucial for enabling competition in the small business and enterprise customer segments, in mobile services and in deployment of services in high-cost areas.

Section 4.2 addresses devices, with a particular focus on set-top boxes. Of the three main categories of broadband devices—mobile devices, personal computing devices and set-top boxes—set-top boxes is the category with the least competition: two manufacturers control more than 90% of the U.S. market and have controlled comparable market shares for many years. Congress recognized the need for change in the set-top box market when it enacted Section 629 of the Telecommunications Act, but the FCC's attempts to meet Congress's objectives have been unsuccessful. As video becomes an increasingly important element of broadband applications, driving usage and adoption, it is crucial that the FCC takes steps that will foster increased innovation in set-top boxes and video navigation devices to bring more competition and choice for consumers.

Section 4.3 addresses applications, focusing on the management of personal data and privacy. The number and variety of applications and content available over broadband connections has exploded over the last few years. Competition within different types of applications and content services must be looked at on a case-by-case basis. However, the importance of digital personal data is a common thread among current and emerging content and application services. Personal data, often aggregated into "digital profiles," are often used to provide consumers with personalized services and to target them with more relevant advertising. These increasingly detailed digital profiles offer both an opportunity and a challenge. The opportunity is to increase the innovations and convenience provided to end-users, who may enjoy better targeted, more customized services and applications, many of them free of charge. The challenge is to enable consumers to take advantage of such innovations while ensuring that they can retain control of their personal data, protect their privacy and manage how the information collected on them is used.

Recommendations

Networks

- The federal government, including the FCC, the National Telecommunications and Information Administration (NTIA) and Congress, should make more spectrum available for existing and new wireless broadband providers in order to foster additional wireless-wireline competition at higher speed tiers.
- The FCC and the U.S. Bureau of Labor Statistics (BLS) should collect more detailed and accurate data on actual availability, penetration, prices, churn and bundles offered by broadband service providers to consumers and businesses, and should publish analyses of these data.
- The FCC, in coordination with the National Institute of Standards and Technology (NIST), should establish technical broadband performance measurement standards and methodology and a process for updating them. The FCC should also encourage the formation of a partnership of industry and consumer groups to provide input on these standards and this methodology.
- The FCC should continue its efforts to measure and publish data on actual performance of fixed broadband services. The FCC should publish a formal report and make the data available online.
- The FCC should initiate a rulemaking proceeding by issuing a Notice of Proposed Rulemaking (NPRM) to determine performance disclosure requirements for broadband.
- The FCC should develop broadband performance standards for mobile services, multi-unit buildings and small business users.
- The FCC should comprehensively review its wholesale competition regulations to develop a coherent and effective framework and take expedited action based on that framework to ensure widespread availability of inputs for broadband services provided to small businesses, mobile providers and enterprise customers.
- The FCC should ensure that special access rates, terms and conditions are just and reasonable.
- The FCC should ensure appropriate balance in its copper retirement policies.
- The FCC should clarify interconnection rights and obligations and encourage the shift to IP-to-IP interconnection where efficient.
- The FCC should move forward promptly in the open proceeding on data roaming.

Devices

- The FCC should initiate a proceeding to ensure that all multi-channel video programming distributors (MVPDs) install a gateway device or equivalent functionality in all new subscriber homes and in all homes requiring replacement set-top boxes, starting on or before Dec. 31, 2012.

- On an expedited basis, the FCC should adopt rules for cable operators to fix certain CableCARD issues while development of the gateway device functionality progresses. Adoption of these rules should be completed in the fall of 2010.

Applications

- Congress, the Federal Trade Commission (FTC) and the FCC should consider clarifying the relationship between users and their online profiles.
- Congress should consider helping spur development of trusted "identity providers" to assist consumers in managing their data in a manner that maximizes the privacy and security of the information.
- The FCC and FTC should jointly develop principles to require that customers provide informed consent before broadband service providers share certain types of information with third parties.
- The federal government, led by the FTC, should put additional resources into combating identity theft and fraud and help consumers access and utilize those resources, including bolstering existing solutions such as OnGuard Online.
- FCC consumer online security efforts should support broader national online security policy, and should be coordinated with the Department of Homeland Security (DHS), the FTC, the White House Cyber Office and other agencies. Federal agencies should connect their existing websites to OnGuard Online to provide clear consumer online security information and direction.
- The federal government should create an interagency working group to coordinate child online safety and literacy work, facilitate information sharing, ensure consistent messaging and outreach and evaluate the effectiveness of governmental efforts. The working group should consider launching a national education and outreach campaign involving governments, schools and caregivers.
- The federal government should investigate establishing a national framework for digital goods and services taxation.

4.1. Networks

Competition in Residential Broadband Markets

Competition is crucial for promoting consumer welfare and spurring innovation and investment in broadband access networks. Competition provides consumers the benefits of choice, better service and lower prices. This section begins by analyzing the available data to assess the current state of competition among wireline broadband services and mobile wireless broadband services, and the competitive dynamics across different broadband technologies. It does not analyze the market power of specific companies or reach definitive conclusions about the current state of competition for residential broadband services. The section then discusses how new technologies and network upgrades present both opportunities and challenges to competition in the near future. It concludes with several recommendations to promote competition and to improve the data the government collects to assess the state of competition in broadband markets in the future.

Competition in industries with high fixed costs

Building broadband networks—especially wireline—requires large fixed and sunk investments. Consequently, the industry will probably always have a relatively small number of facilities-based competitors, at least for wireline service. Bringing down the cost of entry for facilities-based wireline services may encourage new competitors to enter in a few areas, but it is unlikely to create several new facilities-based entrants competing across broad geographic areas.[111] Bringing down the costs of entry and expansion in wireless broadband by facilitating access to spectrum, sites and high-capacity backhaul may spur additional facilities-based competition. Whether wireless competition is sustainable in driving innovation, investment and consumer welfare will depend on the evolution of technology and consumer behavior among many other factors.

The lack of a large number of wireline, facilities-based providers does not necessarily mean competition among broadband providers is inadequate. While older economic models of competition emphasized the danger of tacit collusion with a small number of rivals, economists today recognize that coordination is possible but not inevitable under such circumstances. Moreover, modern analyses find that markets with a small number of participants can perform competitively;[112] however, those analyses do not tell us what degree of competition to expect in a market with a small number of wireline broadband providers combined with imperfect competition from wireless providers.[113] In addition, as the Department of Justice (DOJ) describes the issue, the critical question is not "some abstract notion of whether or not broadband markets are 'competitive'" but rather "whether there are policy levers [around competition policy] that can be used to produce superior outcomes."[114] Given that approximately 96% of the population has at most two wireline providers, there are reasons to be concerned about wireline broadband competition in the United States. Whether sufficient competition exists is unclear and, even if such competition presently exists, it is surely fragile. To ensure that the right policies are put in place so that the broadband ecosystem benefits from meaningful competition as it evolves, it is important to have an ongoing, data-driven evaluation of the state of competition.

New data from the FCC's Form 477 combined with several other sources make possible certain general observations about the state of competition in broadband services today, though additional data are needed to more rigorously evaluate broadband competition.[115, 116]

In general, broadband subscribers appear to have benefited from the presence of multiple providers. Broadband providers have invested in network upgrades to deliver faster broadband speeds and enter new product markets—cable companies providing telephony and telephone companies offering multichannel video—but the data available only provide limited evidence of price competition among providers.

Fixed broadband service

Unlike many countries, the majority of U.S. broadband subscribers do not connect to the Internet via local-access infrastructure owned by an incumbent telephone company. The U.S. cable infrastructure was advanced and ubiquitous enough to allow cable companies to offer broadband access services to large portions of the country, in many cases before the telephone companies. As a result, the U.S. market structure is relatively unique in that people in most parts of the country have been able to choose from two wireline, facilities-based broadband platforms for many years. Approximately 4% of housing units are in areas with three wireline providers (either DSL or fiber, the cable incumbent and a cable over-builder), 78% are in areas with two wireline providers, about 13% are in areas with a single wireline provider and 5% have no wireline provider (see Exhibit 4-A).

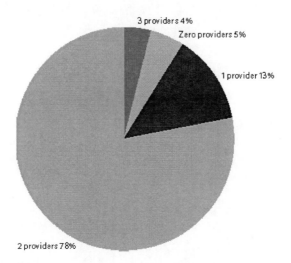

Exhibit 4-A. Share of Housing Units in Census Tracts with 0, 1, 2, and 3 Wireline Providers[117]

These data do not necessarily mean that 82% (78% + 4%) of housing units have two or three competitive options for wireline broadband service—the data used here do not provide adequate information on price and performance to determine if multiple providers present in a given area compete head-to-head.

Additionally, the data show that rural areas are less likely to have access to more than one wireline broadband provider than other areas. The data also show that low-income areas are on average somewhat less likely to have more than one provider than higher-income areas.

There are other types of fixed broadband providers. For instance, satellite-based broadband service is available in most areas of the country from two providers, while hundreds of small fixed wireless Internet service providers (WISPs) offer service to more than 2 million people[118] and Clearwire offers WiMAX service in a number of cities. [119] These providers compete for customers as well, although their services tend to be either more expensive or offer a lower range of speeds than today's wireline offerings. [120]

The presence of a facilities-based competitor impacts investment. Indeed, broadband providers appear to invest more heavily in network upgrades in areas where they face competition. Exhibit 4-B shows that controlling for housing density, household income and state-specific factors that affect supply and demand, providers of broadband over any given wireline technology—Digital Subscriber Line (DSL), cable or fiber—generally offer faster speeds when competing with other wireline platforms. So, for example, available cable speeds are higher in areas in which cable competes with DSL or fiber than in areas where cable is the only option. DSL and fiber show similar results. Available speeds are even higher where three wireline providers compete (e.g., where a cable over-builder is also present). [121]

Indeed, competition appears to have induced broadband providers to invest in network upgrades.[123] Cable and telephone companies invested about $48 billion in capital expenditures (capex) in 2008 and about $40 billion in 2009. While it is very difficult to accurately disaggregate service provider capital expenditures into broadband and other areas, a review of analyst reports at Columbia Institute for Tele-Information (CITI) suggests that of this total, wireline broadband capital expenditures were about $20 billion in 2008 and expected to be about $18 billion in 2009.[124] Companies channeled these investments into network upgrades in recent years, as detailed in Exhibit 4-C. [125]

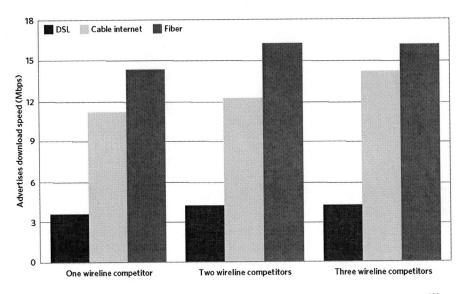

Exhibit 4-B. Average Top Advertised Speed in Areas with 1, 2 and 3 Wireline Competitors[122]

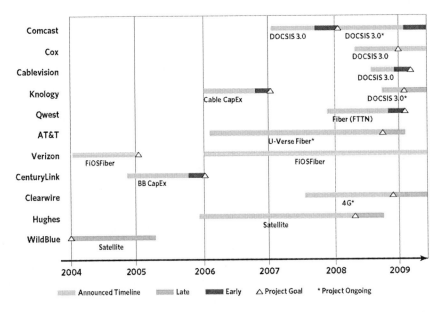

Exhibit 4-C. Select Fixed Broadband Infrastructure Upgrades[126]

Consumers are benefiting from these investments. Top advertised speeds available from broadband providers have increased in the past few years. Additionally, typical advertised download speeds to which consumers subscribe have grown at approximately 20% annually for the last 10 years.[127]

New choices—at new, higher speeds—are becoming available, as well. Clearwire offers download speeds of up to 2 Mbps service in several cities and plans to have its WiMAX service available to about 120 million people by 2011. [128] Two satellite providers plan to launch new satellites in 2011 and 2012, with ViaSat (WildBlue) expecting to advertise download speeds of up to 2–10 Mbps and Hughes Communications planning to advertise download speeds of up to 5–25 Mbps. [129]

In principle, providers can compete on price as well as on service. Unfortunately, the dearth of consistent, comprehensive and detailed price data makes it difficult to evaluate price competition. The data that do exist are imperfect. First, some focus on the price of broadband when not bundled with any other services even though the vast majority of consumers purchase broadband bundled with voice, video or both. [130] Second, sources that have data on bundles do not provide sufficient information to determine the incremental price of the broadband component. Third, broadband providers frequently offer promotions to attract new customers. No data source consistently captures the relevant details of those promotions, including details such as how long the promotional price lasts, the length of the contract the consumer signs to get the promotional price, the price once the promotion expires and any early termination fee. Some international comparisons suggest the number of retail broadband providers may be positively correlated with advertised download speeds, at least at the high end of the market, and with affordability.[131] Others rank the United States high in affordability of broadband, despite the fact that 96% of consumers have two or fewer choices, and suggest that consumers may not be willing to pay as much for high speeds as they are for other functionality. [132]

Nevertheless, the available data can be analyzed to see if they yield consistent results. Merging comprehensive cross- sectional data on prices[133] with Form 477 data makes possible econometric analyses of the effects of competition on prices, controlling for income, density and region-specific factors. These analyses yield some weak evidence that monthly prices are lower when more wireline providers are in a census tract, but the data limitations discussed above make it difficult to draw robust conclusions.

A fundamental question related to competition is how prices paid by consumers evolve as underlying costs change. While the data do not allow us to examine competition in detail, it is possible to examine certain aspects of prices over time. In particular, Greenstein and McDevitt (2010) analyzed about 1,500 broadband contracts[134] to construct price indices (see Exhibit 4-D).[135] The exhibit shows that the price index for standalone nominal prices, adjusted for upload and download speeds, changed modestly between 2006 and 2009 while the index for bundled prices remained relatively constant. [136]

Other data reach similar conclusions. The Internet service provider (ISP) price index compiled by BLS shows a slight increase in Internet service prices between 2007 and 2009.[138] The available time-series data, therefore, show, at best, a small decline in quality-adjusted nominal broadband prices while the econometrics reveal weak evidence that providers compete on prices. One clear conclusion from the analysis, however, is that better data for analyzing price competition would be helpful.

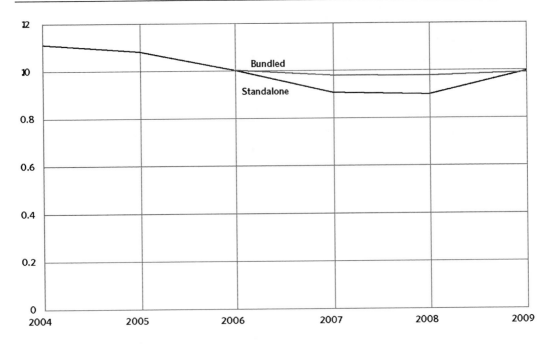

Exhibit 4-D. Price Indices for Broadband Advertised as a Standalone Service and as Part of a Bundle (2006 = 1) [137]

Mobile broadband competition[139]

As discussed in Chapter 3, as of November 2009, according to data from American Roamer, third-generation (3G) wireless service covers roughly 60% of U.S. landmass. [140] In addition, approximately 77% of the U.S. population lived in an area served by three or more 3G service providers, 12% lived in an area served by two, and 9% lived in an area served by one. About 2% lived in an area with no provider (see Exhibit 4-E). [141]

These measures likely overstate the coverage actually experienced by consumers, since American Roamer reports *advertised* coverage as reported by many carriers who all use different definitions of coverage. In addition, these measures do not take into account other factors such as signal strength, bitrate or in-building coverage, and they may convey a false sense of consistency across geographic areas and service providers. [142] As with fixed broadband, most areas without mobile broadband coverage are in rural or remote areas. Nonetheless, the data can help benchmark mobile broadband availability nationwide. In total, while United States service providers are building out mobile broadband coverage, the U.S. is far from having "complete" coverage.

Mobile data users typically receive download speeds ranging from hundreds of kilobits per second to about one megabit per second. [144] Several competing firms offer mobile broadband. In addition to the nationwide service providers AT&T, Verizon, Sprint and T-Mobile (two of which are also leading providers of wireline broadband), new competitors such as Leap Wireless and MetroPCS have emerged in metropolitan areas in recent years. Like wireline broadband providers, these firms may compete along many dimensions including coverage, device selection, roaming and services. Many service providers have focused on network upgrades to 3G services. [145]

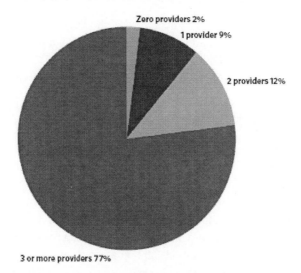

Exhibit 4-E. Share of Population Living in Census Tracts with 0, 1, 2, 3 or More 3G Mobile Providers[143]

As mentioned earlier, identifying broadband-specific capital expenditures is very difficult, but the CITI report indicates that total capital expenditures by major wireless firms were about $21 billion in 2008, of which about $10 billion was for broadband. In 2009 wireless companies were expected to have incurred about $20 billion in capital expenditures, $12 billion of which was for broadband services.[146] While projections should be viewed cautiously, wireless broadband capital expenditures are expected to be about $12 billion in 2010 and increase steadily to $15 billion in 2015 as service providers roll out their 4G services.[147] Mobile broadband services are relatively new and their competitive dynamics are changing rapidly. As new technologies such as High Speed Packet Access (HSPA), WiMAX and Long Term Evolution (LTE) are introduced and rolled out by different carriers, new devices support different uses and consumers turn to different applications.

Wireline-wireless competition

Whether wireless broadband, either fixed or mobile, can compete with wireline broadband is an important question in evaluating the status of broadband services competition. The answer depends on how technology, costs and consumer preferences evolve, as well as on the strategic choices of firms that control wireline and wireless assets,[148] including firms that offer both fixed and mobile broadband.

Consumers' preferences differ depending on how they use their broadband connections and how much they are willing to pay for such use. Some value download speeds more than any other attribute, some value mobility and new converts from dial-up may still even value the simple "always on" connection. A user who values little more than e-mail and browsing news sites has, in principle, many choices—nearly any broadband access technology will do. But a user who streams high-definition video and enjoys gaming probably requires high download and upload speeds and low latency. That user will likely have few choices.

Most consumers' preferences are not so extreme—they tend to value some factors more than others. If a sufficiently large segment of consumers are relatively indifferent about the attributes, performance and pricing of mobile and fixed platforms, then mobile and fixed providers are likely to compete for consumers. Today, however, most consumers who do not value mobility when purchasing broadband, or want high download or upload speeds, face only two choices for their fixed broadband service.[149]

It is not yet clear how that might change. The spectral efficiency of wireless technologies has increased by a factor of roughly 40 or more since the early days of second-generation (2G) wireless (see Exhibit 4-F).[150] These technologies—often deployed for mobile services—can deliver even higher download speeds by replacing mobile devices with fixed terminals. Indeed, terrestrial, fixed wireless access solutions have already been deployed as a substitute for wired access technologies; for example, in the United States by Clearwire with WiMAX and Stelera with HSPA.

Wireless broadband may not be an effective substitute in the foreseeable future for consumers seeking high-speed connections at prices competitive with wireline offers.[152] Given enough spectrum, however, a variety of engineering techniques—including higher transmitter power, high-gain directional antennas and multiple externally mounted antennae—may make wireless a viable price/performance competitor to wired solutions at far higher speeds than are possible today, further increasing consumer choice.

The ongoing upgrade of the wireless infrastructure is promising because of its potential to be a closer competitor to wireline broadband, especially at lower speeds. For example, if wireless providers begin to advertise, say, 4 Mbps home broadband service, wireline providers may be forced to respond by lowering prices of their broadband offerings. This could be true even if wireless services are more expensive, especially if the service is also mobile. Such an outcome is a possibility—for instance, according to CITI,LTE could offer speeds between 4 and 12 Mbps, with sustained speeds of up to 5 Mbps. Further, as with most goods, consumers choose broadband by trading off price and features. Providers offering a product with fewer features may have to reduce prices in order to remain competitive, even if the superior product charges more. Consider, for example, computer monitors. LCD flat-screen monitors were introduced at prices many multiples higher than older and once-standard CRTs. Even though the typical LCD did not offer as clear a picture as the typical CRT, its advantages in terms of weight, the space it took up on a desk, and its rapid technological improvements were such that it quickly put downward price pressure on the already much cheaper CRT.[153]

There is no guarantee, however, that competition will necessarily evolve this way. Technologies, costs and consumer preferences are changing too quickly in this dynamic part of the economy to make accurate predictions. Regardless of how those develop, affordability will remain a principle policy concern. The FCC should therefore carefully monitor affordability of low-end offerings and, if affordability does not improve in light of ongoing wireless upgrades, take further steps beyond those already described in this plan to address the issue.

Potential future issues for fixed broadband competition

Analysts project that within a few years, approximately 90% of the population is likely to have access to broadband networks capable of peak download speeds in excess of 50 Mbps as cable systems upgrade to DOCSIS 3.0. About 15% of the population is likely to be able to choose between two robust high-speed service services—cable with DOCSIS 3.0 and upgraded services from telephone companies offering fiber-to-the-premises (FTTP).

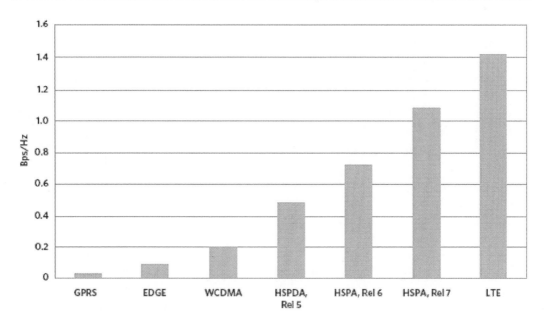

Exhibit 4-F. Evolution of Spectral Efficiency[151]

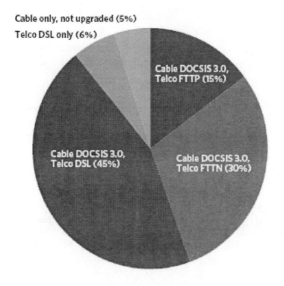

Exhibit 4-G. Projected Share of Households with Access to Various Wireline Broadband Technologies in 2012

These upgrades represent a significant improvement to the U.S. broadband infrastructure, and consumers who value high download and upload speeds will benefit by having a service choice they did not have before the upgrade. The upgrades may, however, change competitive dynamics. Prior to cable's DOCSIS 3.0 upgrade, more than 80% of the population could choose from two reasonably similar products (DSL and cable). Once the current round of upgrades is complete, consumers interested in only today's typical peak speeds can, in principle, have the same choices available as they do today. Around 15% of the population will be able to choose from two providers for very high peak speeds (providers with FTTP and DOCSIS 3.0 infrastructure). However, providers offering fiber-to-the-node and then DSL from the node to the premises (FTTN), while potentially much faster than traditional DSL, may not be able to match the peak speeds offered by FTTP and DOCSIS 3.0.[154] Thus, in areas that include 75% of the population, consumers will likely have only one service provider (cable companies with DOCSIS 3.0-enabled infrastructure) that can offer very high peak download speeds (see Exhibit 4-G).

Some evidence suggests that this market structure is beginning to emerge as cable's offers migrate to higher peak speeds. Exhibit 4-H shows that in 2004 the mean advertised download peak speeds of cable and DSL were similar, and the maximum and minimum advertised peak speeds were identical. By 2009, the mean advertised cable speed was about 2.5 times higher than DSL, while the maximum peak advertised speed was three times higher than DSL.[155] The minimum peak advertised speeds remained identical. While the exhibit does not contain information about demand or uptake of the higher-speed offers, or actual speeds delivered, it shows that the upgrade in network performance for cable companies from DOCSIS 3.0 is likely to continue or accelerate the trend where offers to end-users of traditional DSL cannot keep pace.

As with fixed-mobile substitution, how the evolution of network capabilities affects competition depends on how pricing, consumer demand, technology and costs evolve over time. For example, if users continue to value primarily applications that do not require very high speeds (e.g., speeds in excess of 20 Mbps), and are not willing to pay much for vastly increased speeds,[157] then a provider may not gain much of an advantage by offering those higher speeds. In contrast, if typical users require high speeds and only one provider can offer those speeds, and expected returns to telephone companies do not justify fiber upgrades, then users may face higher prices, fewer choices and less innovation. Because of this risk, it is crucial that the FCC track and compare the evolution of pricing in areas where two service providers offer very high peak speeds with pricing in areas where only one provider can offer very high peak speeds. The FCC should benchmark prices and services and include these in future reports on the state of broadband deployment.

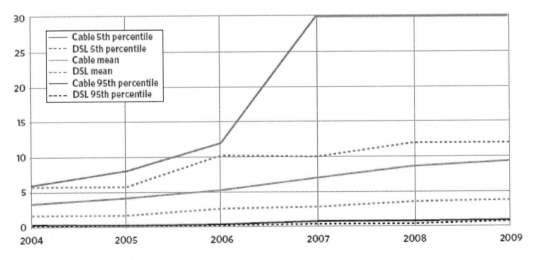

Exhibit 4-H. Broadband Speeds Advertised by Cable and Telco (5th percentile, mean and 95th percentile), 2004–2009[156]

Recommendations

Two sets of recommendations address the current and expected nature of competition in broadband network services in the United States. First, the FCC should take specific steps to make more spectrum available to ease entry into broadband markets and reduce the costs for current wireless providers to offer higher-speed services that can compete with wireline offers for a larger segment of end-users. Second, the FCC and BLS should collect data that enable more detailed analyses of the market and competition and make that data more publicly available to ensure visibility into competitive behavior of firms.[158]

RECOMMENDATION 4.1: The federal government, including the FCC, the National Telecommunications and information Administration (NTIA) and congress, should make more spectrum available for existing and new wireless broadband providers in order to foster additional wirelesswireline competition at higher speed tiers.

Chapter 5 discusses why additional spectrum is crucial to accommodate growing wireless broadband use. Additional spectrum is also critical for increasing competition along two interrelated dimensions.

First, additional spectrum for mobile competitors is likely to enhance mobile competition. Second, more spectrum makes possible faster download speeds, which would allow new and existing companies to use wireless technologies to serve as closer substitutes to fixed broadband providers for consumers seeking more than just low-end plans.

RECOMMENDATION 4.2: The FCC and the U.S. Bureau of Labor statistics (BLS) should collect more detailed and accurate data on actual availability, penetration, prices, churn and bundles offered by broadband service providers to consumers and businesses, and should publish analyses of these data.

- Improve current Form 477 data collection.
- Collect location-specific subscribership data.
- Collect price, switching costs, customer churn and market share information.
- Make more data and FCC analyses publicly available.
- BLS should fully resume its computer and Internet use supplement.

The FCC should revise Form 477 to collect data relevant to broadband availability, adoption and competition. Specifically, it should collect broadband availability data at the census *block* level, by provider, technology and offered speed. Availability for mobile service should be defined in terms of coverage specifications to be determined by the FCC and include information on spectrum used by facilities-based providers. In addition, the FCC should collect broadband service provider ownership and affiliation data and clarify and refine all reporting standards to ensure data consistency and comparability.

To improve its ability to make informed policy decisions and to track deployment, adoption and competition issues, the FCC should transition as quickly as practical to collecting location-specific subscribership data by provider, technology, actual speed and offered speed. Such data would make it possible for the FCC to aggregate the data to any geographic level rather than relying on providers to allocate subscribers by census tract or block. The FCC should also continue to utilize consumer-driven data collection methods, such as voluntary speed tests and broadband unavailability registries.

The FCC is fully cognizant of its obligations under the Electronic Communications Privacy Act (ECPA). To comply with the Act and protect citizens' privacy, the FCC should investigate using a third-party to collect location-specific subscribership data, and aggregate and anonymize it before submitting it to the FCC.

The FCC should collect data on advertised prices, prices actually paid by subscribers, plans, bundles and promotions of fixed and mobile broadband services that have material penetration among users, as well as their evolution over time, by provider and by geographic area.

Collecting information on advertised and promotional prices, rather than only prices current subscribers pay, is very helpful for analyzing competition because advertised prices focus on winning new customers or keeping customers considering switching providers and can offer important insights into how firms compete. In addition, it is important that the FCC collect information about the pricing plans to which customers are actually subscribing. Pricing plans that are available to customers but are not de facto marketed by service providers tend to have more limited competitive impact.

The FCC should also collect information related to switching barriers, such as early termination fees and contract length. To complement this information, the FCC should collect data on customer churn, as well as providers' share of gross subscriber additions.

Finally, the FCC should collect data required to determine whether broadband service is being denied to potential residential customers based on the income of the residents in a particular geographic area.[159]

The data collection should be done in a way that makes possible statistically significant, detailed analyses of at least metropolitan service area (MSA) or rural service area (RSA) levels, thus allowing the FCC to understand the effect of bundles and isolate the evolution of effective pricing and terms for broadband services.

The FCC should have a general policy of making the data it collects available to the public, including via the Internet in a broadband data depository, except in certain circumstances such as when the data are competitively sensitive or protected by copyright. Further, the FCC should implement a process to make additional data that is not accessible by the public available to academic researchers and others, subject to appropriate restrictions to protect confidentiality of competitively sensitive materials.[160]

An analysis of this data should be published and made available through annual existing reports such as the wireless competition report and the 706 report, and through semiannual reports such as the Form 477 data collection. The FCC should investigate if additional methods of providing this data and analyses are necessary.

Finally, BLS should be encouraged to fully resume its computer and Internet use supplement to its current population survey. Better data on adoption and use will facilitate analyses of the effects of competition as well as make it possible to track the effectiveness of adoption programs.

Transparency in the Retail Broadband Market

Collecting better data and allocating spectrum are only the first steps in driving competition. Putting more information in the hands of consumers is a proven method to promote meaningful competition and spur innovation, both of which will generate more and better consumer choices. If customers make well-informed choices, companies will likely invest in new products, services and business models to compete more aggressively and offer greater value.

For example, the U.S. Environmental Protection Agency's miles-per-gallon (mpg) label for cars encouraged automakers to improve fuel economy and design. That in turn helped boost average auto mileage in the United States from less than 15 mpg in 1975 to more than 25 mpg in 1985.[161] Or to take another example, the nutrition label by theU.S. Food and Drug Administration (FDA) has proven both useful and flexible. For example, when the negative health impact of trans fats surfaced, the FDA changed the nutrition label. It supplied the most current and important information to consumers and helped jumpstart the introduction of a wave of healthier food products.[162] With more consumers obtaining information online, the concept of a label should evolve.

Fixed broadband consumers, however, have little information about the actual speed and performance of the service they purchase.[163] Marketing materials typically feature "up to" peak download and upload speeds, although actual performance experienced by consumers is often much less than the advertised peak speed.[164] This disparity confuses consumers and makes it more difficult for them to compare the true performance of different offers. That hinders consumer choice and competition. It also reduces incentives for service providers to invest in better performing networks. Consumers need more information about the speed and overall performance[165] of the services they receive and of competitive offers in their area, and about the gap between actual and advertised speeds and the implications of that difference.

Some providers have added information in advertisements and other communications about what applications different broadband offers will support. But the lack of standards makes it nearly impossible for consumers to compare providers and their offers. For example, describing a specific broadband offer as capable of supporting an application such as video may not be enough to ensure that all consumers clearly understand the capabilities of the offer, as there are many different types of video (e.g., varying standard and high-definition formats and compression techniques).

Four steps must be taken to close this transparency gap.

RECOMMENDATION 4.3: The FCC, in coordination with the National institute of standards and Technology (NIST), should establish technical broadband measurement standards and methodology and a process for updating them. The FCC should also encourage the formation of a partnership of industry and consumer groups to provide input on these standards and this methodology.

The FCC, in coordination with NIST, should determine the technical standards and methodology to measure performance of fixed broadband connections with the objective of giving consumers a more accurate view of the performance of their broadband service. This would include what speeds and quality-of-service metrics should be tracked and how they should evolve with new consumer applications and uses.

The FCC should encourage industry and consumer interest representatives to create a Broadband Measurement Advisory Council (BMAC) to provide input for the measurement of broadband services.[166] The BMAC would focus on the most difficult issues, including where exactly to measure service performance in a network, the timing and frequency of measurements and the standard set of protocols and applications that may be used to establish benchmarks.

The key characteristics to be measured may include (see Exhibit 4-I):

- Actual speeds and performance over the broadband service provider's network (from point 2 to point 5 in Exhibit 4-I) and the end-to-end performance of the service (from point 1 to point 6 in the exhibit). [167]
- Actual speeds and performance at peak use hours. [168]
- Actual speeds and performance achieved with a given probability (*e.g.*, 95%) over a set time period (*e.g.*, one hour) that includes peak use times. [169]
- Actual speeds and performance tested against a given set of standard protocols and applications. [170]

RECOMMENDATION 4.4: The FCC should continue its efforts to measure and publish data on actual performance of fixed broadband services. The FCC should publish a formal report and make the data available online.

The FCC should continue its efforts to measure and report on fixed broadband connections and, similar to the approach taken by the United Kingdom regulator (the Office of Communications, or Ofcom), the FCC should explore contracts with third parties as a means of doing so. [171] These measurement efforts would make data on actual performance easily accessible to all interested parties, especially consumers, and create a mechanism for checking service provider broadband performance claims. The FCC should also use these efforts to conduct pilot projects on different measurement and reporting approaches.

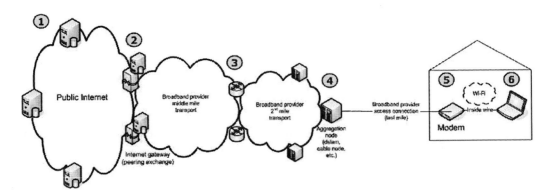

DEFINITIONS

① **Public Internet content:** public Internet content that is hosted by multiple service providers, content providers and other entities in a geographically diverse (worldwide) manner

② **Internet gateway:** closest peering point between broadband provider and public Internet for a given consumer connection

③ **Link between 2nd mile and middle mile:** broadband provider managed interconnection between middle and last mile

④ **Aggregation node:** First aggregation point for broadband provider (e.g. DSLAM, cable node, satellite, etc.)

⑤ **Modem:** Customer premise equipment (CPE) typically managed by a broadband provider as the last connection point to the managed network (e.g. DSL modem, cable modem, satellite modem, optical networking terminal (ONT), etc.)

⑥ **Consumer device:** consumer device connected to modem through internal wire or Wi-Fi (home networking), including hardware and software used to access the Internet and process content (customer-managed)

Exhibit 4-I. Simplified View of Internet Network and Connections

Experience in the United Kingdom, New Zealand, Singapore and elsewhere shows it is possible to provide consumers with information that helps them compare service providers in meaningful ways. [172]

All data should be made available to consumers and interested parties on a public website offering a searchable database. But the process should ensure the privacy of households that voluntarily participate in the measurement study. In addition, the FCC should publish a formal "State of U.S. Broadband Performance" report. This chapter should include detailed information about the actual performance of the country's top broadband service providers in different geographic markets (e.g., by county, city or MSA) and across all the metrics defined by the FCC.

RECOMMENDATION 4.5: The FCC should initiate a rule-making proceeding by issuing a Notice of Proposed Rule-making (NPRM) to determine performance disclosure requirements for broadband.

The FCC should issue an NPRM to determine appropriate disclosure obligations for broadband service providers, including disclosure obligations related to service performance. These obligations should include simple and clear data that a "reasonable consumer" can understand, while providing more detailed disclosure for more interested parties such as tech-savvy consumers, software developers and entrepreneurs designing products for the network.[173]

The purpose of disclosure for consumers is to help foster a competitive marketplace. Consumers need access to information at four different decision-making points in the process: when they are choosing a service provider, when they are choosing a plan, when they are evaluating their billed costs and if and when they decide to change providers. [174]

For broadband today, speed, price and overall performance are important factors in consumer decision-making. Consumers need to understand what broadband speed they actually need for the applications they want to use; how the speeds advertised by a broadband service provider compare to the actual speed a consumer will experience; and what broadband service provider and plan will give them the best value overall. The decision is especially complex because the actual performance of broadband service can vary significantly across geographic areas.

Given these factors, the FCC should look for better ways to improve information availability for consumer decision- making. One example would be to investigate developing or supporting the development by third parties of an online decision-making tool for choosing a broadband ISP, similar to those being developed for cell-phone services.

Some consumers will want a simpler way to gauge performance of different broadband service offers. For them, the FCC should develop a "broadband digital label" that will summarize broadband service performance concisely. Disclosure labels are among the most common tools used to ensure consumers have information about a product or service. They often come in two parts: a simple and clear standard "page 1" and a "page 2" listing more detail. The broadband digital label should take this concept and bring it to the online world. Illustrative examples of the front page of a possible broadband digital label can be found in Exhibit 4-J.

Example 1   Example 2   Example 3

Simplified, clear label with most
critical information

"Star" or index of service as
ranked by third party

Detail that is still clear and focused; list of common
applications and what can be delivered with this service

Exhibit 4-J. Illustrative Broadband Speed and Performance DigitalLabels

In Example 1 in Exhibit 4-J, consumers would know maximum and average upload/download speeds, along with an aggregated quality of service rating incorporating uptime, delay and jitter, as well as a list of standard applications that can be used with that service. Example two includes only actual upload and download speeds and a quality of service rating. Example three, similar to what has been proposed by Cisco and Corning,[175] would create a weighted average "Broadband Quality Index" rating for a service, from zero to five stars. This scoring system would evolve based on input from consumer and industry groups.

The FCC should also consider a broadband service performance disclosure item with the required speeds for different applications. Broadband service providers now claim different required speeds for the same applications in their advertising. A standard and evolving list would help consumers know what they really need—the first step in making an informed decision.

Finally, as noted in the FCC's August 2009 consumer disclosure NOI, consumers need full disclosure of the contractual commitments they are undertaking. These include clear, understandable, and reasonably precise estimates of the likely price of different broadband service offers and plans before they sign-up, as well as all fees and taxes.[176] The FCC should establish appropriate disclosure standards for contractual commitments as part of a rulemaking.

The FCC should conduct consumer research, potentially in collaboration with the FTC, to identify the disclosure obligations that would be most useful for consumers as critical input to a rulemaking proceeding.

RECOMMENDATION 4.6: The FCC should develop broadband performance standards for mobile services, multiunit buildings and small business users.

Mobile

For mobile broadband services, the FCC should create standards of measurement by location, carrier and spectrum band usage as input to a potential future rulemaking. The FCC should maintain and expand initiatives to capture user-generated data on coverage, speeds and performance. The FCC has launched a user-installed, self-testing application on mobile devices that can be used to both aggregate data about mobile broadband and publish the information on a public website. The FCC should continue to work with measurement companies, applications designers, device manufacturers and carriers to create an online database to help consumers make better choices for mobile broadband and spur competition, while ensuring privacy protections. [177]

The FCC should also encourage industry to create more transparent and standard disclosures of coverage, speeds and performance for mobile networks. The FCC should work with industry to identify the unique challenges of mobile disclosure—which requires reporting on speed and performance but also coverage and reliability—to decrease consumer confusion. Standards on disclosure would apply to data disclosed to regulators, to third party aggregators of coverage, and to consumers, with varying levels of detail for different audiences. The FCC should follow the same roadmap as created for fixed broadband disclosures, including the identification of consumer needs, the standardization of technical measurements and the creation of clear and simple consumer disclosure obligations. [178]

Buildings and small business

The FCC should also investigate better ways to improve transparency about the quality of broadband connectivity in residential multi-dwelling buildings and, potentially, in commercial and industrial buildings. The FCC should study the benefits of initiatives such as South Korea's program to institute a voluntary system of building ratings for broadband connectivity.[179] A program in the United States, if created, should carry incentives for developers to put more high-speed connections in new buildings, to upgrade existing structures and to encourage better internal wiring of all buildings, much in the same way that the Leadership in Energy and Environmental Design (LEED) certification program has encouraged developers to incorporate more environmental features into new buildings.

As small and medium-sized businesses (SMBs) use more sophisticated broadband applications, it is important to ensure they have the right performance. Speed, security, reliability and availability requirements may differ greatly from one SMB to another and are often very different from those for consumers. The FCC should determine the appropriate metrics and standards for transparency in SMB broadband to help in purchasing decisions and to encourage innovation among broadband providers.

Competition in Wholesale Broadband Markets

Residential broadband competition—as important as it is—is not the only type of competition we must foster to lay the foundation for America's broadband future. Ensuring robust competition not only for American households but also for American businesses requires particular attention to the role of wholesale markets, through which providers of broadband services secure critical inputs from one another. Because of the economies of scale, scope and density that characterize telecommunications networks, well functioning wholesale markets can help foster retail competition, as it is not economically or practically feasible for competitors to build facilities in all geographic areas. Therefore, the nation's regulatory policies for wholesale access affect the competitiveness of markets for retail broadband services provided to small businesses, mobile customers and enterprise customers.[180]

Unfortunately, the FCC's current regulatory approach is a hodgepodge of wholesale access rights and pricing mechanisms that were developed without the benefit of a consistent, rigorous analytic framework. Similar network functionalities are regulated differently, based on the technology used. Therefore, while networks generally have been converging to integrated, packet-mode, largely-IP networks, regulatory policy regarding wholesale access has followed the opposite trajectory. This situation undermines longstanding competition policy objectives. In some cases it limits the ability of smaller carriers—often those specializing in serving niche markets such as SMBs—to gain access to the necessary inputs to compete.

While facilities such as end-user loops and other point-to- point data circuits often serve as critical inputs to retail broadband services for business, mobile and residential customers, competitors' access to those inputs currently depends on factors that have little bearing on the economics of facilities-based competitive entry. For example, some wholesale access policies vary based on technology—including whether the facility or service operates using a circuit- or packet-based mode or is constructed from copper or fiber—regardless of the economic viability of replicating the physical facility. [181] Similarly, the FCC's wireless roaming policies vary based on the services offered; roaming is only required for voice telephone calls and not mobile data services.[182] As a result, mobile customers may not be able to use all functions of their Smartphone devices when roaming, even in situations where it is technically feasible for all of those functions to work.

In other cases, FCC rules draw distinctions based on the capacity of the facility, or by using various proxies to measure existing or potential competitive entry.[183] The FCC has also been criticized for not collecting better data or monitoring the impact of its current approach to competition. [184] The lack of a consistent analytical framework hinders the FCC's ability to promote competition. Accordingly, the FCC should comprehensively review its current policies and develop a cohesive and effective approach to advancing competition through its wholesale access policies.

RECOMMENDATION 4.7: The FCC should comprehensively review its wholesale competition regulations to develop a coherent and effective framework and take expedited action based on that framework to ensure widespread availability of inputs for broadband services provided to small businesses, mobile providers and enterprise customers.

An effective analytical framework for the FCC's wholesale access competition policies will enable efficient collection of any necessary data, evaluation of current rules and determination of what actions are necessary to achieve the FCC's goals for robust competition in business and consumer markets. The FCC has already taken steps in this direction with regard to the regulation of "special access" services, which encompass a broad array of dedicated, high-capacity transmission services. [185]

Recent filings at the FCC highlight additional dimensions of the FCC's wholesale regulatory framework that deserve attention, including competitive access to local fiber facilities,[186] copper retirement rules and implementation of Section 271 of the Communications Act of 1934 as amended.[187] The FCC should act on these proceedings within the context of rigorous analytic frameworks that establish coherent sets of conditions under which such rules should be applied and appropriately balance the benefits of competitive entry with incentives for carriers to invest in their networks.[188]

RECOMMENDATION 4.8: The FCC should ensure that special access rates, terms and conditions are just and reasonable.

Special access circuits are usually sold by incumbent local exchange carriers (LECs) and are used by businesses and competitive providers to connect customer locations and networks with dedicated, high-capacity links.[189] Special access circuits play a significant role in the availability and pricing of broadband service. For example, a competitive provider with its own fiber optic network in a city will frequently purchase special access connections from the incumbent provider in order to serve customer locations that are "off net."[190] For many broadband providers, including small incumbent LECs, cable companies and wireless broadband providers, the cost of purchasing these high-capacity circuits is a significant expense of offering broadband service, particularly in small, rural communities. [191]

The FCC regulates the rates, terms and conditions of these services primarily through interstate tariffs filed by incumbent LECs. However, the adequacy of the existing regulatory regime in ensuring that rates, terms and conditions for these services be just and reasonable has been subject to much debate. [192]

Much of this criticism has centered on the FCC's decisions to deregulate aspects of these services. In 1999, the FCC began to grant pricing flexibility for special access services in certain metropolitan areas. Since 2006, the FCC has deregulated many of the packet-switched, high-capacity Fast Ethernet and Gigabit Ethernet transport services offered by several incumbent LECs.[193] Business customers, community institutions and network providers regard these technologies as the most efficient method for connecting end-user locations and broadband networks to the Internet.[194]

The FCC is currently considering the appropriate analytical framework for its review of these offerings. [195] The FCC needs to establish an analytical approach that will resolve these debates comprehensively and ensure that rates, terms and conditions for these services are just and reasonable.

RECOMMENDATION 4.9: The FCC should ensure appropriate balance in its copper retirement policies.

Competitive carriers are currently using copper to provide SMBs with a competitive alternative for broadband services. Incumbent carriers are required to share (or "unbundle") certain copper loop facilities, which connect a customer to the incumbent carrier's central office. By leasing these copper loops and connecting them to their own DSL or Ethernet over copper equipment that is collocated in the central office, competitive carriers are able to provide their own set of integrated broadband, voice and even video services to consumers and small businesses. [196]

FCC rules permit incumbents that deploy fiber in their loops to "retire" or remove redundant outside-plant copper facilities after notifying competitive carriers that may be affected. [197] Retirement of these copper facilities affects both existing broadband services and the ability of competitors to offer new services. [198]

There are countervailing concerns, however. Incumbent deployment of fiber offers consumers much greater potential speeds and service offerings that are not generally possible over copper loops. In addition, fiber is generally less expensive to maintain than copper. As a result, requiring an incumbent to maintain two networks—one copper and one fiber—would be costly, possibly inefficient and reduce the incentive for incumbents to deploy fiber facilities.

The FCC should ensure appropriate balance in copper retirement policies as part of developing a coherent and effective framework for evaluating its wholesale access policies generally.

RECOMMENDATION 4.10: The FCC should clarify interconnection rights and obligations and encourage the shift to IP-to-IP interconnection where efficient.

For consumers to have a choice of service providers, competitive carriers need to be able to interconnect their networks with incumbent providers. Basic interconnection regulations, which ensure that a consumer is able to make and receive calls to virtually anyone else with a telephone, regardless of service provider, network configuration or location, have been a central tenet of telecommunications regulatory policy for over a century. For competition to thrive, the principle of interconnection—in which customers of one service provider can communicate with customers of another—needs to be maintained. [199]

There is evidence that some rural incumbent carriers are resisting interconnection with competitive telecommunications carriers, claiming that they have no basic obligation to negotiate interconnection agreements. [200] One federal court agreed with the rural carriers' arguments and concluded that the Act does not require certain rural carriers to negotiate interconnection agreements with other carriers. [201] This decision, which is based on a misinterpretation of the Act's rural exemption and interconnection requirements, has since been followed by several state commissions. [202] Without interconnection for voice service, a broadband provider, which may partner with a competitive telecommunications carrier to offer a voice-videoInternet bundle, is unable to capture voice revenues that may be necessary to make broadband entry economically viable.

Accordingly, to prevent the spread of this anticompetitive interpretation of the Act and eliminate a barrier to broadband deployment, the FCC should clarify rights and obligations regarding interconnection to remove any regulatory uncertainty. In particular, the FCC should confirm that all telecommunications carriers, including rural carriers, have a duty to interconnect their networks. [203] The FCC should also determine what actions it could take to encourage transitions to IP-to-IP interconnection where that is the most efficient approach. [204]

RECOMMENDATION 4.11: The FCC should move forward promptly in the open proceeding on data roaming.

To achieve wide, seamless and competitive coverage, the FCC should encourage mobile broadband providers to construct and build networks. Few, if any, of these networks will provide ubiquitous nationwide service entirely through their own facilities, particularly in the initial stages of construction and in rural areas. In order for consumers to be able to use mobile broadband services when traveling to areas outside their provider's network, their provider likely will need to enter into roaming arrangements with other providers. Roaming arrangements enable a customer to stay connected when traveling beyond the reach of their provider's network by using the network of another provider.

Data roaming is important to entry and competition for mobile broadband services and would enable customers to obtain access to e-mail, the Internet and other mobile broadband services outside the geographic regions served by their providers. For example, small rural providers serve customers that may be more likely to roam in areas outside their providers' network footprints. The industry should adopt voluntary data-roaming arrangements. In addition, the FCC should move forward promptly in its open proceeding on roaming obligations for data services provided without interconnection with the public-switched network. [205]

4.2. Devices

Innovative devices fundamentally change how people use broadband. Smartphones have allowed millions of Americans to use mobile e-mail, browse the Internet on-the-go, and—more recently—to use hundreds of thousands of mobile applications that did not exist a few years ago. Before smartphones, personal computers with graphical user interfaces and growing processing power enabled the emergence of the Web browser, which led to the widespread adoption of the Internet.

Competition, often from companies that were not market leaders, has driven innovation and investment in devices in the past and must continue to do so in the future. When one examines the three main types of devices that connect to broadband service provider networks—mobile devices, computing devices and set-top boxes—one finds that there are many mobile and computing device manufacturers offering hundreds of devices with a dizzying assortment of brands, features and price levels. Whole new device classes, such as tablets, e-book readers and netbooks continue to emerge, shifting firms' market positions and enabling entrants to capture market share. Mobile devices are rapidly incorporating technology such as Global Positioning System, accelerometers, Bluetooth, Wi-Fi, enhanced graphics and multi-touch screens. By any measure, innovation is thriving in mobile and computing devices.

The same is not true for set-top boxes, which are becoming increasingly important for broadband as video drives more broadband usage (see Chapter 3). [206] Further innovation in set- top boxes could lead to:

- Greater choice, lower prices and more capability in the boxes, including applications.[207]
- More competition among companies offering video content (MVPDs). [208]
- Unlimited choice in the content available—whether from traditional television (TV) or the Internet—through an integrated user interface. [209]
- More video and broadband applications for the TV, possibly in conjunction with other devices, such as mobile phones and personal computers (PCs). [210]
- Higher broadband utilization. [211]

Congress wanted to stimulate competition and innovation in set-top boxes and other video navigation devices in 1996 when it added Section 629 to the Communications Act.

Section 629 directed the FCC to ensure that consumers could use commercially available navigation devices to access services from MVPDs.[212] Lawmakers pointed to innovative uses of the telephone network, related to new phones, faxes and other equipment, and said they wanted to create a similarly vigorous retail market for devices used with MVPD services.[213]

The FCC adopted its First Report and Order to implement the provisions of Section 629 in 1998.[214] The order established rules requiring MVPDs to separate the system that customers use to gain access to video programming, called the conditional element, from the device customers use to navigate the programming. Section 629 nominally applies to all MVPDs. The FCC, however, has applied its rules only to cable operators. It either directly exempted other MVPDs, such as satellite TV operators, or implicitly excluded them by taking "no action" against an operator.[215]

Operators and other stakeholders agreed on a proposed solution for cable—called CableCARD —to separate the conditional access element. The CableCARD is about the size of a credit card and roughly similar in function to the Subscriber Identity Module (SIM) card used in mobile phones. Cable operators supply the CableCARD, which is inserted into a set-top box or television set that a consumer buys at a store to authenticate the subscriber. To ensure adequate support for CableCARDs, the FCC required cable operators to use CableCARDs for set-top boxes leased to consumers.

The first devices from third-party manufacturers using CableCARDs hit the retail market in August 2004, six years after the FCC's First Report and Order. Three years later, in July 2007, cable operators began using CableCARDs in their leased set-top boxes.[216]Despite Congressional and FCC intentions, CableCARDs have failed to stimulate a competitive retail market for set-top boxes. The top two cable set-top box manufacturers in North America, Motorola and Cisco, together captured a 95% share of unit shipments over the first three quarters of 2009. That's up from 87% in 2006.[217] A national or global market with relatively low costs of entry, like that for many consumer electronics markets, should support more than two competitors over time.[218] The two companies continue to control both the hardware and the security on the cable set- top box through their proprietary conditional access systems. By contrast, the top two cable set-top box manufacturers in Europe, the Middle East and Asia (EMEA) where open standards are used for conditional access accounted for a market share of approximately 39% between 2006 and the third quarter of 2009.[219] There are 0.5 million CableCARDs deployed in retail devices today,[220] which represents roughly 1% of all set- top boxes deployed in cable homes.[221] Only two manufacturers, TiVo and Moxi, continue to sell CableCARD -enabled set-top boxes through retail outlets.

Other alternatives are starting to emerge. For example, several innovators are attempting to bring Internet video to the TV.[222] Their devices often cannot access traditional TV content that consumers value—content that is not available or difficult to access online. Without the ability to seamlessly integrate Internet video with traditional TV viewing, Internet video devices like Apple TV and Roku have struggled to gain a foothold in U.S. homes.[223]

Retail set-top boxes have been competing on an uneven playing field. The barriers have been well-documented in multiple proceedings[224] and have prompted some companies not to enter the market at all.[225]

BOX 4-1. BROADBAND MODEMS AS AN ANALOG FOR INNOVATION IN SET- TOP BOXES

Broadband modems offer an example of how to unleash competition, investment and innovation in set-top boxes and other video navigation devices for consumer benefit. For stan- dard residential broadband connections, even though there are numerous delivery technologies (including cable, fiber, DSL, satellite and fixed wireless broadband), a customer must use an interface device, such as a cable modem. That device performs all network-specific functions. It also connects via a standardized Ethernet port to numerous devices consumers can buy at the store—including PCs, game consoles, digital media devices and wireless routers. Innovation can happen on either "side" of that device without affecting the other side. Service providers are free to invest and innovate in their networks and the services they deliver. Because the interface device communicates with consumer devices through truly open, widely used and standard protocols, manufacturers can create devices independently from service providers or any related third parties (*e.g.*, CableLabs). For example, PC manufacturers do not need to sign non-disclosure agreements with broadband service providers, license any intellectual property selected or favored by broadband service providers or get approval from any broadband service providers or any non-regulatory certification bodies to develop or sell their PCs at retail or enable consumers to attach them to service provider networks through the interface device.

Establishing an interface device for video networks that serves a similar purpose to modems for broadband networks could spark similar levels of competition, investment and innovation.

To level the field, the FCC should adopt the recommendation that follows. To maximize the likelihood that the recommendation will succeed, it should apply to all MVPDs. Extending the rule to all MVPDs will enable consumer electronics manufacturers to develop products for a larger customer base and allow consumers to purchase retail devices that will continue to function even if the consumer changes providers. Today, four out of the top 10 MVPDs are not cable companies and represent 41% of MVPD subscribers.[226]

RECOMMENDATION 4.12: The FCC should initiate a proceeding to ensure that all multichannel video programming distributors (MVPDs) install a gateway device or equivalent functionality in all new subscriber homes and in all homes requiring replacement set-top boxes, starting on or before Dec. 31, 2012.

To facilitate innovation and limits costs to consumers, the gateway device must be simple. Its *sole* function should be to bridge the proprietary or unique elements of the MVPD network (e.g., conditional access, tuning and reception functions) to widely used and accessible, open networking and communications standards. That would give a gateway device a standard interface with televisions, set-top boxes and other in-home devices and allow consumer electronics manufacturers to develop, market and support their products independently of MVPDs.

The following key principles apply:

- A gateway device should be simple and inexpensive, both for MVPDs and consumers. It should be equipped with only those components and functionality required to perform network-specific functions and translate them into open, standard protocols. The device should not support any other functionality or components.[227]
- A gateway device should allow consumer electronics manufacturers to develop, sell and support network-neutral devices that access content from the network *independently* from MVPDs or any third parties.[228] Specifically, third-party manufacturers should not be limited in their ability to innovate in the user interface of their devices by MVPD requirements. User-interface innovation is an important element for differentiating products in the consumer electronics market and for achieving the objectives of Section 629.

Similar to broadband modems (see Box 4-1), the proposed gateway device would accommodate each MVPD's use of different delivery technologies and enable them to continue unfettered investment and innovation in video delivery. At the same time, it would allow consumer electronics manufacturers to design to a stable, common open interface and to integrate multiple functions within a retail device. Those functions might include combining MVPD and Internet content and services, providing new user interfaces and integrating with mobile and portable devices such as media players and computers. It could enable the emergence of completely new classes of devices, services and applications involving video and broadband.

To ensure a competitive market for set-top boxes, the open gateway device:

- Should use open, published standards for discovering, signaling, authenticating and communicating with retail devices.[229]
- Should allow retail devices to access all MVPD content and services to which a customer has subscribed and to display the content and services without restrictions or requirments on the device's user interface or functions and without degradation in quality (*e.g.*, due to transcoding). [230]
- Should not require restrictive licensing, disclosure or certification. Any criterion should apply equally to retail and operator-supplied devices. Any intellectual property should be available to all parties at a low cost and on reasonable and non-discriminatory terms.[231]
- Should pass video content through to retail devices with existing copy protection flags from the MVPD.[232]

Requiring that the gateway device or equivalent functionality be developed and deployed by the end of 2012 is reasonable given the importance of stimulating competition and innovation in set-top boxes, the extensive public record established in this subject area[233] and the relatively simple architectures proposed to date.[234]

The FCC should establish interim milestones to ensure that the development and deployment of a gateway device or equivalent functionality remains on track. In addition, the FCC should determine appropriate enforcement mechanisms for MVPDs that, as of Dec. 31, 2012, have not begun deploying gateway device functionality in all new subscriber homes and in all homes requiring replacement set-top boxes.

Enforcement mechanisms would be determined with public input as part of the rulemaking proceeding. They could include, for example, issuing fines against non-compliant operators or denying extensions of certain CableCARD waivers like those granted for Digital Transport Adapters (DTAs). The FCC could also reach agreements with operators to provide set-top boxes for free to new customers until a gateway device is deployed.

The FCC should establish up front the criteria for the enforcement mechanisms. The FCC may want, for instance, to grant small operators more time to deploy the gateway device to take account of unique operational or financial circumstances. Transparency in the criteria for the enforcement mechanisms will establish more regulatory certainty in the market and help limit the number of waiver requests.

RECOMMENDATION 4.13: On an expedited basis, the FCC should adopt rules for cable operators to fix certain CableCARD issues while development of the gateway device functionality progresses. Adoption of these rules should be completed in the fall of 2010.

Four factors hinder consumer demand to purchase CableCARD devices and manufacturers' willingness to produce those devices. First, retail CableCARD devices cannot access all linear channels in cable systems with Switched Digital Video (SDV) unless cable operators voluntarily give customers a separate set-top box as an SDV tuning adapter.[235] Second, consumers perceive retail set-top boxes to be more expensive than set-top boxes leased at regulated rates from the cable operator. This perception is partially driven by a lack of transparency in CableCARD pricing for operator-leased boxes and by the bundling of leased boxes into package prices by operators.[236] Third, consumers who buy retail set-top boxes can encounter more installation and support costs and hassles than those who lease set-top boxes from their cable operators.[237] Fourth, the current retail CableCARD device certification process, run through CableLabs, incurs incremental costs of at least $100,000 to $200,000 during product development. The process also currently introduces other negative elements, including complexity, uncertainty and delays.[238]

Specifically, the proposed rules should address the four CableCARD issues. They should:

- Ensure equal access to linear channels for retail and operator-leased CableCARD devices in cable systems with SDV by allowing retail devices to receive and transmit outof-band communications with the cable headend over IP.[239]
- Establish transparent pricing for CableCARDs and operator-leased set-top boxes. Consumers should see the appropriate CableCARD charge, whether they purchase a retail device or lease one from the operator, and they should receive a comparable discount off packages that include the operator-leased set-top box if they choose to purchase one instead. [240]
- Standardize installation policies for retail and operator- leased CableCARD devices to ensure consumers buying CableCARD -enabled devices at retail do not face materially different provisioning hurdles than those using operator- leased set-top boxes.[241]
- Streamline and accelerate the certification process for retail CableCARD devices.[242] For example, the rules could restrict the certification process to cover hardware only,

similar to the certification required for cable-ready TVs, to ensure retail CableCARD devices do not harm a cable operator's network.

Addressing these issues will not require large investments in either headend or customer premise infrastructure.[243]

In fact, fixing these four CableCARD issues will sustain the current retail market for set-top boxes, enable companies that have invested in CableCARD -based products in accordance with current rules to compete effectively until the gateway device is deployed at scale, encourage more innovation until the gateway device is widely deployed and potentially allow for competition in the provision of the gateway device.

4.3. Applications

Over the last 10 years, there has been phenomenal growth in the applications and content available over broadband networks. Whole new markets have emerged, while others have migrated— partially or totally—online. Innovation in applications and content is transforming the way Americans communicate, shop, bank, study, read, work, use maps to find their way as they drive or walk, and are entertained. They have also changed the ways businesses interact with one another and market to their customers. Applications, content and the services they enable are bundled, sold, priced and monetized in many different ways. The nature and intensity of competition in applications and content varies tremendously and must be evaluated on a case-by-case basis.

The collection, aggregation and analysis of personal information are common threads among, and enablers of, many application-related innovations. The data that businesses collect have allowed them to provide increasingly valuable services to end-users, such as customized suggestions for movie rentals or books—often free of charge. These data have also become a source of value to businesses—e.g., as an enabler of more targeted and relevant advertising and increased user loyalty.[244] These data collection and monetization activities are a major driver of innovation for the Internet today and have benefited consumers in many ways.

However, many users are increasingly concerned about their lack of control over sensitive personal data. As aspects of individuals' lives become more "digitized" and accessible through or gleaned from broadband use, the disclosure of previously private, personal information has made many Americans wary of the medium. Innovation will suffer if a lack of trust exists between users and the entities with which they interact over the Internet. Policies therefore must reflect consumers' desire to protect sensitive data and to control dissemination and use of what has become essentially their "digital identity." Ensuring customer control of personal data and digital profiles can help address privacy concerns and foster innovation.

Personal Data, Innovation and Privacy

Historically, many firms have used personal data offline to create consumer profiles that have spawned multibillion dollar industries. The credit rating industry, for instance, tracks personal information including payment history, loan balances and income levels, which it sells to third parties to facilitate critical decisions such as approval of mortgages, loans and credit cards. The credit card industry, advertising industry and telemarketers have always relied on personal profiles of customers to better tailor their products and services. However, the impact has not always been positive for consumers. This fact has led to government actions like the creation of the "do not call" list for telemarketers and the FTC's work on combating fraud and identity theft.

The emergence of broadband and the growing use of the Internet makes aggregation of detailed personal data much easier and more valuable (see Box 4-2). As a result, single firms may be able over time to collect a vast amount of detailed personal information about individuals, including web searches, sites visited, click-stream, e-mail contacts and content, map searches, geographic location and movements, calendar appointments, mobile phone book, health records, educational records, energy usage, pictures and videos, social networks, locations visited, eating, reading, entertainment preferences, and purchasing history.

These data are giving rise to something akin to a "digital identity," which is a major source of potential innovation and opens up many possibilities for better customization of services and increased opportunities for monetization. The value of a targeted advertisement based on personal data can be several times higher than the value of an advertisement aimed at a broad audience. For example, the going rate for some targeted advertising products can be several times the rate for a generic one[245] because consumers can be six times more likely to "click through" a targeted banner advertisement than a non-targeted one.[246] This differential will likely increase as targeting becomes more refined and more capable of predicting preferences, intentions and behaviors.

Firms' ability to collect, aggregate, analyze and monetize personal data has already spurred new business models, products and services, and many of these have benefited consumers. For example, many online content providers monetize their audience through targeted advertising. Whole new categories of Internet applications and services, including search, social networks, blogs and user-generated content sites, have emerged and continue to operate in part because of the potential value of targeted online advertising.[247]

The ability to collect and store increasing amounts of personal data to develop these "digital identities" is accentuated by potential network effects. Firms with more predictive profiles and larger audiences will be able to offer increasingly better-targeted products and services that generate more advertising and consumer usage. This, in turn, enables the firms to collect more and better consumer personal data and develop even more predictive profiles. Those data and profiles are often so valuable for firms that they increasingly offer their products and services free of any monetary charges. Consumers gain access to a valuable service, and businesses gain valuable information.

However, new firms without access to detailed profiles of individual consumers, large audiences or subscriber pools may face competitive challenges as they try to monetize their innovations. They may face competitors offering an inferior service free of charge, and they may not have sufficient information about enough consumers to monetize their "audience" through advertising.

One way to encourage innovation in applications is to give individuals control of their digital profiles.[248] Giving consumers control of their digital profiles and personal data, including the ability to transfer some or all of it to a third party of their choice, may enable the development of new applications and services, and reduce barriers to entry for new firms.

Giving customers increased control of their profiles would also help address growing concerns about privacy and anonymity.

BOX 4-2. ONLINE PERSONAL DATA COLLECTION

Online data collection can be either passive or active. Passive data collection occurs without any overt consumer interaction and generally includes capturing user preferences and usage behavior, including location data from personal mobile devices. The best-known example is the use of "cookies" on a user's computer to capture Internet browsing history.[249] Passive data collection and the sharing of this data among third parties is poorly understood by consumers and often not communicated transparently by websites and applications. Consumers have some tools at their disposal, such as "private" browsing capabilities provided in the latest version of popular Web browsers or tools that allow them to see what passive activity is being captured, but the tools are limited.[250]

Active data collection requires a user to deliberately share personal data—for instance, when completing an online retail transaction or downloading an application on a mobile device. It often includes some disclosure of the use of the data being collected, although disclosures are frequently complex and written for lawyers, limiting how effective they are at conveying information to consumers.[251] Additionally, active data collection disclosure forms can fail to divulge policies on data sharing with third parties; when a consumer enters personal information, it is not clear whether these data might become part of a "digital profile" on a third party site.

Once personal data are collected, either passively or actively, they can be aggregated through third parties. Large firms, with enough interactions with consumers and sufficient information about them, may aggregate the data on their own. Profiles may be simple "contextual" maps, drawing just on immediate actions that consumers take on a page; for instance, someone searching for a flight may see a travel ad generated. Profiles may also be based on complex "behavioral" relationships that are not apparent to consumers; for example, someone may see a more tailored travel offer on that same website based on purchases they made at a retail store a month earlier and on their subsequent spending. These more sophisticated profiles allow for targeting of products to individuals in a predictive fashion.

Privacy and Anonymity

Today, consumers may have limited knowledge (if any) about how their personal data are collected and used. The fiduciary and legal responsibilities of those who collect and use that data are also unclear. Once consumers have shared their data, they often have limited ability to see and influence what data about them has been aggregated or is being used.[252] Further, it is difficult for consumers to regain control over data once they have been released and shared. As a result, privacy concerns can serve as a barrier to the adoption and utilization of broadband. A recent FCC survey showed that almost half of all consumers are concerned about privacy and security online.[253] Clear and strong privacy protections that disclose how and when users can delete or manage data shared with companies will help develop a market for innovative online applications.

Anonymity also must be addressed—both because it can be a positive factor online and because it can be a negative one. Anonymity is critical for allowing Internet users to exercise fundamental rights such as whistleblowing and engaging in activism. However, anonymity could also have negative consequences, such as allowing cybercriminals to go undetected.

Framework for Federal involvement

Several laws grant the FTC, FCC and other agencies regulatory authority over online privacy. The FTC has used its authority to prohibit unfair or deceptive practices and enforce promises made in corporate privacy statements on websites.[254] The FCC, for its part, typically works with the providers of broadband access to the Internet—phone, cable and wireless network providers—and the Communications Act contains various provisions outlining consumer privacy protections.[255] However, existing regulatory frameworks provide only a partial solution to consumer concerns and consist of a patchwork of potentially confusing regulations.[256] For instance, online communications are subject to ECPA,[257] but the privacy protections in ECPA may not apply to the information that websites collect from individual website visitors.[258] The Gramm-Leach-Bliley Act's protections for personal financial data apply only to financial institutions (such as banks, credit institutions and non-bank lenders), even though non-financial institutions (such as data brokers) may possess comparable information not subject to protections.[259] And while traditional telephone and cable TV networks are subject to privacy protections, ISPs operating in an unregulated environment can theoretically obtain and share consumer data through technologies such as deep packet inspection.[260]

In terms of anonymity, communications privacy laws,[261] health privacy regulations[262] and financial privacy laws[263] all prohibit disclosure of some analog to "personally identifiable information." However, defining "personally identifiable information" is not simple. In some cases, a single piece of information could be enough to identify an individual; in other cases, multiple facts might be required. For example, some claim that an aggregate of gender, ZIP code and birth date are unique for about 87% of the U.S. population.[264]

The right to speak anonymously without fear of government reprisal is protected by a number of laws, including federal whistleblower laws[265] and the First Amendment.[2656] The protections for anonymous speech are broad. People who are actually engaging in expressive or political speech are afforded even fuller protections.[267] As a result, anonymity is a complex issue.

As the FTC has stated, existing regulations are not enough in today's rapidly evolving world.[268] However, steps are being taken at the federal level to improve privacy protections, even in the absence of *comprehensive* privacy protections.[269] In particular, the FTC has addressed a wide variety of privacy issues since the 1990s. It has brought enforcement actions against spammers, makers of spyware and those who fail to protect sensitive consumer data. The FTC has also encouraged websites to post privacy policies that describe how personal information is collected, shared, used and secured. Today, nearly all of the top 100 commercial sites post such privacy policies.[270] Several years ago, the FTC launched an initiative to encourage greater transparency and consumer control with respect to online behavioral advertising. As part of that initiative, FTC staff issued a set of "principles" to guide industry self-regulation, including:

- Provide a clear, concise, consumer-friendly, prominent statement about behavioral advertising practices and a choice to consumers about whether to allow the practice.
- Provide reasonable security and have limited data retention.
- Obtain consent for material changes to existing privacy promises.
- Collect sensitive data for behavioral advertising only after obtaining consent from the consumer to receive such advertising.[271]

Following the issuance of these principles, individual companies, industry organizations and privacy groups have taken steps to address the privacy issues raised by behavioral advertising.[272] At the time of this plan's release, the FTC is hosting a series of public roundtables to examine existing privacy frameworks and whether they are adequate to address the vast array of technologies, business models and privacy challenges in today's world.[273] The goal of the roundtables is to explore how best to protect consumer privacy while supporting beneficial uses of information and technological innovation.

BOX 4-3. CRITICAL LEGISLATION— REFORMING THE PRIVACY ACT

This plan contains many recommendations, including some directed to Congress, for how to achieve the Congressional goals of access, affordability, utilization and achieving national purposes. In analyzing barriers to achieving these goals, a recurring theme emerges around privacy and control of personal data. The current legal landscape for how consumers control their personal data, when applied to the online world, may hold back new innovation and investment in broadband applications and content. These applications and content, in turn, are likely the most effective means to advance many of Congress's goals for broadband. New generations of applications and devices in sectors such as health care, energy and education will collect critical data that will help drive the next generation of American innovation, even as they raise important security and privacy considerations.[274]

While it is beyond the scope of this plan to address the details of how the legal landscape should be reformed, it is likely that revising the current Privacy Act to give consumers more control over their personal data and more confidence in the security of their personal data is a positive action Congress could take to improve the broadband ecosystem. Done correctly, this would increase innovation, rather than stifling it, by allowing consumers to transparently understand and choose how their government data are used. Updating the Act for the 21st century reality of digital interaction and seamless content sharing could drive more Americans online, increase their utilization of the Internet and help American businesses and organizations develop deeper and more trusted relationships with their customers and clients.

BOX 4-4. THE FDIC AS AN ANALOG TO TRUSTED "IDENTITY PROVIDERS"

Many government-backed entities have been created to help protect the public interest. The Federal Deposit Insurance Corporation (FDIC) provides one example of how government assists private companies in protecting and better serving consumers. Founded in 1933, the FDIC is an independent agency created by Congress to guarantee the deposits of individuals up to certain levels, thereby increasing trust in the banking system. Since the launch of FDIC insurance on Jan. 1, 1934, no depositor has lost a single cent of insured funds as a result of a failure.[275] The FDIC fulfills its mission:

- By acting as a private entity with the implicit backing of the government but that is fully self-funded through bank insurance payments.
- By creating minimum levels of security for depositors, giving Americans incentives to invest their personal funds in the banking system while limiting risk.
- By providing oversight of banks, assuring depositors that standards for good business and thoughtful risk taking are created and enforced.

Congress could explore the creation of mechanisms similar to those used by the FDIC to foster the emergence of trusted "identity providers" to secure and protect consumer data.

Finally, Congress and NTIA have taken an active interest in privacy and personal data protection. Several congressional committees have held hearings, and members have introduced bills that address various aspects of online privacy, from the brokerage of online information to deep packet inspection. NTIA, as part of its statutory obligation to advise the President, has worked closely with other parts of government on these issues.

RECOMMENDATION 4.14: Congress, the Federal Trade commission (FTC) and the FCC should consider clarifying the relationship between users and their online profiles.

In particular, several questions need to be addressed:

- What obligations do firms that collect, analyze or monetize personal data or create digital profiles of individuals have to consumers in terms of data sharing, collection, storage, safeguarding and accountability?
- What, if any, new obligations should firms have to transparently disclose their use of, access to and retention of personal data?
- How can informed consent principles be applied to personal data usage and disclosures?

RECOMMENDATION 4.15: Congress should consider helping spur development of trusted "identity providers" to assist consumers in managing their data in a manner that maximizes the privacy and security of the information.

Standard safe harbor provisions could allow companies to be acknowledged as trusted intermediaries that properly safeguard information, following appropriately strict guide-lines and audits on data protection and privacy (see Box 4-4). Congress should also consider creating a regime that provides insurance to these trusted intermediaries.[276]

RECOMMENDATION 4.16: The FCC and FTC should jointly develop principles to require that customers provide informed consent before broadband service providers share certain types of information with third parties.[277]

This information should include customers' account and usage information such as patterns of Internet access use and other personally identifiable information. This should not limit the ability of the provider to render reasonable service. Consent to allow sharing of personal information should not be a prerequisite to receiving service.

Identity Theft and Fraud

Identity theft is not a new risk—in fact, it is significantly more common offline than online.[278] However, with increases in electronic communications and online commerce, and the aggregation of information in databases, identity theft has become a growing concern.[279] In 2000, the FTC Consumer Sentinel Network received 31,000 identity theft complaints; by 2008, this number had risen to 314,000.[280] According to the FTC:

> "Credit card fraud (20%) was the most common form of reported identity theft followed by government documents/benefits fraud (15%), employment fraud (15%) and phone or utilities fraud (13%). Other significant categories of identity theft reported by victims were bank fraud (11%) and loan fraud (4%)."

In 2008, the FTC's network collected 1.2 million consumer complaints (up from roughly 900,000 in 2006) involving both online and offline transactions. Fraud and identity theft accounted for nearly 80% of these complaints.[281] Consumer risks like fraud and identity theft create a disincentive for individuals to engage in online transactions, increase the costs of doing business online and create law enforcement challenges.[282] Ensuring growing adoption and utilization of broadband requires that Internet users feel that they can connect and interact safely online.

Recently, fraud has been growing. A separate report by the Internet Crime Complaint Center (IC3) showed a 33.1% increase in fraud from 2007 to 2008.[283] The IC3 found that non-delivered merchandise or payment was, by far, the most reported offense (32.9%) while Internet auction fraud (25.5%) and credit/debit card fraud (9.0%) were also common offenses.

Several federal agencies have authority and responsibility for identity theft. In 1998, Congress passed the Identity Theft and Assumption Deterrence Act, making identity theft a federal crime. By 2002, most states had followed the federal example and enacted identity theft statutes.[284]

The Act called on the FTC to act as a clearinghouse for identity theft complaints and to provide consumer information to potential victims.[285] The FTC has produced several guidebooks with step-by-step information on actions consumers can take if they believe they are victims of identity theft. Those materials are available through the FTC.gov/idtheft website and the OnGuardOnline.gov project.

Beyond existing regulations, the 111th Congress has multiple bills in development that specifically address identity theft and security breaches.[286]

RECOMMENDATION 4.17: The federal government, led by the FTC, should put additional resources into combating identity theft and fraud and help consumers access and utilize those resources, including bolstering existing solutions such as OnGuard Online.

- **Put more resources into OnGuard Online. The federal government should put additional resources into OnGuard Online, ensuring that it is easily accessible to consumers and provides them with information on risks, solutions and who they can contact for further action. Federal agencies should connect their existing online websites to OnGuard Online and direct consumers to its resources.**
- **Maintain and publicize a database of agencies with responsibility. The FTC should maintain and publicize a database of agencies responsible for identity theft and fraud information, with clear information and directions available to consumers.**
- **Continue education efforts around identity theft and fraud. The federal government should continue educational efforts that clarify for consumers and businesses that personal information should only be collected when necessary and that entities should take reasonable measures to protect information from unauthorized access.**
- **Encourage broadband service providers to link to OnGuard Online. All agencies should encourage broadband service providers to link to onGuard online to direct potential victims of identity theft or fraud to necessary resources.**

Consumer Online Security

In 1988, Robert Morris unleashed the Morris Worm on the Internet, bringing approximately 10% of the network to a halt.[287] In response, the Defense Advanced Research Projects Agency set up the first national cybersecurity effort—the CERT Coordination Center at Carnegie Mellon University.[288] Today, the Department of Homeland Security (DHS) leads federal cybersecurity activities supported by numerous efforts such as the FTC's OnGuard Online program and DOJ legal actions. Consumer online security issues such as viruses, spam and malware are closely related to cybersecurity activities.

In October 2009, spam accounted for 87% of all e-mail messages, and 1.9% of these spam messages contained malware.[289] According to the Anti-Phishing Working Group, the number of computers infected with malware viruses rose more than 66% between the fourth quarter of 2008 and the second quarter of 2009, representing more than half of their total sample of scanned computers. The incidence of malware such as password-stealing software directed at banking and financial accounts increased more than 186% in the same period.[290]

DHS is the government agency with primary responsibility for cybersecurity, although the FTC often handles "consumer online security" complaints. DHS, DOJ and the Executive Branch have taken the lead in promoting cybersecurity. Other agencies such as the National Security Agency, the U.S. Department of Defense (DoD), NIST, the National Science

Foundation and the FCC have all had active roles. Recently, these agencies have tried to enable simpler communication to the public about where to go in the case of online security issues, while also detailing strategies for protecting the online environment. [291]

Broadband service providers have an incentive to offer security to customers to protect the network. Some offer antivirus software for free, although installation and control still primarily reside with the consumer. Application providers like Google also help consumers by providing information on vulnerabilities, such as by flagging sites that are security risks. This is a start, but there is a critical need for more consumer education on what threats they face, how to protect their connections and where to turn in case of emergency.

RECOMMENDATION 4.18: FCC consumer online security efforts should support broader national online security policy, and should be coordinated with the Department of Homeland security (DHS), the FTC, the white House Cyber Office and other agencies. Federal agencies should connect their existing websites to OnGuard Online to provide clear consumer online security information and direction.

Child Protection

In the FCC's recent study of broadband adopters and non- adopters, 74% of broadband users strongly agreed that it is important for children to learn how to use the Internet. In fact, technology has already become integral to children's lives.[292] While children can benefit from being online (*e.g.*, through access to novel educational opportunities), they can also be exposed to risks.[293]

Last year's Internet Safety Technical Task Force Report concluded that simply being online does not automatically put youth at risk for online predation.[294] Research also found that "there was no evidence that online predators were stalking or abducting unsuspecting victims based on information they posted at social networking sites."[295]

Still, there is a growing consensus that children need to be taught the critical skills necessary to succeed in an online environment. As stated by the National Academies of Sciences: "Swimming pools can be dangerous for children. To protect them, one can install locks, put up fences and deploy pool alarms. All of these measures are helpful, but by far the most important thing that one can do for one's children is to teach them to swim."[296]

RECOMMENDATION 4.19: The federal government should create an interagency working group to coordinate child online safety and literacy work, facilitate information sharing, ensure consistent messaging and outreach and evaluate the effectiveness of governmental efforts. The working group should consider launching a national education and outreach campaign involving governments, schools and caregivers.

Content and online Copyright Protection

The Internet is revolutionizing the production and distribution of creative works, lowering barriers to entry and enabling far broader and faster access to culture and ideas than previously possible.[297] But the Internet's value as a platform for content— and the ability of online content to drive increased adoption and use of broadband[298]—depends on creators' incentives to create and disseminate their works online, which are in turn at least partly dependent on copyright protection. The Internet must be a safe, trusted platform for the lawful distribution of content. At the same time, copyright protection efforts must not stifle innovation; overburden lawful uses of copyrighted works; or compromise consumers' privacy rights.

The Plan's recommendations regarding content and online copyright protection are limited to a few discrete suggestions regarding educational uses and public media (see Chapters 11 and 15).

Digital Goods and Services Taxation

RECOMMENDATION 4.20: The federal government should investigate establishing a national framework for digital goods and services taxation.

The National Broadband Plan is focused on increasing beneficial use of the Internet, including e-commerce and new innovative business models. The current patchwork of state and local laws and regulations relating to taxation of digital goods and services (such as ringtones, digital music, etc.) may hinder new investment and business models.[299] Entrepreneurs and small businesses in particular may lack the resources to understand and comply with the various tax regimes.

Recognizing that state and local governments pursue varying approaches to raising tax revenues, a national framework for digital goods and services taxation would reduce uncertainty and remove one barrier to online entrepreneurship and investment.

4.4. Competition for Value across the Ecosystem

"The Internet's openness, and the transparency of its protocols, [has] been critical to its success."[300] As the FCC's NPRM on Preserving the Open Internet explains, broadband is a powerful engine for innovation and investment in America in part because the Internet is an open platform, where anyone can communicate and do business with anyone else on a level playing field.[301] The open Internet "ensures that users are in control of the content that they send and receive,"[302] and that inventors and entrepreneurs "do not require the securing of permission" to innovate.[303]

The NPRM notes that these characteristics have made the Internet vibrant, and its continued health and growth—as well as broadband's ability to drive the many benefits discussed in this plan—depend on its continued openness "[B]roadband providers' ability to innovate and develop valuable new services must co-exist with the preservation of the free and open Internet that consumers and businesses of all sizes have come to depend on."[304]

In the latest step in a longstanding effort to ensure these interests remain balanced, the FCC adopted the NPRM on Preserving the Open Internet in October 2009, which launched a rulemaking process that is currently underway.[305] The NPRM asked for public comment on six proposed principles:

1. *Content.* Subject to reasonable network management, a provider of broadband Internet access service may not prevent any of its users from sending or receiving the lawful content of the user's choice over the Internet.

2. *Applications and services.* Subject to reasonable network management, a provider of broadband Internet access service may not prevent any of its users from running the lawful applications or using the lawful services of the user's choice.

3. *Devices.* Subject to reasonable network management, a provider of broadband Internet access service may not prevent any of its users from connecting to and using on its network the user's choice of lawful devices that do not harm the network.

4. *Competitive Options.* Subject to reasonable network management, a provider of broadband Internet access service may not deprive any of its users of the user's entitlement to competition among network providers, application providers, service providers and content providers.

5. *Nondiscrimination.* Subject to reasonable network management, a provider of broadband Internet access service must treat lawful content, applications and services in a nondiscriminatory manner.

6. *Transparency.* Subject to reasonable network management, a provider of broadband Internet access service must disclose such information concerning network management and other practices as is reasonably required for users and content, application and service providers to enjoy the protections specified in this part.

The proposed rules also make clear that the principles would not supersede any obligation or limit the ability of broadband providers to deliver emergency communications or address the needs of law enforcement, public safety or homeland security authorities, consistent with applicable law.

4.5. Transition from a Circuit-Switched Network

Increasingly, broadband is not a discrete, complementary communications service. Instead, it is a platform over which multiple IP-based services—including voice, data and video—converge. As this plan outlines, convergence in communications services and technologies creates extraordinary opportunities to improve American life and benefit consumers. At the same time, convergence has a significant impact on the legacy Public Switched Telephone Network (PSTN), a system that has provided, and continues to provide, essential services to the American people.[306]

Convergence raises a number of critical issues. Consumers benefit from the options that broadband provides, such as Voice over Internet Protocol. But as customers leave the PSTN, the typical cost per line for Plain Old Telephone Service (POTS) increases, given the high fixed costs of providing such service.[307] Between 2003 and 2009, the average cost per line increased almost 20 percent.[308]

Regulations require certain carriers to maintain POTS—a requirement that is not sustainable—and lead to investments in assets that could be stranded.[309] These regulations can have a number of unintended consequences, including siphoning investments away from new networks and services. The challenge for the country is to ensure that as IP-based services replace circuit-switched services, there is a smooth transition for Americans who use traditional phone service and for the businesses that provide it.

This is not the first time the United States has overseen a transition in communications. In the past, the country transitioned mobile service from analog to digital and, more recently, transitioned broadcast television from analog to digital. In each case, government policies helped ensure that legacy regulations and services did not become a drag on the transition to a more modern and efficient use of resources, that consumers did not lose services they needed and that businesses could plan for and adjust to the new standards.

As with earlier transitions, the transition from a circuit- switched network will take a number of years. But to ensure that the transition does not dramatically disrupt communications or make it difficult to achieve certain public policy goals, the country should start considering the necessary elements of this transition in parallel with efforts to accelerate broadband deployment and adoption. As such, the FCC should start a proceeding on the transition that asks for comment on a number of questions, including whether the FCC should set a timeline for a transition and, if so, what the timeline should be,[310] quality of service requirements[311] and safeguarding emergency communications.[312] This proceeding should consider questions of jurisdiction,[313] regulatory structure[314] and legacy voice-specific regulations, including interconnection, numbering and carrier of last resort obligations.[315] It should consider the impact of the transition on employment in the communications industry, particularly given the historic role of the sector in providing high-skill, high-wage jobs.[316] In the proceeding, the FCC should also look at whether there are requirements from other federal entities, such as tax requirements, that would affect the path of the transition.

Finally, a number of recommendations in this plan will affect the path of the transition, including recommendations about universal service and intercarrier compensation (Chapter 8) and recommendations related to access for people with disabilities (Chapter 9). The proceeding should examine how best to proceed with a transition in light of these other recommendations.

4.6. Leveraging the Benefits of Innovation and Investment Internationally

While the National Broadband Plan focuses on developing the domestic broadband ecosystem, broadband policy also unfolds in an interdependent international market full of opportunities and challenges. Global trade in information and communications technology (ICT) is almost $4 trillion and growing. [317] U.S. companies have played a leading role in bringing technologies to market that support a worldwide ICT ecosystem through the development of software, devices, applications, semiconductors and network equipment. This trade and investment is supporting tremendous growth in international Internet traffic, which increased at a compound annual growth rate of 66% over the past five years, supported by a 22% compound annual reduction in international transit port prices over that same period.[318] Further investment and innovation in U.S. broadband networks will provide U.S. businesses and consumers with the infrastructure they need to continue to compete in the rapidly changing ICT market. However, to realize the tremendous promise of a networked world, U.S. leadership and international cooperation are needed to encourage Internet freedom and strengthen cybersecurity.

The United States took a leading role in the global Internet revolution of the 1990s by contributing to the technological and policy developments that enabled the Internet. The breakup of AT&T in the 1980s and the Telecommunications Act of 1996 served as catalysts for the spread of pro-competition policies around the world.[319] In addition, with the adoption of the World Trade Organization's Basic Telecommunications Agreement and Reference Paper in 1996, the world community took steps to adopt important liberalization principles that remain relevant and influential today.[320]

The National Broadband Plan recognizes that making the right policy choices at home that result in domestic market success is essential for the United States to advocate effectively in the debate on policies and practices for the global communications network. The policies contained in the plan form the basic foundations of the U.S. international telecommunications agenda. These principles include support for regulatory frameworks that are pro-competitive, transparent and technology-neutral.

Ubiquitous availability of broadband and universal connectivity enable people and entities in the United States to communicate worldwide, which increases productivity and enables innovation. The National Broadband Plan's emphasis on the promotion of the use of broadband for national priorities, such as education, energy, health care, economic development, e-government, civic engagement and public safety, demonstrates the possibilities for progress that can result from access to broadband. Even for the many people whose access to the global network is limited to mobile phones, there are still innovative examples of how mobile broadband can serve national priorities, such as providing access to health care information through mobile handsets.[321]

Competitive communication policies have facilitated network development around the world. The trends are encouraging, with 1.7 billion Internet users and 4.6 billion mobile phone subscribers in the world today.[322] Mobile networks now constitute the world's largest distribution platform. And today's mobile users will be the next generation of Internet users, as Smartphones enable those with mobile access to experience the benefits of connectivity. But more needs to be done to encourage mobile broadband access. About 40% of the world's population still does not have mobile phones and about three- quarters are not using the Internet.[323]

The United States should continue to support policies that hasten the rollout and uptake of telecommunications technology that bridges the international digital divide. Integrating ICT deployment and utilization into broader regional economic development strategies is as important abroad as it is at home.[324]

Policies that support the uptake of telecommunications technologies not only provide incentives for needed connectivity but also allow U.S. innovations to flourish in a rapidly developing world market. In turn, Americans benefit from a parallel stream of innovations coming from abroad.

As more people gets access to mobile communications services, innovative uses of mobile technology are increasing. But proliferation of mobile phones not only allows people to share more information, it has also spurred innovation and investment in other sectors that would be impossible without global access to broadband. From health care to banking, entrepreneurs have recognized that the commonality and wide distribution of mobile communications devices make them ideal tools for launching a variety of services and applications.

For example, in many developing countries, an entire segment of the population that previously had no access to banks is taking advantage of the convenience and availability of mobile banking. Mobile banking includes a variety of technology and business strategies to leverage mobile communications networks for the provision of transactional and informational financial services. Emerging markets are embracing mobile banking as a more effective means of reaching more people than traditional bricks-and-mortar banks. Access to banking for the previously "unbanked" can have a dramatic impact on individuals, families and small businesses as it increases safety, prevents monetary loss, enables savings and makes business more efficient and successful.[325]

The United States also needs to provide continued leadership to ensure that the Internet will continue to evolve in ways that are cooperative, collaborative and maximally beneficial for the collective community of users, managers and investors. The three primary streams of cooperation—intergovernmental cooperation, cooperation through non-governmental organizations and cooperation through technical bodies—have served the world and the Internet well. The United States needs to provide continued leadership in all of these fora— particularly by working (as recommended in Chapter 5) with the international community, including the ITU, to develop innovative and flexible global spectrum allocation.[326] Global harmonization across spectrum usage, along with international standards-setting, can reduce per-unit costs and lead to increased adoption and usage of the Internet around the world.

Today, as in the 1990s, the changing capabilities of ICT are forcing the world to make critical policy choices. The great achievement of a near-ubiquitous global network is being threatened by curtailed Internet freedom and decreased network security.

The global communications network has created an era in which information is perhaps freer than ever before. Maximizing the benefits of broadband worldwide will require increased attention to policies that promote universal and unrestricted access to the Internet. The United States should lead in efforts to create a global consensus on how to define and guarantee basic rights of openness, access to and creation of information and connection to the global Internet community.

Cybersecurity, as discussed in Chapters 14 and 16,[327] is an important element of the National Broadband Plan. Cybersecurity attacks can be generated from anywhere in the world. The importance of cybersecurity as a policy objective cannot be underestimated. Engaging counterparts in international fora, as appropriate, will be crucial to successfully implementing cybersecurity policies.

5. SPECTRUM

Historically, the federal communications commission (fcc)'s approach to allocating spectrum has been to formulate policy on a band-by-band, service-by-service basis, typically in response to specific requests for service allocations or station assignments. This approach has been criticized for being ad hoc, overly prescriptive and unresponsive to changing market needs. [328] wireless broadband is poised to become a key platform for innovation in the united states over the next decade. As a result, u.s. spectrum policy requires reform to accommodate the new ways that industry is delivering wireless services. These reforms include making more spectrum available on a flexible basis, including for unlicensed and opportunistic uses. Given the length of the spectrum reallocation process, these reforms should reflect expectations of how the wireless world will look 10 years from now. These reforms should ensure that there is sufficient, flexible spectrum that accommodates growing demand and evolving technologies.

Spectrum policy must be a key pillar of U.S. economic policy. The contribution of wireless services to overall gross domestic product grew over 16% annually from 199 2–2007 compared with less than 3% annual growth for the remainder of the economy. [329] Given these growth rates, wireless communications—and mobile broadband in particular—promises to continue to be a significant contributor to U.S. economic growth in the coming decade. Some analysts predict that within five years more users will connect to the Internet via mobile devices than desktop personal computers (PCs). [330]

Disruptive technology transformations happen once every 10 to 15 years. Mobile broadband represents the convergence of the last two great disruptive technologies—Internet computing and mobile communications—and may be more transformative than either of these previous breakthroughs. Mobile broadband is scaling faster and presents a bigger opportunity. This revolution is being led not only by domestic wireless carriers, who are investing billions in network upgrades, but also by American companies such as Amazon, Apple, Intel, Google, Qualcomm and numerous entrepreneurial enterprises that export innovation globally. [331]

Recommendations

Ensure greater transparency concerning spectrum allocation and utilization

- The FCC should launch and continue to improve a spectrum dashboard.
- The FCC and the National Telecommunications and Information Administration (NT IA) should create methods for ongoing measurement of spectrum utilization.
- The FCC should maintain an ongoing strategic spectrum plan including a triennial assessment of spectrum allocations.

Expand incentives and mechanisms to reallocate or repurpose spectrum

- Congress should consider expressly expanding the FCC's authority to enable it to conduct incentive auctions in which incumbent licensees may relinquish rights in spectrum assignments to other parties or to the FCC.
- Congress should consider building upon the success of the Commercial Spectrum Enhancement Act (C SEA) to fund additional approaches to facilitate incumbent relocation.
- Congress should consider granting authority to the FCC to impose spectrum fees on license holders and to NTIA to impose spectrum fees on users of government spectrum.
- The FCC should evaluate the effectiveness of its secondary markets policies and rules to promote access to unused and underutilized spectrum.

Make more spectrum available for broadband within the next 10 years

- The FCC should make 500 megahertz newly available for broadband use within the next 10 years, of which 300 megahertz between 225 MHz and 3.7 GHz should be made newly available for mobile use within five years.
  - The FCC should make 20 megahertz available for mobile broadband use in the 2.3 GHz Wireless Communications Service (WCS) band, while protecting neighboring federal, non-federal Aeronautical Mobile Telemetry (AMT) and satellite radio operations.
  - The FCC should auction the 10 megahertz Upper 700 MHz D Block for commercial use that is technically compatible with public safety broadband services.
  - The FCC should make up to 60 megahertz available by auctioning Advanced Wireless Services (AWS) bands, including, if possible, 20 megahertz from federal allocations.
  - The FCC should accelerate terrestrial deployment in 90 megahertz of Mobile Satellite Spectrum (MSS).
  - The FCC should initiate a rule making proceeding to reallocate 120 megahertz from the broadcast television (TV) bands.

Increase the flexibility, capacity and cost-effectiveness of spectrum for point-to-point wireless backhaul services

- The FCC should revise Parts 74, 78 and 101 of its rules to allow for increased spectrum sharing among compatible point-to-point microwave services.
- The FCC should revise its rules to allow for greater flexibility and cost-effectiveness in deploying wireless backhaul.

Expand opportunities for innovative spectrum access models

- The FCC, within the next 10 years, should free up a new, contiguous nationwide band for unlicensed use.
- The FCC should move expeditiously to conclude the TV white spaces proceeding.
- The FCC should spur further development and deployment of opportunistic uses across more radio spectrum.
- The FCC should initiate proceedings to enhance research and development that will advance the science of spectrum access.

Take additional steps to make U.S. spectrum policy more comprehensive

- The FCC and NTIA should develop a joint roadmap to identify additional candidate federal and non-federal spectrum that can be made accessible for both mobile and fixed wireless broadband use, on an exclusive, shared, licensed and/or unlicensed basis.

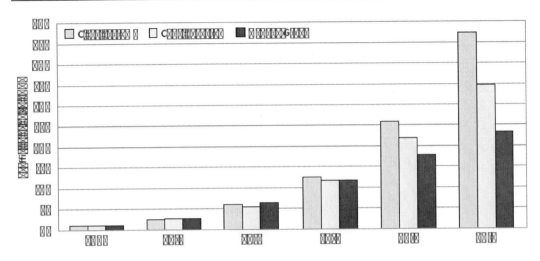

Exhibit 5-A. Forecasted Mobile Data Traffic in North America

- The FCC should promote within the International Telecommunication Union (ITU) innovative and flexible approaches to global spectrum allocation that take into consideration convergence of various radio communication services and that enable global development of broadband services.
- The FCC should take into account the unique spectrum needs of U.S. Tribal communities when implementing the recommendations in this chapter.

5.1. The Growth of Wireless Broadband

The use of wireless broadband is growing rapidly, primarily in the area of mobile connectivity, but also in fixed broadband applications. Key drivers of this growth include the maturation of third-generation (3G) wireless network services, the development of smartphones and other mobile computing devices, the emergence of broad new classes of connected devices and the rollout of fourth-generation (4G) wireless technologies such as Long Term Evolution (LTE) and WiMAX.

3G network services are in full bloom. Data traffic on AT&T's mobile network, driven in part by iPhone usage, is up 5,000% over the past three years,[332] a compound annual growth rate of 268%. Verizon Wireless says it, too, has recently experienced substantial data growth in its network.[333] According to Cisco, North American wireless networks carried approximately 17 petabytes per month in 2009, [334] an amount of data equivalent to 1,700 Libraries of Congress. By 2014, Cisco projects wireless networks in North America will carry some 740 petabytes per month, a greater than 40-fold increase. Other industry analysts forecast large proportional increases (see Exhibit 5-A). [335]

This growth in aggregate traffic is due to increased adoption of Internet-connected mobile computing devices and increased data consumption per device. A recent survey of 7,000 U.S. adults found that smartphone penetration is now at 33% of mobile subscribers across the four largest wireless operators. Penetration rose steadily over the past several quarters.[336] These new devices drive higher data usage per subscriber, as users engage with data- intensive social networking applications and user-generated video content. Advanced smartphones, such as the iPhone, and devices using the Android operating system consume hundreds of megabytes of data per user per month.[337] Laptops using air- cards consume more than a gigabyte per user per month.[338] To put these numbers in perspective, Cisco estimates that smartphones such as the iPhone can generate 30 times more data traffic than a basic feature phone, and that a laptop can generate many times the traffic of a smartphone. [339]

Additionally, experts expect a huge increase in machine- based wireless broadband communications over the next several years, as "smart" devices take advantage of the ubiqui-tous connectivity afforded by high-speed, low-latency, wireless packet data networks.[340] While many of these devices, like smart meters, are expected to consume relatively small amounts of bandwidth, others, such as wireless-enabled cameras, may make use of embedded video and other media that could substantially increase demand for wireless bandwidth. Analysts predict a shift from one device per person to a world where "smart" connected devices greatly outnumber human beings. [341] The aggregate impact of these devices on demand for wireless broadband networks could be enormous.

The rollout of advanced 4G networks using new versions of LTE and WiMAX technologies will also intensify the impact on mobile broadband networks. The next generation of mobile broadband networks will support higher data throughput rates, lower latencies and more consistent network performance throughout a cell site. This will increase the range of applications and devices that can benefit from mobile broadband connectivity, generating a corresponding increase in demand for mobile broadband service from consumers, businesses, public safety, health care, education, energy and other public sector users. Most of the major wireless carriers are building or planning upgrades to 4G technologies (see Exhibit 5-B).

An increase in mobile broadband use raises demand for other wireless services, such as point-to-point microwave back- haul and unlicensed networks, to enhance the overall delivery of broadband. Wireless backhaul transports large quantities of data to and from cell sites, especially in rural areas. Unlicensed services such as Wi-Fi and Bluetooth are important complements to licensed mobile networks and to fixed wireline networks. Most smartphones available today feature Wi-Fi, and users increasingly take advantage of this capability inside homes or businesses where high-speed broadband connectivity is available. According to a November 2008 report from AdMob, 42% of all iPhone traffic was transported over Wi-Fi networks rather than carriers' own networks. [343] Other carriers report similar trends in how their customers use Wi-Fi to complement cellular service.

Growing Spectrum Needs

The growth of wireless broadband will be constrained if government does not make spectrum available to enable network expansion and technology upgrades. In the absence of sufficient spectrum, network providers must turn to costly alternatives, such as cell splitting, often with diminishing returns. If the U.S. does not address this situation promptly, scarcity of mobile broadband could mean higher prices, poor service quality, an inability for the U.S. to compete internationally, depressed demand and, ultimately, a drag on innovation.

The progression to 4G technologies may require appropriately sized bands, including larger blocks to accommodate wider channel sizes. That said, innovative technologies are emerging that take advantage of narrower slices of spectrum, and such complementary approaches provide new opportunities for investment and further technological innovation.

Exhibit 5-B. Selected Announced Upgrades to the U.S. Mobile Broadband Network (Persons Covered) [342]

Technology	Companies	2009	2010	2011	By 2013
LTE	Verizon AT&T MetroPCS Cox		Verizon (100 million) AT&T (trials)	AT&T (start of deployment) Cox (start of deployment) MetroPCS (start of deployment)	Verizon (entire network)
WiMAX	Clearwire/Sprint OpenRange Small wireless Internet service providers (WISPs)	Clearwire (30 million) WISPs (2 million)	Clearwire (120 million)		OpenRange (6 million)

Unlocking the full potential of 4G will require more than a "re-farming" of existing mobile spectrum and deployment using recently released spectrum in the 700 MHz, Advanced Wireless Services (AWS) and 2.5 GHz bands. It cannot focus solely on "last mile" mobile connectivity, but also needs to address other potential network bottlenecks that inhibit speed, including backhaul and other point-to-point applications.

Additional spectrum is also required to accommodate multiple providers in a competitive marketplace, including new entrants and small businesses, as well as to enable wireless services to compete with wireline services. The U.S. Department of Justice (DOJ) aptly summarized: "Given the potential of wireless services to reach underserved areas and to provide an alternative to wireline broadband providers in other areas, the Commission's primary tool for promoting broadband competition should be freeing up spectrum." [344]

Spectrum: The Great enabler

Each of the past three decades has seen a new tranche of mobile spectrum create successive waves of innovation and investment.

In 1983, the FCC allocated the spectrum used to build out the first cellular networks. This spectrum was originally allocated to television channels 70 to 83. Reallocation of the band effectively gave birth to the mobile industry. The spectrum was initially used for analog cellular telephone systems. It constituted the entire spectrum allocation for the cellular industry for a dozen years.

From 1994 to 2000, the FCC auctioned the Personal Communications Service (PCS) spectrum, which made mobile voice communications a mass-market reality and unleashed a tidal wave of innovation and investment. These auctions more than tripled the stock of spectrum for commercial mobile radio services. With spectrum as the catalyst, the mobile industry profoundly changed during this period:

- The number of wireless providers increased significantly in most markets. [345]
- The per-minute price of cell phone service dropped by 50%. [346]
- The number of mobile subscribers more than tripled. [347]

- Cumulative investment in the industry more than tripled from $19 billion to over $70 billion.[348]
- The number of cell sites more than quadrupled, from 18,000 to over 8 0,000[349]
- Industry employment tripled from 54,000 to over 155,000.[350]

That same period saw a rapid uptick in the pace of industry innovation, from the deployment of new wireless technologies, to the introduction of new services such as Short Message Service, to the launch of the first nationwide service plans. As the DOJ explains, "mobile wireless users saw a substantial increase in the variety of pricing plans, lower per-minute prices, the introduction of newer generations of technology, and new features and functionality." [351]

The past decade has seen new spectrum come online in the 700 MHz, AWS and 2.5 GHz bands, providing a foundation for the nation's 4G wireless networks. The history of the 700 MHz band in particular demonstrates the importance of taking active steps to modernize spectrum policies in anticipation of future needs. In 2008, the FCC auctioned spectrum in the 700 MHz band, which was reallocated from the ultra high frequency (UHF) television band as part of America's transition to digital television (DTV). In 1997, the FCC established a ten year transition to digital broadcasting. Congress then modified that to mandate the transition would end when 85% of households owned digital receivers, a milestone that was difficult to measure and did not establish a specific deadline. At that time, this policy did not anticipate the explosion in mobile data that would begin a decade later; but in an effort to ensure a timely transition, Congress eventually accelerated the transition to 2009. In hindsight, setting a definitive transition date unlocked tremendous value for consumers and service providers. The auction garnered over $19 billion, and the spectrum is likely to provide a launch pad for two of the largest 4G network deployments in the coming years.

The Importance of Spectrum Flexibility

The current spectrum policy framework sometimes impedes the free flow of spectrum to its most highly valued uses. The federal government, on behalf of the American people and under the auspices of the FCC and NTIA, retains all property rights to spectrum. [352] In several instances, both agencies assign large quantities of spectrum to specific uses, sometimes tied to specific technologies. In some cases, this approach is appropriate to serve particular public interests that flexible use licenses and market-based allocations alone would not otherwise support. However, because mission needs and technologies evolve, there must be a public review process to ensure that decisions about federal and non-federal use that may have worked in the past can be revisited over time. In general, where there is no overriding public interest in maintaining a specific use, flexibility should be the norm.

In the case of commercial spectrum, the failure to re-visit historical allocations can leave spectrum handcuffed to particular use cases and outmoded services, and less valuable and less transferable to innovators who seek to use it for new services. The market for commercial, licensed spectrum does not always behave like a typical commodities market. Commercially licensed spectrum does not always move efficiently to the use valued most highly by markets and consumers. For example, a megahertz-pop may be worth a penny in one industry context and a dollar in another. Legacy "command and control" rules, high transaction costs and highly fragmented license regimes sometimes preserve outmoded band plans and prevent the aggregation (or disaggregation) of spectrum into more valuable license configurations.

Flexibility of use enables markets in spectrum, allowing innovation and capital formation to occur with greater efficiency. More flexible spectrum rights will help ensure that spectrum moves to more productive uses, including mobile broadband, through voluntary market mechanisms.

Spectrum flexibility, both for service rules and license transfers, has created enormous value. For example, the combined book value of flexible-use licenses held by the four national wireless providers, reflecting the prices paid at auction as well as in mergers and other corporate transactions, is over $150 billion.[353] Some economists estimate that the consumer welfare gains from spectrum may be 10 times the private value to the spectrum holder.[354] If this rule of thumb is true, it suggests that the social value of licensed mobile radio spectrum alone in the United States is at least $1.5 *trillion*.

The process of revisiting or revising spectrum allocations has historically taken 6-13 years, as described in Exhibit 5-C. Deploying networks adds still more time. Therefore, the FCC must maintain a forward-looking perspective as it evaluates reallocations or other rule changes that will make more spectrum available for broadband. In general, a voluntary approach that minimizes delays is preferable to an antagonistic process that stretches on for years. However, the government's ability to reclaim, clear and re-auction spectrum (with flexible use rights) is the ultimate backstop against market failure and is an appropriate tool when a voluntary process stalls entirely.

While flexibility in spectrum use is valuable, flexibility in access to spectrum can be just as important. Creating ways to access spectrum under a variety of new models, including unlicensed uses, shared uses and opportunistic uses, increases opportunity for entrepreneurs and other new market entrants to develop wireless innovations that may not have otherwise been possible under licensed spectrum models. In particular, unlicensed uses—which are technically not allocations *per se*— have enabled innovation in devices at the "edge" of the network. The spectrum novelties of today may become the predominant network technologies of tomorrow. Therefore, allowing technologically flexible access to spectrum is an essential innovation policy that the FCC should continue to develop.

With all of these considerations in mind, the U.S. government should take several actions to address urgent broadband spectrum needs.

Exhibit 5-C. Time Required Historically to Reallocate Spectrum

Band	First Step	Available for Use	Approximate Time Lag
Cellular (Advanced Mobile Phone System)	1970	1981	11 years
PCS	1989	1995	6 years
Educational Broadband Service (EBS)/Broadband Radio Service (BRS)	1996	2006	10 years
700 MHz	1996	2009	13 years
AWS-1	2000	2006	6 years

5.2. Ensuring Greater Transparency Concerning Spectrum Allocation and Utilization

Spectrum policy starts with transparency—disclosure about spectrum allocations, licensing and utilization. Transparency further increases the quality of policymaking by allowing outside parties—including citizens, companies, other government agencies and investors—to engage in the allocation process on an ongoing basis. The FCC and NTIA should create a system for greater transparency on spectrum allocation and utilization.

In the 1990s, the FCC began keeping electronic records of radio licenses and making this information available online. For example, the Universal Licensing System contains data on approximately two million licenses for over 30 different radio services. Nonetheless, it is difficult for stakeholders and the public to access and use these data. Much of the currently available information on spectrum resides in multiple "silos" requiring expert knowledge and interpretation. The complexity of the system and the resulting lack of transparency and usability create impediments to public policy and limit the emergence of new technologies that could employ such data to optimize use of the spectrum automatically.

RECOMMENDATION 5.1: The FCC should launch and continue to improve a spectrum dashboard.

Concurrent with the National Broadband Plan, the FCC is launching a beta release of a spectrum dashboard.[355] This Internet-based software enables user-friendly access to information regarding spectrum bands and licenses, including those that may be suitable for wireless broadband deployment. The initial version includes general information about non-federal use of spectrum bands in the range of 225 MHz to 3.7 GHz as well as more detailed information about bands of particular relevance to broadband.[356]

The spectrum dashboard will allow users to browse spectrum bands more easily, search for spectrum licenses, produce maps and download raw data for further analysis. For the first time, through a single FCC portal, users may access basic information on licenses (e.g., licensee name, contact information, frequency bands) as well as descriptions of allocations. Further, the dashboard includes information not previously available through the FCC website, such as the capability to search for licenses based on commonly recognizable names of companies (e.g., AT&T, T-Mobile, Verizon, etc.) and the amount of spectrum held by licensees on a county-by-county basis for many types of licenses. The screen shot below is illustrative of the spectrum dashboard user interface (see Exhibit 5-D).

The FCC should continue to improve and augment this spectrum dashboard over time, adding more comprehensive data on all bands, including commercial, state and local allocations within one year of the initial launch.[357] The FCC should also implement ongoing improvements to the database that will assist in spectrum policy planning and decision making, promote a robust secondary market in spectrum and improve communications services in all areas of the U.S., including rural, underserved and Tribal areas. Simultaneously, NTIA should develop similar information on federal spectrum operations.[358] This information should be made accessible through common links, with the intent of providing users a comprehensive view of combined FCC and NTIA information.

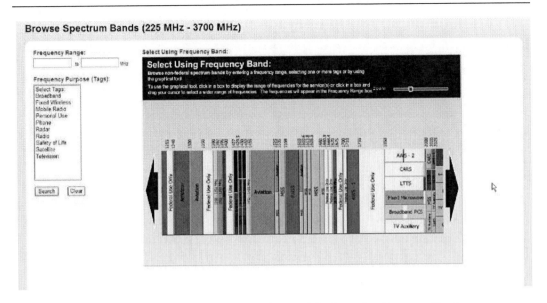

Exhibit 5-D. The Spectrum Dashboard: An Interactive Tool for Browsing Spectrum Bands

RECOMMENDATION 5.2: The FCC and the National Telecommunications and Information Administration (NTIA) should create methods for ongoing measurement of spectrum utilization.

To assist in understanding how, where and when spectrum resources are being used, the FCC and NTIA should develop scientific, statistically valid methods to measure and report the utilization of spectrum bands between 225 MHz and 3.7 GHz.[359] Some studies of spectrum utilization suggest that spectrum goes unused in many places much of the time, although critics assert that larger-scale studies are needed to draw more definitive conclusions.[360] More systematic measurement methods would help to provide a fact base that can inform policymaking, when combined with other forms of analysis.[362]

In the United Kingdom, the independent regulator Ofcom commissioned a study that provided a wealth of insights about spectrum utilization, and demonstrated the practicality of large-scale spectrum measurement.[363] An equivalent study, scaled to the larger scope of U.S. geography, would cost approximately $10–$15 million, and would provide insight into the utilization of spectrum resources with trillions of dollars of social value. Spectrum measurement for this study could use inexpensive frequency scanners installed on postal trucks or other fleet vehicles.

Information on spectrum utilization should be updated annually to provide an accurate snapshot of current use. Results should be made available to the public as an additional layer in the spectrum dashboard.

RECOMMENDATION 5.3: The FCC should maintain an ongoing strategic spectrum plan including a triennial assessment of spectrum allocations.

The recommendations in this chapter form the nucleus of a plan to ensure that spectrum is allocated to support the growth of broadband services and to accommodate new technologies that deliver it. Of course, every plan must evolve to accommodate new circumstances. Therefore, the FCC should maintain and continually update a strategic spectrum plan. Furthermore, the FCC should regularly refresh its analysis of the spectrum market with an assessment of the supply, usage and demand for spectrum, including potential sources of new spectrum. This assessment will draw on data collected from the spectrum dashboard and from spectrum measurement and utilization efforts, as described above in Recommendations 5.1 and 5.2, respectively. The spectrum assessment should be published every three years and should include an assessment of available spectrum and metrics by which to measure potential reallocation to alternative uses.

5.3. Expanding Incentives and Mechanisms to Reallocate or Repurpose Spectrum

The FCC has a variety of methods to manage spectrum pursuant to its authority under the Communications Act. In recent years, Congress has enhanced the FCC's spectrum management abilities by providing additional tools to promote more effective use of spectrum.

For instance, Congress enabled the FCC to develop procedures for assigning hundreds of megahertz more quickly and efficiently by providing the Commission with auction authority in 1993.[364] In 2004, with passage of the CSEA, Congress gave the FCC a powerful mechanism to encourage incumbent federal users to clear spectrum bands so that reallocated spectrum can be made available for commercial use.[365]

While these tools have served their purpose, they may prove insufficient for the spectrum policy challenges ahead. The broadband spectrum needs of the U.S. are growing as it is becoming more difficult to identify large swaths of spectrum—both federal and commercial—that can be reclaimed for auction. In many cases, the traditional auction model is likely to remain the preferred approach. Increasingly, however, the FCC will find itself looking for new ways to move spectrum to more productive uses. Given the practical challenges of reallocation, the FCC needs to create new incentives for incumbent licensees to yield to next-generation users.

RECOMMENDATION 5.4: Congress should consider expressly expanding the FCC's authority to enable it to conduct incentive auctions in which incumbent licensees may relinquish rights in spectrum assignments to other parties or to the FCC.

FCC spectrum licensees often possess certain rights and expectations that can make it difficult, in practice, for the FCC to reclaim and re-license that spectrum for another purpose. Contentious spectrum proceedings can be time-consuming, sometimes taking many years to resolve, and incurring significant opportunity costs. One way to address this challenge is by motivating existing licensees to voluntarily clear spectrum through incentive auctions. Congress should grant the FCC authority to conduct incentive auctions to accelerate productive use of encumbered spectrum.

Incentive auctions can provide a practical, market-based way to reassign spectrum, shifting a contentious process to a cooperative one. In an incentive auction, incumbents receive a portion of the proceeds realized by the auction of their spectrum licenses. This sharing of proceeds creates appropriate incentives for incumbents to cooperate with the FCC in reallocating their licensed spectrum to services that the market values more highly. A market-based mechanism—an auction—determines the value of the spectrum; market-based incentives, such as a share of the revenue received, encourage existing licensees to participate, accelerating the repurposing of spectrum and reducing the cost. Incentive auctions can be especially useful where fragmentation of spectrum licenses makes it difficult for private parties to aggregate spectrum in marketable quantities.

Incentive auctions can come in different forms. For example, in a "two-sided" auction, the FCC could act as a third-party auctioneer for the private exchange of spectrum between willing sellers and buyers, similar to a fine art auction. Alternatively, the FCC could offer a revenue-sharing enhancement to the existing spectrum auction system, in which some portion of revenues generated by an auction are shared between the U.S. Treasury and incumbent licensees who agree to relinquish their licenses.[366]

Incentive auctions present a more efficient alternative to the FCC's overlay auction authority, in which the FCC auctions encumbered overlay licenses and lets the new overlay licensees negotiate with incumbents to clear spectrum. These piecemeal voluntary negotiations between new licensees and incumbents introduce delays as well as high transaction costs as new licensees contend with holdouts and other bargaining problems. Anticipating these delays and negotiating costs, bidders typically pay significantly less for encumbered spectrum. The value of spectrum that must be cleared through such a voluntary process is reduced even more by uncertainty about the final cost of clearing.

Although sharing auction proceeds through incentive auctions means that some funds paid for spectrum will not go to the U.S. Treasury, incentive auctions should have a net-positive revenue impact for a variety of reasons: accelerated clearing, more certainty about costs, and the ability to auction adjacent spectrum that, due to technical rules, is not currently licensed.[367]

RECOMMENDATION 5.5: Congress should consider building upon the success of the Commercial Spectrum Enhancement Act (CSEA) to fund additional approaches to facilitate incumbent relocation.

The CSEA encourages federal incumbents to clear spectrum not being put to its most productive use and facilitates the updating of agency networks for enhanced broadband capabilities.[368] The CSEA establishes a Spectrum Relocation Fund to reimburse federal agencies operating on certain frequencies that have been reallocated to non-federal use.[369] With certain revisions, CSEA could become an even more effective tool for relocating federal incumbents from reallocated spectrum and for developing technological advances that will enable future reallocations of federal spectrum for wireless broadband.

The CSEA funding mechanism was first utilized in connection with the auction of former federal spectrum in the AWS -1 auction, which concluded in September 2006. The auction proceeds attributable to the former federal spectrum amounted to $6.85 billion, or half of the total net winning bids of $13.7 billion. The relocation costs totaled approximately $1 billion.[370] The auction's proceeds thus surpassed relocation costs by nearly $6 billion. At the same time, federal incumbents received modernized systems in other frequency bands. The experience of AWS-1 and CSEA proves that relocation can be a win-win-win: for incumbents, for the U.S. Treasury, and, most importantly, for the American public, which benefits from increased access to the airwaves.

Congress should consider improving the CSEA to ensure that a full range of costs are covered to provide federal agencies adequate incentives and assistance, including up-front planning, technology development and staffing to support the relocation effort. Further, agencies should be compensated for using commercial services and non-spectrum-based operations, in addition to dedicated spectrum-based system deployments. In particular, Congress should revise the CSEA to provide for payments of relocation funds to federal users that vacate spectrum and make use of commercial networks instead of alternative dedicated federal spectrum. Expanding the definition of reimbursable costs to include a federal incumbent's costs incurred to obtain telecommunications services from another existing network will promote agency use of shared commercial infrastructure, thereby freeing federal spectrum to be licensed for broadband deployment.

RECOMMENDATION 5.6: Congress should consider granting authority to the FCC to impose spectrum fees on license holders and to NTIA to impose spectrum fees on users of government spectrum.

In many spectrum bands, the government issues exclusive flexible use licenses that allow licensees to choose what services to offer and to transfer, lease or subdivide their spectrum rights.[372] Many spectrum licensees, however, have inflexible licenses that limit the spectrum to specific uses. These licensees do not incur opportunity costs for use of their spectrum; therefore, they are not apt to receive market signals about new uses with potentially higher value than current uses. The result can be inadequate consideration of alternative uses, artificial constraints on spectrum supply and a generally inefficient allocation of spectrum resources.

One way to address these inefficiencies is to impose a fee on spectrum, so that licensees take the value of spectrum into account.[373] Congress should grant the FCC and NTIA authority to impose spectrum fees, but only on spectrum that is not licensed for exclusive flexible use.[374]

Fees may help to free spectrum for new uses such as broadband, since licensees who use spectrum inefficiently may reduce their holdings once they bear the opportunity cost of spectrum. As the Government Accountability Office noted in a 2006 report to Congress, administrative fees "promote the efficient use of spectrum by compelling spectrum users to recognize the value to society of the spectrum that they use. In other words, these fees mimic the functions of a market."[375] However, it is not clear that the FCC and NTIA at present have authority to impose such fees.[376]

How best to set spectrum fees is a complex question. To be fully effective, fees should reflect the value of the spectrum in its best feasible alternative use, i.e., the opportunity cost. The prices observed from the auction of licenses for comparable spectrum are one indicator, but are imprecise due to differences in the technical characteristics, rules, interference environment and temporal variations in the supply and demand of the spectrum being compared. Recognizing these uncertainties, Ofcom has followed a practice of first setting low fees and then raising them gradually over time in response to observed changes in usage patterns (see Box 5-1). This is a prudent approach that gives users time to adjust to administrative pricing levels.

BOX 5-1. ADMINISTRATIVE INCENTIVE PRICING (AIP) IN THE UNITED KINGDOM

The U.K. has adopted a user fee system called AIP for commercial and government spectrum, including some held by the U.K. Ministry of Defence.[377] A recent Ofcom review of the AIP program concluded that AIP is meeting its objective of providing signals about market value to spectrum users so that they have an incentive to make optimal use of their spectrum.[378] By making the value of spectrum more salient, this pricing system has had its intended impact on government spectrum holders—military holders in particular. For example, spectrum costs are now included in business cases for major programs, long-term spectrum need plans are developed, and some unneeded spectrum has been transferred to other uses.[379]

In addition, a different approach to setting fees may be appropriate for different spectrum users. A fee system must avoid disrupting public safety, national defense, and other essential government services that protect human life, safety, and property and must account for the need to adjust funding through what can be lengthy budgetary cycles.

This year, the Obama Administration requested that Congress grant the FCC authority to impose spectrum fees. The Bush Administration made similar requests from 2001 to 2008.[380] Congress should grant this authority to the FCC and to NTIA.

RECOMMENDATION 5.7: The FCc should evaluate the effectiveness of its secondary markets policies and rules to promote access to unused and underutilized spectrum.

Secondary markets provide a way for some network providers to obtain access to needed spectrum for broadband deployment. While the FCC currently has rules that enable secondary markets, the record is mixed. Some public comments maintain that market-based policies have enabled a wide variety of entities, including non-nationwide providers, to obtain access to spectrum.[381] Others contend that unused or underutilized spectrum is not being made available to smaller providers, especially in rural areas where spectrum goes unused.[382] To ensure that secondary markets are functioning effectively, the FCC should identify and address barriers to more productive allocation and use of spectrum through secondary markets. The FCC should complete its assessment of potential barriers by the end of 2010.

The goal of the FCC's current secondary market policies is to eliminate regulatory barriers that might hinder access to, and permit more efficient use of, valuable spectrum resources.[383] The FCC has expressed concern that existing licensees may not fully utilize or plan to utilize the entire spectrum assigned to them; as a result, a substantial amount of spectrum may be underused, especially in rural areas.[384]

The FCC's policies and rules permit a variety of secondary market transactions: license transfers and assignments, partitioning and disaggregation of licenses, and spectrum leasing.[385] The FCC significantly streamlined the processing of lease transactions in 2003 and 2004.[386] The spectrum leasing policies also permit dynamic leasing arrangements that enable licensees and spectrum lessees to share use of the same spectrum. These arrangements take advantage of more sharing technologies that are possible as a result of innovations and advanced technologies such as cognitive radios.[387]

Preliminary analyses establish that there have been thousands of secondary-market transactions involving mobile broadband licenses over the last several years. These have included license transfers, including partitioning and disaggregation, and spectrum leases,[388] thus providing some evidence that the FCC's policies have enabled "spectrum to flow more freely among users and uses," as envisioned in the Commission's Secondary Markets Policy Statement.[389]

Despite this activity, the pressing spectrum requirements of broadband necessitate the need for a second look. In particular, the FCC should examine additional positive incentives that may assist in the development of secondary markets, such as reducing secondary market transaction costs like lease filing costs, and encouraging and facilitating the use of dynamic spectrum leasing arrangements that harness emerging technologies. The FCC should also consider a more systematic set of incentives, both positive and negative, to ensure productive use of spectrum to address broadband gaps in underserved areas.

5.4. Making More Spectrum Available within the Next 10 Years

RECOMMENDATION 5.8: The FCC should make 500 megahertz newly available for broadband use within the next 10 years, of which 300 megahertz between 225 MHz and 3.7 GHz should be made newly available for mobile use within five years.

In order to meet growing demand for wireless broadband services, and to ensure that America keeps pace with the global wireless revolution, 500 megahertz should be made newly available for mobile, fixed and unlicensed broadband use over the next 10 years. This spectrum would be made available for a variety of licensed and unlicensed flexible commercial uses, as well as to meet the broadband needs of specialized users such as public safety, energy, educational and other important users. Of this amount, 300 megahertz between 225 MHz and 3.7 GHz should be made available for mobile flexible use within five years. The timeline in Exhibit 5-E illustrates a schedule of actions that would fulfill this latter goal.

Exhibit 5-E. Actions and Timeline to Fulfill 300 Megahertz Goal by 2015

Band	Key Actions and Timing	Megahertz Made Available for Terrestrial Broadband
WCS	2010—Order	20
AWS 2/3[390]	2010—Order 2011—Auction	60
D Block	2010—Order 2011—Auction	10
Mobile Satellite Services (MSS)	2010—L-Band and Big LEO Orders 2011—S-Band Order	90
Broadcast TV[391]	2011—Order 2012/13—Auction 2015—Band transition/clearing	120
Total		300

In the bands below 3.7 GHz, 547 megahertz is currently licensed as flexible use spectrum that can be used for mobile broadband.[392] Of this amount, the Cellular and PCS bands compose 170 megahertz and represent the most intensively used spectrum today. The majority of the remaining 377 megahertz was auctioned or rebanded within the past six years and is just now coming online for mobile broadband deployment. This latter portion brought more than a three-fold increase in total spectrum for mobile services and provides a "runway" for the launch of next-generation mobile broadband services.

Looking ahead, operators, regulators and others have attempted to forecast the amount of spectrum that will be needed. Given current trends and future uncertainty, virtually all the major players in the wireless industry have stated on the record that more spectrum is needed.[393] Estimates range from 40 to 150 megahertz *per operator*.[394] In a recent public filing, CTIA summed up the industry-wide need to be approximately 800 megahertz.[395]

Several international organizations have also issued estimates, which vary widely. The ITU released an analysis in 2006 predicting that the total amount of spectrum needed to support mobile broadband in developed countries like the U.S. would be 1,300 megahertz by 2015 and up to 1,720 megahertz by 2020.[396] In the U.K., Ofcom commissioned an analysis of potential spectrum shortages. In the longer term, Ofcom believes that "improvements in spectral efficiency and the move to higher density network architectures will provide sufficient capacity to handle most high-end predictions of future demand." Still, Ofcom warns that "there could still be some limitations due to pressure on spectrum in the 2020 timeframe."[397]

Spectrum forecasts all incorporate a range of assumptions about future network capacity. Demand is difficult to predict due to uncertainties about future devices and user behavior. Supply is also difficult to predict since new technologies can change underlying operating costs, and access to key inputs like backhaul and tower sites can be limited by regulatory and other barriers (see Chapter 6).

In addition, bandwidth supply and demand are co-dependent. More bandwidth begets more data-intensive applications which begets a need for more bandwidth. Indeed, it is this virtuous cycle that has made broadband an innovation growth engine over the past decade— but also makes forecasting difficult.

The forecast of a need to make 300 megahertz available by 2015 reflects a set of reasonable assumptions about the evolution of supply and demand for mobile bandwidth and the resulting cost impact to service providers and their customers. On the demand side, the forecast considers the impact of smart- phones, wireless substitution in broadband, and traffic forecasts by industry experts, all of which incorporate the impact of new applications such as streaming video and cloud computing. On the supply side, the forecast considers expected increases in spectral efficiency from new technologies and increased spatial reuse of spectrum. The forecast also considers the inherent fragmentation in usable channels that is a byproduct of prior spectrum allocations and assignments to competing providers. The forecast suggests that demand growth is likely to outpace advances in technology and network deployment.

Although increased spectrum demands are primarily an urban phenomenon, several factors point to the need to make spectrum available nationwide. A national footprint improves carriers' cost structure, particularly in rural areas, by allowing the use of the same network equipment on a nationwide basis. Additionally, especially for highly propagating lower bands, increased availability of spectrum provides sufficient capacity to serve very large rural areas with a single cell, thereby further reducing the cost of rural deployments.

Three considerations further support the 300 megahertz goal. First, the accelerating nature of industry analyst demand forecasts makes clear that it is not a question of *if* the U.S. will require 300 megahertz of spectrum for mobile broadband, but *when*. Second, the use of flexible mechanisms such as incentive auctions to meet the need for more spectrum ensures that the market will self-correct if the forecast proves to be inaccurate. If the U.S. needs more than 300 additional megahertz for mobile broadband, prices for spectrum will go up and market mechanisms will help move spectrum to mobile broadband use. On the other hand, if the market demands less than that amount, prices may fall and less bandwidth will be made available for mobile broadband. Third, because there are ways to free up spectrum by delivering existing services more efficiently (rather than eliminating them altogether), the risk of overestimating spectrum needs is much lower than the risk of underestimating them.

This discussion focuses on *availability* of spectrum for mobile broadband. The FCC has a number of tools at its disposal to make spectrum usable for broadband, including changing allocations and modifying service, technical and auction rules. For some bands, reallocation may be the appropriate action. However, for others, reallocation may not be practical given international agreements and other constraints. In these situations, making spectrum available for broadband means taking steps appropriate to the specific circumstances of individual bands. It means working within the authority of the FCC or NTIA to remove legacy constraints that limit the usefulness of a band for appropriate broadband services and applications.

Increasing spectrum availability does not necessarily imply a traditional spectrum auction. In instances where the government is able to reclaim spectrum, a traditional auction will be the most appropriate and efficient method of reallocation. In other cases, the most expedient path to repurposing spectrum to broadband may be to use incentive auctions or to take other steps to energize the secondary markets for a particular band.

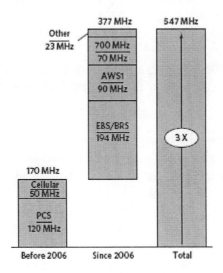

Exhibit 5-F. Spectrum Baseline

Ultimately, the cost of not securing enough spectrum may be higher prices, poorer service, lost productivity, loss of competitive advantage and untapped innovation. It would not be wise for America to bet its mobile future on a strategy of "demand reduction." As noted above, it can take many years to make spectrum available for new uses. With only 50 megahertz currently in the FCC pipeline, now is the time to act. Specifically, the following spectrum bands should be prioritized for reallocation or other rule changes in order to make progress toward the five-year, 300-megahertz goal.

RECOMMENDATION 5.8.1: The FCC should make 20 megahertz available for mobile broadband use in the 2.3 GHz Wireless Communications Service (WCS) band, while protecting neighboring federal, non-federal Aeronautical Mobile Telemetry (AMT) and satellite radio operations.

The Commission established the 2.3 GHz WCS band in 1997.[398] At that time, the FCC adopted strict operating parameters to protect operations in the adjacent Satellite Digital Audio Radio (SDARS) band. Certain WCS technical rules, particularly the out-of-band emission (OOBE) limits, largely preclude the provision of mobile broadband services in the spectrum. Based on an extensive record,[399] the FCC should revise certain technical rules, including the WCS OOBE limits, to enable robust mobile broadband use of the 2.3 GHz WCS spectrum, while protecting federal, non-federal AMT and satellite radio operations in the neighboring SDARS band.

Since the FCC first auctioned the WCS spectrum in 1997, a number of new and robust wireless telecommunications technologies have been successfully introduced, including Time Division Duplex and Orthogonal Frequency Division Multiplexing[400] technologies. Such dynamic technologies, coupled with the exploding demand for broadband services, suggest that the WCS spectrum may provide fertile ground for the provision of high-value mobile broadband services to the public. The same frequency band is currently being used in South Korea and other countries to deploy mobile WiMAX service today. Accordingly, the FCC should accelerate efforts to ensure that the WCS spectrum is used productively for the benefit of all Americans.

RECOMMENDATION 5.8.2: The FCC should auction the 10 megahertz Upper 700 MHz D Block for commercial use that is technically compatible with public safety broadband services.

The FCC should auction the Upper 700 MHz D Block for commercial use with limited technical requirements that would ensure technical compatibility between the D Block and the adjacent public safety broadband spectrum block and would enable, but not obligate, the licensee to enter into a spectrum-sharing partnership with the neighboring Public Safety Broadband Licensee (PSBL). Due to its favorable propagation characteristics and the emergence of a 4G technology ecosystem in the 700 MHz band, the D Block is likely to have high value for the delivery of commercial mobile broadband services. Our recommendation is intended to unlock this value while supporting the simultaneous development of public safety broadband capability through equipment development, roaming and priority access, pursuant to the recommendations described in Chapter 16.

The D Block consists of 10 megahertz (2x5 megahertz) that did not receive a winning bid in the 700 MHz auction held in 2008. The original rules required the D Block licensee to enter into a public-private partnership with the PSBL to build a public safety broadband network. The absence of meaningful bidding activity indicated that the public safety obligations as designed were not commercially viable. The approach recommended in Chapter 16 would allow for a voluntary partnership between public safety broadband spectrum holders and commercial partners, including the D Block licensee(s). Limited technical requirements on the D Block can help maximize the number of partners available to public safety, while also maximizing the commercial potential of the spectrum.

Specifically, the D Block should be auctioned with the following rules:

- The D Block licensee(s) must use a nationally standardized air interface. The emerging consensus in the public safety community is that the LTE family of standards is most appropriate.[401] A standardized air interface will ensure that the D block will be technically capable of supporting roaming and priority access by public safety users of the neighboring public safety broadband block.
- The FCC should initiate a proceeding to enable authorized state, local and federal public safety users to have rights to roaming and priority access for broadband service on commercial networks subject to compensation, as described in Chapter 16. Before the D Block is auctioned, it must be clear that D Block licensee(s) are required to provide such roaming and priority access to public safety users.
- D Block licensee(s) must develop and offer devices that operate both on the D Block and the neighboring public safety broadband block, with a path toward scale production of components and devices that can utilize both blocks, in order to stimulate the public safety broadband equipment "ecosystem."[402]
- The D Block license should be subject to commercially reasonable buildout requirements. The Commission should also consider the use of incentives to promote additional deployment by the D Block licensee(s) for the benefit of rural citizens and for public safety agencies.

The FCC should promptly take steps needed to implement these recommendations.

RECOMMENDATION 5.8.3: The FCC should make up to 60 megahertz available by auctioning Advanced Wireless Services (AWS) bands, including, if possible, 20 megahertz from federal allocations.

The FCC should move expeditiously to resolve the future of the spectrum already allocated for AWS. The AWS-2 and AWS3 allocations consist of the following bands:

- *AWS-2 "H" Block.* Total of 10 megahertz at 1915–1920 MHz paired with 1995–2000 MHz.
- *AWS-2 "J" Block.* Total of 10 megahertz at 2020–2025 MHz paired with 2175–2180 MHz.
- *AWS-3 Band.* Twenty megahertz unpaired at 2155– 2175 MHz.

The FCC proposed rules for AWS-2 spectrum in 2004 and sought comment on AWS-3 spectrum in 2007. Potential synergies exist between the AWS-3 band and spectrum currently allocated to federal use at 1.7 GHz. There are a number of countries that have allocated spectrum in the 1710–1780 MHz band for commercial use[403] and devices already exist in the international market for that spectrum. Consequently, pairing the AWS-3 band with spectrum from the 1755–1780 MHz band has the potential to bring benefits of a global equipment ecosystem to this band.

NTIA, in consultation with the FCC, should conduct an analysis, to be completed by October 1, 2010, of the possibility of reallocating a portion of the 1755–1850 MHz band to pair with the AWS-3 band. NTIA has commented that, "the Administration supports exploring both commercial and government spectrum available for reallocation."[404] If there is a strong possibility of reallocating federal spectrum to pair with the AWS3 band, the FCC, in consultation with NTIA, should immediately commence reallocation proceedings for the combined band. If, at the end of this inquiry, there is not a strong possibility of reallocation of federal spectrum, the FCC should proceed promptly to adopt final rules in 2010 and auction the AWS-3 spectrum on a stand-alone basis in 2011.

The AWS-2 "J" block also has potential synergies with AWS-3 and with the adjacent MSS S-Band. If developments in those other bands warrant, the FCC should integrate the J Block into one or the other of the band plans in order to maximize the broadband potential of the spectrum. For example, it may make sense to group the J Block with contiguous S-Band spectrum if the AWS-3 band is paired with federal spectrum, or to group the J Block with the AWS-3 band if there is no reallocation of federal spectrum.

RECOMMENDATION 5.8.4: The FCC should accelerate terrestrial deployment in 90 megahertz of Mobile Satellite Spectrum (MSS).

The FCC should build on past efforts to enable terrestrial deployment in MSS bands. The MSS allocation consists of a significant amount of bandwidth with propagation characteristics suitable for mobile broadband. The FCC should take actions that will optimize license flexibility sufficient to increase terrestrial broadband use of MSS spectrum, while preserving market-wide capability to provide unique mission- critical MSS services.

MSS is a radio communication service involving transmission between mobile earth stations and one or more space stations. MSS can provide mobile communications, from a handheld device such as a smartphone, in areas where it is difficult or impossible to provide coverage using terrestrial base stations, such as in remote or rural areas and non-coastal maritime regions, and at times when coverage may be unavailable from terrestrial-based networks, such as during hurricanes and other natural disasters. For this reason, MSS has a unique role in our communications infrastructure, and the preservation of sufficient spectrum for MSS incumbent users is important for ensuring continuity of mission-critical communications services.

The FCC first allocated spectrum for MSS in 1986. Since then, the Commission has allocated spectrum in four bands to MSS: the Little LEO Band, the L-Band, the S-Band, and the Big LEO band. The latter three MSS bands are capable of supporting broadband service, and several public comments have identified MSS as a potential focal point for a broadband spectrum strategy.[405] Exhibit 5-G provides a snapshot of the current broadband-capable MSS bands.

Exhibit 5-G. Broadband- Capable MSS Bands

MSS Band	Allocated Bandwidth	Bandwidth Usable for Terrestrial Broadband	Licensees	Subscribers[406]
L-band	Two 34-megahertz blocks at 1525–1 559 MHz, 1626.5–1660.5 MHz[407]	40 megahertz	SkyTerra	18,235
			Inmarsat	254,000
S-band	Two 20-megahertz blocks at 2000– 2020 MHz, 2180–2200 MHz	40 megahertz	DBSD (ICO)	—
			TerreStar	—
Big LEO	Two 16.5-megahertz block at 1610– 1626.5 MHz, 2483.5–2500 MHz,	10 megahertz	Globalstar	382,313
			Iridium	359,000

The FCC adopted rules in February 2003 that allow MSS operators to construct and operate Ancillary Terrestrial Components (ATCs) in their licensed spectrum. Although satellites permit nationwide coverage, satellite links are limited without line-of-sight transmission, particularly in urban areas and inside buildings. The ATC rules allow MSS providers to deploy terrestrial networks to enhance coverage in areas where the satellite signal is attenuated or unavailable.

When it enacted the ATC rules, the FCC stated that it would "authorize MSS ATC subject to conditions that ensure that the added terrestrial component remains ancillary to the principal MSS offering."[408] In this regard, the FCC adopted gating criteria that require MSS operators to satisfy certain requirements prior to using ATC. Specifically, the FCC requires MSS licensees to provide substantial satellite service, including satisfying geographic and temporal coverage requirements, maintaining spare satellites, and offering commercial service to the public for a fee. In addition, MSS licenses must integrate MSS and ATC services, including, notably, a requirement that all ATC handsets must have a satellite communications capability.

No licensee is operating a live commercial ATC network at this time, although Globalstar, SkyTerra, DBSD, and recently Terrestar have been authorized to provide ATC services. So far, the ATC gating criteria have made it difficult for MSS providers to deploy ancillary terrestrial networks, as well as to establish partnerships with wireless providers or other well-capitalized potential entrants. Requiring full satellite coverage prior to initiation of ATC forces MSS licensees to incur substantial costs and obligations to provide satellite services before integrated ATC can be deployed. Several MSS licensees have sought waivers of the ATC requirements in an effort to create a more cost-effective framework for terrestrial deployment.[409] Some critics of the ATC rules consider the added costs to be appropriate, given the fact that the terrestrial rights were never assigned through competitive bidding.

Looking forward, commercial and technological developments suggest that the potential exists for increased deployment of ATC networks and possible inclusion in consumer devices. In recent months, multiple providers have unveiled business partnerships with terrestrial-based providers and equipment manufacturers, indicating that the MSS industry might be ready to deploy ATC networks with updated business plans that appeal to mass-market consumers.[410] In addition, satellite technology continues to advance, with the development of larger satellite antennas designed to work with smaller terrestrial mobile handsets that more closely resemble mass-market mobile devices. However, until these technical advances are market-tested, it is premature to conclude that the current ATC regime will succeed in deploying terrestrial broadband networks and attracting commercial interest.

From the standpoint of promoting broadband through increased use of the MSS spectrum, the FCC can take action to accelerate terrestrial deployments in the MSS bands. At the same time, the FCC must take care to ensure that the MSS market continues to provide public safety and government users with mission-critical satellite capabilities. To this end, the FCC should seek to ensure that these actions to introduce greater flexibility in the MSS spectrum do not interfere with non-ATC MSS operations, or with the ability of MSS providers to supply emergency "surge capacity" when authorized by the FCC, especially in light of the important role these licensees play in ensuring public safety.

Specifically, the FCC should take the following actions to promote more productive use of MSS spectrum:

- The FCC and other government agencies should work closely with L-Band licensees and foreign governments to accelerate efforts to rationalize ATC-authorized L-Band spectrum to make it usable for broadband ATC service.
- The FCC should add a primary "mobile" (terrestrial) allocation to the S-Band, consistent with the international table of allocations, which will provide the option of flexibility to licensees to provide stand-alone terrestrial services using the spectrum. Exercise of this option should be conditioned on construction benchmarks, participation in an incentive auction, or other conditions designed to ensure timely utilization of the spectrum for broadband and appropriate consideration for the step-up in the value of the affected spectrum.
- The FCC should grant licensees flexibility under the ATC regime in the 2.4 GHz Big LEO band, already being used for terrestrial broadband deployments, to make this spectrum permanently suitable for terrestrial broadband service, subject to appropriate safeguards to promote the public interest.

The FCC should initiate proceedings on these recommendations immediately.

RECOMMENDATION 5.8.5: The FCC should initiate a rule-making proceeding to reallocate 120 megahertz from the broadcast television (TV) bands, including:[411]

- **Update rules on TV service areas and distance separations and revise the Table of Allotments to ensure the most efficient allotment of six-megahertz channel assignments as a starting point.**
- **Establish a licensing framework to permit two or more stations to share a six-megahertz channel.**
- **Determine rules for auctions of broadcast spectrum reclaimed through repacking and voluntary channel sharing.**
- **Explore alternatives—including changes in broadcast technical architecture, an overlay license auction, or more extensive channel sharing—in the event the preceding recommendations do not yield a significant amount of spectrum.**
- **Take additional measures to increase efficiency of spectrum use in the broadcast TV bands.**

The spectrum occupied by broadcast television stations has excellent propagation characteristics that make it well-suited to the provision of mobile broadband services, in both urban and rural areas. Enabling the reallocation of a portion of this spectrum to broadband use in a way that would not harm consumers overall has the potential to create new economic growth and investment opportunities with limited potential impact on broadcast business models. Consumers would retain access to free, over-the-air television. Reallocation would focus primarily on major markets where the broadcast TV bands are most congested and the need for additional spectrum for broadband use will be greatest.[412] Moreover, the FCC should study and develop policies to ensure that its longstanding goals of competition, diversity, and localism are achieved. Changes to the TV broadcast spectrum need to be carefully considered to weigh the impact on consumers, the public interest, and the various services that share this spectrum, including low-power TV, wireless microphones and prospective TV white space devices. While the FCC has performed initial analyses to consider the viability of various options, further work will be required and all options must be examined through rulemaking.

Over-the-air television continues to serve important functions in our society. It delivers free access to news, entertainment and local programming, and provides consumers an alternative video service to cable or satellite television.[413] It is the only such service to a segment of the population that either cannot afford paid television or broadband services or cannot receive those services at their homes currently. Over-the-air television also serves numerous public interests, including children's educational programming, coverage of community news and events, reasonable access for federal political candidates, closed captioning and emergency broadcast information. Through broadcast television, the FCC has pursued longstanding policy goals in support of the Communications Act, such as localism and diversity of views. Finally, emerging broadcast applications, such as mobile DTV and data casting, may provide an opportunity to take advantage of the relative efficiencies of point-to-multipoint and point-to-point architectures in order to deliver various types of content in the most spectrum-efficient ways.

Because of the continued importance of over-the-air television, the recommendations in the plan seek to preserve it as a healthy, viable medium going forward, in a way that would not harm consumers overall, while establishing mechanisms to make available additional spectrum for flexible broadband uses.

The need for such mechanisms is illustrated by the relative market values of spectrum for alternative uses. For example, the market value for spectrum used for over-the-air broadcast TV and the market value for spectrum used for mobile broadband currently reveal a substantial gap.[414] In 2008, the FCC held an auction of broadcast TV spectrum in the 700 MHz band recovered as part of the DTV Transition. That auction resulted in an average spectrum valuation for mobile broadband use of $1.28 per megahertz-pop.[414] The TV bands have propagation characteristics similar to those of the 700 MHz band. However, the market value of these bands in their current use ranges from $0.11 to $0.15 per megahertz-pop.[415] Other attempts to size the current economic value of spectrum for over-the-air television using alternative methods have resulted in comparable megahertz-pop valuations.[416] While there are other possible valuation methods that could result in further variations, this analysis illustrates the order of magnitude of the gap.

This gap in economic value between spectrum used for wireless broadband and spectrum used for over-the-air broadcast television reflects in part the long-term market trends in both industries. Demand for mobile broadband services is growing rapidly with the introduction of new devices (e.g., smartphones, netbooks) and with 3G and 4G upgrades of mobile networks. The mobile broadband industry is expected to continue to drive innovation, job growth and investment through the next decade.

Over-the-air broadcast television, on the other hand, faces challenging long-term trends. The percentage of households viewing television solely through over-the-air broadcasts steadily declined over the last decade, from 24% in 1999 to 10% in 2010.[417] Since 2005, broadcast TV station revenues have declined 26%,[418] and overall industry employment has declined as well.[419]

The gap in economic value also reflects two characteristics of broadcast TV licensing constraints. First, since broadcast TV requires channel interference protections, only a fraction of the total spectrum allocated to broadcast TV is currently being used directly by stations.[420] Second, as a universally available, free over-the-air medium, television broadcasting has long been required to fulfill certain public interest and technical requirements. It is important to allow television broadcasting to continue to fulfill these obligations to local communities, while at the same time utilizing less spectrum, thus freeing up additional airwaves for mobile broadband. This could yield more service to local communities overall—broadcast television that consumers have always received along with more and better mobile broadband connectivity.

The FCC should initiate a rulemaking proceeding to real-locate 120 megahertz from the broadcast TV bands. The proceeding should pursue four sets of actions in parallel to achieve this objective. In addition, the FCC should take a fifth set of actions to increase efficiency of spectrum use in the broadcast TV bands.

1. **Update rules on TV service areas and distance separations and revise the Table of Allotments to ensure the most efficient allotment of 6 megahertz channel assignments as a starting point.**

Changes to the current broadcast TV technical rules and channel assignments could reduce the amount of spectrum allocated to its use without impacting the bandwidth of any individual station. First, updating the technical rules defining TV service areas and required distance separations between stations may enable stations to operate at currently prohibited spacing on the same or adjacent channels without increasing interference to unacceptable levels.[421] Second, the FCC may be able to "repack" channel assignments more efficiently to fit current stations with existing 6 megahertz licenses into fewer total channels, thus freeing spectrum for reallocation to broadband use.

Repacking alone could potentially free up to 36 megahertz of spectrum from the broadcast TV bands.[422] If the repacking takes place in conjunction with updated technical rules and some or all of the additional recommendations below, the amount of spectrum recovered could be substantially greater.[423]

2. **Establish a licensing framework to permit two or more stations to share a 6 megahertz channel.**

With the appropriate regulatory structure in place, broadcasters could combine multiple TV stations onto single six-megahertz channels. The current broadcast TV rules provide each licensee a six-megahertz channel that is capable of transmitting data at a rate of 19.4 Mbps.

Television stations broadcast their primary video signal either in high definition (HD), requiring approximately 6–17 Mbps, or in standard definition (SD), requiring approximately 1.5–6 Mbps.[424]

Two stations could generally broadcast one primary HD video stream each over a shared six-megahertz channel.[425] Some stations are already broadcasting multiple HD streams simultaneously today and claim to deliver "spectacular" signal quality that "consistently satisfies [their] discerning viewers." [425] Alternatively, more than two stations broadcasting in SD (not HD) could share a six-megahertz channel. Numerous permutations are possible, including dynamic arrangements whereby broadcasters sharing a channel reach agreements to exchange capacity to enable higher or lower transmission bit rates depending on market-driven choices. [426] The FCC should ensure that the framework it adopts retains carriage rights for the primary signal of each station with a modified license to share a six-megahertz channel.[427] The FCC also should address any potential concerns regarding anti-competitive behavior or media ownership consolidation arising from such arrangements.

To date, although there are examples of individual stations employing these techniques to broadcast multiple HD streams or signals from two major broadcast networks, there are no examples of two or more stations combining transmissions to share a single channel. Television stations will need to consider their desire to multicast additional video streams, such as digital side channels and mobile DTV streams, relative to the possible sharing of channels. Multicasting mobile DTV streams and digital side channels requires additional bandwidth to ensure reception quality. Stations are just now beginning to deploy such services, and it is not yet clear whether they will be widely accepted or how they might affect the ability of stations to share channels.

3. Determine rules for auctions of broadcast spectrum reclaimed through repacking and voluntary channel sharing.

The FCC should conduct an auction of some or all of the nationwide, contiguous spectrum recovered through the repacking described above and through decisions by stations to voluntarily relinquish some or all of their bandwidth. Stations would receive a share of the proceeds from the spectrum they directly contribute to the auction.[428] By this time, Congress would need to have authorized the FCC to conduct such an incentive auction and share proceeds. Stations could choose to share channels voluntarily under the regulatory framework established for channel sharing described above in order to participate in the incentive auction. Following the auction, stations continuing to broadcast over the air would receive channel assignments according to a new Table of Allotments, modified licenses if they are sharing a channel with other stations, and reimbursement from auction winners for any expenses incurred as a result of repacking.

The preference is to establish a voluntary, market-based mechanism to effect a reallocation, such as the incentive auctions described previously in this chapter. To date, markets have only operated within the broadcast TV allocation and license regime—e.g., ownership of TV stations changing hands, stations going out of business and returning licenses for reissue, or stations leasing bandwidth for other broadcast uses. Additional market mechanisms could broaden choices for both incumbent and would-be licensees and facilitate movement of spectrum to flexible broadband use. Market trends and legal and regulatory developments could affect the outcome of these auctions, including the demand trajectory for mobile broadband services, the financial condition of broadcast TV stations, the resolution of Cablevision's must-carry challenge in the Supreme Court,[429] and the outcome of the FCC's quadrennial review of broadcast ownership rules.

The voluntary, market-based reallocation should be implemented in a way that will have limited long-term impact on consumers overall, broadcast business models and the public interest, including the FCC's goals with respect to competition, diversity and localism. Moreover, the substantial benefits of more widespread and robust broadband services would outweigh any impact from reallocation of spectrum from broadcast TV.

Consumers would continue to receive over-the-air television. Some over-the-air consumers would lose reception from one or more stations as a result of stations voluntarily going off the air, choosing to share channels with other stations (and thus change their service area), or experiencing loss in service area due to increased interference following a repacking. Others might gain reception from one or more stations as a result of changes to service areas. In addition, over-the-air consumers would need to reorient antennas or rescan their TVs, as they did following the DTV Transition in June 2009.

There are several actions the FCC should take to mitigate the impact on over-the-air consumers. First, as a matter of policy, the FCC should ensure that consumers in rural areas and smaller markets retain service and are not significantly impacted by these changes. The reallocation mechanisms are most likely to be in the country's largest, most densely populated markets, where the greatest demand for spectrum and the greatest congestion within the broadcast TV bands coincide. Consumers in these markets tend to have a relatively large number of alternatives to view television content—a median of 16 over-the-air full-power television stations, over-the-air low- power stations and digital multicast channels, at least three to four multichannel video programming distributors (MVPDs), and a growing amount of broadband Internet video content, increasingly delivered to the TV (see Chapter 3).

Second, in all markets, the FCC should seek to ensure that longstanding policy goals under the Communications Act are to be met, such as localism, viewpoint diversity, competition and opportunities for new entrants to participate in the industry, including women and members of minority groups.

Finally, the FCC should explore through rulemaking proceedings appropriate compensation mechanisms and levels to retain free television service for those consumers who meet the criteria established. For example, these consumers could become eligible for a "lifeline" video service from MVPDs, consisting of all over-the-air television signals in their market. These mechanisms could be coordinated with the provision of broadband service for unserved and underserved populations. Congress would determine the criteria and compensation mechanisms, if necessary, and allocate the funding (e.g., from auction proceeds). In all areas, the incentives provided by the incentive auction, the focus of reallocation mechanisms only where needed, and ongoing FCC vigilance would ensure that decisions made by broadcasters and the FCC itself do not adversely affect particular communities of American consumers.[430]

Under the recommended voluntary approach, some broadcasters moving channel assignments would need to replace transmission equipment (with reimbursement) and adjust transmission parameters to match previous coverage areas. Any impact on a broadcast TV station's revenue or business model would result from a decision that station chose to make regarding participation in the incentive auction. Broadcast TV stations derive their revenue primarily based on "eyeballs," or the size and composition of viewership on their primary video signal.[431] Stations gain viewers through distribution reach and the appeal of their programming.[432] The reallocation mechanisms described above could have a negative impact on reach for some stations, but would most likely affect reach in a neutral to positive way overall.[433] The effect on programming appeal would depend on the choices broadcasters make as a result of an incentive auction and on the importance of and impact on picture quality to viewers. Based on analyses of programming and signal throughput, as well as case examples, two stations could each broadcast a primary video stream in HD simultaneously over the same channel without causing material changes in the current consumer viewing experience.[434] As a result of neutral impacts on both reach and programming appeal of stations' primary signals, the impact of a voluntary, market- based reallocation on current revenue streams for stations that continue broadcasting over-the-air could be minimal.

The voluntary incentive auction would give stations another variable to consider in choosing the type of primary video signal to broadcast over-the-air, HD or SD, and in pursuing new business models enabled by the DTV Transition: multicasting and mobile DTV. Stations could balance these choices, based on projected market demand for these services, against the market value of bandwidth for other uses, such as wireless broadband.

Multicasting additional digital sub-channels can generate advertising, leasing or subscription revenue. To date, stations have launched approximately 1,400 multicast channels, or fewer than one per station on average.[435] The revenue generated by such services has been modest thus far and is forecast to remain so in the near term—0.9% of revenue for broadcast TV stations in 2010, projected to rise to 1.5% of revenue in 2011.[436]

The second newly emerging business model, mobile DTV, could serve as a potential evolution path for broadcasters to fixed/mobile and broadcast/broadband convergence. In particular, broadcasting popular video content to mobile devices may help offload growing video streaming traffic from mobile point- to-point broadband networks.[437] As of July 2009, approximately 70 broadcast stations serving 28 markets had announced plans to begin mobile broadcasting through the Open Mobile Video Coalition. The business model for mobile DTV is uncertain, with forecasts and comparisons to domestic and international examples representing varying points of view.[438] Many entities are pursuing the delivery of television content to mobile devices, but the method of delivery that will be favored by consumers and be successful in the market has yet to be determined.

By preserving over-the-air television as a healthy, viable medium, while reallocating spectrum from broadcast TV bands to flexible mobile broadband use, the recommendations in this plan seek to protect longstanding policy goals and public interests served by overthe-air television and further support those served by broadband use. In particular, all stations that broadcast a primary video signal would continue to serve existing public interest requirements.

Depending on the particular mechanisms pursued and on the individual choices of TV stations, the reallocation mechanisms could impact the number and diversity of broadcast "voices" in a community or market. As noted above, these effects would primarily take place in major markets, where the number and diversity of local community voices are the highest. The FCC should implement these mechanisms consistently with its policies supporting competition, localism, and diversity, and with the outcome of the current quadrennial review of broadcast ownership rules. In particular, the FCC should study the potential impact on minority and women ownership of TV stations. Recommendations in the plan to create a public interest media trust fund (see Chapter 15) will fortify public media across platforms, further bolstering viewpoint diversity and localism in communities throughout the country.

4. **Explore alternatives—including changes in broadcast technical architecture, an overlay license auction or more extensive channel sharing—in the event the preceding recommendations do not yield a significant amount of spectrum.**

If the FCC does not receive authorization to conduct incentive auctions, or if the incentive auctions do not yield a significant amount of spectrum, the FCC should pursue other mechanisms.[439] Through a rule-making proceeding, it should consider other approaches, potentially including:

- *Transition to a cellular architecture on a voluntary or involuntary basis.* With a cellular architecture, stations would broadcast television service over many low-powered transmitters that collectively provide similar coverage to the current architec- ture with one high-powered transmitter. Cellularizing the architecture could reduce or eliminate the need for channel interference protections that result in only a fraction of the total spectrum allocated to broadcast TV being used directly by stations.[440] A cellular architecture could also facilitate broadcasters' offerings of converged broadcast/broadband services. The FCC has approved Distributed Transmission Systems/ Single Frequency Networks (DTS/SFN), using multiple transmitters operating on a single channel, as one alternative for a cellular architecture.[441] Other alternatives are possible, such as a Multi-Frequency Network (MFN).[442] Moving to a cellular architecture would be expensive, take a long period of time, and potentially introduce substantial operational challenges for broadcasters. The potential spectrum dividend is unknown at this point, but could be very high.[443] Though stations could voluntarily move to a cellular architecture on individual bases, such moves would achieve greater overall spectrum efficiency if they are conducted in a coordinated manner by all stations in major markets. DTS/SFN and MFN are cutting-edge technologies that need to be developed further to evaluate their viability and the various trade-offs. The FCC should encourage and closely monitor their development.
- *Auction of overlay licenses.* Under its current authority,[444] the FCC could auction overlay, flexible-use licenses with secondary rights in the broadcast TV bands. Overlay auction winners would negotiate with broadcast TV stations and other licensed users to clear their respective bands.[445] Proceeds from the overlay auction would go to the U.S. Treasury but could be significantly lower than the proceeds of an incentive auction, primarily due to greater uncertainty over the amount and timing of spectrum recovered.[446]
- *More extensive channel sharing of two or more broadcast TV stations on a single six-megahertz channel.* Under this alternative, the FCC would modify licenses to require channel sharing where necessary.
- *Other innovative solutions that may emerge.* Stations would not share in auction proceeds under these alternatives, but they should receive reimbursement from auction winners for any relocation or other transition expenses incurred.

5. Take additional measures to increase efficiency of spectrum use in the broadcast TV bands.

In addition to the above, the following recommendations would enable more efficient use of the broadcast TV spectrum:

- *Full-power TV spectrum fees.* If authorized by Congress, the FCC should consider assessing spectrum fees on commercial, full-power broadcast TV licensees as part of a broader review of broadcast ownership rules and public interest obligations.[447]
- *Low power DTV transition.* The FCC should establish a deadline to achieve the DTV transition of low-power TV (LPTV) stations by the end of 2015 or after the reallocation of spectrum from the broadcast TV bands is complete.[448] In addition, the FCC should grant similar license flexibility to LPTV stations post-DTV transition as full-power stations have, allow LPTV stations to use certain technologies (such as mask filters) to enable more efficient channel allotments, and authorize LPTV stations to participate in incentive auctions.
- *Very high frequency (VHF) reception issues.* The FCC should pursue additional options to address VHF reception issues, such as increased power limits or adoption of enhanced antenna and receiver standards.[449] Without these measures, VHF stations may continue to request channel reassignments to the UHF band, complicating efforts to reallocate spectrum from that band to mobile broadband use.
- *Trust fund for public media.* Congress should consider legislation to establish an endowment to fund public interest media from auction proceeds or spectrum fees (see Chapter 15).

The recommendations in this section depending on the extent to which that are implemented, might not significantly affect other current or future occupants of the broadcast TV bands, notably land mobile radio system (LMRS) operators, wireless microphone users, and TV white spaces devices. LMRS operators would continue to operate under existing licenses in channels 14–20 in certain major metropolitan areas. The FCC should complete rulemaking proceedings on the above steps for which it currently has authority as soon as practicable, but no later than 2011, and should conduct an auction of some or all of the reallocated spectrum in 2012. If Congress grants the FCC authority to conduct incentive auctions prior to the auction in 2012, then the FCC should delay any auction of reallocated broadcast TV spectrum until 2013. This delay would allow time to complete rulemaking proceedings on a voluntary incentive auction. All reallocated spectrum should be cleared by 2015. Though aggressive by historical standards, this timeline would bring additional mobile broadband capacity to market when it may be most needed.

5.5. Increasing the Flexibility, Capacity and Cost-Effectiveness of Spectrum for Point-to -Point Wireless Backhaul Services

Many wireless providers increasingly rely on microwave for backhaul, especially in rural areas. Therefore, the FCC should take steps to ensure that sufficient microwave spectrum is available to meet current and future demand for wireless backhaul, especially in the prime bands below 12 GHz. As a starting point, the FCC is considering revisions to its Part 101 rules permitting operation of wider channels in the Upper 6 GHz Band, and faster activation of links on additional channel pairs in the 23 GHz Band. The FCC should take further actions to enhance the flexibility and speed with which companies can obtain access to spectrum for use as wireless backhaul, which is critical to the deployment of wireless broadband and other wireless services

Backhaul costs currently constitute a significant portion of a cellular operator's network operating expense. With 4G deployments, this burden will become more acute as the demand for backhaul capacity increases. When fiber is not proximate to a cell site, microwave backhaul can often provide a cost- effective substitute for data rates up to 600 Mbps. Further, in certain remote geographies, microwave is the only practical high-capacity backhaul solution available. Policies that facilitate microwave usage for backhaul will lower the cost of 4G deployment and increase 4G availability in rural America. As with all wireless communications, operators' ability to use microwave depends on availability of spectrum and the distance of the link itself. In general, spectrum below 12 GHz is preferred for long-link backhaul because of rain-fading effects at higher frequencies.[450]

Although microwave backhaul is a point-to-point service, interference with other systems may occur in the beam contour as well as in side lobes near the radiating antenna. Therefore, frequency coordination is required to ensure sufficient spectral and geographic reuse to maintain a high level of service reliability. [451] In practice, this can create a scarcity of useful backhaul spectrum in high-traffic locations. This scarcity will only be exacerbated as the increase in broadband traffic drives greater use of microwave services.

RECOMMENDATION 5.9: The FCC should revise Parts 74, 78 and 101 of its rules to allow for increased spectrum sharing among compatible point-to-point microwave services.

The FCC should commence a proceeding to examine Parts 74, 78 and 101 of its rules and opportunities to increase sharing of spectrum bands currently used for Mobile Broadcast Auxiliary Service (BAS) and Mobile Cable TV Relay Service (CARS) with microwave services. Such sharing appears feasible as BAS and CARS have started to migrate to Internet protocol (IP)-based communications, making the traffic that is carried on these links fundamentally the same as that on common carrier microwave links. Increased sharing would have the practical effect of increasing the supply of backhaul-suitable spectrum in the prime frequencies below 12 GHz.[452] In the course of this review, the FCC should consider making below-1 GHz "white spaces" spectrum available for backhaul in very rural areas where it otherwise may go unused, to the extent that such use is consistent with Recommendation 5.8.5 above and the ongoing white spaces proceeding.

RECOMMENDATION 5.10: The FCC should revise its rules to allow for greater flexibility and cost-effectiveness in deploying wireless backhaul.

The FCC's Part 101 microwave rules are intended to enable a high level of service reliability, but they may also limit deployment flexibility in coverage- or capacity-limited situations. Therefore, the FCC should commence a proceeding to update these rules to reduce the cost of backhaul in capacity-limited urban areas and range-limited rural areas. In particular, the proceeding should revise rules consistent with the following:

- *Greater spatial reuse of microwave frequencies, particularly in urban areas.* Public comment has raised the possibility that rule changes could enable more efficient use of spectrum, particularly in the area immediately surrounding a microwave station.[453] Such changes, it is claimed, could dramatically increase the ability to use spectrum for backhaul in high-congestion areas, especially urban areas. The FCC, in the context of a larger Part 101 proceeding, should expeditiously consider whether the proposal merits changes to the existing rules.
- *Modification of minimum throughput rules, particularly in rural areas.* The FCC should consider modifying rules on minimum data throughput for each authorized microwave channel when the benefits are clear. Several parties have noted the potential benefits of using adaptive modulation in rural areas to expand the range of backhaul systems.[454] Adaptive modulation is a technique whereby the data rate is dynamically adjusted based on channel conditions at any moment in time. All of these changes could potentially reduce operational costs, particularly in rural areas where microwave backhaul is essential to providing broadband service.
- *Restrictions on antenna size.* The tower lease costs for mounting antennas can constitute up to 40% of the total cost of microwave ownership.[455] These lease costs are directly related to the size of the antenna. Smaller antennas may also "cost less to manufacture and distribute, are less expensive to install because they weigh less and need less structural support, and cost less to maintain because they are less subject to wind load and other destructive forces."[456] Current rules on antenna sizes are designed to maximize the use of microwave spectrum while avoiding interference between operators. It is important to ensure these standards are up-to-date in order to maximize the cost-effectiveness of microwave services.
- *Use of higher frequencies.* Technology has historically been the most important factor limiting the use of higher frequencies. Every successive decade has seen that limit pushed higher. This does not mean that differences in propagation factors at higher frequencies can be ignored. Systems using higher frequencies will need to adopt new architectures and technologies, appropriate to the frequency and the application, as has every past innovative radio application. It must be emphasized that the use of higher frequencies is "compatible and synergistic" with the new wireless paradigms, rather than the new paradigms evolving as forced responses to the necessity of using higher frequencies. Simultaneously, it is important to be mindful of the implications for network engineering of systems operating at higher frequencies, and the impact of those implications on the economic viability of those systems. This Part 101 proceeding should commence in 2010.

5.6. Expanding Opportunities for Innovative Spectrum Access Models

Advances in technology hold much promise for enabling new modes of efficient spectrum access. Many of these advances have led to the development of innovative uses and, ultimately, can complement more conventional licensed approaches. It is important to create a spectrum environment that provides plenty of room for experimentation and growth of new technologies to ensure that the next great idea in broadband spectrum access is first developed and deployed in the U.S.

The FCC and NTIA have made progress in making spectrum available and open to the development and evolution of new technologies. The FCC's decision not to dictate a technological standard for PCS licenses ultimately contributed to the development and widespread commercialization of the CDMA technology now widely in use by 3G networks. Similarly, the creation of the flexible Part 15 rules allowed for the growth and proliferation of unlicensed devices, particularly in the 2.4 GHz Industrial, Scientific and Medical (ISM) band. More recently, the FCC has taken steps to allow innovative spectrum access models in the white spaces of the digital television spectrum bands and in the 3.65 GHz band. Notably, and not coincidentally, innovation sometimes occurs in bands that conventional wisdom had at one time considered to be "junk" spectrum.

In June 2006, the FCC concluded a rulemaking allowing commercial users to employ opportunistic sharing techniques to share 355 MHz of radio spectrum with incumbent federal government radar system operators. Using Dynamic Frequency Selection detect-and-avoid algorithms, commercial interests are now able to operate Wireless Access Systems in the radio spectrum occupied by preexisting radar systems. Opportunistic sharing arrangements offer great potential to meet an increasing market demand for wireless services by promoting more efficient use of radio spectrum.[457]

The FCC and NTIA can take significant steps toward ensuring that the next generation of spectrum access technology can take root in the next few years.

RECOMMENDATION 5.11: The FCC, within the next 10 years, should free up a new, contiguous nationwide band for unlicensed use.

As the FCC seeks to free up additional spectrum for broadband, it should make a sufficient portion available for use exclusively or predominantly by unlicensed devices. This would enable innovators to try new ideas to increase spectrum access and efficiency through unlicensed means, and should enable new unlicensed providers to serve rural and unserved communities. Such an approach would represent a departure from the way the FCC has treated most unlicensed operations in the past. Unlicensed operations are typically overlays to licensed bands, with intensive unlicensed use emerging in some bands (e.g., the 2.4 GHz band) over a long period of time. However, targeting bands for unlicensed use could yield important benefits.

The FCC's Part 15 rules[458] permit unlicensed devices to operate on any spectrum except spectrum specifically designated as restricted.[459] This widespread access to spectrum comes with a trade-off—unlicensed devices must generally operate at very low power levels and on a sufferance basis with respect to any allocated service. In particular, they are subject to the conditions that they cause no harmful interference and must accept interference that may be caused by other operations in the band, including licensed operations.[460] Ever since such unlicensed operation under these rules has been allowed, developers have found ways to provide for a wide variety of devices that perform an assortment of applications that serve consumers. These innovations continue to evolve and proliferate, and include not only garage-door openers, key fobs to open car doors, and Bluetooth headsets, but also the increasingly important deployment of Wi-Fi access points.

The innovations spurred by unlicensed device usage have occurred because of benefits associated with such usage, including low barriers to entry and faster time to market, that have reduced costs of entry, spurred innovation and enabled very efficient spectrum usage. Taken together, these benefits have allowed many communities, entrepreneurs and small businesses to rapidly deploy broadband systems. Often, as has been the case for many WISPs, this has occurred in rural or previously underserved communities.

As mentioned previously, unlicensed and licensed broadband networks can complement one another in important ways. For instance, with the availability of Wi-Fi networks in many locations that enable users to take much of their data off of a licensed network, users benefit by obtaining much faster service while licensed providers have less congestion and can deliver a better overall quality of service. Near-field communications devices operating under the unlicensed provisions are being integrated into cell phones to facilitate electronic transactions. ZigBee and other unlicensed devices are being integrated with Smart Grid applications on licensed wireless systems. Providing additional spectrum for unlicensed use will only amplify these and other complementary benefits by allowing carriers to optimize their networks for mobile use in areas where Wi-Fi is not available or not practical.

RECOMMENDATION 5.12: The FCC should move expeditiously to conclude the TV white spaces proceeding.

The FCC should move expeditiously to resolve pending petitions for reconsideration in the TV white spaces proceeding (ET Docket No. 04-186). This proceeding has introduced a new approach to gaining access to spectrum through use of a database and cognitive radio techniques. The approach to spectrum access used in this proceeding could conceivably be expanded and extended to other spectrum on either a licensed or unlicensed basis.

Industry has demonstrated the promise of and potential for use of the TV white space spectrum. For example, TV white space devices have been used to provide broadband service to a school in rural Virginia and are currently being used for demonstration of a wireless broadband network in Wilmington, North Carolina.

The development of rules for TV white space devices has taken several years. Industry has invested heavily in this process by offering prototype devices that were submitted to the FCC for testing in an open process that included laboratory and field tests. The FCC should complete the final rules for TV white space devices in order to accelerate the introduction of new innovative products and services. As the FCC considers other changes to the TV broadcast spectrum, it should also evaluate the impact on the viability of use of TV white spaces.

RECOMMENDATION 5.13: The FCC should spur further development and deployment of opportunistic uses across more radio spectrum.

Using existing allocations more intelligently is another way to provide for growth in data services. Public comment has suggested that "opportunistic" or "cognitive" technologies can significantly increase the efficiency of spectrum utilization by enabling radios to access and share available spectrum dynamically.[461] These technologies could allow access to many different frequencies across the spectrum chart that may not be in use at a specific place and time and could do so without harming other users' operations or interests. Given the upside potential of these technologies, the FCC and NTIA should take steps to expand the environment in which new, opportunistic technologies can be developed and improved.[462]

Opportunistic spectrum use involves a spectrum-agile radio that can operate on spectrum determined to be unused and available at any moment in time over a given transmission path. That determination can be made through devices that effectively sense available spectrum or consult a database containing that information. Thus, the radio would be able to access available spectrum on a dynamic basis as the opportunity presents itself.[463] Many entities are conducting research or taking part in standardization efforts aimed at continued development. Much of this research is still in its early stages and some barriers must be overcome before the technology gains wide acceptance.[464] The FCC should take two actions to accelerate the development of opportunistic use technologies and expand access to additional spectrum.

First, the FCC should allow opportunistic radios to operate on spectrum currently held by the FCC (such as in certain license areas where spectrum was not successfully auctioned). The availability of such unauctioned spectrum in multiple bands could provide a technical "sandbox" for the creation of, and innovation in, cognitive technologies (including frequency hopping) that take advantage of the ability to operate in different frequency bands dispersed throughout the radio spectrum. Use of a geo-location database that enables opportunistic devices to identify this available spectrum, as discussed below, could be helpful in the development and future deployment of such technologically sophisticated devices.

Second, the FCC should initiate a proceeding that examines ways to extend the geo-location database concept, currently being implemented in the TV bands, to additional spectrum bands that are made available for access by opportunistic radios.[465] As described above, the FCC adopted rules which permit unlicensed devices to access TV white spaces after checking a database to determine which channels are available for use. In the TV bands, the development of an effective database is possible because TV stations, as well as other facilities that must be protected, generally are fixed and known, so that locating the specific protection zone around these facilities is relatively straightforward. It is possible to extend this concept for opportunistic use to other frequency bands where the behavior of stations is well understood and predictable.[466] In addition, devices that operate under this database approach may serve effectively as "listening posts" to measure and report usage of the spectrum back to the database. These reports could improve the opportunistic use of the selected frequencies without causing harmful interference.

The FCC should determine which particular frequency bands should be identified for opportunistic use and what specific information may need to be included in the relevant database. Such determination should also include whether and to what extent the FCC should exclude LPTV band devices in the border areas with Mexico and Canada, including the Tribal lands in those areas, and whether to allow higher power fixed operations in rural areas, which often include Tribal lands. For example, some frequency bands are used for satellite and fixed microwave operations. Similar to TV, microwave stations are fixed and can be protected fairly easily. Protecting satellite use is more complicated, but it is possible if earth station locations can be found through a database search. Moreover, the spectrum dashboard could eventually provide a data resource to enable a more generalized geo-location system, particularly if supplemented with data on spectrum construction and usage (see Recommendations 5.1 and 5.2).

RECOMMENDATION 5.14: The FCC should initiate proceedings to enhance research and development that will advance the science of spectrum access.

A robust research and development pipeline is essential to ensuring that spectrum access technologies continue to evolve and improve. As described in Chapter 7, the FCC should start a rule-making process to establish more flexible experimental licensing rules. Additionally, the National Science Foundation, in consultation with the FCC and NTIA, should fund wireless research and development that will advance the science of spectrum access.

5.7. Taking Additional Steps to Make U.S. Spectrum Policy More Comprehensive

RECOMMENDATION 5.15: The FCC and NTIA should develop a joint roadmap to identify additional candidate federal and non-federal spectrum that can be made accessible for both mobile and fixed wireless broadband use, on an exclusive, shared, licensed and/or unlicensed basis.

As noted elsewhere in this plan, additional spectrum is needed for wireless broadband use. While the plan identifies specific bands that can partially meet this need, access to additional spectrum will still be required in the future. NTIA and the FCC, as co-managers of the spectrum, should develop a plan by October 1, 2010 to identify additional federal and non-federal spectrum that can be made accessible for wireless broadband use.

In developing a national spectrum policy, this plan makes recommendations for reallocating or repurposing several non-federal spectrum bands for wireless broadband use. This plan also recommends that the FCC should coordinate with NTIA on the possible reallocation of certain federal spectrum in the 1755–1850 MHz band. Certain recommendations apply to both non-federal and federal spectrum, such as providing for increasing opportunistic use of the spectrum. However, these steps alone are insufficient. All of the non-federal and federal spectrum, not just certain bands, must be closely examined for possible reallocation.

NTIA and FCC staff have held initial discussions to identify additional candidate federal spectrum bands that might be considered for reallocation, sharing or opportunistic use to help meet the spectrum needs for wireless broadband. These discussions are not sufficiently advanced to identify specific bands at this time. However, this process should continue and be accelerated.

Any reallocation or repurposing of federal spectrum is a complex process. Federal spectrum is used to support national security and public safety applications that must be protected and preserved. Many federal systems have unique capabilities that cannot be easily replaced with off-the-shelf equipment operating in other spectrum, which means it may not be possible to gain access to the spectrum for many years. As in the case of the reallocation of the federal spectrum at 1710–1755 MHz to AWS-1, federal users may require access to non-federal spectrum to accommodate displaced systems.

Given these complexities and timing considerations, it is vital to develop a well-defined and ongoing process to ensure that all spectrum is examined for additional opportunities.

RECOMMENDATION 5.16: The FCC should promote within the international Telecommunication union (ITU) innovative and flexible approaches to global spectrum allocation that take into consideration convergence of various radio communication services and enable global development of broadband services.

As the FCC participates in international organizations like the ITU and regional organizations such as the Inter-American Telecommunication Commission of the Organization of American States, it should promote innovative approaches to spectrum allocation to ensure maximum flexibility for advanced communications services that will enable global broadband services.

In addition to multilateral and regional organizations, the FCC also participates with other U.S. government agencies, such as the U.S. Department of State and NTIA, in bilateral meetings where spectrum issues and approaches to broadband deployment are discussed. In all of these fora, the FCC should ensure that innovative approaches to spectrum allocation are considered and supported.

For example, an item on the agenda for consideration at the ITU's World Radiocommunication Conference in 2012 (WRC12) calls for taking appropriate action with a view to enhancing the international regulatory framework and the international spectrum framework (Agenda Item 1.2). The primary objective of this agenda item is to examine international radio allocation and associated regulatory procedures to meet the demands of current, emerging and future radio technologies, while also taking into account existing services and spectrum usage.

The introduction of many new wireless technologies and applications, especially in consumer products, has spurred growing interest in reviewing spectrum management practices. Consumers want to use many applications offered on wireline and fixed radio communication systems on mobile terminals. The next generation of mobile terminals encompasses multiple radio communication services functions (e.g., fixed, mobile, broadcasting and even radio determination) that provide for voice, data and video as well as positioning (i.e., convergence).

The ITU's Radio Regulations, however, may not be sufficiently flexible to accommodate these technological changes. Therefore, the FCC and the U.S. government should consider whether alternatives are necessary to accommodate advancements in technologies, particularly those that allow many radio communication services to be implemented in the same terminal or handset.

RECOMMENDATION 5.17: The FCC should take into account the unique spectrum needs of U.S. Tribal communities[467] when implementing the recommendations in this chapter.

Some Tribes have successfully used wireless infrastructure to deliver broadband connectivity to their communities. Increasing Tribal access to and use of spectrum would create additional opportunities for Tribal communities to obtain broadband access. Through the following actions, the FCC should evaluate its policies and rules to address obstacles to greater use of spectrum on Tribal lands, including access to spectrum by Tribal communities:

- *Spectrum dashboard.* Facilitating access to the FCC's spectrum dashboard described in Recommendation 5.1 will be critical to helping Tribal communities use spectrum or identify non-Tribal parties that hold licenses to serve Tribal lands.[468] To enhance Tribal access to such information, future iterations of the spectrum dashboard should include information identifying spectrum allocated and assigned in Tribal lands. If the FCC conducts spectrum utilization studies in the future, those studies should identify Tribal lands as distinct entities.

- *Tribal Land Bidding Credit.* Since 2000, the Commission has administered a Tribal Land Bidding Credit (TLBC) program to provide incentives to wireless telecommunications carriers to serve Tribal lands.[469] The FCC should revisit the TLBC program to determine whether it can be modified to facilitate Tribal access to spectrum in Tribal lands and better promote deployment of communications services to Tribal communities.

- *Tribal priority.* The FCC has established a Tribal priority in the threshold analysis stage of the FM radio allotment and AM radio licensing processes.[470] Recognizing that the statutory and regulatory procedures for licensing wireless services are different in some respects from those applicable to broadcast stations, the FCC should consider expanding any Tribal priority policy to include the process for licensing fixed and mobile wireless licenses covering Tribal lands, potentially considering geographic carve-out license areas for Tribal lands.

- *Build-out.* The FCC should consider providing additional flexibility and incentives for the build-out of facilities serving Tribal lands. For example, if a licensee has fulfilled its construction requirement but has failed to provide service to Tribal lands, the FCC should consider alternative mechanisms to facilitate Tribal access to such unused spectrum. These mechanisms might include developing rules for re-licensing the unused spectrum to the Tribal community for the provision of services, mandating partitioning or disaggregation of the spectrum, and encouraging the use of secondary market mechanisms for the purpose of deploying services to Tribal areas.[471]

- *White spaces.* The FCC should move expeditiously to resolve pending petitions for reconsideration in the TV white spaces proceeding. Among other issues, this proceeding should determine whether and to what extent the FCC should exclude LPTV band devices in the border areas with Mexico and Canada, including the Tribal lands in those areas. Further, the FCC should proceed to consider higher-power fixed operations in rural areas, which often include Tribal lands.

6. INFRASTRUCTURE

Just as wireless networks use publicly owned spectrum, wireless and wired networks rely on cables and conduits attached to public roads, bridges, poles and tunnels. Securing rights to this infrastructure is often a difficult and time-consuming process that discourages private investment. Because of permitting and zoning rules, government often has a significant role in network construction. Government also regulates how broadband providers can use existing private infrastructure like utility poles and conduits. Many state and local governments have taken steps to encourage and facilitate fiber conduit deployment as part of public works projects like road construction. similarly, in November 2009, the Federal communications commission (Fcc) established timelines for states and localities to process permit requests to build and locate wireless equipment on towers.[472]

While these are positive steps, more can and should be done. Federal, state and local governments should do two things to reduce the costs incurred by private industry when using public infrastructure. First, government should take steps to improve utilization of existing infrastructure to ensure that network providers have easier access to poles, conduits, ducts and rights-of-way. Second, the federal government should foster further infrastructure deployment by facilitating the placement of communications infrastructure on federally managed property and enacting "dig once" legislation. These two actions can improve the business case for deploying and upgrading broadband network infrastructure and facilitate competitive entry.

Recommendations

Improving utilization of infrastructure

- The FCC should establish rental rates for pole attachments that are as low and close to uniform as possible, consistent with Section 224 of the Communications Act of 1934, as amended, to promote broadband deployment.
- The FCC should implement rules that will lower the cost of the pole attachment "make-ready" process.
- The FCC should establish a comprehensive timeline for each step of the Section 224 access process and reform the process for resolving disputes regarding infrastructure access.
- The FCC should improve the collection and availability of information regarding the location and availability of poles, ducts, conduits and rights-of-way.
- Congress should consider amending Section 224 of the Act to establish a harmonized access policy for all poles, ducts, conduits and rights-of-way.
- The FCC should establish a joint task force with state, Tribal and local policymakers to craft guidelines for rates, terms and conditions for access to public rights-of-way.

Maximizing impact of federal resources

- The U.S. Department of Transportation (DOT) should make federal financing of highway, road and bridge projects contingent on states and localities allowing joint deployment of conduits by qualified parties.
- Congress should consider enacting "dig once" legislation applying to all future federally funded projects along rightsof-way (including sewers, power transmission facilities, rail, pipelines, bridges, tunnels and roads).
- Congress should consider expressly authorizing federal agencies to set the fees for access to federal rights-of-way on a management and cost recovery basis.
- The Executive Branch should develop one or more master contracts to expedite the placement of wireless towers on federal government property and buildings.

6.1. Improving Utilization of Infrastructure

The cost of deploying a broadband network depends significantly on the costs that service providers incur to access conduits, ducts, poles and rights-of-way on public and private lands.[473] Collectively, the expense of obtaining permits and leasing pole attachments and rights-of-way can amount to 20% of the cost of fiber optic deployment.[474]

These costs can be reduced directly by cutting fees. The costs can also be lowered indirectly by expediting processes and decreasing the risks and complexities that companies face as they deploy broadband network infrastructure.

The FCC has already begun to take important steps in this direction with policies that will speed the deployment of wireless equipment on towers. With regard to other infrastructure such as utility poles, the FCC has authority to improve the deployment process and should use that authority. Lowering the costs of infrastructure access involves every level of government; active consultation among all levels of government will be needed to put in place pro-deployment policies such as joint trenching, conduit construction and placement of broadband facilities on public property.

RECOMMENDATION 6.1: The FCC should establish rental rates for pole attachments that are as low and close to uniform as possible, consistent with section 224 of the communications Act of 1934, to promote broadband deployment.

As Exhibit 6-A shows, the rental rates paid by communications companies to attach to a utility pole vary widely—from approximately $7 per foot per year for cable operators to $10 per foot per year for competitive telecommunications companies to more than $20 per foot per year for some incumbent local exchange carriers (ILECs).[475] The impact of these rates can be particularly acute in rural areas, where there often are more poles per mile than households.[476] In a rural area with 15 households per linear mile, data suggest that the cost of pole attachments to serve a broadband customer can range from Annual Pole Rates Vary Conside $4.54 per month per household passed (if cable rates are used) to $12.96 (if ILEC rates are used). If the lower rates were applied, and if the cost differential in excess of $8 per month were passed on to consumers, the typical monthly price of broadband for some rural consumers could fall materially.[477] That could have the added effect of generating an increase—possibly a significant increase—in rural broadband adoption.

Different rates for virtually the same resource (space on a pole), based solely on the regulatory classification of the attaching provider, largely result from rate formulas established by Congress and the FCC under Section 224 of the Communications Act of 1934, as amended ("the Act").[479] The rate structure is so arcane that, since the 1996 amendments to Section 224, there has been near-constant litigation about the applicability of "cable" or "telecommunications" rates to broadband, voice over Internet protocol and wireless services.[480]

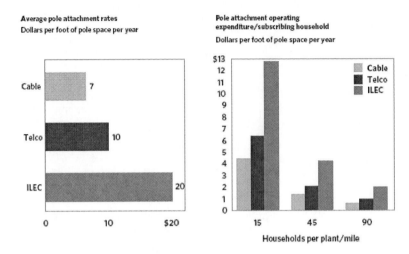

Exhibit 6-A. Annual Pole Rates Vary Considerably by Provider Type[478]

To support the goal of broadband deployment, rates for pole attachments should be as low and as close to uniform as possible. The rate formula for cable providers articulated in Section 2 24(d) has been in place for 31 years and is "just and reasonable" and fully compensatory for utilities.[481] Through a rulemaking, the FCC should revisit its application of the telecommunications carrier rate formula to yield rates as close as possible to the cable rate in a way that is consistent with the Act.

Applying different rates based on whether the attacher is classified as a "cable" or a "telecommunications" company distorts attachers' deployment decisions. This is especially true with regard to integrated, voice, video and data networks. This uncertainty may be deterring broadband providers that pay lower pole rates from extending their networks or adding capabilities (such as high-capacity links to wireless towers). By expanding networks and capabilities, these providers risk having a higher pole rental fee apply to their entire network.[482]

FCC rules that move toward low rates that are as uniform as possible across service providers would help remove many of these distortions. This approach would also greatly reduce complexity and risk for those deploying broadband.

RECOMMENDATION 6.2: The FCC should implement rules that will lower the cost of the pole attachment "make- ready" process.

Rearranging existing pole attachments or installing new poles—a process referred to as "make-ready" work—can be a significant source of cost and delay in building broadband networks. FiberNet, a broadband provider that has deployed 3,000 miles of fiber in West Virginia, states that "the most significant obstacle to the deployment of fiber transport is FiberNet's inability to obtain access to pole attachments in a timely manner."[483]

Make-ready work frequently involves moving wires or other equipment attached to a pole to ensure proper spacing between equipment and compliance with electric and safety codes. The make-ready process requires not only coordination between the utility that owns the pole and a prospective broadband provider, but also the cooperation of communications firms that have already attached to the pole. Each attaching party is generally responsible for moving its wires and equipment, meaning that multiple visits to the same pole may be required simply to attach a new wire.

Reform of this inefficient process presents significant opportunities for savings. FiberNet commented that its make- ready charges for several fiber runs in West Virginia averaged $4,200 per mile and took 182 days to complete, [484] but the company estimates that these costs should instead have averaged $1,000 per mile.[485] Another provider, Fibertech, states that the make-ready process averages 89 days in Connecticut and 100 days in New York, where state commissions regulate the process directly.[486]

Delays can also result from existing attachers' action (or inaction) to move equipment to accommodate a new attacher, potentially a competitor.[487] As a result, reform must address the obligations of existing attachers as well as the pole owner.

An evaluation of best practices at the state and local levels reveals ample opportunities to manage this process more efficiently. Yet, absent regulation, pole owners and existing attachers have few incentives to change their behavior.

To lower the cost of the make-ready process and speed it up, the FCC should, through rulemaking:

- Establish a schedule of charges for the most common categories of work (such as engineering assessments and pole construction).
- Codify the requirement that gives attachers the right to use space- and cost-saving techniques such as boxing or extension arms where practical and in a way that is consistent with pole owners' use of those techniques.[488]
- Allow prospective attachers to use independent, utility- approved and certified contractors to perform all engineering assessments and communications make-ready work, as well as independent surveys, under the joint direction and supervision of the pole owner and the new attacher.[489]
- Ensure that existing attachers take action within a specified period (such as 30 days) to accommodate a new attacher. This can be accomplished through measures such as mandatory timelines and rules that would allow the pole owner or new attacher to move existing communications attachments if the timeline is not met.
- Link the payment schedule for make-ready work to the actual performance of that work, rather than requiring all payment up front.

These cost-saving steps can have an immediate impact on driving fiber deeper into networks, which will advance the deployment of both wireline and wireless broadband services.

RECOMMENDATION 6.3: The FCC should establish a comprehensive timeline for each step of the section 224 access process and reform the process for resolving disputes regarding infrastructure access.

There are no federal regulations addressing the duration of the entire process for obtaining access to poles, ducts, conduit and rights-of-way. While the FCC in the past has recognized that "time is critical in establishing the rate, terms and conditions for attaching," current FCC rules only require that a utility provide a response to an application within 45 days.[490] The FCC does not have any deadlines for subsequent steps in the process, which can drag on for months if not years.[491] This causes delays in the deployment of broadband to communities and anchor institutions.[492]

Several states, including Connecticut and New York, have established firm timelines for the entire process, from the day that a prospective attacher files an application, to the issuance of a permit indicating that all make-ready work has been completed.[493] Timelines speed the process considerably in states where they have been implemented,[494] thus facilitating the deployment of broadband.

The FCC should establish a federal timeline that covers each step of the pole attachment process, from application to issuance of the final permit. The federal timeline should be implemented through a rulemaking and be comprehensive and applicable to all forms of communications attachments.[495] In addition, the FCC should establish a timeline for the process of certifying wireless equipment for attachment.[496]

The FCC also should institute a better process for resolving access disputes. For large broadband network builds, the pole attachment process is highly fragmented and often involves dozens of utilities, cable providers and telecommunications providers in multiple jurisdictions. Yet there is no established process for the timely resolution of disputes.[497]

The FCC has the authority to enforce its pole attachment rules, but today it generally attempts to informally resolve attachment disputes through mediation. This process has significant flaws. Under the current system of case-by-case adjudication, the attacher always bears the burden of bringing a formal complaint.[498] The formal dispute rules also do not provide for compensation dating from the time of the injury, so attachers have minimal incentive to initiate costly formal pole attachment cases that may linger for years.

Also, because time is often of the essence during the make- ready process, methods for resolving disputes over application of individual safety and engineering standards may be neces - sary. Informal local procedures and mediation may sometimes result in satisfactory settlements, but they do not create precedents for what constitutes a "just and reasonable" practice under Section 224 of the Act.

In revising its dispute resolution policies, the FCC should consider approaches that not only speed the process but also provide future guidelines for the industry. Institutional changes, such as the creation of specialized fora and processes for attachment disputes, and process changes, such as target deadlines for resolution, could expedite dispute resolution and serve the overarching goal of lowering costs and promoting rapid broadband deployment. The FCC also could use its authority under Section 224 to require utilities to post standards and adopt procedures for resolving safety and engineering disagreements and encourage appropri- ate state processes for resolving such disputes. Finally, awarding compensation that dates from the denial of access could stimulate swifter resolution of disputes.

RECOMMENDATION 6.4: The FCC should improve the collection and availability of information regarding the location and availability of poles, ducts, conduits and rights-of-way.

There are hundreds of private and public entities that own and control access to poles, ducts, conduits and rights-of-way, and an even greater number of parties that use that infrastructure. Accurate information about pole owners and attachments is critical if there is to be a timely and efficient process for accessing and utilizing this important infrastructure.[499]

The FCC should ensure that attachers and pole owners have the data they need to lower costs and accelerate the buildout of broadband networks.

Consistent with its current jurisdiction under Section 224, the FCC should ensure that information about utility poles and conduits is up-to-date, readily accessible and secure, and that the costs and responsibility of collecting and maintaining data are shared equitably by owners and users of these vital resources. For example, data could be collected systematically as in Germany, which is mapping fiber, ducts and conduits and is planning to coordinate these data with information about public works and infrastructure projects.[500] Existing industry efforts to collect and coordinate data could be expanded and made more robust.[501] In addition, the participation of all pole owners subject to Section 224 and attaching parties in any such database effort could be regulated and streamlined. These databases should be easily searchable, identify the owner of each pole and should contain up-to-date records of attachments and make-ready work that has been performed. For conduits and ducts, any database should note whether there is space available. Whichever methods are used, data must be regularly updated, secure and accessible in order to further the FCC's efforts to ensure that broadband providers have efficient access to essential infrastructure information.

RECOMMENDATION 6.5: Congress should consider amending section 224 of the Act to establish a harmonized access policy for all poles, ducts, conduits and rights-of-way.

Even if the FCC implemented all of the recommendations related to its Section 224 authority, additional steps would be needed to establish a comprehensive national broadband infrastructure policy. As previously discussed, without statutory change, the convoluted rate structure for cable and telecommunications providers will persist. Moreover, due to exemptions written into Section 224, a reformed FCC regime would apply to only 49 million of the nation's 134 million poles.[502] In particular, the statute does not apply in states that adopt their own system of regulation and exempts poles owned by co-operatives, municipalities and non-utilities.[503]

The nation needs a coherent and uniform policy for broadband access to privately owned physical infrastructure. Congress should consider amending or replacing Section 224 with a harmonized and simple policy that establishes minimum standards throughout the nation—although states should remain free to enforce standards that are not inconsistent with federal law. The new statutory framework could provide that:

- All poles, ducts, conduits and rights-of-way be subject to a regulatory regime addressing a minimum set of criteria established by federal law.
- All broadband service providers, whether wholesale or retail, have the right to access pole attachments, ducts, conduit and rights-of-way based on reasonable rates, terms and conditions.
- Infrastructure access be provided within standard timelines established by the FCC, and that the FCC has the authority to award damages for non-compliance.
- The FCC has the authority to compile and update a comprehensive database of physical infrastructure assets.

RECOMMENDATION 6.6: The FCC should establish a joint task force with state, Tribal and local policymakers to craft guidelines for rates, terms and conditions for access to public rights-of-way.

Because local, state, Tribal and federal governments control access to important rights-of-way and facilities, a comprehensive broadband infrastructure policy necessarily requires a coordinated effort among all levels of government.

There is wide diversity among state and local policies regarding access to and payment for accessing public rightsof-way. Many jurisdictions charge a simple rental fee. Other jurisdictions use other compensation schemes, including per-foot rentals, one-time payments, in-kind payments (such as service to public institutions or contributions of fiber to city telecommunications departments) and assessments against general revenues.[504] Some jurisdictions calculate land rental rates based on local real estate "market value" appraisals.

Many states have limited the rights-of-way charges that municipalities may impose, either by establishing uniform rates (Michigan) or by limiting fees to administrative costs (Missouri).[505] Other states, including South Carolina, Illinois and Florida, do not allow municipalities to collect rightsof-way fees directly; instead, the state compensates local governments for the use of their rights-of-way with proceeds from state-administered telecommunications taxes.

Broadband service providers often assert that the expense and complexity of obtaining access to public rights-of-way in many jurisdictions increase the cost and slow the pace of broadband network deployment.[506] Representatives of state and local governments dispute many of these contentions.[507] However, nearly all agree that there can and should be better coordination across jurisdictions on infrastructure issues.[508]

Despite past efforts by the National Telecommunications and Information Administration (NTIA) and the National Association of Regulatory Utility Commissioners (NARUC),[509] a coordinated approach to rights-of-way policies has not taken hold. There are limits to state and local policies; Section 253 of the Communications Act prohibits state and local policies that impede the provision of telecommunications services while allowing for rights-of-way management practices that are nondiscriminatory, competitively neutral, fair and reasonable.[510] However, disputes under Section 253 have lingered for years, both before the FCC and in federal district courts.[511]

In consultation and partnership with state, local and Tribal authorities, the FCC should develop guidelines for public rights-of-way policies that will ensure that best practices from state and local government are applied nationally. For example, establishing common application information and inspection protocols could lower administrative costs for the industry and governmental agencies alike. Fee structures should be consistent with the national policy of promoting greater broadband deployment. A fee structure based solely upon the market value of the land being used would not typically take into account the benefits that the public as a whole would receive from increased broadband deployment, particularly in unserved and underserved areas. In addition, broadband network construction often involves multiple jurisdictions. The timing of the process and fee calculations by one local government may not take into account the benefits that constituents in neighboring jurisdictions would receive from increased broadband deployment. The cost and social value of broadband cut across political boundaries; as a result, rights-of-way policies and best practices must reach across those boundaries and be developed with the broader public interest in mind.

To help develop this consistent rights-of-way policy, the FCC should convene a joint task force of state, local and Tribal authorities with a mandate to:

- Investigate and catalog current state and local rights-ofway practices and fee structures, building on NTIA's 2003 compendium and the 2002 NARUC Rights-of-Way Project.
- Identify public rights-of-way and infrastructure policies and fees that are consistent with the national public policy goal of broadband deployment and those that are inconsistent with that goal.[512]
- Identify and articulate rights-of-way construction and maintenance practices that reduce overall capital and maintenance costs for both government and users and that avoid unnecessary delays, actions, costs and inefficiencies related to the construction and maintenance of broadband facilities along public rights-of-way.[513]
- Recommend appropriate guidelines for what constitutes "competitively neutral," "nondiscriminatory" and "fair and reasonable" rights-of-way practices and fees.
- Recommend a process for the FCC to use to resolve disputes under Section 253. Creating a process should expedite resolution of public rights-of-way disputes in areas either unserved or underserved by broadband.

The FCC should request that the task force make its recommendations within six months of the task force's creation. These recommendations should then be considered by the FCC as part of a proceeding that seeks industry-wide comment on these issues.

6.2. Maximizing Impact of Federal Resources

Federal government can also play an important role in directly lowering the costs of future infrastructure deployment. The federal government has already made efforts to simplify access to federal rights-of-way under President George W. Bush,[514] and to improve access to federal government facilities for wireless services under President William J. Clinton.[515] However, policies have generally taken a permissive approach, simply allowing the federal government to take steps, rather than requiring that those steps be taken.

RECOMMENDATION 6.7: The U.S. Department of Transportation (DOT) should make federal financing of highway, road and bridge projects contingent on states and localities allowing joint deployment of conduits by qualified parties.

RECOMMENDATION 6.8: Congress should consider enacting "dig once" legislation applying to all future federally funded projects along rights-of-way (including sewers, power transmission facilities, rail, pipelines, bridges, tunnels and roads).

Although pushing fiber deeper into broadband networks considerably improves the performance and reliability of those networks, deploying a mile of fiber can easily cost more than $100,000 (see Exhibit 6-B). The largest element of deployment costs is not the fiber itself, but the placement costs associated with burying the fiber in the ground (or attaching it to poles in an aerial build). These placement costs can, in certain cases, account for almost three-quarters of the total cost of fiber deployment. Running a strand of fiber through an existing conduit is 3–4 times cheaper than constructing a new aerial build.[516]

Substantial savings can be captured if fiber builds are coordinated with other infrastructure projects in which the right-of-way (e.g., road, water, sewer, gas, electric, etc.) is already being dug. For example, the city of San Francisco has a "trench once" policy, in which a 5-year moratorium is placed on opening up a road bed once the trench along that road bed has been closed.[518] San Francisco uses a notification process to ensure that other interested parties have the opportunity to install conduits and cabling in the open trench.[519] The city of Boston has implemented a "Shadow Conduit Policy," in which the first company to request a trench takes a lead role, inviting other companies to add additional empty (or "shadow") conduits for future use by either the city of Boston or a later entrant.[520] The city of Chicago seeks to "inexpensively deploy excess conduit when streets are opened for other infrastructure and public works projects."[521] In the Netherlands, a committee in the city of Amsterdam similarly coordinates digging and trenching activities between the public and private sector.[522]

These policies have clear benefits, as shown by the case of Akron, Ohio. When Akron was deploying facilities and conduit to support its public safety network, it shared those facilities with OneCommunity, a northeast Ohio public-private partner ship that aggregates demand by public institutions and private broadband service providers. As a result of that coordination, those same facilities and conduits now support health care institutions, schools and Wi-Fi access in Akron.[523] Similarly, along Interstate 91 in western Massachusetts, collaboration among the Massachusetts Department of Transportation, the Massachusetts Broadband Institute and the federal DOT is resulting in the installation of 55 miles of fiber optic cable with 34 interconnection points.[524]

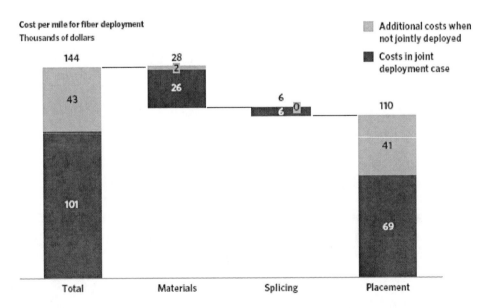

Exhibit 6-B. Joint Deployment Can Materially Reduce the Cost of Fiber Deployment[517]

DOT should implement "joint trenching" and conduit policies to lower the installation costs for broadband networks.[525] At a minimum, states and localities undertaking construction along rights-of-way that are partially or fully financed by DOT should be required to give at least 90 days' notice before projects begin. This would allow private contractors or public entities to add conduits for fiber optic cables in ways that do not unreasonably increase cost, add to construction time or hurt the integrity of the project. Opportunities for joint trenching and conduit deployment are varied, from construction of Intelligent Transportation Systems alongside interstates to building and maintenance of recreational rail trails.[526] As a result, information about potential joint trenching and conduit deployment opportunities should be available and accessible to prospective broadband network providers whenever government engages in an infrastructure project, subject to security precautions.

Congress also should consider enacting "dig once" legislation to extend similar joint trenching requirements to all rights-ofway projects (including sewers, power transmission facilities, rail, pipelines, bridges, tunnels and roads) receiving federal funding.

RECOMMENDATION 6.9: Congress should consider expressly authorizing federal agencies to set the fees for access to federal rights-of-way on a management and cost recovery basis.

RECOMMENDATION 6.10: The Executive Branch should develop one or more master contracts to expedite the placement of wireless towers on federal government property and buildings.

The federal government is the largest landowner in the country—650 million acres, constituting nearly one-third of the land area of the United States.[527] The federal government's General Services Administration (GSA) also owns or leases space in 8,600 buildings nationwide.[528] To effectively deploy broadband, providers often need to be able to place equipment on this federally controlled property, or to use the rights-ofway that pass through the property.

Based on an August 1995 executive memorandum by President Clinton,[529] GSA developed guidelines to allow wireless antennas on federal buildings and land.[530] Additionally, since 1989, GSA has run the National Antenna Program to facilitate wireless tower placement on federal government buildings.[531] On more than 1,900 buildings administered by GSA, there are currently antennas covered by approximately 100 leases that result in millions of dollars in revenue for the Federal Buildings Fund annually.[532] For each of the leases managed by GSA, market rent is charged, and the leases are tightly crafted to cover rooftop space, specific equipment and technology.

Even given this progress, the federal government can do more to facilitate access to its rights-of-way and facilities that it either develops or maintains. In many instances, federal law currently requires that rental fees for rights-of-way controlled by federal agencies be based upon the market value of the land. As a result, these fees are often much higher than the direct costs involved.[533] To facilitate the development of broadband networks, Congress should consider allowing all agencies to set the fees for access to rights-of-way for broadband services on the basis of a direct cost recovery approach, especially in markets currently underserved or unserved by any broadband service provider.

The Executive Branch should also develop one or more master contracts for all federal property and buildings covering the placement of wireless towers. The contracts would apply to all buildings, unless the federal government decides that local issues require non-standard treatment. In the master contracts, GSA should also standardize the treatment of key issues covering rooftop space, equipment and technology. The goal of these master contracts would be to lower real estate acquisition costs and streamline local zoning and permitting for broadband network infrastructure.

While reducing the prices for leases on government property may reduce fees paid to governments at the local, state and federal levels, the decline in prices may also greatly increase the number of companies that acquire leases on government property. In any case, the increased deployment of broadband will stimulate investment and benefit society.

7. RESEARCH AND DEVELOPMENT

In the 1970s, research funded by the Defense Advanced Research projects Agency and later the National Science Foundation (NSF) was an important part of the development of the Internet. In the late 20th century, American companies led in the development of digital switching technologies, optical communications, cellular communications, Internet hardware and Internet applications. Federal investments in research and development, coupled with private firms' innovative research and product development, have led to the robust broadband ecosystem users enjoy today. such investments have also made possible the creation of multibillion-dollar companies that are global leaders in networking, search and other Internet-based businesses. [534]

This R&D activity drove innovation and productivity gains, which aided economic growth. The National Research Council found that in the case of information technology (IT), "The unanticipated results of research are often as important as the anticipated results," "The interaction of research ideas multiplies their impact," and "Past returns on federal investment in IT research have been extraordinary for both United States society and the United States economy." [535]

America's top research universities and laboratories continue this R&D effort today in their experiments with very fast 1 Gbps networks (gigabit networks). For example, Case Western Reserve University in Cleveland, with 40 institutional partners, vendors and community organizations, is planning a University Circle Innovation Zone in the economically impoverished area around the university to provide households, schools, libraries and museums with gigabit fiber optic connections. [536] Case Western expects this network to create jobs in the community and spawn software and service development for Smart Grid, health, science and other applications, as well as foster technology, engineering and mathematics education services. [537]

The private sector continues to invest in high-speed networks, as revealed in several recent announcements during the course of the National Broadband Plan proceeding. Google has announced a plan to provide 50,000 to 500,000 consumers in a small number of test communities with gigabit connections. [538] And Cisco Systems is deploying a telemedicine pilot solution to 15 medical sites in California to spur e-health application development. [539]

All of these efforts aim to accelerate the pace of innovation by placing next-generation technology in the hands of individuals and entrepreneurs, and allowing them to discover the best uses for it. Very fast networks may lead to unanticipated discoveries that will change how people connect, work, learn, play and contribute online.

The federal government must continue to do its part to foster the development of research networks and wireless testbeds through a clear R&D funding agenda that is focused on broadband networks, equipment, services and applications. These efforts should include expanding access to ultra-highspeed connectivity through regulatory policy and direct action in communities where the federal government has a long-term presence, such as Department of Defense (DoD) installations.

The broadband ecosystem—networks, devices and applications—has benefited from research breakthroughs in a broad variety of areas such as networking, software, semiconductors, material sciences, applied mathematics, construction and engineering. Advancement in all these fields and many others is essential for continued innovation and improvement. For U.S. companies to continue to be leaders in high-value areas of the global broadband ecosystem, they must continue to generate and benefit from scientific innovation.

Although measuring the effects of R&D is difficult, studies find that firms earn 20% to 30% returns on their investments. [540] R&D returns to society are even higher as innovators beyond original research teams are able to access research and take work in new directions[541] The gap between R&D returns for private companies and those for society presents a challenge for funding and conducting R&D. [542]

Government can help fill the R&D investment gap by funding research that would yield net benefits to society but that would not earn sufficient returns to be privately profitable. [543] This approach should include funding for direct research, for R&D at universities and other institutions, and for subsidizing private R&D through mechanisms such as the R&D tax credit. [544] Alongside direct funding, the government can take an active role in creating new next-generation applications and uses by linking DoD locations with ultra-high-speed broadband connectivity.

The federal government needs to create a clear agenda and priorities for broadband-related R&D funding, focused on important research that would not be conducted absent government intervention. The government can also promote R&D through regulatory policies allowing increased use of government resources. Examples include establishing research centers or allowing access to spectrum in order to evaluate new technologies in ways that theoretical studies and simulations do not support.

Recommendations

- The government should focus broadband R&D funding on projects with varied risk-return profiles, including a mix of short-term and long-term projects (e.g., those lasting 5 years or longer).
- Congress should consider making the Research and Experimentation (R&E) tax credit a long-term tax credit to stimulate broadband R&D.

- The federal government should provide ultra-high-speed broadband connectivity to select DoD installations to enable the development of next-generation broadband applications.
- The National Academy of Sciences and the National Academy of Engineering (National Academies) should develop a research road map to guide federal R&D funding priorities.
- NSF should establish an open, multi-location, interdisciplinary research center for broadband, addressing technology, policy and economics. Center priorities should be driven by the agenda identified in the National Academies research road map.
- NSF, in consultation with the Federal Communications Commission (FCC), should consider funding a wireless testbed for promoting the science underlying spectrum policymaking and a testbed for evaluating the network security needed to provide a secure broadband infrastructure.
- The FCC should start a rulemaking process to establish more flexible experimental licensing rules for spectrum and facilitate the use of this spectrum by researchers.

Some high-risk, high-return R&D initiatives or projects requiring sustained, long-term collaboration across highly diverse fields may be underfunded by the private sector. Federal research funding should close any potential gaps due to private sector risk- reward expectations or inability to coordinate and cooperate.

RECOMMENDATION 7.1: The government should focus broadband R&D funding on projects with varied risk-return profiles, including a mix of short-term and long-term projects (e.g., those lasting 5 years or longer).

In September, the White House Office of Science and Technology Policy (OSTP) found that, in regards to R&D policy, "[a] short-term focus has neglected fundamental investments."[545] The National Research Council's report, *Renewing U.S. Telecommunications Research*, states, "Long-term, fundamental research aimed at breakthroughs has declined in favor of shorter-term, incremental and evolutionary projects whose purpose is to enable improvements in existing products and services. This evolutionary work is aimed at generating returns within a couple of years to a couple of months and not at addressing the needs of the telecommunications industry as a whole in future decades."[546]

Similarly, in FCC workshops, researchers repeatedly noted that, like industry funding, federal funding is now focused more on short-term work than on long-term fundamental research projects.[547]

The academic community also noted the lack of funding for research that has a high probability of failure, even when success would lead to significant advances in technology. Researchers have indicated that the current review process for government research grants takes a conservative approach to project review and more risky projects are rarely funded.[548]

RECOMMENDATION 7.2: Congress should consider making the Research and Experimentation (R&E) tax credit a longterm tax credit to stimulate broadband R&D.

A number of economic studies have shown that R&D tax incentives are a cost-effective way to spur private sector research and investment. These types of tax incentives may help move the United States toward the goal of developing and building world-class broadband networks.

The Research and Experimentation tax credit, established in the 1980s, stimulated about $2 billion in research per year while costing about $1 billion in lost tax revenue. [549] Bronwyn Hall has estimated that a permanent 5% R&E tax credit would lead to a permanent increase in R&D spending of 10% to 15%. Similarly, Klassen, Pittman and Reed have found that R&D tax incentives stimulate $2.96 of additional R&D investment for every dollar of lost tax revenue. [550]

The long-term R&E tax credit applies broadly across and will benefit many industries.

RECOMMENDATION 7.3: The federal government should provide ultra-high-speed broadband connectivity to select DoD installations to enable the development of next-generation broadband applications.

The nation's military installations "are the platforms from which America's military capability is generated, deployed and sustained." [551] These installations house, train, educate and support tens of thousands of service personnel and their families. [552] There is no doubt that the nation's military personnel deserve to have access to the latest technology, the most resilient and cost-effective methods of communications and services, and ultra-high-speed broadband connectivity.

As a start, DoD, in consultation with OSTP, should consider expanding the deployment of ultra-high-speed connectivity to a select number of DoD installations in a manner consistent with the missions and operational requirements of the Armed Forces.

DoD installations are ideal communities for ultra-highspeed broadband due to their scale and the variety of services they provide to military personnel and their families. Expanded access to ultra-high-speed connectivity will further enable educational applications such as advanced distance learning. In addition, base personnel will have greater access to distance learning content from military staff colleges to better prepare the them to be the next generation of officers, while enhanced distance post-secondary offerings can smooth the transitions of those looking for new careers in civilian life.

Typical base medical facilities treat thousands of soldiers, retirees and their families every year. Next-generation health applications, such as high-definition video consultations and continuous remote monitoring of patients, can improve quality of care for these patients.

Bases are also intense users of energy. DoD is the nation's single largest energy user, accounting for nearly 1% of all energy consumed by the United States in FY2006. [553] Broadband capability and advanced information services allow deployment of Smart Grid and smart meter technologies. If deployed on military installations, these technologies would facilitate improved power management that will reduce energy consumption, allow for incorporating more renewable generation on site and enable new continuity of operations capabilities like micro-grids. [554]

Because of bases' large population under the age of 25, including families and children, increased access to ultra-highspeed Internet would act as a catalyst for the development of increasingly sophisticated applications that would support military personnel and their families. Indeed, as these applications evolve, DoD installations would be showcases for advanced educational, training and other uses of broadband.

The first step in implementing this idea should be a task force led by DoD, with consultation from OSTP. This task force should make recommendations on installation selection, level of connectivity and potentially, next-generation applications—both commercial and military—that could be deployed to these installations. The task force must consider a variety of requirements in order to prevent adverse operational impact to force readiness. These requirements include information assurance, integration and governance with existing commercial and DoD networking capability, non-federal spectrum availability, identification of funding sources and a cost-benefit analysis. In selecting the initial sites, the task force should also explore whether this program should work in conjunction with DoD's existing "green bases" effort. DoD would of course retain operational control of the project to ensure that the technology and services deployed are consistent with the missions of the Armed Forces, and may terminate the project at any time based on mission impacts, capabilities delivered and cost.

RECOMMENDATION 7.4: The National Academy of Sciences and the National Academy of Engineering (National Academies) should develop a research road map to guide federal broadband R&D funding priorities.

The National Academies, which gather committees of experts across scientific and technological endeavors to offer advice to the federal government and the public, [555] should take the lead in developing a research road map to guide federal broadband R&D funding priorities. The road map should identify gaps, critical issues, competitive shortfalls and key opportunities in areas associated directly or indirectly with broadband networks, devices or applications. It should leverage the input of public and private stakeholder communities. Additionally, the President's Council of Advisors on Science and Technology, an advisory group of the nation's leading scientists and engineers, as well as the FCC's Technology Advisory Committee might play key advisory roles. [556]

Input from the Broadband Research Public Notice and Workshop[557] identified the following potential research priorities, which are summarized as input to the National Academies:

- *Breakthroughs in network price/performance.* Increasing price/performance and lowering unit costs fuel the computer industry. Research is needed to enable similar price/ performance improvements in wired and wireless networks to make truly high-speed broadband more affordable. Closing gaps to achieve these breakthroughs may require research in networking, materials science, optics, semiconductors, electromagnetism, construction engineering and other fields.
- *Communications research to support national purposes.* In the Recovery Act, Congress defined key national purposes that broadband should support. Multi-disciplinary, government-funded communications research may be required to ensure progress in accessibility, health care, energy management, education and public safety networks.
- *Social science and economic research on broadband adoption and usage.* Lack of adoption is a larger barrier to universal broadband than lack of availability. Moreover, usage and acceptance of broadband varies greatly across population segments and the sources of this variation are not well understood. Social science and

economic research may help explain the reasons underlying broadband non-adoption, as well as network evolution and its impact on the user.

- *Secure, trustworthy and reliable broadband infrastructure.* The vast complexity of today's networks has created massive vulnerabilities to security at the same time that society has become increasingly dependent on these networks. Research is needed to improve the trustworthiness, security and reliability of these networks, the devices that attach to them and the software and applications they support. This is critical to continued growth of networks and applications.
- *Broadband network measurement and management.* Research is needed to provide the tools to measure network operations and to gain a better understanding of the Internet's "health."

Enabling new service models. Continued exponential improvements in processing power and storage, coupled with broadband networking, are enabling both new applications and more cost-effective means of providing those applications. Research is needed to support development of new architectures and operational breakthroughs in emerging issue areas like cloud computing, content distribution networks, content centered networks, network virtualization, social applications and online personal content—as well as topics of study that remain nascent.

RECOMMENDATION 7.5: NSF should establish an open, multi- location, interdisciplinary research center for areas related to broadband, addressing technology, policy and economics. center priorities should be driven by the agenda identified in the National Academies research road map.

Creating new technologies often involves interdisciplinary collaboration. In networking, for example, scientists in fields such as dynamic spectrum access, robust wireless networking and applications might need to work together to develop breakthrough solutions. [558]

The NSF should consider establishing an interdisciplinary research center for broadband networking, devices, applications and enabling technologies. Such a center could be modeled on the Engineering Research Centers (ERCs) that the NSF established in 1984. ERCs are partnerships among universities, technology-based industries and the NSF that focus on integrated engineering systems and produce technological innovations that strengthen the competitive position of industry. They currently operate in a number of fields such as biotechnology, energy and microelectronics. The NSF funds each ERC for 10 years, and most centers become self-sustaining. [559]

Only 2 of the existing ERCs touch on broadband networking, and their current research is limited to optical technologies and integrated microsensor networks. The NSF should establish a broadband networking research center in partnership with the FCC. The involvement of the FCC, as the government's expert agency on telecommunications, would help assure that the ERC agenda includes topics that are relevant to broadband policy.

The research center could illustrate what can be accomplished by connecting multiple, geographically dispersed physical research centers through very-high-speed optical wavelength networking. Examples of such connectivity include Internet2 and National LambdaRail in the United States and SURFnet in the Netherlands. [560] As a platform for research and innovation, the center ought to collaborate with private research centers, academic research networks and the gigabit community testbeds referenced above that are being constructed by industry and the non-profit sector. The center should practice open research, and the networks connecting these locations should adhere to open network principles as defined by the FCC. [561]

The research center should be broadly interdisciplinary so that it can address not only the technical issues raised by broadband, but also the economic and policy issues it raises. Researchers should include not only technologists such as engineers, computer scientists and physicists, but also economists and other social scientists. Bringing together a large number of diverse researchers should allow the center to work on projects of a larger scale than is typical under NSF grants.

RECOMMENDATION 7.6: NSF, in consultation with the FCC, should fund both a wireless testbed for promoting the science underlying spectrum policymaking and a testbed for evaluating the network security needed to provide a secure broadband infrastructure.

Spectrum (along with fiber) will be critical to the effective operation of future communications networks. However, there is uncertainty about how spectrum can be most efficiently and innovatively used in such networks. Wireless testbeds could be valuable tools to develop the science to support modern spectrum policy principles, which could guide FCC rulemaking on spectrum matters. For example, today there is uncertainty about how best to establish technical rules for exclusive spectrum, unlicensed spectrum and shared spectrum. Wireless testbeds can permit empirical assessment of radio systems and the complex interactions of spectrum users, which are nearly impossible to assess through simulation or analytical methods. As a result, they can reveal a great deal about how sharing can best be facilitated, how spectrum rights might be established, and the impact of dynamic spectrum access radios on existing and future communications services.

A request for proposal should be made to build and assess a network testbed that is sufficiently secure. With sensitive information about almost all Americans available in computerized databases and with the recent growth of electronic commerce, cybersecurity has become a vital issue. Many of the tools exist for building secure networks, but from an end-to-end systems perspective, difficult problems remain to be solved (particularly those that cross technical and non-technical disciplines). [562]

RECOMMENDATION 7.7: The FCC should start a rulemaking process to establish more flexible experimental licensing rules for spectrum and facilitate the use of this spectrum by researchers.

For the most part, spectrum is lightly used outside major urban areas. This holds true for prime frequency bands such as 800 MHz cellular and 1850–1990 MHz Personal Communications Services. In non-prime frequency bands such as those above 20 GHz, use may be modest even in major urban areas and limited or nonexistent in most other areas. Allowing research organizations such as universities greater flexibility to temporarily use fallow spectrum can promote more efficient and innovative communications systems.

Currently, there are restrictions on market trials conducted under experimental authorizations. [563] The FCC, building on relevant ideas from the Wireless Innovation Notice of

Inquiry,[564] should evaluate whether regulatory restrictions should be relaxed to permit research organizations to conduct broader market studies. Similarly, such organizations could be permitted to operate experimental stations without individual coordination of frequencies, conditioned on not causing harmful interference to authorized stations. Such a program could allow the FCC to work cooperatively with research organizations to identify topics and frequency bands for further study and to learn about new wireless technologies.

To facilitate the use of spectrum by researchers, the FCC should work with the National Telecommunications and Information Administration (NTIA) to identify underutilized spectrum that may be suitable for conducting research activities. It should also conduct workshops with NTIA to advance research activities involving spectrum use.

PART II—INCLUSION

Equality of opportunity is a fundamental principle of american democracy. For too long, the geographic limitations of one's life have determined access to many critical resources—employment, schools and services. Too often, we can predict the outcome of children's lives by the ZIp code in which they live.[565] people are shut out from economic and social opportunity by blighted neighborhoods, lack of sustainable employment and failing schools—excluded from making informed choices about their family's future.

Access to broadband is the latest challenge to equal opportunity, but it also offers new and innovative avenues to achieve it. Broadband can be a platform for significant economic, cultural and social transformation, overcoming distance and transcending the limitations of one's physical surroundings. Americans can use broadband to take online classes and read digital textbooks. They can utilize broadband to make and maintain community connections and obtain information about their health care. They can use broadband to bank, shop and apply for jobs. In these many ways, broadband can help create opportunity.

Yet approximately 100 million people in the United States do not use broadband at home.[566] Some of these Americans do not see the need for the technology; they may not value the extra speed broadband delivers or do not think it is relevant to their day-to-day lives. And some will never choose to subscribe to broadband, just as a small percentage of Americans do not see the need for television or telephone service.

But for others, lack of broadband is not a simple choice. More than 14 million Americans do not have access to broadband infrastructure that can support today's applications. Some cannot afford broadband service or the cost of a computer. Some lack the basic skills needed to take advantage of broadband. Still others may only get service via satellite.

The cost of this digital exclusion is large and growing. For individuals, the cost manifests itself in the form of lost opportunities. As more aspects of daily life move online and offline alternatives disappear, the range of choices available to people without broadband narrows. Digital exclusion compounds inequities for historically marginalized groups. People with low incomes, people with disabilities, racial and ethnic minorities, people living on Tribal lands and people living in rural areas are less likely to have broadband at home. Digital exclusion imposes inefficiencies on our society as one-third of Americans carry out tasks by means that take more time, effort and resources than if they had used broadband. Since government agencies must maintain both offline and online systems for transactions, many government services are not as effective or efficient as they could be. [567]

Like the costs of poverty, it is difficult to quantify the costs of digital inequality. It is certain, however, that people will not experience the promised benefits of broadband—increased earning potential, enhanced connections with friends and family, improved health and a superior education—without a connection.

Some of the recommendations in Part I of this plan (Innovation and Investment) discussed improving the economics of deploying and upgrading networks, both in unserved and served areas. More spectrum for wireless broadband, reducing the cost and complexity of access to utility poles and rights-ofway, ensuring fair prices in the wholesale market for backhaul service and implementing policies to stimulate broadband demand will ultimately push the network farther into unserved areas. Unfortunately, this will not finish the job of connecting people to broadband, since many areas of the country are just too expensive to serve without government support.

Part II (Inclusion) makes recommendations to ensure that any American who wants to subscribe to broadband can get the service. Chapter 8 sets a path to providing broadband to all Americans by extending the network through public investment in privately owned infrastructure. Chapter 9 examines the barriers many Americans face in adopting broadband—such as cost, digital literacy and relevance—and considers specific programs to reduce these barriers.

At stake is the equality of opportunity on which America was built. The nation needs to provide everyone with the opportunity to join the world that broadband is helping reshape.

8. AVAILABILITY

Everyone in the united states today should have access to broadband services supporting a basic set of applications that include sending and receiving e-mail, downloading web pages, photos and video, and using simple video conferencing. [568]

Ensuring all people have access to broadband requires the Federal Communications Commission (FCC) to set a national broadband availability target to guide public funding. An initial universalization target of 4 Mbps of *actual* download speed and 1 Mbps of *actual* upload speed, with an acceptable quality of service for interactive applications, would ensure universal access. [569]

This represents a speed comparable to what the typical broadband subscriber receives today, and what many consumers are likely to use in the future, given past growth rates. [570] While the nation aspires to higher speeds as described in Chapter 2, it should direct public investment toward meeting this initial target.

A universalization target of 4 Mbps download and 1 Mbps upload is aggressive. It is one of the highest universalization targets of any country in the world. Many nations, such as South Korea and Finland, have already adopted short-term download targets around 1 Mbps (see Exhibit 8-A). Over time, these targets, both in the United States and abroad, will continue to rise.

Exhibit 8-A. Universalization Goals in Selected Countries[571]

Country	"Universal" availability target (download)	Type of speed	Date
United States	4 Mbps	Actual	2020
South Korea	1 Mbps (99%)	Actual	2008
Finland	1 Mbps	Actual	2009
Australia	0.5 Mbps	Unspecified	2010
Denmark	0.5 Mbps	Unspecified	2010
Ireland	1 Mbps	Unspecified	2010
France	0.5 Mbps	Unspecified	2010
Germany	1 Mbps	Unspecified	2010
United Kingdom	2 Mbps	Unspecified	2012
Australia	12 Mbps	Unspecified	2018

It is possible the speed requirements for the most common applications will grow faster than they have historically. But it is also possible compression technology or shifts in customer usage patterns will slow the growth of bandwidth needs. To account for this uncertainty, the FCC should review and reset this target for public investment every four years.[572]

Recommendations

The FCC should conduct a comprehensive reform of universal service and intercarrier compensation in three stages to close the broadband availability gap.

Stage One: Lay the foundation for reform (2010–2011)

- The FCC should improve Universal Service Fund (USF) performance and accountability.
- The FCC should create the Connect America Fund (CAF).
- The FCC should create the Mobility Fund.
- The FCC should design new USF funds in a tax-efficient manner to minimize the size of the gap.
- Throughout the USF reform process, the FCC should solicit input from Tribal governments on USF matters that impact Tribal lands.
- The FCC should take action to shift up to $15.5 billion over the next decade from the current High-Cost program to broadband through common-sense reforms.
- The FCC should adopt a framework for long-term intercarrier compensation (ICC) reform that creates a glide path to eliminate per-minute charges while providing carriers an opportunity for adequate cost recovery, and establish interim solutions to address arbitrage.
- The FCC should examine middle-mile costs and pricing.

BOX 8-1. NATIONAL BROADBAND AVAILABILITY TARGET

Every household and business location in America should have access to affordable broadband service with the following characteristics:

- *Actual* download speeds of at least 4 Mbps and *actual* upload speeds of at least 1 Mbps
- An acceptable quality of service for the most common interactive applications

The FCC should review and reset this target every four years.

Stage Two: accelerate reform (2012–2016)

- The FCC should begin making disbursements from the CAF.
- The FCC should broaden the universal service contribution base.
- The FCC should begin a staged transition of reducing per- minute rates for intercarrier compensation.

Stage Three: Complete the transition (2017–2020)

- The FCC should manage the total size of the USF to remain close to its current size (in 2010 dollars) in order to minimize the burden of increasing universal service contributions on consumers.
- The FCC should eliminate the legacy High-Cost program, with all federal government funding to support broadband availability provided through the CAF.
- The FCC should continue reducing ICC rates by phasing out per-minute rates for the origination and termination of telecommunications traffic.

Accelerating broadband deployment

- To accelerate broadband deployment, Congress should consider providing optional public funding to the Connect America Fund, such as a few billion dollars per year over a two to three year period.

Congress should consider providing other grants, loans and loan guarantees

- Congress should consider expanding combination grant- loan programs.
- Congress should consider expanding the Community Connect program.
- Congress should consider establishing a Tribal Broadband Fund to support sustainable broadband deployment and adoption on Tribal lands, and all federal agencies that upgrade connectivity on Tribal lands should coordinate such upgrades with Tribal governments and the Tribal Broadband Fund grant-making process.

Government should facilitate Tribal, state, regional, and local broadband initiatives

- Congress should make clear that state, regional and local governments can build broadband networks.
- Federal and state policies should facilitate demand aggregation and use of state, regional and local networks when that is the most cost-efficient solution for anchor institutions to meet their connectivity needs.
- Congress should consider amending the Communications Act to provide discretion to the FCC to allow anchor institutions on Tribal lands to share broadband network capacity that is funded by the E-rate or the Rural Health Care program with other community institutions designated by Tribal governments.
- The federal government and state governments should develop an institutional framework that will help America's anchor institutions obtain broadband connectivity, training, applications and services.

8.1. The Broadband Availability Gap

Setting a target clarifies where the United States should focus its resources to universalize broadband. At present, there are 14 million people living in seven million housing units[573] that do not have access to terrestrial broadband infrastructure capable of meeting the National Broadband Availability Target.[574]

This broadband availability gap is greatest in areas with low population density.[575] Because service providers in these areas cannot earn enough revenue to cover the costs of deploying and operating broadband networks, including expected returns on capital, there is no business case to offer broadband services in these areas. As a result, it is unlikely that private investment alone will fill the broadband availability gap. The question, then, is how much public support will be required to fill the gap.

An FCC analysis finds that the level of additional funding required is approximately $24 billion (present value in 2010 dollars) as described in Exhibit 8-B.[576]

Exhibit 8-B presents the broadband availability gap in greater detail. Initial capital expenditures ("initial capex") are the incremental investments required to *deploy* networks that can deliver the targeted level of service to everyone in the United States; this covers new networks and upgrades of existing networks. "Ongoing costs" are the incremental costs that must be incurred to *operate* those networks. They include the cost of replacing old or outdated equipment, access to middle-mile transport and other continuing costs such as customer service, marketing and network operations.

"Revenue" includes all incremental revenue generated as a result of deploying the networks that meet the National Broadband Availability Target, whether the revenue comes from the sale of voice, data or, in limited cases, multichannel video services.

Cash Flows Associated With Broadband Availability Gap

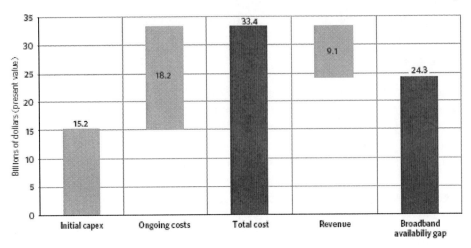

Exhibit 8-B. The Present Value (in 2010 Dollars) of the Broadband Availability Gap is $24 Billion[577]

Adding initial capex and continuing costs and subtracting revenue yields a gap of approximately $24 billion.[578]

This estimate is based on a number of key assumptions:

- First, the gap was calculated based on the economics of terrestrial technologies only, although a variety of technologies and architectures were considered. While satellite is capable of delivering speeds that meet the National Broadband Availability Target,[579] satellite capacity can meet only a small portion of broadband demand in unserved areas for the foreseeable future.[580] Satellite has the advantage of being both ubiquitous and having a geographically independent cost structure, making it particularly well suited to serve high-cost, low-density areas. However, while satellite can serve *any* given household, satellite capacity does not appear sufficient to serve *every* unserved household. In addition, the exact role of satellite-based broadband and its impact on the total cost of universalizing access to broadband depends on the specific disbursement mechanism used to close the broadband availability gap.

- Second, this calculation assumes that, whenever possible, a market-based mechanism will be used to select which providers receive support (as discussed in Section 8.3), and that there is competitive interest in receiving a subsidy to extend broadband to an unserved area. But it is impossible to know precisely how and whether this will occur until the details of the distribution mechanism are defined.

- Third, the estimated gap does not assume that currently announced fourth-generation (4G) wireless buildouts will provide service that meets the target without investments incremental to the planned commercial builds. Fourth- generation technology holds great promise and will likely play a large role in closing the broadband availability gap if speed and consumer satisfaction are comparable to traditional wired service, such as that provided over Digital Subscriber Line (DSL) or cable modem. If buildouts occur as announced, about five million of the seven million unserved housing units will have 4G coverage.[581] However, in order to provide actual download speeds of 4 Mbps or more, it may be necessary for providers to make investments that are incremental to their planned commercial builds. The FCC will revisit this issue as this new technology is implemented.

- Fourth, the estimated gap does not include any amounts necessary to support companies that currently receive universal service support for voice and already offer broadband that meets the National Broadband Availability Target. Some federal USF amounts indirectly support broadband, and going forward will do so directly. Nor do the estimates take into account the impact on existing recipients of support if other providers receive support to build out broadband in an area where the current provider has a carrier of last resort obligation.

- Fifth, there are a number of recommendations throughout this plan that may lower the cost of entering or operating in currently unserved areas, or that could increase or decrease potential revenues. The calculation does not include the impact of any of these recommendations. To the extent these recommendations are implemented, they may change the overall gap. The analysis also does not take into account any available federal, state, regional, Tribal, local or other funding sources that could help close the gap.

The support needs of different geographic areas are distinct and depend on many factors, including the existing network infrastructure and household density. In some areas, subsidizing all or part of the initial capex will allow a service provider to have a sustainable business. Elsewhere, subsidizing initial capex will not be enough; service providers will need support for continuing costs. Support for one-time deployment or upgrades will likely be enough to provide broadband to 46% of the seven million unserved housing units. Closing the gap for the remaining 54% of housing units will probably require support for both one-time and recurring costs.

Moreover, serving the 250,000 housing units with the highest gaps accounts for $14 billion of the broadband availability gap. As Exhibit 8-C depicts, this represents less than two-tenths of 1% of all housing units in the United States. The average amount of funding per housing unit to close the gap for these units with terrestrial broadband is $56,000.[582]

8.2. Closing the Broadband Availability Gap

Closing the broadband availability gap requires financial support from federal, state and local governments. This section will discuss the current state of government support for infrastructure deployment and will make recommendations for targeting this support more directly to close the availability gap.

The federal government spends nearly $10 billion annually on grants, loans and other subsidy programs that support communications connectivity; in 2010, the American Recovery and Reinvestment Act (Recovery Act) provided an additional $7.2 billion in one-time funding (see shaded rows Exhibit 8-D). Historically, much of this funding has supported voice service in certain areas of the country, but more recently it also has been used to modernize networks to deliver broadband as well. While this funding has improved broadband infrastructure in the

U.S., federal efforts have not been coordinated to meet the universal broadband goals of Congress.

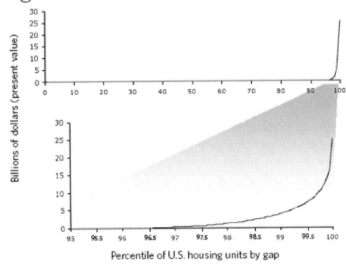

Exhibit 8-C. The Most Expensive Unserved Housing Units Represent a Disproportionate Share of the Total Gap[583]

Exhibit 8-D. Existing Sources of Federal Support for Communications Connectivity[584]

Agency	Program	Description	Annual funding amount
Federal Communications Commission	Universal Service Fund	Provides funding for companies serving high-cost areas, low-income consumers, rural health care providers, and schools and libraries.	$8.7 billion (FY2010)
National Tele-communications and Information Administration	Broadband Technology Opportunities Program	Grant program to promote deployment and adoption of broadband throughout the country, particularly in unserved and underserved areas. Priority in the second Notice of Funding Availability (NOFA) will be given to middle-mile broadband infrastructure projects that offer new or substantially upgraded connections to community anchor institutions, especially community colleges.	$4.7 billion (one-time ARRA)—includes at least $2.5 billion for infrastructure, $250 million for adoption, and $200 million for public computing centers.

Exhibit 8-D. (Continued)

Agency	Program	Description	Annual funding amount
Rural Utilities Service	Broadband Initiatives Program	Loan, loan guarantee and grant program to increase broadband penetration and adoption, primarily in rural areas. Priority in the second NOFA will be given to last-mile projects, and middle-mile projects involving current RUS program participants.	$2.5 billion (one-time ARRA)—includes at least $2.2 billion for infrastructure.
Rural Utilities Service	Telephone Loans and Loan Guarantees Program	Provides long-term, direct and guaranteed loans to qualified organizations, often telephone companies, to support invest-ment in broadband-capable telephone networks.	$685 million
Rural Utilities Service	Rural Broadband Access Loans and Loan Guarantees Program	Provides loans and loan guarantees to eligible applicants—in cluding telephone companies, municipalities, non-profits and Tribes—to deploy broadband in rural communities.	$298 million
Institute of Museum and Library Services	Library Services and Technology Act Grants	Provides funds for a wide range of library services including installation of fiber and wireless networks.	$164 million
Multiple agencies	Other programs[585]	Multiple purposes	$49 million
Total			$17.1 billion

Nearly half of the funding appropriated in 2010 to support greater connectivity comes from the Recovery Act, which Congress passed in February 2009. Congress appropriated $7.2 billion to create the Broadband Telecommunications Opportunities Program (BTOP) at the U.S. Department of Commerce and the Broadband Initiatives Program (BIP) at the U.S. Department of Agriculture. BTOP "makes available grants for deploying broadband infrastructure in unserved and underserved areas in the United States, enhancing broadband capabilities at public computer centers, and promoting sustainable broadband adoption projects."[586] BIP "extend[s] loans, grants and loan/grant combinations to facilitate broadband deployment in rural areas."[587]

Awards under BTOP and BIP are ongoing, and many projects should help meet the goal of providing universal broadband access. For instance, the ION Rural Broadband Initiative will add middle-mile connectivity for 70 rural communities in upstate New York, and Project Connect South Dakota will provide a cash infusion to add 140 miles of back- haul service and 219 miles of middle-mile connections to an existing fiber optic network.[588]

Through the Broadband Data Improvement Act mapping process, the FCC may be able to improve its estimate of the gap. But it is impossible to know with precision how much the

BTOP and BIP programs will contribute to closing the gap before all of the funds are awarded.

In any event, BTOP and BIP alone will not be sufficient to close the broadband availability gap. Other government support is required to complete the task of connecting the nation to ensure that broadband reaches the highest-cost areas of the country. Closing the broadband availability gap and connecting the nation will require a substantial commitment by states and the federal government alike. This commitment must include initial support to cover the capital costs of building new networks in areas that are unserved today, as well as ongoing support for the operation of newly built networks in areas where revenues will be insufficient to cover ongoing costs.

8.3. Universal Service

Universal service has been a national objective since the Communications Act of 1934, in which Congress stated its intention to "make available, so far as possible, to all the people of the United States... a rapid, efficient, Nation-wide, and world-wide wire and radio communication service with adequate facilities at reasonable charges." [589]

The current federal universal service programs were created in the aftermath of the Telecommunications Act of 1996 at a time when only 23% of Americans had dial-up Internet access at home, and virtually no one had broadband. [590] While the federal USF and earlier programs have played a critical role in the universalization of voice service in the last century, the current USF was not designed to support broadband directly, other than for schools, libraries and rural health care providers. [591]

Exhibit 8-E. The Federal Universal Service Fund[594]

Program	Description	FY 2010 disbursements (projected)
High Cost	Ensures that consumers in all regions of the nation have access to and pay rates for telecommunications services that are reasonably comparable to those in urban areas.	$4.6 billion
Low Income (Lifeline and Link-Up)	Provides discounts that make basic, local telephone service affordable for low-income consumers.	$1.2 billion
Schools and Libraries (E-rate)	Subsidizes telecommunications services, Internet access and internal connections to enable schools and libraries to connect to the Internet.	$2.7 billion
Rural Health Care	Provides reduced rates to rural health care providers for telecommunications and Internet access services and, on a pilot basis, support for infrastructure.	$214 million
Total		**$8.7 billion**

In 2010, the federal USF is projected to make total outlays of $8.7 billion through four programs (see Exhibit 8-E).[592] The High- Cost program, which subsidizes telecommunications services in areas where costs would otherwise be prohibitively high, will spend $4.6 billion. E-rate, which supports voice and broadband connectivity for schools and libraries, will spend $2.7 billion.[593] The Low Income program, which subsidizes the cost of telephone service for low-income people, will spend $1.2 billion, and the Rural Health Care program, which supports connectivity for health care providers, will spend $214 million.

At least 21 states have high-cost funds that collectively distribute over $1.5 billion. [595] Thirty-three states have a state low-income program, nine states have a state subsidy program for schools and libraries, and at least 27 states support state telehealth networks. [596] A number of states have established specific programs to fund broadband deployment. [597] Some states provide tax credits for investment in broadband infrastructure. [598]

The remainder of this section will discuss how the current federal High-Cost program should be modernized to shift from supporting legacy telephone networks to directly supporting high-capacity broadband networks. The federal Low Income program provides critical support to low-income households and will be discussed in Chapter 9. The Rural Health Care and E-Rate programs provide important support for broadband to critical institutions like schools, libraries and health care facilities, and will be addressed in Chapters 10 and 11.

Accelerating the pace of investment in broadband networks in high-cost areas will also require consideration of related policy issues that affect the revenue streams of existing carriers. The ICC system provides a positive revenue stream for certain carriers, which in turn affects their ability to upgrade their networks during the transition from voice telephone service to broadband service. In rural America USF and ICC represent a significant portion of revenues for some of the smallest carriers—i.e., 60% or more of their regulated revenues. [599] The rules governing special access services also affect the economics of deployment and investment, as middle-mile transmission often represents a significant cost for carriers that need to transport their traffic a significant distance to the Internet backbone. For that reason, the FCC needs to consider the middle mile in any discussion of government support to high-cost areas. [600]

USF and ICC regulations were designed for a telecommunications industry that provided voice service over circuit-switched networks. State and federal ratemaking created implicit subsidies at both the state and federal levels and were designed to shift costs from rural to urban areas, from residential to business customers, and from local to long distance service.

Unfortunately, the current regulatory framework will not close the broadband availability gap. A comprehensive reform program is required to shift from primarily supporting voice communications to supporting a broadband platform that enables many applications, including voice. This reform must be staged over time to realign these systems to support broadband and minimize regulatory uncertainty for investment.

The goal of reform is to provide everyone with affordable voice and broadband. The reforms must be achieved over time to manage the impact on consumers, who ultimately pay for universal service. The FCC should target areas that are currently unserved, while taking care to ensure that consumers continue to enjoy broadband and voice services that are available today. Given that USF is a finite resource, the FCC should work to maximize the number of households that can be served quickly, focusing first on those areas that require lower amounts of subsidy to achieve that goal, and over time addressing those areas that are the hardest to serve, recognizing that the subsidy required may decline in the future as technology advances and costs decline. Ongoing support should be provided where necessary.

Sudden changes in USF and ICC could have unintended consequences that slow progress. Success will come from a clear road map for reform, including guidance about the timing and pace of changes to existing regulations, so that the private sector can react and plan appropriately.

Stage One of this comprehensive reform program starts with building the institutional foundation for reform, identifying funding that can be shifted immediately to jumpstart broadband deployment in unserved areas, creating the framework for a new Connect America Fund and a Mobility Fund, establishing a long-term vision for ICC, and examining middle-mile costs and pricing (see Chapter 4). In Stage Two, the FCC will begin disbursements from the CAF and Mobility Fund, while implementing the first step in reducing intercarrier compensation rates and reforming USF contribution methodology. Stage Three completes the transformation of the legacy High-Cost program, ends support for voice-only networks and completes reforms on ICC.

Before going into the details of this plan, it is important to consider the unique characteristics of each system in more detail.

The High-Cost Program

The High-Cost program ensures that consumers in all parts of the country have access to voice service and pay rates for that service that are reasonably comparable to service in urban areas. The program currently provides funding to three groups of eligible telecommunications carriers (ETCs) (see Box 8-2). In 2009, approximately $2 billion went to 814 rate-of-return carriers, $1 billion to 17 price-cap carriers and $1.3 billion to 212 competitive eligible telecommunications carriers (competitive ETCs). [601]

The current High-Cost program is not designed to universalize broadband. While some companies receiving High-Cost support have deployed broadband-capable infrastructure to serve most of their customers, [602] others have not. Carriers receiving High-Cost support are not required to provide any households in their service area with some minimal level of broadband service, much less provide such service to *all* households in their service area.

In addition, the High-Cost program only supports certain components of a network, such as local loops and switching equipment, but not other components necessary for broadband, like middle-mile infrastructure that transports voice and data traffic to an Internet point of presence. As a result, the amount of support provided is not appropriately sized for the provision of broadband in high-cost areas.

Because broadband is not a supported service, today there is no mechanism to ensure that support is targeted toward extending broadband service to unserved homes. Today, roughly half of the unserved housing units are located in the territories of the largest price-cap carriers, which include AT&T, Verizon and Qwest, while about 15% are located in the territories of mid-sized price-cap companies such as CenturyLink, Windstream and Frontier. [603] While current funding supports phone service to lines served by price-cap carriers, the amounts do not provide an incentive for the costly upgrades that may be required to deliver broadband to these customers. [604]

In addition, current oversight of the specific uses of High-Cost support is limited. While some states require both incumbents and competitive ETCs to report on their use of funding for network infrastructure projects, [605] many states do not. [606] There is no uniform framework at the federal level to track the progress of any infrastructure deployment, broadband-capable or not, that is subsidized through the use of federal funds.

Box 8-2. High-Cost Program Recipients

Rate-of-Return Carriers—Incumbent telephone companies that are given the opportunity to earn an 11.25% rate of return on their interstate services.

Price-Cap Carriers—Incumbent telephone companies that may only raise interstate rates on the basis of a formula that considers expense growth and a productivity growth factor.

Competitive ETCs—Competitive wireline and wireless providers that are certified by a state utility regulator or the FCC to receive funds from the High-Cost program based on the level of support provided to the incumbent in a given area.

While the High-Cost program has made a material difference in enabling households in many high-cost areas of America to have access to affordable voice service, it will not do the same for broadband without reform of the current system.

Intercarrier Compensation

ICC is a system of regulated payments in which carriers compensate each other for the origination, transport and termination of telecommunications traffic. For example, when a family in Philadelphia calls Grandma in Florida, the family's carrier usually pays Grandma's carrier a per-minute charge, which may be a few cents a minute, for terminating the call. Estimates indicate that this system results in up to $14 billion in transfers between carriers every year. [607]

The current per-minute ICC system was never designed to promote deployment of broadband networks. Rather, ICC was implemented before the advent of the Internet when there were separate local and long distance phone companies. Local companies incurred a traffic-sensitive cost to "switch" or connect a call from the long distance company to the carrier's customer. The per-minute rates charged to the long distance carrier were set above cost and provided an implicit subsidy for local carriers to keep residential rates low and promote universal telephone service. [608] ICC has not been reformed to reflect fundamental, ongoing shifts in technology and consumer behavior, and it continues to include above-cost rates. The current ICC system is not sustainable in an all-broadband Internet Protocol (IP) world where payments for the exchange of IP traffic are not based on per-minute charges, but instead are typically based on charges for the amount of bandwidth consumed per month.

The current ICC system also has fundamental problems that create inefficient incentives. First, terminating rates are not uniform despite the uniformity of the function of terminating a call, which leads to unproductive economic activity. Rates vary from zero to 35.9 cents per minute,[609] depending on the jurisdiction of the call, the type of traffic[610] and the regulatory status of the terminating carrier.[611] Rate differences lead to arbitrage opportunities such as phantom traffic, in which traffic is masked to avoid paying the terminating carrier intercarrier compensation entirely, and/or redirected to make it appear that the call should be subject to a lower rate.[612] Such behavior leads to disputes and underpayment to the terminating carrier.

Most ICC rates are above incremental cost, which creates opportunities for access stimulation, in which carriers artificially inflate the amount of minutes subject to ICC payments. For example, companies have established "free" conference calling services, which provide free services to consumers while the carrier and conference call company share the

ICC revenues paid by interexchange carriers.[613] Because the arbitrage opportunity exists, investment is directed to free conference calling and similar schemes for adult entertainment that ultimately cost consumers money,[614] rather than to other, more productive endeavors.

Broadband providers have begun migrating to more efficient IP interconnection and compensation arrangements for the transport and termination of IP traffic. Because providers' rates are above cost, the current system creates disincentives to migrate to all IP-based networks. For example, to retain ICC revenues, carriers may require an interconnecting carrier to convert Voice over Internet Protocol (VoIP) calls to time-division multiplexing in order to collect intercarrier compensation revenue. While this may be in the short-term interest of a carrier seeking to retain ICC revenues, it actually hinders the transformation of America's networks to broadband.[615]

ICC may be stalling the development of the broadband ecosystem in other ways as well. For example, there are allegations that regulatory uncertainty about whether or what intercarrier compensation payments are required for VoIP traffic,[616] as well as a lack of uniform rates, may be hindering investment and the introduction of new IP-based services and products.[617]

Moreover, fewer terminating minutes ultimately mean a smaller revenue base for intercarrier compensation. According to FCC data, for example, total minutes of use of incumbent carriers decreased from 567 billion minutes in 2000 to 316 billion minutes in 2008, a drop of 56%.[618] Price-cap carriers have no means of increasing per-minute rates to offset these declines. Even rate-of-return carriers, who are permitted to increase per-minute rates so they have the opportunity to earn their authorized rate of return, acknowledge that the current system is "not sustainable" and could lead to a "death spiral" as higher rates to offset declining minutes exacerbate arbitrage and non-payment.[619] As the small carriers recognize, revenues are also decreasing due to arbitrage and disputes over payment for VoIP traffic.[620]

The continued decline in revenues and free cash flows at unpredictable levels could hamper carriers' ability to implement network upgrade investments or other capital improve- ments. Any consideration of how government should provide supplemental funding to companies to close the broadband availability gap should recognize that ICC revenue is an important part of the picture for some providers.

Special Access Policies

High-capacity dedicated circuits are critical inputs in the provision of fixed and mobile broadband services in rural America. Special access circuits connect wireless towers to the core network,[621] provide fiber optic connectivity to hospitals and health centers,[622] and are sometimes the critical broadband link that traverses up to 200 miles between a small town and the nearest Internet point of presence.[623] The law requires that the rates, terms and conditions for these circuits be just and reasonable.[624]

The rates that firms pay for these critical middle- and second-mile connections have an impact on the business case for the provision of broadband in high-cost areas. Small local exchange carriers, wireless firms and small cable companies typically purchase these connections from other providers. It may well be the case that the cost of providing these circuits is so high that there is no private sector business case to offer broadband in some areas, even if the rates, terms and conditions are just and reasonable.

High-Cost funds today are generally distributed on the basis of loop and switching costs and not the cost of middle-mile transport of voice traffic. Because data traffic is aggregated on backhaul facilities, per-customer middle-mile costs will increase significantly as consumers and businesses use their broadband connections more. [625]

It is not clear whether the high costs of middle-mile connectivity in rural areas are due solely to long distances and low population density, [626] or also reflect excessively high special access prices as some parties have alleged. [627] The FCC is currently examining its analytic framework for regulating special access services generally (see Chapter 4). Because of the link between middle- and second-mile costs and special access policies, the FCC's review of its special access policies should be completed in concert with other aspects of this reform plan.

Comprehensive Reform

As federal and state regulators have recognized, the federal USF must be modernized to support the advanced broadband networks and services of the future—and must be modernized quickly, in a way that will accelerate the availability of broadband to all Americans. [628] Closing the broadband availability gap requires comprehensive reform of the USF High-Cost program, as well as consideration of ICC and an examination of special access costs and pricing. These actions should be consistent with a set of guiding principles:

- *Support broadband deployment directly.* The federal government should, over time, end all financial support for networks that only provide "Plain Old Telephone Service" (POTS) and should provide financial support, where necessary and in an economically efficient manner, for broadband platforms that enable many applications, including voice.[629]
- *Maximize broadband availability.* USF resources are finite, and policymakers need to weigh tradeoffs in allocating those resources so that the nation "gets the most bang for its buck." The objective should be to maximize the number of households that are served by broadband meeting the National Broadband Availability Target.[630]
- *No flash cuts.* New rules should be phased in over a reasonable time period. Policymakers must give service providers and investors time to adjust to a new regulatory regime.[631]
- *Reform requires federal and state coordination.* The FCC should seek input from state commissions on how to harmonize federal and state efforts to promote broadband availability.[632]

These guiding principles will inform a long-term plan for reform that will unfold over a decade (see Exhibit 8-F). This plan balances the need to direct more capital to broadband networks, particularly in high-cost areas, while recognizing the significant role that the private sector plays in broadband deployment.

One variable that will impact the pace of broadband availability is the time it will take to implement various reforms. The proposed reforms on the timeline presented could enable the buildout of broadband infrastructure to more than 99% of American households by 2020. Any acceleration of this path would require more funding from Congress, deeper cuts in the existing USF program or higher USF assessments, which ultimately are borne by consumers. While this plan makes the best use of the assets the country currently has to advance the availability of broadband, a more aggressive path is available if Congress so chooses.

Before discussing the reforms in Stage One to advance broadband availability, we address administrative reforms to improve the management and oversight of USF.

RECOMMENDATION 8.1: The FCC should improve universal service Fund (USF) performance and accountability.

The Universal Service Administrative Company (USAC), a not-for-profit subsidiary of the National Exchange Carrier Association (NECA), serves as the day-to-day administrator of USF, working under FCC direction. As part of its overall effort to make the FCC more open and transparent, data- driven and a model of excellence in government, the FCC is reviewing its oversight of the funds it administers to determine whether changes are necessary to improve efficiency and effectiveness. USF is part of that review and includes oversight and management of USAC and all of the universal service programs. While there is no doubt that federal universal service programs have been successful in preserving and advancing universal service, it is vital to ensure that these public funds are administered appropriately.

To provide stronger management and oversight of the program, the FCC already has begun to implement a number of changes:

- The FCC has moved oversight of the audit program to the Office of Managing Director and has directed USAC to revise its audit approach.
- The FCC has implemented a new Improper Payments Information Act (IPIA) assessment program that is tailored to cover all four USF disbursement programs, measure the accuracy of payments, evaluate the eligibility of applicants, test information obtained by participants, and ensure a reasonable cost while meeting IPIA requirements.
- The FCC has implemented a new compliance audit program for all four USF disbursement mechanisms and contributors. This audit program takes into account such factors as program risk elements and size of disbursements. This audit program is also conducted at a reasonable cost in relation to program disbursements and reduces unnecessary burdens on beneficiaries.

These new assessment and audit programs will reduce the cost of USF-related audits going forward and will be more efficient. These changes will also help deter fraud, waste and abuse and identify levels of improper payments.

As the FCC reforms its USF support and disbursement mechanisms after the release of the National Broadband Plan, it should also ensure that any future enhancements to the USF program have accountability and oversight provisions built in from the outset. The FCC should also examine its Memorandum of Understanding with USAC to ensure that it reflects programmatic changes and evaluate whether any modifications to its existing relationship with USAC are necessary.[633]

Across the four USF programs, there is a lack of adequate data to make critical policy decisions regarding how to better utilize funding to promote universal service objectives. For instance, recipients of USF funding currently are not required to report the extent to which they use the funding they receive to extend broadband-capable networks. As the FCC moves forward on the reforms in the plan, it should enhance its data collection and reporting to ensure that the nation's funds are being used effectively to advance defined programmatic goals.

Stage One: Laying the Foundation for Reform (2010–2011)

The FCC should create a Connect America Fund to address the broadband availability gap in unserved areas and provide any ongoing support necessary to sustain service in areas that already have broadband because of previous support from federal USF. The FCC should create a fast-track program in CAF for providers to receive targeted funding for new broadband construction in unserved areas. In addition, the FCC should create a Mobility Fund to provide one-time support for deployment of 3G networks (used for both voice and data) to bring all states to a minimum level of 3G availability which will improve the business case for investment in the rollout of 4G in harder to serve areas.

In Stage One, a series of actions will identify initial funds to be shifted from the current High-Cost program to the CAF and Mobility Funds. The FCC also should establish a glide path to long-term ICC reform, while taking interim steps to address phantom traffic and access stimulation to provide the industry a greater degree of revenue stability and predictability. Because middle- and second-mile connectivity is a key cost component for broadband service providers in high-cost areas, the FCC should also examine the rates for high-capacity circuits to ensure they are just and reasonable.

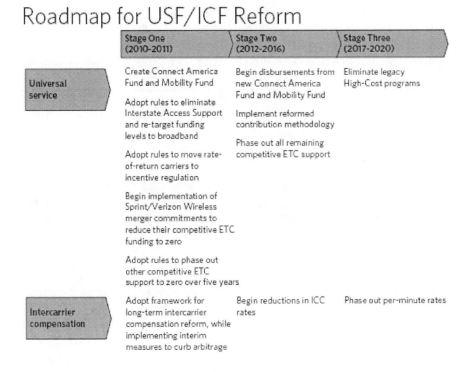

Exhibit 8-F. Roadmap for USF/ICC Reform

Throughout the USF reform process, the FCC should solicit input from Tribal governments on USF matters that impact Tribal lands.[634]

RECOMMENDATION 8.2: The FCC should create the Connect America Fund (CAF).

The FCC's long range goal should be to replace all of the legacy High-Cost programs with a new program that preserves the connectivity that Americans have today and advances universal broadband in the 21st century. CAF will enable all U.S. households to access a network that is capable of providing both high-quality voice-grade service and broadband that satisfies the National Broadband Availability Target. There are many issues that will need to be addressed in order to fully transition the legacy programs into the new fund. The FCC should create an expedited process[635], however, to fund broadband infrastructure buildout in unserved areas with the USF savings identified below.

As a general roadmap, CAF should adhere to the following principles:

- *CAF should only provide funding in geographic areas where there is no private sector business case to provide broadband and high-quality voice-grade service.*[636] CAF support levels should be based on what is necessary to induce a private firm to serve an area. Support should be based on the net gap (i.e., forward looking costs less revenues).[637] Those costs would include both capital expenditures and any ongoing costs, including middle-mile costs, required to provide high-speed broadband service that meets the National Broadband Availability Target.[638] Revenues should include all revenues earned from broadband-capable network infrastructure, including voice, data and video revenues,[639] and take into account the impact of other regulatory reforms that may impact revenue flows, such as ICC, and funding from other sources, such as Recovery Act grants.[640] The FCC should evaluate eligibility and define support levels on the basis of neutral geographic units such as U.S. Census-based geographic areas, not the geographic units associated with any particular industry segment.[641]

  In targeting funding to the areas where there is no private sector business case to offer broadband service, the FCC should consider the role of state high-cost funds in supporting universal service and other Tribal, state, regional and local initiatives to support broadband. A number of states have established state-level programs through their respective public utility commissions to subsidize broadband connections, while other states have implemented other forms of grants and loans to support broadband investment.[642] As the country shifts its efforts to universalize both broadband and voice, the FCC should encourage states to provide funding to support broadband and to modify any laws that might limit such support.[643]

- *There should be at most one subsidized provider of broadband per geographic area.*[644] Areas with extremely low population density are typically unprofitable for even a single operator to serve and often face a significant broadband availability gap. Subsidizing duplicate, competing networks in such areas where there is no sustainable business case would impose significant burdens on the USF and, ultimately, on the consumers who contribute to the USF.

- *The eligibility criteria for obtaining support from CAF should be company- and technology-agnostic so long as the service provided meets the specifications set by the FCC.* Support should be available to both incumbent and competitive telephone companies (whether classified today as "rural" or "non-rural"), fixed and mobile wireless providers, satellite providers and other broadband providers, consistent with statutory requirements.[645] Any broadband provider that can meet or exceed the specifications set by the FCC should be eligible to receive support.

- *The FCC should identify ways to drive funding to efficient levels, including market-based mechanisms where appropriate, to determine the firms that will receive CAF support and the amount of support they will receive.*[646] If enough carriers compete for support in a given area and the mechanism is properly designed, the market should help identify the provider that will serve the area at the lowest cost.

- *Recipients of CAF support must be accountable for its use and subject to enforceable timelines for achieving universal access.* USF requires ongoing adjustment and re-evaluation to focus on performance-based outcomes. The recipients of funding should be subject to a broadband provider-of-lastresort obligation.[647] The FCC should establish timelines for extending broadband to unserved areas. It should define operational requirements and make verification of broadband availability a condition for funding. [648] The subsidized providers, should be subject to specific service quality and reporting requirements, including obligations to report on service availability and pricing. Recipients of funding should offer service at rates reasonably comparable to urban rates.[649] The FCC should exercise all its relevant enforcement powers if recipients of support fail to meet FCC specifications.

RECOMMENDATION 8.3: The FCC should create the Mobility Fund.

As discussed in Chapter 3, both broadband and access to mobility are now essential needs, and America should have healthy fixed and mobile broadband ecosystems. Based on past experience with mobile wireless, it is not clear that government intervention will be necessary to enable a robust mobile broadband ecosystem in most parts of the country. According to American Roamer, 3G wireless networks, used for both voice and data, cover 98% of the population in the United States— more people than are passed by terrestrial broadband.[650]

However, some states have materially lower 3G deployment than the national average. For example, 77% of Alaska's population is covered by 3G networks, and a mere 71% of West Virginia's population is covered by 3G networks. [651]

This lack of coverage is even more significant if one considers that 3G infrastructure will be used in many cases to enable the rollout of 4G networks. U.S. companies will soon embark on 4G buildouts, expecting to reach at least 94% of the U.S. population by 2013. [652] The 4G footprint is likely to mirror closely the 3G footprint, because providers will use their existing infrastructure as much as possible. But how much this build will ultimately cost, and exactly which parts of the country it will cover, or not cover, remains unclear.

Timely, limited government intervention to expand the availability of 3G networks would help states with 3G buildout below the national standard to catch up with the rest of the nation and improve the business case for 4G rollout in harderto-serve areas. In addition, expanding 3G coverage would benefit public safety users to the extent that public safety agencies use commercial services. It would benefit public safety by establishing more cell sites that could be used for a 4G public- private broadband network, serving commercial as well as public safety users.

The FCC should create a Mobility Fund to provide one-time support for deployment of 3G networks, to bring all states to a minimum level of 3G (or better) mobile service availability.[653] The FCC should select an efficient method, such as a market- based mechanism, for supporting mobility in targeted areas.

RECOMMENDATION 8.4: The FCC should design new USF funds in a tax-efficient manner to minimize the size of the gap. [654]

In certain circumstances, the Department of Treasury's Internal Revenue Service treats governmental payments to private parties for the purpose of making capital investments to advance public purposes as contributions to capital under section 118 of the U.S. Internal Revenue Code. Such treatment allows recipients to exclude the payments from income, but reduces depreciation deductions in future years. The Department of Treasury recently issued a ruling that BTOP grants to corporations that are restricted solely to the acquisition of capital assets to be used to expand the business and that meet a five- part test would be excluded from income as a nonshareholder contribution to capital under section 118(a). [655] Ultimately, the impact of taxes incurred may depend on the specific details of how the support is distributed, as well as the profitability of the service providers that receive support.

Box 8-3. Tribal Input

The United States currently recognizes 564 American Indian Tribes and Alaska Native Villages (Tribes).[656] Tribes are inherently sovereign governments that enjoy a special relationship with the U.S. predicated on the principle of government-to-government interaction. This government-to-government relationship warrants a tailored approach that takes into consideration the unique characteristics of Tribal lands in extending the benefits of broadband to everyone.

Any approach to increasing broadband availability and adoption should recognize Tribal sovereignty, autonomy and independence, the importance of consultation with Tribal leaders, the critical role of Tribal anchor institutions, and the community- oriented nature of demand aggregation on Tribal lands. [657]

RECOMMENDATION 8.5: Throughout the USF reform process, the FCC should solicit input from Tribal governments on USF matters that impact Tribal lands.

In recognition of Tribal sovereignty, the FCC should solicit input from Tribal governments on any proposed changes to USF that would impact Tribal lands. Tribal governments should play an integral role in the process for designating carriers who may receive support to serve Tribal lands.[658] The ETC designation process should require consultation with the relevant Tribal government after a carrier files an ETC application to serve a Tribal land. It should also require that an ETC file a plan with both the FCC (or state, in those cases where a carrier is seeking ETC designation from a state) and the Tribe on proposed plans to serve the area.

RECOMMENDATION 8.6: The FCC should take action to shift up to $15.5 billion over the next decade from the current High-cost program to broadband through commonsense reforms. [659]

In Stage One, the FCC should identify near-term opportunities to shift funding from existing programs to advance the universalization of broadband. These targeted changes are designed to create a pathway to a more efficient and targeted funding mechanism for government support for broadband investment, while creating greater certainty and stability for private sector investment.

While these shifts could move as much as $15.5 billion (present value in 2010 dollars) into new broadband programs, they are not risk-free. Shifting identified funds to support broadband could have transitional impacts that will need to be carefully considered. To the extent the FCC does not realize the full amount of savings described below, it will need to identify additional opportunities for savings in Stage Two in order to achieve the National Broadband Availability Target, unless Congress chooses to provide additional public funding for broadband to mitigate some of hese risks.

First, the FCC should issue an order to implement the voluntary commitments of Sprint and Verizon Wireless to reduce the High-Cost funding they receive as competitive ETCs to zero over a five-year period as a condition of earlier merger decisions. [660] Sprint and Verizon Wireless received roughly $530 million in annual competitive ETC funding at the time of their respective transactions with Clearwire and Alltel in 2008. Their recaptured competitive ETC funding should be used to implement the recommendations set forth in this plan. This represents up to $3.9 billion (present value in 2010 dollars) over a decade.

Second, the FCC should require rate-of-return carriers to move to incentive regulation. As USF migrates from supporting voice telephone service to supporting broadband platforms that can support voice as well as other applications, and as recipients of support increasingly face competition in some portion of their service areas, [661] how USF compensates carriers needs to change as well.

Rate-of-return regulation was implemented in the 1960s, when there was a single provider of voice services in a given geographic area that had a legal obligation to serve all customers in the area and when the network only provided voice service. Rate-of-return regulation was not designed to promote efficiency or innovation; indeed, when the FCC adopted price- cap regulation in 1990, it recognized that "rate of return does not provide sufficient incentives for broad innovations in the way firms do business."[662]. In an increasingly competitive marketplace with unsubsidized competitors operating in a portion of incumbents' territories, permitting carriers to be made whole through USF support lessens their incentives to become more efficient and offer innovative new services to retain and attract consumers.

Conversion to price-cap regulation would be revenue neutral in the initial year of implementation, assuming that amounts per line for access replacement funding known as Interstate Common Line Support (ICLS) would be frozen (consistent with existing FCC precedent). [663] Over time, however, freezing ICLS would limit growth in the legacy High-Cost program on an interim basis, while the FCC develops a new methodology for providing appropriate levels of CAF support to sustain service in areas that already have broadband. [664] This step could yield up to $1.8 billion (present value in 2010 dollars) in savings over a decade.

The amount of interim savings achieved by freezing ICLS support during the CAF transition is dependent on the timing of the conversion to price caps and carrier behavior before the conversion. There is some chance that rate-of-return carriers could accelerate their investment before conversion to price caps to lock in higher support per line. Depending on the details of implementation, such a spike in investment activity could result in further broadband deployment that would narrow the broadband availability gap, but could increase the overall size of the fund.

Third, the FCC should redirect access replacement funding known as Interstate Access Support (IAS) toward broadband deployment. [665] Incumbent carriers received roughly $457 million in IAS in 2009. [666] When the FCC created IAS in 2000, it said it would revisit this funding mechanism in five years "to ensure that such funding is sufficient, yet not excessive." [667] That re-examination never occurred. Now, in order to advance the deployment of broadband platforms that can deliver high-quality voice service as well as other applications and services, the FCC should take immediate steps to eliminate this legacy program and re-target its dollars toward broadband. This could yield up to $4 billion (present value in 2010 dollars) in savings over a decade.

Freezing ICLS and refocusing IAS could have distributional consequences for existing recipients; individual companies would not necessarily receive the same amount of funding from the CAF as they might otherwise receive under the legacy programs. As the FCC considers this policy shift, it should take into account the impact of potential changes in free cash flows on providers' ability to continue to provide voice service and on future broadband network deployment strategies.

Fourth, the FCC should phase out the remaining legacy High-Cost support for competitive ETCs. [668] In 2008, the FCC adopted on an interim basis an overall competitive ETC cap of approximately $1.4 billion, pending comprehensive USF reform. [669] As the FCC reforms USF to support broadband, it is time to eliminate ongoing competitive ETC support for voice service in the legacy High-Cost program.

In some areas today, the USF supports more than a dozen competitive ETCs that provide voice service, [670] and in many instances, companies receive support for multiple handsets on a single family plan. Given the national imperative to advance broadband, subsidizing this many competitive ETCs for voice service is clearly inefficient. [671] The FCC should establish a schedule to reduce competitive ETC support to zero over five years, which will be completed in Stage Two. In order to accelerate the phase-down of legacy support, the FCC could immediately adopt a rule that any wireless family plan should be treated as a single line for purposes of universal service funding. [672] As competitive ETC support levels are reduced, this funding should be redirected toward broadband. This could yield up to $5.8 billion (present value in 2010 dollars) in savings over a decade.

Depending on the details and timing of implementation, these actions collectively will free up to $15.5 billion (present value in 2010 dollars) in funding from the legacy High-Cost program between now and 2020. In addition to funding the CAF, the savings identified should be used to implement a number of USF and ICC recommendations in this plan. Approximately $4 billion (present value in 2010 dollars) will go to a combination of activities including the new Mobility Fund, potential revenue replacement resulting from intercarrier compensation reform, expanding USF support for health care institutions up to the existing cap, enabling E -rate funding to maintain its purchasing power over time, and conducting pilots for a broadband Lifeline program. The remaining amount, up to $11.5 billion (present value in 2010 dollars), can be expressly targeted to supporting broadband through the CAF so that no one is left behind.

RECOMMENDATION 8.7: The FCC should adopt a framework for long-term intercarrier compensation (ICC) reform that creates a glide path to eliminate per-minute charges while providing carriers an opportunity for adequate cost recovery, and establish interim solutions to address arbitrage.

During Stage One, the FCC should establish a framework for phased reform of ICC to eliminate current distortions that are created by recovering fixed network costs through per-minute rates for the origination and termination of traffic. The FCC also should provide carriers the opportunity for adequate cost recovery.

The first step of the staged reform should move carriers' intrastate terminating switched access rates to interstate terminating switched access rate levels in equal increments over a period of two to four years. [673] The FCC has authority to establish a new methodology for ICC, but Congress could make explicit the FCC's authority to reform intrastate intercarrier rates by amending the Communications Act in order to reduce litigation and expedite reform. Following the intrastate rate reductions, the framework should set forth a glide path to phase out per-minute charges by 2020.

To offset the impact of decreasing ICC revenues, the FCC should permit gradual increases in the subscriber line charges (SLC) and consider deregulating the SLC in areas where states have deregulated local rates.[674]

The FCC should also encourage states to complete rebalancing of local rates to offset the impact of lost access revenues. Even with SLC increases and rate rebalancing, some carriers may also need support from the reformed Universal Service Fund to ensure adequate cost recovery. When calculating support levels under the new CAF, the FCC could impute residential local rates that meet an established benchmark.[675] Doing so would encourage carriers and states to "rebalance" rates to move away from artificially low $8–$12 residential rates that represent old implicit subsidies to levels that are more consistent with costs.[676]

As part of comprehensive ICC reform, the FCC should adopt interim rules to reduce ICC arbitrage. The FCC should, for example, prohibit carriers from eliminating information necessary for a terminating carrier to bill an originating carrier for a call. Similarly, the FCC should adopt rules to reduce access stimulation and to curtail business models that make a profit by artificially inflating the number of terminating minutes. The FCC also should address the treatment of VoIP traffic for purposes of ICC.

RECOMMENDATION 8.8: The FCC should examine middle- mile costs and pricing.

As discussed above, the cost of second- and middle-mile connectivity has a direct impact on the cost of providing broadband service in unserved areas of the country. As a result, there is a direct link between whether the FCC's policies regarding the rates, terms and conditions of special access services are effective and the funding demands that will be placed on the new CAF. It may be the case that the cost of providing these circuits in areas supported by CAF is so high that there is no private sector business case to offer broadband services, even if the rates, terms and conditions are just and reasonable. An examination of middle-mile costs and pricing should occur in concert with the comprehensive USF/ICC reform program.

Stage Two: Accelerating Reform (2012–2016)

In Stage Two, the FCC will need to take further steps and answer a number of questions in order to accelerate reform of the High-Cost program and ICC. Some have proposed other ways that current High-Cost funding could be shifted towards broadband without having a deleterious effect on existing network deployment or operations. [677] The FCC should examine the potential costs and benefits of additional ways to shift funding from the legacy High-Cost program to the CAF.

Implementation decisions in Stage Two will impact the speed with which broadband service is available throughout the United States and the overall cost of filling the broadband availability gap. Two critical issues will be to determine what ongoing support is necessary to sustain areas that already meet the National Broadband Availability Target due to current USF subsidies, and how rights and responsibilities should be modified when the incumbent is not the broadband providerof-last-resort for a particular geographic area. [678]

During this phase, the FCC will begin distributing support from CAF, with an initial focus on extending broadband to unserved areas. Intrastate rates for ICC will be lowered over several years to interstate levels, and competitive ETC support will be phased out. The FCC should also stabilize USF for the future by expanding the USF contribution base.

RECOMMENDATION 8.9: The FCC should begin making disbursements from the CAF.

Once the FCC completes rulemakings to establish the parameters of the new CAF, it should begin to distribute CAF funding to discrete geographic areas that contain unserved households. The FCC potentially could focus first on those states that have a higher absolute number or percentage of unserved housing units per capita, or those states that provide matching funds for broadband construction.

RECOMMENDATION 8.10: The FCC should broaden the universal service contribution base.

Today, federal universal service funding comes from assessments on interstate and international end-user revenues from telecommunications services and interconnected VoIP services. Service providers typically pass the cost of these assessments on to their customers.

The revenue base for universal service contributions—telecommunications services—has remained flat over the last decade, even though total revenues reported to the FCC by communications firms grew from $335 billion in 2000 to more than $430 billion in 2008. [679] Broadband-related revenues are projected to grow steadily over time. [680]

Service providers are increasingly offering packages that "bundle" voice and broadband and deliver them over the same infrastructure. Assessing only telecommunications services revenues provides incentives for companies to characterize their offerings as "information services" to reduce contributions to the fund.

There is an emerging consensus that the current contribution base should be broadened, though with differing views on how to proceed. Some parties urge the FCC to expand the contribution base to include broadband revenues, [681] while others urge the FCC to assess broadband connections through a hybrid numbers- and connections-based approach. [682] Some parties suggest that the FCC should explore some method of assessing entities that use large amounts of bandwidth. [683] Some suggest that broadband should not be assessed because that would lessen broadband adoption, or that residential broadband should be exempted. [684]

As the FCC establishes the CAF, it also should adopt revised contribution methodology rules to ensure that USF remains sustainable over time. Whichever path the FCC ultimately takes, it should take steps to minimize opportunities for arbitrage as new products and services are developed and remove the need to continuously update regulation to catch up with technology and the market.

RECOMMENDATION 8.11: The FCC should begin a staged transition of reducing per-minute rates for intercarrier compensation.

The comprehensive ICC reforms adopted in Stage One should be implemented in Stage Two. The FCC should begin by reducing intrastate rates to interstate rate levels in equal increments over a period of time. The FCC should also implement interim solutions to address arbitrage, which will help offset revenue losses from the reduction in intrastate rates.

The FCC should continue the staged reduction of per- minute rates adopted as part of the comprehensive ICC reform. After reducing intrastate rates, the FCC could, for example, reduce interstate rates to reciprocal compensation rate levels for those carriers whose interstate rates exceed their reciprocal compensation rates, and reduce originating access rates in equal increments. Doing so would transition all ICC terminating rates to a uniform rate per carrier, which is an important step to eliminate inefficient economic behavior. The rate reduction in a staged approach will give carriers adequate time to prepare and make adjustments to offset the lost revenues.

Stage Three: Completing the Transition (2017–2020)

In Stage Three, the FCC should complete the transition with an emphasis on measurement and adjustment. To the extent there remain a small number of households that still do not have service meeting the National Broadband Availability Target, the FCC should consider alternative approaches to extend service to those areas.

RECOMMENDATION 8.12: The FCC should manage the total size of the USF to remain close to its current size (in 2010 dollars) in order to minimize the burden of increasing universal service contributions on consumers.

Unrestrained growth of the USF, regardless of reason, could jeopardize public support for the goals of universal service. [685]

The USF has grown from approximately $4.5 billion in 2000 to a projected $8.7 billion in 2010. [686] Portions of the USF are already capped, and with the implementation of the interim competitive ETC cap for the High-Cost program in 2008, the only significant parts of the fund that remains uncapped are the Low Income program and a part of the High-Cost program that provides access replacement funding (ICLS) to small, rate-of-return carriers.

The FCC's Low Income program has grown significantly in the last year, [687] in large part due to the efforts of companies to create targeted offerings for Lifeline recipients. Since Low

---

Income support comes from an uncapped fund for which eligibility is determined by need, future demand for Low Income support will likely depend on many factors, including the state of the economy, the efficacy of outreach efforts, the level of subsidy provided, the price elasticity of demand among low-income households, the number and type of eligible service offerings and the evolution of consumer demand.

The FCC needs to proceed with measured steps to assure that as it advances the nation's broadband goals, it does not increase the USF contribution factor, which is already at a public historic high. Unless Congress chooses to provide additional public funding to accelerate broadband deployment, the FCC should aim to keep the overall size of the fund close to its current size (in 2010 dollars), while recognizing that the uncapped parts of USF may continue to grow due to factors outside the scope of this plan.[688] As the FCC implements the recommendations of the plan, it should evaluate innovative strategies to leverage the reach of existing governmental support programs and evaluate whether to adjust the relative proportion of supply-side versus demand-side subsidies over time.

RECOMMENDATION 8.13: The FCC should eliminate the legacy High-Cost program, with all federal government funding to support broadband availability provided through the CAF.

By 2020, the "old" High-Cost program will cease operations, and service providers will only receive support for deployment and provision of supported services (i.e., broadband that offers high-quality voice) through the CAF.

The FCC should set a deadline for recipients of USF to offer supported services. As noted above, based on current terrestrial technology, providing broadband to the 250,000 housing units with the highest gaps accounts for approximately $14 billion of the total investment gap, which represents an average cost of $56,000 per housing unit to serve the last two-tenths of 1% of all housing units.

The FCC should consider alternative approaches, such as satellite broadband, for addressing the most costly areas of the country to minimize the contribution burden on consumers across America. The FCC could consider means-tested consumer subsidies for satellite service. Another approach would be to provide a limited waiver of the requirement to offer broadband to providers that demonstrate that it is economically or technically infeasible to upgrade a line to offer broadband service,[689] while ensuring that consumers are able to continue to receive the high-quality voice service that they enjoy today.

RECOMMENDATION 8.14: The FCC should continue reducing ICC rates by phasing out per-minute rates for the origination and termination of telecommunications traffic.

The elimination of per-minute above-cost charges should encourage carriers to negotiate alternative compensation arrangements for the transport and termination of voice and data traffic. Given that there may be market power for terminating traffic, the FCC should carefully monitor compensation arrangements for IP traffic as the industry transitions away from per-minute rates, particularly in areas where there is little or no competition, to ensure that such arrangements do not harm the public interest.[690]

In summary, this roadmap for comprehensive universal service and ICC reform over the next decade represents a critical first step to ensure that all people in the United States have access to affordable broadband. To begin turning this roadmap into reality, the FCC will embark on a series of rulemakings to seek public comment and adopt rules to implement this reform. Although these proceedings will need to make specific decisions on implementation details, this plan sets forth a clear vision for the end state we seek to achieve as a nation— preserving the connectivity that Americans have today and advancing universal broadband in the 21st century.

Achieving this vision will not happen automatically. Indeed, significant changes to the existing regulatory structure will need to be made, including adjustments to existing USF support mechanisms to redirect funding away from supporting single-purpose voice telephone networks and toward supporting integrated, multifunctional broadband platforms in a more efficient manner. Additional capital must be directed toward broadband infrastructure. The plan sets forth a pathway to shift up to $15.5 billion (present value in 2010 dollars) over the next decade from the existing USF High-Cost program to broadband, with up to $11.5 billion specifically focused on broadband deployment in unserved areas. By implementing this plan as written, broadband will be available to more than 99% of the people in the United States by 2020.

This plan is not without risk. The baseline estimates that form the foundation for this plan are subject to a number of assumptions, most notably relating to the timing and outcome of regulatory proceedings. [691] The timing of some shifts such as implementation of the voluntary commitments from Sprint and Verizon Wireless to give up their competitive ETC funding is known, while the timing of other changes that could yield savings is not.

The FCC's ability to shift funds from existing programs to broadband assumes that shifting the identified money from voice service to broadband will not negatively impact company operations or future deployment strategies.

The gap estimates assume that the FCC implements an effective market-based mechanism to determine who should receive support and the level of that support, and that the market-based mechanism is designed in a way to target support first to those areas that require only support for new construction. The estimates also assume that the market mechanism will fund the areas requiring the least amount of support first, thus connecting the most housing units as quickly as possible. In some areas of the country, however, the number of interested parties may be insufficient to implement a market-based mechanism, and the FCC therefore may need to use an alternative approach to drive subsidies to efficient levels.

The plan does not estimate the amount of support that may be necessary to sustain broadband service in those areas where it already is available. The estimates focus on the investment gap to make broadband capable of delivering high-quality voice universally available in unserved areas. While the FCC will initially target CAF funding toward unserved areas, the objective over time is to develop a mechanism that supports the provision of affordable broadband and voice in all areas, both served and unserved, where governmental funding is necessary. The amount of support ultimately required for those areas that currently are served through the receipt of universal service subsidies will depend on many factors, including the evolution of market demand, the precise distribution mechanism selected, and the achievement of efficiencies in an IP-based network. To the extent an incumbent rate-of-return company is not the designated broadband provider-of-last-resort for its entire territory, for instance, the FCC would need to determine how changing support levels would impact service to consumers and how to address the costs of past network investments.

The fact that many questions remain to be answered should not stop the nation from starting down the road to universal broadband. There will be ample opportunity to adjust in the years ahead.

Accelerating Broadband Deployment

Active management of the entire USF program by the FCC as described in this plan is the best way to mitigate these risks going forward. To speed deployment, provide the FCC greater flexibility, and ensure significant capital available for broadband, Congress should act.

RECOMMENDATION 8.15: To accelerate broadband deployment, Congress should consider providing optional public funding to the Connect America Fund, such as a few billion dollars per year over a two to three year period.

If Congress were to provide such funding in a timely manner, it would enable the FCC to achieve more quickly the objectives set forth in the plan for universal broadband, without having to obtain such funding through the current USF contribution mechanism. Since consumers and businesses bear both the USF contribution burden and the general tax burden, additional public funding would draw money for deployment from the same parties that contribute today, but potentially with less relative impact on vulnerable populations that may have lower broadband adoption rates than the general population.[692] Additional funding would allow the country to achieve the National Broadband Availability Target faster and ease the glide path for implementing other reforms in this plan by removing regulatory uncertainty over USF and ICC revenue streams potentially available for further broadband deployment. In addition, in the event additional funding becomes available, whether through new government funding or careful management of existing funds, that funding could be used to build upon lessons learned from successful Lifeline broadband pilots and expand innovations in the E-rate and other programs to support community institutions (see Chapters 9 and 11).

Although the plan sets forth a vision to achieve universal broadband, no one can accurately foresee every potential market dynamic between now and 2020, nor would it be possible for the plan to accurately predict how private sector investment may occur in the future. The precise timing to achieve universal availability will depend on multiple variables, many of which are beyond the control of regulators. Technology, markets and the industry can and will change. One thing that we can reliably predict is that the world in 2020 will be different than what we envision today. But the fact that the FCC may need to make mid-course corrections along the way does not change the overarching national policy imperative—the need for a connected, high-performance America. For the nation to achieve this goal, the steps outlined in this plan must be taken promptly.

8.4. Other Government Actions to Promote Availability

Other Federal Financing

Congress should also consider measures to provide greater flexibility to the Rural Utilities Service (RUS) and other agencies in order to provide additional financing solutions to advance broadband availability.

RECOMMENDATION 8.16: Congress should consider expanding combination grant-loan programs.

Most existing funding mechanisms for telecommunications infrastructure, such as those run by RUS, are designed to provide funds via loans, loan guarantees or grants. Recovery Act funding and RUS's Farm Bill Broadband Program and Distance Learning Program have allowed some combinations. To optimize use of taxpayer dollars, more funding should be directed to such combinations. By allowing agencies like RUS to structure funding as combinations of loans, grants and guarantees, [693] they can select the most efficient use of taxpayer dollars while simultaneously providing service providers a one-stop financing solution.

RECOMMENDATION 8.17: Congress should consider expanding the Community Connect program.

The Community Connect program, administered by RUS, is intended to provide funding for broadband to communities that are otherwise unserved. The program had $13.4 million in funding available in 2009, [694] while demand for program funding runs into the hundreds of millions of dollars, principally from communities that are too small to attract interest from private capital. To meet the needs of such communities, Congress should consider expanding the Community Connect program (both in size and in the scope of its eligibility criteria) to be more inclusive in serving such communities.

RECOMMENDATION 8.18: Congress should consider establishing a Tribal Broadband Fund to support sustainable broadband deployment and adoption in Tribal lands, and all federal agencies that upgrade connectivity on Tribal lands should coordinate such upgrades with Tribal governments and the Tribal Broadband Fund grant-making process.

Tribal lands face unique connectivity challenges (see Box 8-4). Grants from a new Tribal Broadband Fund would be used for a variety of purposes, including bringing high-capacity connectivity to Tribal headquarters or other anchor institutions, deployment planning, infrastructure buildout, feasibility studies, technical assistance, business plan development and implementation, digital literacy, and outreach. [695] In addition, a portion of the fund should be allocated to provide small, targeted grants on an expedited basis for Internet access and adoption programs. [696] The fund should be administered by NTIA in consultation with the FCC and the Bureau of Indian Affairs.

In order to provide state-of-the-art services to Tribal communities and promote the deployment of high-capacity infrastructure on Tribal lands, Congress should consider providing ongoing public funding for federal facilities serving Tribal lands in order to upgrade and maintain their broadband infrastructure. Telecommunications infrastructure at federal facilities located on Tribal lands frequently has limited broadband capacity. [697]

Consistent with Recommendation 6.8, which encourages government entities to actively seek out and leverage "dig once" coordination opportunities, all federal agencies that upgrade network connectivity on Tribal lands should coordinate such upgrades with Tribal governments and the Tribal Broadband Fund grant-making process to exploit opportunities for joint trenching, laying of conduit or construction of additional fiber optic facilities. [698]

BOX 8-4. BROADBAND ON TRIBAL LANDS

Available data, which are sparse, suggest that less than 10% of residents on Tribal lands have broadband available. [699] The Government Accountability Office noted in 2006 that "the rate of Internet subscribership [on Tribal lands] is unknown because no federal survey has been designed to capture this information for Tribal lands."[700] But, as the FCC has previously observed, "[b]y virtually any measure, communities on Tribal lands have historically had less access to telecommunications services than any other segment of the population." [701]

Many Tribal communities face significant obstacles to the deployment of broadband infrastructure, including high build-out costs, limited financial resources that deter investment by commercial providers and a shortage of technically trained members who can undertake deployment and adoption planning.[702] Current funding programs administered by NTIA and RUS do not specifically target funding for projects on Tribal lands and are insufficient to address all of these challenges.[703] Tribes need substantially greater financial support than is presently available to them, and accelerating Tribal broadband deployment will require increased funding. [704]

Tribal, State, Regional and Local Broadband Initiatives

In addition to Tribal, federal, and state efforts to support broadband deployment, local governments and regions often organize themselves to support deployment in their communities. According to recent market research, as of October 2009, there were 57 fiber-to-the-premises (FTTP) municipal deployments, either in operation or actively being built, in 85 towns and cities in the United States. These deployments collectively serve 3.4% of the FTTP subscribers in North America. [705]

Not all government-sponsored networks serve consumers directly. Several government-sponsored entities, such as NOANet in the Pacific Northwest and OneCommunity in Ohio, are major providers of backhaul capacity in areas that benefit community institutions and local broadband service providers. Their networks are often "constructed" by patching together and opening up to wider use fiber and other connections that might originally have been built for single-purpose institutional needs, such as the needs of government offices and local transportation. By offering up that existing capacity to wider use, including the service provider community, these efforts can benefit an entire community, not just one institution. [706]

While it is difficult to measure the impact of many local efforts, these efforts should be encouraged when they make sense. However, 18 states have passed laws to restrict or explicitly prohibit municipalities from offering broadband services. Some states, like Nebraska, have outright bans on municipalities offering any wholesale or retail broadband service. Other states, such as South Carolina and Louisiana, set conditions that make municipal broadband both harder to deploy and more costly for consumers. [707] In addition, restrictions on the use of institutional networks can substantially impede the ability of local and regional authorities to utilize that infrastructure to benefit the broadband needs of the community as a whole. Restricting these networks in some cases restricts the country's ability to close the broadband availability gap, and should be revisited.

RECOMMENDATION 8.19: Congress should make clear that Tribal, state, regional and local governments can build broadband networks.

Local entities typically decide to offer services when no providers exist that meet local needs. These local entities do so only after trying to work with established carriers to meet local needs.[708] This experience is similar to how some municipalities responded in the early part of the 20th century, when investor- owned electric utilities left rural America in the dark while they electrified more lucrative urban centers. Public and cooperatively owned power utilities were created to fill the void. More than 2,800 public and co-op operators still provide electricity to 27% of Americans today.[709] Many of these same rural areas now face similar challenges attracting private investment to connect civic institutions, businesses and residences to highspeed data networks. In some areas, local officials have decided that publicly–owned communications services are the best way to meet their residents' needs (see Box 8-5).

Municipal broadband has risks. Municipally financed service may discourage investment by private companies. Before embarking on any type of broadband buildout, whether wired or wireless, towns and cities should try to attract private sector broadband investment. But in the absence of that investment, they should have the right to move forward and build networks that serve their constituents as they deem appropriate.

RECOMMENDATION 8.20: Federal and state policies should facilitate demand aggregation and use of state, regional and local networks when that is the most cost-efficient solution for anchor institutions to meet their connectivity needs.

Government policy often limits the ability of schools, hospitals and other community institutions to serve as community broadband anchors. FCC universal service policies and the policies of other grant-making agencies frequently drive institutions to use dedicated, single-purpose networks that are not available for broader community use, resulting in a situation in which "[c]ommunity residents working in healthcare or education often have unlimited access to the Internet while other rural residents are left with no access."[710] These restrictions make it difficult to expand and share broadband with other community institutions in the most cost-effective way.

This problem is especially acute in rural areas and Tribal lands where broadband may only be available and affordable to residents and small businesses in a community if the fiber optic infrastructure in that town is shared not only by commercial users but also by the local hospital, government office and school system. [711] Because broadband networks—particularly fiber optic networks—demonstrate large economies of scale, bulk purchasing arrangements for forms of connectivity like second-mile and middle-mile access can drive down the per-megabit cost of such   access considerably. As a result, policy restrictions that impede the ability of school networks funded by E -rate to share capacity with hospitals funded by the Rural Health Care program, or the public safety system which may be funded by state and other federal sources, drive up the cost of connectivity for those institutions and for others in the community.[712]

Box 8-5. Community Broadband in Rural America

Bristol, Va., provides a good example of the potential of community broadband in rural America. This small town, which also operates the local electric utility, initially deployed a fiber optic network to connect its government, electric utility and school buildings. Local businesses and residents expressed interest in connecting to this high-speed network, so Bristol made plans to build a fiber-to-the-premises network. After overcoming a series of state legislative barriers and legal challenges by incumbent providers offering slower services, Bristol launched a FTTP service. Today 62% of Bristol's residents and businesses subscribe to the service despite competition from the incumbent telephone company and cable.

At least 30 states have established state networks operated by public agencies or the private sector to aggregate demand among schools, universities, libraries, and state and local government agencies to reduce costs.[713] Better collaboration among government agencies could reduce the potential for waste of federal resources and maximize available federal funding for broadband-related community development projects. Federal and state policy should not preclude or limit networks that serve one category of institution from serving other institutions and the community as a whole.[714] The FCC should explore creative solutions to help schools, libraries and health care providers reduce their broadband-related costs by aggregating demand with other community institutions so that they can purchase the maximum amount of broadband with their USF dollars. For instance, the FCC should remove barriers to the shared use of state, regional, Tribal, and local networks by schools, libraries and health care providers when such networks provide the most cost-efficient choice for meeting broadband needs.[715]

Because community anchor institutions are large—if not the largest—potential consumers of broadband in even the smallest of towns, adopting these recommendations will not only expand broadband options for the institutions themselves but also will improve availability in the community as a whole.

RECOMMENDATION 8.21: Congress should consider amending the Communications Act to provide discretion to the FCC to allow anchor institutions on Tribal lands to share broadband network capacity that is funded by the E-rate or the Rural Health care program with other community institutions designated by Tribal governments.

In recognition of the unique challenges facing Tribal communities, Congress should consider amending the Communications Act to provide discretion to the FCC to define circumstances in which schools, libraries and health care providers that receive funding from the E -rate or Rural Health Care program may share broadband network capacity that is funded by the E-rate or the Rural Health Care program with other community institutions designated by Tribal governments. [716]

RECOMMENDATION 8.22: The federal government and state governments should develop an institutional framework that will help America's anchor institutions obtain broadband connectivity, training, applications and services.

Earlier in this chapter, the plan proposes a path to ensure that homes in high-cost areas have access to broadband, largely by reforming the High-Cost program and intercarrier compensation. In other chapters, the plan proposes reforms to USF to improve connectivity to schools, libraries and health care providers. Government should take additional steps to enable these and other community institutions to better utilize their connectivity to provide a better quality of life for all people.

One approach to ensure connectivity for facilities that serve public purposes is to give a non-profit institution the mission and capability to focus on serving the broadband needs of public institutions, including health clinics, community colleges, schools, community centers, libraries, museums, and other public access points. In the past, the connectivity needs of research institutions have been met by non-profit research and education (R&E) networks such as Internet2 and National LambdaRail. R&E networks played a central role in the development and growth of the Internet itself through ARPANET and later NSFNET. Today, similar R&E networks provide high-speed (10 Mbps-1 Gbps) connectivity to 66,000 community anchor institutions. [717] But more can be done—it is estimated that only one-third of anchor institutions have access to an R&E network today. [718] This model should be expanded to other community institutions.

A group of R&E networks, including Internet2 and the National LambdaRail, with the support of the National Association of Telecommunications Officers and Advisors and the Schools, Health and Libraries Broadband Coalition, have proposed that the federal government and state governments create a non-profit coordinating entity, the "Unified Community Anchor Network," that would support and assist anchor institutions in obtaining and utilizing broadband connectivity. [719] Expanding the R&E network model to other anchor institutions would offer tremendous benefits. Many community institutions lack the institutional resources to undertake the many tasks necessary to maximize their utilization of broadband. Facilitating collaboration on network design and how best to utilize applications to meet public needs could result in lower costs and a far more efficient and effective utilization of broadband by these institutions.

Working with the R&E and non-profit community, the federal government and state governments should facilitate the development of an institutional framework that will help anchor institutions obtain broadband connectivity, training, applications and services. One method of implementation would be to establish federal and state coordinators and consortia of anchor institutions. These coordinators would help secure connectivity and would also provide hands-on experience and capacity in the building and running of networks.[720] A coordinating entity also could have a national procurement role in negotiating bulk equipment and connectivity purchase agreements, acting as a sophisticated buyer, which would then be available to community institutions.[721] There also could be a platform for interconnected networks to share resources and applications and provide training opportunities. Coordinating and building common resources and capacity in this manner at the national and state levels would lower the overall costs of building and running anchor institutional networks.

9. ADOPTION AND UTILIZATION

While 65% of americans use broadband at home, the other 35% (roughly 80 million adults) do not. [722] some segments of the population—particularly low-income households, racial and ethnic minorities, seniors, rural residents and people with disabilities—are being left behind. As Exhibit 9-A demonstrates, some communities are significantly less likely to have broadband at home. Half of all Hispanics do not use broadband at home, while 41% of African Americans do not. only 24% of Americans with less than a high school diploma use broadband at home, and the adoption rate for those with annual household incomes less than $20,000 is only 40%.

If history is a guide, adoption rates will continue to rise. [724] Broadband adoption reached 50% in 2007, up from 12% at the end of 2002 and 32% in early 2005. [725] But gaps will likely persist with certain segments of the population continuing to lag the national average.

Consider the history of telephone adoption. Traditional telephone service reached saturation around 1970, when 93% of households subscribed. At that point, roughly 20% of African Americans and Hispanics did not have telephone service. By 1985, households earning less than $10,000 per year still lagged those earning $40,000 or more by nearly 19 percentage points; by 2008, they continued to trail by almost 9 percentage points.[726] As described in Chapter 8, government action through the Universal Service Fund ultimately contributed to telephone adoption to near universal levels.

Exhibit 9-A. Broadband Adoption Among Certain Demographic Groups[723]*

Demographic group	Current adoption rates, by %
National average	65
Low income (under $20,000/year)	40
Less educated (no high school degree)	24
Rural Americans	50
Older Americans (65+)	35
People with disabilities	42
African Americans	59
Hispanics	49

*The sample size of the FCC Survey, though the largest survey of non-adopters to date, is too small to make statistically reliable broadband adoption estimates for certain population subgroups, particularly racial and ethnic minorities. Data released by National Telecommunications and Information Administration (NTIA) from the U.S. Census Bureau's Current Population Survey Internet and Computer Use Supplement offer some insight into computer and Internet use by less numerous population subgroups. In particular, NTIA reports 67% of Asian Americans have broadband at home while 43% of American Indians/Alaska natives (living on and off Tribal lands) report having broadband at home. See NTIA, DIGITAL NATION: 21st Century America's Progress Toward Universal Broadband Internet Access (2010), available at http://www.ntia. doc.gov/reports/2010/NTIA_ internet_use_report_Feb2010.pdf.

Absent action, broadband adoption rates will continue to be uneven. Even if broadband reaches saturation in coming years, the aggregate adoption number may mask troubling differences along socioeconomic and racial and ethnic lines. If broadband adoption follows the trajectory of telephone adoption, one in four African Americans and one in three Hispanics could still be without broadband service at home even when an overwhelming majority of Americans overall have it.

To understand broadband adoption trends, many questions must be answered. Who chooses not to adopt, and why? What is the appropriate role for government in general, and the federal government in particular, to spur sustainable adoption? How can stakeholders such as state, local and Tribal leaders, non-profit community partners and private industry support the goals of bringing all citizens online and maximizing their utilization of broadband applications?

The following recommendations outline targeted investments the United States should consider in order to increase adoption levels. Federal action is necessary but needs to be taken in partnership with and in support of state, local and Tribal governments, corporations and non-profits.

Recommendations

Address cost barriers to broadband adoption and utilization

- The Federal Communications Commission (FCC) should expand Lifeline Assistance (Lifeline) and Link-Up America (Link-Up) to make broadband more affordable for low- income households.
- The FCC should consider free or very low-cost wireless broadband as a means to address the affordability barrier to adoption.

Address digital literacy barriers to broadband adoption and utilization

- The federal government should launch a National Digital Literacy Program that creates a Digital Literacy Corps, increases the capacity of digital literacy partners and creates an Online Digital Literacy Portal.

Address relevance barriers to broadband adoption and utilization

- The National Telecommunications and Information Administration (NTIA) should explore the potential for public-private partnerships to improve broadband adoption by working with other federal agencies.
- Public and private partners should prioritize efforts to increase the relevance of broadband for older Americans.
- The federal government should explore the potential of mobile broadband access as a gateway to inclusion.
- The private sector and non-profit community should partner to conduct a national outreach and awareness campaign.

Address issues of accessibility for broadband adoption and utilization

- The Executive Branch should convene a Broadband Accessibility Working Group (BAWG) to maximize broadband adoption by people with disabilities.
- The FCC should establish an Accessibility and Innovation Forum.
- Congress, the FCC and the U.S. Department of Justice (DOJ) should consider modernizing accessibility laws, rules and related subsidy programs.

Expand federal support for regional broadband capacity- building, program evaluation and sharing of best practices

- Federal support should be expanded for regional capacity-building efforts aimed at improving broadband deployment and adoption.
- Congress and federal agencies should promote third- party evaluation of future broadband adoption programs.
- NTIA should establish a National Broadband Clearinghouse to promote best practices and information sharing.

Coordinate with Tribes on broadband issues

- The Executive Branch, the FCC and Congress should make changes to ensure effective coordination and consultation with Tribes on broadband-related issues.

9.1. Understanding Broadband Adoption

On Feb. 23, 2010, the FCC published the results of its first Broadband Consumer Survey. This national survey of 5,005 adult Americans focused on non-adopters and the issues they face in adopting broadband. While many surveys track broadband adoption, this survey is one of the first efforts to oversample non-adopters. [727] This section builds off these survey results to develop a set of programs to improve the adoption and utilization of broadband services, focusing on the barriers faced by non-adopters.

Barriers to Adoption and Utilization

The 35% of adults who do not use broadband at home generally are older, poorer, less educated, more likely to be a racial or ethnic minority, and more likely to have a disability than those with a broadband Internet connection at home. The FCC survey identified three major barriers that keep non-adopters from getting broadband:

Cost. When prompted for the main reason they do not have broadband, 36% of non-adopters cite cost. Almost 24% of non- adopters indicate reasons related to the cost of service—15% point to the monthly service cost, and 9% say they do not want the financial commitment of a long-term service contract or find the installation fee too high. For 10% of non-adopters, the cost of a computer is the primary barrier. The additional 2% cite a combination of cost issues as the main reason they do not adopt. [728]

Digital Literacy. About 22% of non-adopters cite a digital literacy-related factor as their main barrier. This group includes those who are uncomfortable using computers and those who are "worried about all the bad things that can happen if [they] use the Internet." [729]

Relevance. Some 19% of non-adopters say they do not think digital content delivered over broadband is compelling enough to justify getting broadband service. Many do not view broadband as a means to access content they find important or necessary for activities they want to pursue. Others seem satisfied with offline alternatives. These non-adopters say, for instance, the Internet is a "waste of time." [730]

An important and cross-cutting issue is accessibility for people with disabilities. Some 39% of all non-adopters have a disability, much higher than the 24% of overall survey respondents who have a disability. [731] It is not a surprise that non-adopters include a disproportionately high share of people with disabilities. Americans with disabilities share many characteristics with other non-adopters (i.e., both groups are older and have lower incomes than adopters), but having a disability may be an independent factor contributing to lower levels of broadband adoption at home. [732] For example, some of the other impediments that people with disabilities face include:

- Devices often are not designed to be accessible for people with disabilities. [733]
- Assistive technologies are expensive (Braille displays, for example, can cost between $3,500 and $15,000). [734]
- Services, including emergency services, are not accessible. [735]
- Web pages and new media applications cannot be accessed by a person using a screen reader. [736]
- Internet-based video programming does not have captions or video descriptions offering an account of what is on the screen. [737]

Despite these barriers, ways that non-adopters use other forms of information and communications technology (ICT) bodes well for the future of broadband adoption. Some non- adopters have a positive view of the benefits of ICT; they buy and use such technology, even though they have not purchased broadband. For example, 80% of non-adopters have satellite or cable premium television, 70% have cell phones and 42% have at least one working computer at home. [738]

In addition to using ICT, many non-adopters have positive attitudes about the Internet. Fifty-nine percent of non- adopters strongly agree with the statement "the Internet is a valuable tool for learning;" 54% strongly agree that "it is important for children to learn to use the Internet;" and 37% strongly agree people can be more productive if they learn to use the Internet. This level of ownership of and interest in technology indicates that many non-adopters may be inclined to subscribe to broadband. [739]

Overcoming Barriers to Adoption and Promoting Utilization

The recommendations in this chapter address both adoption and utilization. "Adoption" refers to whether a person uses a broadband service at home or not; "utilization" refers to the intensity and quality of use of that connection to communicate with others, conduct business and pursue online activities. Research indicates that "differentiated use"—different levels of intensity and varied complexity of activities one pursues online—can affect the kind of offline benefits users experience.[745] Adoption is necessary for utilization, but utilization is necessary to extract value from a connection.

BOX 9-1. BROADBAND MEANS OPPORTUNITY

Broadband is a platform for social and economic opportunity. It can lower the geographic barriers and help minimize socioeconomic disparities—connecting people from otherwise disconnected communities to job opportunities, avenues for educational advancement and channels for communication. Broadband is a particularly important platform for historically disadvantaged communities including racial and ethnic minorities, people with disabilities and recent immigrants. For example:

- In Santa Barbara County, Calif., a parent reads an email from her child's teacher. Although seemingly unexceptional, this event is actually quite remarkable because the teacher and the parent do not speak the same language. Using a donated foreign language translation program, a refurbished computer, heavily discounted Internet access and training provided through the local school system, this mother can now converse with her child's teacher for the first time.[740] The Computers for Families (CFF) program is a partnership between the Santa Barbara County Education Office and Partners in Education, a group of county business and education leaders that brings together the technological and educational resources to allow hundreds of families to benefit from the power of computers and the Internet.[741]

- Three in 10 families headed by a single mother live below the federal poverty line.[742] In 2001, to address the barriers that low-wage workers face in attaining skills, training and education, the New Jersey Department of Labor piloted a workforce development program in which single, working mothers received a computer, Internet access and online- skills training. The program had a 92% completion rate. Participants saw average annual wage increases of 14%, and several enrolled in community college, college programs and other educational offerings. All the women reported that they would not have completed a training program if it were not available at home—just one more demonstration of how online learning equalizes access to education and skills training.[743]

- In Tribal lands in Southern California, broadband helps bridge the physical distance between Tribal residents. Although 18 designated Tribal lands are located in the region, they are geographically separated and often isolated. In 2005, with a grant from Hewlett-Packard, the Southern California Tribal Chairmen's Association (SCTCA) launched the Tribal Digital Village. The initiative brought communications infrastructure, training and online content together. Because of the broadband provided via this initiative, the SCTCA was able to start its first for-profit business, Hi-Rez Printing.[744]

While cost is the leading barrier to adoption, nearly two- thirds of non-adopters note that something else keeps them from getting broadband at home.[746] In addition to cost, lack of digital skills, irrelevance of online content and inaccessible hardware and software often work together to limit adoption.[747] For non-adopters to find broadband valuable enough to subscribe, they need a basic knowledge of how to find and use trustworthy, substantive content.[748] Similarly, if broadband costs fall because of lower prices or subsidies, consumers might be more willing to try it, in spite of doubts about its relevance or their own abilities to use it.

There is also an important social dimension to broadband adoption that cannot be overlooked. The primary incentive for broadband adoption is communication—two-way communication through e-mail, social networking platforms, instant messaging or video-chatting.[749] People find broadband relevant when the communities they care about are online, exchanging information and creating content.[750] Once online, individuals will stay online if they continue to find information and broadband applications that are useful and relevant to their lives and when the people around them do the same.[751] E-mailing friends and family is difficult if they do not also have e-mail.

Ultimately, broadband adoption and utilization are not about owning a specific piece of technology or subscribing to a service but about making the Internet work for people. Getting people online is a critical first step, but the goal must be to *keep* people online through sustainable efforts that promote utilization and help each user derive value from the Internet in his or her own way.

Federal Efforts

Historically, the federal government has supported Internet adoption through efforts that are part of broader programs. For example, the Community Connect program, run by the Rural Utilities Service, has granted more than $39 million to fund broadband infrastructure investment in 67 rural communities.[752] This program requires communities that apply to create a Community-Oriented Connectivity Plan, which must include a state-of-the-art community center that provides free Internet access to residents with the goal of facilitating economic development and enhancing educational and health care opportunities in rural communities.[753]

To take another example, from 1994-2004, NTIA's Technology Opportunity Program (TOP)[*] made 610 matching grants to Tribal, state and local governments, as well as health care providers, schools, libraries and non-profits, for self-sustaining adoption programs. The grants totaled $233.5 million and leveraged $313.7 million in local matching funds.[754]

TOP emphasized how ICT could be efficiently and innovatively deployed. While this program often promoted broadband, broadband was not its central focus. TOP has not been funded since 2004, but many grantees have maintained operations with other funds. In this way, projects such as Austin Free Net, which provides technology training and access to residents of East Austin, Texas, and the Mountain Area Information Network, a community network for western North Carolina, continue to serve their communities.[755]

The American Recovery and Reinvestment Act of 2009 (Recovery Act), in addition to funding broadband deployment, marked the first large-scale federal broadband adoption effort. A minimum of $450 million within NTIA's Broadband Technology Opportunities Program (BTOP)[*] was set aside for sustainable broadband adoption programs and public computing centers.[756]

---

[*] BTOP and TOP are distinct programs. BTOP was created and funded by the Recovery Act.

Thus far in the first round of BTOP funding awards, $15.9 million have been allocated for six public computer center projects and $2.4 have been for three sustainable broadband adoption projects.[757] The recipients include:

- Fast-Forward New Mexico, which will offer eight training courses on basic computer literacy, Internet use and e-commerce while providing outreach to Spanish-, Navajo- and Pueblo-speaking populations. [758]

- The Spokane Broadband Technology Alliance in the state of Washington, [759] which will train 12,000 individuals and 300 small businesses in courses ranging from basic computer skills to advanced multimedia production, e-commerce and online business applications. The training will take place at public libraries and other area sites.

- The Los Angeles Computer Access Network, which received $7.5 million to upgrade and expand 188 public computing centers that provide free access to broadband Internet. [760]

Additional awards are expected as this program continues.

State and Local Efforts

While the federal government has provided important financing for Internet adoption efforts, Tribal, state and local governments are often in the best position to identify barriers and circumstances unique to their communities.[761]

The Minnesota Ultra High-Speed Broadband Task Force final report provides an example of a state-level strategy to address adoption. Issued in November 2009, the report recommends that the state government promote adoption through general outreach and education and specific policies directed toward people who are not connected to the Internet for financial or other socioeconomic reasons. [762] To boost broadband adoption and utilization, the report suggested programs to make computers more affordable, including creating a clearinghouse of used computers, expanding the Minnesota Computers for Schools program and establishing a support mechanism to provide assistance for the cost of monthly broadband service for low-income consumers. The plan also suggested that the state explore a variety of partnerships to increase adoption and utilization. [763]

Local leaders can play an important role by building on existing social programs and partnering with community organizations that non-adopters already rely on as trusted sources of information. [764] They can tailor adoption efforts to address language barriers, lack of credit, low basic literacy levels and other issues faced by non-adopters.

Cities can also play a role. For instance, the City of Seattle has developed a number of initiatives to promote a "technology healthy community." In 2000, the City's Department of Information Technology and the City's Citizens Telecommunications and Technology Advisory Board, with the non-profit Sustainable Seattle, launched the Information Technology Indicators project. Through this project, the City identified a set of goals for a technology healthy community and indicators to track their progress.[765] Using these indicators, the city saw its broadband adoption rate grow from 18% in 2000 to 74% in 2009.[766]

Over the past several years, Seattle has taken a number of steps to address gaps in access, digital literacy and content. The City also has a number of ongoing digital inclusion initiatives including: The Bill Wright Technology Matching Fund which funds community-driven technology projects; promoting public access terminals in public places; Puget SoundOff, a youth-driven online portal to promote civic engagement and digital skills[767]; and, Seniors Training Seniors in Technology, a peer education program helping seniors learn basic computer and Internet skills.[768]

The point is that there is no "one-size-fits-all answer. States and municipalities across the country are working on specific efforts to increase adoption and utilization of broadband. Through local action, coupled with federal support, the US can connect people with technology to improve their lives.

Guiding Principles for Broadband Adoption and Utilization

Creating the conditions necessary to promote broadband adoption and increase utilization requires a range of activities. The federal government has a role in providing support to people with low incomes, ensuring accessibility, funding sustainable community efforts, convening key stakeholders and measuring progress. Tribal, state and local governments can develop and implement specific programs to meet their unique needs. Non-profits and philanthropic organizations often work cooperatively with government, focusing on issues important in their communities. Private industry also has a stake; businesses stand to gain because new adopters can become skilled customers and employees.

All stakeholders should work together on broadband adoption issues, guided by a set of consistent principles:

- *Focus on the barriers to adoption.* Successful efforts address multiple barriers to adoption simultaneously. They combine financial support with applications and training that make broadband connectivity more relevant for non-adopters. Relevance, in turn, boosts the technology's perceived value and affordability.[769]

- *Focus on broadband in the home.* While libraries and other public places are important points of free access that help people use online applications, home access is critical to maximizing utilization.[770] Broadband home access can also help rural, low-income, minority and other communities overcome other persistent socioeconomic or geographic disparities.[771]

- *Promote connectivity across an entire community.* New users adopt broadband to stay in touch with others.[772] In addition, people are more likely to adopt and use broadband if the people they care about are online[773] and if they see how broadband can improve their quality of life in key areas such as education, health care and employment.[774]

- *Promote broadband utilization.* Promoting access and adoption are necessary steps, but utilization is the goal. People must be able to use broadband to efficiently find information or use applications to improve their lives.[775] A connection is just the beginning.

- *Plan for changes in technology.* Adoption programs have to evolve with technology. Both the trainers and the equipment they use to serve non-adopters must employ up-to- date technology and applications.

- *Measure and adjust.* Measurement and evaluation are critical to success because they allow programs to make adjustments on an ongoing basis.[776]

- *Form partnerships across stakeholder groups.* Promoting adoption requires federal commitment, state, local and Tribal action, industry partnership and support from nonprofits and philanthropic organizations. Sustainable broadband adoption and use will require efforts from all partners.

9.2. Addressing Cost Barriers to Broadband Adoption and Utilization

As mentioned, some 36% of non-adopters cite a financial reason as the main reason they do not have broadband service at home. Nearly a quarter cite service-related concerns, while one in 10 says that the cost of getting a computer is too high.

To address this barrier directly, the FCC's Lifeline and Link-Up programs—which focus on support for telephone service—should be expanded to include broadband support.

RECOMMENDATION 9.1: The Federal Communications Commission (FCC) should expand Lifeline Assistance (Lifeline) and Link-up America (Link-up) to make broadband more affordable for low-income households.

- **The FCC and states should require eligible telecommunications carriers (ETCs) to permit Lifeline customers to apply Lifeline discounts to any service or package that includes basic voice service.**
- **The FCC should integrate the expanded Lifeline and Link-up programs with other state and local e-government efforts.**
- **The FCC should facilitate pilot programs that will produce actionable information to implement the most efficient and effective long-term broadband support mechanism.**

Forty percent of adults with household incomes less than $20,000 have broadband at home, compared to 93% with household incomes greater than $75,000.[777] Many people with low incomes simply cannot afford the costs associated with having a broadband connection at home. To make broadband more affordable and overcome some of the barriers that have kept the penetration rate for these households low, the FCC should extend low-income universal service support to broadband.

The FCC created Lifeline Assistance and Link-Up America in the mid-1980s to ensure that low-income Americans could afford traditional local telephone service. Lifeline lowers the cost of monthly service for eligible consumers by providing support directly to service providers on behalf of consumer households. Link-Up provides a onetime discount on the initial installation fee for telephone service. Enhanced support is available for Tribal lands. The programs helped increase low-income telephone subscriber- ship from 80.1% in 1984 to 89.7% in 2008.[778] The FCC expects to distribute approximately $1.4 billion in low-income support during calendar year 2010.[779]

Approximately 7 million of an estimated 24.5 million eligible households (less than 29%) participated in Lifeline in 2008.[780] Statewide participation rates vary dramatically; some states have participation rates of more than 75% and others have rates less than 10%.[781]

There are several reasons for this variance across states. They include different consumer technology preferences; restrictions on consumers' ability to apply the Lifeline discount to certain types of services; lack of service options; lack of information about the program; and differences in funding levels, enrollment procedures, eligibility criteria and outreach and awareness efforts.[782]

While the FCC establishes default eligibility criteria for Lifeline and Link-Up, states that provide additional state-funded discounts can determine their own eligibility requirements.[783] Some states, such as Florida, rely on the federal default eligibility criteria. Others, like Vermont, use more liberal criteria so that more people are eligible for support. Many states allow the discount to be used on any basic voice service— including voice service bundled with other services—as well as packages that include optional features such as caller ID or call waiting. In other states, consumers are limited to specific Lifeline-branded service offerings. Finally, some states play a more active role in managing eligibility certification, outreach and verification, while others leave the burden to service providers.

Lifeline discounts apply only to service (not customer premises equipment) offered by participating ETCs. Each eligible household is entitled to a discount on only one voice line, either fixed or mobile.

The FCC and states should require eligible telecommunications carriers (ETCs) to permit Lifeline customers to apply Lifeline discounts to any service or package that includes basic voice service. By clarifying that Lifeline consumers can apply the current Lifeline discount to any offering that includes voice and data service, the FCC and states can help low-income consumers benefit from the same discounts provided through bundled service offerings that are affordable to wealthier households in the United States. Many of these bundled offerings include broadband services. Letting consumers apply their Lifeline discounts to bundled offerings will help make broadband more affordable.

Likewise, as low-income support is extended to cover broadband, the FCC should ensure that consumers are free to apply Lifeline discounts to any service offering or package containing a broadband service that meets the standards established by the FCC.[784]

The FCC should also integrate the expanded Lifeline and Link-Up programs with other state and local e-government efforts. Under the current Lifeline program, ETCs are responsible for consumer outreach and confirming consumer eligibility. Under this model, multiple service providers collect and maintain personal consumer information to determine eligibility.[785] Requiring providers to conduct outreach and verify eligibility may add to existing disincentives to serving historically under- served, low-income populations.[786] This, in turn, affects consumer awareness of and participation in these programs.

State social service agencies should take a more active role in consumer outreach and in qualifying eligible end-users. Agencies should make Lifeline and Link-Up applications routinely available and should discuss Lifeline and Link-Up when they discuss other assistance programs. The FCC should continue to develop and provide educational and outreach materials for use in these efforts.

Furthermore, the FCC should encourage state agencies responsible for Lifeline and Link-Up programs to coordinate with other low-income support programs to streamline enrollment for benefits. Unified online applications for social services, including the low-income programs, and automatic enrollment for Lifeline and Link-Up based on other means-tested programs are potential examples of such efforts.[787] For example, following its introduction of an automatic enrollment process, the state of Florida has seen increased Lifeline participation.[788] The FCC should also work with the states and providers to clarify obligations and identify best practices for outreach, certification and verification of eligibility. As part of these efforts, and in conjunction with Universal Service Administrative Company (USAC) reform efforts outlined in Chapter 8, the FCC should also consider whether a centralized database for online certification and verification is a cost-effective way to minimize waste, fraud and abuse.

The broadband marketplace is much more complex than the traditional world of voice telephony that the existing Lifeline program was designed to support. To make broadband more affordable, the low-income support program should expand provider eligibility to include any broadband provider selected by the consumer—be it wired or wireless, fixed or mobile, terrestrial or satellite—that meets minimum criteria to be established by the FCC.[789] Doing so will maximize consumer choice and stimulate innovation in serving low-income users.[790]

As the FCC designs a Lifeline broadband program, it should consider its recent experience with expanding Lifeline to non-facilities-based prepaid wireless providers. That change substantially increased participation in Lifeline and likely made telephone service more available to people who are less likely to subscribe to wireline voice services. As noted in Chapter 8, increased participation (associated with extending support to prepaid mobile) is one of the factors that led USAC to project a 38% year-over-year increase in low-income disbursements for calendar year 2010.[791] Extending government support to prepaid mobile service has created additional complexities when it comes to eligibility and verification.

To ensure Universal Service Fund (USF) money is used efficiently, the FCC should begin the expansion of Lifeline to broadband by facilitating pilot programs that will experiment with different program design elements. The pilots should determine which parameters most effectively increase adoption among low-income consumers by examining the effects of:

- Different levels of subsidy and/or minimum-payment requirements for consumers.
- A subsidy for installation (equivalent to Link-Up).
- A subsidy for customer premises equipment (CPE) such as aircards, modems and computers.
- Alternative strategies for integrating Lifeline into other programs to encourage broadband adoption and digital literacy. For instance, when signing up for Lifeline, new subscribers could be provided with packets of information that include sources of refurbished computers and digital literacy courses.[792] Additionally, they could receive information about Lifeline from organizations offering digital literacy courses or refurbished computers.

The FCC should also consider the unique needs of residents on Tribal lands.

The FCC should explore ways to conduct the pilots through competitive processes that would encourage providers to test alternative pricing and marketing strategies aimed at maximizing adoption in low-income communities.[793] Upon completion of the pilot programs, the FCC should report to Congress on such issues as whether CPE subsidies are a cost-effective way to increase adoption. After evaluating the results by looking at outputs such as total cost per subscriber, subscriber increases and subscriber churn rate, the FCC should begin full-scale implementation of a low-income program for broadband.

RECOMMENDATION 9.2: The FCC should consider free or very low-cost wireless broadband as a means to address the affordability barrier to adoption.

Another option that can reduce the affordability barrier is the use of special spectrum rules as an inducement to provide a free (or very low-cost), advertising-supported service. The FCC could develop rules for one or more spectrum bands requiring licensees to provide a free or very low-cost broadband service tier. This service would act as a complement to the Lifeline Program.

A free broadband service requirement would be similar to the way in which America currently provides universal access to video services. The FCC provides spectrum for broadcast television stations on the condition they offer a free service in the public interest. As a result, all Americans have access to a free, over-the-air video service: broadcast television, in most instances, supported by advertising. Broadcast television provides all Americans a basic package of news, information and other programming. This free service offers fewer channels and less choice in programming than paid services offer. Indeed, the difference in offerings is so great that despite the financial differences between free and $49, which is the average monthly price of a multichannel video subscription, more than 86% of American households subscribe to a paid service.[794]

The FCC could take a similar approach to broadband: license spectrum through an auction, conditioned on the offering of a free or very low cost broadband service. This free or very-low cost service would provide sufficient connectivity for a basic package of broadband applications.[795] As with broadcast television, the consumer would still need to purchase a device that could be used to access the service. Depending on the specific details of implementation, a free or very low-cost service may be unlikely to compete with paid services that offer greater capabilities.

The FCC should consider both the likely costs and benefits of this program. If undertaken, many more consumers who cannot afford any broadband or Internet service would have access to 21st century communications infrastructure—especially important as public-interest media content, including local news and information, is increasingly provided online. In addition, upon becoming operational, such a service could reduce the assessment of USF contributions needed to support a Lifeline broadband service. However, costs of this approach would include lower auction revenues (due to the conditions placed on use of the spectrum) and the opportunity cost of using the spectrum for other purposes.

The FCC would need to ensure that consumers actually receive the benefits of the free (or very low-cost) broadband program—for example, ensuring that devices tuned to the applicable frequency band(s) are widely available at an affordable price and acceptable bandwidth levels, and that sufficient capacity is reserved for the service. Historically, free advertising-supported telecommunications services have not had the same success as free over-the-air television services. But they might meet with more success if an appropriate business model can be identified.

Decisions about the use of spectrum for a particular purpose should be reached with special attention paid to whether a suitable band is available for this purpose. These decisions should be reached at the same time that the Lifeline pilot programs are launched.

9.3. Addressing Digital Literacy Barriers to Broadband Adoption and Utilization

Tasks that experienced users take for granted—using a mouse, navigating a website or creating a username and password—can be daunting for new or less experienced users of the Internet. As described earlier, 22% of non-adopters cite digital literacy as their main barrier to broadband adoption. This group includes people who are uncomfortable using computers and those "worried about all the bad things that can happen if [they] use the Internet."[796]

Digital literacy is an evolving concept. Though there is no standard definition, digital literacy generally refers to a variety of skills associated with using ICT to find, evaluate, create and communicate information. It is the sum of the technical skills and cognitive skills people employ to use computers to retrieve information, interpret what they find and judge the quality of that information. It also includes the ability to communicate and collaborate using the Internet—through blogs, self-published documents and presentations and collaborative social networking platforms. Digital literacy has different meanings at different stages of a person's life. A fourth grader does not need the same skills or type of instruction as a 45-year-old trying to re-enter the job market. Digital literacy is a necessary life skill, much like the ability to read and write.

The recommendations in this section will help all Americans to develop basic digital skills, lowering barriers to broadband adoption and utilization.

RECOMMENDATION 9.3: The federal government should launch a National Digital Literacy Program that creates a Digital Literacy corps, increases the capacity of digital literacy partners and creates an Online Digital Literacy Portal.

- **Congress should consider providing additional public funds to create a Digital Literacy corps to conduct training and outreach in non-adopting communities.**
- **Congress, the Institute of Museum and Library Services (IMLS) and the office of Management and Budget (OMB) should commit to increase the capacity of institutions that act as partners in building the digital literacy skills of people within local communities.**
  - **Congress should consider providing additional public funds to iMLs to improve connectivity, enhance hardware and train personnel of libraries and other community-based organizations (CBOS).**
  - **OMB consulting with IMLs should develop guidelines to ensure that librarians and CBOs have the training they need to help patrons use next-generation e-government applications.**
- **Congress should consider funding an online Digital Literacy Portal.**

An independent study commissioned by the FCC and conducted by the Social Science Research Council used qualitative research techniques to examine broadband adoption and use in context, particularly in low-income communities. The report draws on focus groups, interviews and group conversations with non-adopters, librarians, community organizers, teachers, human service workers, health professionals, AmeriCorps volunteers and others involved in supporting digital literacy and broadband use in their communities.[797]

The report highlights the important role of communities in supporting digital literacy: Non-adopters and new users often rely on the assistance of others to get online or get one-on-one support when they use the Internet. As the FCC Survey and a recent survey by the Joint Center for Political and Economic Studies found, these are most often family and friends, or trusted intermediaries like librarians and social service providers.[798] Very rarely, however, is it someone's only job to provide technical assistance or training in their community.[799]

The federal government should ensure that all citizens have access to the online and offline resources they need to develop basic digital literacy by launching a National Digital Literacy Program.[800] Such a program would have three closely related parts: the creation of a Digital Literacy Corps, a commitment to increasing the capacity of local institutions that act as partners in building digital literacy and the creation of an Online Digital Literacy Portal.

Creating A Digital Literacy Corps

Many digital literacy training programs, both in the United States and abroad, rely on face-to-face training provided by trusted resources within local communities.[801] Whether using intergenerational training that allows youth committed to community service to train senior citizens,[802] peer-to-peer training that enhances connections among seniors or youth[803] or mentoring models under which skilled college graduates reach out to underprivileged citizens,[804] these programs have helped non-adopters become more comfortable with technology while also fostering volunteers' commitment to community service and increasing their confidence.

Efforts to date have provided valuable lessons; a national program can build on these successful models and ensure the scale needed to address digital literacy barriers. To address this national need, Congress should consider providing additional public funding for NTIA to create a Digital Literacy Corps. In collaboration with the Corporation for National and Community Service (CNCS), NTIA should design and administer a Corps that builds on recognized best practices for both national service and technology learning.

NTIA and CNCS can explore best-practice models for building and managing the Corps, leveraging lessons learned from existing programs like AmeriCorps, Senior Corps and Learn and Serve America. CNCS can also leverage its own experience with the digital television transition, during which it made sure that AmeriCorps members were in communities across the country helping individuals become more comfortable with unfamiliar technology.

CNCS can provide additional lessons on how to build the national scale and operational capabilities (including recruitment, training and technical assistance) to support locally based efforts to provide face-to-face assistance for individuals who need help acquiring digital skills.[805] CNCS's history of helping people of all ages who are interested in serving their communities while learning valuable life skills will help ensure that Corps members receive appropriate training through programs that rely on best practices to adapt to the needs of each community.

This training should ensure that Corps members gain a sufficient understanding of digital literacy and learn how to teach relevant lesson plans. It should also be designed to improve Corps members' own digital literacy skills, as well as other professional skills that can enhance future career prospects.

The Corps should target segments of the population that are less likely to have broadband at home, including low-income individuals, racial and ethnic minorities, senior citizens, people with disabilities, those with lower education levels, people in rural communities, those on Tribal lands and people whose primary or only language is not English.

Efforts should be made to recruit members with foreign language skills who can work in communities where the primary language spoken is not English. Research indicates the dearth of non-English online content and the lack of comfort with English are correlated with low levels of broadband adoption. Just 20% of Hispanics who chose to take the FCC survey in Spanish have broadband at home. For these non-adopters, perceived irrelevance of broadband and lack of digital skills are the primary barriers to adoption.[806] One-on-one digital skills training in a user's native language with accompanying content can begin to alleviate the effects of cultural or linguistic isolation.

Some Corps members might be based out of urban schools where they could work with teachers, staff and administrators to create digital literacy lesson plans and integrate digital skills into the teaching of other subjects (see Box 9-2). Other members might work with broader social service programs to provide digital literacy training as part of a workforce development program. Still other members could incorporate demonstration projects into training activities in rural areas to show the relevance of broadband technology to rural non-adopters and to encourage people to invest time in digital skills training.

Corps members will help non-adopters overcome discomfort with technology and fears of getting online while also helping people become more comfortable with content and applications that are of immediate and individual relevance. For example, Corps members might help people research health information, seek employment, manage finances and engage with or utilize government services.

Beyond their service terms, former Corps members would bring technology teaching skills back to their own communities, magnifying the impact of the program. As happens in numerous CNCS programs currently, Corps members would build other basic work skills: time management, team leadership, planning, contingency management and critical thinking. For example, 90% of AmeriCorps members reported learning new skills as part of their service, and, of those members, nearly all of those members (91%) said they use those skills in their education or career pursuits following the program.[807]

BOX 9-2. A MODEL FOR A DIGITAL LITERACY CORPS

In 42 locations across the city of Chicago, a group of young people is helping others unlock the potential of information communication technology. These young volunteers, mostly in their 20s, are CyberNavigators who, in conjunction with librarians in the Chicago Public Library system, help patrons with everything from basic computer instruction to advanced computer troubleshooting.

These young people teach classes aimed at the beginning computer user—Internet Basics, Mouse Skills and Introduction to e-mail—to support adults trying to enter the workforce after an extended absence. For example, CyberNavigators work with job seekers to update their résumés, set up e-mail accounts, post résumés online and email potential employers.

The CyberNavigators provide one-on-one instruction, at times roaming the library to help users as necessary. Many speak a language other than English, enabling them to better assist a broader group of residents.[808]

Increasing the Capacity of Community Partners

For millions of Americans, libraries and other public computing centers are important venues for free Internet access. Libraries are established institutions where non-adopters know they can access the Internet, but community centers, employment offices, churches and other social service offices play increasingly important roles. Low-income Americans and racial and ethnic minorities, in particular, rely on public institutions and community access centers for Internet access. Over half (51%) of African Americans and 43% of Hispanics who use the Internet do so at a public library.[809]

But public computing centers provide more than just free access to the Internet. They provide supportive environments for reluctant and new users to begin to explore the Internet, become comfortable using it and develop the skills needed to find, utilize and create content.[810] Patrons of these centers overwhelmingly express the value of the personnel who staff them and can offer one-on-one help, training or guidance.[811]

Researchers from the SSRC have found that community- based organizations, such as libraries and non-profits, are key institutions in underserved and non-adopting communities— often providing Internet access, training and support services even when those activities fall outside their traditional missions.[812] While the challenges and opportunities they face vary, these libraries and other community partners are critical to improving digital proficiency in communities.[813]

The United States has more than 16,000 public libraries, 99% of which provide free Internet access. Ninety-one percent of libraries overall and 97% of libraries serving high-poverty areas report offering formal training classes in general computer skills, and 93% offer classes in general Internet use.[814]

However, many libraries lack the computer equipment to meet the needs of today's patrons. Eight in 10 libraries report hardware shortages that produce waiting lists during part or all of the day. More than 80% of libraries enforce time limits on use; 45% of libraries enforce time limits ranging from 31 minutes to 60 minutes,[815] which is not enough time to complete many popular and highly useful tasks such as the mathematics review course for the General Educational Development (GED) tests, which can take up to 150 minutes.[816] In addition, other CBOs such as community centers, churches and local non-profits lack resources to maintain their own computers, technical support and Internet access (see Box 9-3).[817]

Providing Resources for Digital Literacy Partners

Libraries and other CBOs need additional resources to continue to serve as access points and partners in achieving the country's digital literacy goals. IMLS administers the Library Services and Technology Act (LSTA) program which funds the long-standing Library Grants to States Program[818] and Native American Library and Museum Services grants. From 2003 to 2008, these programs distributed over $800 million in federal grants to states and territories. Professionals across the country credit LSTA with helping libraries improve technology, engage the public and establish new models for serving their communities. The State Library of Maryland, for example, reports that funds distributed through the program have "impacted [their] ability to stay on the leading edge of technology and in the delivery of resources."[819] The recommended allocation could enhance connectivity, hardware and personnel training at these community anchor institutions.

IMLS should develop guidelines for public access technology based on populations served and organization size. These guidelines would help libraries and CBOs assess their needs for public access workstations, portable devices and bandwidth. IMLS should work with these organizations to develop guidelines and review them annually to reflect changing technology and practices.

BOX 9-3. COMMUNITY-BASED ORGANIZATIONS AS TRUSTED RESOURCES FOR DIGITAL LITERACY

The Centro Cultural serves as a link between the digital world and the rural community of Moorhead, Minn. A community center with a public computer lab, the Centro connects community members with online resources—such as jobs, scholarships and online civic engagement opportunities—that directly affect their lives. The staff has demonstrated success in reaching out to low-income, high-risk youth about the opportunities that exist on the Internet.

Owing to its popularity and the diverse populations it serves, the Centro has experienced higher than expected demand. During the last year, it has seen an increase in its electricity bills and expenses for maintaining equipment and has had to hire a full-time employee to run the lab. In working with refugees and recent immigrant youth, the Centro Cultural has found that it is difficult to provide all of the resources needed to make their broadband experience meaningful. For example, keyboards become a barrier when users do not speak English. Centro staff members have recognized that accessing the Internet in an environment that is multicultural and multilingual creates a more meaningful experience for users of diverse cultural and linguistic backgrounds.

After public access technology guidelines are developed, Congress should consider providing additional public funds to expand organizational training and capacity—with a matching requirement and minimum percentage set aside for organizations other than libraries. These funds would enhance connectivity, hardware and personnel training at libraries and other public access points and shorten the wait for broadband access at those sites.

Training the Personnel of Digital Literacy Partners

As government services increasingly go online, libraries shoulder responsibility for helping people learn how to use these online services.[820] Eighty percent of libraries report that they help patrons use e-government applications. However, some librarians say they have been overwhelmed by patrons seeking help with government services and online programs, including applications for digital television converter box coupons, Federal Emergency Management Agency forms following Hurricane Katrina and Medicare Part D paperwork. These librarians also say that they did not receive suitable training or information from the agencies that provided the e-government solutions.[821]

OMB should consider developing guidelines to help federal agencies develop e-government services that take into account the role of public libraries and CBOs as delivery points. OMB should consult with IMLS to develop the guidelines. Agencies should work with

IMLS to develop online tutorials for using government websites and toolkits for librarians who help patrons use online government services.

Creating an Online Digital Literacy Portal

Every American should have access to free, age-appropriate content that imparts digital skills. This content should be available in a user's native language and should meet the accessibility requirements applicable to federal agencies under Section 508 of the Rehabilitation Act.

To achieve this, the Federal Trade Commission (FTC), FCC, U.S. Department of Education and NTIA should launch an Online Digital Literacy Portal. Congress should consider providing public funds to support this effort, and these agencies should partner with the technology industry and education sector to approve or create high-quality online lessons that users can access and use at their own pace. The collaboration between the agencies and non-government partners should be similar to the efforts that have produced the online safety resources available through OnGuardOnline.gov.[822] Offline resources will be important complements to this online content. They should be made available for printing or ordering and distributed by libraries, CBOs and other organizations.

This collaborative model has been successful in programs such as the U.S. Department of Housing and Urban Development (HUD) Community Outreach Partnerships Program, which brings institutions of higher education and community partners together to revitalize communities. Historically Black Colleges and Universities (HBCUs), Hispanic-Serving Institutions Assisting Communities (HSIACs) and Tribal Colleges and Universities (TCUs) serve critical roles educating members of minority communities in the United States. [823] In addition to their educational missions, through the Community Outreach Partnerships Program, these organizations provide links to community employment assistance, child care, health care information, fair housing assistance, job training, youth programs and other services. As crucial community institutions and trusted sources of information, HBCUs, HSIACs and TCUs could also serve as offline ambassadors to promote digital literacy and other national digital priorities.

Executive Branch agencies such as HUD and NTIA shouldalso use existing relationships—for example, with Neighborhood Networks and Public Computing Center grant recipients—to distribute outreach materials associated with the Online Digital Literacy Portal. E-rate recipients should also be encouraged to promote the portal. Chapter 11 details how recipients of E-rate funds could use their facilities to allow community members to build digital literacy skills through after-hours access to school computing labs.

The Online Digital Literacy portal should be evaluated after two years to assess its impact. The evaluation should consider, among other metrics, the total number of individuals accessing the portal, the number of individuals from specific target populations accessing the portal and the effectiveness of different offline resources in promoting the portal.

9.4. Addressing Relevance Barriers to Broadband Adoption and Utilization

As mentioned, 19% of non-adopters say they do not think digital content delivered over broadband is compelling enough to justify getting broadband service. [824] Many Americans may not feel broadband can help them achieve specific purposes and do not view online resources as helpful to their lives. [825] Others seem satisfied with offline alternatives. These respondents say, for example, that the Internet is a "waste of time." [826] The country has a unique opportunity to spur adoption by making broadband content relevant to these non-adopters.

Many federal agencies, from HUD to the Social Security Administration (SSA), already administer programs that support disconnected Americans, including people with low incomes and senior citizens. These agencies can serve as advisers and channels for outreach, training and information to link the populations they serve with the digital world.

This effort will require more than federal action. The federal government should support the public-private partnership model to implement these programs at the local level; private, non-profit and community-based entities should work together to draw people online, particularly those that under adopt. Using targeted, culturally relevant messaging and trusted community intermediaries, these groups should work together to inform their communities about the tangible benefits of broadband.

Finally, while the recommendations in this section focus primarily on boosting adoption of fixed Internet at home or at public access points, this plan recognizes that Internet use on handheld devices may be a gateway for home broadband adoption. Further investigation into consumer use of wireless devices is necessary.

RECOMMENDATION 9.4: The National Telecommunications and Information Administration (NTIA) should explore the potential for public-private partnerships to improve broadband adoption by working with other federal agencies.

NTIA should consider supporting public-private partnerships of hardware manufacturers, software companies, broadband service providers and digital literacy training partners to improve broadband adoption and utilization by working with federal agencies already serving non-adopting communities. Congress should consider providing additional public funds, or NTIA should use existing funds to support these partnerships.

Getting people online and connected to technology means engaging non-adopters where they are. Low-income and other vulnerable populations—groups that make up a disproportionate share of non-adopters of broadband—may already receive government services or participate in ongoing public programs. To bring non-adopters online, these agencies should integrate broadband connectivity into their goals, services and operations (see Box 9-4).

These partnerships would support the communities hit hardest by poverty. Participants would be eligible to receive discounted technology products, reduced-priced service offerings, basic digital literacy training and ongoing support. In addition, these partnerships would offer customized training, applications and tools. Government agencies could facilitate and help qualify participants to receive technology products and inspire people to use the Internet. Agencies could advise industry and non-profit partners how to make broadband service important to people's lives, while simultaneously making agency operations more efficient.

For example, a public-private partnership program specifically targeting people living in HUD-subsidized housing could reach more than nine million low-income people including nearly four million school-aged children, more than 1.4 million older Americans and nearly one million households headed by people with disabilities.[827] HUD households, including those on Tribal lands, are often located in areas of concentrated poverty with limited educational and employment opportunities. [828]

While families with school-age children generally have higher-than-average levels of broadband adoption, families with annual income less than $20,000, such as the ones living in HUD housing, are less likely than higher-earning families to have broadband service in the home. [829] Children from low-income families that cannot afford broadband devices or services are at a disadvantage relative to their connected peers. Recent surveys have found that 71% of teens say the Internet has been the primary source for recent school projects; 65% of teens go online at home to complete Internet-related homework. [830]

Similar partnerships, working with SSA, could benefit the seven million children and adults with disabilities who receive Supplemental Security Income (SSI) under the program run by the SSA to provide financial assistance to these Americans.[831] Like HUD, SSA programs would combine contributions from private and non-profit partners to create and fund broad solutions that open the way for SSI recipients to receive a similar package of discounted hardware and broadband service, as well as access to relevant software, training and applications.

Initially, HUD, SSA, the U.S. Department of Education and the U.S. Department of Agriculture are high-impact agencies for partnership programs to target. But interactions with other agencies could provide future opportunities for partnerships to reach non-adopters.

RECOMMENDATION 9.5: Public and private partners should prioritize efforts to increase the relevance of broadband for older Americans.

The broadband adoption rate for Americans over the age of 65 is 35%—well below the national average. The average age of people who identify relevance as their main barrier to getting online is 61. [832] The lag in broadband adoption is particularly acute for older African Americans and Hispanics. Just 21% of African American senior citizens and 23% of Hispanic seniors have broadband. This means that roughly 1.2 million African American and Hispanic seniors do not have broadband at home. [833]

While cost and lack of comfort with technology are almost certainly impediments to older Americans adopting broadband, data indicate that relevance is an issue as well. Experience has shown that older Americans will adopt broadband at home when exposed to its immediate, practical benefits and after receiving focused, hands-on training (see Box 9-5). [834]

The FCC should work with the National Institute on Aging (NIA) to conduct a survey of older Americans to more clearly identify barriers to their adoption of broadband technol- ogy. The survey should particularly focus on relevance and skills. Service providers, other federal agencies and non-profit agencies that serve as trusted information sources can work together to develop government initiatives, broadband service offerings, online tools and content that give people a reason to be online, a low-cost way to do it and an easy way to do the things they need to do.

In addition, the FCC and NIA should work together to identify how to best target adoption programs to older Americans. These programs should address the social infrastructure that supports adoption, including family members and others who care for older Americans, and organizations that serve as trusted sources of information. This work should focus on incorporating the needs of older Americans into the implementation of other recommendations in this section, such as the National Digital Literacy Program, the Best Practices Clearinghouse and any programs to improve broadband affordability for low-income populations.

One way to increase the relevance of broadband for older Americans is to highlight how broadband can improve their access to health care information and services. Broadband enables telemedicine solutions like videoconferencing and remote monitoring, which allow for better health management, lower health care costs and effective aging-in-place programs (see Chapter 10). Numerous initiatives, led by partnerships among the medical community, the private sector and the academic and research community, are underway. [835]

BOX 9-4. USING BROADBAND TO CREATE STRONGER COMMUNITIES IN WASHINGTON, D.C.

Engaging people where they live has already proven to be a successful program model, as demonstrated by the example of Edgewood Terrace, a mixed-income housing complex in northeast Washington, D.C. Through a joint effort, the Community Preservation and Development Corporation, HUD and the U.S. Department of Commerce's TOP initiative developed a strategy to create a stronger community using broadband.

Each of Edgewood Terrace's 792 residences is wired for broadband. But connections are only one part of the overall strategy for this community. Edgewood Terrace's 2,400 network-registered residents use subsidized devices to connect to the Internet and to a specially tailored intranet known as the EdgeNet. The EdgeNet gives residents free e-mail accounts and access to an online forum which residents use to exchange community information and news. Community empowerment staff members have worked with residents to create training classes on community issues.

Beyond the walls of the housing complex, project partners use broadband to connect residents with social services, counseling, financial and educational resources. The community operates learning centers where residents take instructional classes. In one course, the Career and Skills Enhancement Program, students receive information technology (IT) training, skills training and assistance using the Internet to search for jobs. Other courses focus on career preparation and building digital skills (for youth) or health IT (for seniors).

Edgewood Terrace residents and the community have experienced direct benefits as a result of these harmonized efforts. School attendance is up, graduates of IT skills training courses have seen an increase in their average incomes and community residents report feeling more engaged. Community members are using broadband as a tool to accomplish shared goals and create a more involved neighborhood.

The example of Edgewood Terrace makes clear that using existing agency channels and relationships to incorporate broadband into people's lives can have a transformative impact on traditionally under- served communities.

In addition, the private sector, in collaboration with nonprofits that serve older Americans, could launch a competition to invite development of applications that enhance the social benefits of broadband for older Americans. Social networking tools can help older adults to reconnect, to stay connected with others or to expand their social network to people they could never have met in person without traveling.[836] Research shows that social networking can help prevent depression[837] and provide information resources, feedback and support. [838] Despite these benefits, older adults rarely use popular social networking websites such as Facebook and MySpace, [839] which were designed for younger, more tech-savvy users. A competition to encourage the development of "entry-level" social networking applications for older Americans could induce innovators to direct their attention to the needs of this community and encourage older Americans to adopt other broadband applications in the future.

RECOMMENDATION 9.6: The federal government should explore the potential of mobile broadband access as a gateway to inclusion.

Although home broadband adoption (of wireline or fixed wireless technology) is lower for African Americans and Hispanics, these groups are relatively heavier users of mobile Internet. Although African Americans and Hispanics are as likely as other demographic groups to own a cell phone (86% do), they are more likely to have ever accessed the Internet on a mobile handheld device.[840] This handheld access may or may not be high-speed; it is difficult to determine in a survey whether participants' access occurs over 3G networks. Research also indicates that handheld online access is often a supplementary access path rather than a substitute. [841]

BOX 9-5. A WEB PORTAL FOR SENIOR CITIZENS

The Brooklyn, N.Y., nonprofit Older Adults Technology Services (OATS) encourages older adults to use information technology to enhance their quality of life. In addition to specially targeted training methods and device support, OATS has developed a model to engage older adults with information technology by aggregating useful, trustworthy information.

SeniorPlanet is a Web portal for older adults. It promotes health, wellness and quality-of-life improvements. Developed by OATS in 2006, SeniorPlanet is a grassroots digital community seeded with trusted resources and improved by users. The site includes a forum for resource exchanges, an events calendar and user-created blogs. Through SeniorPlanet, a person can register to attend a seminar on Internet safety, ask a technology question, create and share content or find information about legal services in the New York area.

As broadband technology and devices continue to evolve, mobile broadband applications may become important gateways to broadband.[842] The FCC should conduct an in-depth examination of consumer mobile use with particular focus on Americans with lower broadband adoption rates—low-income households, people with lower education levels, seniors, non- English speakers and rural Americans. Any study should also consider mobile use among racial and ethnic minorities that tend to have higher than average use of the mobile Internet.

The results of the study will give developers, community leaders and private industry insight into potential opportunities to use mobile Internet to support individuals and communities.

RECOMMENDATION 9.7: The private sector and non-profit community should partner to conduct a national outreach and awareness campaign.

How people perceive the Internet shapes how they use it. People with strong concerns about potential hazards online reported engaging in a narrower range of activities online than users without those worries. [843] For broadband to be beneficial to their lives, consumers need to be aware of both the benefits of broadband as a means for solving everyday problems and of ways to manage potential hazards. While digital literacy training supports this goal, it is important to explicitly demonstrate the relevance of broadband to people's lives in order to create comfort and familiarity with technology in communities. [844]

Leading media, broadband providers and other technology companies should partner with national non-profits with strong ties to underserved communities to conduct a nationwide outreach and awareness campaign. [845]

The campaign should specifically target key segments of non-adopters such as the elderly, low-income Americans, ethnic and racial minorities and rural Americans. Its messaging should communicate to audiences and their families, in a culturally relevant way, why broadband matters.[846] The campaign's media strategy should include public service announcements and local broadcast messages, but should also focus on printed materials and other resources for local media outreach. In addition to creating targeted, culturally relevant outreach information and materials, the campaign should make media and other resources available in multiple languages so that they are accessible by non-adopters whose primary or only language is not English.

Although the federal government may not directly coordinate the campaign, the FCC and other actors from federal, Tribal, state and local government should work with the partnership to ensure that existing government outreach efforts communicate consistent messages (when possible). The FCC's Consumer Advisory Committee should also monitor the campaign and report back to the FCC on the campaign's effectiveness and private sector's level of engagement with the campaign.

9.5. Addressing Issues of Accessibility for Broadband Adoption and Utilization

Broadband-enabled applications create unique opportunities for people with disabilities. To allow Americans with disabilities to experience the benefits of broadband, hardware, software, services and digital content must be accessible and assistive technologies must be affordable.

In order to achieve this goal, the federal government must become a model for accessibility. Further, the federal government must promote innovative and affordable solutions to ensure that people with disabilities have equal access to communications services and that they do not bear disproportionate costs to obtain that access.

RECOMMENDATION 9.8: The Executive Branch should convene a Broadband Accessibility Working Group (BAWG) to maximize broadband adoption by people with disabilities.

The Executive Branch should convene a working group to coordinate federal efforts to maximize broadband adoption by people with disabilities. The BAWG also should work to make the federal government itself a model of accessibility. Members of the BAWG would bring together representatives from the Executive Branch including the departments of Agriculture, Commerce, Defense, Education, Health and Human Services, Justice, Labor and Veterans Affairs; the Access Board; the FCC; the FTC; the General Services Administration; the National Council on Disability and the National Science Foundation.

The BAWG would take on several important tasks:

- Ensure the federal government complies with Section 508 of the Rehabilitation Act.[847] Under Section 508 of the Rehabilitation Act, federal agencies must "develop, procure, maintain and use" electronic and information technologies that are accessible to people with disabilities—unless doing so would cause an "undue burden."[848] The record indicates that the government's efforts with respect to procurement and website accessibility need improvement.[849] Section 508 requires the U.S. Office of the Attorney General to submit a biennial report to the President and Congress providing information on agency compliance and making recommendations.[850] The Attorney General prepared an interim report in 2000; prospectively, the Attorney General should carry out his statutory duty of submitting a biennial report to the President and Congress providing information on agency compliance with Section 508 and making recommendations.[851] The BAWG should work with the Executive Branch to conduct an ongoing and public assessment of the degree to which agencies are complying with Section 508. The BAWG should also survey federal agencies to determine how they could apply Section 508 requirements to grant recipients and licensees.

- Coordinate policies and develop funding priorities across agencies. The BAWG should work to identify and modify program restrictions that prevent new and efficient technologies from being funded. [852] It also should explore whether any public funding should be used for the development and operation of new software enhancements that could support a network-based delivery system for assistive technologies to allow users to "call up interface features or adaptations that they need anytime, anywhere and on any device that they encounter."[853]

- Prepare a report on the state of broadband accessibility in the United States within a year after the BAWG is created and biennially thereafter. This chapter should consider broadband adoption, barriers and usage among people with disabilities and incorporate the results from questions included in FCC surveys conducted pursuant to the Broadband Data Improvement Act.[854] It should also analyze the root causes of the relatively low broadband adoption rate by people with disabilities and make specific recommendations to address these problems.

RECOMMENDATION 9.9: The FCC should establish an Accessibility and innovation Forum.

The Accessibility and Innovation Forum could allow manufacturers, service providers, assistive technology companies, third-party application developers, government representatives and others to learn from consumers about their needs, to share best practices and to demonstrate new products, applications and assistive technologies. The forum could hold workshops to share and discuss breakthroughs by technologists, engineers, researchers and others that promote accessibility. The Chairman of the FCC, in conjunction with the forum, could also present an annual Accessibility and Innovation Award recognizing innovations by industry, small business, individuals and public-private partnerships that have made the greatest contribution to advancing broadband accessibility. The forum could have an ongoing web presence to allow participants to share information about public and private accessibility efforts and discuss accessibility barriers and inaccessible products.

RECOMMENDATION 9.10: Congress, the FCC and the U.S. Department of Justice (DOJ) should modernize accessibility laws, rules and related subsidy programs.

Accessibility laws, regulations and subsidy programs should be updated to cover Internet Protocol (IP)-based communications and video-programming technologies.[855] To do so:

- The FCC should ensure services and equipment are accessible to people with disabilities. The FCC should extend its Section 255 rules[856] to require providers of advanced services[857] and manufacturers of end-user equipment, network equipment and software used for advanced services to make their products accessible to people with disabilities.[858] Further, the FCC should extend its Hearing Aid Compatibility rules to all devices that provide voice communications via a built-in speaker and are typically held to the ear, to the extent that it is technologically feasible.[859] Finally, the FCC should open a proceeding to implement a standard for reliable and interoperable real-time text any time that Voice over Internet Protocol is available and supported.[860]
- The federal government should ensure the accessibility of digital content. The DOJ should amend its regulations to clarify the obligations of commercial establishments under Title III of the Americans with Disabilities Act[861] with respect to commercial websites. The FCC should open a proceeding on the accessibility of video programming distributed over the Internet, the devices used to display such programming and related user interfaces, video programming guides and menus.[862] Congress should consider clarifying the FCC's authority to adopt video description rules.[863]
- The FCC should materially support assistive technologies to make broadband more usable for people with disabilities. Congress should consider authorizing the FCC to use Universal Service Funds to provide assistive technologies that would enable individuals who are deaf or blind to access broadband services (up to $10 million per year)[864] and to provide funding for competitive awards to be given to developers of innovative devices, components, software applications or other assistive technologies that promote access to broadband (up to $10 million per year). As part of its ongoing

reform efforts,[865] the FCC should issue a Notice of Proposed Rulemaking on whether to establish separate subsidy programs to fund broadband services and assistive technologies under the Telecommunications Relay Services (TRS) program.[866] The FCC should also determine whether additional Internet Protocol-enabled TRS services, such as Video Assisted Speech-to-Speech Service,[867] could benefit people with disabilities.

9.6. EXPANDING FEDERAL SUPPORT FOR REGIONAL BROADBAND CAPACITY-BUILDING, PROGRAM EVALUATION AND SHARING OF BEST PRACTICES

Over the past decade several Tribal, state and local governments have developed broadband adoption and deployment strategies. The federal government has an important role in supporting these complementary state and local efforts and encouraging the "partnership of the public and private sectors in the continued growth of broadband services and information technology for residents and businesses."[868]

Building sustainable efforts to support Tribal, state and local initiatives requires sufficient financial, technical and information resources. The federal government can bolster these efforts by providing additional funding for regional capacity- building and by investing in program evaluation, identification of best practices and facilitation of information sharing among stakeholders across the country[869]

RECOMMENDATION 9.11: Federal support should be expanded for regional capacity-building efforts aimed at improving broadband deployment and adoption.

Many states have shown leadership by developing digital inclusion policies and programs. For example, California, Georgia, Illinois, Kentucky, Maine, Massachusetts, Minnesota and New York have created broadband offices. These offices are building state-level plans, supporting local programs and leading broadband initiatives aligned with the states' economic development, education and health care goals. The federal government can use these strong state programs to achieve national broadband objectives by relying on states to be local advocates for national programs that boost awareness about broadband and ICT.

Some state programs have taken advantage of unique funding opportunities. California, for example, imposed merger conditions on telecommunications providers to establish the California Emerging Technology Fund, which helps fund local efforts to bring broadband to unserved and underserved communities within the state. [870] However, not all states have been able to develop and consistently fund state-level programs. Additional federal support of state efforts can encourage state and local initiatives.

In 2008, the Broadband Data Improvement Act (BDIA) recognized this opportunity.[871] BDIA established a state grant program, eventually funded by the Recovery Act, to begin to ensure all residents and businesses had affordable access to broadband and to promote state efforts to improve technology literacy, computer ownership and broadband use.[872]

Initial grants allocated a per-state maximum of $500,000 over the course of five years for strategic planning; many states have used these grants to create state broadband task forces or hire dedicated broadband staff.[873] States can use additional funding to continue the work begun under these initial planning grants and establish state and local adoption programs envisioned by the legislation.

NTIA should provide additional funding to support ongoing grants aligned with Section 106 of BDIA. The Recovery Act made $350 million available to NTIA to fund the state data-gathering and development goals set in BDIA. NTIA has currently assigned only a portion of these funds; the remainder should be obligated to state-level organizations in 2010. To ensure long-term sustainable efforts, states that have designated an outside entity should be encouraged to include state agency oversight of the planning. These state-level organizations should:[*]

- Complete strategic planning based on gap analysis of broadband availability, adoption and the existing capacity of local support organizations.[874]
- Establish programs to improve computer ownership andInternet access in unserved and underserved areas.[875]
- Provide technical expertise to local institutions, non-profits and governments to develop deployment and adoption- related initiatives.[876]
- Work with the private sector to create public-private partnerships to access infrastructure, technical expertise, training and program funding.
- Accelerate broadband application usage in key areas like government, education and health care.[877]
- Gather state and local benchmark data to determine program success over time.[878]
- Coordinate and enhance volunteer and non-profit programs that provide digital literacy and small business broadband training.[879]

If Congress makes additional funding available under BDIA, it should consider amending BDIA to make Tribes eligible to receive funding. In addition, if BDIA is amended, Congress should consider allowing NTIA to require that new state funding award recipients re-grant a portion of their total award to local and regional broadband programs. Congress also should consider allowing local, community and non-profit entities to apply independently for this new funding in the event that any state, territory or the District of Columbia fails to designate an eligible entity.

RECOMMENDATION 9.12: Congress and federal agencies should promote third-party evaluation of future broadband adoption programs.

Better measurement is widely recognized as necessary for understanding the costs, benefits and efficiency of different adoption programs. But little progress has been made.[880] More systematic evaluation is required to make the most of the federal government's broadband investment.[881] Most adoption programs spend their money on program activities, rather than measuring results. This is an understandable choice in the short run. But in the long run it has left the country with a limited understanding of what works and what does not.[882] The government needs to invest in detailed evaluations of how adoption programs actually influence broadband adoption and use. Such evaluations should also assess the

---

[*] Each of the following is consistent with the uses outlined by BDIA.

impact of adoption programs on educational achievement and literacy as well as cost effectiveness.

Future federal appropriations for broadband adoption should include specific requirements and funding for third-party evaluation and assessment. Each grant should include funding for program evaluation, with additional funding to conduct in- depth assessments and longitudinal program assessment.

Program evaluation should not use a single methodology or type of data collection; evaluations will differ depending on project type and intended outcomes. But evaluations must provide a clear framework against which programs can be measured. They should define what makes a person a broadband "adopter" and track costs per incremental adopter. Further, evaluation should be a basic part of planning a project and adjusting that project when necessary. Evaluations should be designed to track progress and results at the program level, the organizational level and the community level. Longitudinal assessments should sample outcomes across program types.

RECOMMENDATION 9.13: NTIA should establish a National Broadband clearinghouse to promote best practices and information sharing.

In addition to detailed evaluation, practitioners, including the federal government, need better information sharing. A National Broadband Clearinghouse would promote best practices and collaboration among those involved in programs aimed at boosting broadband adoption and utilization. NTIA should work with the FCC, Tribal, state and local governments, regulators, CBOs and the private sector to create, maintain and market a nationally recognized online clearinghouse for best practices. It should serve as a resource for all parties involved in establishing broadband services—providers, Tribal, state and local governments and non-profits. NTIA should establish standards for managing the clearinghouse's online information. NTIA should also provide the clearinghouse with relevant content, including results and data collected during an evaluation of its own programs. States and other entities receiving federal broadband funding from NTIA would be expected to contribute content.

As part of the clearinghouse, NTIA should create a National Broadband Data Warehouse to serve as a central repository for broadband consumer data that exist across government agencies. NTIA's BTOP program rightly places strict reporting requirements on grant recipients in order to gather important performance data. To make the most of these data, they should be included in the warehouse. To the extent possible, the warehouse should provide data in standard and interoperable formats.

Those managing the clearinghouse should conduct outreach efforts and promote the online clearinghouse and its services. They also should encourage community members and broadband users to submit and update information that could be shared online and to develop a review system to ensure the content's quality and usefulness. If necessary, Congress should consider providing additional public funds to support development and management of the clearinghouse and a program of regional outreach, events and field-based data collection.

9.7. Coordinating with Tribes on Broadband Issues

Developing and executing a plan to ensure that Tribal lands have broadband access and that Tribal communities utilize broadband services requires regular and meaningful consultation with Tribes on a government-to-government basis, as well as coordination across multiple federal departments and agencies.

To facilitate effective Tribal consultation and streamline coordination across federal entities on broadband-related issues, the following changes are recommended:

RECOMMENDATION 9.14: The Executive Branch, the FCC and Congress should consider making changes to ensure effective coordination and consultation with Tribes on broadband related issues.

- **The Executive Branch should establish a Federal-Tribal Broadband initiative through which the federal government can coordinate both internally and directly with Tribal governments on broadband-related policies, programs and initiatives.**
- **The FCC should increase its commitment to government-to-government coordination with Tribal leaders.**
- **Congress and the FCC should consider increasing Tribal representation in telecommunications planning.**
- **Federal agencies should facilitate Tribal access to broadband funding opportunities.**
- **The FCC and congress should support technical training and development on Tribal lands.**
- **The federal government should improve the quality of data on broadband in Tribal lands.**

Government-to-Government Coordination and Consultation Tribal governments must interact with multiple federal agencies and departments on a wide range of programs. Because broadband is a critical input to the achievement of goals in many areas, including education, health care, public safety and economic development, the federal government should establish a Federal-Tribal Broadband Initiative to coordinate both internally and directly with Tribal governments on broadband- related policies, programs and initiatives. The initiative will include elected Tribal leaders or their appointees and officials from relevant federal departments and agencies.

The FCC should create an FCC-Tribal Broadband Task Force consisting of senior FCC staff and elected Tribal leaders or their appointees to carry out its commitment to promoting government-to-government relations.[883] The task force will assist in developing and executing an FCC consultation policy, ensure that Tribal concerns are considered in all proceedings related to broadband and develop additional recommendations for promoting broadband deployment and adoption on Tribal lands. The FCC should also create an FCC Office of Tribal Affairs to consult regularly with Tribal leaders, to develop and drive a Tribal agenda in coordination with other FCC bureaus and offices and to manage the FCC-Tribal Broadband Task Force.

Further, the Secretary of Agriculture should complete the department's ongoing consultation process with Tribes and implement provisions of the 2008 Farm Bill relating to substantially underserved trust areas for all broadband funding programs. [884]

In addition, Congress should consider amending the Communications Act to establish a Tribal seat on the USF Joint Board. The FCC should establish a Tribal seat on the USAC Board of Directors.

Technical Training for Tribes

Congress should consider additional annual funding for the FCC to expand the Indian Telecommunications Initiatives' Tribal workshops and roundtables to include sessions on education, technical support and assistance with broadband initiatives.[885] In order to help Tribes acquire technical knowledge and expertise, Congress should also consider additional annual funding to allow Tribal representatives to participate in FCC University training programs at no cost.

Improving Data on Tribal Lands

The FCC should identify methods for collecting and reporting broadband information that is specific to Tribal lands, working with Tribes to ensure that any information collected is accurate and useful. In the interim, the FCC should immediately coordinate discussions between broadband providers and Tribal governments to develop a process for Tribes to receive information about services on Tribal lands. In addition, NTIA should provide BDIA planning and mapping grantees with guidance on how to work with Tribes to obtain data about Tribal lands, and ensure that Tribal governments have the opportunity to review mapping data about Tribal lands and offer supplemental data or corrections.[886] Congress should also consider allowing NTIA to provide separate grants to Tribes or their designees for any purpose permitted under the BDIA, including future planning and mapping projects on Tribal lands.

PART III— NATIONAL PURPOSES

Why is it that some parts of the u.s. economy have greatly improved their performance through the use of technology while others lag far behind?

Why is it that banks have moved their data and transactions online over the past decade, but hospitals collect and disseminate data just as they did 20 years ago?

Why is it that printed newspapers are disappearing, but a high school student's backpack contains the same 25 pounds of textbooks it did decades ago?

Why is it that many jobs are posted online, but too many Americans—particularly in low-income and minority communities—lack the access or skills to see those postings?

Why is it that a football helmet allows a coach and his quarterback to communicate, but first responders from different jurisdictions still cannot communicate at the scene of a disaster?

The private sector offers some hints to the answers to these questions. In their book *Wired for Innovation*, Massachusetts Institute of Technology professors Erik Brynjolfsson and Adam Saunders[887] explore why certain companies benefit from the use of information technology while other similarly situated companies do not. They find that companies only realize the benefits of technology if they also change their fundamental processes and develop a "digital culture."[888] Technology alone is not enough.

The 1990 paper "The Dynamo and the Computer"[889] reveals more clues. In the paper, Stanford professor Paul David tries to explain why major technological innovations in the 1980s had not yet shown up in productivity statistics by the start of the 1990s.

Part of the answer was a "diffusion lag."[890] It takes time for a new technical system to replace an existing technical system. For example, in the early 1900s "the transformation of industrial processes by the new electric power technology was a long-delayed and far from automatic business."[891] Factories didn't reach 50% electrification until four decades after the first central power station opened.[892]

This lag was due in part to the unprofitability of replacing "production technologies adapted to the old regime of mechanical power derived from water and steam."[893] In other words, the problem was not getting electricity—it was reengineering factories designed and optimized for the steam era to embrace the potential benefits of electric power.

Similarly, today some sectors suffer a diffusion lag. The world, the economy and our lifestyles are all moving from analog to digital. Yet some sectors—particularly health care, education, energy, public safety and government generally— have not adapted their processes to take advantage of the modern communications era. Today's diffusion lag precludes the country from realizing the improvements broadband can bring in key national priority areas.

To help America realize world-leading high performance, Congress directed that the National Broadband Plan include a "plan for use of broadband infrastructure and services in advancing consumer welfare, civic participation, public safety and homeland security, community development, health care delivery, energy independence and efficiency, education, worker training, private sector investment, entrepreneurial activity, job creation and economic growth and other national purposes."[894]

Each of these priorities is unique—each faces different challenges, offers different opportunities and demands a different response. As great as the differences are among these national purposes, certain themes are common. For example, there are connectivity requirements for institutions and for relevant functions. Yet in many cases today's connectivity levels are insufficient for current use, let alone the needs of potential future applications. In addition, the right incentives to motivate the use of broadband are critical, yet incentive structures are often hampered by entrenched interests and even deeper entrenched ways of thought.

Across all these priorities, broadband enables the free and efficient exchange of information. Doctors can understand the needs of their patients better and faster by exchanging electronic health records, which improves the quality of care and reduces costs. Smart meters for energy can arm consumers and businesses with information to reduce energy consumption and unlock new opportunities for energy entrepreneurship. Citizens can have better visibility into and involvement in policymaking.

Broadband also removes barriers of time and space. A patient can be monitored at home 24 hours a day, seven days a week. The elderly and frail can avoid frequent trips to the doctor's office that might expose them to illness. A brilliant physics teacher can engage students in classrooms across the country. A working mother can advance her career by taking a job training course at her convenience. A small business in rural America can transact efficiently with customers and suppliers worldwide at any time.

Finally, broadband allows for aggregation of information. With sophisticated data storage, transfer and mining techniques, medical researchers can develop new treatments that improve medical practice. Similarly, teachers can analyze the impact of particular instructional strategies on student progress toward specific learning objectives. The chapters that follow include recommendations that aim to unlock the value of personal data for new applications and research, while taking into account privacy considerations.

In addition to these common themes, several common recommendations span these national priorities.

The connectivity needs of institutions that may further national purposes are varied, and no single solution fits all. But collaboration and coordination between these institutions has significant potential to meet connectivity requirements. Government policy can promote and facilitate that collaboration.

In the past, many institutions have used a collaborative model to achieve connectivity. The Internet2 Project was established in 1996 by 34 university researchers to better support the unique needs of the research community like data mining, medical imaging and particle physics. This partnership and others like it (e.g., National LambdaRail) have emerged to provide the unique capabilities that our nation's top institutions require.

Unfortunately, the job of connecting all of our institutions is not complete. The proposed Unified Community Anchor Network (UCAN) (see Chapter 8) and other networks like it would extend the collaborative model favored by many of our research institutions for the benefit of our other community institutions such as rural health clinics and community colleges. UCAN would enable more demand aggregation and sharing, remove barriers to entry and support efforts to and empower all of our community institutions that need connectivity. [895]

Additionally, national priorities should not be restricted by caps on bandwidth. Broadband usage patterns and pricing models are evolving rapidly. In some cases, fixed and mobile broadband service providers have put in place volume caps that have differential impact on users; in other cases, they have offered specific plans that charge on a usage basis. Such pricing schemes may raise policy issues, but it is premature for this plan to address them, as there are a wide variety of methods by which they can be implemented.

If ISPs adopt volume caps or usage-based pricing as the model for how broadband should be priced, the FCC should ensure that such decisions do not inhibit the use of broadband for public purposes such as education, health care, public safety, job training and general government uses.

It is critical that the country move now to enact the recommendations in this part of the plan in order to accelerate the transformation that broadband can bring in areas so vital to the nation's prosperity. Diffusion of new technologies can take time, but the country does not have time to spare. There are students to inspire, lives to save, resources to conserve and people to put back to work. Integrating broadband into national priorities will not only change the way things are done, but also the results that can be achieved for Americans.

10. HEALTH CARE

Improving americans' health is one of the most important tasks for the nation. Health care already accounts for 17% of u.s. gross domestic product (GDp); by 2020, it will top 20%.[896] America is aging—by 2040, there will be twice as many Americans older than 65 as there are today—and health care costs will likely increase as a consequence.

Rising costs would be less concerning if there were results. But Americans are not healthy. Sixty-one percent of American adults are overweight or obese, which often leads to medical complications.[897] Chronic conditions, which already account for 75%[898] of the nation's health care costs, are increasing across all ages.[899] The nation has 670,000 new cases of congestive heart failure every year, many of them fatal.[900] And too often the care itself causes harm. One and a half million Americans are injured every year because of prescription drug errors,[901] while a person dies every six minutes from an infection developed after arriving at a hospital.[902]

In addition, the United States has a health care supply problem. The country is expected to have a shortage of tens of thousands of physicians by 2020.[903] An aging physician workforce that is nearing retirement and working fewer hours exacerbates the situation.[904] Supply will be further strained if previously uninsured Americans enter the care delivery system.

Another significant problem plaguing the nation's health care system is the fact that there are health disparities across different ethnic groups. "African Americans, for example, experience the highest rates of mortality from heart disease, cancer, cerebrovascular disease, and HIV/AIDS than any other U.S. racial or ethnic group. Hispanic Americans are almost twice as likely as non-Hispanic whites to die from diabetes. Some Asian Americans experience rates of stomach, liver and cervical cancers that are well above national averages."[905] Further exacerbating this problem, members of ethnic groups are less likely than whites to have health insurance, have more difficulty getting health care and have fewer choices in where to receive care.[906]

Broadband is not a panacea. However, there is a developing set of broadband-enabled solutions that can play an important role in the transformation required to address these issues. These solutions, usually grouped under the name health information technology (IT), offer the potential to improve health care outcomes while simultaneously controlling costs and extending the reach of the limited pool of health care professionals. Furthermore, as a major area of innovation and entrepreneurial activity, the health IT industry can serve as an engine for job creation and global competitiveness.

This chapter's recommendations aim to encourage maximum utilization of these solutions. In its traditional role, the FCC would evaluate this challenge primarily through a network connectivity perspective. However, it is the ecosystem of networks, applications, devices and individual actions that drives value, not just the network itself. It is imperative to focus on adoption challenges, and specifically the government decisions that influence the system in which private actors operate, if America is to realize the enormous potential of broadband-enabled health IT.

This chapter has five sections. Section 10.1 reviews the potential value that broadband-enabled health IT solutions can unlock. Section 10.2 offers an overview of current health IT utilization in America, reviews recent federal government actions to enhance utilization of health IT and highlights outstanding challenges.

Sections 10.3–10.5 provide recommendations concerning four critical areas in which the government should take action to help unlock the value of broadband and health IT: better reimbursement, modern regulation, increased data capture and utilization and sufficient connectivity.

Recommendations

Create appropriate incentives for e-care utilization

- Congress and the Secretary of Health and Human Services (HHS) should consider developing a strategy that documents the proven value of e-care technologies, proposes reimbursement reforms that incent their meaningful use and charts a path for their widespread adoption.

Modernize regulation to enable health IT adoption

- Congress, states and the Centers for Medicare & Medicaid Services (CM S) should consider reducing regulatory barriers that inhibit adoption of health IT solutions.
- The FCC and the Food and Drug Administration (FDA) should clarify regulatory requirements and the approval process for converged communications and health care devices.

Unlock the value of data

- The Office of the National Coordinator for Health Information Technology (ONC) should establish common standards and protocols for sharing administrative, research and clinical data, and provide incentives for their use.
- Congress should consider providing consumers access to— and control over—all their digital health care data in machine-readable formats in a timely manner and at a reasonable cost.

Ensure sufficient connectivity for health care delivery locations

- The FCC should replace the existing Internet Access Fund with a Health Care Broadband Access Fund.
- The FCC should establish a Health Care Broadband Infrastructure Fund to subsidize network deployment to health care delivery locations where existing networks are insufficient.
- The FCC should authorize participation in the Health Care Broadband Funds by long-term care facilities, off- site administrative offices, data centers and other similar locations. Congress should consider providing support for for-profit institutions that serve particularly vulnerable populations.

- To protect against waste, fraud and abuse in the Rural Health Care Program, the FCC should require participating institutions to meet outcomes-based performance measures to qualify for Universal Service Fund (USF) subsidies, such as HHS's meaningful use criteria.
- Congress should consider authorizing an incremental sum (up to $29 million per year) for the Indian Health Service (IHS) for the purpose of upgrading its broadband service to meet connectivity requirements.
- The FCC should periodically publish a Health Care Broadband Status Report.

10.1. The Promise of Health It and the Role of Broadband

Health IT plays a key role in advancing policy priorities that improve health and health care delivery. Priorities set forth by HHS include the following: [907]

- Improving care quality, safety, efficiency and reducing disparities
- Engaging patients and families in managing their health
- Enhancing care coordination
- Improving population and public health
- Ensuring adequate privacy and security of health information

Health IT supports these priorities by dramatically improving the collection, presentation and exchange of health care information, and by providing clinicians and consumers the tools to transform care. Technology alone cannot heal, but when appropriately incorporated into care, technology can help health care professionals and consumers make better decisions, become more efficient, engage in innovation, and understand both individual and public health more effectively.

Analysis of information gathered through health IT can provide a basis for payment reform. Payors, providers and patients are focusing increasingly on value. However, data to measure the effectiveness of prevention and treatment on individual and population-wide bases are lacking. This hampers attempts to shift from a volume-focused system that pays for visits and procedures to a value-based regime that rewards cost-effective health improvements. [909]

10-1. EXPLANATION OF REFERENCED TERMS[908]

Health IT	Information-driven health practices and the technologies that enable them. Includes billing and scheduling systems, e-care, EHRs, telehealth and mobile health.
E-Care	The electronic exchange of information—data, images and video—to aid in the practice of medicine and advanced analytics. Encompasses technologies that enable video consultation, remote monitoring and image transmission ("store-and-forward") over fixed or mobile networks.
EHR	An electronic health record is a digital record of patient health information generated by one or more encounters in any care delivery setting. Included in this information are patient demographics, progress notes, diagnoses, medications, vital signs, medical history, immunizations, laboratory data and radiology reports.
Telehealth	Often used as a synonym for e-care, but includes non-clinical practices such as continuing medical education and nursing call centers.
Mobile Health	The use of mobile networks and devices in supporting e-care. Emphasizes leveraging health-focused applications on general-purpose tools such as smartphones and Short Message Service (SMS) messaging to drive active health participation by consumers and clinicians.

Broadband is necessary for these transformations in three ways. First, it enables efficient exchange of patient and treatment information by allowing providers to access patients' electronic health records (EHRs) from on-site or hosted locations. Second, it removes geography and time as barriers to care by enabling video consultation and remote patient monitoring. Third, broadband provides the foundation for the next generation of health innovation and connected-care solutions.

Broadband and Electronic Health Records

Physicians report that electronic health records improve patient care in many ways. [910] The e-prescribing component of EHRs helps avert known drug allergic reactions and potentially dangerous drug interactions, while facilitating the ordering of laboratory tests and reducing redundancy and errors. EHRs also provide easier access to critical laboratory information and enhance preventive care. For example, influenza and pneumonia vaccination reminders displayed to clinicians during a patient visit could play a part in saving up to 39,000 lives a year. [911]

According to one study often cited, electronic health record systems have the potential to generate net savings of $371 billion for hospitals and $142 billion for physician practices from safety and efficiency gains over 15 years. [912] Potential savings from preventing disease and better managing chronic conditions could double these estimates. [913]

Hosted EHR solutions tend to be more affordable and easierto-manage alternatives for small physician practices and clinics. In certain settings, they cost on average 20% less than on-site solutions, reduce the need for internal IT expertise and provide timely updates to clinical decision-support tools (e.g., drug interaction references and recommended care guidelines). [914]

Broadband and Video Consultation

Video consultation is especially beneficial for extending the reach of under-staffed specialties to patients residing in rural areas, Tribal lands and health professional shortage areas (HPSAs). [915] For example, the American Heart Association and American Stroke Association recommend use of video consultation technology for stroke patients to help overcome the dearth of neurologists and to make decisions about whether to deliver the life-saving, clot-busting drug known as tPA (see Box 10-2). [916]

10-2. "Stroke Victim Makes Full Recovery—Thanks o E-Care" [917]

At only 49 years of age, Beverly suffered a stroke. Her best friend drove her to St. Luke's Hospital, which has a video link to the stroke center at Massachusetts General Hospital ("Mass General"), 75 miles away. Minutes after her arrival, St. Luke's emergency department staff assessed her symptoms, ordered a brain scan and called Mass General.

A Mass General stroke specialist activated a video link through which he could see Beverly on a gurney at St. Luke's. He had to determine whether she was having a stroke and, if so, what caused it. A hemorrhage could require emergency brain surgery, whereas a clot could be treated with tPA, which must be administered within the first three hours of stroke onset. The wrong diagnosis could provefatal.

The specialist conducted a neurologic exam over the video link while receiving critical vital signs and lab values. He determined a clot was the cause and figured out when the stroke started by asking her yes/no questions to which she could nod her responses.

Beverly received tPA right at the three-hour deadline. An ambulance took her to Mass General and at the end of the hour-long ride, the nurse recalled being shocked at Beverly's recovery—"We were literally pulling into Mass General, and I said, 'Beverly, how are you?' And she said, 'I'm fine!'" It was as if all the symptoms weregone.

"Wow! I can talk!" the nurse remembers Beverly exclaiming. "'Wow, if it's that medicine, it really worked!'"

In addition to increasing access to otherwise unavailable care, video consultations combined with store-and-forward technologies (e.g., sending images to a specialist at night, as opposed to obtaining a diagnosis during a patient's visit) [918] could lead to significant cost savings from not having to transport patients. Avoiding costs from moving patients from correctional facilities and nursing homes to emergency departments and physician offices, or from one emergency department to another, could result in $1.2 billion in annual savings. [919]

Video consultation and remote access to patient data may also be critical during pandemic situations. If hospitals are at capacity or if isolation protocols are necessary to prevent the spread of infection, these technologies can help health care providers assist more patients and help patients avoid public areas.

Broadband and Remote Patient Monitoring

Remote patient monitoring enables early detection of health problems, usually before the onset of noticeable symptoms. Earlier detection allows earlier treatment and, therefore, better outcomes. For example, after an initial hospitalization for heart failure, 60% of patients are readmitted at least once within six to nine months.[920] If a congestive heart failure patient has a common problem indicator, such as increase in weight or a change in fluid status, a monitoring system instantly alerts the clinician who can adjust medications, thereby averting a hospital readmission. Estimates indicate that remote monitoring could generate net savings of $197 billion over 25 years from just four chronic conditions. [921]

Mobile Broadband and the Future of Health

Mobile health is a new frontier in health innovation. This field encompasses applications, devices and communications networks that allow clinicians and patients to give and receive care anywhere at any time. Physicians download diagnostic data, lab results, images and drug information to handheld devices like PDAs and Smartphones; emergency medical responders use field laptops to keep track of patient information and records; and patients use health monitoring devices and sensors that accompany them everywhere.[923] Through capabilities like these, mobile health offers convenience critical to improving consumer engagement and clinician responsiveness.

BOX 10-3. "HOW HEALTH IT SAVES VETERANS AFFAIRS BILLIONS EACH YEAR"[922]

The Veterans Health Administration (VHA) coordinates the care of 32,000 veteran patients with chronic conditions through a national program called Care Coordination/Home Telehealth (CCHT). CCHT involves the systematic use of health informatics, e-care and disease management technologies to avoid unnecessary admission to long-term institutional care. Technologies include videophones, messaging devices, biometric devices, digital cameras and remote monitoring devices.

CCHT led to a 25% reduction in the number of bed days of care and a 19% drop in hospital admissions. At $1,600 per patient per year, it costs far less than the VHA's home-based primary care services ($13,121 per year) and nursing home care rates ($77,745 on average per patient per year).

Based on the VHA's experience, e-care is an appropriate and cost-effective way to manage chronic care patients in urban and rural settings. Most importantly, it enables patients to live independently at home.

Innovations in mobile medicine include new modalities of non-invasive sensors and body sensor networks.[924] Mobile sensors in the form of disposable bandages and ingestible pills relay real-time health data (e.g., vital signs, glucose levels and medication compliance) over wireless connections.[925] Sensors that help older adults live independently at home detect motion, sense mood changes and help prevent falls.[926] Wireless body sensor networks reduce infection risk and increase patient mobility by eliminating cables; they also improve caregiver effectiveness. Each of these solutions is available today, albeit with varying degrees of adoption.

Mobile medicine takes remote monitoring to a new level. For example, today's mobile cardiovascular solutions allow a patient's heart rhythm to be monitored continuously regardless of the patient's whereabouts.[927] Diabetics can receive continuous, flexible insulin delivery through real-time glucose monitoring sensors that transmit data to wearable insulin pumps.[928]

Advances in networked implantable devices enable capabilities that did not seem possible a few years ago. For example, micropower medical network services support wideband medical implant devices designed to restore sensation, mobility and other functions to paralyzed limbs and organs.[929] These solutions offer great promise in improving the quality of life for numerous populations including injured soldiers, stroke victims and those with spinal cord injuries. Human clinical trials of networked implantable devices targeting an array of conditions are expected to begin at the end of 2010.[930]

Mobile and networked health solutions are in their infancy. The applications and capabilities available even two years from now are expected to vary markedly from those available today. Some will be in specialized devices; others will be applications using capabilities already built into widely available mobile phones, such as global positioning systems and accelerometers. Networked implantable devices stand to grow in sophistication and broaden the realm of conditions they can address. These solutions represent a glimpse into the future of personal and public health—an expanded toolkit to achieve better health, quality of life and care delivery.

10.2. The Need for Action: Maximizing Health It Utilization

Limited Health IT Utilization

The United States is not taking full advantage of the opportunities that health IT provides. It lags other developed countries in health IT adoption among primary health care providers (see Exhibit 10-A).

The United States ranks in the bottom half (out of 11 countries) on every metric used to measure adoption, including use of electronic medical records (10th), electronic prescribing (10th), electronic clinical note entry (10th), electronic ordering of laboratory tests (8th), electronic alerts/prompts about potential drug dose/interaction problems (8th) and electronic access to patient test results (7th).

Adoption rates for e-care are similarly low. A Joint Advisory Committee to Congress found that less than 1% of total U.S. provider locations use e-care. Approximately 200 e-care networks connect only 3,000 providers across the country; typically, the networks are used on a limited basis.[932] A 2008 American Hospital Association survey found that for each of six conditions, only 2–12% of hospitals use Internet-enabled monitoring devices (fixed and mobile), covering 4–8% of relevant patient populations for each condition.[933] Only 17% of home-care agencies use remote monitoring solutions in their practices.[934]

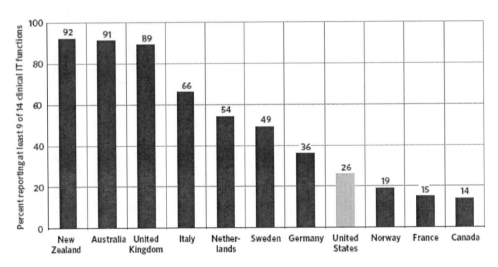

Exhibit 10-A. International Comparison of Electronic Health Adoption[931]

Significant Government Action

The federal government has launched a set of major health IT initiatives to overcome some of the barriers preventing the use of technology, with the goal of transforming America's health care. The largest step by far is a $19 billion net investment to incent the meaningful use of certified EHR technology.

This action is transformative for two reasons: the investment is substantial, and the funding mechanism is focused on measurable outcomes, not inputs. Physicians can earn up to $44,000 in extra Medicare payments from 2011 to 2015 if they become meaningful users of EHRs; hospitals can collect an initial bonus and an extra payment each time a Medicare patient is discharged.[935] There is a similar scheme for Medicaid providers. Rather than provide physicians grants to purchase software, computers and broadband, a set of outcomes such as e- prescribing, data exchange and capturing quality measurements defines "meaningful use."[936] Participants determine the best way to achieve those outcomes. To further adoption, incentives give way to penalties for those that fail to meaningfully use EHRs by 2015.

It is important to recognize the radical change in this approach. The health care delivery system has been dogged for years by criticism that incentives are not aligned to outcomes. The meaningful use mechanism is an attempt, supported by an enormous federal investment and the threat of financial penalties, to develop a new incentive model.

In addition to these incentives, more than $2 billion has been allocated to help the EHR transition succeed. A nationwide network of Regional Extension Centers is being launched to support physician practices as they adopt EHRs; states are being supported to develop policies and technologies that facilitate trusted health information exchange among providers and institutions; and more than a dozen Beacon Communities are being funded to showcase the program's potential, while providing important outcome data and implementation lessons.

All these actions were authorized by the Health Information Technology for Economic and Clinical Health (HITECH) Act, which was part of the American Recovery and Reinvestment Act of 2009.[937] The HITECH Act provisions were designed to improve individuals' health and the performance of the health care system. They focus on four basic goals: define meaningful use, encourage and support the attainment of meaningful use through incentives and grant programs, bolster public trust in electronic information systems by ensuring their privacy and security and foster continued health IT innovation.[938] The HITECH Act is implemented by two agencies within HHS: ONC and CMS.

Despite government actions, three gaps remain: adoption, information utilization and connectivity. These gaps must be filled to accelerate the benefits of broadband. Many fall outside the FCC's traditional purview. For those areas—adoption and data utilization—this chapter highlights some of the most pressing issues and offers high-level recommendations for moving the country forward. Hopefully Congress and the federal agencies responsible for these issues can use these ideas as a starting point or to reinforce efforts underway.

10.3. Closing the Broadband –Enabled Health IT Adoption Gap

Create Appropriate incentives for Health IT Utilization

A key barrier to greater broadband-enabled health IT adoption is misaligned incentives.[939] Those who benefit most from use of these technologies are often not the same as those who shoulder the implementation costs. Providers are expected to pay for equipment and training and adjust to altered workflows. These costs often outweigh the direct benefits they can reasonably expect to gain in terms of reimbursement for services facilitated by health IT.[940] As a result, hospitals and physicians cite funding and unclear investment returns as major barriers to electronic health record adoption.[941]

Instead, it is payors and patients who reap most of the direct benefits of health IT.[942] For example, the federal government— as the payor for veterans' health care—saves money by using a robust e-care program to avoid hospital admissions and expensive home-based care.[943] If a private hospital had implemented a similar program, it might have lost money—forgoing revenue earned through admissions and home-based care services.[944]

The health IT industry has long looked to the country's largest payor, CMS, to lead the way in correcting this incentive imbalance. If CMS were to pay providers more for using effective health IT solutions, all sides would benefit: providers could practice 21st century medicine without losing money; patients could receive 21st century care and achieve better health outcomes; and CMS could save money over time.

Unfortunately, the fee-for-service reimbursement mechanism is not an effective means for realizing health IT's benefits. Fee-for-service rewards providers for volume, and more reimbursement under such a model exposes CMS to the risk of higher costs absent demonstrated health improvements.[945] Coupled with budget neutrality restrictions, it is difficult for CMS to incent broader health IT adoption under this scheme.

HHS's meaningful use approach addresses the incentive misalignment problem for EHRs by moving to outcomes-based reimbursement. Outcomes-based reimbursement alleviates the incentive problem by tying payments to proven, measurable expenditure reductions and health improvements.[946] However, no such systematic solution has been offered for e-care. Currently, CMS only reimburses about $2 million in telehealth services[947] from a budget that exceeds $300 billion.[948]

RECOMMENDATION 10.1: Congress and the Secretary of Health and Human services (HHS) should consider developeing a strategy that documents the proven value of e-care technologies, proposes reimbursement reforms that incent their meaningful use and charts a path for their widespread adoption.

HHS is moving toward outcomes-based reimbursement to stimulate EHR adoption and is well positioned to do the same for e-care. A clearly articulated e-care strategy will accomplish two main purposes:

- Marshal support from Congress, states and the health care community to drive e-care use
- Provide the health IT industry with a clear understanding of the federal government's policies toward e-care

In crafting an e-care strategy, HHS should consider developing new payment platforms to drive adoption of applications proven to be effective. It should also support evaluation of nascent e-care technologies through pilots and demonstration projects. In the course of this effort, HHS should look for opportunities to broaden reimbursement of e-care under the current fee-for-service model. After a reasonable timeframe, Congress should consider convening a panel to review HHS's recommendations and taking action to ensure these technologies' wider adoption. The National Broadband Plan recommends including the following steps as part of this initiative:

1. **HHs should identify e-care applications whose use could be immediately incented through outcomes-based reimbursement.** In its recommendations to Congress, HHS should prioritize e-care applications that it believes are proven to warrant reimbursement incentives. Using the same rigor applied to meaningful use of

EHRs, HHS should define these applications' use cases, data requirements and associated outcomes (expenditure reductions and health improvements). Models such as the VHA's e-care pilot, for instance, could be codified into concrete use cases and criteria for gauging outcomes. These could then be translated into CMS reimbursement incentives for demonstrating meaningful use of the technologies and achieving specified outcomes.

Future iterations of the meaningful use program could offer one means for implementing these reimbursement changes. Draft 2013 and 2015 meaningful use standards require EHRs to be capable of leveraging certain e-care technologies. However, as currently worded, these requirements will not address modifying reimbursement to incent e-care utilization.[949]

2. **When testing new payment models, HHs should explicitly include e-care applications and evaluate their impact on the models. Where proven and scalable, these alternative payment models would provide an additional solution for incenting e-care.** Several alternative payment models have been proposed by the Medicare Payment Advisory Commission and through the health care legislative process. Tests of these models, which are in various stages of implementation, offer an ideal venue for understanding the role e-care can play in outcomes-based reimbursement. Tests include Acute Care Episode Demonstration,[950] Medicare Medical Home Demonstration,[951] Independence at Home, Patient-Centered Medical Home, Accountable Care Organization pilots and Bundled Payment pilots.[952] These pilots and demonstration projects could include an explicit objective to identify e-care use cases and evaluate their effect on health outcomes and expenditure reductions. For instance, in an Independence at Home pilot, remote monitoring could be evaluated as a tool at sample participant sites to understand its impact on quality, data capture and cost savings.

3. **For nascent e-care applications, HHs should support further pilots and testing that review their suitability for reimbursement.** HHS should champion e-care technology pilots where additional data are needed to evaluate their value. HHS has a number of testing mechanisms that it should use to prove the system-wide potential of e-care. Where possible, major pilots of e-care should be designed to adhere to HHS standards for program design, data capture and other requirements for reimbursement decisions and payment model reform. HHS should collaborate in design stages with parties conducting pilots and provide additional funding when its criteria create extra administrative cost.

There are a number of opportunities for HHS to pursue further pilots:

- HHS should make e-care pilots and demonstration projects a top priority across the agency, including the Health Services Resources Administration, the Substance Abuse and Mental Health Services Administration, IHS,[953] NIH and the Agency for Healthcare Research and Quality. HHS-funded projects should be designed with the objectives of understanding use cases, measuring outcomes and determining optimal payment methodologies to produce efficient, high-quality care.

- HHS should collaborate with federally administered providers of care (e.g., VHA, IHS and the Bureau of Prisons) that can act as role models and testbeds for health IT use. For future programs similar to VHA's e-care program (see Box

10-3), HHS should become involved early on to ensure that programs are designed appropriately to inform reimbursement decisions and payment model reform.

- Large-scale private pilots of e-care such as the Connected Care Telehealth Program in Colorado[954] and the Community Partnerships and Mobile Telehealth to Transform Research in Elder Care[955] should similarly consult with HHS and share valuable lessons learned. For pilots that meet HHS's data collection standards, Congress should consider tax breaks or other incentives. For example, Medicare Advantage plan administrators could receive tax credits for testing e-care within their Medicare populations.

The FCC should use data from e-care pilots to update its understanding of health care institutions' broadband requirements. Pilots showcasing emerging technologies that will be used more widely in the subsequent 10 years will be good opportunities to test the network demands of those technologies. Updated use requirements should be coupled with periodically updated reviews of the country's state of connectivity (both wired and wireless) to give the public and other government agencies a better understanding of potential health care broadband gaps. (See Section 10.5 for further recommendations on the FCC's role in monitoring health care broadband.)

4. **As outcomes-based payment reform is developed, CMS should seek to proactively reimburse for e-care technologies under current payment models.** While outcomes-based reimbursement is the optimal payment model for realizing the potential of e-care, it will be years before payment reform transforms the U.S. health care delivery system. In the meantime, CMS should proactively seek means for reimbursing e-care under the current fee-for-service model. This might include the following:

- Collaborating with physicians, researchers, vendors and government stakeholders to design tests that will prove system-wide expenditure reduction under CMS's fee-forservice model.
- Widening coverage for currently reimbursed use cases wherethey have been proven to reduce system-wide expenditures.
- Providing feedback to the community of physicians, researchers and vendors who are trying to enact solutions. Through greater decision-making transparency, CMS could provide critical information that allows that community to target its efforts where they matter most.
- Incenting Medicare Advantage plans to invest rebates (the difference between the established price of care for enrollees and the benchmark for care in that county, of which 75% must be invested as mandatory, health-related supplemental benefits) in the adoption of e-care technologies. Incentives should stipulate tracking health outcomes and expenditure reductions associated with use of these technologies (in compliance with HHS's tracking guidelines).
- Incenting Home Health Agencies reimbursed through CMS to use e-care technologies where CMS believes the technologies will create better health outcomes and reduced expenditures, while requiring participants to track impact associated with the supported technologies.

Physician associations and vendors have recommended areas where they believe expanded reimbursement of e-care, under the current fee-for-service model, will reduce overall CMS expenditures while expanding access to care.[956] As long as the fee-for-service model is the standard, the onus remains on these stakeholders to meet CMS's criteria to expand reimbursement. Examples such as the Veterans Affairs program are less relevant in this case because they operate under a closed payment system. However, CMS's review board should ensure it fully analyzes the system-wide benefits of e-care when making reimbursement decisions.

Modernize Regulation to Enable Health IT Adoption

There is a wide range of problems around the legal and regulatory framework that underpins the use of health IT.[957] Outdated laws and regulations inhibit adoption, and regulatory uncertainty deters investments in both innovation and utilization.

RECOMMENDATION 10.2: Congress, states and the centers for Medicare and Medicaid services (CMS) should consider reducing regulatory barriers that inhibit adoption of health iT solutions.

Several rules have not kept up with technology changes and inhibit adoption of e-care and other health IT solutions. They include the following:

- *Credentialing and privileging.* CMS should revise standards that make credentialing and privileging overly burdensome for e-care; such standards conflict with the goal of expanding access to care. A hospital is not allowed to use the decisions of another hospital as the basis for credentialing and granting privileges; rather, hospitals must conduct their own assessments. For e-care, this means the site where a patient is located (the originating site) may not rely on the site where the physician is located (the distant site) for credentialing and privileging the doctor prescribing care and must instead follow the same process used to credential and privilege any other physician on staff.[958] It can be expensive and time-consuming for originating sites to identify and grant privileges to all the physicians treating its patients via e-care, and they often lack the in-house expertise to privilege specialists. It also creates an undue burden on remote physicians to maintain privileges at numerous additional hospitals and limits the pool of experts a hospital may access. The additional complexity and expense from these standards inhibit e-care. CMS should engage the e-care community and other experts to explore national standards or processes that facilitate e-care while protecting patient safety and ensuring accountability for care.
- *State licensing requirements.* States should revise licensing requirements to enable e-care. State-by-state licensing requirements limit practitioners' ability to treat patients across state lines. This hinders access to care, especially for residents of states that do not have needed expertise in-state. For example, the national ratio for developmental-behavioral pediatricians is 0.6 per 100,000 children; 27 states fall below that level.[959] The increase in autism-spectrum condition diagnoses creates greater demand for this scarce subspecialty. The nation's governors and state legislatures could collaborate through such groups as the National Governors Association, the National Conference

of State Legislatures and the Federation of State Medical Boards to craft an interstate agreement.[960] If states fail to develop reasonable e-care licensing policies over the next 18 months, Congress should consider intervening to ensure that Medicare and Medicaid beneficiaries are not denied the benefits of e-care.

- *E-prescribing.* Congress and states should consider lifting restrictions that limit broader acceptance of electronic prescribing, a technology that could eliminate more than two million adverse drug events and 190,000 hospitalizations, as well as save the U.S. health care system $44 billion per year.[961] One set of rules that needs to be addressed relates to the ban on e-prescribing of controlled substances such as certain pain medications and antidepressants. Drug Enforcement Administration rules require doctors to maintain two systems: a paper-and-fax-based system for auditing controlled substances and an electronic system for other drugs. The complexity of dual systems is at best an inconvenience and at worst an impediment to adoption.[962] Although a pilot to test e-prescribing of controlled substances is pending, stricter security requirements may prove too burdensome and inhibit adoption. Furthermore, the solution for e-prescribing controlled substances must be compatible with EHRs certified to meet meaningful use criteria. Failure to resolve security protocols and interoperability issues for controlled substances may further delay widespread adoption of e-prescribing.

RECOMMENDATION 10.3: The FCC and the Food and Drug Administration (FDA) should clarify regulatory requirements and the approval process for converged communications and health care devices.

The use of communications devices and networks in the provision of health care is increasing. Smartphones have become useful tools for many physicians managing patient care on the go. Medical devices[963] increasingly rely on commercial wireless networks to relay information for patient health monitoring and decision support. Some examples of the convergence between communications and medicine include the following:

- Mobile applications that help individuals manage their asthma, obesity or diabetes
- A Smartphone application that displays real-time fetal heartbeat and maternal contraction data allowing obstetricians to track a mother's labor
- An iPhone application that presents images for clinicians making appendicitis diagnoses
- Wearable wireless patch-like sensors that transmit health data over commercial wireless networks to practitioners, caregivers and patients

These and other products cover a broad range of health IT solutions. At one end, general-purpose communications devices such as smartphones, videoconferencing equipment and wireless routers are regulated solely by the FCC when not created or intended for medical purposes. At the other, medical devices including life-critical wireless devices such as remotely controlled drug-release mechanisms are regulated by the FDA. However, the growing variety of medical applications that leverage communications networks and devices to transmit information or to provide decision support to both clinicians and consumers presents challenges to the current federal regulatory regime. Potential lack of clarity about the appropriate regulatory approach to these convergent technologies threatens to stifle innovation, slow application approval processes and deter adoption.

The FCC and the FDA should collaborate to address and clarify the appropriate regulatory approach for these evolving technologies. As part of this process, the FCC and the FDA should seek formal public input within the next 120 days and hold a workshop with representatives from industry and other relevant stakeholders to examine real case studies. Through this joint, transparent process the agencies should seek to answer questions such as: "Which components of a health solution present risk that must be regulated?" "How can the process for introducing products to the market be improved?" and "What are the characteristics needed for 'medical-grade' wireless?" After public input is received, the agencies should offer joint guidance to address these and other relevant questions.

The FCC and the FDA are committed to working together to facilitate innovation and protect public health in the continued development of safe and effective convergent devices and systems.

10.4. Unlocking the Value of Data

Data are becoming the world's most valuable commodity. In multiple sectors—including finance, retail and advertising—free-flowing and interoperable data have increased competition, improved customer understanding, driven innovation and improved decision-making. Fortune 500 companies such as Google and Amazon have based their business models on the importance of unlocking data and using them in ways that produce far-reaching changes.

In personal finance, for example, individuals can share their data from multiple bank accounts, credit cards and brokerage accounts with trusted third parties. These parties provide personalized services that benefit consumers, such as credit card recommendations that tailor reward programs to a customer's spending patterns.

The advanced use of data in health care offers immense promise in many areas:

- *Better treatment evaluations.* Therapeutic drugs are not tested across all relevant populations. For example, pharmaceutical companies do not conduct widespread tests of new drugs on children for ethical and practical reasons. But increasingly, physicians are treating them with medications that were designed for adults. This may be the right treatment, but, too often, no one knows. The federal government, recognizing the need for better data in comparing treatment options, has recently allocated $1.1 billion toward comparative effectiveness research.[964] Health IT can further this priority. By using applications to collect and analyze the existing data, which today are locked in paper charts, physicians and researchers can evaluate the efficacy and side effects of treatments from disparate groups of patients in order to develop best practices.
- *Personalized medicine.* Many therapeutic drugs are indiscriminately applied to vast populations without sufficient understanding of which treatments work better or worse on certain people. Genomic research produces huge amounts of data that,

when combined with clinical data, could enable development of better targeted drugs. Such drugs could improve outcomes and reduce side effects.

- *Enhanced public health.* Accurately measuring health status, identifying trends and tracking outbreaks and the spread of infectious disease at a population level are extremely difficult. Health IT enables widespread data capture which in turn allows better real-time health surveillance and improved response time to update care recommendations, allocate health resources and contain population-wide health threats.

- *Empowered consumers.* Consumers are too often passive recipients of care, not accessing, understanding or acting upon their own data. Health IT applications that provide easy access and simplify vast amounts of data empower consumers to proactively manage their health. Empowered consumers better grasp their health needs, demand high-quality services and make informed choices about treatment options.

- *Improved policy decisions.* Innovation in health care delivery systems and payment models is stifled by the lack of suitable interoperable data. The prevailing health care payment model mainly pays for volume of services rendered rather than quality of services provided. However, the right data will help make outcomes-based reimbursement possible by allowing consumers, payors and providers to understand the impact of various prevention and treatment options.

Digital health care in America is at an inflection point. The HITECH Act should vastly improve both the capture of interoperable clinical data and consumer access to such data. Nevertheless, a number of barriers prevent the advanced use of data to make Americans healthier for less money. First, not all types of health data are uniformly captured and interoperable. Second, government regulations continue to limit consumer access to personal health data.

RECOMMENDATION 10.4: The Office of the National Coordinator for Health information Technology (ONC) should establish common standards and protocols for sharing administrative, research and clinical data, and provide incentives for their use.

Digital health data are difficult to collect and aggregate. Such data generally are held in proprietary "siloed" systems that do not communicate with one another and therefore cannot be easily exchanged, aggregated or analyzed. The meaningful use incentives for electronic health records will greatly increase the capture of interoperable clinical health information. However, the inability of researchers to access clinical data in standard format and in a secure manner hinders clinical breakthroughs. Performing research across an amalgamation of all types of health care data will remain a challenge absent uniform data standards.

Coordinated standards and protocols will likely increase innovation and discovery within basic science research, clinical research and public health research, helping alleviate many failings of the health care system. The analysis of combined genomic, clinical and real-time physiological data (often captured wirelessly) could help researchers better understand the interplay of genetics and the environment. This could result in personalized interventions based on associations between people and their surroundings, leading to better outcomes.

Combined administrative and clinical data could be an invaluable tool for shifting to an outcomes-based reimbursement system, as well as providing the ability to build statistical models outlining the economic and clinical effects of novel health policy prior to implementation.

The vision is to enable a continuously learning and adaptive health care system that ubiquitously collects information, aggregates it and allows real-time analysis and action. Extending data interoperability to administrative and research data is possible without creating a centralized database controlled by government or private sector actors. But significant administrative, privacy, technology and financial concerns must be resolved in order to empower decentralized solutions. ONC is best positioned to convene a group of experts across the public and private sectors to address these difficult issues and develop a path forward. While developing new versions of meaningful use, ONC should move to extend data interoperability standards.

RECOMMENDATION 10.5: Congress should consider providing consumers access to—and control over—all their digital health care data in machine-readable formats in a timely manner and at a reasonable cost.

BOX 10-4. DATA ADVANCE MEDICINE AND PUBLIC HEALTH[965]

The Framingham Heart Study (FHS), which focused on cardiovascular disease (CVD), illustrates how widespread data capture, aggregation, sharing and analysis can transform medicine.

By 1948, when the study was initiated, cardiovascular disease had become an epidemic in the U.S. Death rates for CVD had been on the rise for half a century, but little was known about the causes of heart disease and strokes. The study began with a group of more than 5,200 men and women who provided detailed medical histories and underwent physical exams, lab tests and lifestyle interviews every two years since joining the study. The data were initially painstakingly captured in written form. Today the study spans three generations of participants, totals nearly 15,000 lives and the data are available online.

FHS is cited as the seminal study in understanding heart disease. The data collected made possible fundamental changes in its knowledge base and treatment. For example, FHS led to the identification and quantification of CVD risk factors—high blood pressure, high blood cholesterol, smoking, obesity, diabetes and physical inactivity. CVD risk factors are now an integral part of modern medical curricula and have facilitated the development of novel therapeutics and effective preventive and treatment strategies in clinical practice. FHS has led to the publication of approximately 1,200 research articles in leading journals.[966]

Broadband will enable the capture of digital health information for all diseases, from patients across the country. Wider availability and analysis of such rich data will allow similar studies for numerous other conditions and populations to be conducted easier and faster. This could broadly transform understanding of disease risk factors and treatment options.

There are too many barriers between consumers and their health data, including administrative, diagnostic, lab and medication data. For example, in Alabama it can take up to 60 days to receive medical records and cost $1 per page for the first 25 pages of those records.[967] The Health Insurance Portability and Accountability Act (HIPAA) gave individuals the right to access their protected health information, and the recent HITECH Act broadened this right by allowing individuals to obtain a copy of their records digitally within 96 hours of the provider obtaining the information. Both were important steps. However, depending on the nature of the data, there are barriers preventing consumer access. Lab results, for example, may only be released to "authorized persons," which often excludes the patient, despite their requests. In contrast, consumers can access their prescription medication lists from their treating physicians or individual pharmacies that have patient portals, but not from e-prescribing intermediaries that aggregate much of this data. The latter is not a regulatory problem; rather, it is due to a lack of incentives for payors, pharmacy benefit managers and pharmacies to allow e- prescribing intermediaries to disseminate the information.

But it is *consumers'* data. A troubling statistic is that patients are not informed of approximately 7% of abnormal lab results.[968] Consumers armed with the right information could do a better job managing their own health, demanding higher quality services from their providers and payors and making more informed choices about care.[969] With seamless access to their raw health data including lab data and prescriptions, consumers could plug the information into specialized applications of their choice and get personalized solutions for an untold number of conditions (see Box 10-5).

Innovation within this space is occurring from the ground up and it is impossible to predict the potential of future applications. What is certain is that in order to maximize innovation and further personalization of health care, consumers must be able to have access to all their health care data and the right to provide it to third-party application developers or service providers of their choice.[970] Congress should consider updating HIPAA, with suitable exceptions,[971] to include consumers as "authorized persons" of their digital lab data. In a similar vein, barriers relevant to all other forms of health data should be examined and removed.

10.5. Closing the Health IT Broadband Connectivity Gap

Characterizing and Sizing the Gap

Research is scarce on health care providers' broadband connectivity needs and the ability of the country's infrastructure to meet those needs. This plan is one of the first attempts to quantify both. A number of challenges that prevented earlier study are relevant to this analysis. Pricing data, for instance, are proprietary and fluctuate widely according to a number of variables, making it difficult to quantify an aggregate price curve. Databases of practice locations bear inconsistent category classifications and often overlap (e.g., a small hospital may also be called a rural health clinic; a small health clinic may also be called a medium-sized physician office). Despite these shortcomings, this analysis is necessary to inform health care policy changes related to broadband, including the effort underway to reform the FCC's Rural Health Care Program.

BOX 10-5. ASTHMAMD: A CASE STUDY IN THE POWER OF CONSUMER HEALTH DATA

A newly released smart- phone application offers a glimpse of the potential when consumers enter even a small amount of data.[972] AsthmaMD helps patients manage their asthma by inputting a number of parameters, including current medications, and attack timing and severity. Users can opt to share their data anonymously with the service. The data are aggregated and analyzed with the aim of better understanding the disease, as well as providing specific personalized solutions for the consumer. For example, the application can help users better understand the effectiveness of different medications for asthma management and offer insights into specific triggers for that individual's attacks (*e.g.*, pollen, dust, exercise). The application also can track the consumer's precise location and the timing of their asthma activity, which can be correlated with local pollutant count, adverse weather changes and different types of pollutants. In addition, it can alert users with higher risks of an attack in real time if it detects users with a similar asthma history reporting asthma issues. Ultimately it could send live Twitter streams showing geographic areas with asthma flare-ups in real time.

Health Care Providers' Broadband Needs

Health care providers' broadband needs are largely driven by the rapidly increasing amount of digital health-related data that is collected and exchanged. A single video consultation session can require a symmetric 2 Mbps connection with a good quality of service.[973] There is a wide range of requirements to support EHRs and medical imaging. Exhibit 10-B shows the variation in file sizes for common health care file types. Over the next decade, physicians will need to exchange increasingly large files as new technologies such as 3D imaging become more prevalent.

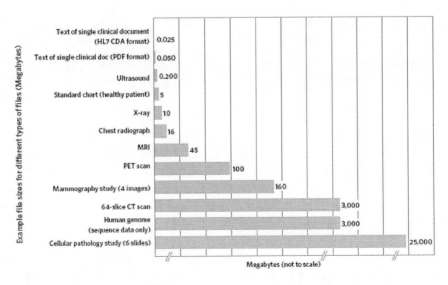

Exhibit 10-B. Health Data File Sizes[974]

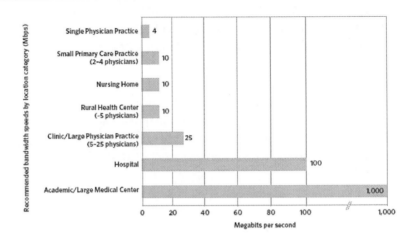

Exhibit 10-C. Required Broadband Connectivity and Quality Metrics (Actual)[975]

The connectivity needs of different health delivery settings will vary depending on their type (e.g., tertiary care center versus primary care physician practice) and their size. In addition, applications that integrate real-time image manipulation and live video will stimulate demand for more and better broadband[976] because these applications have specific requirements for network speeds, delay and jitter. Exhibit 10-C shows an estimate of the required minimum connectivity and quality metrics to support deployment of health IT applications today and in the near future at different types of health delivery settings. Although some delivery settings currently function at lower connectivity and quality, those levels are straining under increasing demand and are unable to support needs likely to emerge in the near future.[977]

Most businesses in the United States, physician offices included, have two choices of broadband service categories: mass-market "small business" solutions[978] or Dedicated Internet Access (DIA),[979] such as T-1 or Gigabit Ethernet service. DIA solutions include broader and stricter Service Level Agreements (SLAs) by network operators. DIA services are substantially more expensive than mass-market packages. For example, in Los Angeles, 10 Mbps Ethernet service with an SLA averages $1,044/month,[980] while Time Warner Cable's similar mass-market package, Business Class Professional, which offers 10 Mbps download speeds 2 Mbps upload speeds, is approximately $400/month.[981]

Connectivity Gap: Small Providers (Four or Fewer Physicians)

In general, smaller providers can achieve satisfactory health IT adoption with mass-market "small business" packages of at least 4 Mbps for single physician practices and 10 Mbps for two-tofour physician practices, even though these solutions may not provide business-grade quality-of-service guarantees.[982] Since most small physician offices do not provide acute care services, they do not require the same degree of instant and guaranteed responsiveness that large practices and hospitals require.

Based on the requirements listed above, an estimated 3,600 out of approximately 307,000 small providers face a broadband connectivity gap. The gap is particularly wide among providers in rural areas (see Exhibit 10-D). In locations defined as rural by the FCC, approximately 7% of small physician offices are estimated to face a connectivity gap. In contrast, across all locations, only approximately 1% of physician offices face a connectivity gap.[983]

Connectivity Gap: Medium & Large Providers (Five or More Physicians)

Larger physician offices, clinics and hospitals face connectivity barriers of a different nature. Because of their size and service offerings, these providers often cannot rely on mass-market broadband and must usually purchase DIA solutions. DIA pricing is determined on a case-by-case basis depending on factors such as capacity, type and length of the connection; type of service provider; and type of facility used. It often varies significantly by geography. Exhibit 10-E illustrates how widely DIA prices fluctuate in urban areas.

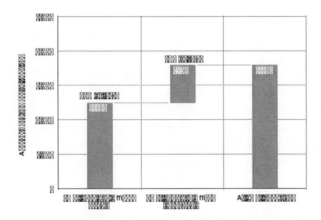

Exhibit 10-D. Estimate of Small Physician Locations Without Mass- Market Broadband Availability[984]

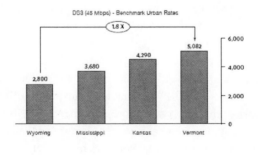

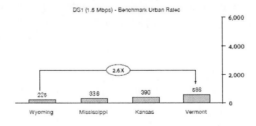

Exhibit 10-E. Wide Fluctuations in Dedicated Internet Access Prices[986] (Monthly Service Cost in Dollars)

For two large physician offices seeking to capitalize on meaningful use incentives, a disparity of more than $27,000 per year[985] in broadband costs puts one at a disadvantage to the other, negates a significant portion of the incentives and may prove an insurmountable obstacle to EHR adoption.

Rural and Tribal areas are likely to face even greater price inequities. There are more than 2,000 rural providers participating in the FCC's Telecommunications Fund, and their broadband prices average three times the price of urban benchmarks.[987]

Connectivity Gap: Federally Funded Providers

Several federally funded providers[988] have insufficient connectivity. For example, 92% of IHS sites purchase a DIA connection of 1.5 Mbps or less.[989] These bandwidth constraints prevent IHS providers from achieving full adoption of video consultation, remote image diagnostic and EHR technology. Similarly, federal subsidy recipients such as Federally Qualified Health Centers,[990] Rural Health Clinics[991] and Critical Access Hospitals[992] face challenges in securing broadband solutions relative to the rest of the country. Exhibit 10-F shows the FCC's estimate of these providers' mass-market broadband gaps. It is important to note that these gaps in mass-market broadband do not preclude locations from purchasing DIA solutions. Nearly every IHS location purchases DIA broadband. However, the fact that such high percentages of federally funded providers are located outside the mass-market footprint means that they face significantly higher prices.

Federally funded providers have a direct impact on the government's costs and serve health care populations for whom the government assumes responsibility; the federal government should improve their connectivity and make them models of harnessing health IT to ensure better health (see Recommendation 10.10).

Connectivity Gap: Next Phase of Analysis

Understanding the state of broadband connectivity for health care providers is a new but important area of analysis. There is more to be done, especially as the need for better data continues to grow. As nascent health IT applications become more prevalent and the importance of wireless connectivity grows, an up-to-date understanding of broadband use cases and connectivity levels will be invaluable. Immediate efforts should be made to quantify the price disparity problem on a more granular level. Similarly, the levels—and costs—of broadband that providers purchase warrant further analysis.

The FCC should play an ongoing role in serving this knowledge base via the Health Care Broadband Status Report proposed in Recommendation 10.11. This information is important not only to policymakers and regulators, but to the health IT industry and the health care provider community. These groups are also invested in understanding the role broadband plays in health care delivery and should participate in shaping this body of research.

Reform the Rural Health Care Program

The recommendations throughout this plan will have a tremendous impact on health care institutions, particularly the consumers and small providers that will likely be using mass-market solutions. However, because of health care's role in the lives of consumers and its importance to the national economy, it is critical to retain a dedicated set of programs within the Universal Service Fund (USF) to help spur broadband adoption by health care providers.

The FCC's Rural Health Care Program as currently structured, however, is not meeting the country's needs.

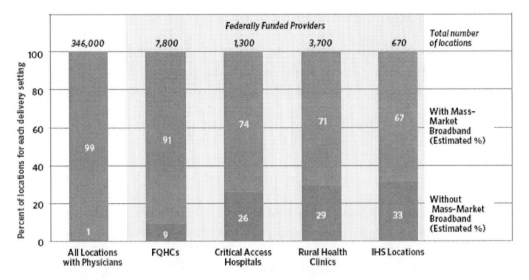

Exhibit 10-F. Estimated Health Care Locations Without Mass-Market Broadband Availability[993]
(Percent of locations for each delivery setting)

In 1997, the FCC implemented the directives of the Telecommunications Act of 1996 by creating a Rural Health Care Program, funded through the USF.[994] The program provides three types of subsidies to public and nonprofit health care providers. First, the program subsidizes the rates paid by rural health care providers for telecommunications services to eliminate the rural/urban price difference within each state (via the Telecommunications Fund).[995] Second, to support advanced telecommunications and information services the program provides a 25% flat discount on monthly Internet access for rural health care providers and a 50% discount for health care providers in states that are entirely rural (via the Internet Access Fund).[996] Lastly, the FCC adopted a three- year program that provides support for up to 85% of the costs associated with deploying broadband health care networks in a state or region (the Pilot Program).[997] The Pilot Program funds one-time capital costs for network deployment, as well as recurring capital and operational costs over five years.

Problems with the Current Program

As previous sections demonstrate, many health care providers have difficulty accessing broadband services because they are located in areas that lack sufficient infrastructure or areas where broadband service is significantly more expensive. Less than 25% of the approximately 11,000 eligible institutions are participating in the program,[998] and many are not acquiring connections capable of meeting their needs.[999] In 2009, 82% of Telecommunications Fund spending supported connections of 4 Mbps or less,[1000] which is a minimum for single physician practices that are using a robust suite of broadband-enabled health IT. That speed is increasingly insufficient for the clinics and hospitals that are the major participants in the program.

Thousands of eligible rural health care providers currently do not take advantage of this program. Some claim that this is because the subsidy is too low and the application process is too complex to justify participation.[1001] Large gaps in broadband access and price disparities for broadband services suggest that change is needed in the support program. Statutory restrictions that limit support to public and non-profit entities and program rules that limit support to rural entities should be reexamined. Many deserving health care providers, such as urban health clinics and for-profit physician offices that function as the safety net for the country's care delivery system, should become eligible for funding under the program.[1002] In rural areas alone, for-profit eligibility restrictions exclude more than 70% of the 38,000 health care providers; many face the same disadvantages in securing broadband as the eligible providers.[1003]

The Pilot Program represents an important first step in extending broadband infrastructure to unserved and underserved areas, and ensuring that health care providers in rural areas and Tribal lands are connected with sophisticated medical centers in urban areas. Over 35% of projects have received funding commitments to date. Much of this progress has come in the last 12 months. Extensive outreach from the FCC and efforts of program participants have resulted in funding commitment letters for 22 projects, for a total of $44.5 million.[1004] To ensure that each program participant has ample time to finalize its project, the FCC has extended the deadline for funding commitment submissions. It should continue to assist participants to ensure networks are built as quickly and effectively as possible.

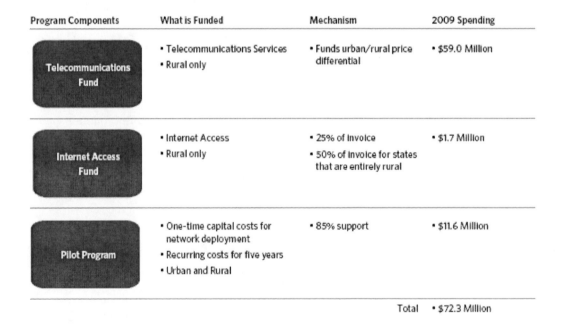

Program Components	What is Funded	Mechanism	2009 Spending
Telecommunications Fund	• Telecommunications Services • Rural only	• Funds urban/rural price differential	• $59.0 Million
Internet Access Fund	• Internet Access • Rural only	• 25% of Invoice • 50% of Invoice for states that are entirely rural	• $1.7 Million
Pilot Program	• One-time capital costs for network deployment • Recurring costs for five years • Urban and Rural	• 85% support	• $11.6 Million
		Total	• $72.3 Million

Exhibit 10-G. 2009 Rural Health Care Program Spending

Despite the FCC's efforts to date, many health care providers remain under-connected. The FCC's programs are in need Exhibit 10-G: 2009 Rural Health Care Program Spending of improved performance measures to assess their impact on broadband services and, more importantly, patient care. The FCC should take a fresh look and evaluate how it can improve the Rural Health Care Program to ensure that funds are used efficiently and appropriately to address the adoption and deployment challenges outlined above. In doing so, lessons learned from the Rural Health Care Pilot Program should be incorporated into this examination.

RECOMMENDATION 10.6: The FCC should replace the existing Internet Access Fund with a Health Care Broadband Access Fund.

The Health Care Broadband Access Fund should support bundles of services, including bundled telecommunications, broadband and broadband Internet access services for eligible health care providers. This program would replace the existing underutilized Internet Access Fund. Health care providers eligible to participate in the new program should include both rural and urban health care providers, based on need. The FCC should develop new discount levels based on criteria that address such factors as:

- Price discrepancies for similar broadband services between health care providers.
- Ability to pay for broadband services (i.e., affordability).
- Lack of broadband access, or affordable broadband, in the highest HPSAs of the country.
- Public or safety net institution status.[1005]

To allow health care providers to afford higher bandwidth broadband services, the subsidy support amount under the Health Care Broadband Access Fund should be greater than the current 25% subsidy support under the Internet Access Fund. In addition, support should better match the costs of services for disadvantaged health care providers. To better encourage participation, the FCC should also simplify the application process and provide clarity on the level of support that providers can reasonably expect, while protecting against potential waste, fraud and abuse.

After approximately three years of data collection for the new Health Care Broadband Access Fund, the FCC should examine, based on the success of that program, whether the Telecommunications Fund program needs to be adjusted.

RECOMMENDATION 10.7: The FCC should establish a Health Care Broadband Infrastructure Fund to subsidize network deployment to health care delivery locations where existing networks are insufficient.

Many health care providers are located in areas that lack adequate physical broadband infrastructure. Specifically, as demonstrated by the overwhelming interest in the Pilot Program, the FCC was able to identify and begin addressing the lack of access to appropriate broadband infrastructure throughout the nation. The FCC should permanently continue this effort by creating a Health Care Broadband Infrastructure Fund, incorporating lessons learned from administering the Pilot Program. In particular, the Pilot Program has enabled the FCC to obtain valuable data on how to better target support to deploy health care networks where the need is most acute. The following recommendations are based on preliminary lessons from the Pilot Program.

The FCC should establish demonstrated-need criteria to ensure that deployment funding is focused in those areas of the country where the existing broadband infrastructure is insufficient. For example, demonstrated-need criteria could include any combination of the following:

- Demonstration that the health care provider is located in an area where sufficient broadband is unavailable or unaffordable. The forthcoming BDIA broadband map should be a factor in determining availability.
- A financial analysis that demonstrates that network deployment will be significantly less expensive over a specified time period (e.g., 15–20 years) than purchasing services from an existing network carrier.
- Certification that the health care provider has posted for services under the Telecommunications Funds and/or the Internet Access Fund (or the new Health Care Broadband Access Fund) for an extended period of time (for example, six to 12 months) and has not received any viable proposals from qualified network vendors for such services. The FCC should also:
- Require that program participants pay no less than a minimum percentage of all eligible project costs, such as the 15% match requirement used in the Pilot Program. The match contribution requirement aligns incentives and helps ensure that the health care provider values the broadband services being developed and makes financially prudent decisions regarding the project.
- Facilitate efficient use of USF-funded infrastructure. For the Pilot Program, the FCC has required that any excess capacity (broadband capacity in excess of the amount required for the eligible health care providers) must be paid for by the health care provider or a third party, at fair share.[1006] Fair share has been defined as a proportionate share of all costs, including trenching and rights-of-way. In instances where excess capacity will be used by other USF-eligible institutions, the FCC should allow the excess capacity to be paid for by those institutions at incremental cost rather than fair share. The FCC should also explore ways to encourage joint applications between eligible health care providers and other USF-qualifying institutions, such as schools and libraries.
- Simplify the community buildout fair share rules so nonUSF-eligible institutions can accurately and efficiently estimate their proper share of network deployment costs and join the infrastructure projects. It is in a community's best interest when public, non-profit and private institutions share infrastructure costs and bring broadband to more of the community. The FCC should define, early in the process, permissible ways in which excess capacity can be deployed and allocated to non-USF-eligible institutions.
- Maintain existing criteria utilized in the Pilot Program, including requirements that projects are sustainable, create statewide or regional networks and leverage existing network technology. Moreover, the FCC should continue to allow (but not require) the connection of networks to proprietary nationwide backbones that link government research institutions and academic, public and private health care providers that house significant medical expertise.
- Simplify program application and administration. For example, the FCC should allow some limited funding of project administration costs for network design and project planning.

The FCC should set a target for how much yearly support should go to infrastructure versus ongoing support. Based on the benefits these programs can deliver to American health care, the FCC should plan to spend up to the current annual cap and then consider additional funding if the need exists and funds can be made available.

RECOMMENDATION 10.8: The FCC should authorize participation in the Health Care Broadband Funds by long-term care facilities, off-site administrative offices, data centers and other similar locations. congress should consider providing support for for-profit institutions that serve particularly vulnerable populations.

The term "health care provider" has been interpreted narrowly, excluding, for example, nursing homes, hospices, other long-term care facilities, off-site administrative offices and health information data centers.[1007] The FCC should re-examine that decision in light of trends in the health care landscape and expand the definition to include, where consistent with the statute, those institutions that have become integral in the delivery of care in the United States. The expanded definition of eligible health care providers should explicitly include off-site administrative offices of eligible health care providers, long-term care facilities, data centers used for health care purposes and owned (directly or indirectly) by eligible health care providers, dialysis centers and skilled nursing facilities.

The FCC should periodically look to the ONC (e.g., every two years) to determine whether the definition of institutions eligible for funding as an eligible health care provider should be changed while the health IT landscape evolves.

In addition, Congress should consider expanding the definition of health care providers eligible for USF funding to include certain for-profit entities.[1008] Under the Communications Act, eligibility for funding under the Rural Health Care Program is limited to public or nonprofit entities.[1009] Not supporting private and for-profit health care providers has a significant impact on some important components of the health delivery system that serve needy populations. In rural areas, for example, private physician clinics can be the most critical—and sometimes the only—health care delivery location in the community. The power of digitized patient records is most valuable when all providers, including private physicians, are connected.

Including for-profit locations will require appropriate limitations to ensure that money from USF is targeted to health care providers that serve particularly vulnerable populations. For instance, funding for health IT in the Recovery Act is available to private physicians that either bill Medicare or have patient volumes consisting of at least 30% Medicaid beneficiaries (20% for pediatricians).[1010] This methodology could provide Congress a template to consider for expanding USF eligibility.

RECOMMENDATION 10.9: To protect against waste, fraud and abuse in the Rural Health care Program, the FCC should require participating institutions to meet outcomes-based performance measures to qualify for USF subsidies, such as HHS's meaningful use criteria.

The FCC should align its health care program with other federal government criteria intended to measure the efficient use of health IT, such as the meaningful use criteria being developed by HHS.[1011] This will help ensure the FCC's programs encourage physicians and hospitals to not only deploy networks or purchase broadband services, but to use them in a way that improves the country's health delivery system. For example, participants in the FCC programs should be required to achieve meaningful use certification for EHRs, after a certain period of support (e.g., three years).

The FCC should work with HHS (and other relevant agencies) and seek comment from the public to determine which outcome metrics (e.g., coordination with Regional Extension Centers, remote monitoring of chronic patients) should be utilized to assess its programs' impact on broadband usage and the delivery of medicine at participating locations. For metrics that are deemed particularly difficult to attain, the FCC should consider offering additional support to those health care providers that are most successful in utilizing broadband services to improve the lives of their patients.

By following the path Congress laid out in the HITECH Act and re-focusing federal investments away from process and toward outcomes (specifically meaningful use of health IT), the FCC can contribute to an important transformation of federal spending. Most importantly, it can ensure that the program funds not just wires, but health. Also, it can allow the FCC to give program participants more authority over project administration as long as they are achieving well-defined objectives. The FCC should evaluate and improve upon its oversight (e.g., competitive bidding, audits and investigations) to ensure that funds are being used to further the statutory purposes of universal service and doing the most to impact broadband usage and the delivery of medicine while minimizing waste, fraud and abuse.

RECOMMENDATION 10.10: Congress should consider providing an incremental sum (up to $29 million per year) for the indian Health service for the purpose of upgrading its broadband service to meet connectivity requirements.

The Indian Health Service offers a unique opportunity for Congress to consider taking action. There is a clear need for broadband—many IHS sites are extremely remote and Tribal lands generally have low broadband penetration rates (see Exhibit 10-F). Since IHS is an integrated system that directly impacts the federal government's bottom line,[1012] taxpayers stand to realize the savings and efficiency improvements promised by best-practice health IT utilization across IHS. IHS can serve as a testbed for forward-looking health IT use, much as VHA does with its CCHT program.[1013]

Congress should consider providing additional public funding for IHS locations that currently have insufficient levels of broadband connectivity. IHS estimates that the annual expenditure to upgrade its broadband service is $29 million.[1014] New funding should be contingent on a competitive process that ensures efficient use of funds and clear goals tied to the meaningful use of health IT, as outlined in the proposed reforms for the Health Care Broadband Access Fund. Where new infrastructure needs to be deployed, it should be deployed in a way that maximizes value for the surrounding communities, providing low-cost, high-speed infrastructure where it did not previously exist.

After one year of administering the IHS funding, Congress should consider doing the same for other federally funded providers with a connectivity gap. Where Congress does not act directly, these networks of providers should remain a high priority for the FCC's reformed Health Care Broadband Access and Infrastructure programs.

RECOMMENDATION 10.11: The FCC should periodically publish a Health Care Broadband Status Report.

Health IT is in its infancy. The private sector innovations and public programs described in this chapter are merely an overview of the explosion in activity. While the National Broadband Plan lays the path forward, it will be critical for the FCC to play a more prominent and sustained role in evaluating broadband infrastructure and in supporting the nation's health transformation. The health care connectivity analysis should serve as a starting point for measuring the health care connectivity problem and assessing the effectiveness of potential solutions.

The FCC should publish a Health Care Broadband Status Report every two years. It should discuss the state of health care broadband connectivity, review health IT industry trends, describe government programs and make reform recommendations. For the FCC's programs, these analyses should be coupled with a dedicated effort to assess their impact on broadband usage and health care delivery at participating locations. The Rural Health Care Program has improved access to quality medical services, but the FCC lacks comprehensive information to determine how funding actually changes behavior. In conjunction with HHS, which has experience evaluating the effectiveness of clinical programs, the FCC should look for better ways to test the impact of the Health Care Broadband Access and Health Care Broadband Infrastructure funds. For instance, the FCC could conduct the following tests:

- Determine how unsupported health care providers differ from supported providers in the utilization of e-care.
- Assess the impact of changing the level of broadband subsidies to a targeted community and determine if there is an increased use of broadband and health IT as a result of such subsidies.
- Explore whether including funding for training would lead to better broadband utilization and improved care.
- Evaluate the impact the program is having on vulnerable populations, such as the elderly, racial and ethnic minorities or low-income rural and urban communities, to understand whether targeted efforts would be more effective.

Through these mechanisms, the FCC should develop a culture of testing and learning. Working in conjunction with participants, policymakers and industry leaders, the FCC should seek to continuously evaluate the impact of its programs and change direction when they do not meet expectations. To ensure sufficient support for these tests, the FCC should allocate a portion of the existing funding cap (e.g., $5 million) for innovative ideas or programs that can evaluate existing efforts or improve upon them in the future. These actions could also help reduce waste, fraud and abuse, because program effectiveness could be continuously monitored, with rules and administration adjusted as necessary.

As technologies rapidly evolve, so too do expectations for health IT adoption in America. Supporting health IT requires further analysis of complex issues and the development of solutions to address them. The work ahead will be most successful if it combines the efforts of government, industry and the health care community.

11. EDUCATION

The united states has some of the best schools and research universities in the world and produces top professionals in every industry. The public education system has effectively developed a workforce for the industrial age, and its graduates have helped the united states become the most prosperous nation in the world.

However, the demands of the new information-based economy require substantial changes to the existing system. American businesses have pointed to a widening gap between the skills of graduates and modern workforce demands.[1015] The U.S. Department of Labor predicts "occupations that usually require a postsecondary degree or award... to account for nearly half of all new jobs from 2008 to 2018."[1016] The 21st century workplace requires both a better-educated and a differently educated work force.[1017]

While some U.S. Students perform extremely well, the educational system as a whole faces huge challenges. Thirty-two percent of all public school students and nearly 50% of African American and Hispanic students fail to graduate from high school.[1018] A significant gap in achievement persists, with African American and Hispanic students trailing white students of the same age by two to three years.[1019] Measured against international benchmarks, the United States lags significantly behind other advanced nations in preparing its students, particularly in math and science (see Exhibit 11-A).[1020]

Researchers have been studying these outcomes for years and have identified several factors that need to be addressed. These include a scarcity of well-trained teachers in key areas such as science, technology, engineering and mathematics (STEM),[1021] inequitable distribution of highly qualified teachers[1022] and a deficit of well-trained principals and administrators.[1023] In addition, there is widespread inability to engage students in learning,[1024] a lack of standards and assessments that measure learning effectively[1025] and insufficient access to timely, individualized content for students.[1026] Exacerbating these challenges are limited organizational transparency and accountability and the inability of teachers and principals to share best practices, content and strategies to improve achievement.[1027] The escalating cost of education, measured against overall results, is also a critical issue.[1028]

Four core assurances drive the U.S. Department of Education's strategy to address these challenges:

- Making progress toward rigorous college- and career-ready standards and high-quality assessments that are valid and reliable for all students, including English-language learners and students with disabilities.
- Establishing pre-kindergarten to college and career data systems that track progress and foster continuous improvement.
- Making improvements in teacher effectiveness and in the equitable distribution of qualified teachers for all students, particularly those most in need.
- Providing intensive support and effective interventions for the lowest-performing schools.[1029]

PISA rankings show United States trailing other OECD countries

Average PISA mathematics score, 2006		Average PISA science score, 2006	
Higher quality Finland	548	Korea	552
Korea	547	Finland	548
Netherlands	531	Canada	527
Switzerland	530	New Zealand	522
Canada	527	Netherlands	519
Japan	523	Australia	516
New Zealand	522	Switzerland	514
Belgium	520	Belgium	511
Australia	520	Japan	511
Denmark	513	Ireland	509
Czech Republic	510	Sweden	505
Iceland	506	Denmark	504
Austria	505	Poland	502
Germany	504	Germany	499
Sweden	502	Austria	498
Ireland	501	Czech Republic	496
France	496	United Kingdom	495
United Kingdom	495	Iceland	495
Poland	495	France	492
Slovak Republic	492	Norway	487
Hungary	491	Hungary	487
Luxembourg	490	Luxembourg	485
Norway	490	Slovak Republic	479
Spain	480	**24th** United States	474
25th United States	474	Spain	470
Portugal	466	Portugal	469
Italy	462	Italy	465
Greece	459	Greece	459
Lower quality Turkey	424	Turkey	436
Mexico	406	Mexico	408
Average=498		Average=494	

Note: Results are for all OECD countries; OECD partner countries not included. Differences may not be statistically significant.
Source: OECD

Exhibit 11-A. Programme for International Student Assessment (PISA) Rankings Show the United States Trailing Other Organisation for Economic Cooperation and Development (OECD) Countries

Broadband can be an important tool to help educators, parents and students meet major challenges in education. The country's economic welfare and long-term success depend on improving learning for all students,[1030] and broadband-enabled solutions hold tremendous promise to help reverse patterns of low achievement.

With broadband, students and teachers can expand instruction beyond the confines of the physical classroom and traditional school day. Broadband can also provide more customized learning opportunities for students to access high-quality, low-cost and personally relevant educational material.[1031] And broadband can improve the flow of educational information, allowing teachers, parents and organizations to make better decisions tied to each student's needs and abilities. Improved information flow can also make educational product and service markets more competitive by allowing school districts and other organizations to develop or purchase higher-quality educational products and services.

This chapter is arranged in three sections. Section 11.1 contains recommendations to help improve online learning opportunities, both inside and outside the classroom. Section 11.2 recommends ways to gather and provide information that fosters innovation. Section 11.3 recommends changes to the E-rate program—which offers schools and libraries discounted telecommunications services, Internet access and internal connections to improve the broadband infrastructure available to schools.

Recommendations

Support and promote online learning

- The U.S. Department of Education, with support from the National Institute of Standards and Technology (NIST) and the Federal Communications Commission

(FCC), should establish standards to be adopted by the federal government for locating, sharing and licensing digital educational content by March 2011.

- The federal government should increase the supply of digital educational content available online that is compatible with standards established by the U.S. Department of Education.
- The U.S. Department of Education should periodically reexamine the digital data and interoperability standards it adopts to ensure that they are consistent with the needs and practices of the educational community, including local, state and non-profit educational agencies and the private sector.
- Congress should consider taking legislative action to encourage copyright holders to grant educational digital rights of use, without prejudicing their other rights.
- State accreditation organizations should change kindergarten through twelfth grade (K–12) and post-secondary course accreditation and teacher certification requirements to allow students to take more courses for credit online and to permit more online instruction across state lines.
- The U.S. Department of Education and other federal agencies should provide support and funding for research and development of online learning systems.
- The U.S. Department of Education should consider investment in open licensed and public domain software alongside traditionally licensed solutions for online learning solutions, while taking into account the long-term effects on the marketplace.
- The U.S. Department of Education should establish a program to fund the development of innovative broadband- enabled online learning solutions.
- State education systems should include digital literacy standards, curricula and assessments in their English Language Arts and other programs, as well as adopt online digital literacy and programs targeting STEM.
- The U.S. Department of Education should provide additional grant funding to help schools train teachers in digital literacy and programs targeting STEM. States should expand digital literacy requirements and training programs for teachers.

Unlock the value of data and improve transparency

- The U.S. Department of Education should encourage the adoption of standards for electronic educational records.
- The U.S. Department of Education should develop digital financial data transparency standards for education. It should collaborate with state and local education agencies to encourage adoption and develop incentives for the use of these standards.
- The U.S. Department of Education should provide a simple Request for Proposal (RFP) online "broadcast" service where vendors can register to receive RFP notifications from local or state educational agencies within various product categories.

Modernize educational broadband infrastructure

- The FCC should adopt its pending Notice of Proposed Rulemaking (NPRM) to remove barriers to off-hours community use of E-rate funded resources.

- The FCC should initiate a rulemaking to set goals for minimum broadband connectivity for schools and libraries and prioritize funds accordingly.
- The FCC should provide E-rate support for internal connections to more schools and libraries.
- The FCC should give schools and libraries more flexibility to purchase the lowest-cost broadband solutions.
- The FCC should initiate a rulemaking to raise the cap on funding for E-rate each year to account for inflation.
- The FCC should initiate a rulemaking to streamline the E-rate application process.
- The FCC should collect and publish more specific, quantifiable and standardized data about applicants' use of E-rate funds.
- The FCC should work to make overall broadband-related expenses more cost-efficient within the E-rate program.
- Congress should consider amending the Communications Act to help Tribal libraries overcome barriers to E-rate eligibility arising from state laws.
- The FCC should initiate a rulemaking to fund wireless connectivity to portable learning devices. Students and educators should be allowed to take these devices off campus so they can continue learning outside school hours.
- The FCC should award some E-rate funds competitively to programs that best incorporate broadband connectivity into the educational experience.
- Congress should consider providing additional public funds to connect all public community colleges with high-speed broadband and maintain that connectivity.

11.1. Supporting and Promoting Online Learning

Broadband breaks down traditional barriers so that teaching and learning happen in new ways.

A student attending a rural school that does not offer an Advanced Placement (AP) calculus course can receive instruction online from a teacher in a different part of the state or even the country. That teacher, who is online because of her passion for the subject and because of her demonstrated ability to teach it, might not only provide lectures but may also use instant messaging and e-mail to communicate with the student. The teacher also might steer the student toward interactive tools that let students practice on their own. And the teacher might even pique the student's curiosity by using video showing how calculus applies to real-world examples such as a major league baseball player hitting a home run or how Isaac Newton developed calculus to understand gravity and the motion of the planets.

A student with a strong interest in Roman history might take an online class that includes video of an archaeologist demonstrating Roman glassmaking techniques. Outside of school hours, the student might monitor a blog the archaeologist writes while working on a dig and might e-mail the archaeologist questions and comments.

As these examples illustrate, broadband offers tremendous potential to improve education. Thanks in large part to the $2.25 billion per year in support provided by the E-rate program, virtually every school in the country has Internet access. However, computer and Internet access alone do not produce greater student achievement.[1032] Access needs to be combined with appropriate online learning content, systems and teacher training and support.[1033]

Carnegie Mellon University's Open Learning Initiative has shown that online learning, when "blended" with in-person instruction, can dramatically reduce the time required to learn a subject while greatly increasing course completion rates (see Exhibit 11-B).[1034]

There is strong evidence that online learning classes do not sacrifice quality of instruction for convenience and efficiency. For example, students attending Florida Virtual Schools (FLVS) earned higher AP scores and outscored the state's standardized assessment average by more than 15 percentage points in grades 6 through 10 (see Exhibit 11-C). [1035]

Students at Oregon Connections Academy met or exceeded state achievement averages,[1036] and students in the Florida Virtual Academy (unrelated to FLVS) have consistently outscored state test averages.[1037] In its first year, the Missouri Virtual Instruction Program showed significantly improved achievement for its students compared with the same students' achievement in the same subject the previous year; greater percentages of these students scored 3 or higher on AP exams than their peers.[1038]

Some school districts are finding that online systems can help with high dropout rates as well.[1039] Aldine Independent School District in Texas was able to reach at-risk students and get them to take classes online that earned school credit. Salem-Keizer School District in Oregon has re-enrolled more than 50% of dropouts and at-risk students through its online Bridge Program annually. At FLVS, 20% of the program's students enrolled to earn remedial credit. The passing rate of students taking makeup courses was 90%.[1040] In addition to dropout prevention, online systems provide flexibility to students who cannot be in school for health, child care, work or other reasons.[1041]

Teachers also benefit from online professional learning communities, lesson development websites and certified professional development opportunities. This allows them to fulfill their learning requirements in more flexible and diverse ways. A 2005 Texas study found the Online Post-Baccalaureate Program was just as successful as traditional teacher preparation programs and was more successful in attracting more diverse candidates in terms of race and gender. It also was more successful in recruiting science and math teachers.[1042]

But there are still major barriers to realizing the full potential of online learning:

- There is a limited pool of high-quality digital content that is easily found, bought, accessed and combined with other content to allow teachers to customize classroom materials to their students' needs.
- Students often have trouble obtaining course credit for online classes, and teachers licensed in one state may not be able to teach online courses in another.[1043]
- Students and teachers may lack the digital literacy skills necessary to make use of broadband tools. [1044]

The following recommendations, which expand digital content and online learning systems and promote digital literacy, will help address these barriers.

Expanding Digital Educational Content

The federal government can address the first barrier through three steps. First, it should define and adopt standards for finding and sharing digital educational content as well as licensing educational material for digital use. Teachers, students and other users should be able to easily find, purchase, access and combine any digital resources meeting the standards. Second, government should take steps to create a pool of digital educational resources meeting the U.S. Department of Education's standards. Third, government should encourage authors and private sector organizations to contribute their material within these standards.

RECOMMENDATION 11.1: The U.S. Department of Education, with support from the National institute of standards and Technology (NIST) and the Federal communications commission (FCC), should establish standards to be adopted by the federal government for locating, sharing and licensing digital educational content by March 2011.

As with the music industry[1045] and, increasingly, with video[1046] and books,[1047] broadband can generate new models for creation, publication and distribution of educational resources. Greater flexibility in the way content can be accessed can have a direct impact in the classroom. For example, it allows for differentiated instruction that can help students with variable levels of subject-area mastery by providing more tailored learning opportunities.[1048] A strong reader can be given more challenging material rather than wait while the rest of the class catches up. A weaker reader can be given material more appropriate to his level without holding back the rest of the class. Teachers can more easily select materials that fit the specific needs of different students. Digital content standards can help make that possible by offering a much wider choice of content than typically found in traditional printed curricular materials.

While digital content is available currently, there are significant challenges to finding, buying and integrating it into lessons. Content is not catalogued and indexed in a way that makes it easy for users to search. It is also hard for teachers to find content that is most relevant and suitable for their students. Even if one finds the right content, accessing it in a format that can be used with other digital resources is often difficult or impossible. And if the desired content is for sale, the problem is even harder because online payment and licensing systems often do not permit content to be combined. These three problems—finding, sharing and license compatibility— are the major barriers to a more efficient and effective digital educational content marketplace. These barriers apply to organizations that want to assemble diverse digital content into materials for teachers to use, as well as to teachers who want to assemble digital content on their own.

Digital content standards will make it possible for teachers, students and other users to locate the content they need, access it under the appropriate licensing terms and conditions, combine it with other content and publish it. This way, a teacher preparing a presentation on greenhouse gas emissions could easily find and combine National Aeronautics and Space Administration (NASA) pictures and videos on the impact of global warming on the polar ice caps with U.S. Department of Energy (DOE) graphs on fossil fuel consumption and a text-book chapter on clean energy sources.

The U.S. Department of Education should select standards for digital educational content after consulting with other government agencies, the educational community and the private sector. Once the standards are selected, the federal government should ensure all educational content it develops or sponsors is compatible with those standards. The following recommendation lays out specific steps the U.S. Department of Education can take to achieve this.

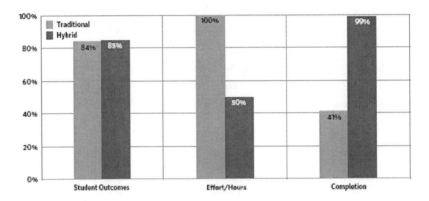

Exhibit 11-B. Carnegie Mellon Online Learning Initiative

RECOMMENDATION 11.2: The federal government should increase the supply of digital educational content available online that is compatible with standards established by the U.S. Department of Education.

- **The Executive Branch should make digital educational resources they own available online in a format compatible with the standards defined in Recommendation 11.1.**
- **Whenever possible, federal investments in digital educational content should be made available under licenses that permit free access and derivative commercial use and should be compatible with the standards defined in Recommendation 11.1.**
- **The U.S. Department of Education should encourage vendors that sell paper-based educational materials to sell digital versions or provide digital rights independent of rights on printed materials; whenever possible this content should be aligned with the standards defined in Recommendation 11.1.**

Many federal agencies own and develop new educational content. Making this content available online—in accordance with standards that allow for discovery, sharing and license compatibility—has two effects. It benefits end-users as it makes it easier for them to use the content. And it may encourage third parties such as universities, publishers and individuals to ensure the digital resources they own and produce comply with the same standards.

Millions of digital learning resources already are available under open and commercial licenses. Publishers of digital content include NASA, DOE, the Corporation for Public Broadcasting, universities nationwide,[1049] large publishing houses and authors.[1050] By providing greater access to a broad set of educational content, the federal government can give teachers and schools more tools to address their instructional challenges. This also can create business opportunities for companies to develop new educational solutions without the costs of re-creating educational content that already exists.

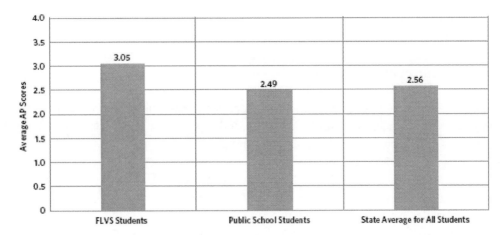

Exhibit 11-C. Florida Virtual Schools Students Taking Distance- Learning Courses Get Higher AP Scores

In addition, the U.S. Department of Education should provide grants and other incentives to vendors to offer their materials in digital formats compatible with the standards it adopts. The ultimate goal of such incentives is to provide more choice for customers and a more competitive market. The Department could use incentives and other strategies to help identify and make available the highest-quality and most relevant digital content to educators so that teachers can find what they need with less effort and have a greater impact in the classroom.

RECOMMENDATION 11.3: The U.S. Department of Education should periodically reexamine the digital data and interoperability standards it adopts to ensure that they are consistent with the needs and practices of the educational community, including local, state, and non-profit educational agencies and the private sector.

Recommendation 11.2 above could lead to the creation of a large enough pool of digital educational content to catalyze the private sector to adopt the same set of standards or standards that are compatible with those chosen by the federal government. Whether or not this will in fact occur is not certain. Because of the quickly changing nature of this space, it is also possible that in the future the private sector will develop and adopt standards that are fundamentally different from those chosen by the U.S. Department of Education in the near term.

Therefore, in addition to evolving its standards definitions and implementations to take into account incremental market and technology changes, the U.S. Department of Education should set a specific timeline to re-examine its overall choice for digital educational content (e.g., every 5 years). This reexamination should take into account both the success and effectiveness of the chosen standards and the evolution of digital educational content in broader contexts such as local, state, non-profit and commercial content.

RECOMMENDATION 11.4: Congress should consider taking legislative action to encourage copyright holders to grant educational digital rights of use, without prejudicing their other rights.

New broadband-enabled solutions are transforming how teachers and students use content and media. But copyright law must keep pace as new technologies and media are developed. In part due to a lack of clarity regarding what uses of copyrighted works are permissible, current doctrine may have the effect of limiting beneficial uses of copyrighted material for educational purposes, particularly with respect to digital content and online learning. In addition, it is often difficult to identify rights holders and obtain necessary permissions. As a result, new works and great works alike may be inaccessible to teachers and students. For instance, a film containing archival and documentary footage of Martin Luther King, Jr. and the struggle to end segregation could no longer be shown or distributed because of the expense and legal complications of license renewals related to "orphan works" (copyrighted works whose owners are difficult or impossible to identify).[1051] Teachers seeking to use Beatles lyrics to promote literacy, employing music as a cultural bridge, could not afford the $3,000 licensing fee charged by the rights holders. [1052] Text-to-speech features for the Amazon Kindle e-book reader were shut off because of a copyright dispute–While both parties to the dispute raised legitimate concerns, several universities chose not to provide the device to students. That, in turn, slowed the adoption of lower-cost e-textbooks and eliminated a useful tool for the visually impaired. [1053] Penalties for copyright infringement can be substantial,[1054] but the boundaries between permissible and impermissible uses of copyrighted works in educational contexts—particularly with respect to digital content and online learning—are not always clear. That produces a chilling effect on teachers, schools, and school districts, which limits the use of cultural works for educational purposes.

Increasing voluntary digital content contributions to education from all sectors can help advance online learning and provide new, more relevant information to students at virtually no cost to content providers. Congress should consider ways for educators to interact with their students using new educational content contributed by the public in the following ways:

- *Update TEACH Act.* Congress could consider updating the TEACH Act[1055] to better allow educators and students to use content for educational purposes in distance and online learning environments without prejudicing the other rights of copyright holders.
- *New Copyright Notice.* Congress could consider directing the Register of Copyrights to create additional copyright notices to allow copyright owners to authorize certain educational uses while reserving their other rights (see Exhibit 11-D).
- *Facilitate Licensing.* Congress could consider providing a statutory framework to facilitate identification of copyright holders and securing of permissions in an efficient and cost-effective way, while retaining existing protections for educational uses without exceeding permissible exceptions and limitations under copyright law.

Expanding Online Learning Systems

Effective broadband-based solutions exist. But they often are deployed only in limited ways for various reasons, including regulatory barriers, market forces, limited resources and capacity constraints. Many promising ideas and applications have been developed in ways that do not foster wide-scale use and adoption or integration into the classroom. The following recommendations propose steps to bring online learning opportunities to scale.

RECOMMENDATION 11.5: State accreditation organizations should change kindergarten through twelfth grade (K–12) and post-secondary course accreditation and teacher certification requirements to allow students to take more courses for credit online and permit more online instruction across state lines.

Exhibit 11-D. Proposed Copyright Notice Permitting Free Educational Use

Educational opportunities in the United States are distributed inequitably, usually because of unequal access to high-quality teachers and curricula.[1056] Online learning can help reduce such disparities.

In a survey of more than 10,000 school districts, 70% of respondents saw distance learning[1057] as important for delivering courses not otherwise available in their schools; 60% cited AP courses. Forty percent cited distance learning as a way to provide certified teachers when not enough are available for face-to-face instruction.[1058] Rural and high-poverty schools often have difficulty placing highly qualified teachers in every classroom.[1059] Rural districts, in particular, strongly identify distance learning as important for meeting the needs of their students, who do not always have access to specialized teachers.[1060] These schools, as well as charter and small schools, have difficulty affording teachers for advanced classes because of limited budgets and programming flexibility.[1061]

Despite the benefits of distance learning, students often have trouble obtaining course credit for online classes. Also, teachers licensed in one state may not be able to teach online courses in another.[1062] Although many states and districts offer make-up courses online, very few virtual schools are able to grant high school diplomas.[1063]

It is unusual for a teacher certified in one state to be allowed to teach in another without recertification. If a teacher experienced in a specific subject is available in one state but the student is enrolled in a different state, current regulations can make it difficult and sometimes impossible for the student to obtain course credit. Additionally, many states have course hour requirements that make it challenging to obtain course credit from online solutions that do not track "seat time" in the same way as traditional classes.

While states need to change their requirements, the U.S. Department of Education should help states work together to achieve the national goal of improving online education opportunities.

RECOMMENDATION 11.6: The U.S. Department of Education and other federal agencies should provide support and funding for research and development of online learning systems.

Online learning systems too often are deployed without effective research and development strategies. Moreover, designs are often not improved over time based on quantitative data.[1064] Because online learning can take place "anytime, anywhere," research has proved to be more difficult than for in-class instruction.[1065] The federal government can help by supporting, requiring and publishing data on the effective—and ineffective—aspects of online learning systems.

As online learning systems are deployed, research must be designed to measure their effectiveness—including "real- time, interaction-level data on how [students] are learning to inform further course revisions and improvements."[1066] The U.S. Department of Education and state governments can play a key role in this process by using field research and other data to highlight the most promising systems.

RECOMMENDATION 11.7: The U.S. Department of Education should consider investments in open licensed and public domain software alongside traditionally licensed solutions for online learning solutions, while taking into account the long-term effects on the marketplace.

Cost is a significant problem for online learning solutions: Utah's state government said that it "lack[s] affordable digital asset management systems that will be able to take full advantage of public repositories of information such as that made available from the PBS Digital Learning Library and the vast treasure trove of online content yet to be harvested from other public repositories like the National Archives and the Smithsonian Institution."[1067] Traditionally, licensed commercial products can cost 10–13% more in total cost of ownership than open-source equivalents, while delivering equivalent capability.[1068] Although adopting open-source software has unique risks, it can also offer significant benefits when implemented appropriately.

Some federal and state agencies have already found open- source software to be cost-effective across a wide array of applications. The Department of Defense determined in 2006 that it was inadvertently increasing its own software costs "by not enabling internal distribution" of open-source technologies.[1069] By funding development of innovative educational software applications under open-source licenses, the U.S. Department of Education may, in some cases, be able to accelerate the deployment of new technologies until they are mature enough to be resold by the educational vendor community.

Where suitable commercial online learning products are already available, it may be cheaper to buy product licenses rather than develop new open licensed solutions. However, open licensed investments can offer an additional strategy that can be pursued alongside licensing to strengthen the solutions available to the educational market. Ensuring that private capital continues to enter the educational online learning market needs to be an important consideration when the federal government considers open licensing strategies.

RECOMMENDATION 11.8: The U.S. Department of Education should establish a program to fund development of innovative broadband-enabled online learning solutions.

Currently, the educational technology market suffers from "a classic market failure... that discourages private industry from heavily investing in basic research to exploit emerging information technologies for learning. . . This situation requires a federal research investment to do for learning what the National Science Foundation does for science, the National Institutes of Health does for health and what the Defense Advanced Research Projects Agency (DARPA) does for defense." [1070] Education markets, however, are "notoriously difficult to enter [because] they are highly fragmented and often highly political."[1071]

Government investment in other sectors has helped fill gaps in private investment.[1072] For example, federal funding for research in broadband technologies has encouraged numerous innovations, creating billions of dollars of economic value.[1073]

Several examples exist of government funding of innovation in education. The American Graduation Initiative bill proposes $50 million over 10 years to finance an Online Skills Laboratory (OSL) to develop innovative learning solutions for Community Colleges. OSL's proposed focus on solutions that are free for use and resale will help ensure that the innovations that emerge can be used widely. The U.S. Department of Education's Race to the Top and Investing in Innovation funds are also good examples. But these programs have limited funding cycles. Attention and funding must be given over an extended period to ensure that the best ideas, products and businesses survive to become marketable and sustainable.

Establishing such an "ARPA-ED" [1074] educational broadband investment fund with a longer lifetime—eight years, for example—to make seed loans and grants to early-stage education companies or nonprofits can help stimulate sector-wide progress.

Promoting Digital Literacy

In an increasingly digital world, literacy must be defined more broadly to include fluency in digital skills and information. Digital literacy is "the ability to find, evaluate, utilize, and create information using digital technology."[1075] Additional skills include "the ability to read and interpret media (text, sound, images), to reproduce data and images through digital manipulation and to evaluate and apply new knowledge gained from digital environments."[1076] It can include the ability to analyze and reflect critically on digital media.[1077] Digital citizenship and safety are often included in definitions of digital literacy as well. A detailed consideration of digital literacy can be found in Chapter 9 of this plan. The following recommendations address strategies to promote digital literacy for educators and students.

RECOMMENDATION 11.9: State education systems should include digital literacy standards, curricula and assessments in their English Language Arts and other programs, as well as adopt online digital literacy and programs targeting STEM.

Digital literacy skills are required to take full advantage of online learning systems[1078] and future job opportunities. But students and teachers often lack such skills.[1079] While today's students may be competent with some technology, they are far from expert when it comes to locating and using information.[1080] Internet skill levels and usage rates among young people in the European Union now exceed those of their peers in the United States.[1081]

Many U.S. Students can handle computer keyboards and wireless devices, but digital literacy involves more than the ability to use a device. Students must be able to analyze problems so they can determine what information is needed to perform an academic or work task; access, assimilate, organize and analyze the information; interpret the information; conduct research; and effectively communicate their understanding and interpretation of the information to others.[1082] Integrating digital literacy into existing subject areas such as English Language Arts allows for these skills to be used and developed in a practical manner, without taking time away from other subjects by creating stand-alone courses. Students must also understand their ethical responsibilities online and know how to stay safe while using advanced broadband technologies.[1083] To succeed in the 21st century workplace, students must be digitally proficient at developing, advancing and applying their own knowledge and skills within virtually any field or profession.[1084]

BOX 11-1. ONLINE LEARNING CAN SUPPORT INVESTMENT IN STEM

Expertise in STEM will be critical to maintaining the United States' competitive edge in the 21st century.[1085] A critical shortage of highly qualified math and science teachers, particularly in low- income urban school districts and rural districts, threatens this competitive edge.[1086] Providing access to more online learning systems, coursework and materials in STEM can improve opportunities for students who are interested in working in these areas but lack local, high-quality learning opportunities.[1087] The Executive Office of the President recently announced a $250 million public-private investment for STEM teacher recruitment, professional development and the use of innovative teaching methods such as online learning. This is an excellent example of the kind of investment that should be made in this area.[1088] In addition, improved online solutions for professional development of teachers can help train new teachers and give existing teachers new techniques and resources for instruction in these fields.

RECOMMENDATION 11.10: The U.S. Department of Education should provide additional grant funding to help schools train teachers in digital literacy and programs targeting STEM. States should expand digital literacy requirements and training programs for teachers.

Achieving digital literacy goals for students means teachers also must be digitally literate (see Box 11-1). While teacher use of technology continues to grow, most teachers still do not use technology in their classrooms for many key activities.[1089] Teachers without digital literacy skills find it difficult to incorporate online learning solutions into instruction. Similarly, it is hard for students who lack such skills to engage with the systems to learn.[1090]

Teachers report that teaching online requires different skills than teaching in a bricks-and-mortar classroom.[1091] Students also need training in online learning methods. Consequently, teachers need training both as online instructors and in teaching methods that combine online and face-to-face learning.[1092] Online courses at the secondary level often serve younger-than-average students seeking access to accelerated courses in math or science that are not available in their regular schools. Online courses also serve older-than-average students needing a slower pace and more individualized attention.[1093] This variability in students' skills, combined with the geographical distribution that occurs in an online environment, provides additional challenges for which teachers must prepare.

11.2. Unlocking the Power of Data and Improving Transparency

Ideally, a teacher would have real-time access to accurate information about each student's mastery of skills, course grades, test scores and progress over time. Other pertinent information would include the student's behavior and learning style, his or her prior experiences in school and more. As students transfer among multiple classrooms during the year—something more likely to happen with at-risk children—the same information would be available as soon as the child walks through the door. In addition, if an issue arose that was outside a teacher's experience—for instance, providing alternative teaching strategies for an individual student—the teacher would have instantaneous access to online information about the issue and, perhaps, to experts and colleagues who could offer advice.

In addition to benefiting individual students and teachers, the creation of a large-scale pool of electronic educational records could potentially transform education. Anonymized records with detailed data on schools, educators and students would allow educators to determine in a fact-based fashion what works and when, and what the actual costs and benefits are of different practices. It would allow researchers to learn from the best practices and brightest ideas of every great teacher and principal in America. It would help educators determine when improved educational outcomes are a consequence of practices and techniques that are transferable to different contexts or due to factors not directly associated with educational practices.

At the moment, however, schools run on a patchwork of proprietary data systems that make sharing meaningful information about students slow and difficult. Disjointed administrative systems and processes currently keep schools, school systems, colleges and universities from conducting fast, efficient transfers of student data and related information.[1094] Consequently, teachers often have only bare-bones information about their students. "Only 37 percent of all teachers reported having electronic access to achievement data for the students in their classrooms in 2007."[1095] This results in a situation where "a significant proportion of teachers still do not have access to the data necessary for making instructional decisions."[1096] Any design of electronic educational records should account for parent and student privacy and rights to control their information, as well as the need for schools and researchers to share data.

Schools suffer from other data issues, too. They lack adequate market data about vendors, products and services, making purchases of technology and resources inefficient.[1097] The difficulty in obtaining overall market data means federal and state policies are not always informed by up-to-date information about what products and services are in use, which product categories are growing quickly and where rapid turnover in product choices might indicate underlying problems that policy could address.

The recommendations that follow address a number of the barriers preventing the free and efficient flow of information in education.

RECOMMENDATION 11.11: The U.S. Department of Education should encourage the adoption of standards for electronic educational records.

- **The U.S. Department of Education should support and accelerate the adoption of electronic educational records capability among states and local education**

- **agencies. it should also set standards for sharing this information so data can be transferred across states.**
- **The U.S. Department of Education should support any secure authentication strategy developed by the Federal Chief Information Officer that permits private, decentralized identification of educational agencies, students and their data records.**
- **The U.S. Department of Education should recommend to congress updates to student data privacy and protection laws that would improve online educational services.**

The health care and education sectors face similar problems: Just as educators lack important information about students' histories, doctors and nurses are often in the dark about the needs of new patients who arrive for treatment for the first time. These patients may have long, complicated histories of symptoms and treatments, many of which may not be readily apparent without careful interview and diagnosis. And the risks of missing an important issue are severe. The federal government is making significant investments in electronic health records (see Chapter 10). [1098] The federal government should also encourage development of electronic educational records to allow schools to support each student with a more complete digital picture.

Information in an electronic educational record could include student demographic and academic information as well as course history, student work, attendance and health data. Electronic educational records also could include information about teachers, schools, curriculum and other administrative data. Currently, these data often are stored in a variety of systems across a school or district and sometimes are available only on paper.

Data stored in these systems typically cannot be transferred from one system to another. This means it is expensive and time-consuming to look at all the different data together. Consequently, it can be difficult or impossible to analyze data for trends about what kind of instruction seems to be producing the best results. The inability to share data in a standardized form also makes it hard to identify students requiring special attention, especially those who change schools frequently.

Complete pictures of student performance need to be available to teachers, principals, districts, states, the federal government, research communities and colleges and universities.[1099] More effective tools and standards are needed to create a national network of data systems to manage and transfer data between organizations while maintaining student privacy.

The U.S. Department of Education, along with a number of states, independent standards groups and other organizations, have been working toward developing educational data-sharing solutions for more than a decade. [1100] The U.S. Department of Education is currently working on a National Educational Data Model, which is a critical step toward data sharing and interoperability. The Schools Interoperability Framework Association, IMS Global Learning Consortium and others continue to advance important technical standards. Numerous components remain undeveloped. And many of the existing incentives for local education agencies and states to adopt electronic educational records are insufficient to justify the cost and risk associated with implementation. A more comprehensive solution is required. The U.S. Department of Education is positioned to convene the necessary stakeholders to develop an effective national solution that accommodates the different needs of the educational agencies across the country.

The federal government needs to:

- Develop standards for electronic educational records and the ability to share this information through interoperability.
- Encourage state and local adoption of electronic education records consistent with these standards.
- Integrate digital authentication.
- Strengthen and modernize privacy and protection laws.

Working toward the goal of national educational data sharing, the U.S. Department of Education should convene stakeholders to adopt the standards by implementing them in ways that make it easier for schools to satisfy reporting requirements or by funding projects that help vendors test and implement the standards in their products.

Privacy and data protection laws for students and their families need to be modernized to reap the full benefit of improved information flow about student performance while still fully protecting student data. For example, organizations offer tutoring and supplemental services to students, but the legal status of the data they collect is unclear. Issues include whether parents and regulators have the same rights to the data as they have with school records. A relatively small change in the law to allow parents to combine data from outside sources with school data would provide a richer picture of students' learning needs so all providers can support them effectively. There may also be cases in which fine-grained levels of privacy control are appropriate. For example, students should be able to select and share their best work with other educational institutions, the military or future employers from within their digital portfolios or other materials linked to electronic educational records.

RECOMMENDATION 11.12: The U.S. Department of Education should develop digital financial data transparency standards for education. It should collaborate with state and local education agencies to encourage adoption and develop incentives for the use of these standards.

The public education system is highly decentralized, with total annual spending of hundreds of billions of dollars.[1101] Escalating expenditures in education have not resulted in improvement in student gains.[1102] Public education finances are a matter of public record. But it is difficult—if not impossible—to aggregate this information because it is stored in a distributed manner across thousands of county, district and regional administrative agencies. As a result, decisions about how to invest resources in education are often made without the benefit of understanding what investments have the greatest impact.

The benefits of improving access to these financial data over the Internet could be significant. State and local education agencies, academic researchers and others could more easily gather and analyze financial data to inform resource allocation decisions at the school, district, state and national levels, as well as research and policy questions about the educational impact of financial decisions. In addition, the availability of school expenditure data in machine-readable format may motivate the development of new applications and tools for school communities, districts and other support organizations to help them manage finances more effectively.

In some circumstances, making financial information— including product pricing— easier to access, compare and analyze can lead to tacit price collusion among competing providers and to overall higher prices.[1103] Delaying publication of these data, or aggregating them in ways that still allow meaningful and actionable tracking and comparison, could help reduce the chances that collusion will occur while still providing the benefits of making financial data more accessible. In developing standards and procedures for collecting and sharing educational financial data in digital form, the U.S. Department of Education should determine the appropriate level of aggregation for financial data collection[1104] and amount of time that should elapse between expenditure and publication, based on trends in market pricing.

RECOMMENDATION 11.13: The U.S. Department of Education should provide a simple Request for Proposal (RFP) online "broadcast" service where vendors can register to receive RFP notifications from local or state educational agencies within various product categories.

In addition to financial data transparency standards for education, the federal government can provide RFP notification services—similar to RSS feeds on the traditional Internet— where vendors could register to receive notifications of new RFPs and where local educational agencies (LEAs) could transmit their RFPs when they want to receive maximum exposure and bidding for a purchasing contract. [1105] This would make it easier for LEAs to find vendors with products or services they want to purchase. Past RFPs could be stored in a central repository as they are posted, providing useful historical data.

This product pricing information database and RFP broadcast service could together give many LEAs the opportunity to improve their ability to find and acquire the best product or service at the best price.

11.3. Modernizing Educational Broadband Infrastructure

Congress directed the FCC in 1996 to provide discounts on telecommunications and other services "to elementary schools, secondary schools and libraries for educational purposes"[1106] and authorized the FCC to support broadband services as part of that program.[1107] In response, the FCC developed the Schools and Libraries universal service support mechanism (also known as E-rate), which offers schools and libraries the chance to receive telecommunications services, Internet access and internal connections at a discounted rate. Thousands of schools and libraries have received billions of dollars since the E-rate program began 12 years ago.

As a result, Internet access is nearly universal in the nation's schools and libraries. Today, about 97% of public schools have access to the Internet.[1108] In classrooms, more and more students have access to Internet-connected computers, and 94% of instructional rooms have at least some Internet access.[1109] In addition, in-school use of the Internet and technology by students and teachers is growing rapidly.[1110] Public schools are connected to a district network 92% of the time. Types of connections from schools to districts include direct fiber (55%), T-1 or DS1 lines (26%) and wireless connections (16%).[1111]

Eighty-four percent of districts have district-wide networks. These districts have connections to Internet service provider(s) via T-1 or DS1 lines (42%), direct fiber (37%), wireless connections (18%), broadband cable (13%) and T-3 or DS3 lines (12%). Direct fiber connections are found in a larger percentage of city districts than in suburban, town or rural districts (62% versus 49%, 46% and 24%, respectively). More rural districts than city districts report T-1 or DS1 connections (51% versus 18%).[1112]

However, inadequate connectivity speeds and infrastructure issues are frequently reported,[1113] and bandwidth demands are projected to rise dramatically over the next few years.[1114] Moreover, there is pent-up demand in schools and communities for access to more broadband content and tools. This demand has not been met in part because applicants require greater bandwidth to use these tools; E-rate provisions do not always support the latest strategies for deploying broadband networks (which have evolved significantly since 1996); the application process is cumbersome; and the E-rate program is oversubscribed.[1115]

Additionally, many schools will need significant upgrades to meet projected broadband bandwidth demands in the future.[1116] Online educational systems are rapidly taking learning outside the classroom, creating a potential situation where students with access to broadband at home will have an even greater advantage over those students who can only access these resources at their public schools and libraries. The E-rate program needs to be updated and strengthened to ensure the rapid growth of online learning and data sharing in education are not limited by insufficient bandwidth.

This section recommends a number of changes to the E-rate program to address these challenges and the opportunities presented by new broadband-enabled technologies.

Three key goals should drive modernization of the E-rate program:

- Improve flexibility, deployment and use of infrastructure
- Improve program efficiency
- Foster innovation

Improve Flexibility, Deployment and Use of Infrastructure

RECOMMENDATION 11.14: The FCC should adopt its pending Notice of Proposed Rulemaking (NPRM) to remove barriers to off-hours community use of E-rate funded resources.

Currently, FCC rules require schools seeking support under the E-rate program to certify that services funded by E-rate "will be used solely for educational purposes."[1117] Schools are the site of many community activities. Use of school networks should be permitted when such activities do not interfere with the educational use of the network. Moreover, such access should be available free of charge because the school's excess capacity is otherwise unused. For example, adult job-training programs by community nonprofits are currently discouraged from using school network facilities because of network cost-sharing requirements—even though night-time programs would have no impact on students' network use. Schools should have the option to use their broadband resources in this way. Numerous organizations have cited the benefits these changes would bring to schools and communities.[1118]

The FCC recently approved an order to temporarily waive the rules dealing with these barriers, and it should adopt its pending NPRM to implement this recommendation.

RECOMMENDATION 11.15: The FCC should initiate a rule-making to set goals for minimum broadband connectivity for schools and libraries and prioritize funds accordingly.

All schools and libraries should provide sufficient broadband Internet access to their students and patrons. Setting minimum service goals for schools and libraries can help ensure adequate services to all communities. Minimum service goals for schools and libraries should not be set based on speed and quality of service alone. Factors including the number of peak active users as well as the type and quantity of broadband services consumed should be factored into defining these minimum service goals. The minimum service goals for schools and libraries should be adjusted regularly (every three to five years) because broadband bandwidth requirements change frequently. [1119]

Some schools and libraries need help making the transition to broadband. Data from the Universal Service Administrative Company (USAC) for FY2009 show the E-rate program received at least 200 requests for funding for dial-up access to the Internet. The FCC should investigate the reasons behind those funding requests. For example, the FCC should explore whether those schools and libraries lack access to the physical infrastructure necessary for broadband, whether it is simply an issue of funding and/or whether they lack the other resources, such as hardware, to make the best use of faster connectivity speeds. The FCC should also examine whether there are economic and social characteristics of the communities relevant to those 200 requests that are common. For example, do they tend to be communities with a large percentage of residents that are lower-income? The FCC should determine if there are other communities that may have similar characteristics and may need this funding.

Once the barriers to access and adoption have been identified, the FCC should develop strategies to address those barriers. For example, the FCC could give additional funding to or place a higher priority on schools and libraries using dial-up so that they could transition to broadband services. Such a plan could also be used to upgrade schools and libraries with low-tier broadband services.

RECOMMENDATION 11.16: The FCC should provide E-rate support for internal connections to more schools and libraries.

The E-rate program provides two "priorities" for discounting telecommunications services. Priority 1 is for external telecommunications connections and Priority 2 is for internal connections and wiring. While the E-rate program has always been able to fund all Priority 1 requests, Priority 2 funding requests have exceeded the E -rate program's cap in every year but one during the program's existence. In the past 10 years, only the neediest schools and libraries have received funding for the internal connections necessary to utilize increased broadband capacity, and the vast majority of requests for internal connections have gone unfunded. For example, in funding year 2007, applicants requested more than $2 billion for internal connections and internal connections maintenance but only $600 million was authorized for funding. Only schools or libraries at a discount level of 81% or higher received funding.

The result is that the vast majority of schools and libraries, while receiving discounts to help pay for broadband services, do not receive funds for the internal infrastructure necessary to utilize increased broadband capacity. In order to ensure that schools and libraries have robust broadband connections and the capability to deliver that capacity to classrooms and computer rooms, the FCC should develop ways that Priority 2 funding can be made available to more E-rate applicants.

RECOMMENDATION 11.17: The FCC should give schools and libraries more flexibility to purchase the lowest-cost broadband solutions.

Numerous E -rate applicants have provided input in the National Broadband Plan record, asserting that current E-rate rules do not always make it possible for them to acquire the lowest-cost, highest-value broadband available to them. Applicants should be able to acquire the lowest-cost broadband service, whether it is a fully leased or a mixed lease/own solution. For instance, the current ineligibility of dark fiber prevents applicants from pursuing lower-cost mixed lease/ own strategies for broadband infrastructure. Allowing funding for ownership or leasing of dark fiber and associated communications equipment could allow recipients to use locally underutilized commercial or governmental capacity to provide lower-cost, high-value broadband instead of leased services currently eligible for E-rate discounts. The FCC should reexamine specific E-rate rules that appear to limit the flexibility of applicants to craft the most cost-effective broadband solutions based on the types of broadband infrastructure, services and providers available in their geographic areas.

For example, the Mukilteo School District in the state of Washington reports that it currently uses dark fiber (without support from E -rate) at a cost of $0.0009/student/Mbps/month, which is 1/300th of the cost charged by a telecommunications carrier for a similar E-rate-approved service (costing $0.27/student/Mbps/month).[1120] The district indicates its costs include maintenance and service level agreements providing equivalent service to an E-rate-eligible service. Similarly, the Council of Great City Schools noted the flexibility to lease dark fiber from providers and own the related equipment would permit "the most cost-effective pricing" for schools and libraries.[1121] The state of Wisconsin said E-rate should prefer the most cost-effective solution.[1122] Other commenters expressed support for giving recipients more flexibility to use dark fiber as part of their broadband solutions. These organizations also said participants need more flexibility to reduce the overall cost of broadband, increase bandwidth and participate in local and regional networks using dark fiber.[1123]

The E-rate program already has a three-year amortization rule for "special construction" fees that E-rate applicants pay carriers that construct infrastructure to serve them. This is done to avoid front-loading the E-rate fund with expenses tied to such long-lasting projects. Extending this rule to situations where recipients receive funding for broadband solutions that may involve ownership or mixed lease/ownership of network components—such as the need to purchase equipment to light leased dark fiber—could reduce the short-term impact on the fund.

Improve Program Efficiency

RECOMMENDATION 11.18: The FCC should initiate a rule- making to raise the cap on funding for E-rate each year to account for inflation.

The current program's annual spending has fallen by about $650 million in inflation-adjusted dollars since the program began.[1124] It also is significantly oversubscribed, leaving most internal wiring requests unmet each year. Annual funding applications consistently have exceeded the cap by nearly a two-to-one margin. Some applicants do not apply for internal wiring (Priority 2) funding because they know from experience the cap is reached before many Priority 2 requests are funded.[1125] The E-rate program should be indexed to the inflation rate to prevent continued depreciation.[1126]

RECOMMENDATION 11.19: The FCC should initiate a rule- making to streamline the E -rate application process.

The FCC has reduced administrative burdens on applicants over the past several years. However, procedural complexities still exist, sometimes resulting in applicant mistakes and the imposition of unnecessary administrative costs. These complexities also may deter eligible entities from even applying for funds in the first place. The FCC should continue to protect the E-rate program from waste, fraud and abuse. However, straightforward modifications to the program can improve the administration, allocation and disbursement of funds while still ensuring that funding is used for its intended purpose.

Some existing application requirements may be unduly burdensome and also may result in applicants duplicating their efforts in order to meet other federal or state requirements. The FCC can ease burdens on applicants for Priority 1 services that enter into multiyear contracts. Applications for small amounts could be streamlined with a simplified application similar to the "1040EZ" form the Internal Revenue Service makes available for some taxpayers. The FCC should also work with other relevant federal agencies, including the U.S. Department of Education and the Department of Agriculture, to streamline requirements between agencies and ensure that schools and libraries do not have to duplicate work because of uncoordinated deadlines or other requirements that differ only slightly. [1127]

RECOMMENDATION 11.20: The FCC should collect and publish more specific, quantifiable and standardized data about applicants' use of E -rate funds.

Currently, USAC obtains from applicants applying for E-rate funding certain basic information about their Internet connectivity but does not analyze the responses in the aggre-gate.[1128] As a result, the FCC lacks comprehensive knowledge of the different types or capacities of broadband services that are supported through the E-rate program. The collection of this type of information from E -rate program participants will enable the FCC to determine how the E-rate program can better meet applicants' needs. Therefore, the FCC should modify the relevant FCC forms to determine more accurately how schools and libraries connect to the Internet, their precise levels of connectivity and how they use broadband. The collection of this additional information will enable the FCC to continue to improve the management and design of the program as network technologies and uses change in the future.

RECOMMENDATION 11.21: The FCC should work to make overall broadband-related expenses more cost-efficient within the E -rate program.

The FCC should encourage schools and libraries to use state, regional, Tribal and local networks to increase school and library purchasing power.[1129] It should support the establishment of state, regional, Tribal and local networks through the E-rate program. In addition, better collaboration among state and federal programs, including the FCC's Rural Health Care Program, could reduce the potential waste of federal resources and maximize available federal funding for broadband-related projects.[1130] The FCC should explore creative solutions to aid schools and libraries in reducing their broadband-related costs so that they can purchase the maximum amount of broadband for their limited dollars. For example, the FCC could establish a website that facilitates an exchange of information among federal agencies, state networks and schools and libraries so that the state networks can provide consulting support and share best practices for efficient technological solutions for broadband needs. The same website could also allow state networks to collaborate and share information with federal agencies so that federal funding for broadband projects can be better utilized.[1131]

RECOMMENDATION 11.22: Congress should consider amending the communications Act to help Tribal libraries overcome barriers to E-rate eligibility arising from state laws.

Current eligibility requirements for the E -rate program prevent Tribal libraries in some states from qualifying for E-rate funding.[1132] Under the Communications Act, a library can be eligible for E -rate funding only if it is eligible for assistance from a state library administrative agency under the Library Services and Technology Act (LSTA). LSTA has two types of library grants that primarily relate to governmental entities: one for states and one for federally recognized Tribes and organizations that primarily serve and represent Native Hawaiians. To be eligible for E -rate funds, a Tribal library must be eligible for state LSTA funds and not just Tribal LSTA funds. However, some states preclude Tribal libraries from being eligible to receive state LSTA funds, thus making Tribal libraries in those states ineligible for E-rate funding. Congress should consider amending the Act to allow Tribal libraries to become eligible for E-rate funding if they are eligible to receive funding from either a state library administrative agency or a Tribal government under the LSTA.[1133] The FCC should also explore ways to remove technical barriers that may prevent some Tribal libraries from receiving E-rate support.

Foster Innovation

RECOMMENDATION 11.23: The FCC should initiate a rule- making to fund wireless connectivity to portable learning devices. Students and educators should be allowed to take these devices off campus so they can continue learning outside school hours.

Online learning can occur anytime, anywhere. Research shows that home use of computers and broadband technologies for learning can be a significant factor in boosting math and reading achievement.[1134] Use of computers and broadband at home for educational purposes has also been shown to motivate students and to increase the relevance of content during school hours—ultimately improving student achievement.[1135]

E-rate should support online learning by providing wireless connectivity to portable learning devices so students[1136] can engage in learning while not at school. Restricting student access to network services while on school grounds is becoming increasingly indefensible given the new educational opportunities presented by cloud-based desktops, smartphones, tablet PCs, netbooks and other highly portable solutions. Demand for wireless services in education is rapidly growing, and students without off-campus access to online educational services will be increasingly left behind in terms of skills, experience and confidence in their online capabilities.

Where applicant-managed hardware can use wireless services off campus, E-rate should provide appropriate Priority 1 discounts for those services. Potentially high demand for this service should be accounted for in the program design to ensure equitable overall distribution of E-rate funds. For example, providing a limited amount of funding for wireless services within a pilot program could help determine demand levels and cost-effectiveness.[1137]

RECOMMENDATION 11.24: The FCC should award some E-rate funds competitively to programs that best incorporate broadband connectivity into the educational experience.

Competitive programs are an effective strategy in government and philanthropy to stimulate new ideas, reward the best applicants, spread new ideas and make efficient use of scarce resources. E-rate is designed to provide telecommunications services to all schools and libraries. It is also intended to ensure that advanced services are deployed and improved over time. By rewarding innovative ideas, the E-rate program can encourage more strategic integration of broadband into education by applicants as well as recognize and potentially spread best practices among applicants. Broadband-enabled solutions are demonstrating new pathways for innovation and research in education.[1138] According to Philip R. Regier, Dean of Arizona State University's Online and Extended Campus program, the education system is "at an inflection point in online education"[1139] with large increases in use and improvements in quality expected in the near future.

The U.S. Department of Education is encouraging similar innovation in education with its Race to the Top and Investing in Innovation programs. A competitive component to E-rate could foster similar innovative applications for use of broadband networks nationwide. Importantly, competitions should be designed to offer funding opportunities both to smaller institutions with fewer resources to develop competitive applications and larger institutions with the ability to undertake larger programs.

Providing Connectivity to Community Colleges

RECOMMENDATION 11.25: Congress should consider providing additional public funds to connect all public community colleges with high-speed broadband and maintain that connectivity.

Community colleges are anchor institutions for training a highly skilled 21st century workforce. Providing broadband connectivity to these institutions will help provide better services to students.[1140] As of 2007, according to the Integrated Postsecondary Education Data System, there were 1,138 public two-year institutions in the United States.[1141] These institutions operated an estimated 3,439 distinct campuses. Only 16% of these public community college campuses currently have highspeed broadband connections comparable to those of American research universities.[1142]

Access to high-quality broadband connectivity and innovative online technologies will allow community colleges to extend their reach even further. They can offer powerful learning opportunities to even broader audiences. With adequate funding and innovative technology development, community colleges can offer college credit for online courses for advanced high school students; offer specialized science and technology online learning experiences in subjects where there are too few specialized K–12 teachers; support adult students through personalized career and technical programs while working around the needs of their jobs and families; and extend continuing education programs by offering diverse, quality content to the public to foster job skills, community development and personal growth.

Community colleges with broadband connectivity and quality online instructional programs serve as learning and career development centers for the K-12 community and for local citizens. Community colleges also play integral roles in educating Americans about math and science and preparing students for their future careers as teachers. Forty percent of teachers have taken a math or science course at a community college, and 44% of science and engineering graduates attended a community college as part of their postsecondary education. Twenty percent of teachers begin their postsecondary education at community colleges. [1143]

The most recent Notice of Funding Availability from the Department of Commerce related to the Broadband Technology Opportunities Program created an opportunity for community colleges to obtain funding to upgrade their levels of connectivity. After such funding is determined, Congress should evaluate whether additional action is warranted for community colleges.

12. ENERGY AND THE ENVIRONMENT

America depends on reliable and affordable access to diverse sources of energy. The $1.2 trillion u.s. energy industry powers the rest of the economy, making possible a good quality of life and strong economic productivity.[1144]

U.S. prosperity and national security, as well as the health of the planet, require a national transition to a low-carbon economy and reduced dependence on foreign oil. Congress has demonstrated significant resolve in jump-starting this transition, devoting more than $80 billion in the American Recovery and Reinvestment Act of 2009 (Recovery Act) to clean energy and efficiency investments. [1145] Americans have mounted solar panels on their roofs, weatherized their homes, installed efficient light bulbs and traded their "clunkers" for vehicles that get higher gas mileage. But the U.S. economy still runs mostly on domestic fossil fuels and imported oil.

Broadband and advanced communications infrastructure will play an important role in achieving national goals of energy independence and efficiency. Broadband-connected smart homes and businesses will be able to automatically manage lights, thermostats and appliances to simultaneously maximize comfort and minimize customer bills. New companies will emerge to help manage energy use and environmental impact over the Internet, creating industries and jobs. Televisions, computers and other devices in the home will consume just a fraction of the power they use today, drawing energy only when needed. Large data centers, built and managed to leading energy efficiency standards, will be located near affordable and clean energy sources. Finally, broadband connectivity in vehicles will power the next generation of navigation, safety, information and efficiency applications while minimizing driver distraction. Next-generation safety systems will alert drivers to hazards, helping to avoid accidents and saving lives. In the process, broadband and information and communication technologies (ICT) can collectively prevent more than a billion metric tons of carbon emissions per year by 2020. [1146]

The path to reliable, affordable and clean energy will require ingenuity and hard work from legions of scientists, entrepreneurs and green-collar workers, as well as the participation of every American. Consumers and businesses will need easy access to information about the type, amount and price of energy to make informed decisions about their consumption. The price of electricity will also have to better reflect the cost of providing power, which can skyrocket during critically hot days.

Broadband alone cannot solve the country's energy and environmental challenges, but it will be an important part of the solution.

This chapter is divided into four sections. The first two focus on how broadband and advanced communications can make the greatest impact on energy and the environment: as the foundation of a smarter electric grid and as a platform for innovation in smart homes and buildings, especially if utilities unlock energy data. The third section highlights how industry and the federal government can improve the energy efficiency and environmental impact of ICT usage. The fourth explores how broadband and advanced communications can make transportation safer, cleaner and more efficient.

Recommendations

Integrate broadband into the Smart Grid

- As outlined in Chapter 16, the Federal Communications Commission (FCC) should start a proceeding to explore the reliability and resiliency of commercial broadband communications networks.
- States should reduce impediments and financial disincentives to using commercial service providers for Smart Grid communications.
- The North American Electric Reliability Corporation (NERC) should clarify its Critical Infrastructure Protection (CIP) security requirements.
- Congress should consider amending the Communications Act to enable utilities to use the proposed public safety 700 MHz wireless broadband network.

- The National Telecommunications and Information Administration (NTIA) and the FCC should continue their joint efforts to identify new uses for federal spectrum and should consider the requirements of the Smart Grid.
- The U.S. Department of Energy (DOE), in collaboration with the FCC, should study the communications requirements of electric utilities to inform federal Smart Grid policy.

Unleash innovation in smart homes and smart buildings

- States should require electric utilities to provide consumers access to, and control of, their own digital energy information, including real-time information from smart meters and historical consumption, price and bill data over theInternet. If states fail to develop reasonable policies over the next 18 months, Congress should consider national legislation to cover consumer privacy and the accessibility of energy data.
- The Federal Energy Regulatory Commission (FERC) should adopt consumer digital data accessibility and control standards as a model for states.
- DOE should consider consumer data accessibility policies when evaluating Smart Grid grant applications, report on the states' progress toward enacting consumer data accessibility and develop best practices guidance for states.
- The Rural Utilities Services (RUS) should make Smart Grid loans to rural electric cooperatives a priority, including integrated Smart Grid-broadband projects. RUS should favor Smart Grid projects from states and utilities with strong consumer data accessibility policies.

Accelerate sustainable ICT

- The FCC should start a proceeding to improve the energy efficiency and environmental impact of the communications industry.
- The federal government should take a leadership role in improving the energy efficiency of its data centers.

12.1. Broadband and the Smart Grid

The United States is undertaking a massive communications and information technology buildout to produce the Smart Grid, which the National Institute of Standards and Technology (NIST) defines as the "two-way flow of electricity and information to create an automated, widely distributed energy delivery network."[1147]

The vision is to build a modern grid that enables energy efficiency and the widespread use of both renewable power and plug-in electric vehicles, reducing the country's dependence on fossil fuels and foreign oil. This grid will intelligently detect problems and automatically route power around localized outages, making the energy system more resilient to natural disasters and terrorist attacks. It will keep bills low and minimize greenhouse gas emissions.

Realizing the promise of the Smart Grid will require the addition of two-way communications, sensors and software to the electrical system, both in the grid and in the home. Communications are fundamental to all aspects of the Smart Grid, including generation, transmission, distribution and consumption.

The Energy Independence and Security Act of 2007 (EISA) made modernizing the grid national policy, and the Recovery Act devoted $4.5 billion to accelerating standardization and deployment of the Smart Grid. The Electric Power Research Institute estimates that the U.S. will spend $165 billion over the next 20 years building the Smart Grid. [1148]

The Smart Grid is a national priority for several reasons. It will increase the reliability of the electric grid, more efficiently integrate renewable generation, reduce peak demand and support the widespread adoption of electric vehicles.

First, as the current patchwork grid has become more interconnected and complex, reliability has become more critical. Power blackouts cost the nation as much as $164 billion per year. [1149] The Smart Grid could prevent many blackouts by sensing problems and routing power around them (see the story of the 2003 blackout in Box 12-2).

Second, to combat climate change, national and state energy policies increasingly encourage the development of generation assets—such as solar, wind and nuclear—that emit fewer greenhouse gases. But renewable power can be intermittent; clouds can mask the sun and wind can stop blowing without warning. The country will need greater intelligence in the grid and viable energy-storage solutions in order to meaningfully displace fossil fuel generation. Renewable power and distributed generation will also drive the need for greater communication because they will transform the one-way power system into a sophisticated two-way system, where homes, vehicles and buildings sometimes draw power from the grid and sometimes contribute power to it. [1150] A recent study by the Pacific Northwest National Laboratory estimates the Smart Grid can reduce greenhouse gas emissions from electricity generation by as much as 12% by 2030, which is equivalent to taking 65 million of today's cars off the road. [1151]

12-1. UNITED STATES ENERGY FLOW (PETAJOULES, 2007)[1152]

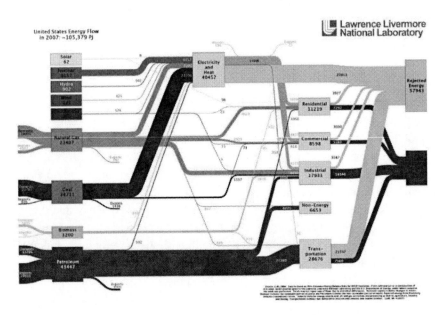

The national energy balance sheet reveals a number of pertinent facts. First, coal-fired power plants generate almost half of our electricity and are responsible for nearly two billion metric tons of greenhouse gas emissions per year—equivalent to the emissions of the entire transportation industry.[1153]

Greenhouse gas emissions from coal, and to a lesser extent natural gas and oil, explain why the electric power industry is the single largest contributor to U.S. greenhouse gas emissions.[1154]

Second, although there has been explosive growth in solar, wind and biomass power in recent years, renewable generation still provides a small amount of our generating capacity. Third, the current electricity system, from generation to end-user, wastes vast sums of energy; for example, a light bulb receives less than half of the energy contained in a piece of coal. Finally, the U.S. transportation sector is almost wholly reliant on oil, more than half of which is imported.

Third, it is important to shift energy usage away from the cripplingly expensive times of peak demand. To meet those peaks, utilities build and maintain power plants that only run for hours per year. In New England, for example, 15% of the total generating capacity is needed less than 1% of the time— fewer than 90 hours per year.[1155] As a result, state regulators are increasingly looking to change the structure of retail rates— which are mostly flat today—to time-varying or dynamic rates that better reflect the cost of supplying power. A smarter grid is necessary to communicate those prices to consumers and help them manage their energy use. According to a recent FERC report, dynamic pricing and better demand-side engagement can reduce peak demand by as much as 20% by 2019, limiting the need to build expensive new power plants.[1156]

Fourth, a smarter grid is necessary if America wants to lead in the shift toward vehicle electrification. Almost all of the global automakers are developing plug-in hybrid electric or full electric vehicles, and, if successful in the market, these vehicles have the potential to reduce U.S. dependence on foreign oil by half and decrease greenhouse gas emissions of the light- duty vehicle fleet by 27%.[1157] Without a Smart Grid, widespread adoption of electric vehicles would require the construction of many more power plants. A 2008 study illustrates the challenge: California's grid has enough spare capacity to charge a fleet of more than 10 million plug-in electric hybrids at night without requiring new plants. But if drivers plugged in the same 10 million vehicles at the end of the workday, California would require 10 gigawatts of new capacity (see Exhibit 12-A). According to a DOE study, the U.S. has enough existing capacity to power 73% of its light-duty vehicle fleet once a smarter grid is in place that can charge vehicles entirely at off-peak times.[1158]

Smart meters, which are located at customers' homes and provide two-way communications with their utility, will play a major role in the Smart Grid. FERC estimates that the number of smart meters deployed will rise from eight million today to 80 million in 2019.[1160]

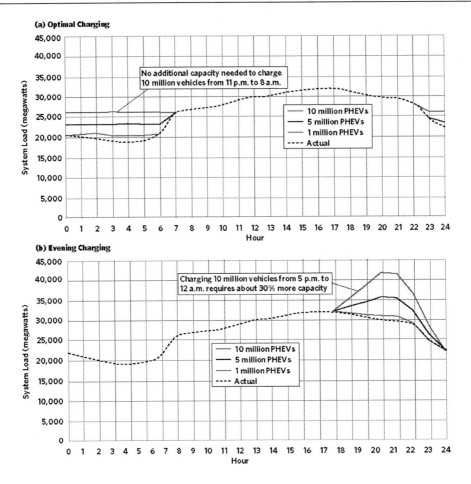

Exhibit 12-A. California Independent System Operator (ISO) System Load Profiles in Various Plug-in Hybrid Electric Vehicle (PHEV) Deployment Scenarios[1159]

Smart meters, however, are just one part of the effort to modernize the electric system. The Smart Grid also includes new and legacy applications in the generation, transmission and distribution systems, including Supervisory Control and Data Acquisition systems, outage management systems, energy management systems and a host of new sensing technologies, such as synchrophasors (see Box 12-2). These systems allow utilities to operate the grid more efficiently, safely and reliably. They also allow grid operators to detect, prevent and recover from faults, helping avoid blackouts. But they require communications networks capable of operation during and immediately following disasters.

Today, the more than 3,000 electric utilities in the United States use a variety of networks, including wired and wireless, licensed and unlicensed, private and commercial, fixed and mobile, broadband and narrowband. Traditionally, electric utilities build private networks to support applications with a high level of reliability, such as those for grid control and protection. These systems have operated separately from commercial networks, often utilizing privately owned, proprietary narrowband solutions.

BOX 12-2. THE 2003 NORTHEAST BLACKOUT AND SYNCHROPHASORS

On Aug. 14, 2003, a high- voltage power line in Ohio failed after contact with an overgrown tree. When a grid alarm system also failed, a cascading set of faults traveled throughout eight northeastern states and southeastern Canada over the next two hours, as transmission system operators tried to determine the cause and full extent of the problem. In total, more than 50 million people lost power, trapping some in elevators and leaving vulnerable populations at home without air conditioning.

According to Secretary of Energy Steven Chu, a smarter grid could have prevented the blackout, which cost the nation an estimated $6-10 billion.[1161]

A key finding of the U.S.- Canada Power System Outage Task Force was that network operators did not have the right data and tools in place to view, analyze and control grid events as they quickly deteriorated. First, each operator only had visibility in his or her own control area. The grid, however, is heavily interconnected across regions and so operators must be able to see the status of the grid beyond their area to make appropriate adjustments in response to grid events. Second, only limited real-time, time-coded, synchronized energy data was available in 2003, preventing operators from quickly seeing the cascading events even within their own areas.

Advanced grid sensors, called synchrophasors, would have given those grid operators sufficient visibility to prevent the spread of the blackout. Synchrophasors measure voltage, current and frequency 30 times or more per second, compared with once every four seconds for legacy systems. Given higher bandwidth and low latency requirements, these advanced sensors are often connected with utility fiber networks. Synchrophasors improve wide-area visibility and control, allowing grid operators to track real-time grid conditions, observe emerging problems and take actions to protect system reliability. The high granularity of the data can also facilitate: 1) better post-disturbance analysis, 2) improved system utilization, and 3) better analysis of the integration of renewable power into the grid.

Along with industry, the Recovery Act is funding the deployment of synchrophasors across the country's electric transmission system. The funds will help pay for the installation of nearly 900 synchrophasors, improving reliability, security and visibility of the entire electric transmission system.[1162] In the future, synchrophasors will extend throughout the distribution grid, transmitting data over wide-area broadband networks.[1163]

However, current narrowband solutions are not able to support the growing number of endpoints requiring connectivity in the modern electric grid,[1164] and many utilities believe that solutions using unlicensed spectrum will be suboptimal for mission-critical control applications.[1165]

The amount of data moving across Smart Grid networks is modest today but is expected to grow significantly because the number of devices, frequency of communications and complexity of data transferred are all expected to increase. [1166] Various parties have attempted to estimate bandwidth requirements; none expect existing narrowband communications will be sufficient. Sempra Energy has found that it will require "pervasive mobile coverage of at least 100 kbps to all utility assets and customer locations."[1167] Similarly, DTE Energy believes it will require connectivity of 200-500 kbps to support pole-mounted distribution devices.[1168] And, as Southern California Edison points out, "the history of new technology deployments shows that performance and bandwidth needs were underestimated at early stages."[1169]

Commercial networks are not available in all areas where utilities have assets and provide service.[1170] Commercial data networks are less commonly used for mission-critical control applications, in part because they have historically been unable to ensure service continuity during emergency situations, which is a fundamental requirement for utility control networks. The record indicates that commercial wireless data networks can become congested or may fail completely because of a lack of power backup or path redundancy.[1171]

In summary, the lack of a mission-critical wide-area broadband network capable of meeting the requirements of the Smart Grid threatens to delay its implementation.[1172]

The country should pursue three parallel paths. First, existing commercial mobile networks should be hardened to support mission-critical Smart Grid applications. Second, utilities should be able to share the public safety mobile broadband network for mission-critical communications. Third, utilities should be empowered to construct and operate their own mission-critical broadband networks. Each approach has significant benefits and tradeoffs, and what works in one geographic area or regulatory regime may not work as well in another. Rather than force a single solution, these recommendations will accelerate all three approaches.

RECOMMENDATION 12.1: As outlined in chapter 16, the Federal communications commission (FCC) should start a proceeding to explore the reliability and resiliency of commercial broadband communications networks.

Commercial broadband networks, and wireless broadband networks in particular, can serve more mission-critical and wide-area utility communications needs as service providers adopt measures to improve the reliability and resiliency of these networks during emergency scenarios. Because 97.8% of Americans are already covered by at least one 3G network,[1173] a hardened commercial wireless data network could serve as a core part of the Smart Grid.

The benefits of a more reliable commercial broadband network are much broader than enabling the Smart Grid alone. A more reliable network would also benefit homeland security, public safety, businesses and consumers, who are increasingly dependent on their broadband communications, including their mobile phones. Today, more than 22% of households in America do not subscribe to fixed-line telephone service.[1174]

RECOMMENDATION 12.2: States should reduce impediments and financial disincentives to using commercial service providers for smart Grid communications.

Commercial wireless networks are often suitable and widely used for many Smart Grid applications, particularly metering and routine sensing systems. In certain situations, compared with private networks, commercial networks may provide substantially similar network performance at an equal or lower total cost of ownership.[1175] A commercial network that can ensure service continuity would be capable of supporting additional mission-critical applications. However, many large utilities have economic disincentives to use commercial networks and may be making suboptimal choices. As rate-ofreturn regulated utilities, they typically earn guaranteed profits on the assets they deploy—including private communications networks—but only receive cost recovery if they use commercial networks.

Public utility commissions (PUCs) must ensure that utilities' incentives do not lead them to make suboptimal communications and technology decisions. State regulators should carefully evaluate a utility's network requirements and commercial network alternatives before authorizing a rate of return on private communications systems. Consistent with EISA,[1176] PUCs should also consider letting recurring network operating costs qualify for a rate of return similar to capitalized utility-built networks. California is currently considering this question.[1177]

In many states, electric utility incentives are still oriented toward deploying assets and selling more power, not selling less or cleaner power.[1178] This thorny structural problem is outside the scope of the National Broadband Plan, despite its explicit Congressional mandate to address energy efficiency. However, a national strategy to support the growth of the Smart Grid must recognize that many large electric utilities have inherent financial incentives to deploy regulator-approved communications systems but have mixed-to-poor incentives to use these systems to deliver energy more efficiently. There are meaningful exceptions: Box 12-3 illustrates an example of a U.S. utility working collaboratively with customers to reduce peak load and to encourage energy efficiency.

BOX 12-3. THE IDAHO POWER COMPANY: A CASE STUDY[1179]

The Idaho Power Company, which serves more than 485,000 customers in the state, has had some of the lowest electricity prices in the nation due to its heavy reliance upon cheap hydroelectric power. The impact of a statewide drought and the 2000-01 Western energy crisis led prices to spike tenfold, and the Idaho Public Utilities Commission put in place an aggressive set of energy efficiency programs to reduce price volatility and help lower customer bills.

The utility instituted a demand response and direct load control program, supported by broadband and other communications technologies, that compensates homeowners, farmers and businesses for reducing their electricity use during periods of peak demand. Homeowners receive a $7 credit if the utility can automatically cycle their air conditioners. Farmers, who require a significant amount of electricity to pump water to irrigate their fields, can earn rewards if they cut their irrigation time by up to 15 hours a week.

In addition, Idaho Power offers rebates for attic insulation, advertises to promote consumer-oriented energy efficiency products and runs energy-saving classes for customers. Since state regulators have decoupled the company's profits from how much energy it sells, the utility has new incentives to get its customers to reduce their energy use.

These measures have led to a 5.6% drop in the state's peak power demand and have saved more than 500,000 MWh of energy since 2002, equivalent to eliminating the energy used by 5,000 homes over the intervening eight years. In addition, some customers have seen reductions of as much as 30% in their electricity bills.

RECOMMENDATION 12.3: The North American Electric Reliability corporation (NERC) should clarify its critical infrastructure Protection (CIP) security requirements.

NERC, the organization under FERC's authority responsible for the reliability of the bulk power system, should revise its security requirements to provide utilities more explicit guidance about the use of commercial and other shared networks for critical communications. In future versions of the CIP standard, NERC should clarify whether such networks are suitable for grid control communications. NERC should also clarify how its CIP requirements will coexist with NIST's cybersecurity standards. The perceived ambiguity on CIP requirements appears to be slowing utility decision-making and stifling the deployment of some Smart Grid applications on commercial networks. [1180]

RECOMMENDATION 12.4: Congress should consider amending the communications Act to enable utilities to use the proposed public safety 700MHz wireless broadband network.

The wide-area network requirements of utilities are very similar to those of public safety agencies. Both require near- universal coverage and a resilient and redundant network, especially during emergencies. In a natural disaster or terrorist attack, clearing downed power lines, fixing natural gas leaks and getting power back to hospitals, transportation hubs, water treatment plants and homes are fundamental to protecting lives and property. Once deployed, a smarter grid and broadband-connected utility crews will greatly enhance the effectiveness of these activities.

Congress should consider amending the Communications Act to enable utilities to use the public safety wireless broadband network in the 700MHz band, subordinated to the communications of Section 337-defined public safety services. Jurisdictions that are licensees or lessees of the public safety 700MHz broadband spectrum should be allowed to enter into agreements with utilities on uses and priorities. At the sole discretion of the public safety licensee, utilities should also be able to purchase services on a public safety network, contribute capital funds and infrastructure or even be the operator of a joint network. These statutory changes should create more options for the construction and operation of a public safety wireless broadband network. Although the network will take years to build, carrying critical traffic from multiple users can help lower costs for all.

Several examples already exist of networks that are being shared successfully by public safety entities and utilities. SouthernLINC, a subsidiary of the Southern Company, provides commercial wireless service in the Southeast and voice communications for Southern Company itself. Because the network was built to very high reliability standards, almost a quarter of SouthernLINC's customers are public safety and other public agencies. Another example is the Nevada Shared Radio System, which is jointly operated by two Nevada utilities and the Nevada Department of Transportation (the Nevada State Patrol is also a customer). [1181]

RECOMMENDATION 12.5: The National Telecommunications and Information Administration (NTIA) and the FCC should continue their joint efforts to identify new uses for federal spectrum and should consider the requirements of the Smart Grid.

Many large utilities plan to build their own private wireless broadband networks to support their mission-critical Smart Grid applications.[1182] Traditionally, utilities have not participated in broadband spectrum auctions because the geographic boundaries and regulatory requirements of these licenses have been incompatible with utility business models and service territories.[1183] Utilities report they are limited by their lack of access to suitable wireless broadband spectrum[1184] and that lack of a nationwide band to build an interoperable Smart Grid will slow the nation's progress toward greater energy independence and energy efficiency.[1185] Several vendors do provide private wireless solutions in licensed spectrum, but in various bands, protocols and speeds.[1186]

Identifying a nationwide band in which Smart Grid networks could operate would speed deployment of a standardized and interoperable broadband Smart Grid.[1187] Establishing a nationwide band would also promote vendor competition and lower equipment costs.[1188]

NTIA and the FCC should specifically explore possibilities for coordination of Smart Grid use in appropriate federal bands. Any new broadband network built in the identified spectrum should be required to meet standards of interoperability, customer data accessibility, privacy and security. Use of this spectrum should not be mandated, so that legacy systems are not stranded and that commercial, other shared networks and unlicensed wireless networks can be used where appropriate.

RECOMMENDATION 12.6: The U.S. Department of Energy (DOE), in collaboration with the FCC, should study the communications requirements of electric utilities to inform Federal smart Grid policy.

Understanding the evolving communications requirements of electric utilities will help DOE develop informed Smart Grid policies for the nation. As an input to this plan, the FCC solicited public comment on Smart Grid technologies, and a number of utilities filed detailed responses. However, many utilities declined to comment, and others understandably declined to reveal confidential or sensitive information in public filings.

DOE, in collaboration with the FCC, should conduct a thorough study of the communications requirements of electric utilities, including, but not limited to, the requirements of the Smart Grid. Building upon the FCC's recent efforts, DOE should collect data about utilities' current and projected communications requirements, as well as the types of networks and communications services they use.

12.2. Unleashing Innovation in Smart Homes and Buildings

One of the most important and cost-effective ways to meet national energy goals is to encourage energy efficiency in homes and businesses—but end-users need better information in order to maximize energy and cost savings.

Today, most Americans receive an electricity bill—via paper or an electronically delivered PDF—12 times a year after the energy use occurs. They do not know the price of electricity, the source of the power or the amount of power needed to run each of their appliances. Most Americans know how much gasoline they need for a week's worth of commuting, yet almost no one knows how much electricity it takes to run a load of laundry, turn on an additional flat-screen television or cool a home an extra two degrees.

Smart meters help change this equation because they generate real-time data. In addition to their other operational capabilities such as automated meter reading and remote power monitoring, smart meters can record or transmit three types of information:

- Historical energy consumption data (e.g., "How much power did I use yesterday, last month and last winter?")
- Real-time data (e.g., "How much power am I using right now?")
- Price and demand response data (e.g., "What is the price of electricity right now?")

In dozens of consumer trials, Advanced Metering Infrastructure (AMI) technologies combined with time-based pricing tariffs have led to reductions of both peak demand and total energy consumption. A recent study of 15 utility pilots by the Brattle Group found that time-based or dynamic pricing of electricity resulted in a drop of peak demand between 3% and 20%, depending on how the pricing was set up. Adding technologies such as two-way programmable communicating thermostats, in-home energy displays and two-way load control systems drove the drop in peak demand to between 27% and 44%.[1189] When people see just how expensive electricity is when demand peaks on a hot summer day, they find ways to conserve energy or defer their usage. This not only saves consumers money, but also greatly cuts costs for the utilities, given that the plants brought on line to meet peak demand are easily the highest-cost producers. A drop in peak demand also helps the environment because it helps prevent the need for new fossil-fueled power plants.

Even without price incentives, simply providing consumers better information about their energy use has been shown to reduce total consumption by 5–15%,[1190] equating to savings of $60 –180 per year for the average American household.[1191] Making better information widely available would result in billions of dollars in savings per year by consumers and businesses.

Real-time energy consumption and price data also create an opportunity for consumers to select from a growing number of products and services that can help save energy. General Electric, for example, is developing refrigerators that automatically wait until power is less expensive before they run a defrost cycle or make ice.[1192] Whirlpool plans to have one million Smart Grid-compatible clothes dryers available by 2011 and has announced that by 2015 all of its appliances will be able to connect to a Smart Grid.[1193] Programmable communicating thermostats and energy displays like those made by Tendril, EnergyHub and others can show consumers how much they have spent to date and can automatically adjust the temperature based on a customer's desired energy spending amount and level of comfort.[1194] Google and Microsoft, among others, have released Internet-based visualization tools that help consumers get a better handle on their energy use. [1195]

For commercial and industrial customers, innovative software companies are already finding ways to deliver real value from energy data. Minnesota-based Verisae, for example, remotely monitors and manages its customers' assets, such as a grocery chain's freezers, over the Internet. Analyzing detailed data, Verisae can identify opportunities for its customers to invest in energy-efficiency improvements that maximize return on investment. Verisae can even identify when assets require maintenance, preventing costly failures and extending equipment life.[1196] And as explained in Box 12-4, Massachusetts-based EnerNOC uses real-time energy data and secure communications over the Internet to create a virtual power plant made up of commercial and industrial customers who earn money by temporarily reducing their loads during critical peaks.[1197]

BOX 12-4. A VIRTUAL POWER PLANT

Downtown Boston is home to one of the country's largest power plants. But instead of nuclear fuel rods or massive piles of coal, this plant is powered only by sophisticated software, broadband Internet and companies that are willing to reduce their energy use on demand.

The idea behind this "virtual power plant," run by Massachusetts-based EnerNOC, is simple. Typically, when electricity demand rises above supply, utilities must either generate more electricity or buy additional power from other suppliers on the grid. Financially, EnerNOC functions like an extra power plant during these peaks. But instead of generating additional electricity, EnerNOC provides the grid a temporary reduction in demand (the service is called demand response in industry parlance). EnerNOC partners with more than 3,000 commercial and industrial customers who are willing to temporarily reduce their power consumption. These businesses, from grocery stores to factories, reduce energy demand by dimming noncritical lights in a warehouse or by temporarily suspending an energy-intensive industrial process.

EnerNOC needs two things to make this virtual power plant work: broadband Internet and access to its customers' real-time energy consumption data to verify they are really curtailing load when called upon. Many customers already have Internet connections and where they lack connectivity, EnerNOC can install commercial wireless data modems. But getting a customer's real-time energy information can be an onerous process, often involving a meter upgrade. As more residential, commercial and industrial customers upgrade to smart meters, the number of customers that can participate in such virtual power plants will expand, but only if these customers and their vendors have access to real-time digital energy information.

Broadband is essential to realizing the full potential of smart homes and buildings.[1198] Pervasive Internet connectivity brings innovative competitors, technologies and business models to energy management systems, from sophisticated building management systems to simple home thermostats. Internet connectivity to stand-alone energy displays, multipurpose security and home automation systems, televisions, computers and smartphones enables consumers to see more information (e.g. weather conditions, energy prices, bills-to-date) and make smarter decisions about energy use. Broadband allows consumers to monitor and control their home energy use from the convenience of a mobile phone.

However, broadband by itself is not sufficient to unleash the full innovation potential of smart homes and buildings. The country also needs open standards and customer data accessibility policies.

Standards are critical to the Smart Grid. For example, the faster NIST can accelerate market convergence toward a small number of appliance communications standards, the sooner manufacturers can offer smart appliances that communicate with the rest of the smart home. Standards will help ensure that the Smart Grid is "plug-and-play," encouraging innovation by giving companies a large potential market for devices and applications and providing customers with the ability to use any of them to take advantage of the grid. The NIST standards development process should continue to draw on lessons from the Internet. Open standards are critically important—Internet Protocol being a prime example. In addition, security and privacy should be fundamental to both network architectures and everyday business processes.

Despite the wide variety of potential uses for the energy information created by smart meters, these data are not yet available to customers. One study of a number of large utilities found that of the almost 17 million meters being planned or deployed by the respondents, there were clear plans to provide customer access to the data only 35% of the time. Furthermore, less than 1% of the respondents' customers have real-time access to their energy data today. [1199]

A national Smart Grid policy should encourage tens of thousands of entrepreneurs to innovate—using new technologies and business models—to create a wide variety of in-building energy management and information services. Making energy data available to customers and their authorized third parties, while employing open and nonproprietary standards, is the best way to unleash this vast potential for innovation. [1200] The history of the Internet illustrates how entrepreneurs can develop disruptive applications, attract investment capital and compete to deliver value to customers—thereby driving innovation, economic growth and job creation (see Box 12-5).

BOX 12-5. ENERGY MANAGEMENT APPLICATIONS[1201]

It is a blistering hot summer day. You have just arrived at work and realize that you forgot to turn off your home air conditioning, which is blowing full blast. In the past there was nothing you could do until you returned home. But today there are new mobile applications ("apps") that allow you to take action anytime, anywhere.

There are already dozens of apps on smartphones, computers and other devices dedicated to home energy measurement and management. Companies such as Visible Energy, Control4 and many others offer apps that let you monitor your energy consumption and control your lights, security system, entertainment system and thermostat from the comfort of your living room couch or a remote location.

These applications are not just for early adopters with high-end home automation systems. Socially minded or cost-conscious consumers who want to better track their energy use can use online sites like Microsoft's Hohm and Google's PowerMeter.

A national broadband plan in 2010 cannot fully anticipate how Americans will use energy in 2050. Perhaps energy generation (and storage) will be much more distributed by then, with the grid functioning mostly as an intelligent broker between net-zero buildings exchanging power. Maybe energy transactions, not just energy management and efficiency, will be the next killer application of the Internet. The federal government need not know the answer in 2010; rather, it should use a combination of incentives, rules and standards to foster an open marketplace where the best ideas, technologies and entrepreneurs can compete for investment capital and customers.

RECOMMENDATION 12.7: states should require electric utilities to provide consumers access to, and control of, their own digital energy information, including real-time information from smart meters and historical consumption, price and bill data over the internet. if states fail to develop reasonable policies over the next 18 months, congress should consider national legislation to cover consumer privacy and the accessibility of energy data.

Consumers, and their authorized third parties, must be able to get secure, non-discriminatory access to energy data in standardized, machine-readable formats. Customers should have access to their data in the same granular form in which it is collected, and in as close to real-time as possible. Innovative companies—from large service providers to small startups—and utilities should be able to compete on a level playing field to provide a wide variety of home and building energy information and management services.

PUCs should mandate data accessibility as a part of Smart Grid rate cases, especially smart meter deployments. Consistent with EISA, these policies should mandate secure consumer accessibility to real-time energy consumption data, time-series consumption and billing data and dynamic price data.[1202] Regulators should also require regulated utilities to adopt business processes that clearly articulate the methods by which consumers can authorize and de-authorize third-party access. Regulators should also strongly consider requiring distribution utilities to provide consumers' generation mix and emissions data in as close to real time as possible. [1203]

Several state PUCs and legislatures have already started to require customer access to energy data. The California PUC has recently ruled that its major investor-owned utilities must provide customers access to their usage and price data by the end of 2010 and must provide real-time access by the end of 2011.[1204] The Pennsylvania legislature has required all large utilities to create a plan for deploying AMI systems with customer data access capabilities. In Texas, the PUC has established a common data portal in which customers, utilities, electricity retailers and third parties will be able to securely access and exchange digital energy information over the Internet.

States and utilities should not wait for full smart meter deployments to take these steps. Though smart meters will provide increased data resolution, digital access to simple monthly consumption data has many benefits. Historical usage and bill information lets consumers analyze their energy usage over time, evaluate prospective energy-efficiency measures and even compare their consumption against similarly sized houses. Better access to utility bill data also lets new buyers of homes or buildings factor energy efficiency information into their purchase decisions.

With reasonable privacy protections, the federal government should be granted limited access to utility bills from homes receiving federal energy efficiency funds to better evaluate the government's energy efficiency programs, such as weatherization. Energy consumption data, when aggregated, can be very useful to a wide variety of public policy and economics researchers. States should consider how third parties might get access to anonymized datasets for research purposes, with strict privacy protections.

By the end of 2010, every state PUC should require its regulated investor-owned utilities to provide historical consumption, price and bill data over the Internet, in machine-readable, standardized formats. By the end of 2011, every investor-owned utility should develop and implement this capability.

While a handful of states are moving quickly to develop pro- innovation energy data policies, a number of states are moving too slowly, or not at all. Congress should monitor the issue and should consider national legislation if states fail to act. America's energy and environmental challenges are too important to wait.

RECOMMENDATION 12.8: The Federal Energy Regulatory commission (FERC) should adopt consumer digital data accessibility and control standards as a model for the states.

RECOMMENDATION 12.9: DOE should consider consumer data accessibility policies when evaluating smart Grid grant applications, report on states' progress toward enacting consumer data accessibility, and develop best practices guidance for states.

The federal government should promote consumer accessibility to digital energy information. Although retail energy services are regulated at a state level, FERC and DOE should encourage consumer data accessibility and control. As FERC begins its rulemaking to adopt NIST standards, it should also include NIST standards focused on consumer data access to provide states a model on which to base their own Smart Grid rulemakings. FERC should also encourage wholesale market entities—independent system operators/regional transmission network organizations—to provide information on generation mix and emissions data as close to real time as possible at a system level. In future versions of its Smart Grid Systems Report, [1205] DOE should specifically provide updates on the progress of each state in enacting strong consumer data accessibility policies. DOE should also develop a set of best practices for states by publishing a set of model energy data policies.

RECOMMENDATION 12.10: The Rural utilities service (RUS) should make smart Grid loans to rural electric cooperatives a priority, including integrated smart Grid-broadband projects. RUs should favor smart Grid projects from states and utilities with strong consumer data accessibility policies.

The U.S. Department of Agriculture's Rural Utilities Service can play an important role in modernizing the operations of the rural electric cooperatives that own and operate 42% of the nation's distribution infrastructure.[1206] In FY2009, RUS disbursed 209 electric loans and loan guarantees totaling \$6.6 billion, giving it a total loan portfolio of \$40 billion.[1207] Similar to the directive in EISA, RUS should ensure that electric cooperatives have considered investment in qualified Smart Grid systems before undertaking investment in less sophisticated grid technologies.

12.3. Sustainable Information and Communications Technology

ICT industries account for 120 billion kilowatt-hours (kWh) of electricity use annually—approximately 3% of all U.S. electricity.[1208] They are responsible for 2.5% of the national greenhouse gas emissions, and their emissions share is forecast to grow three times faster than those from other sectors of the economy.[1209] The growth in energy usage and resulting emissions can be divided into three components: increased penetration and usage of personal computers (PCs) and peripherals, growing demand for communications services and rapid growth of data centers.

PCs and peripherals made up approximately 3.3% of residential and commercial electricity use in the U.S. in 2005,[1210] a share that is expected to grow to approximately 4.7% by 2011.[1211] This growth will be driven by the increased penetration and usage of devices such as mobile phones, netbooks and video-game consoles. Simple behavioral changes can lessen the impact of these devices. For example, one study found that 60% of all desktop PCs remain fully powered during nights and weekends.[1212]

A new standard for a universal charging solution for mobile phones, recently approved by the International Telecommunication Union, will cut standby power consumption in half. The drop will occur because the same highly energy-efficient charger will be used for all future handsets, regardless of their make or model. The change will eliminate up to 21.8 million tons of greenhouse gas emissions a year and reduce by up to 82,000 tons annually the chargers that need to be produced, shipped and subsequently discarded.[1213]

Communications networks also can be made more efficient. Approximately 0.8% of U.S. electricity is consumed by the telecommunications industry.[1214] Emissions related to mobile networks, in particular, are expected to increase from 10.5 million metric tons of greenhouse gases in 2008 to 11.2 million metric tons in 2013 under a business-as-usual scenario.[1215] But the large service providers are not sitting still. They recognize that reducing the energy intensity of their operations will not only help the planet but also reduce costs and maximize their profits. To take one example, Sprint has audited all of its facilities and installed building automation systems and Web-based meter-information systems, leading to a 9% annual energy savings (~23 million kWh) and preventing 21,400 tons of CO_2 emissions per year.[1216] The company has also installed hydrogen fuel cells and solar power at a number of its cell tower sites.[1217]

Data centers accounted for 1.5% of U.S. electricity consumption in 2006, and demand is expected to double by 2011.[1218] Demand will rise in large part because of the rapid increase in the need for data processing and storage of electronic information, compounded by data center servers' very low utilization rates and inefficient cooling systems.[1219]

The largest efficiency opportunities for data centers can be achieved through virtualization, a technique that lets a single server be treated as though it is multiple machines. This means that servers do not need to be dedicated to specific purposes and can be used wherever processing power is needed. At the moment, only 5-15% of server capacity in a typical data center is being used at any one time, but virtualization can significantly increase that figure.[1220] Such increased efficiency can reduce a data center's greenhouse gas emissions by an average of 27%.[1221] Better temperature monitoring and control devices, as well as reducing a data center's reliance on air conditioning, can cut emissions by 18%.[1222] Lastly, locating data centers in areas where a high proportion of baseload power is generated from low- carbon sources can lead to significant emissions reductions.[1223]

RECOMMENDATION 12.11: The FCC should start a proceeding to improve the energy efficiency and environmental impact of the communications industry.

The FCC should start a Notice of Inquiry to study how the communications industry could improve its energy efficiency and environmental impact. This proceeding should examine such topics as data center energy efficiency, the use of renewable power for communications networks and the steps that communications companies can take to reduce their carbon emissions. The proceeding should also study how service providers can impact the energy usage of peripherals in the home, including mobile phone chargers.

RECOMMENDATION 12.12: The federal government should take a leadership role in improving the energy efficiency of its data centers.

The federal government owns and operates approximately 10% of the nation's data centers and servers.[1224] Research suggests that data centers can cut their electricity use by up to 45% by adopting best practices in energy efficiency.[1225] Federal agencies should take measures to improve the energy efficiency of their data centers in accordance with President Obama's Oct. 5, 2009, Executive Order 13514 that promotes environmental stewardship (including "implementing best management practices for energy-efficient management of servers and Federal data centers") and the announced 28% greenhouse gas emissions reduction target set for the federal government by 2020.

Specifically, the federal government should set a goal of earning the government's ENERGY STAR for all eligible data centers it operates. A first step toward this goal should be metering the energy use in all federal data centers as soon as practicable. This will enable data centers to receive an ENERGY STAR rating upon the U.S. Environmental Protection Agency's release of the data center Portfolio Manager in June 2010. By metering their data centers and using the rating tool, departments and agencies will be able to measure their progress toward earning the ENERGY STAR, which will be given to the top 25% of energy-efficient facilities. With limited national security exceptions, agencies should post their data center efficiency ratings online so the public can track the government's progress. In addition, all new federal data centers should be designed to earn the ENERGY STAR. Finally, DOE should consider and report on whether the government can go beyond ENERGY STAR savings, and if so, how.

12.4. Smart Transportation

The transportation industry is the second-largest consumer of energy, a primary reason for the country's reliance on oil and the sector that is the second-highest emitter of greenhouse gases.[1226] Broadband and advanced communications infrastructure will play an important role in modernizing various transportation systems by making them safer, cleaner and more efficient.

Broadband and other information and communications technologies can reduce emissions by enabling more efficient driving. Adding communications technologies to vehicles and to key infrastructure, such as traffic signals, can help reduce the amount of time spent on the road. Drivers can optimize routes based on real-time traffic conditions, and commercial operators can plan more efficient routes and supply chain logistics. Communications can also enable potential future transportation policies such as congestion pricing and performance-based mileage standards, which would cut traffic and encourage drivers to be as efficient as possible. Collectively, information and communications technologies can eliminate as much as 440 million metric tons of greenhouse gas emissions from transportation by 2020.[1227]

Automakers are increasingly building wireless communications into vehicles, for safety, navigation, entertainment and productivity. OnStar, a service offered by General Motors, uses an embedded cellular connection to provide emergency alert services and diagnostics that can improve a vehicle's performance and gas mileage. Vehicle communications can also come from a driver's personal mobile phone; Ford's SYNC service, for example, allows drivers to use their wireless phones to provide in-vehicle connectivity for a variety of entertainment, communications and safety applications.

While the number of vehicles with broadband is small today, all U.S. automakers have begun offering integrated or aftermarket-compatible solutions that presage eventual mass-market, in-vehicle broadband adoption. Whatever its form factor or application, in-vehicle broadband is likely to contribute to the growing need for commercial broadband spectrum.

The benefits of broadband-connected vehicles will be great, but the risks of increased driver distraction must be proactively addressed. The addition of new technologies in the vehicle must be coupled with a commitment by individuals, families and automakers to use and deploy these technologies responsibly, in a manner that minimizes driver distraction. Solving these challenges will require coordinated leadership from industry, government and consumer groups. Solutions must be pursued before these applications are widely deployed, rather than as an afterthought.

The federal government has already swung into action. The U.S. Department of Transportation (DOT) held a distracted driving summit and launched Distraction.gov, the federal government's official website for distracted driving—currently featuring Oprah Winfrey's campaign against distracted driving. The FCC held a workshop exploring technologies that could play a role in reducing the risk of distracted driving. DOT and the FCC have also launched an interagency collaboration on distracted driving, focused on consumer outreach and on technological approaches to the problem. The federal government should continue to work with industry to safely incorporate the next generation of in-vehicle communications technology.

Broadband can also encourage the use of alternatives to automobile transportation. Route-planning applications make public transportation easier to use, and in-vehicle broadband can make mass transit more attractive. For example, intercity bus companies cite broadband as one factor increasing ridership since 2006.[1228] Several companies offer free Wi-Fi to passengers, a feature Megabus credits with attracting new riders to its Boston- New York City service, which saw ticket sales rise 67% in 2009.[1229]

As discussed in Chapter 13, broadband itself provides an alternative to transportation and travel, through Web conferencing, telecommuting and videoconferencing. Already, many companies are minimizing emissions and saving costs by avoiding air travel, and telecommuters are saving time and gas by working from home.

Advanced communications systems also have the potential to help reduce the nation's tens of thousands of automobile fatalities each year. [1230] For example, imagine a driver needs to suddenly brake while traveling on a busy highway. An ad hoc vehicle-to-vehicle communications system could allow cars following several vehicles to be alerted of the danger almost as soon as the first car's driver pushed the brake pedal. This would give more drivers a critical opportunity to prevent a high-speed, rear-end collision—a common cause of highway fatalities. In 1999, the FCC allocated 75 MHz of spectrum in the 5.850–5.925 GHz band for these types of specialized Intelligent Transportation Systems (ITS) applications. The transportation industry envisioned using dedicated short- range communication (DSRC) protocols to communicate between vehicles (vehicle-to-vehicle) and roadway infrastructure (vehicle-to-infrastructure). Despite promising tests, these networks have not been deployed.

For some ITS applications, such as vehicle-to-vehicle collision avoidance, DSRC technology may be required because it provides extremely low latency communication between vehicles. However, these applications require a critical mass of vehicles with the technology to deliver real benefits. Practically speaking, this means DOT would need to mandate the technology in new vehicles or otherwise encourage adoption, possibly by implementing a consumer information program through the New Car Assessment Program. DOT has committed to making a decision on its approach by 2013.

Whatever the ultimate decision, the country need not wait for deployment of DSRC technology to begin aggressively developing and deploying smart transportation applications. In the 10 years since the FCC allocated spectrum for ITS applications, commercial wireless data networks have been built to cover much of the country's roadways. These networks and Internet-hosted applications are capable of delivering many of the efficiency, mobility and sustainability applications envisioned in ITS. DOT should explore ways to leverage commercial wireless data networks and the Internet to achieve its goals.

13. ECONOMIC OPPORTUNITY

Broadband is becoming a prerequisite to economic opportunity for individuals, small businesses and communities. Those without broadband and the skills to use broadband-enabled technologies are becoming more isolated from the modern American economy.

This is due in part to the rapidly changing nature of work in the digital age. Sixty-two percent of American workers rely on the Internet to perform their jobs.[1231] The Bureau of Labor Statistics forecasts that jobs depending on broadband and information and communication technologies (ICT) —such as computer systems analysts, database administrators and media and communications workers—will grow by 25% from 2008–2018, 2.5 times faster than the average across all occupations and industries.[1232]

The benefits that flow to the regions, workers and businesses that adopt and use broadband can be seen across the country. Diller, Neb., population 287,[1233] is home to Blue Valley Meats, which has seen its business grow more than 30% and its employee ranks double over the last five years, thanks in large part to the creation of a website to extend its product reach.[1234] In Youngstown, Ohio, located in the country's hard-hit "rust belt," the Youngstown Business Incubator is fostering companies such as Turning Technologies, recognized by *Inc.* magazine as one of America's fastest-growing software firms.[1235] In post-Katrina New Orleans, entrepreneurs are using the Web to serve other small businesses with online marketplaces and customized reservations systems. These new firms are contributing to a flourishing tech community in the Crescent City.[1236]

Braodband and the Internet make it possible for small businesses to reach new markets and improve their business processes. They have also become a critical pathway for individuals to gain skills and access careers. And it is a core infrastructure component for local communities seeking to attract new industries and skilled work forces. As a result, small businesses, workers, and communities must have the broadband infrastructure, training and tools to participate and compete in a changing economy. Broadband can help every community. Unfortunately, certain communities such as African Americans, Hispanics and rural Americans face low adoption rates, further limiting the potential benefits of broadband (see Chapter 9).

This chapter contains recommendations to extend the benefits of broadband, and the economic opportunities broadband creates, to more communities. Section 13.1 discusses the impact of broadband on small businesses and entrepreneurship. The section recommends ways to accelerate small business adoption and use of broadband applications by expanding application training and entrepreneurship mentoring programs, while giving businesses access to improved broadband network performance information.

Section 13.2 reviews how broadband connectivity and Web- based applications can help the American workforce build skills and find jobs in more effective ways. This section also recommends the virtual delivery of job training and employment assistance programs.

Section 13.3 explores ways to promote telework among American employees.

Section 13.4 focuses on community development, where broadband availability can be a key element of an integrated approach to regional economic development. This section recommends online tools for regional development managers, more efficient and effective uses of federal resources for regional growth, and expanded technology transfer efforts within local universities.

Additionally, Chapters 8 and 9 of the plan explore how broadband access and adoption by minority populations can further economic opportunities for all, particularly through initiatives such as expanding Universal Service Fund support for low-income and rural communities, and launching a Digital Literacy Corps.

Recommendations

Support entrepreneurship and America's small and medium-sized businesses

- Small Business Administration (SBA) resource partner programs should provide enhanced information technology (IT) applications training.
- Current federal small and medium enterprise (SME) support programs should use broadband and online applications to scale their services and give small businesses access to a virtual nationwide network of experts.
- The government should develop a public-private partnership to provide technology training and tools for small disadvantaged businesses (SDBs) and SMEs in low-income areas.
- Congress should consider additional funds for the Economic Development Administration (EDA) to bolster entrepreneurial development programs with broadband tools and training.

Deliver high quality federally-supported job training and placement services virtually

- The Department of Labor (DOL) should accelerate and expand efforts to create a robust online platform that delivers virtual employment assistance programs and facilitates individualized job training.

Remove barriers and promote telework within the federal government

- Congress should consider eliminating tax and regulatory barriers to telework.
- The federal government should promote telework internally.

Enable local and regional economic development

- The federal government should develop regional and community broadband benchmarks for use as a central component within economic development planning and programs.
- EDA should create an easy-to-use, dynamic online information center that gives regional development managers access to integrated federal, state, local and Tribal data.
- The National Science Foundation (NSF) should use its technology transfer grants to spur regional innovation and development as well as greater collaboration across universities.

13.1. Supporting Entrepreneurship and America's Small Businesses

Broadband can provide significant benefits to the next generation of American entrepreneurs and small businesses—the engines of job creation and economic growth for the country. Small and medium enterprises (SMEs) —businesses with fewer than 500 employees—employ more than half of America's private sector workers and create roughly 64% of net new private sector jobs each year.[1237] As of 2006, there were almost 5.4 million firms employing less than 20 people in the U.S. and an additional 20.8 million nonemployer firms.[1238] Of that total, approximately 7.6 million firms were owned by women and 4.6 million firms were owned by minorities.[1239] In the last 10 years, minority-owned businesses have accounted for more than half of the two million new businesses started in the United States, and created 4.7 million jobs.[1240] Home-based businesses and entrepreneurs also have a profound effect on the economy, employing more than 13 million people in the United States in 2008.[1241]

Small businesses have been particularly important in high-tech industries. They currently hire roughly 40% of all high-tech workers,[1242] and account for a majority of the more than 1.2 million new jobs generated by the growth of the Internet during the last 10–15 years.[1243] Moreover, telecommunications has proven to be a particularly successful sector for women- and minority-owned businesses. For instance, in 2002, the more than 6,000 women-owned businesses in the telecom sector generated revenues of more than $7 billion. That works out to $1.1 million in revenues per business, far more than $145,000 in revenues per women-owned business in the economy overall.[1244]

Broadband and broadband-dependent applications allow small businesses to increase efficiency, improve market access, reduce costs and increase the speed of both transactions and interactions. By using Web-based technology tools, 68% of businesses surveyed boosted the speed of their access to knowledge, 54% saw reduced communications costs and 52% saw increased marketing effectiveness.[1245] However, many small businesses have a knowledge gap about how best to utilize broadband tools, leaving potential productivity gains unrealized. Though private sector options exist for training and educating small businesses, those options are currently insufficient. Targeted government support can help small businesses achieve an optimum level of broadband use.

The Benefits of Broadband for SMEs

The conduct of key business activities such as communication, collaboration, process enhancements and transactions is made easier by use of broadband applications such as online conferencing, social networking, cloud-based business software and e-commerce. Perhaps chief among the benefits of broadband for business is that it allows small businesses to achieve operational scale more quickly. Broadband and associated ICTs can help lower company start-up costs through faster business registration and improved access to customers and suppliers. Broadband also gives SMEs access to new markets and opportunities by lowering the barriers of physical scale and allowing them to compete for customers who previously turned exclusively to larger suppliers.[1246] E-commerce solutions eliminate geographic barriers to getting a business's message and product out to a broad audience. However, small businesses are not fully capitalizing on these opportunities. An estimated 60 million Americans go online every day to find a product or service;[1247] but only 24% of small businesses use e-commerce applications to sell online.[1248] The large majority of small businesses are missing an opportunity to level the playing field versus their larger rivals.[1249]

Supporting IT and Application Adoption among SMEs

The benefits described above are most compelling when broadband is supported with significant investment in IT hardware, software and services and material improvement in business processes.[1250] Even technologically lagging firms in the small and midsize space recognize that broadband is a key part of a firm's basic IT infrastructure. Yet IDC, a research firm, indicates that roughly half of small and midsize firms say that they are cautious when it comes to investing in new IT.[1251] Other small businesses voice skepticism about select broadband applications either because of a perceived lack of applicability or uncertain profitability.[1252] In addition, small businesses often identify important problems that IT applications can help address but do not link those problems to available solutions. For example, IDC Research shows that approximately 33% of SMEs identify "strengthening customer service and support" as a key spending priority, but only about 10% cite as a priority "improving customer relationship management tools" which are specifically designed to help in this area. [1253]

To address these challenges, many small businesses rely on outsourced support when selecting and implementing broadband applications. Applications training and online tutorials are widely available from private application providers such as salesforce.com, Google and Amazon. However, despite these resources, private sector support mechanisms are not sufficient to address the full range of SME training and education needs for a number of reasons:

- Particularly in economically disadvantaged, rural or remote areas, direct application training and integration services are often too expensive or unavailable for many small businesses.
- Small businesses already pay significantly more per employee for broadband and communications services,[1254] making it difficult to afford additional training and support services within a limited IT budget.
- Existing support and training initiatives typically target IT staff, omitting the broad range of other employees who can benefit from broadband applications.
- Many service providers and vendors do not provide direct support for SMEs.
- Service providers and vendors that do target SMEs (such as value-added resellers) are often small themselves and have limited capabilities to support SME broadband and IT needs.

While private sector options exist—particularly as suppliers place emphasis on the SME market—public programs may in certain cases be valuable for addressing these gaps, particularly for rural businesses and those in economically disadvantaged areas. There are some select programs that offer dedicated training to these areas, such as the Louisiana Business & Technology Center Mobile Classroom, which provides seminars, workshops and training programs for small businesses and entrepreneurs in rural communities.[1255] However, these programs are uncommon and should be augmented by other dedicated public efforts.

BOX 13-1. UNITED KINGDOM TRANSFORMATIONAL ICT PROGRAM

The United Kingdom is one of the few nations emphasizing assisting SMEs in the adoption and use of ICTs. In the 2009 Digital Britain report, the Department for Business, Innovation & Skills (BIS) announced £23 million for a three-year pilot program of business support interventions for SMEs to assist them to exploit advanced ICT to transform their business processes. This focus on business support recognizes that the key obstacle to ICT use is SME understanding of the benefits of broadband applications, rather than connecting these businesses to broadband. As a result, BIS has made ICT education and training key priorities to help SME growth. To support this effort, BIS has created the Transformational ICT program. The program has six components that address supply and demand of ICTs for business:

1. Seminars for business owners that demonstrate benefits of ICTs.
2. Assessment of IT challenges for businesses that go through the seminars.
3. Training assistance for employers to increase skills in key areas related to ICT, through the "Train to Gain" program.
4. Assistance for implementing key technology purchases, as well as funding for specialist support.
5. Certification of business service and equipment suppliers to provide guidance on business purchasing decisions.
6. Collaboration with third parties such as financial institutions and insurers to address business needs in a coordinated fashion.[1256]

RECOMMENDATION 13.1: Small Business Administration (SBA) resource partner programs should provide enhanced information technology (IT) applications training.

Many businesses currently receive a range of assistance from federally sponsored small business support programs, including help with business planning, application usage, finance and marketing. These training efforts, often initiated by the SBA and administered through Small Business Development Centers (SBDCs)[1257] and Women's Business Centers (WBCs),[1258] may or may not include broadband or IT content, depending on both the goals of the program and the entity in charge.

The SBDCs can be an effective conduit for serving small business needs, reaching more than 600,000 business clients annually and helping create more than 12,000 new small businesses in 2009.[1259] Congress should consider ways to leverage existing assistance provided through those programs to focus training on advanced IT and broadband applications. The budget for upgrading existing SBDC lead centers to receive technology accreditation as Small Business Technology Development Centers (SBTDCs) is estimated at $1 million annually, including costs for supporting 10–12 sub-centers each. This pilot program would create as many as 12 new SBTDCs and 180 sub-centers. This budget reflects the typical scope of technology training initiatives within the SBDCs.

Congress could also consider ways to support technology training among women entrepreneurs through the WBCs. The 110 WBCs currently reach a broad client base that typically includes low-income women, first generation immigrant populations, Native Americans and veterans. These funds will be used to develop a curriculum tailored to women entrepreneurs on the value of broadband-based programs and applications, such as online marketing, financial management, Web 2.0 tools and other online based services. SBA would design this training curriculum to be scalable in addressing the needs of entrepreneurs at all stages of development.

The training programs should include an entry-level "Broadband 101" course to give small businesses an introduction to how to capitalize on broadband connectivity, as well as more advanced applications for IT staff. In addition, SME IT training should include resources for non-IT staff, such as how to use e-commerce tools for sales, streamline finance with online records or leverage knowledge management across an organization. The Manufacturing Extension Program, which provides manufacturing companies with services focused on business and process improvements, is one example of a government initiative external to the SBA that has incorporated IT and technology training effectively. In scaling the training program, SBA should also identify outside consultants and private vendors from a variety of communities to help develop curricula and support the creation of a shared online directory to leverage these experts and training courses across locations.

Given that 19% of Americans speak a language other than English at home,[1260] SBA should also encourage its SBDCs and WBCs to support more staff and volunteer trainers who can speak a language other than English to ensure that small business digital skills are made available to all Americans.

RECOMMENDATION 13.2: Current federal small and medium enterprise (SME) support programs should use broadband and online applications to scale their services and give small businesses access to a virtual network of experts.

In addition to the SBDC and WBC networks, the SBA's portfolio of tools to help entrepreneurs includes programs such as the Veterans Business Outreach Centers and the Service Corps of Retired Executives (SCORE). Collectively, these programs help thousands of entrepreneurs and small businesses by delivering free and low-cost training and one-on-one mentoring and counseling support.[1261] Broadband tools and connectivity can further boost the effectiveness of these programs. The Small Business Committees in the House of Representatives and the Senate have already turned their attention to this issue, recommending areas where broadband and the Internet can help the SBA's resource partners.

All of these programs, with the backing of the SBA, should undergo a two-step assessment to identify how broadband can make them more effective:

- Identify locations and mentors with sufficient broadband connectivity and collaboration tools to enable them to participate in an online network.
- Identify counselor strengths and availability for distance mentoring. SCORE is already prepared to deploy this system; its current online system for pairing individuals with e-mail mentors tracks individual mentor competencies.[1262]

The SBDC network, WBCs and the Veterans Business Outreach Centers would need to undergo similar assessments.

Some of these programs have significant scale already. Today, more than 10,500 SCORE volunteers provide counseling to small businesses at more than 800 locations.[1263] Nearly 1,000 SBDCs nationwide offer training and one-on-one mentoring for small businesses.[1264] Yet many of the SBA partner programs remain constrained by a shortage of brick-and-mortar resource centers, as well as mentors, particularly in rural areas. Moreover, these partner programs must serve a growing and diverse range of businesses. Nationwide, there is an average of 6,500 SMEs per SBDC, with nine states having more than 10,000 SMEs per SBDC.[1265]

Tools such as webinars and online training courses, provided by the SBA's existing Small Business Training Network, can potentially provide an effective platform for these efforts. Similarly, adoption of videoconferencing and distance mentoring practices can allow these programs to move beyond networks defined by the location of the mentors to networks defined by the expertise of the mentors. One private sector model is Cisco's internal Specialist Optimization Access and Results (SOAR) program. SOAR allows Cisco employees to leverage experts from different locations through tools such as unified communications and collaboration (including Web conferencing and videoconferencing), customer reference databases, expertise locators, virtual demos and online communities for specialists.[1266] The effectiveness of the SBA partner programs can be similarly improved through the use of these tools.

To fully implement next-generation technology within its operations, the SBA should also appoint a broadband and emerging IT coordinator. This individual would ensure that SBA programs maintain the requisite broadband expertise, tools and training courses to serve small businesses.

RECOMMENDATION 13.3: The government should develop a public-private partnership to provide technology training and tools for small disadvantaged businesses (SDBs) and SMEs in low-income areas.

Small businesses represent a crucial source of economic development and growth in low-income areas. They comprise 99% of establishments and 80% of total employment in inner cities and economically challenged areas.[1267] They also account for roughly 5.6 million self-employed workers in rural areas.[1268] Broadband can serve as a transformational force not just for these businesses, but also for their surrounding communities.[1269] Too often, however, businesses in low-income areas—even when they have broadband—lack the necessary tools, expertise and resources to take full advantage of the technology. These businesses can benefit from digital literacy and assistance in fundamental online business activities such as website construction, URL registration and use of social media.[1270]

Existing support programs within the SBA, such as SCORE, already help businesses address general training needs, including business planning, identifying sources of capital and improving business efficiency. Assistance with broadband and emerging technologies should be added to the list. Although SCORE is currently positioned to offer a minimum level of technology tools and training to small businesses, these needs are not currently part of the program's core focus. However, SCORE is attempting to increase its support of small businesses in low- income areas and small disadvantaged businesses[1271] by expanding its technology expertise and coordinating with local partners.

The SBA and SCORE should enter into a public-private partnership with private communications and technology firms to better address the broadband and technology needs of the small businesses that they serve, with a particular focus on SDBs and small businesses in low-income areas. The partner firms should provide applications, training materials, support services and skills expertise. In addition, SCORE and SBA should work to include SDBs as partners in this effort, to provide both technical expertise and insight on training small businesses across a wide range of rural and urban communities. Contributions by private firms to the partnership should include:

- "How to" training for key activities such as digital literacy, e-commerce, online collaboration, search optimization, cybersecurity, equipment use and Web 2.0 tools.
- Technical and professional support for hardware, software and business operations.
- Licenses for business applications such as document creation, antivirus and security software, and online audio- and videoconferencing.
- Website development and registration.
- Basic communications equipment, such as low-cost personal computers and wireless routers.
- "Train the trainer" assistance to prepare SCORE volunteers.
- Funding contributions.

SCORE should provide program coordination while disseminating these new resources through its nationwide network of business counselors and mentors. In doing so, SCORE should coordinate with local community organizations through its chapters in low-income areas to assist with small business implementation and use.[1272] This effort ties into SCORE's existing plans to double its volunteer base over the next seven years and reorient its volunteer corps to include more full-time trainers who have the technology expertise that small businesses require.[1273] As SCORE expands its volunteer base, it should partner with local educational institutions and graduate programs to recruit young students with business and technology expertise as volunteer trainers. This would create a high-impact service opportunity for young Americans and enable SCORE to cultivate new volunteers who can mentor local businesses over the long term.

The majority of the resources for this program will come as donations of time, money, materials and intellectual property from the collection of private partners and participating foundations. The SBA and SCORE should also coordinate with the Minority Business Development Agency at the U.S. Department of Commerce and the FCC's Office of Communications Business Opportunities to help reach the target small business populations. Congress could consider leveraging the federal investment in SCORE through the SBA's Office of Entrepreneurship Education to integrate content and support rollout of this effort.

RECOMMENDATION 13.4: Congress should consider additional funds for the Economic Development Administration (EDA) to bolster entrepreneurial development programs with broadband tools and training.

Existing entrepreneurial development efforts focus on providing assistance in the following areas: funding, business plans, market testing, mentoring, connections with peer entrepreneurs and training courses.[1274] Broadband applications increasingly are becoming necessary components of this curriculum, as e-commerce, online marketing and website design skills are critical to business success. Yet too often they are not part of the core mandate of these efforts. Moreover, broadband is allowing individuals in dispersed or rural areas (where high-growth entrepreneurs may be an untapped resource)[1275] to access these entrepreneurial development resources through tools such as online collaboration software, knowledge sharing, online mentoring communities, webinar platforms and videoconferencing.

Successful entrepreneurial development programs have been built around a small group of high-growth entrepreneurs, with an emphasis on hands-on mentorship and strong community support. Today, a few such examples of micro-focused programs exist at the state level, including JumpStart in Ohio, KTEC PIPELINE in Kansas, Innovation Works in Pennsylvania and Innovate Illinois. Based on initial evidence showing the effectiveness of these programs, they should be considered models for new entrepreneurial development programs.

In areas with existing state-level entrepreneurial development programs, the federal government can augment state and non-profit funding to help increase the scale and reach of these programs. This can be done through grants earmarked for broadband communications tools. Additionally, EDA should encourage these existing programs to add broadband-centered training courses focused on online marketing and sales, website design and business process applications.

Congress should consider funding to create parallel entrepreneurial development programs that include broadband tools and training in areas not covered by existing programs. Each pilot would have a $3 million annual budget—reflective of the annual budget for those programs currently in place—funded roughly one-third each from federal sources, state and local economic development agencies, and private entrepreneurial support organizations. Ten million dollars in federal funding for this effort, with equal matching funds from state/local and private entities would create 10 new support organizations in areas where EDA identifies the greatest needs. These new programs should have an emphasis on broadband communications tools and training. Federal funds for the pilot program should be granted through a competitive process similar to the U.S. Department of Education's Race to the Top Fund, which will ensure that communities with innovative approaches, strong community support for entrepreneurial development and the appropriate tools to achieve success will receive adequate funding for their programs.

13.2. Job Training and Workforce Development

Jobs increasingly require new skills. Today, the average worker will hold more than 10 different jobs during their prime working years, and the duration of the average job often remains short even as workers approach middle age.[1276] Most new jobs today require some level of post-secondary education or professional credentials, but 88 million working adults either have low literacy skills, limited English proficiency or no postsecondary educational credential.[1277]

A changing economy, supported by workers taking on jobs that require more skills, demands better training—training that evolves in real time to meet shifting workforce needs. Broadband-enabled job training and search platforms can scale training to reach the greatest possible number of people and do so at a lower cost and in a more flexible manner. Decades of research have found that using technology-based instruction for vocational training reduces the cost of that training by about a third, while increasing the effectiveness of instruction by a third and using a third less time.[1278]

Numerous employment assistance solutions targeting various demographic groups exist in the public and private sectors. DOL delivers services through the federally supported workforce development system that help low-income, low-skilled Americans find jobs. These Americans face unique barriers—including low literacy, an absence of digital skills, lack of social networks to connect to opportunities and difficulty accessing traditional training resources due to geography, disability, family responsibilities and other constraints. These groups traditionally depend almost entirely on government assistance to obtain career guidance, employment information and job training funding.

However, the current workforce development system is fragmented[1279] and relies heavily on bricks-and-mortar facilities to deliver services.[1280] This physical infrastructure makes it difficult to adjust to changes in demand, resulting in inconsistent supply, quality and information distribution. DOL-operated One-Stop Career Centers faced heavy demand in the wake of the 2008–2009 recession, but served only a fraction of the unemployed due to a lack of capacity—in some cases serving 10% or less of a region's unemployed.[1281] The challenge of scaling the physical infrastructure of the workforce system is particularly critical during a recession with widespread impact. For instance, in New York City, according to a July 2009 study, 26% of low-income Latinos and 18% of low-income African Americans reported losing their jobs due to the recession, meaning this problem is more acute in certain communities.[1282] In addition, skills of One-Stop personnel differ from center to center, creating inequity in the types of information and services customers receive. Delivering services online through a scalable platform would expand the reach of One-Stops to everyone who has access to the Internet. Additionally, adopting content and service standards would ensure every participant receives consistent high-quality service.

Broadband-enabled solutions also address time, information and technology barriers faced by disadvantaged Americans seeking jobs and training. The "anytime, anywhere" nature of an online environment allows people who have daytime responsibilities to participate in programs during evenings and off-hours. For those without home access to a computer, the 16,000 libraries across the country along with other community access points will help ensure increased access to career tools. Minority groups are often particularly reliant on public Internet access points; a 2002 study found that 13% of African American and 12% of Hispanic households used the Internet in a public library in a single month, compared with

8% of white households.[1283] Moreover, 83% of African-Americans and 68% of Hispanics have used their broadband connection to search or apply for a job online, compared to a national average of 57%.[1284] Recommendations in Chapter 9 to expand free Internet access at community anchor institutions will help bolster the effectiveness of online workforce development tools.

Innovative online career tools make available a wealth of information and technology to which low-income Americans may not otherwise have had access. Encouraging workforce participation in online job training could also yield long-term cost savings and better outcomes.[1285] The National Skills Coalition estimates that an increase in any level of post-secondary education could increase output per capita, increase annual federal tax revenues and reduce use of public programs such as food stamps, Medicaid and Temporary Assistance for Needy Families.[1286]

Building a workforce system that allows individuals to seek training more easily and effectively is a significant step in preparing the workforce for future jobs. DOL's Employment and Training Administration is spearheading several efforts to introduce new technology solutions to the workforce development community, including development of a virtual One-Stop. In December 2009, DOL launched the Tools for America's Job Seekers Challenge, in which the country's workforce community sampled and ranked numerous companies' online job search and career advancement tools.[1287]

RECOMMENDATION 13.5: The Department of Labor (DOL) should accelerate and expand efforts to create a robust online platform that delivers virtual employment assistance programs and facilitates individualized job training.

Creating a broadband-enabled job training and search tool for disadvantaged Americans is of paramount importance to keeping the workforce competitive and ensuring that Americans can earn family-supporting wages. This tool could help participation in job search and training programs among low-income, low-skilled Americans for whom private sector options may not be sufficiently accessible or compre- hensive. Developing this online One-Stop platform effectively would involve several steps—termed versions 1.0, 2.0 and 3.0. Each successive iteration of the tool would feature increased functionality, starting with making resources currently available through off-line One-Stops available online and later offering dynamic features that allow users to discover careers with growth potential in their region. Ultimately, those careers could be mapped to the training required to qualify.

The recommended platform would help unemployed individuals who are motivated to search and train for jobs but who do not know about the existing universe of federally supported employment assistance programs. It would tell them how to access state, local and Tribal programs, which careers are within their reach, which careers have high chances of upward mobility, whether their credentials are competitive with other applicants for the same jobs, where to find job training and how to pay for job training.

The platform's version 1.0 should deliver many of the programs that One-Stops currently deliver. One-Stops operate under a sequential delivery model in which customers must participate in Core Services to be considered eligible to receive Intensive and Training Services. The end-users of the platform would qualify for different levels of service and advance automatically from one level of service to the next until services or eligibility have been exhausted. Encouraging customers with basic levels of digital literacy to use the platform would allow One-Stop counselors to provide more in-person assistance to people who will benefit from additional attention.

Version 2.0 of the platform should offer basic skills training, intermediate digital literacy training and English as a second language coursework. The Council of Economic Advisors has found that "employers currently bemoan the lack of basic skills in the U.S. workforce, and individuals without such skills have a hard time adapting to the ever-changing U.S. workplace."[1288] Mechanisms should be put in place for private employers to offer real-time input on tailoring basic skills training to meet the needs of available jobs in the future. Over time, this platform would allow collaboration with community colleges to deliver interactive certificate-bearing online training modules as envisioned for the Online Skills Laboratory.[1289]

In version 3.0 of the platform, DOL would transform the way One-Stops deliver job training services by launching an algorithmic, long-term career planning and job training tool. Through the platform, users should be able to:

- Assess levels of digital literacy, basic literacy and English proficiency, then review recommended training opportunities to address any basic skills deficiencies.
- Evaluate job skills and work experience.
- Learn about growth industries and other labor market trends by region.
- Access detailed information about professions.
- Chart pathways to advance within professions of interest, including understanding specific professional certifications required to pursue and advance within each career path.
- Search for jobs on a national level rather than at the state or community level.
- Build a resume, write a cover letter and obtain interview preparation assistance.
- Apply for jobs, store pertinent application documents and track progress of job applications through a personal dashboard.
- Obtain detailed information on necessary job training opportunities, providers and costs, then use the information to apply for federal and state funding for these opportunities.

Research shows that unemployed workers who receive re-employment services land a job and exit unemployment insurance approximately one week sooner than those who do not receive such services. This results in cost savings for DOL, the federal government and society.[1290]

In this third phase of development, the platform should serve as a medium through which the workforce development community—non-profit, public and private players—can share best practices, initiate sector partnerships and track long-term program participant outcomes through a high-level dashboard. State-to-state collaboration might generate programs that multiple states could offer together. With better tracking capabilities, the federal government could adjust funding for programs more easily by investing in proven successes while pulling funds from programs producing poor results.

To develop the various versions of this tool, DOL should award "prize" funding to private sector firms that compete to build this employment assistance and job training platform. DOL should work to promote these funding opportunities among SDBs to ensure that there is strong participation across a wide range of eligible firms. DOL should also oversee product development and set relevant data, content and formatting standards. DOL should consider any cost savings that might come from collaborating with the U.S. Department of Energy, which is creating a virtual training software platform focused initially on training materials for weatherization jobs, but that may include advanced functionality that could be used to enhance other training content. DOL has allocated $20 million for its virtual One-Stop project. Additional funding for the platform should be considered in discussions related to reauthorization of the Workforce Investment Act. The platform's ongoing annual maintenance costs should be budgeted to provide quality control, customer service and academic support, on top of technology development costs.

13.3. Promoting Tele Work

Soon after the September 11 attacks, letters containing anthrax spores were sent to Congress, forcing members of Congress and their staffs to work from the Government Accountability Office (GAO) building. This displaced GAO analysts from their offices. But thanks to their government-issued laptop computers, more than 1,000 analysts were able to continue working remotely, maintaining the continuity of operations.[1291]

Telework has broader implications than mere continuity of operations. Jeffrey Taggart, a resident of Des Moines, Iowa, has multiple mental and physical disabilities that make working in an office difficult, if not impossible. However, thanks to the Internet, Taggart makes a living from home as a customer service professional.[1292]

Such stories are increasingly common as home broadband access has become more widespread. From 2003 to 2008, the number of teleworkers in America increased by 43% to 33.7 million people.[1293] One survey estimates that 14% of retirees, 31% of homemakers and 29% of adults with disabilities would be willing to join the workforce if given the option to telework. Making telework a more widespread option would potentially open up opportunities for 17.5 million individuals.[1294] Moreover, the average American spends more than 100 hours per year commuting; 3.5 million people spend more than 90 minutes commuting to work each way every workday. Telework allows workers to be more productive by eliminating their daily commuting time. And it gives workers greater flexibility to handle family responsibilities, attend school full time and perform more community service.[1295] This is particularly important for those living in rural areas as it can enable these workers to more effectively compete for jobs located elsewhere and perform those jobs via telework.[1296]

Telework solutions also help the environment. Every additional teleworker reduces annual CO_2 emissions by an estimated 2.6–3.6 metric tons per year.[1297] Replacing 10% of business air travel with videoconferencing would reduce carbon emissions by an estimated 36.3 million tons annually.[1298]

RECOMMENDATION 13.6: Congress should consider eliminating tax and regulatory barriers to telework.

Tax and regulatory policy may prevent some employees from teleworking more regularly. Many teleworkers live in a different state from where their firm is located. This can sometimes result in double taxation issues that end up discouraging telework. Most states tax telecommuters based on the percentage of time worked within that state. However, some states tax the full income of nonresident teleworking employees of companies based in their state unless they are working at home "for the convenience of the employer," a category that telework advocates claim is nearly impossible to prove.[1299] Since teleworkers are technically working in their home state as well, this opens them up to potential double taxation. There is pending federal legislation to ban states from taxing nonresidents on work done outside the state.[1300] Congress should consider addressing this double taxation issue that is preventing telework from becoming more widespread.

RECOMMENDATION 13.7: The federal government should promote telework internally.

The federal government employs more than 2.6 million civilians and more than one million uniformed military personnel.[1301] As of 2008, 102,900 federal employees actively teleworked, a 9% increase from 2007.[1302] Key institutions are beginning to support telework within the government. The U.S. Office of Personnel Management announced a new telework plan for federal employees in April 2009, including a Telework Managers Council that would develop standards and review agency telework policies.[1303] However, more can be done to increase the use of telework within the government.

BOX 13-2. VIRTUAL ENGLISH TEACHERS IN POWELL, WYOMING

In late January 2009, the city of Powell, Wyo. (population 5,524),[1301] finished an ambitious municipal fiber network, which provides fiber-to-the-premises to 95% of households in the community.[1302] The project spurred the growth of new business opportunities in Powell, including the hiring of more than 100 certified English teachers by Wyoming-based Eleutian to teach conversational English to South Korean students using videoconferencing.[1303] Eleutian was able to attract $1.5 million in venture funding from Skylake Incuvest, a South Korean venture capital fund. Eleutian's CEO said that Powell's fiber project was "critical" to hiring the teachers, noting: "Without fiber-tothe-home like Powell [has], we would not be able to offer home-based jobs in Powell."[1304]

Agencies must develop guidelines for managers of teleworking employees. According to the American Electronics Association, "The most daunting challenges to widespread adoption [of telework] are cultural, not technical."[1304] Giving managers guidelines on best practices for managing teleworking employees will help overcome manager resistance and alleviate any stigma associated with telework as a viable alternative work arrangement. The Telework Managers Council should review agency-developed guidelines in the course of reviewing telework plans and should promulgate best practices to the wider federal, state and local government communities.

Agencies should also evaluate and deploy, where economically attractive, a unified communications platform, including instant messaging, Web conferencing, videoconferencing, voice and a unified message center for all methods of communication. In addition, the federal government should evaluate the impact of videoconferencing to replace travel and improve government efficiency. The General Services Administration should oversee the initial deployment of advanced videoconferencing technologies to overcome cultural resistance to telework and determine whether it should be implemented more broadly.

13.4. Local and Regional Economic Development

The benefits of broadband and its centrality to economic life make it an essential element of local and regional economic development in the 21st century. Broadband enables regions and industries to compete globally, from rural farmers marketing their products nationwide to start-up companies along Massachusetts's Route 128 corridor achieving dramatic break-throughs in biotechnology that are attracting global attention. Looking ahead, communities without broadband infrastructure will find it more difficult to attract investment and IT-intensive jobs, particularly because they face growing national and international competition. The story of one community in rural Georgia proves to be today's norm rather than the exception. After losing its local textile manufacturing base, the community tried to attract once-outsourced customer services jobs for those left jobless. A major airline expressed interest in developing a customer call center but ultimately passed for one basic reason: The community lacked adequate broadband infrastructure.[1305]

Local economic developers should view broadband as a part of local infrastructure development and should incorporate it into local economic development strategies. The federal government can also leverage broadband to facilitate better integration of its diverse investments in localities. The Brookings Institution estimates that \$76 billion in federal funding for local and regional economic development was scattered across 14 agencies comprising 250 separate programs.[1306] This fragmentation makes the need for regional integration of broadband investments into local economic development investments even more critical. Broadband-enabled tools can help federal and local policymakers and citizens get a clearer, more transparent view of these disparate funding streams.

RECOMMENDATION 13.8: The federal government should develop regional and community broadband benchmarks for use as a central component within economic development planning and programs.

- **The U.S. Department of commerce and U.S. Department of Agriculture (USDA) should ensure that regions integrate broadband infrastructure into local economic development.**
- **To support local community benchmarking, the Department of Housing and urban Development (HUD) and USDA should integrate technology assessments into the Empowerment Zone (EZ), Enterprise community (EC) and Renewal community (RC) programs.**

Broadband infrastructure and a digitally skilled workforce are essential for a region to attract new jobs and investment. One way for communities to determine the level of broadband utilization in their local economy is to develop a set of broadband metrics that can be used to benchmark their performance against communities nationally. For communities with high levels of broadband use, this will help demonstrate the integration of broadband into the local economy, while attracting new private-sector investments. For communities with below-average use, community benchmarking can be an important tool for local planners to set broadband policy goals while ensuring that broadband programs effectively target gaps left by the private sector.

These benchmarks should include the following metrics:

- *Access.* The share of community or region with access to broadband services
- *Adoption.* Broadband adoption rates by local residents, businesses and institutions
- *Usage.* Applications used by local residents, businesses and institutions

These benchmarking efforts should be divided between larger regions that are served by a common network—focusing on broadband access and adoption—and smaller neighborhoods and communities, where benchmarking would focus on usage by local residents, businesses and institutions. Focus at the regional and community level would help ensure that the benchmarking program would serve the needs of regional or local policymakers. This effort would also help to coordinate federal support for technology planning and economic development, which would lead to more focused investments, as well as cost savings as projects are implemented.

Under the Recovery Act, both the National Telecommunications and Information Administration (NTIA) and USDA's Rural Utilities Service (RUS) were given the responsibility to disburse $7.2 billion for broadband adoption and deployment.[1307] In making future disbursements beyond Recovery Act funding, both NTIA and RUS should review how broadband projects integrate into local economic strategies. NTIA and RUS should partner with EDA to develop both broadband and economic development benchmarking metrics that can be integrated into regional development strategies. These efforts could include existing federally supported economic development planning efforts developed by local groups, such as workforce development boards, community colleges and other institutions. Strategies could include a combination of plans for attracting new businesses and industries, plans for local workforce training and development, and measures for improving local digital literacy

One way to implement regional broadband benchmarking is by expanding EDA's Comprehensive Economic Development Strategy (CEDS) process to include a technology assessment. A CEDS is developed by a local strategy committee that includes public officials, community leaders and local business leaders, among others.[1308] The CEDS process requires local input concerning strengths and weaknesses of the region and requires a plan of action to address issues such as transportation infrastructure, environmental impact and workforce development. Currently, each economic development district or region eligible for EDA grant funding must complete a CEDS plan at least once every five years to remain eligible for program grants.[1309] Moving forward, the CEDS process should require a plan for promoting the use of technology regionally along with an assessment and benchmarking of local broadband resources. Such measurements would help regions determine how attractive their technology infrastructure is for businesses and how equipped their local workforce is to fill new jobs.

HUD and USDA's Empowerment Zone, Enterprise Community and Renewal Community programs encourage the revitalization of impoverished urban and rural communities through economic, physical and social investments.[1310] As part of their administration of Enterprise Communities, Empowerment Zones, Renewal Communities and HOPE VI developments, HUD and USDA should incorporate technology as a critical input into the communities that they support. These programs should include a community technology assessment that measures availability, price and adoption of broadband services. HUD and USDA should also require community plans to set goals for increasing adoption and use of broadband for local development.

Residents of areas currently receiving, or eligible to receive, federal redevelopment assistance pay more for broadband and have lower maximum speeds available to them. There is some evidence broadband prices tend to be higher in low-income rural areas than similarly populated areas with higher median incomes.[1311] Enterprise Zones, Empowerment Zones, Enterprise Communities and Renewal Communities have broadband penetration rates of 56%, below the national average of 61% across all Census tracts according to FCC's 2009 Form 477 data.[1312] Thirty-four percent of these areas have average penetration rates below 30%.[1313] (Penetration rates in Enterprise Zones, which tend to be in more densely populated areas, only match the national average.)

BOX 13-3. CONNECTING BROADBAND WITH OTHER INFRASTRUCTURE TO CREATE JOBS AND OPPORTUNITY IN RURAL VIRGINIA

Planning commissions in rural southwest Virginia accelerated job growth by combining broadband deployment with new economic development projects to take full advantage of broadband's benefits. These commissions deployed fiber efficiently by coordinating its deployment with trenching for water or sewer lines, forming the groundwork for a regional broadband network in an area previously unserved due to the high cost of deployment. In addition, localities supported broadband infrastructure by upgrading other key economic development infrastructure assets. For example, the town of Lebanon converted an old strip mall to serve as a job-training center to deliver high school equivalency courses and train workers for IT-related jobs. These efforts helped the community attract new employers and create new jobs. The Lenowisco Planning District Commission reported 1,200 new jobs, $55 million in new private investments and $35 million in new payroll as a result of the region's broadband network. Its sister planning organization, the Cumberland Plateau Planning District Commission, reported 1,100 new jobs, $60 million in private investments and $40 million in new payrolls. The regional networks, which were designed to serve schools, incubators and health care providers, helped attract new employers, such as Northrop Grumman and CGI, to rural southern Virginia, enabling job opportunities that did not exist in the area before.[1314]

Though geographic characteristics limit deployment of some higher-speed technologies, fewer businesses in EZ/EC/ RC areas and census tracts with HOPE VI developments have access to the highest cable and DSL speeds, even when controlling for population density.[1314] Opportunities for growth in community broadband connectivity exist in these zones, and communities should leverage existing support for broadband infrastructure deployment, last-mile connectivity and sustainable Internet adoption efforts. Including ICT in strategic plans will enable EZs/ECs/RCs to use grant funds for community technology initiatives in support of economic development.[1315]

RECOMMENDATION 13.9: EDA should create an easy-touse, dynamic online information center that gives regional development managers access to integrated federal, state local Tribal data.

To help local economic developers in regions and localities support more competitive clusters, the EDA should build an online information center for regional economic development data.[1316] This information center would have three components:

- It would continuously update a distributed database containing key economic development indicators[1317] at the local, regional and state level, and it would allow users to custom- define regions (comprised of multiple localities or counties) for analysis.
- It would offer a searchable online database of federal funding programs that can be used by local developers and matched to their local conditions and industries. This tool would help address the fragmentation and complexities of the grant process.
- It would provide an interactive map of current and previous grantees across programs, which would include all completed impact assessments and grantee contact information.

An easy-to-use online resource could help regions identify central "clusters" of industries that provide a competitive advantage, attract skilled labor and reduce company operating costs. These clusters could create spillover effects of formal and informal networks of information sharing as firms participate in what one paper called the "social structure of innovation."[1318] Collectively, federal agencies have data on employment, education, traded goods, patents and more. The national information center could bring together these data sources to present a broader picture of how individual communities are performing economically.

The information center would also include an algorithmic tool to match federal grant programs to local conditions and industries. This capability should start with EDA's funding streams and expand over time to include 26 federal grant-making agencies.[1319] Further, the center would have information to help grantees understand what projects others in their region are pursuing. And it would have impact assessments from prior federal grants, to help regions learn from past projects and make the development process more sustainable.

Congress should consider providing public funding for the creation and operation of a Regional Information Center, as part of EDA's Regional Innovation Cluster Initiative. The information center will gather, analyze and distribute regional economic data, as well as promote best practices in economic development.

RECOMMENDATION 13.10: The National science Foundation (NSF) should use its technology transfer grants to spur regional innovation and development as well as greater collaboration across universities.

Technology transfer grants can accelerate regional innovation by supporting existing research facilities and improving coordination among local universities, development managers and the business community. NSF is launching a university innovation grant program to support the technology commercialization process through several pilot university programs. Each grant would support the creation of an innovation center that provides proof-of-concept funding and mentoring to accelerate the creation of spin-off companies.[1320]

However, smaller colleges and universities may find it difficult to apply for innovation grants because of limited connectivity, exacerbating the divide between large and small institutions. In 2007, the 50 research universities that spent the most on R&D each had an average annual research budget of nearly $550 million, representing (in total) more than 55% of all university research and development (R&D) spending.[1321] In contrast, the next 613 universities averaged just $36 million each, accounting for the remaining 45% of university R&D spending.[1322]

To assist smaller universities in applying for these grants, NSF should encourage consortia of these universities to pool their R&D resources, technology transfer staff and mentoring and research networks into a single innovation center. Supporting these university consortia could catalyze technology commercialization and drive regional economic development. It could also provide benefits to a wider range of higher education institutions, including Historically Black Colleges and Universities, regional campuses, and liberal arts colleges.

In addition, NSF should offer support for broadband networks between consortium partners and other institutions that receive the innovation grants. This approach would allow smaller universities to create a critical mass of researchers and technologies, helping attract private-sector support. In addition, it would create an online network of expertise from the participating universities, helping academic institutions adopt best practices for technology transfer management while allowing local businesses to tap into a larger pool of resources to address their innovation challenges. NSF is already supporting these universities with the Experimental Program to Stimulate Competitive Research (EPSCoR), which provides up to $6 million in grants for broadband infrastructure for universities. By starting a new effort coordinating its EPSCoR broadband infrastructure grants and its university innovation grants, NSF can allow consortia to access funds not just for connectivity but also for technology transfer and innovation.

By creating a shared communications network, these consortia would also give researchers and university spin-offs access to resources like grid computing, cloud-based applications, telepresence networks and connections to academic research networks such as the Internet2 Network. In a recent survey of Internet2 universities, all members reported research networks with connections of 100 Mbps or higher, with 76% planning on expanding their connections to 10 Gbps or higher over the next five years.[1323] By contrast, universities that conduct research but lack doctoral programs were twice as likely as universities with doctoral programs to have connection speeds below 100 Mbps.[1324] To help address this issue, groups of universities that are not connected to an academic network should be given funding priority to expand their connectivity infrastructure.

14. GOVERNMENT PERFORMANCE

Americans can check their bank accounts, communicate with customer service repre-sentatives and do their shopping anytime, anywhere by using applications enabled by broadband. Americans now expect this level of service from their government and are often disappointed with what they find. while some bright spots exist around filing taxes and paying parking tickets, these are the exception, not the rule. Government has fallen behind the private sector in using broadband to deliver services, and it is time to catch up.[1325]

From city hall to the U.S. Capitol, government can better serve the American people by relying more on broadband. The implications are enormous.

The federal government can use broadband to increase the efficiency of its own internal operations. And it can use its size and purchasing power to help state and local governments and communities deploy more broadband capability.

Consider also the impact on low-income families. At the moment, many Americans do not receive all the benefits for which they are eligible. The reasons are many, including the complexity of determining eligibility, as well as lengthy and repetitive applications. Integrating and streamlining processes can help low-income Americans receive all the safety-net benefits for which they qualify, and that has had a demonstrable effect on bettering their chances of getting out of poverty.[1326] Meanwhile, government services will operate more efficiently with the paperwork reduction that broadband technology allows. And when caseworkers assigned to these families spend fewer hours filling out paperwork, they can become more personally involved in helping their clients.

Broadband, in short, can change the way government serves the public. This chapter makes recommendations to accelerate this change. Section 14.1 focuses on how the government can take action to improve deployment of broadband in local communities. Section 14.2 proposes ways that broadband can improve government performance and service delivery. It also makes recommendations related to strengthening cybersecurity.

Recommendations

Improve connectivity through government action

- Federal government agencies and departments should serve as broadband anchor tenants for unserved and underserved communities.
- When feasible, Congress should consider allowing state and local governments to get lower service prices by participating in federal contracts for communications services.
- The Office of Management and Budget (OMB) should review and coordinate federal grants that have a broad-band connectivity requirement. Federal government grant funding should not limit or permit limitations on the use of federally funded facilities or services for broadband deployment, except when technology solutions cannot ensure privacy or security of data.
- The Executive Branch and Congress should consider using federal funding to encourage cities and counties to gather information on initiatives enabled by broadband in ways that allow for rigorous evaluation and lead to an understanding of best practices.

Enhance internal government efficiency

- OMB should develop a vision and strategy to guide agencies on cloud computing.
- OMB and the Federal Chief Information Officers (CIO) Council should develop a competition to annually recognize internal efforts to transform government using broadband- enabled technologies.
- The Executive Branch should create an interagency working group, comprised of the senior grants officials from each agency, to implement guidelines and requirements for interagency coordination of grants and to improve Grants.gov to make it easier for applicants to use.
- The Federal CIO Council should accelerate agency adoption of social media technologies for internal use.

Strengthen cybersecurity

- The Executive Branch, in collaboration with relevant regulatory authorities, should develop machine-readable repositories of actionable real-time information concerning cybersecurity threats in a process led by the White House Cybersecurity Coordinator.
- The federal government should take an active role in developing public-private cybersecurity partnerships.
- The Executive Branch should expand existing and develop additional educational programs, scholarship funding, training programs and career paths to build workforce capability in cybersecurity.
- The Executive Branch should develop a coordinated foreign cybersecurity assistance program to assist foreign countries in the development of legal and technical expertise to address cybersecurity.
- The FCC should work with Internet service providers (ISPs) to build robust cybersecurity protection and defenses into networks offered to businesses and individuals without access to cybersecurity resources. ISPs that participate in this program should receive technical assistance from the federal government in securing their networks.
- OMB should accelerate technical actions to secure federal government networks.

Improve service delivery

- OMB and the Federal CIO Council should develop a single, secure enterprise-wide authentication protocol that enables online service delivery.
- The Executive Branch should establish MyPersonalData.gov as a mechanism that allows citizens to request their personal data held by government agencies.
- Congress should consider re-examining the Privacy Act to facilitate the delivery of online government services and to account for changes in technology.
- The federal government should undertake a series of efforts to improve the delivery of government services online.

- The Executive Branch's review of the Paperwork Reduction Act should aim to enable government to solicit input to improve government services.
- The White House Office of Science and Technology Policy (OSTP) should develop a five-year strategic plan for online service delivery.
- The federal government should improve the delivery of means-tested benefits to low-income Americans.

14.1. Improving Connectivity through Government Action

The federal government spends billions of dollars annually on broadband connections for its office buildings and facilities throughout the United States and provides billions more in funding for programs that have a broadband communications component. The government does not, however, leverage that spending in a coordinated way to improve broadband connectivity and access within local communities. In many cases, doing so would have a nominal incremental cost, but the impact on communities, especially those that are unserved or underserved, could be transformative.

Government can help in the deployment of broadband by serving as an anchor tenant in unserved and underserved communities, by leveraging the purchasing power of the federal government to provide lower prices for broadband communications services for state and local governments and by coordinating federal grants with a broadband connectivity requirement.

RECOMMENDATION 14.1: Federal government agencies and departments should serve as broadband anchor tenants for unserved and underserved communities.

State and local governments have expressed a strong desire to share broadband communications infrastructure deployed by the federal government to extend broadband connectivity to state and local agencies as well as unserved and underserved communities.[1327] In response to Section 414 of the Transportation, Treasury, Independent Agencies, and General Government Appropriations Act of 2005,[1328] the President directed federal departments and agencies to deploy redundant communications links for all facilities.[1329] Implementation efforts did not account for the potential spillover benefits to people and businesses in unserved or underserved communities that are allowed to tap into the high-speed connection to the Internet that the government secured for its facilities. In the future, when deploying redundant links, the federal government should consult with local communities and use those links to extend broadband access to the unserved and underserved.

RECOMMENDATION 14.2: When feasible, congress should consider allowing state and local governments to get lower service prices by participating in federal contracts for communications services.

The federal government is one of the largest buyers of products and services in the country, especially when it comes to information technology (IT). Since passage of the E - Government Act of 2002,[1330] state and local government entities have been authorized to leverage the bulk purchasing power of the federal government to purchase a wide variety of information technology hardware, software and services. Use of that authority has increased every year, and state and local governments have saved millions of dollars. Purchasing authority is, however, restricted to items found on the General Services Administration (GSA)'s IT Schedule 70.

In 2007, GSA negotiated a 10-year, $68 billion telecommunications and network services contract to provide voice, IP, wireless, satellite and IP-centric services to 135 federal agencies operating out of 191 countries, at rates that are 10-40% lower than in previous contracts. This contract, called Networx, includes a provision that allows state and local governments to utilize the contract if federal law is changed to allow the practice.

Congress should consider allowing state and local governments to take advantage of Networx and other communications contracts to enable cost savings and encourage broadband deployment.

RECOMMENDATION 14.3: The Office of Management and Budget (OMB) should review and coordinate federal grants that have a broadband connectivity requirement. Federal government grant funding should not limit or permit limitations on the use of federally funded facilities or services for broadband deployment, except when technology solutions cannot ensure privacy or security of data.

In certain cases, well-intentioned grant programs require that money be spent on broadband connections even though a review of other projects would show that spending to be redundant.[1331] Sometimes, a broadband connection already exists. In other cases, multiple grants may be used to build multiple connections. For example, grants for primary and secondary education networks and grants for rural health care networks often call for the development of independent networks, even though one would suffice.[1332] Coordination at the OMB level would greatly reduce inefficiencies in federally-financed broadband rollouts.

RECOMMENDATION 14.4: The Executive Branch and Congress should consider using federal funding to encourage cities and counties to gather information on initiatives enabled by broadband in ways that allow for rigorous evaluation and lead to an understanding of best practices.

Examples abound of potentially powerful initiatives including IBM's Smart Cities,[1333] Cisco's Connected Communities[1334] and Google's proposed 1 Gbps fiber-to-the-home "broadband testbed."[1335] These initiatives use broadband connections to try to solve some of today's most challenging public policy problems in areas such as transportation, health care, education, public safety and government services. Dubuque, Iowa, is reducing water and electricity use by deploying sensors connected via broadband. Alameda County, California, has implemented an integrated data warehouse for social services that saves $11 million a year by reducing duplicative work and improving detection of fraud. Unfortunately, information on projects like these is not collected systematically.

Federal broadband grant programs can fill the gap by including reporting requirements for recipients.[1336] Gathering the information will not only help the federal government set priorities when issuing grants but also will assist local governments in identifying best practices across the nation.

Executive Branch agencies should run these initiatives like pilot programs and evaluate their success against pre-established benchmarks. This would help inform the next set of Congressional actions to promote widespread adoption of the techniques that prove successful with the pilots.

14.2. Improving Government Performance

Innovative applications of broadband have transformed the private sector, creating countless new ways of collaborating with partners and interacting with customers. Government, however, has not kept pace.

A poll of U.S. citizens by the Pew Research Center for the People & the Press found that in 2007, 62% agreed that government is usually inefficient and wasteful, up from 53% in 2002.[1337] This gap may be widening in part because the private sector has raised expectations that government has not met. While customers increasingly can go online to interact with private companies, the public still mostly deal with government via mail or in person, standing in line. While companies have made it easy for customers to find what they want, the government has been slow to adopt technological efficiencies to speed citizen service and eliminate its siloed structure.[1338]

Smarter use of broadband can facilitate a vast change in government. Like private companies, government can make its services available 24 hours a day, seven days a week, 365 days a year. Broadband-enabled online services can create paths across government's bureaucratic silos so that someone wanting to access unemployment benefits can deal with the local government and the federal government at the same time. Broadband holds the potential to move all government forms online, eliminating paperwork. Broadband allows for online tutorials for simple government services, which can help free government employees to focus on the most complicated cases. And broadband can increase efficiency by increasing the speed and depth of cooperation across departments and across different levels of government.

Enhance Internal Government Efficiency
In government, historically siloed institutions have bred siloed systems that are inefficient. Through strategic use of broadband-enabled technologies, the federal government has the opportunity to become a model of efficiency and performance.

RECOMMENDATION 14.5: OMB should develop a vision and strategy to guide agencies on cloud computing.[1339]

During the past decade, federal spending on information technology has grown substantially. On IT infrastructure alone, the federal government spends $20 billion per year.[1340] The number of federal government data centers has more than doubled over the last 10 years from 493 to more than 1,200.[1341]

Cloud computing has the potential to at least slow the growth in federal spending while increasing efficiency. A study by Booz Allen Hamilton estimates that an agency that migrates its infrastructure to a public or private cloud can achieve savings of 50-67%.[1342] For example, the District of Columbia recently moved toward using a commercial cloud computing solution for its mail, calendar, instant messaging, word processing and spreadsheet needs. The cost was only $50 per user per year; the District's previous solution for enterprise e-mail alone cost $96 per user per year.[1343]

The federal government has already launched a number of limited cloud computing initiatives, with positive results. Electronic payroll systems have been consolidated from 26 systems to four shared-service provider centers; this will result in estimated savings of more than $1 billion during the next 10 years.[1344] Apps.gov has allowed agencies to nimbly procure software and information technology services from GSA's Schedule 70[1345] and deploy these solutions in the cloud. Agencies such as the U.S. Department of Defense (DoD) and the Central Intelligence Agency are also moving forward on internal cloud solutions for sensitive data.[1346] The Rapid Access Computing Environment functions as an internal cloud for DoD, allowing for certification of applications that meet proper security standards within 40 days, half the time of the noncloud-based method.[1347]

Despite these successes, federal government IT executives harbor concerns about security and privacy. These concerns have some merit, but the risks can be mitigated through technology and policy solutions.[1348] Because the risks many federal agencies face are the same, they would benefit from a community approach. OMB should develop a coordinated vision and strategy that touches upon the security and privacy policy concerns that must be resolved as the government moves to deploy cloud computing.

RECOMMENDATION 14.6: OMB and the Federal Chief Information Officers (CIO) council[1349] should develop a competition to annually recognize internal efforts to transform government using broadband-enabled technologies.

Federal government employees often generate ideas for innovation and efficiency within government, yet many of their ideas go unnoticed or unheralded. The federal government has taken initial steps to celebrate innovation and efficiency by launching the Securing Americans Value and Efficiency Award, a monthlong contest that allowed every federal employee to submit ideas for how government can save money and perform better. The program received more than 38,000 suggestions.[1350] The winning innovation was an idea to eliminate the waste of medications in VA hospitals.[1351] This innovation has been included in the President's FY2011 budget, and agencies have been directed to implement many other recommendations resulting from the contest.[1352] Expanding upon this, OMB and the Federal CIO Council should create a competition focused on transforming government operations using broadband-enabled applications.

RECOMMENDATION 14.7: The Executive Branch should create an interagency working group, comprised of the senior grants officials from each agency, to implement guidelines and requirements for interagency coordination of grants and to improve Grants.gov to make it easier for applicants to use.

During FY2009, the federal government awarded more than $1 trillion in grants.[1353] Using broadband-enabled online services in the grant process can improve how the federal government implements its policies and programs.

Grants.gov was set up as a central portal for grants across the federal government to make the grants application process easier, but it has not succeeded on many metrics.[1354] On average, federal government websites earn a satisfaction score of 75/100, but Grants.gov scores only 56/100.[1355] Potential applicants must download forms to complete applications offline. There is no system for generating feedback about Grants.gov, limiting the ability to improve it.[1356]

The proposed interagency working group should be empowered to recommend improvements to Grants.gov. Also, Grants.gov should allow tagging, or the labeling of grants, to make searches (especially of broadband grants) easier. This would enable the public to use USASpending.gov to gain a crosscutting view of all federal broadband expenditures while reducing the burden on applicants searching for grants.

The grant process should also be improved to require grantors to certify that any project requiring broadband has sufficient connectivity or that the funds from the grant would pay for that connectivity. Oversight for this process should rest with the interagency group.

RECOMMENDATION 14.8: The Federal CIO council should accelerate agency adoption of social media[1357] technologies for internal use.

Social media technologies provide the federal government another platform to spur innovation and collaboration. For example, the National Academy of Public Administration uses a wiki to synthesize interview data. This simple collaborative tool has reduced data analysis time by nearly 15%.[1358]

The private sector has come to recognize the efficiency gains and other benefits of social media within the workplace.[1359]

The federal government has not made widespread use of these tools despite evidence that federal government employees embrace the use of social media to make their organizations more efficient and effective. The Transportation Security Administration (TSA) uses a social media platform called IdeaFactory that allows its 43,000 officers to securely share ideas for improving their workplace and performance. TSA employees have submitted more than 9,000 ideas, generating more than 39,000 comments.[1360] More than 40 ideas from IdeaFactory have been implemented, including changes to standard operating procedures.[1361] The DoD has also embraced social media platforms to enhance internal efficiency, with 87% of DoD workers using these tools at work.[1362]

Many agencies continue to have concerns about social media and block employee access to outside websites such as YouTube, Facebook and Wikipedia.[1363] The Federal CIO Council has expressed concerns that these social technologies and tools could be susceptible to cyber attacks.[1364] Still, there are clear benefits to adopting social media platforms for internal or cross-agency collaboration, and the Federal CIO Council should address concerns and accelerate adoption of these platforms (see Box 14-1).

Strengthen Cybersecurity

According to the Preamble to the United States Constitution, the federal government must "provide for the common defence" (*sic*). The United States has evolved dramatically since its founding, and one of the most significant changes that has marked the 21st century is the country's reliance upon the Internet in all sectors of society—from individuals to government to the economy at large.

BOX 14-1. THE INTELLIGENCE WIKI

In 2006, members of the Intelligence Community formally launched the social media site Intellipedia to help solve information-sharing problems.[1365] The effort has been well-received and is used by the Intelligence Community to share information classified up to "Top Secret." It now has more than 900,000 pages and 100,000 users who make 5,000 page edits every day.[1366] Using Intellipedia, officials can quickly learn about new topics, scrutinize information and ensure it is up-to-date and complete.

The global, borderless nature of the Internet has also led to the emergence of new categories of threats that can come from anyone, anywhere in the world, at any time. Protecting the Internet and providing for cybersecurity is both an economic and national security challenge and collectively, one of the most serious challenges of the 21st century.[1367] How the federal government approaches and provides cybersecurity will be critical to the continuing evolution of the Internet in the United States.

The recommendations that follow apply to the federal government's approach to cybersecurity. Specific recommendations relating to the FCC and cybersecurity can be found in Chapter 16.

RECOMMENDATION 14.9: The Executive Branch, in collaboration with relevant regulatory authorities, should develop machine-readable repositories of actionable real-time information concerning cybersecurity threats in a process led by the white House cybersecurity coordinator.

The federal government recognizes that no operational mechanism currently exists for the United States to provide a "coordinated and unified effort to detect, prevent, mitigate, and carry out a real-time response to significant cyber issues affecting the Nation."[1368] Recent real[1369] and simulated events[1370] demonstrate that responding to a cyberattack in real time is complex. Every second counts. Cyber threat detection, prevention, mitigation and response require coordinated action by public and private entities. In addition, traditional approaches to cybersecurity, including intrusion-detection systems and antivirus software, are ineffective against new rapidly evolving threats.[1371] As a result, new methods are required to facilitate a coordinated response.

To begin addressing this challenge, the Executive Branch should develop machine-readable repositories containing actionable real-time information concerning cybersecurity threats (including signatures for viruses, spam, IP address blacklists and other indicators). By delivering information faster and in a more useful fashion, the Executive Branch will become an active partner in the public-private battle to protect cyberspace. These repositories will further facilitate timely interaction with both the private sector and international partners.

RECOMMENDATION 14.10: The federal government should take an active role in developing public-private cybersecurity partnerships.

- **The Executive Branch should develop protocols and incentives for establishing public-private cybersecurity partnerships with all major industry sectors. These** protocols would enable sharing of cybersecurity information, threats, and incidents in a non-attributable manner, and would provide an existing channel for government to communicate actionable cybersecurity information to the private sector.

- The Executive Branch and the small Business Administration should work together to develop a cybersecurity resource program, in conjunction with state and local governments, to develop cybersecurity partnerships for small and medium enterprises (SMEs) that are not covered by cybersecurity partnerships developed for major industry sectors.

Cybersecurity continues to be a concern for the private sector in the United States, which relies on robust intellectual property protection to undergird its competitiveness. As a result, private sector networks in the United States, where most of its intellectual property resides, have been a major target for attacks, and despite the significant resources that the private sector devotes to cybersecurity, there have been a number of successful attacks on its networks. Recent victims of well-publicized cyber attacks include Google[1372] and the U.S. oil industry.[1373]

Due to the diffuse nature of cyberattacks, sharing of information is critical when responding to, mounting sufficient defenses against and remediating attacks. However, businesses are often reluctant to share information, either with other private sector entities or the government, due to worries about the potential disclosure of such an attack and related concerns about corporate liability, despite the fact that the resources necessary to successfully respond often exceed those of individual private sector organizations.

The public and private sectors must work together to overcome these challenges to ensure the security of the Internet. Information Sharing and Analysis Centers (ISACs), which convene a representative industry body to interact with the federal government on cybersecurity issues full-time, are good models for the kind of collaboration that is needed. Today, ISACs exist for the financial services sector (FS-ISAC), the information technology sector (IT-ISAC), and state and local governments (the Multi-State ISAC, or MS-ISAC). To ensure that ISACs for other industry sectors are effective, ongoing communication and actionable information will be required from both industry participants and the federal government.

SMEs often have fewer resources to dedicate to cybersecurity than large businesses in major industrial sectors. However, despite limited resources, cybersecurity is no less important to small and medium businesses. Recognizing both resource constraints and the importance of cybersecurity, the Executive Branch and the Small Business Administration should develop a cybersecurity resource program, in conjunction with state and local governments, through the MS-ISAC.

The effectiveness of public-private partnerships depends on ongoing communication and actionable information from both industry sector participants and the federal government. To ensure that this occurs, protocols and incentives should be developed for the sharing of cybersecurity information, threats and incidents in a non-attributable manner.

RECOMMENDATION 14.11: The Executive Branch should expand existing and develop additional educational programs, scholarship funding, training programs, and career paths to build workforce capability in cybersecurity.

Cybersecurity is a rapidly evolving field, requiring specialized training and expertise. The importance of this field to the economy, competitiveness and national security underscores the need to build a robust and capable workforce with the skills to sustain it. The federal government has an additional challenge in retaining skilled IT security officials because training and career advancement opportunities are limited.[1374] However, the quality of professionals in the field of cybersecurity is mixed, with current training insufficient to meet the needs of either the public or private sectors.[1375]

Immediately following the launch of Sputnik, governments in both the United States and Western Europe were deeply concerned about the growing quantity and quality of scientists and engineers in the Soviet Union. One of the major policy actions to address this concern was education and training in basic science, laying the groundwork for the United States' Apollo mission to go to the moon. Similarly, to meet the security challenges of the present day, a new professional cybersecurity workforce needs to be cultivated. The Executive Branch should expand existing and develop additional educational programs, scholarship funding, training programs and career paths to build workforce capability in cybersecurity. The Executive Branch should increase its current funding for these efforts.

RECOMMENDATION 14.12: The Executive Branch should develop a coordinated foreign cybersecurity assistance program to assist foreign countries in the development of legal and technical expertise to address cybersecurity.

The Internet knows no geographic boundaries, and threats and attacks emanating from cyberspace can come from anywhere at any time. The volume of cyberattacks originating internationally continues to grow.[1376] To respond to these attacks effectively, a global response involving both the U.S. and foreign governments is necessary.[1377] Although the U.S. government has been working to address cyber incidents through legal and policy actions and public-private partnerships many foreign countries lack either the legal framework or the capacity to respond in a similar manner.

To address this challenge, as it has done in cases of counternarcotics and human trafficking, the federal government must work collaboratively with international partners to address detection, prevention, mitigation and response with respect to cybersecurity. The International Criminal Investigative Training Assistance Program at the Department of Justice is an example of one program that works with foreign governments to develop professional and transparent legal institutions, with a focus on protecting human rights, combating corruption and reducing the threat of transnational crime and terrorism.[1378]

Each federal government agency[1379] with expertise should work collaboratively with its counterpart agencies in foreign governments to nourish the worldwide development of legal and technical cybersecurity expertise. In 1999, the U.S. led a similar collaborative effort to develop global expertise in telecom regulation, leading to the publication of *Connecting the Globe: A Regulator's Guide to Building a Global Information Community*.[1380] A similar effort should be undertaken by the United States government in cybersecurity, bringing multiple countries together to share information on best practices.

RECOMMENDATION 14.13: The FCC should work with internet service providers (ISPs) to build robust cybersecurity protection and defenses into networks offered to businesses and individuals without access to cybersecurity resources. ISPs

that participate in this program should receive technical assistance from the federal government in securing their networks.

Protecting computers and other devices from new and evolving threats found on the Internet is a full-time activity that occurs 24 hours a day, seven days a week. Most Fortune 500 companies spend millions of dollars annually on specialized staff and technology supporting cybersecurity efforts to protect their corporate computers and networks. Smaller businesses and individuals, however, may have limited or even no cybersecurity protection.

ISPs have taken some steps to provide cybersecurity resources to small business and residential customers. For example, Comcast has provided a commercial antivirus and security software suite for free to customers since 2005[1381] and will alert customers if their computers are infected with botnets, viruses or other online threats.[1382] But these efforts only offer incomplete protection at best, since antivirus and security software may miss up to 80% of previously unknown Internet threats and attacks.[1383]

As cybersecurity becomes increasingly specialized and technologically complex, it is no longer reasonable to expect that small business and individuals can engage in self-help when it comes to cybersecurity. By having ISPs take a more proactive role in securing their networks, Internet security can be enhanced, especially since the top 23 ISPs in the United States represent over 75% of all U.S. Internet subscribers.[1384] Building upon efforts already taken by ISPs, the FCC should work with ISPs to build robust cybersecurity protection and defenses into networks offered to business and individuals. Participation by end-users would be voluntary: ISPs could offer a choice to subscribers between a network with built-in cybersecurity protection or a network with no cybersecurity protection. The FCC should identify ways that the federal government can provide ongoing technical assistance to secure these networks as an incentive for participation in this program.

RECOMMENDATION 14.14: OMB should accelerate technical actions to secure federal government networks.

Under the Federal Information Security Management Act (FISMA), OMB, through the Federal Chief Information Officer (CIO), has responsibility for securing all federal networks, except those under the purview of DoD and the Intelligence Community. OMB has undertaken a number of technical efforts to secure its networks. The Federal Desktop Core Configuration, a common platform for end-user computers, has been rolled out throughout the federal government and incorporates a standard information security configuration developed by the National Institute of Standards and Technology (NIST) in collaboration with DoD and the Department of Homeland Security (DHS).[1385] The Trusted Internet Connections initiative is reducing the number of federal government Internet connections from over 8,000 connections down to approximately 50, and then deploying security solutions—including antivirus, firewall, intrusion detection, and traffic monitoring—on the remaining connections.[1386]

In addition to these initiatives, further steps can be taken to bolster the federal government's cybersecurity efforts. The Federal CIO should accelerate technical steps to secure these networks and better position the federal government to react swiftly to new attack vectors. Particularly, the Federal CIO should speed the implementation of Internet Protocol Version 6 throughout federal government computer networks as a step towards implementing Internet Protocol Security and computer security at the network level. The Federal CIO should also accelerate efforts to securing the Internet's routing system.

OMB recently automated the FISMA data collection process, reducing the burden on agencies for FISMA compliance. Automating the data collection process will also allow the Federal CIO to more readily ensure FISMA compliance and improve existing benchmarks towards outcomes-based metrics so that federal agencies are taking all steps necessary to secure federal government IT networks.[1387] Moving towards outcomes-based metrics is vital to securing the nation's critical infrastructure.

Improve Service Delivery

Americans can have a high-performance government that delivers many services online. But to realize this vision, technical and structural barriers must be addressed, including finding secure ways to establish identity and share information across agencies. Many government services rightly require identity authentication, such as presentation of a driver's license when applying for a U.S. passport. Additionally, government agencies must be able to share information across departments, with appropriate privacy safeguards, in order to reduce the burden on the public requesting government services.

In addition to removing these barriers, the government can improve service delivery by leveraging broadband-based tools to support the improvement, integration and modernization of federal government processes.[1388] Low-income Americans accessing government benefits and services must navigate a fragmented world. They deal with multiple agencies and a host of forms. They typically must make in-person visits. A U.S. Government Accountability Office (GAO) report found that a family seeking to apply for the 11 largest means-tested benefits programs—including Temporary Assistance for Needy Families (TANF), food stamps, Medicaid and school meals— would have to complete six to eight applications and visit as many as six government offices. The process often requires many unpaid hours away from work and lengthy commutes.[1389] A government employee on the other side of the desk spends hours per day entering data into antiquated systems that do not allow the kind of data sharing that could save money, improve productivity, reduce error rates and improve outcomes.

RECOMMENDATION 14.15: OMB and the Federal CIO Council should develop a single, secure enterprise-wide authentication protocol that enables online service delivery.

A robust, secure authentication protocol would enable new online government services as well as improvements to existing online government services, like online passport applications and electronic receipt of benefits. Such a system would enable a single sign-on so that individuals could access their college loan and tax information without creating multiple digital identities.

The federal government has released a strategy for development of secure authentication services for federal employees called the Federal Identity, Credential, and Access Management (ICAM) Roadmap.[1390] In addition, the federal government has moved forward with limited implementation of an OpenID[1391] pilot to provide public services requiring the lowest assurance level, or "little or no confidence in the asserted identity's validity."[1392] Consider that a webmail account has some security and is associated with some identity, but because it is simple to claim any name one wishes, there is "little or no confidence" that an email from "John Doe" is indeed from a person named John Doe. OpenID enables simple applications such as using existing credentials (for example, with a webmail account) to provide individual customized Web-page functionality[1393] for the National Institutes of Health (NIH) and other agencies. NIH is also currently testing applications with higher levels of identity assurance that draw on information from providers like Equifax and PayPal.[1394]

A secure authentication protocol would allow the federal government to use broadband to deliver a greater set of government services online to the American people,[1395] but efforts to improve authentication are limited. Even the ICAM Roadmap offers minimal guidance because it focuses primarily on secure authentication as a cybersecurity issue. The Roadmap says little about services for the public and provides no metrics for measuring the delivery of services.

To address these gaps, OMB and the Federal CIO Council should take the lead in developing a flexible, secure government-wide authentication protocol that covers all levels of identity assurance, from the most secure to the least, and that facilitates the deployment of the next generation of online government services. There is support for a federated scheme with OMB and the Federal CIO Council setting standards.[1396] The Federal CIO Council should also revise the ICAM Roadmap to include performance metrics related to government delivery of services to the public.

RECOMMENDATION 14.16: The Executive Branch should establish MyPersonalData.gov as a mechanism that allows citizens to request their personal data held by government agencies.

The federal government holds data on many of its citizens, and the Privacy Act contains provisions for giving people access to it and letting them correct it.[1397] As currently implemented, this is a manual and costly process, and it is not easy for citizens to get access to their information online. Were citizens able to securely authenticate their identity online, they could easily verify the information (and correct any errors), thereby increasing its value.[1398] Therefore, the Executive Branch should create and maintain MyPersonalData.gov. This tool and corresponding website would serve as an interface so citizens could access the data about them held by federal agencies.

For example, MyPersonalData.gov could allow taxpayers to create tax returns by importing data submitted to the Internal Revenue Service by employers and financial institutions into tax forms. This would save individuals time and money in the preparation of their taxes.[1399]

RECOMMENDATION 14.17: Congress should consider re-examining the Privacy Act to facilitate the delivery of online government services and to account for changes in technology.

The Privacy Act is the legal framework for how the federal government handles personal data and information, but it does not address how private third parties handle personal data and information. Its limitations in dealing with the issues that arise with data in electronic databases are well-recognized.[1400]

The Privacy Act also provides no guidance on new technologies that have privacy implications, such as the use of persistent cookies on websites.[1401] Congressional changes to the Act could allow agencies to significantly reduce the administrative burden on students applying for financial aid if agencies are allowed to share personal information with each other given appropriate privacy safeguards such as the permission of the person securely authenticated online.

RECOMMENDATION 14.18: The federal government should undertake a series of efforts to improve the delivery of government services online.

- **OMB should benchmark federal government websites against the private sector and hold agencies accountable for making improvements on an annual basis.**
- **OMB should modernize the Advance Planning Document (APD) process to encourage state governments to develop enterprise-wide solutions.**
- **The Federal web Managers council should promulgate web standards and templates to make the federal web presence easier to navigate, easier to recognize and accessible to people with disabilities.**
- **OMB should deploy a portion of the E – Government Fund to facilitate replication of leading best practices.**
- **The results of these efforts should be included in oMB's annual E-Government Report to congress.**

Though some government websites show great promise, many are still built from a siloed, agency-centric perspective, with insufficient focus on developing websites and portals that are integrated, user-friendly and consumer-centric. Though more than 75% of Internet users have visited a government website,[1402] reports consistently show that public sector websites lag the private sector.[1403] Additionally, the government has failed to meaningfully integrate lessons learned from best practices of leading online government services into its operations. Notable exceptions include the new U.S. Citizenship and Immigration Services (USCIS) portal, which allows applicants to check their immigration status instantly along with typical wait times,[1404] and the Open Government Initiative (see Exhibit 14-A and Box 14-2).[1405] At the state and local government level, the eCityGov Alliance, comprised of nine cities in the state of Washington, is a successful effort to share best practices and offer cross-government online services.[1406] The problem is that the successes are isolated. Not enough has been done to share lessons learned so that other efforts can benefit from the successes.

Sharing best practices can particularly improve the provision of benefits for low-income individuals by state governments. Millions of federal dollars are spent annually on IT that supports these services, and the APD process allows states to obtain approval for the portion of the costs of acquiring new online systems that the federal government contributes. The current system contains important mechanisms to hold states accountable for making smart choices about what systems are developed, but it may also encourage siloed systems, which might add greater costs for later integration as well as biasing states against migrating to solutions that could be more cost-effective in the long term. To address this gap, OMB should work with relevant agencies to modernize the APD process to encourage governments to develop enterprise-wide solutions.

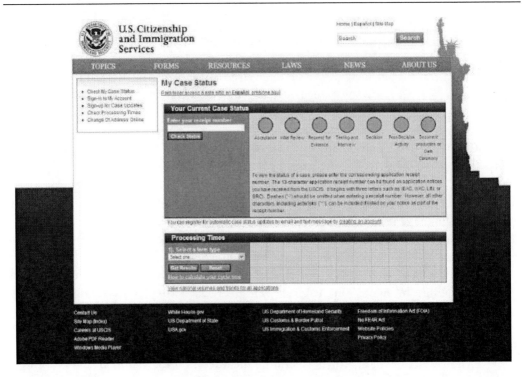

Exhibit 14-A. The U.S. Citizenship and Immigration Services Dashboard

Because public sector websites lag the private sector in usability and design, the Federal Web Managers Council should benchmark the design and usability of government websites against leading industry best practices.

OMB should continually recommend specific improvements that agencies should make, highlight best practices in its annual E- Government Report to Congress and deploy the E - Government Fund to help replicate best practices across the federal government.

BOX 14-2. U.S. CITIZENSHIP AND IMMIGRATION SERVICES OFFERS ONLINE ACCESS

Until recently, when an individual filed an application for citizenship with the U.S. Citizenship and Immigration Services (USCIS), the applicant had no knowledge of his case status. USCIS has recently revamped its website to allow applicants to use an identifying number and immediately check a case status online. Applicants can receive alerts about changes in status via text message and e-mail updates. Most importantly from the applicant's perspective, the whole system is more transparent because wait times and changes in status are clearly documented.

RECOMMENDATION 14.19: The Executive Branch's review of the Paperwork Reduction Act should aim to enable government to solicit input to improve government services.

The Paperwork Reduction Act is a barrier to implementing many best practices.[1407] For example, the Act precludes surveying Web users to improve an agency's Web presence without undertaking an onerous survey-approval process that could take months. One federal employee commented, "[The Paperwork Reduction Act] imposes a burden to obtain any user-generated input ... The result is that we often don't go to the trouble."[1408] The director of USA.gov, the online gateway to the federal government, has stated that the Act needs to be reexamined for the new media world.[1409]

The Executive Branch has begun work on updating the 15-year-old Paperwork Reduction Act.[1410] This review should aim to enable the government to engage in a two-way conversation with the public.

RECOMMENDATION 14.20: The White House Office of Science and Technology Policy (OSTP) should develop a five- year strategic plan for online service delivery.

Since the release of the Quicksilver plan for deployment of 24 Presidential-level E - Government initiatives in 2002,[1411] there has been no subsequent government-wide effort to develop a strategic plan for online federal government services. OMB currently submits an annual E-Government Report to Congress pursuant to the E - Government Act,[1412] but this is an historical summary, not a forward-looking strategic vision.

It is clear that Americans want the opportunity to conduct simple transactions with the federal government online.[1413] OSTP should develop a strategic plan, updated every two years, that addresses issues such as accessibility (including issues raised in the Attorney General's biennial report on Section 508 compliance), benefits administration, alternative platforms, and state and local government partnerships.

RECOMMENDATION 14.21: The federal government should improve the delivery of means-tested benefits to low-income Americans.

- **OMB should enhance Partner4solutions.gov, a platform for improving service delivery of government means- tested benefits, to include a database of government, non-profit and private tools.**
- **OMB should convene a summit in 2010 of state government cios, local health and human services leaders and technology innovators to focus on using technology to modernize benefit services.**

Integrating and streamlining processes through the use of broadband can help low-income Americans receive all the safety-net benefits for which they qualify, demonstrably bettering their chances of getting out of poverty. A 2002 Urban Institute report found that getting access to both Supplemental Nutrition Assistance Program benefits (or food stamps) and Medicaid increases the likelihood of job retention for those leaving TANF. Twenty percent of former recipients who secured both benefits returned to welfare, compared with 51% of those who did not secure both benefits. In our current system, many poor people do not receive all the benefits they need or for which they are eligible. Just over half of those eligible for food stamps receive them. Two- thirds of those eligible for Medicaid or the State Children's Health Insurance Program receive it. One-third of those eligible for TANF receive these benefits. Many cite confusion over eligibility and difficulty of application as major barriers.[1414]

Many states have started to experiment with a continuum of changes that leverage the Internet. ACCESS NYC uses online calculators that screen residents for 35 benefits in seven languages. Other states have set up "one-stop" online applications for multiple sets of benefits. Still others have gone to large-scale systems integration. Moving toward a modernized, integrated online benefits system would improve service delivery, reduce access barriers and drive efficiency.

A recently-launched federal program, the Partnership Fund for Program Integrity, has begun helping state and local governments find innovative ways to improve benefits programs. It should be used to encourage the move to "one-stops" for online applications. Instead of merely aggregating application forms that will ultimately need to be printed, grantees should move toward electronic signatures, full electronic submission and pre-population of fields based on applications for other benefits, which would save clients time and agencies money. These systems could potentially include secure document imaging and storage. A 2007 GAO report notes that Florida's document management and imaging system lets caseworkers retrieve electronic case records in seconds, compared with as long as 24 hours for paper case files.[1415]

Partner4Solutions.gov is a platform for improving service delivery in this space. It should develop a database of online benefits tools from state, local governments and non-profits, functioning as an Apps.gov of the benefits world. Where applicable, the database should include prices (because they can vary so widely). For example, the cost of purchasing or developing a pre-screening tool—an online set of questions to give families a sense of the range and amount of benefits for which they are eligible—costs $15,000 to $5 million.[1416]

Finally, numerous state and local governments are working on initiatives to utilize broadband and online service delivery to improve the administration of benefits programs. Although many best practices are being developed, these efforts are occurring independently of each other. To address this gap, OMB should convene a summit in 2010 of state government CIOs, local health and human services leaders, and technology innovators so they can focus on using technology to modernize benefit services. This summit would have three goals: to develop a shared time horizon for moving toward integrated online platforms for key programs for low-income Americans; to showcase and share available data on costs and benefits of current state tools as well as external innovations such as the Annie E. Casey Foundations' Casebook, a Web 2.0 tool for child welfare case management; and to develop a shared set of best practices that states can use to improve service delivery.

15. CIVIC ENGAGEMENT

Civic engagement is the lifeblood of any democracy and the bedrock of its legitimacy. Broadband holds the potential to strengthen our democracy by dramatically increasing the public's access to information and by providing new tools for Americans to engage with this information, their government and one another. Increasingly our national conversation, our sources for news and information and our knowledge of each other will depend upon broad-band. The transition to new information technologies and services can open new doors to enhance America's media environment, but with traditional sources of news and information journalism under severe stress in the current media and economic environments, we confront serious challenges to ensure that broadband is put to work to strengthen our democracy.

Civic engagement starts with an informed public, and broadband can help by strengthening the reach and relevance of mediated and unmediated information.

Broadband can enable government to share unmediated information more easily with the American people. Providing more information and data to the public about the processes and results of government can strengthen the citizenry and its government.

Broadband can also empower citizens to engage their government through new broadband-enabled tools. Broadband has already increased access to information and revolutionized the way citizens interact with each other. Companies such as YouTube enable the distribution of "user-generated content" over the Internet; YouTube now supports monthly more than 120 million viewers watching more than 10 billion videos.[1417] More than 80% of U.S. adults who are online use social media at least once per month, and half of them participate in social networks such as Facebook.[1418] Today, out of the 36% of Americans involved in a civic or political group, more than half of them (56%) use digital tools to communicate with other group members.[1419] Government must take advantage of these trends and adopt broadband-enabled tools to encourage citizens to communicate with government officials more often and in richer ways—and to hold these officials more accountable.

Building the infrastructure for America's democracy has been a challenge since the birth of this nation. The Founders worried about it long ago. In 1787, when talking about newspapers—the broadband of its time—Thomas Jefferson wrote:

"The basis of our governments being the opinion of the people, the very first object should be to keep that right; and were it left to me to decide whether we should have a government without newspapers or newspapers without a government, I should not hesitate a moment to prefer the latter. But I should mean that every man should receive those papers, and be capable of reading them."[1420]

More than two centuries ago, Jefferson was addressing *deployment*—getting newspapers out ubiquitously—and *adoption*— ensuring people read, recognizing the value of knowledge, and making use of the information infrastructure. Although our technology may change, our democratic challenge remains the same.

Recommendations

Create an open and transparent government

- The primary legal documents of the federal government should be free and accessible to the public on digital platforms.
- Government should make its processes more transparent and conducive to participation by the American people.
- All data and information that the government treats as public should be available and easy to locate online in a machine-readable and otherwise accessible format in a timely manner. For data that are actionable or time-sensitive in nature, the Executive

Branch should provide individuals a single Web interface to manage e-mail alerts and other electronic communications from the federal government.

- All responses to Freedom of Information Act (FOIA) requests by Executive Branch and independent agencies should be made available online at www.[agency].gov/foia.
- The Executive Branch should revise its Data Quality Act guidance to encourage agencies to apply the Act more consistently and facilitate the re-publishing of government data.

Build a robust digital media ecosystem

- Congress should consider increasing funding to public media for broadband-based distribution and content.
- Congress should consider amending the Copyright Act to provide for copyright exemptions to public broadcasting organizations for online broadcast and distribution of public media.
- The federal government should create and fund Video.gov to publish its digital video archival material and facilitate the creation of a federated national digital archive to house public interest digital content.
- Congress should consider amending the Copyright Act to enable public and broadcast media to more easily contribute their archival content to the digital national archive and grant reasonable non-commercial downstream usage rights for this content to the American people.

Expand civic engagement through social media

- The Federal Chief Information Officers (CIO) Council should accelerate the adoption of social media technologies that government can use to interact with the American people.

Increase innovation within government

- The White House Office of Science and Technology Policy (OSTP) should create an Open Platforms Initiative that uses digital platforms to engage and draw on the expertise of citizens and the private sector.
- The Executive Branch and independent agencies should expand opportunities for Americans with expertise in technological innovation to serve in the federal government.

Modernize democratic processes

- Federal, state and local stakeholders should work together to modernize the elections process by addressing issues such as electronic voter registration, voting records portability, common standards to facilitate data exchanges across state borders and automatic updates of voter files with the most current address information.

- The Department of Defense (DoD) should develop a secure Internet-based pilot project that enables members of the military serving overseas to vote online.

15.1. Creating an Open and Transparent Government

Open and transparent governance is central to democratic values. In order for government to be accountable to the public, it must share the results of its policies with the public as well as the processes by which those results are achieved. Ultimately, democracy rests on the ability of the people to evaluate the performance of their government in order to make informed electoral decisions.

RECOMMENDATION 15.1: The primary legal documents of the federal government should be free and accessible to the public on digital platforms.

- **For the Executive Branch and independent agencies, this should apply to all executive orders and other public legal documents.**
- **For congress, this should apply to all votes, as well as proposed and enacted legislation.**
- **For the Judicial Branch, this should apply to all judicial opinions.**

Every person who is subject to the laws of this country should have free access to those laws online.[1421] Online legal documents should be appropriately digitally watermarked to preserve their integrity. For the Executive Branch and independent agencies, this means publishing all executive orders and other public legal documents on the Internet and in an easily accessible, machine- readable format. For the Legislative Branch, this means that Congress should publish all votes, as well as proposed and enacted legislation, in a timely manner, online and in a machine-readable and otherwise accessible format.[1422]

Finally, all federal judicial decisions should be accessible for free and made publicly available to the people of the United States. Currently, the Public Access to Court Electronic Records system charges for access to federal appellate, district and bankruptcy court records.[1423] As a result, U.S. federal courts pay private contractors approximately $150 million per year for electronic access to judicial documents.[1424] While the E - Government Act has mandated that this system change so that this information is as freely available as possible, little progress has been made.[1425] Congress should consider providing sufficient funds to publish all federal judicial opinions, orders and decisions online in an easily accessible, machine-readable format.

RECOMMENDATION 15.2: Government should make its processes more transparent and conducive to participation by the American people.

- **For the Executive Branch, independent agencies, congress and state and local government, all government meetings, public hearings and town hall meetings should be broadcast online.**

- **Congress should consider allowing the American public to track and comment on proposed legislation online.**

In addition to Recommendation 15.1 to make final documents open and transparent to the public, government processes should also be made open and transparent. As a guiding principle, the Knight Commission has declared, "the public's business should be done in public."[1426] Public hearings and town hall meetings are among the most direct and frequent opportunities for the public to engage in their democracy. Video streaming of government meetings expands access to the government by eliminating geographic limitations and allowing for "time shifting," so that a person who is unable to watch a meeting in real time (because they are at work, for example) can still watch the proceedings and provide feedback.[1427] That is why federal, state and local governments should require that all public agency meetings and hearings be streamed over the Internet.[1428] Additionally, these events should offer closed-captioning services to increase accessibility for persons with disabilities and, to the extent practical, enable individuals to ask questions online.[1429]

Congress should consider enabling the American people to electronically track and comment on proposed legislation from anywhere in the U.S.[1430] Tools to enable greater civic participation are already being implemented in some states. For example, in New York the State Senate empowers individuals not only to see bills that have been proposed, but to comment on them.[1431] The Sunlight Foundation has experimented with the use of this tool at a federal level.[1432] Congress should consider offering a similar tool to more actively engage the American people.

RECOMMENDATION 15.3: All data and information that the government treats as public should be available and easy to locate online in a machine-readable and otherwise accessible format in a timely manner. For data that are actionable or time-sensitive in nature, the Executive Branch should provide individuals a single web interface to manage email alerts and other electronic communications from the federal government.

Information enables citizens to monitor inefficiency, waste, fraud and abuse and hold their government accountable. It also empowers the public to more actively participate in government processes and decision-making.[1433] That is why all public information should be easily accessible online and should be posted in real time, whenever possible.[1434]

For government at all levels to be more open, it must provide more information online in open formats.[1435] Data.gov shows the demand for such information. A Web portal that offers an index of data generated by government agencies in machine-readable formats, Data.gov received more than 47 million visits in its first seven months of existence.[1436] Data.gov has also received national and international recognition, providing a model for transparency that cities and nations around the world are looking to emulate.[1437] By publishing all public data online, government can empower the private sector to innovate. In some instances, this is already taking place. As an example, the city of San Francisco launched DataSF.org, publishing more than 100 data feeds and enabling the public to create new applications. These include applications to show individuals crime data and health inspection scores for restaurants.[1438]

Despite this progress, most efforts are far from comprehensive. Even Data.gov contains only a small amount of the data that the federal government possesses.[1439] One survey found that only half of the states provided at least 12 of 20 types of information online in areas that are important to the public. These types of information were selected based on their relevance to people's lives and their usefulness in holding the government accountable. They include financial disclosure reports, audit reports, nursing home and child care center inspection reports and building inspection reports.[1440]

For data that are actionable or time-sensitive in nature, the Executive Branch should provide individuals a single Web interface to manage e-mail alerts and other electronic communications from the federal government. Currently, individual agencies manage e-mail communications and alerts independently in a variety of ways. Developing a single Web interface will simplify individuals' access to alerts and other communications from the federal government.

RECOMMENDATION 15.4: All responses to Freedom of Information Act (FOIA) requests by Executive Branch and independent agencies should be made available online at www.[agency].gov/foia.

FOIA ensures a fair and equitable process through which the public can access information about their government.[1441] However, agencies often do not consider the usability of the information they provide to the American people in response to FOIA requests. For example, the U.S. Customs and Immigration Service (USCIS) received nearly 80,000 FOIA requests in 2008, but in the 60% of cases where the requester asked for the information electronically, USCIS mailed a CD, rather than providing the data online.[1442] Additionally, there are no guidelines regarding the format in which this underlying data should be delivered.

That is why all Executive Branch and independent agencies should make all responses to FOIA requests available online in each agency's FOIA Reading Room. Once records are released pursuant to a FOIA request, they are in the public domain. Agencies are currently required to make frequently requested records (generally defined as records requested three or more times) available on their websites. Nevertheless, agencies have not proactively posted materials likely to be the subject of FOIA requests on their websites, nor have they made records released pursuant to a FOIA request routinely available on their websites. Even initial FOIA determinations by agencies are often not routinely available on agency websites. The U.S. Department of Justice should issue further guidance stating that all records (and not just frequently requested records) released pursuant to a FOIA request (which exclude any information subject to a FOIA exemption) should be posted in an agency's Electronic Reading Room to preempt repeat requests. Doing so would eliminate repetitive FOIA requests, make more agency records accessible to the public and significantly drive down the costs (approximately $338 million per year[1443]) of proces sing FOIA requests.

RECOMMENDATION 15.5: The Executive Branch should revise its Data Quality Act guidance to encourage agencies to apply the Act more consistently and facilitate the republishing of government data.

The federal government should eliminate unnecessary internal barriers to making data available to the public. That is why the Executive Branch should revise its guidance regarding the Data Quality Act. This legislation's purpose is to "ensure and maximize the quality, objectivity, utility and integrity of information" disseminated by the federal government to the public.[1444] Unfortunately, the Act often impedes the release of data. For example, current administration of the Act requires data owners to certify the quality of their datasets before they can be published on Data.gov—even if the data are already publicly available on an agency's website. In practice, this re-certification imposes a burden that keeps data off Data.gov. That burden should be removed. In addition, the U.S. Government Accountability Office (GAO) has noted that the Act is often implemented inconsistently and inefficiently.[1445] These issues have led to confusion regarding what types of data can be posted and the process for posting it.[1446]

15.2. Building a Robust Digital Media Ecosystem

America's communities require a media ecosystem that provides the educational, news and other content necessary to inform the citizenry and to sustain our democracy. Just as communities depend on individuals to create and maintain communities, individuals rely on trusted media intermediaries to connect them with relevant and accurate information so they can make informed decisions in their daily lives.[1447] Today, traditional media and journalism institutions, which serve as essential watchdogs over both the public and private sectors, face significant challenges.

These challenges are well documented. Newspapers are shutting down at an astonishing rate, local television (TV) news stations are laying off reporters and as a consequence statehouses and other governmental institutions are drawing fewer and fewer journalists to cover the news. Between 2001 and 2009, newspapers laid off an estimated 14,000 journalists, 25% of their workforce.[1448] TV news shows eliminated 1,200 people in 2007 alone,[1449] and radio newsrooms shed 16% of their staff in 2008.[1450] Such a drastic contraction in the news media means fewer checks on government and other powerful institutions, more corruption and injustice going unreported and less information being made available to citizens. Whether uncovering the horrific abuse of veterans at a Veterans Affairs hospital or informing the public of toxic chemicals in toys, professional journalism at its best arms citizens and consumers with the information they need to hold leaders accountable and to improve their own communities and the quality of their lives.

The contraction of traditional professional journalism has prompted concern from a wide variety of independent analysts and groups that the United States may end up with fewer "informed communities." The Pew Project for Excellence in Journalism recently stated that business trends in the media were "chilling,"[1447] and a 2009 report from the Columbia Graduate School of Journalism observed that "accountability journalism, particularly local accountability journalism, is especially threatened by the economic troubles that diminished so many newspapers."[1448] A shrinking of journalistic capacity could mean fewer checks on government and other powerful institutions, more scandals and injustices that go unreported, less information available to citizens and less civic engagement.

At the same time, all is not bleak. The popularity and accessibility of the Internet have already led to the development of some creative and experimental media. In San Diego and Minneapolis, journalists created Voice of San Diego and MinnPost, respectively, to fill some of the gaps created by contracting newspapers.[1449] The American Standard covers state government and politics, and ProPublica provides high quality investigative reporting that many news outlets can no longer afford on their own.[1450] Some organizations have enlisted journalism students; others are experimenting with "pro-am" journalism—professionals and amateurs collaborating via the Internet. The spread of broadband can fuel ever more creative uses of technology, including new ways of gathering, explaining and distributing news and information. Never before have the barriers to add one's voice to the civic dialogue been so low. We should seek ways to harness some of these same digital forces that, in part, disrupted old models of journalism to bring creative solutions for restoring American journalism to both large and small communities.

There are differing views about how these negative and positive developments net out. Some feel that private and non-profit sector innovations will fully replace the loss in traditional journalism and, in some cases, improve upon it.[1451] Others, however, are concerned about the state of traditional media in America and believe that these problems may extend to new forms of Internet-based media as well.[1452] For example, these observers argue that the proliferation of choices on the Internet should not obscure the reality that even most online news originates with traditional journalistic organizations.[1453] They suggest, too, that excessive private sector media industry consolidation, coupled with misdirected public sector policies, has inflicted serious harm on traditional news and information media and that special vigilance must be taken to avoid similar outcomes for new media.[1454] The FCC understands the importance of these lines of inquiry and the need to address these questions expeditiously.

These questions will be studied by the FCC's new project on the Future of Media and Information Needs of Communities in a Digital Era (see www.fcc.gov/futureofmedia and GN Docket No. 10-25).[1455] The FCC will move expeditiously to determine what actions are needed to ensure that all citizens have access to vibrant, local and diverse sources of information and news that enable them to enhance their lives, communities and democracy. This project will review trends in the provision of local news by local TV stations, radio and other media in the context of the Internet and evolving economic conditions. The project will hold workshops, seek public input and release a report this year.

Though the Future of Media project is in an early stage, two points should be clear. First, broadband technology can only make a valuable contribution to our civic dialogue if everyone has access to it. As the Internet increasingly becomes the standard platform for receiving information, those who do not have high-speed access to the Internet will be left completely out of the civic dialogue. The media they used to rely on (often inexpensively) will be increasingly weakened if not better fortified for the transition, while salutary alternatives will be only available to the well-wired.

Second, public media will play a critical role in the development of a healthy and thriving media ecosystem. Public media plays a vital and unique role in our democracy, informing individuals and leading our public conversation as well as building cohesion and participation in our communities.[1456] This strength comes from its ability to create connected and informed communities, empower citizens to hold their government accountable and enable people to actively participate in government processes and decision-making.[1457] And at a time of increasing skepticism, cynicism and distrust of institutions, public media has earned and maintained the trust of the American people. According to a 2007 Roper opinion poll, nearly half of all Americans trust the Public Broadcasting Service (PBS) "a great deal"; this is more than trust commercial television or newspapers.[1458] This trust reinforces the critical role that public media plays in American democracy.

This trust enables public media to provide tremendous educational resources to America's families. Last year, after more than 4,000 episodes, Sesame Street celebrated its 40th year on the air.[1459] This is a remarkable testament to public media and to its educational programming. Fittingly, last year public television also launched a tremendous resource for the broadband age in the form of the PBS KIDS preschool video player. During the first month alone, more than 87 million streams of educational content were delivered across PBS KIDS sites.[1460] Providing rich public media content on new digital multimedia platforms will help ensure that another generation of kids will grow up with Sesame Street and other great public television content. Public media's past is a tremendous success story that our communities and our nation should celebrate, and it has already begun developing its 21st century digital identity in myriad ways. This is evidenced by the work of PBS and National Public Radio (NPR) as well as individual public television and radio stations, all of which are playing important roles in communities across the country. For example, Boston public television station WGBH has developed the Teachers' Domain, a free collection of more than 2,000 standards-based digital resources covering diverse content for students and teachers. This collection offers video, audio, articles, lesson plans and student-oriented activities for more than 333,000 registered users.[1461] Additionally, Philadelphia's WHYY radio station has partnered with the Philadelphia Daily News to produce a multimedia civic engagement blog that solicits essays from Philadelphians about their city.[1462]

These examples demonstrate how broadband can bring public media into the digital age and help public media achieve its full potential. But there is more work to do if its future is to be as successful as its past. Public media has historically focused on broadcasting, with its capacity constraints and one- way limitations.[1463] Today, public media is at a crossroads.[1464] It is predominantly structured around broadcast-based communications, both legally and in practice, presenting a challenge in the digital age. That is why public media must continue expanding beyond its original broadcast-based mission to form the core of a broader new public media network that better serves the new multi-platform information needs of America.[1465] To achieve these important expansions, public media will require additional funding.[1466]

RECOMMENDATION 15.6: Congress should consider increasing funding to public media for broadband-based distribution and content.

If public media is to continue playing an important role in supporting civic engagement with online content, it will need expanded support. Public broadcasting is financed by a combination of annual federal appropriations, federal grants, state and local funds and private donations; it receives less than 20% of its funding from the federal government.[1467] As broadband adoption and utilization continue to grow, public media will require greater and more flexible funding to support new digital platforms.[1468]

As one avenue for the funding of online content, Congress should consider creating a trust fund for digital public media that is endowed by the revenues from a voluntary auction of spectrum licensed to public television. By doing so, Congress can increase public media's role by expanding the resources directed to the digital public media ecosystem without diminishing station operations. As discussed in Chapter 5, this plan recommends a process by which commercial television broadcasters may contribute some or all of their spectrum allocation to an auction in the 2012-2013 time period. Non-commercial broadcasters should also be allowed to participate in such an auction on a completely voluntary basis. Stations that contribute some (e.g., half) of their licensed spectrum would then share channels and transmission facilities with other public television stations who also contributed a portion of their spectrum allocation. These stations would not go off the air and would still broadcast their primary streams under their on-air call letters. In addition, these stations would remain direct FCC licensees as they are today, and would continue receiving all the benefits of being a direct FCC licensee, such as must- carry rights.

Congress should consider dedicating all the proceeds from the auctioned spectrum contributed by public broadcasters to endow a trust fund for the production, distribution, and archiving of digital public media.

There would be multiple benefits to public television stations who participate in this auction. First, it could provide significant savings in operational expenses to stations that share transmission facilities. Second, 100% of proceeds from the public television spectrum auction would be used to fund digital multimedia content. The proceeds should be distributed so that a significant portion of revenues generated by the sale of spectrum go to public media in the communities from which spectrum was contributed.

RECOMMENDATION 15.7: Congress should consider amending the copyright Act to provide for copyright exemptions to public broadcasting organizations for online broadcast and distribution of public media.

Creating a robust digital public media ecosystem requires changes to copyright law as well. Congress passed special copyright exemptions for public broadcasting in the 20th century, but these provisions no longer fulfill their original purpose. Current licensing practices make it difficult for public broadcasters to produce and distribute the highest quality programming. These exemptions should be updated to facilitate the distribution of the highest quality programming on 21st century digital platforms.[1469]

RECOMMENDATION 15.8: The federal government should create and fund video.gov to publish its digital video archival material and facilitate the creation of a federated national digital archive to house public interest digital content.

RECOMMENDATION 15.9: Congress should consider amending the copyright Act to enable public and broadcast media to more easily contribute their archival content to a digital national archive and grant reasonable noncommercial downstream usage rights for this content to the American people.

The federal government should facilitate the creation of a federated national archive for digital content. Creating such an archive will require tackling digital rights challenges and coordinating among multiple stakeholders. As part of this federated archive, the Executive Branch should create Video.gov, which would be modeled after Data.gov. This platform would house the federal government's public digital video content, current and historical, and would make it accessible and available to the public. All agencies should be encouraged to release as much video content as possible onto Video.gov. The Executive Branch should also work closely with Congress to ensure that the Library of Congress participates in this effort. Additionally, Congress should consider making a one-time appropriation to fund the creation of this federated collection of national digital archives.

Public and broadcast media are critical to creating a robust national digital archive. Today, public media and much of broadcast media sit on a wealth of America's civic DNA in the form of millions of hours of historical news coverage of wars, elections and daily life. This archival content could provide tremendous educational opportunities for generations of students and could revolutionize how we access our own history (see Box 15-1).

BOX 15-1. NPR'S OPEN APPLICATION PROGRAMMING INTERFACE (API): A MODEL FOR A NATIONAL DIGITAL ARCHIVE

In July 2008, NPR launched an Open API. The API framework provides mediated access to almost 15 years of NPR-produced content to NPR member stations. This allows NPR's member stations to curate NPR content. For example, WBUR in Boston re-launched its website, using the API to mix local and national news stories. Third parties can also consume and share NPR content (non- commercially) using the API. Opening this cache of data for non-commercial use led to the development of both an iPhone application (app) and an Android app. Both apps were not developed by NPR, but rather by supporters and programmers who used the API to build them. This open framework is an example of the kind of digital archive that would significantly expand access to rich content.

These opportunities will only be realized if several challenges are addressed.[1470] For example, public television has attempted to launch such a digital video archive but has run into difficulties obtaining necessary clearances from holders of intellectual property rights. To address this issue, Congress should consider amending the Copyright Act to enable public and broadcast media to more easily contribute their archival content to a digital national archive. In addition to clearing these upstream rights for submission into a digital national archive, the amendment to the Copyright Act should grant the public reasonable non-commercial downstream usage rights to all materials deposited into the archive. This would ensure that archival content is open and accessible.[1471] Any such amendment to the Copyright Act should take into account the interests of affected copyright holders.

15.3. Expanding Civic Engagement through Social Media

Government must also improve the quality and number of points at which the American people can contact their government by implementing social media tools, providing opportunities for outside experts to increase innovation within government, and empowering citizens to engage in the democratic process in a digital age.

RECOMMENDATION 15.10: The Federal CIO council should accelerate the adoption of social media technologies that government can use to interact with the American people.

Just as the internal use of social media tools can enhance the performance of government, social media presents a tremendous opportunity for Americans to provide meaningful input into their democracy. Americans use these tools in their daily lives and are more likely to interact with government officials and agencies if these tools make it easier.

Recent growth in adoption of social media has been dramatic. According to the Pew Internet & American Life Project, 35% of American adult Internet users have a profile on an online social networking site. That is four times as many as three years ago. These tools are likely to become even more prevalent over the coming years as the 65% of American teens that are online use social networks to engage with their government.[1472] In order to maintain effective contact with the American people, government will need to adopt these tools.

The government should view social media technologies not as pilot projects or add-ons, but as tools central to the achievement of its mission. Government should adopt a variety of new media tools across many areas—from those primarily used to communicate to those that enable more intensive participation.

While adoption of these tools has been uneven, there are many success stories (see Box 15-2). The Centers for Disease Control and Prevention (CDC) utilizes social media platforms to provide access to credible, science-based health information. Between April 22, 2009, and Dec. 6, 2009, the CDC had more than 2.6 million views of H1N1 podcasts, more than three million views of H1N1-related YouTube videos, and more than 37 million views of H1N1-related media feeds.[1473] The Transportation Security Administration (TSA) has also achieved success with social media, launching a blog in 2008 to give travelers the opportunity to ask questions and raise concerns.[1474] TSA's blog has had more than one million hits and has resulted in improvements like educating screeners about certain computers and translating regulations into easy-tounderstand language.[1475]

BOX 15-2. BROADBAND-ENABLED DIPLOMACY: CITIZEN-TO-CITIZEN ENGAGEMENT AS AN EXAMPLE OF 21ST CENTURY STATECRAFT

Government can also use new technologies to reach people around the world. On Nov. 13, 2009, the U.S. Embassy in Beijing launched pages on two leading social networking portals in China.[1479]

Social media tools are also connecting individuals across nations and regions. The U.S. Department of State recently announced the creation of a "Virtual Student Foreign Service." This program creates "dorm-room diplomats" by matching American college students with embassies and college students in other nations to build transnational relationships and cultural understanding through digital citizen-to-citizen diplomacy.[1480] The State Department has also used Skype videoconferencing capabilities to connect students in Massachusetts to students in Afghanistan, enabling the Afghan students' first face-to-face conversations with Americans.[1481] Broadband-based diplomacy will only become more important in the years to come.

The FCC has also made extensive use of social media tools, regularly communicating with its more than 330,000 Twitter followers (the third most of any federal agency) and actively engaging the public.[1476] So far, individuals have submitted more than 450 ideas to the FCC, which have generated more than 7,500 comments and over 37,000 votes, all online.[1477] The FCC has also posted more than 175 entries on its 4 blogs, which have generated more than 11,000 comments.[1478]

Government can use social media in innovative ways to engage individuals on a state and local level as well. Spartanburg County, S.C., and the town of Cary, N.C., have used social networking to engage residents, soliciting ideas and feedback concerning local community projects.[1482] The state of New York has released a series of Web-based tools to engage residents in the state's budgetary challenges, including an online calculator that allows individuals to create their own proposal to balance the budget.[1483] Maine has engaged residents in the budgeting process through a similar online budget-balancing tool as well.[1484]

15.4. Increasing Innovation in Government

Beyond transparency, government should leverage broadband to experiment with new ideas and technologies to extend opportunities for engagement.

RECOMMENDATION 15.11: The White House Office of Science and Technology Policy (OSTP) should create an open Platforms initiative that uses digital platforms to engage and draw on the expertise of citizens and the private sector.

- This initiative should create open expert and peer review platforms to bring outside expertise to government.
- This initiative should create open problem-solving plat-forms, including competitions, to bring innovative solutions to government.
- This initiative should create open grantmaking platforms to improve the grantmaking process and enable greater innovations in grantmaking.

Although progress has been uneven, there are examples of innovative collaboration throughout the government. As part of the development of the Open Government Initiative, OSTP solicited comments online through a public brainstorming blog, a wiki and a collaborative drafting tool.[1485] To build on this progress, OSTP and the Office of Management and Budget (OMB) should launch and manage an initiative to develop open platforms that increase participatory governance.[1486] These include open peer review and open expert network platforms that enable subject matter experts to volunteer to review policies under consideration and brainstorm policy ideas with each other. The federal government has already taken steps to empower citizen experts. In 2007, the U.S. Patent and Trademark Office launched its Peer-to-Patent program, a groundbreaking Internet-based program in which expert volunteers assist the federal government with reviewing patent applications. Within the first year, Peer-to-Patent attracted more than 2,000 reviewers, and 93% of patent examiners surveyed said that they would welcome examining another patent application with public participation.[1487] This kind of knowledge-sharing platform can reduce the cost of policymaking and improve government performance.

RECOMMENDATION 15.12: The Executive Branch and independent agencies should expand opportunities for Americans with expertise in technological innovation to serve in the federal government.

Because many of the best ideas come from outside government, OSTP and the FCC should create an Innovation Corps and an Innovation Corps to ensure that new ideas continue to flow to the federal government. An FCC-operated Innovation Corps of volunteers would serve as a think tank for technologists from inside and outside government who would volunteer to design and develop platforms and applications for all levels of government. An OSTPadministered Innovation Fellows program could be structured similarly to the White House Fellows program.[1488] It would place leading private sector experts and innovators throughout the federal government for one year.

15.5. Modernizing the Democratic Process

More Americans engage in democratic election processes than in any other civic act. By bringing the elections process into the digital age, government can increase efficiency, promote greater civic participation and extend the ability to vote to more Americans.

The current paper-based system for voter registration can include multiple steps: collecting information on paper forms, manually entering handwritten data onto voter lists and offering third-party groups the opportunity to distribute, collect and submit handwritten registration cards. These practices result in a system that is often inaccurate and cumbersome, with large numbers of registration forms inundating election offices prior to each election. One recent study estimates that voter registration problems resulted in more than two million voters being unable to vote in the 2008 general election. The problems are even worse for members of the military serving overseas; service members are more than twice as likely to face registration problems as the general public.[1489] According to an Overseas Vote Foundation survey, nearly a quarter (23.7%) of experienced overseas voters had questions or problems when registering to vote in 2008.[1490] Maintaining this poorly functioning system is costly in terms of dollars as well as votes. A study of voter registration costs in Oregon found that in 2008 voter registration alone cost taxpayers more than four dollars per vote, with an ultimate bill of almost nine million dollars.[1491]

RECOMMENDATION 15.13: Federal, state and local stakeholders should work together to modernize the elections process by addressing issues such as electronic voter registration, voting records portability, common standards to facilitate data exchanges across state borders and automatic updates of voter files with the most current address information.

Government can improve the voting system by modernizing voter registration to increase efficiency and decrease confusion. This change would also increase accessibility for those who have difficulty with current voter registration processes, such as people living in rural areas and on Tribal lands and disabled populations who have difficulty traveling or face other accessibility challenges.[1492] These recommendations will not provide instant solutions, but they are important steps toward creating a more rational system.

The first step must be to modernize the voter registration process. Arizona, Kansas and Washington already permit citizens to complete and submit voter registration applications online.[1493] In Utah, the Governor's Commission on Strengthening Democracy published a final report in December 2009 recommending that all citizens of Utah be allowed to register to vote online.[1494]

Common standards will assist in making voting records portable so that these records update whenever citizens change party affiliation, marital status or move. Several states have already begun to adopt common standards to facilitate data exchange across state borders.[1495] Delaware has implemented a new eSignature system that requires every visitor to the Division of Motor Vehicles to register to vote, update their registration or decline to do so. Delaware's system immediately downloads updated data directly into voters' files, eliminating the need for data entry and reducing the possibility for human error. The eSignature program saved Delaware $200,000 annually, and it can save other states money as well.

Local governments have also reaped benefits from modernizing voter registration. In Maricopa County, Arizona, paper registration forms cost at least 83 cents each to process, while online registration costs an average of only three cents.[1496] In Travis County, Texas, the County Tax Office implemented an Internet-based application that allows citizens to register to vote online, reducing citizen calls by 30% and walk-ins by 40%. Voter fraud was also minimized by using wireless devices to instantly confirm voter eligibility.[1497]

The federal government has taken steps in this direction as well. The Military and Overseas Voter Empowerment Act, recently passed by Congress, requires that states (beginning with the 2010 general election) establish procedures to allow voters covered by the Uniformed and Overseas Citizens Absentee Voting Act to electronically request voter registration applications and absentee ballot applications for federal elections.[1498]

While this is a positive step, empowering citizens to register to vote online would remove additional obstacles.[1499]

RECOMMENDATION 15.14: The Department of Defense (DoD) should develop a secure internet-based pilot project that enables members of the military serving overseas to vote online.

According to the Overseas Voter Foundation, more than half (52%) of the military serving overseas who tried to vote were unable to do so because their ballots were late or never arrived.[1500] Based on a survey of seven states by the Congressional Research Service, an average of more than 25% of military and overseas ballots were returned as undeliverable, lost or rejected in the 2008 election.[1501]

The federal government has demonstrated clear intent to address these issues. In 2002, the Help America Vote Act established the Election Assistance Commission to serve as a national clearinghouse, develop voluntary guidelines and study new technologies related to voting. The National Defense Authorization Act for FY2005 mandated the creation of a secure Internet-based electronic voting pilot after the Election Assistance Commission establishes Internet voting guidelines. In 2007, the GAO built on this momentum, recommending that the Election Assistance Commission work with major stakeholders such as the DoD to create an action plan to address security and privacy issues and develop a timeframe for developing Web-based absentee voting guidelines.

In the meantime, other groups are taking important steps forward. The Overseas Voter Foundation and the Pew Center on the States have developed an online tool to give U.S. military personnel and other citizens living overseas easier access to Federal Write-in Absentee Ballots. This tool was implemented prior to the 2008 election and yielded positive results, receiving 4.5 million visitors in 2008 and registering almost 90,000 voters.[1502] Several states, including Minnesota and Ohio, have launched similar tools.[1503]

Some states have already made significant progress on these issues. In September 2008, Arizona launched a Web-based voting system that allows the military and overseas citizens to vote online, with completed ballots uploaded directly to the Secretary of State's website. It has been approved by the Department of Justice and uses "industry standard, 128-bit encryption technology to ensure security, privacy and the overall integrity of the ballot." At least five other states, including Missouri, Florida, Colorado, Montana and Washington, permitted some version of electronic voting (via e-mail or a secure online system) in the 2008 general election.[1504]

16. PUBLIC SAFETY

Safety and security are vital to America's prosperity. Broadband can help public safety personnel prevent emergencies and respond swiftly when they occur. Broadband can also provide the public with new ways of calling for help and receiving emergency information.

A cutting-edge public safety communications system uses broadband technologies:

- To allow first responders anywhere in the nation to send and receive critical voice, video and data to save lives, reduce injuries and prevent acts of crime and terror.
- To ensure all Americans can access emergency services quickly and send and receive vital information, regardless of how it is transmitted.
- To revolutionize the way Americans are notified about emergencies and disasters so they receive information vital to their safety.
- To reduce threats to e-commerce and other Internet-based applications by ensuring the security of the nation's broadband networks.

Unfortunately, the United States has not yet realized the potential of broadband to enhance public safety. Today, first responders from different jurisdictions and agencies often cannot communicate during emergencies. Emergency 911 systems still operate on circuit-switched networks. Similarly, federal, Tribal, state and local governments use outdated alerting systems to inform the public during emergencies.

The United States also faces threats to the resiliency and cybersecurity of its networks. As the world moves online, America's digital borders are not nearly as secure as its physical borders.

The country must do better. In a broadband world, there is a unique opportunity to achieve a comprehensive vision for enhancing the safety and security of the American people. Careful planning and strong commitment could create a cutting-edge public safety communications system to allow first responders anywhere in the nation to communicate with each other, sending and receiving critical voice, video and data to save lives, reduce injuries and prevent acts of crime and terror.

Broadband can also make 911 and emergency alert systems more capable, allowing for better protection of lives and property. For example, with broadband, 911 call centers (also known as public safety answering points or PSAPs) could receive text, pictures and videos from the public and relay them to first responders. Similarly, the government could use broadband networks to disseminate vital information to the public during emergencies in multiple formats and languages.

Finally, well-structured and well-protected broadband networks could reduce threats to Internet-based applications. The proliferation of Internet Protocol (IP)-based communications requires stronger cybersecurity. Disasters and pandemics can lead to sudden disruptions of normal IP traffic flows. As a result, broadband communications networks must be held to high standards of reliability, resiliency and security.

The recommendations in this chapter are designed to realize this vision.

Recommendations

Promote public safety wireless broadband communications

- Create a nationwide interoperable public safety wireless broadband communications network (public safety broad-band network).
- Survey public safety broadband wireless infrastructure and devices.
- Ensure that broadband satellite service is a part of any emergency preparedness program.
- Preserve broadband communications during emergencies.

Promote cybersecurity and the protection of critical broadband infrastructure

- The Federal Communications Commission (FCC) should issue a cybersecurity roadmap.
- The FCC should expand its outage reporting requirements to broadband service providers.
- The FCC should create a voluntary cybersecurity certification regime.
- The FCC and the Department of Homeland Security (DHS) should create a cybersecurity information reporting system (CIRS).
- The FCC should expand its international participation and outreach.
- The FCC should explore network resilience and preparedness.
- The FCC and the National Communications System (NCS) should create priority network access and routing for broadband communications.
- The FCC should explore broadband communications' reliability and resiliency.

Encourage innovation in the development and deployment of Next Generation 911 (NG 911) networks and emergency alert systems

- The National Highway Traffic Safety Administration (NHTSA) should prepare a report to identify the costs of deploying a nationwide NG 911 system and recommend that Congress consider providing public funding.
- Congress should consider enacting a federal regulatory framework.
- The FCC should address IP-based communications devices, applications and services.
- The FCC should launch comprehensive next-generation alert system inquiry.
- The Executive Branch should clarify agency roles on the implementation and maintenance of a next-generation alert and warning system.

16.1. Promoting Public Safety Wireless Broadband Communications

RECOMMENDATION 16.1: Create a public safety broadband network.

- **Create an administrative system that ensures access to sufficient capacity on a day-to-day and emergency basis.**
- **Ensure there is a mechanism in place to promote interoperability and operability of the network.**
- **Establish a funding mechanism to ensure the network is deployed throughout the united states and has necessary coverage, resiliency and redundancy.**
- **Conform existing programs to operate with the public safety broadband network.**

The country has long recognized the potential for broadband technologies to revolutionize emergency response wireless mobile communications. This technology will give first responders new tools to save American lives. The country needs a public safety broadband network that allows first responders to communicate with one another. A three-pronged approach will allow the speedy deployment, operation and continued evolution of such a network.

First, an administrative system must ensure that users of the public safety broadband spectrum have the capacity and service they require for their network and can leverage commercial technologies to capture economies of scale and scope. There are significant benefits, including cost efficiencies and improved technological advancement, if the public safety community can increasingly use applications and devices developed for commercial wireless broadband networks. Ultimately, this system must be flexible, allowing public safety entities to forge incentive-based partnerships with commercial operators and others. [1505]

This system will allow the public safety community to realize the benefits of commercial technologies, which will reduce costs and ensure the network evolves. However, leveraging commercial broadband will not be sufficient to develop a truly interoperable nationwide network that meets public safety standards. To ensure the necessary resiliency, capacity and redundancy, the public safety community should be able to roam and obtain priority access on other commercial broadband networks. Commercial operators will need to be compensated at a reasonable rate for this service.

Past efforts to create a public safety narrowband interoperable voice network have failed. Data suggest that many public safety radio systems lack basic interoperability. They also suggest that most jurisdictions that have improved their systems still only have an "intermediate" level of interoperability at best—not the advanced level of interoperability that is required for truly seamless communications in the event of a major emergency.[1506] The public safety broadband network offers a new opportunity to achieve advanced interoperability now.

In addition to a strong administrative system, the FCC should also create an Emergency Response Interoperability Center (ERIC) to ensure that these applications, devices and networks all work together, so that first responders nationwide can communicate with one another seamlessly. In addition, the Federal Emergency Management Agency (FE MA) should undertake a survey to track progress on broadband interopera- bility for the public safety community. ERIC will set the course for interoperability immediately and ensure it is maintained. Focusing on interoperability from the beginning should help the public safety broadband network to overcome the difficulties faced by other earlier voice efforts.

Finally, a grant program will be designed to provide federal support to local efforts in order to fund the capital and ongoing costs of the public safety broadband network. The grant program must provide public safety network operators with long-term support and enough flexibility to form appropriate partnerships with systems integrators and other vendors to en- sure the public safety broadband network is deployed properly.

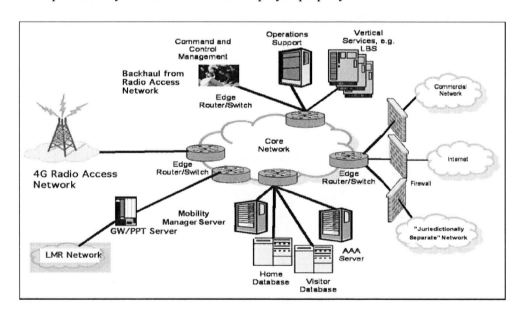

Exhibit 16-A. Public Safety Broadband Network Architecture[1507]

BOX 16-1. REALIZING THE PROMISE OF BROADBAND TO IMPROVE EMERGENCY MEDICAL RESPONSE

Cardiologist Richard Katz knows the life-saving potential of broadband. During an FCC field hearing at Georgetown University Medical Center, the George Washington University (GWU) professor of medicine vividly detailed how wireless broadband technologies can help him provide emergency medical care. A "smart band-aid" attached to an accident victim's chest or wrist can detect vital signs and wirelessly transmit this information to Dr. Katz over GWU's mVisum network. He can receive electrocardiograms of "pristine" quality on his cell phone. And he can use his phone to access patient medical records and disseminate emergency messages and alerts. In short, broadband technologies allow Dr. Katz to integrate aspects of medical care, improving his ability to offer assistance during a disaster or other emergency.

Administrative System

In 1997, Congress directed the FCC to provide public safety agencies with spectrum in the 700 MHz band, considered prime spectrum for public safety communication. In 2007, the FCC adopted rules to promote the construction, deployment and operation of a nationwide and seamless wireless 700 MHz public safety broadband network[1508] by creating a mandatory partnership between the public safety community and the private licensee of a 700 MHz commercial spectrum allocation known as the "D block." The FCC subsequently held an auction in which the D block spectrum failed to attract a required minimum bid. There are many possible reasons for this failure.[1509]

The FCC should overcome past challenges by encouraging, though not requiring, incentive-based partnerships to ensure success. The FCC should encourage network solutions that reduce costs and should provide options for the public safety community to leverage commercial networks, private networks or both.[1510] These rules should also provide the public safety community with more competitive choice among commercial partners. In addition, once the new network is able to support "mission critical" voice communications, the FCC should evaluate the spectrum requirements necessary to ensure adequate capacity for that use, as well as for existing networks. Ultimately, a more flexible set of rules should allow a better balance between the needs of the public safety community and the companies that will partner to build this network.

In more detail, this administrative system should include:

- *An opportunity to enter flexible spectrum-sharing partnerships with commercial operators.* The public safety community must be able to partner with commercial operators and others (such as systems integrators) to lower the costs of building the network and encourage its evolution. Unlike the previous approach that focused solely on the D block, an incentive-based partnership model that addresses not just the D block, but commercial wireless spectrum more broadly, will provide enhanced flexibility and the benefits of economies of scale. Such partnerships should be subject to interoperability requirements set forth by ERIC. Public safety licensees should also be able to allow non-public safety partners to use their spectrum on a secondary basis—that can be preempted—through leasing or similar mechanisms.

Partners could include critical infrastructure users such as utilities connecting to the Smart Grid. [1511] However, any revenues received by a public safety entity for such use must be used to build or improve the public safety broadband network.

- *Public safety access to roaming and priority access on commercial networks.* To improve the capacity of public safety networks during emergencies, the FCC should begin a rule- making to require commercial mobile radio service providers to give public safety users the ability to roam on commercial networks in 700 MHz and potentially other bands. The public safety community should have this ability both in areas where public safety broadband wireless networks are unavailable and where there is currently an operating public safety network but more capacity is required to respond effectively to an emergency.

  The rulemaking also should stipulate that, when a public safety broadband wireless network is at capacity or unavailable, authorized public safety users should get priority access on commercial networks, including all networks using the 700 MHz band and potentially other networks as well. The licensee(s) should be able to obtain priority access under terms similar to those required in today's Wireless Priority Service (WPS). But, unlike WPS, this capacity should be available for state and local first responders as well as National Security/Emergency Preparedness (NS/EP) communications. In addition, the priority access framework should take advantage of the additional access and prioritization capabilities of 4G wireless technologies. Unlike today's circuit-switched cellular networks, 4G wireless networks can give public safety data immediate priority without waiting for commercial capacity to be freed up. Commercial operators should receive reasonable compensation for public safety priority access and roaming capabilities on their networks.

- *Licensing the D block for commercial use, with options for public safety partnership.* The FCC should quickly license the D block for commercial use, while implementing several requirements for the D block licensee(s) to maximize options for partnerships with public safety. First, the FCC should require the D block licensee(s) and the public safety broadband licensee(s) each to operate their networks using the same air interface technology standard. The emerging consensus of the public safety community and carriers is that 700 MHz networks will use the Long Term Evolution (LTE) family of standards. The FCC should consider designating this standard. [1512] A consistent air interface creates a greater likelihood of interoperability between the public safety and commercial D block networks. It will facilitate roaming between networks to improve coverage and access for public safety and commercial customers. In addition, a consistent air interface will encourage a larger number of potential users and allow public safety entities to benefit from commercial economies of scale that otherwise would not exist. Before the D block is auctioned, it must be clear that any D block licensee(s) will be required to provide roaming and WPS-like priority access with reasonable compensation.

  Second, it is critical to develop commercial devices that can operate across 3GPP Band 14 in its entirety. (Band 14 in the 700 MHz band includes the D block and the public safety broadband spectrum.) Accordingly, the FCC should require the D block licensee(s), and potentially other 700 MHz commercial licensees, to develop and offer devices capable of providing service using all 700 MHz Band 14 spectrum and identify a path toward the large-scale production of such devices. Commercial devices should allow the public safety community access to better and less expensive options for use in the public safety spectrum, and will facilitate access to spectrum blocks where the D block licensee and the public safety licensee enter into a shared network partnership. The FCC should explore other ways to encourage the deployment of public safety devices that transmit across the entire broadband portion of the 700 MHz band (i.e., Band 12, Band 13, Band 14 and Band 17).

- *Liability protection for commercial partners.* A federal statute provides wireless, Voice over Internet Protocol (VoIP) and other emergency communications providers with immunity or liability protection for carriage of public safety communications that is not less than the immunity or liability protection given to local exchange carriers.[1513] Commercial licensees should have similar liability protection for public safety communications when, for example, public safety licensees are roaming or using priority access on commercial networks or on shared networks supporting both commercial and public safety communications.
- *Leveraging purchasing power.* The FCC, working with other federal agencies, should explore other cost-saving measures for the buildout of public safety broadband networks. ERIC and DHS should work with the General Services Administration (GSA) to provide rate schedules that public safety entities can use to access commercial nationwide broadband networks and to obtain equipment for their networks. This would generate immediate cost savings and provide an important cost benchmark. In addition, state, Tribal and local governments can help lower costs. Infrastructure sharing can also reinforce network reliability and service continuity among commercial networks, particularly carriers entering into incentive-based partnerships with public safety organizations.

ERIC

The FCC should create ERIC under the umbrella of the Public Safety and Homeland Security Bureau immediately. ERIC will develop common standards for interoperability and operating procedures to be used by the public safety entities licensed to construct, operate and use this nationwide network. To establish a common vision, ERIC must exist before any licensees begin construction of such a network. This will ensure that government, public safety and the communications industry move away from creating and supporting fragmented public safety networks for broadband wireless communications.[1514]

ERIC will establish a baseline for the seamless exchange of public safety wireless broadband communications on a nationwide, interoperable basis from the start of the network's development. This is crucial to allow responders from varying jurisdictions and disciplines to communicate with one another when they converge at an emergency, or when incidents span several jurisdictions. Similarly, first responders must have access to common applications in any situation or location.[1515] To ensure success and leverage existing expertise, ERIC should be chartered to work closely with DHS's Office of Emergency Communications (OEC). Close coordination will enable ERIC to complement OEC's mission of creating standard operating procedures and governance to ensure that public safety communications flow over a seamless network. ERIC also should have a public safety advisory body to ensure appropriate consultation.[1516]

The FCC's FY2011 budget proposes $1.5 million in funding to establish ERIC and support initial staffing requirements. As ERIC and the proposed broadband networks mature, about $5.5 million will be necessary each year starting in FY2012 for ERIC to be fully functional.[1517] These additional funds will allow the FCC to partner with the National Institute of Standards and Technology (NIST) to develop appropriate standards and to maintain ERIC's expertise. The funds will also ensure adequate staffing to address the three core functions of ERIC: network engineering, network technical operations and network governance. In addition, Congress should consider providing DHS $1 million of public funding in FY2011, as proposed in its budget, and each year thereafter. The funding will help DHS to coordinate ERIC with OEC and relevant DHS entities, and enhance OEC outreach to Tribal, state and local agencies.

At a minimum, ERIC should:

- Adopt technical and operational requirements and procedures to ensure a nationwide level of interoperability; this should be implemented and enforced through FCC rules, license and lease conditions and grant conditions.
- Adopt and implement other enforceable technical, interoporability and operational requirements and procedures to address, at a minimum, operability, roaming, priority access, gateway functions and interfaces and interconnectivity of public safety broadband networks.
- Adopt authentication and encryption requirements for common public safety broadband applications and network use.
- Coordinate the interoperability framework of regulations, license requirements, grant conditions and technical standards with other entities (e.g., the public safety broadband licensee(s), DHS, NIST and the National Telecommunications and Information Administration).

ERIC should also work with DHS and the public safety community to ensure that the public safety broadband network and public safety narrowband wireless networks can communicate with one another seamlessly. ERIC's public safety advisory committee[1518] will provide input from the public safety community on ERIC's proposed actions.

ERIC should work with NIST's Public Safety Communications Research Program to ensure that it collaborates in its work on research, development, testing, evaluation and standards with both the public safety community and industry. No federal laboratory facilities exist to independently test and demonstrate public safety 700 MHz broadband technologies. Creating a neutral host facility will allow all stakeholders to work together to develop a nationwide seamless public safety wireless broadband network and ensure that commercial broadband standards can meet public safety's specific requirements. This will help make networks and equipment compatible for public safety use.

NIST has announced that it is moving forward with development of a demonstration 700 MHz public safety broadband network in FY2010. Congress should consider allocating long-term public funding to continue this and other programs that support the new public safety network.

Grant Program

Development of a nationwide public safety broadband network through incentive-based partnerships will make Americans safer and more secure.[1519] A grant program will give public safety its own "hardened" broadband wireless access network; ensure that the most vulnerable areas of the United States have the coverage they require; provide public safety with additional capacity and resiliency via access to nearby commercial spectrum; ensure that the emergency response community has the tools it requires; and optimize the effective use of resources.

As shown in Exhibit 16-B, a multi-pronged approach will provide public safety with greater dependability, capacity and cost savings. First, the hardened network will provide reliable service throughout a wide area. Second, since emergency responders will be able to roam on commercial networks, capacity and resiliency will improve, at a reasonable cost. Third, localized coverage will improve through the use of fixed microcells—like those that provide indoor coverage in skyscrapers—and mobile microcells, which can be placed in fire trucks, police cars and ambulances. Fourth, equipment can be retrieved from caches and used during a disaster when infrastructure is destroyed or insufficient or unavailable. Grants to support the public safety broadband network should be distributed by a single agency to streamline operations, reduce costs and ensure that grants are made in a consistent manner. The grants should only fund projects that comply with ERIC requirements and should be made for the following four purposes:

- Construction of a public safety 700 MHz broadband network that involves partnerships and uses commercial infrastructure, the public safety infrastructure or both through incentive-based partnerships.
- Coverage of the rural areas within the network's geography.
- Hardening of the existing commercial network and new sites that operate as part of the public safety network (including covering non-recurring engineering costs for priority broadband wireless).[1520]
- Development of an inventory of deployable capability for the 700 MHz public safety band.

A single grant-making agency, in coordination with ERIC, should structure the funding to ensure the network is built efficiently. The grant-making agency should have flexibility to limit the time that a grant recipient has to spend any granted funds. It should also ensure that the money spent is accounted for through reporting and auditing requirements. The grant-making agency should encourage grant recipients to enter into infrastructure-sharing agreements, where appropriate, with entities deploying broadband networks with support from other grant programs. Such arrangements should be reviewed annually, and any savings they generate should be taken into account when allocating funds for each program.

The public safety broadband network requires a substantial investment. Using a 99% population coverage model,[1521] deployment of this network will require as much as $6.5 billion in capital expenditure in 2010 dollars over a 10-year period,which can be reduced through efficiency measures such as state and local programs and USF.[1522] Initial public funding for the capital requirement should commence in a timely manner to enable the public safety network to benefit from the planned build-outs of the private 4G wireless broadband networks, which are scheduled to begin in 2010. Congress should consider providing the bulk of these funds in the second to fifth years of the network's construction.

Ongoing costs, including operating expense and appropriate network improvement costs are expected to rise from zero at the beginning of FY2011 to a peak of as much as $1.3 billion per year in year 10 of the capital build program, following a substantial ramp-up that coincides with the network's expansion.[1523]

The total present value of the capital expenditure and ongoing costs over the next 10 years is approximately $ 12–16 billion. State and local governments could contribute funds to cover some of these costs, and there may be additional cost-saving methods that reduce this estimate—such as sharing federal infrastructure, working with utilities, or use of state and local tower sites to improve coverage. This undertaking is also expected to produce a significant number of long-term U.S. jobs. [1524]

It is essential that the United States establish a long-term, sustainable and adequate funding mechanism to help pay for the operation, maintenance and upgrade of the public safety broadband network. America's safety depends on it. Congress should consider creating such a funding mechanism in FY2011, but in any event, no later than FY2012. Recognizing that Americans will obtain substantial benefits from the creation of this network, imposing a minimal public safety fee on all U.S. broadband users would be a fair, sustainable and reasonable funding mechanism. The fee should be sufficient to support the operation and evolution of the public safety broadband network.

It is essential that the public safety community has the funds to operate, maintain and improve this network. All U.S. broadband users will benefit from this network. Spreading nominal costs among them will ensure that this country's emergency responders have access to critical communications capabilities when and where they need them. [1525]

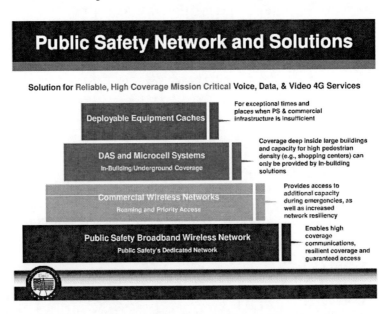

Exhibit 16-B. Public Safety Network and Solutions

Congress should consider authorizing the FCC to impose or require the imposition of such a fee or other funding means. Congress should also consider enabling FCC the to implement or authorize mechanisms to collect, manage, audit and support the grant-making agency's disbursement of these funds. Receipts would fund the grant-making agency's program for public safety broadband operations and evolution. Strict conditions must be established to prohibit any diversion of these funds by state and local governments, and require adherence to ERIC-developed standards. The grant-making agency should be authorized to determine how to best allocate these funds to ensure an appropriate balance among urban, suburban and rural users and to require grant recipients to account for the funds they receive. And it should distribute the funds in a way that also enables the evolution of the network.

Existing Programs

In emergencies, the federal government uses an FCC-developed system called Project Roll Call to determine the operational status of wireless and broadcast communications (including public safety communications) and to help emergency managers restore operations when necessary. However, the system is not designed to operate in a 700 MHz broadband spectrum environment. Deployment of a new broadband public safety network will require a redesign of Project Roll Call and the procurement of new equipment to operate over the new spectrum. These efforts will give the federal government the capability it needs to rapidly restore public safety broadband communications in a disaster or emergency. Accordingly, Congress should consider providing an additional $6.9 million no later than FY 2012—and $1.9 million of public funding on a recurring annual basis—to the FCC for the design and acquisition of enhanced Roll Call systems.

RECOMMENDATION 16.2: Survey public safety broadband wireless mobile infrastructure and devices.

There is a lack of detailed information about state and local deployments of public safety broadband networks, infrastructure and equipment. FEMA, working with Regional Emergency Communications Coordination working groups, periodically collects data on narrowband systems.[1526] But there is no systematic study of public safety wireless broadband communications networks. Documentation of deployment and use of broadband by the state, Tribal and local public safety community, including the status of interoperability, will help in evaluating programs that support this technology.

Accordingly, Congress should consider providing public funding of $3.75 million per year for three years (for a total of $11.3 million) to allow FEMA to expand its data collection and survey efforts with states and territories. Providing federal, Tribal, state and local governments with up-to-date information on public safety broadband capabilities can help target grants to fill broadband gaps. [1527]

Exhibit 16-C. Selection of Proposed Broadband Applications and Services for the Public Safety Broadband Network

Public Safety Spectrum Trust	• Remote access to criminal databases • High-speed file downloads • Distribution of surveillance video feeds to on-scene personnel
The National Association of State EMS Officials	• Medical-quality video • Multiple vital signs transmission • Real-time resource tracking (e.g., of ambulances) • Secure transmission of patient records
National Public Safety Telecommunications Council	• Intelligence gathering • Automated inspections • Environmental monitoring • Traffic management
AT&T	• Location-based services • Messaging • Virtual private networking
Telcordia	• Real-time command and control • Logistics and decision support
District of Columbia	• Real-time identity management and credentialing • Interoperability with computer-aided dispatch systems, emergency operation centers and voice systems

RECOMMENDATION 16.3: Ensure that broadband satellite service is a part of any emergency preparedness program.

Technical factors can affect broadband service during disasters, but it is vital that broadband networks operate reliably and have redundant capabilities in an emergency. A way to ensure this is to use existing broadband mobile and fixed satellite services in an affected area in the event of a disaster or crisis. Satellites can serve as a communications option and a critical source of redundancy, particularly when terrestrial infrastructure is unavailable. Satellite services may be even more important as a method of communication in the first few hours or days of a disaster, should terrestrial-based services be damaged or destroyed—providing unique value for public safety purposes. Already, several state, local and federal agencies use broadband satellite service applications for public health, continuity of government and disaster preparedness activities.[1528]

Federal agencies should recommend the use of broadband fixed and mobile satellite service for emergency preparedness and response activities, as well as for national security, homeland security, continuity and crisis management.[1529] These recommendations should be issued when the agencies offer emergency preparedness and response information guidelines to the emergency response community, or when they develop plans and programs on emergency response. The U.S. Government Accountability Office (GAO) should issue a report on the current and future capability of satellite broadband to provide necessary service during an emergency.

RECOMMENDATION 16.4: Preserve broadband communications during emergencies.

Current law bars for-profit entities, such as hospitals, broadcasters and service providers, from receiving federal assistance to maintain or restore communications—including broadband and broadcast services—immediately following a disaster. However, certain for-profit communications entities provide vital services that ensure public safety. Hospitals, for example, provide public health information, while broadcasters distribute important information and warn the public of impending dangers. The inability to maintain or restore broadband service may prevent hospitals and public health officials from sharing time-sensitive information. Loss of power or broadband connectivity also could prevent broadcasters from distributing health information to the public on a timely basis. [1530] Without federal efforts to maintain and quickly restore broadband and broadcast services, the most vulnerable residents could be cut off from essential services such as NG 911, alerts and warnings, including Emergency Alert System (EAS) messages.

Accordingly, Congress should consider amending the Stafford Act to permit limited federal assistance during a disaster to private, for-profit entities—including health care providers, broadcasters and communications service providers—to maintain or restore public safety-related critical communications services (e.g., public warning and alerts, law enforcement, fire, medical, search and rescue, PSAPs and other emergency services) during a major disaster. The Federal Coordinating Officer or Federal Resource Coordinator at the Joint Field Office (JFO)—or, prior to establishment of a JFO, the Operations Section Chief at the National Response Coordination Center—should be authorized to decide whether to grant requests for such federal assistance. [1531] To prevent abuse, requests should be granted only for services related to operational issues and only for a limited duration, such as 30 days. [1532] These statutory and regulatory changes should be made effective prior to the start of the 2010 hurricane season in June, because of the possibility of frequent and large-scale weather-related disasters.

16.2. Promoting Cybersecurity and Protecting Critical Infrastructure

Improving Cybersecurity

Communications providers have experienced frequent attacks on critical Internet infrastructure. A variety of state and non-state entities has demonstrated the ability to steal, alter or destroy data and to manipulate or control systems designed to ensure the functioning of portions of our critical infrastructure. Additional safeguards may be necessary to protect our nation's commercial communications infrastructure from cyberattack. Such safeguards could promote confidence in the safety and reliability of broadband communications and spur adoption.

RECOMMENDATION 16.5: The FCC should issue a cybersecurity roadmap.

Admiral Mike McConnell, former Director of National Intelligence, said recently that "the United States is fighting a cyber-war today, and we are losing."[1533] He noted that "to the extent that the sprawling U.S. economy inhabits a common physical space, it is in our communications networks."[1534] The country needs a clear strategy for securing the vital communications networks upon which critical infrastructure and public safety communications rely. Within 180 days of the release of this plan, the FCC should issue, in coordination with the Executive Branch, a roadmap to address cybersecurity. The FCC roadmap should identify the five most critical cybersecurity threats to the communications infrastructure and its end users. The roadmap should establish a two-year plan, including milestones, for the FCC to address these threats.

RECOMMENDATION 16.6: The FCC should expand its outage reporting requirements to broadband service providers.

Today the FCC currently does not regularly collect outage information when broadband service providers experience network outages. This lack of data limits our understanding of network operations and of how to prevent future outages. The FCC should initiate a proceeding to extend FCC Part 4 outage reporting rules to broadband Internet service providers (ISPs) and interconnected VoIP providers. Such reports will allow the FCC, other federal agencies and, as appropriate, service providers to analyze information on outages affecting IP-based networks. The information also will help prevent future outages and ensure a better response to actual outages. The timely and disciplined reporting of network outages will help protect broadband communications networks from cyberattacks, by improving the FCC's understanding of the causes and how to recover. This will help improve cybersecurity and promote confidence in the safety and reliability of broadband communications.[1535]

RECOMMENDATION 16.7: The FCC should create a voluntary cybersecurity certification program.

Many Internet users apparently do not consider cybersecurity a priority. Nearly half of all businesses in the 2009 Global State of Information Security Study reported that they are cutting budgets for information security initiatives. A 2008 Data Breach Investigations Report concluded that 87% of cyber breaches could have been avoided if reasonable security controls had been in place.[1536] The FCC should explore how to encourage voluntary efforts to improve cybersecurity.

The FCC should begin a proceeding to establish a voluntary cybersecurity certification system that creates market incentives for communications service providers to upgrade their network cybersecurity. The FCC should examine additional voluntary incentives that could improve cybersecurity as and improve education about cybersecurity issues, and including international aspects of the issues. A voluntary cybersecurity certification program could promote more vigilant network security among market participants, increase the security of the nation's communications infrastructure and offer end- users more complete information about their providers' cybersecurity practices. In this proceeding, the FCC should consider all measures that will promote confidence in the safety and reliability of broadband communications.[1537]

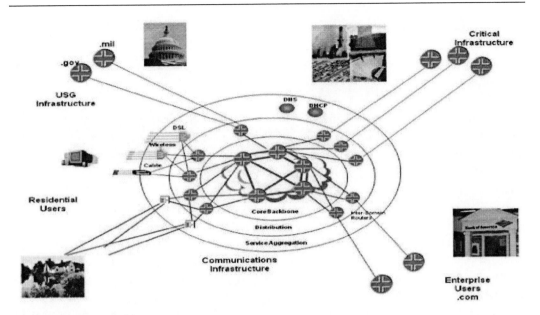

Exhibit 16-D. The Cyber World

RECOMMENDATION 16.8: The FCC and the Department of Homeland security (DHS) should create a cybersecurity information reporting system (CIRS).

The FCC, other government partners and ISPs lack "situational awareness" to allow them to respond in a coordinated, decisive fashion to cyber attacks on communications infrastructure. The FCC and DHS's Office of Cybersecurity and Communications together should develop an IP network CIRS to accompany the existing Disaster Information Reporting System. CIRS will be an invaluable tool for monitoring cybersecurity and providing decisive responses to cyberattacks.

CIRS should be designed to disseminate information rapidly to participating providers during major cyber events. CIRS should be crafted as a real-time voluntary monitoring system for cyber events affecting the communications infrastructure. The FCC should act as a trusted facilitator to ensure any sharing is reciprocated and that the system is structured so ISP proprietary information remains confidential.

RECOMMENDATION 16.9: The FCC should expand its international participation and outreach.

The FCC should increase its participation in domestic and international fora addressing international cybersecurity activities and issues. It should also engage in dialogues and partnerships with regulatory authorities addressing cybersecurity matters in other countries. This should include outreach to foreign communications regulators and international organizations about elements of the National Broadband Plan (see Chapter 4 which discusses international outreach). The FCC should also continue to review other nations' and organizations' cybersecurity activities so it is better aware of those activities as they relate to U.S. domestic policies. And it should continue to participate in domestic initiatives that relate to cybersecurity activities in the international arena.

Critical Infrastructure Survivability

RECOMMENDATION 16.10: The FCC should explore network resilience and preparedness.

Simultaneous failure of or damage to several IP network facilities or routers could halt traffic between major metropolitan areas or between national security and public safety offices. Because many companies colocate equipment, damage to certain buildings could affect a large amount of broadband traffic, including NG 911 communications. The FCC should begin an inquiry into the resilience of broadband networks under a set of physical failures—either malicious or non-malicious—and under severe overload. This will allow the FCC to assess the ability of next-generation public safety communications systems to withstand direct attacks and to determine if any actions should be taken in this regard.

This proceeding should also examine commercial networks' preparedness to withstand overloads that may occur during extraordinary events such as bioterrorism attacks or pandemics. DHS has developed pandemic preparedness best practices for network service providers, but adherence to these voluntary standards is not tracked. For example, a surge in residential broadband network use during a pandemic or other disaster could hinder network performance for critical users and applications by hindering the flow of time-sensitive medical and public health information over public networks. This proceeding will give the FCC insight into pandemic preparedness in commercial broadband networks. In addition, it will yield important information about the susceptibility of such networks to severe overloads and how network congestion on residential-access networks—particularly in the "last mile"—may undermine public safety communications and 911 access during a pandemic or other large-scale event.[1538]

RECOMMENDATION 16.11: The FCC and the National Communications System (NCS) should create priority network access and routing for broadband communications.

Broadband users in the public safety community have no system of priority access and routing on broadband networks. Such a system is critical to protect time-sensitive, safety-of-life information from loss or delay due to network congestion. While technical work is under way to allow the creation of such a system, no corresponding set of FCC rules exists to support it. The FCC and the National Communications System (NCS) should leverage their experience with the Government Emergency Telecommunications Service (GETS) and the WPS to jointly develop a system of priority network access and traffic routing for national security/emergency preparedness (NS/EP) users on broadband communications networks. The Executive Branch should consider clarifying a structure for agency implementation and delineating responsibilities and key milestones; the order should be consistent with national policies already in existing presidential documents.The FCC and NCS should jointly manage this program.

RECOMMENDATION 16.12: The FCC should explore standards for broadband communications reliability and resiliency.

For years, communications networks were designed and deployed to achieve "carrier-class" reliability. As the communications infrastructure migrates from older technologies to broadband technology, critical communications services will be carried over a communications network that may or may not be built to these high standards. The potential decline in service reliability is a concern for critical sectors, such as energy and public safety, and for consumers in general. The FCC should begin an inquiry proceeding to gain a better understanding of the reliability and resiliency standards being applied to broadband networks. The proceeding should examine the standards and practices applied to broadband infrastructure at all layers, from applications to facilities. Its objective should be to determine what action, if any, the FCC should take to bolster reliability of broadband infrastructure.

16.3. Leveraging Broadband Technologies to Enhance Emergency Communications with the Public

The Move to Next Generation 911

The nation's 911 system is evolving toward supporting NG911, which will integrate the core functions and capabilities of Enhanced 911 (E911) while adding new 911 capabilities in multiple formats, such as texting, photos, video and e-mail. NG911 also will integrate entities involved in emergency response beyond the PSAP (see Exhibit 16-E.). This will vastly improve the quality and speed of response, giving all callers—including people with disabilities—equal service. The possibility of sending video and photographs to the PSAP will transcend language barriers and provide eyewitness-quality information to give first responders the most relevant information at the scene of an emergency. NG911 will provide a more interoperable and integrated emergency response capability for PSAPs, first responders, hospitals and other emergency response professionals.

The four fundamental purposes of NG911 are to:

- Replace the E911 system while retaining its core functions, such as automatic location information and automatic number identification.
- Add capabilities to support 911 access in multiple formats for all types of originating service providers, application developers and device manufacturers.
- Increase system flexibility, redundancy and efficiency for PSAPs and 911 governing authorities.
- Add capabilities to integrate and interoperate with entities involved in emergency response beyond the PSAP.

Broadband will make it possible for PSAPs to push and pull video, images, medical information, environmental sensor transmissions and a host of other data through shared databases and networks. This will make it easier for the public—including persons with disabilities—to access 911 services. Users will be able to transmit voice, text or images to PSAPs from a variety of broadband-capable devices.

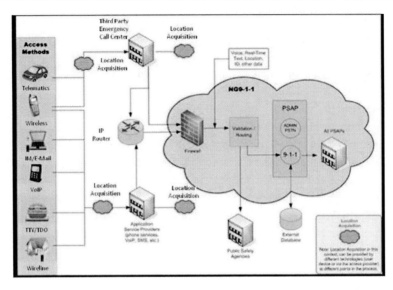

Exhibit 16-E. Call Flow in NG911[1539]

Using Broadband to Bridge the Gap to NG911

Many in the public safety community lack access to broadband services.[1540] Some PSAPs are located in areas where broadband communications are unavailable.[1541] Many PSAPs cannot afford broadband connectivity, and existing grant programs are not focused on long-term funding activities. Further, regulatory roadblocks have hindered NG911 deployment. A more efficient transition needs to be developed to support these services.

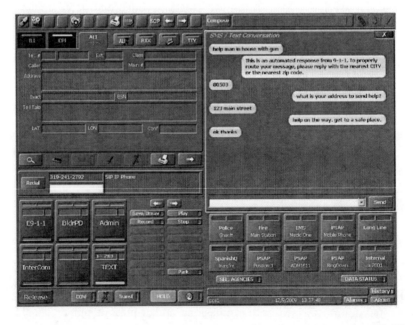

Exhibit 16-F. NG911 Will Enable the Public to Access 911 Through Text Messaging (SMS) and Other Formats

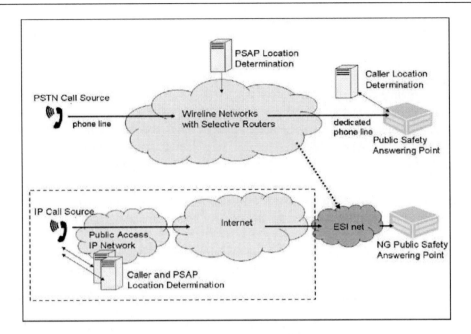

Exhibit 16-G. Physical Architectures of Current and Next- Generation 911

The transition from the legacy 911 system to NG911 has begun. Public safety and industry standards organizations have reached a consensus on NG911 technical architecture to meet demands posed by new forms of technology and methods of communication. The U.S. Department of Transportation (DOT) has published a transition plan for NG911 migration.[1542] Several states and localities have begun deploying NG911. At least one ongoing live test of 911 texting is underway[1543] (see Exhibit 16-F).

Yet financial and regulatory barriers hinder NG911 implementation. Grant programs that support NG911 are uncoordinated and limited in scope. Inconsistent, overlapping and outdated state and federal regulations have slowed NG911 development.

RECOMMENDATION 16.13: The National Highway Traffic Safety Administration (NHTSA) should prepare a report to identify the costs of deploying a nationwide NG911 system and recommend that congress allocate public funding.

The lack of coordinated funding is a significant roadblock for NG911 deployment. Several agencies administer existing grant and loan programs without any central coordination or uniform criteria.[1544] Moreover, limited information has been developed on the potential cost of NG911 implementation. Though DOT estimated in mid-2008 that the total cost of implementing and operating a nationwide NG911 system over the next 20 years would be $82 to $87 billion,[1545] the country requires a more detailed and targeted report to help Congress develop a grant program. A NHTSA analysis should determine detailed costs for specific NG911 requirements and specifications, and specify how costs would be broken out geographically or allocated among PSAPs, broadband service providers and third-party providers of NG911 services. The NHTSA report should also address the current state of NG911 readiness among PSAPs and how differences in PSAP access to broadband across the country may affect costs.

> ## BOX 16-2. IOWA 911 CALL CENTER BECOMES FIRST TO ACCEPT TEXTS[1546]
>
> An emergency call center in Black Hawk County, Iowa, became the first in the nation to accept text messages sent to "911" in August 2009. "Ithink there's a need to get out front and get this technology available," Black Hawk County police chief Thomas Jennings told the Associated Press.
>
> Black Hawk County's system is designed so people with speech and hearing impediments can text 911 for emergency services. It eliminates the cumbersome process of having a deaf person using a keyboard to write a message, which is then delivered via a relay center to the operator answering the call. An added advantage is that 911 operators can text back.
>
> While voice communication is still the primary method for 911 communications, this new wave of Next Generation 911 capability is just one example of the way the nation is modernizing its 911 system to better serve the public.

It is critical that the NG911 system is developed in a way that most effectively ensures Americans can access 911 systems anytime and anyplace. (see Exhibit 16-G for differences between the architecture of current legacy 911 and NG911 systems.) Further, the NG911 system must be able to quickly communicate caller-generated information to first responders. U.S. policy on NG911 should focus on fosteringrapid transition from analog, voice-centric 911 and emergency communications systems to a broadband-enabled, IP-based emergency services model.

Congress should consider providing public funding for NHTSA to analyze the costs of deploying a nationwide NG911 system. The report should be completed by Dec. 1, 2011. It should include a technical analysis and cost study of different delivery platforms—such as wireline, wireless and satellite—and an assessment of the architectural characteristics, feasibility and limitations of NG911 delivery. The report also should include an analysis of the needs of persons with disabilities and should identify standards and protocols for NG911 and for incorporating VoIP and "Real Time Text" standards.[1547] The report should be a resource for Congress as it considers creating a coordinated, long-term funding mechanism for NG911 deployment and operation, accessibility, application development, equipment procurement and training. This analysis is essential to identify funding requirements for the implementation of NG911.

RECOMMENDATION 16.14: Congress should consider enacting of federal NG911 regulatory framework.

Federal and state regulations that focus on legacy 911 systems have hampered NG911 deployment.[1548] Many rules were written when the technological capabilities of NG911 did not exist.[1549] Congress should consider establishing a federal legal and regulatory framework for development of NG911 and the transition from legacy 911 to NG911 networks. This framework should remove jurisdictional barriers and inconsistent legacy regulations and provide legal mechanisms to ensure efficient and accurate transmission of 911 caller information to emergency response agencies. Without such a comprehensive framework and a funding mechanism, it is unlikely all Americans will receive the benefits of NG911 in the near term.

The legislation should recognize existing state authority over 911 services but require states to remove regulatory roadblocks to NG911 development. It should also give the FCC the authority to implement a NG911 federal regulatory framework, eliminate outdated 911 regulations at the federal level and preempt inconsistent state regulations. This legislation should be coordinated with the NHTSA report to ensure federal regulation of NG911 is consistent.

Congress should also consider steps to curtail Tribal, state and local use of 911 funds for purposes other than 911. In the FCC's "Report to Congress on State Collection and Distribution of 911 and Enhanced 911 Fees and Charges" for the year ending Dec. 31, 2008, some states reported that 911/ E911 funds collected at the state level are or may be used, at least in part, to support non-911 and E911 programs.

Congress should also consider amending and reauthorizing the ENHANCE 911 Act and restoring the E911 Implementation Coordination Office (ICO) with appropriate funding. ICO can build upon its prior work with wireless and IP-enabled 911 services and help ensure NG911 is deployed in an interoperable and reliable fashion.

RECOMMENDATION 16.15: The FCC should address IP-based NG911 communications devices, applications and services.

The FCC is considering changes to its location accuracy requirements and the possible extension of Automatic Location Identification (ALI) requirements to interconnected VoIP services.[1550] The FCC should expand this proceeding to explore how NG911 may affect location accuracy and ALI.

The current 911 system will also need to be re-evaluated as broadband-based communications continue to proliferate. The 911 system mainly provides a voice-centric communications platform between the public and 911 operators. However, the deployment of different types of communications, devices, applications and services has meant consumers are changing their expectations about how they can access 911. Many consumers, for example, already have come to expect they may send non-voice communications, such as short text messages and multimedia messages, to PSAPs. But PSAPs typically cannot receive such communications. The national strategy for NG911 deployment should be designed to meet future consumer expectations.

New broadband-based devices and applications may not offer the traditional voice and "call" capabilities that wireless or VoIP phones do today. Thus, consumers may assume they can reach PSAPs via various IP-based communications modes. Non-voice methods of communicating with 911 would have the added benefit of promoting accessibility to 911 for non- English-speaking persons and persons with disabilities. Thus, the FCC should initiate an additional proceeding to address how NG911 can accommodate communications technologies, networks and architectures beyond traditional voice-centric devices. It should also explore how public expectations may evolve in terms of the communications platforms the public would rely upon to request emergency services.

BOX 16-3. EMERGENCY ALERT SYSTEM SAVES LIVES IN AMERICAN SAMOA[1551]

On Sept. 29, 2009, an 8.1 magnitude earthquake triggered a tsunami in American Samoa—the biggest earthquake of that year. KKHJ, the primary station in American Samoa's Emergency Alert System, issued 2 EAS alerts— one after the earthquake hit and a second when waters in Pago Pago Harbor began to rise. This EAS alert warned residents to evacuate the area. Upon receiving the alert, a pastor from the village of Amanave rang his church bells, providing a further warning to locals to evacuate the area. Although more than 180 people perished in the earthquake and tsunami, the early warning system is credited with saving lives.

Moving Toward Next-Generation Alerting

Building on today's emergency alerting technology, FEMA has taken steps to develop an Integrated Public Alert and Warning System (IPAWS) that will lead to a next-generation public alert and warning system.[1552] The IPAWS vision is to build and maintain an effective, reliable, integrated, flexible and comprehensive system that allows Americans to receive alert and warning information through as many communication pathways as possible.[1553] But in a September 2009 report, GAO identified a number of challenges with IPAWS implementation, including some related to the inclusion of new technologies,[1554] stakeholder coordination[1555] and technical issues.[1556] States and localities need additional resources to upgrade their alerting operations to effectively access IPAWS. Further, the federal government should disseminate information about IPAWS development and deployment.

RECOMMENDATION 16.16: The FCC should launch a comprehensive next-generation alert system inquiry.

The FCC should quickly begin a proceeding exploring all issues for developing a multiple-platform, redundant next- generation alert system. Next-generation alerting should include delivery of emergency alerts throughout the nation via broadband. The inquiry should consider Emergency Alert System (EAS) and Commercial Mobile Alert Service (CMAS) developments, as well as FEMA's development of IPAWS. It also should consider all potential multiplatform technologies, including the use of emergency alerts via video programming on the Internet. The inquiry should determine how best to ensure all Americans can receive timely and accurate alerts, warnings and critical information about emergencies regardless of the communications technologies used.

The FCC has not yet begun a wide-ranging inquiry into next-generation alerting. Such an inquiry can bridge the gap from the current EAS and CMAS systems to a comprehensive next-generation alerting system by detailing an implementation strategy. Such a proceeding should be initiated.

Next-generation technologies will transform the information delivery capabilities of both EAS and CMAS. They can also increase the effectiveness of alerts during emergencies. Emergency managers could provide alerts to communities now served poorly—such as persons with disabilities and non-English speakers—and provide improved alert file "trails" containing valuable information, such as full-motion videos of radar-tracked storm systems.

Emergency alerts in Internet video format would allow emergency alert originators to reach people who are not, at the time, listening to broadcast radio and television or other current sources of alerts. Providing alternative methods for distributing emergency alerts to all Americans will save lives. However, the systems that assemble, manage and transmit alerts will need to be upgraded to accommodate broadband.

The system should alert the public of emergencies through all possible means of communications. In the event of a tornado, for example, alerts would be broadcast on local media outlets, sent to wireless and wireline phones within the affected area, posted on Internet feeds and websites sites, and issued through any other communication outlet serving the affected area. That would ensure the public is informed of an emergency and has the information it needs to protect itself. The FCC's inquiry should focus primarily on how to develop such a system.

FEMA's development of IPAWS should help ensure that a ubiquitous alert transmission system is available to accommodate multiple alert platforms and participation by all federal, state, Tribal, local and private sector alert stakeholders. There also needs to be a comprehensive evaluation of the ability of alert managers to participate in IPAWS when launched.

A comprehensive inquiry will allow the FCC to obtain input on the alerting system's future and to form a new regulatory framework for next-generation alerting. This inquiry should focus on the wide-ranging technical, legal and policy issues associated with this new multi-platform system. The proceeding should analyze the developing IPAWS architecture to evaluate the ability of IPAWS to support a broadband-based, next generation alert system. The inquiry also should examine the needs of state, Tribal and local emergency alert originators in utilizing the next-generation alerting system; what assistance, if any, the FCC and its federal partners should provide to address those needs; and what actions the FCC and federal partners should take to ensure the system's timely development and deployment.

RECOMMENDATION 16.17: The Executive Branch should clarify agency roles on the implementation and maintenance of a next-generation alert and warning system.

The Executive Branch through an interagency policy council or through a directive should take action by executive order, federal interagency policy committee or other formal means, to clarify the responsibilities of each federal agency in the implementation, maintenance and administration of next generation alerting systems. This action should also set milestones, benchmarks and necessary actions for implementation and establish a system of accountability among the federal agencies responsible for emergency alerting.

17. IMPLEMENTATION AND BENCHMARKS

This plan is in beta, and always will be. Like the Internet itself, this plan will always be changing—adjusting to new developments in technologies and markets, reflecting new realities and evolving to realize previously unforeseen opportunities.

The plan is both a "noun" and a "verb."[1557] Of course, the "noun"—the March 2010 version of this plan—will be forever available, preserved deep in caches and crawled by search engines. The "verb," though, will be forever alive—updated regularly and driven by new data, analysis and scenarios that the "noun" could not foresee.[1558]

Implementation of this National Broadband Plan requires a long-term commitment to measuring progress and adjusting programs and policies to improve performance. It requiresperiodic assessments of where the country stands in broadband deployment, adoption and utilization; in competition across networks, devices and applications; and in how effectively national priorities embrace the power of broadband.

But evaluation is not an excuse for paralysis. Actions and their results matter most to capturing the opportunities broadband presents.

This plan recommends significant action by the Federal Communications Commission (FCC), the Executive Branch and Congress and a strong partnership among all broadband stakeholders. Federal action is necessary, but state, local and Tribal governments, corporations and community-based organizations must all do their part to build a high-performance America.

Recommendations

- The Executive Branch should create a Broadband Strategy Council to coordinate the implementation of National Broadband Plan recommendations.
- The FCC should quickly publish a timetable of proceedings to implement plan recommendations within its authority, publish an evaluation of plan progress and effectiveness as part of the annual Section 706 Advanced Services Inquiry, create a Broadband Data Depository and continue to utilize Broadband.gov as a public resource for broadband information.
- The FCC should publish a Broadband Performance Dashboard with metrics designed to track broadband plan goals.

17.1. Implementation

More than 20 other nations have published national broadband plans. Their implementation efforts highlight the importance of a long-term commitment and coordination across multiple institutions.

International Lessons

Many countries have depended on long-term and high-level coordination and collaboration efforts across government to implement their broadband plans. For example, in the mid- 1990s South Korea created a durable structure for long-term broadband policy planning by passing a law requiring publication of a national broadband strategy every five years (along with annual implementation plans).[1559] Since then, South Korea has published three master plans, some with multiple versions.[1560] The statutory obligation to produce new plans every five years has ensured that successive political administrations have made broadband a national priority.

South Korea's Prime Minster chairs the Informatization Promotion Committee (IPC), the entity responsible for implementing South Korea's broadband plans.[1561] The IPC's membership includes 24 ministerial-level representatives, thereby fostering intragovernmental coordination.[1562] Member ministries submit annual implementation plans to the IPC for approval.[1563]

Japan provides another example of successful long-term implementation. Japan created an IT Strategy Headquarters to oversee the execution of its broadband strategies, beginning with the e-Japan Strategy of 2001.[1564] Japan's Prime Minister chairs the IT Strategy Headquarters. It also is composed of ministers across agencies with responsibility for broadband policy.[1565] The IT Strategy Headquarters conducts an annual review of broadband policy priorities and directs the implementation of plan recommendations by government agencies, local governments and independent institutions.[1566]

The United Kingdom has also established a high-level coordinating body to implement broadband strategy. In June 2009, the U.K. government published *Digital Britain*, its first broadband plan.[1567] Soon after *Digital Britain* was published, the U.K. government published an implementation plan providing for a cross-agency coordination staff and a dedicated legislative affairs group.[1568] The implementation plan also created a Programme Board, responsible for policy proposals, monitoring progress and ensuring value for the public's financial investment.[1569] Recognizing the importance of keeping stakeholders and the public informed of plan progress, the U.K. government also periodically releases implementation updates.[1570]

RECOMMENDATION 17.1: The Executive Branch should create a Broadband strategy council to coordinate the implementation of National Broadband Plan recommendations.

The FCC is the focus of approximately half of the plan's recommendations and the National Telecommunications and Information Administration (NTIA) the President's advisor for telecommunications policy, has responsibility for many actions in the plan. Most of the remaining proposals are directed at other Executive Branch agencies. The Executive Branch should create an entity accountable to ensure implementation across, and foster effective coordination among the multiple agencies targeted by specific recommendations and engage senior-level officials in these efforts.

This proposed Broadband Strategy Council (BSC) could include senior officials from the White House Office of Science and Technology Policy, the National Economic Council and the Office of Management and Budget. The BSC's membership could also include high-level personnel drawn from the FCC, NTIA and other agencies with key roles in implementing plan recommendations.[1571] The BSC could also rely on the President's Council of Advisors on Science and Technology for external input and support.

Charter of the Broadband Strategy Council
This plan contains recommendations directed at more than 20 agencies. To ensure timely and effective implementation, the BSC should be given direct responsibility for managing the execution of the plan's recommendations to the Executive Branch.

The President could require that Executive Branch departments and agencies submit project plans to the BSC on proposed steps to implement plan recommendations. Additionally, the BSC could track recommendations requiring congressional action with the FCC and legislative affairs offices in the Executive Branch.

Today, the responsibility for broadband-related government policy and programs is spread across many federal agencies as well as state, Tribal and local governments. Successful implementation of the recommendations in this plan will intensify the need for coordination among these actors. The BSC should create a forum for relevant agencies to discuss broadband policy, assign responsibility for joint duties, share best practices and coordinate broadband funding so that broadband-related government spending has maximum economies of scale and maximum impact.[1572]

RECOMMENDATION 17.2: The FCC should quickly publish a timetable of proceedings to implement plan recommendations within its authority, publish an evaluation of plan progress and effectiveness as part of the annual section 706 Advanced services inquiry, create a Broadband Data Depository, and continue to utilize Broadband.gov as a public resource for broadband information.

The FCC is responsible for implementing approximately half of the plan's recommendations. It should quickly publish a timetable of proceedings for implementing broadband plan recommendations directed to the FCC.

Additionally, given the evolving nature of the broadband ecosystem, the National Broadband Plan should be periodically reviewed and revised to reflect new realities. The FCC should conduct a National Broadband Plan strategy review as part of its annual Section 706 Advanced Services Inquiry. The review should analyze plan progress and effectiveness, and, if necessary, recommend strategic and tactical adjustments that will help America meet plan goals. This review should also track the implementation of plan recommendations.

FCC data collection and analysis efforts are essential to understanding the effectiveness of plan policies and the progress being made toward plan goals. The plan includes recommendations to improve the quality and transparency of this process.

The FCC should also create a Broadband Data Depository on the Internet to give researchers and the public better access to the FCC's data. This will help the FCC serve its essential role as a source of independent data on broadband deployment, adoption and usage in America. The FCC should have a general policy of making the data it collects available to the public, ideally over the Internet, except in certain circumstances such as when the data are competitively sensitive, protected by copyright or classified. Additionally, the FCC should have a separate process for allowing researchers access to non-public data, subject to certain restrictions.[1573]

The FCC should also continue to utilize Broadband.gov, which has been a successful Web portal for communicating with the public in an open and interactive fashion about the development of the National Broadband Plan. Going forward, this website should serve as a source for tracking the implementation of the plan. It should also serve as a consumer resource for information about broadband. In addition to hosting the Broadband Performance Dashboard (see Recommendation 17.3), Broadband.gov should contain updates on the progress made in implementing each recommendation, links to the National Broadband Map, access to broadband quality tests and surveys, details on how to obtain computer literacy education, and links to third-party resources from which consumers can purchase broadband.

17.2. Benchmarking

Measuring the effects of a broadband plan over time is a critical challenge. This plan recommends that the FCC track and report several important broadband indicators: how many people and businesses have access to broadband, how many subscribe, what speeds they get, how much they pay and what they do with it.[1574]

In the same spirit as these recommendations, other countries have expanded their broadband data compilation and dissemination efforts to provide more information to policy-makers and consumers. These efforts include collecting and publishing richer information about the extent of broadband deployment, utilization and pricing through broadband map-ping,[1575] usage surveys,[1576] pricing portals[1577] and broadband quality of service measurements.[1578]

In preparing the National Broadband Plan, the FCC used existing resources such as data from Broadband Deployment Form 477, which was recently updated to include census tract-level data. The FCC created a broadband deployment model, conducted surveys of residential and business broadband consumers and performed a detailed consumer preference analysis of consumers' willingness to pay for broadband services. The FCC has also developed tools and mobile applications to collect address level and location-based data on actual delivered speed over fixed and mobile broadband networks.

Nevertheless, as recommended in Chapter 4, the FCC needs to collect more detailed and accurate data on actual broadband availability, penetration, pricing and network performance in order to accurately benchmark progress toward plan goals.[1579] Only with these data inputs can the FCC publish a Broadband Performance Dashboard.

RECOMMENDATION 17.3: The FCC should publish a Broadband Performance Dashboard with metrics designed to track broadband plan goals.

The FCC should publish a Broadband Performance Dashboard to supplement the improved data collection process recommended in the plan. This dashboard should display key progress indicators aligned with plan goals, enable the public to understand important broadband performance metrics and clearly communicate plan progress and effectiveness. The dashboard should be updated regularly and provide data metrics that, track the broadband performance goals detailed in Chapter 2. The sample dashboard (see Exhibit 17-A) details the metrics that the FCC should collect and analyze in order to track progress towards plan goals.

While these fundamental broadband indicators are important, it is equally important to know how broadband affects the very core of the economy: innovation, productivity and the way people live and work. Measures like broadband availability and adoption, while enormously important, cannot provide that kind of information.

The problem is that it is difficult, if not impossible, to know how new technologies, like broadband, will ultimately integrate themselves into the economy. Measurement bias against new technologies by conventional indices makes this even more challenging.[1580] Nobel

Laureate Robert Solow famously quipped more than a decade ago that "you can see the computer age everywhere but in the productivity statistics."[1581] Indeed, it was not until well after companies began using computers that it was possible to statistically attribute any productivity effects to computers or information technology.

Exhibit 17-A. Broadband Goals and Performance Dashboard Sample

Goals for 2020 (see Chapter 2)	Metrics	Sources
At least 100 million U.S. homes should have affordable access to world-class actual download speeds of at least 100 megabits per second and actual upload speeds of at least 50 megabits per second	The nationwide, and per provider, average actual upload and download speeds of broad- band networks	FCC network performance measurements and provider disclosures (See Recs. 4.4–4.6.)
	Number of households with access to broad-networks with sufficient speed	Future revisions to Form 477 data (See Rec. 4.2.)
	The nationwide, and per provider, minimum price for a broadband subscription with sufficient speed	Future revisions to Form 477 data (See Rec. 4.2.)
The United States should lead the world in mobile innovation, with the fastest and most extensive wireless networks of any nation.	MHz of spectrum released since 2010	FCC self-reporting
	The nationwide, and per pro-vider, average actual upload and download speeds of mobile broadband networks, by geographic area	FCC network performance measurements and provider disclosures (See Recs. 4.4–4.6.)
	Percentage of population covered by 3G and 4G services	Future revisions to Form 477 data (See Rec. 4.2)
	Percentage of Americans that subscribe to mobile broadband services, both overall and per socio-economic and demographic groups	FCC consumer surveys (Broadband Data Improvement Act (BDIA) mandated survey)
Every American should have affordable access to robust broadband service and the means and skills to subscribe if they so choose.	Percentage of households with access to broadband networks with sufficient speed	Future revisions to Form 477 data (See Rec. 4.2.)
	The nationwide, and per provider, minimum price for a broadband subscription with sufficient speeds	Future revisions to Form 477 data (See Rec. 4.2.)
	Percentage of Americans that subscribe to broadband services, both overall and by socio- economic and demographic group	Future revisions to Form 477 data, FCC Consumer Surveys (See Rec. 4.2), and mandated survey[1582]
	Percentage of Americans with sufficient digital literacy skills	FCC consumer surveys (BDIA mandated survey)

Every American community should have affordable access to service of at least 1 gigabit per second to anchor institutions such as schools, hospitals and government buildings.	Average actual upload and download speeds of broadband networks	FCC network performance measurements and provider disclosures (See Recs. 4.4–4.6.)
	Deployment of networks with sufficient speed	Future revisions to Form 477 data (See Rec. 4.2.)
	Percentage of communities with sufficient access to broadband	Future revisions to Form 477 data (See Rec. 4.2.)
	The nationwide, and per provider, minimum price for an institutional broadband subscription with sufficient speeds	Future revisions to Form 477 data (See Rec. 4.2.)
To ensure the safety of the American people, every first	Percentage of first responders using the	Federal Emergency Management Agency survey (See Rec. 16.1)
responder should have access to a nationwide, wireless, interoperable broadband public safety network.	nationwide public safety network	
To ensure that America leads in the clean energy economy, every American should be able to use broadband to track and manage their real-time energy consumption.	Percentage of American homes that have smart electric meters capable of communicat- ing real-time energy information to consumers	Federal Energy Regulatory Commission metering assessment and the Department of Energy Smart Grid Systems Report[1583]

If this broadband plan is effective, we will see rapid progress in terms of increased adoption, especially by currently disadvantaged groups; faster speeds; transitions to electronic medical health records and Smart Grids; and better incorporation of broadband into education and government. But none of those are ends in themselves. Broadband access by more people opens up new opportunities for them, helping them to unleash their potential. Faster broadband speeds and better broadband quality improve incentives for entrepreneurs to innovate. And savings realized by incorporating broadband into existing areas like education and health care represent resources that can be newly invested elsewhere.

Thus, if we succeed, not only will the indicators that we currently measure improve, but we will also see improvement in other areas of the economy and will need to derive new indicators to measure changes in industries and activities that do not yet exist.

17.3. The Legal Framework for the FCC's Implementation of the Plan

The plan sets out a strategic vision for America, establishing national goals regarding broadband and recommending specific policies to achieve those goals. It does not reach conclusions about or explore in detail the many legal issues that will be relevant to the FCC's implementation of the plan. These will be addressed through notice-and-comment rule-makings the FCC will conduct following the plan. A variety of parties have, however, offered thoughts on the proper legal framework for the FCC's plan implementation. The following section provides the relevant background and summarizes these comments.

Historically, the FCC treated broadband transmission as a common carrier service subject to the statutory requirements set forth by Title II of the Communications Act.[1584] Facilities-based carriers that provided "enhanced" or "information" services—remote computer applications that allow subscribers to access, modify, or interact with information—were required to offer on a common carrier basis the underlying transmission function known as a "basic" service.[1585]

Beginning in 2002, the FCC adopted a series of orders classifying broadband Internet access services as information services subject to the FCC's general jurisdiction under Title I of the Communications Act.[1586] Although the Act does not establish specific rules for providers of information services, the Supreme Court has held that the Communications Act gives the FCC "ancillary authority" to regulate matters that fall within its general jurisdiction but are not directly addressed by the substantive provisions of the Act.[1587] In *NCTA v. Brand X*, the United States Supreme Court held that the FCC's conclusion that cable modem service providers offer only an information service was a reasonable interpretation of an ambiguous statute.[1588] The Commission then applied a similar analysis to Internet access provided via Digital Subscriber Line (DSL),[1589] broadband over power line,[1590] and wireless broadband technologies,[1591] classifying all of these as information services. These broadband services are not subject to the requirements Congress established for common carrier services, unless the provider chooses to offer broadband transmission as a standalone telecommunications service.[1592]

Comments in the record include competing views on the appropriate legal framework for implementing plan recommendations that involve broadband Internet access services. One approach would involve Congress enacting legislation to direct or enable the FCC to implement specific plan recommendations. Absent Congressional action, however, parties discuss two alternative approaches to plan implementation.

The first suggested approach is to rely on ancillary authority under Title I when promulgating most of the recommended rules and regulations regarding broadband. Some parties believe that Title I and the doctrine of ancillary authority, together with various other provisions of the Act addressing such matters as spectrum, cable television, and universal service, provide the FCC sufficient authority to advance broadband deployment and adoption, including to establish direct support for broadband under the Universal Service Fund's High Cost, Lifeline and Link-Up programs;[1593] to ensure privacy protections regarding sharing of consumers' personal information;[1594] and to promote accessibility for people with disabilities.[1595] Others have expressed doubts about the adequacy of Title I to support FCC efforts to advance broadband goals.[1596]

Some commenters have suggested a second approach, in which the FCC would implement certain plan recommendations under its Title II authority, after classifying broadband services as telecommunications services. These commenters believe such an approach would provide a sounder legal basis for establishing direct support for rural broadband under the Universal Service Fund's High Cost program and broadband access under the Lifeline and Link-Up programs;[1597] requiring enhanced disclosures of broadband speed, performance and pricing;[1598] and other plan recommendations, including ensuring privacy protections regarding sharing of consumers' personal information.[1599] Commenters further note that classifying broadband services as telecommunications services would not require the application of all requirements of Title II to broadband.[1600] Congress gave the FCC "forbearance authority" in section 10 of the Act. Consistent with the comments, this forbearance authority would permit the FCC to narrowly tailor its use of Title II to advance the policies described above without imposing additional regulatory burdens. To the degree that wireless-based broadband is a common carrier service, section 332 of the Act grants similar authority to forbear.[1601] Other parties, however, believe that reverting to Title II to implement the plan would be unwise policy, contending that Title II is an ill- fitting, over-regulatory legal framework for broadband Internet access services.[1602]

The FCC will consider these and related questions as it moves forward to implement the plan.

17.4. Conclusion

This plan is premised on the potential of broadband to improve lives today and for generations.

But broadband alone will not solve America's problems. It cannot guarantee that the United States will lead the world in the 21st century. It cannot promise that the U.S. and other nations will conquer crippling inequality. It cannot ensure that the U.S. bestows the best job, education, health care, public safety and government services on every American.

Broadband is a critical prerequisite, though, to solutions to many of America's problems. It can open up ways for American innovators and entrepreneurs to reassert U.S. leadership in some areas and extend it in others. It can unlock doors of opportunity long closed by geography, income and race. It can enable education beyond the classroom, health care beyond the clinic and participation beyond the town square.

In 1938, President Roosevelt travelled to Gordon Military College in Barnesville, Georgia, to speak at the dedication of a local utility. "Electricity is a modern necessity of life, not a luxury," the President told the audience, "That necessity ought to be found in every village, in every home and on every farm in every part of the wide United States."[1603]

He added, "Six years ago, in 1932, there was such talk about the more widespread and the cheaper use of electricity." But words did not matter until the country, "reduced that talk to practical results."[1604]

Broadband, too, is a modern necessity of life, not a luxury. It ought to be found in every village, in every home and on every farm in every part of the United States.

There has long been talk of the widespread and affordable use of broadband. This plan is a transition from simple chatter to the difficult but achievable reality of implementation. It is a call to action for governments, businesses and non-profits to replace rhetoric with targeted, challenging actions.

It is time again to reduce talk to practical results.

APPENDIX A. BTOP PROGRESS ASSESSMENT

In addition to directing the FCC to develop a plan to ensure that all Americans have access to broadband, Congress also directed the FCC to evaluate the progress of projects supported by grants under the National Telecommunications and Information Administration (NTIA)'s Broadband Technology Opportunities Program (BTOP). This section considers the program so far and makes recommendations for future evaluation—as BTOP has only just funded some projects.

This plan acknowledges the substantial investment BTOP is making to improve connectivity and advance the adoption of broadband. Chapters 8 and 9 make specific mentions of this important program and how it likely will improve the broadband ecosystem. Careful evaluation of BTOP investments will provide valuable insights into the effectiveness of different funding mechanisms, project structures and technologies for future investments.

Recommendations

- Ensure that assessment tracks program outcomes, not only execution.
- Develop measures that specify outcomes to be assessed.
- Create a panel of experts from the academic and research community to advise on assessment approaches.
- Employ longitudinal design in assessing programs where possible.

Background

The American Recovery and Reinvestment Act (Recovery Act) appropriated $7.2 billion to fund programs to promote the adoption and deployment of broadband. NTIA was charged with using $4.7 billion of these funds to create BTOP which funds three types of programs:

- Infrastructure projects that aim to deploy broadband infrastructure in unserved and underserved areas.
- Projects that enhance the capacity of public computing centers (PCCs).
- Efforts to support the sustainable adoption of broadband service by users.

Infrastructure projects are set to receive the bulk of this funding. With regard to the latter two types of programs, Congress specifically stated that NTIA should spend $250 million on "innovative programs that encourage sustainable adoption of broadband services" and spend at least $200 million "to upgrade technology and capacity at public computing centers, including community colleges and public libraries."[1605]

Funds are being disbursed in two rounds. Applications for the first round were due Aug. 14, 2009. As of mid-February 2010, the BTOP program had awarded $597 million in grants:

- $547 million for infrastructure projects;
- $42 million for PCC projects; and
- $8 million for sustainable adoption programs.[1606]

Applications for the second round of funding were due on March 15, 2010. The Recovery Act directs that all funds be awarded by Sept. 30, 2010.

Programs Funding infrastructure Deployment

BTOP infrastructure grants are intended to promote community and economic development by connecting community anchor institutions—such as public schools, universities, libraries, and community colleges—to high-speed infrastructure. Many funded grantees promote connectivity in the middle mile.[1607] By solving the middle-mile problem, the hope is to foster investment in "last mile" facilities to provide service to individuals and institutions that need it.

Most grantees leverage in-kind or financial contributions, not relying solely on BTOP support to complete projects.

Public Computing Centers

Grants for PCCs will provide funding for additional computers for institutions such as public housing developments, typically with the goal of offering training and access for community members. The FCC recently announced a grant for the Housing Authority of San Bernardino, Calif. which aims to serve 350 additional users per week. On a larger scale, a grant awarded to the New York State Education Department intends to serve an additional 50,000 users per week system-wide and provide access to job-search resources 24 hours per day, seven days per week. Both these grants are intended to serve additional users and make a difference in their employment prospects.[1608]

Sustainable Adoption Grants

Grants intended to foster and sustain adoption often focus on the community level. A grant to the West Virginia Future Generations Graduate School funds a community-based approach to promote adoption among low-income and rural residents of the state.[1609] This particular project creates a partnership between fire and emergency rescue squads and the community. The squads will use computers that will also be made available to the public. At the same time, they will promote outreach about and awareness of the Internet's potential to members of the community—adopters and non-adopters alike. Training programs will build capacity and confidence with the Internet and, it is hoped, foster at-home adoption.

Assessing BTOP

BTOP was designed as a short-term investment in broadband infrastructure, broadband adoption and job creation. At the same time, Congress charged the FCC with developing a long-term plan for increasing accessibility, affordability and utilization of broadband, as well as a plan to use broadband to serve designated national purposes—a charge that led to the creation of this plan. In addition to deploying infrastructure and providing resources to communities, BTOP-funded projects can serve as testbeds. Examining projects funded under BTOP can help answer these questions:

- What leads individuals and communities to adopt broadband?
- What quantifiable difference does broadband make in communities?
- What is the impact of broadband on economic development in communities?
- How does the "broadband experience" vary by community, demographics and institutions?

Congress did not allocate funds to assess BTOP's effectiveness. It did allocate $10 million to the U.S. Department of Commerce's Office of Inspector General for oversight and auditing of the program. Such oversight and auditing activities are important, but they focus on execution of the program. Assessing program impacts on a community or on individuals or groups is different.[1610]

The plan makes the following recommendations for assessing the BTOP program, some of which may require action by NTIA and some of which may require that NTIA coordinate with the research community:

RECOMMENDATION A.1: Ensure that assessment tracks program outcomes, not only execution.

Recommendations for how to assess BTOP must take into account the program's multiple goals (as discussed above). BTOP infrastructure grants have a primary goal of making broadband service more available, typically with a secondary goal of promoting economic development. Moreover, BTOP grants for sustainable adoption have the goal of bolstering adoption rates among individuals.

Any assessment should at a minimum determine whether a grantee carried out the project funded by its grant in the time horizon specified. This kind of assessment can be completed in a relatively short period of time.

Thereafter, the assessment should focus on whether the grant had a meaningful impact in the context for which funding was specified. This is a longer-term undertaking and recognizes that the proper basis to assess a program that promises to fund infrastructure is not simply to determine whether the grantee in fact built the infrastructure. The first step in this assessment must be to ascertain whether the grant itself was responsible for the new infrastructure, or whether the infrastructure would likely have been built anyway within a reasonable time period. While it is impossible to know this with any certainty, assessors could identify control groups against which to measure the potential for this result. Such control groups might include projects (or areas) that were not funded and, if possible, geographically or socioeconomically similar areas that submitted no BTOP applications.

Once control groups are identified, assessors should measure whether the infrastructure built with BTOP grant money fostered economic growth, how additional adoption impacted users' lives or other relevant metrics. Similarly, a PCC project with a goal of placing more computers at a specific site should not be considered successful simply if it increases the number of computers at a particular location. Instead, the success of a PCC project depends, instead, on its precise impacts—whether those additional computers helped more people go online for the first time, allowed computer users to spend more productive time online and materially improved a users' lives. In assessing these impacts, NTIA should develop measures that determine the grantees' cost of adding new adopters.

RECOMMENDATION A.2: Develop measures that specify outcomes to be assessed.

Assessing outcomes requires well-defined measures for programs. An infrastructure program may seek to foster economic growth or better connectivity among particular institutions. Whatever the goal, common measures across individual grants are necessary for proper evaluation of the BTOP program as a whole. The process of developing metrics should be done in coordination with other government-wide initiatives to promote broadband infrastructure and adoption.

RECOMMENDATION A.3: Create a panel of experts from the academic and research community to advise on assessment approaches.

The Recovery Act's funding of broadband investment and adoption promotion has prompted some academic researchers to explore how effective such investments have been in other contexts.[1611] There is little empirical evidence on the impact of demand-side adoption programs, and evidence on infrastructure investments is thin as well. As researchers explore the limits of the current assessment literature, a discussion has developed about the kind of evidence, metrics and methods needed to undertake rigorous assessment. NTIA should take advantage of this discussion by convening an expert panel and having the panel coordinate with other experts within the government.

RECOMMENDATION A.4: Employ longitudinal design in assessing programs where possible.

When feasible, assessments should compare outcomes from the beginning of an award's life to a date in the future. Proper assessment of newly connected anchor institutions in an infrastructure grant would take a baseline reading of the institutions' characteristics at the time the grant is made and at periodic intervals time periods into the future. The characteristics to be measured will depend on specification of proper metrics.

Longitudinal design takes into account the fact that the impacts of BTOP grants are likely to unfold over a longer time horizon than the period of the grant itself. The impact of a sustainable adoption grant on an individual who may have passed through a training program can only be determined at some point *after* the individual has completed the program. Similarly, the proper way to determine the impact of an infrastructure grant is to compare conditions at some point (or several points) beyond completion of deployment of the infrastructure.

Finally, assessment approaches should take into consideration the context of programs under study. Infrastructure projects may have fewer measurement challenges than programs which more directly affect users. If so, program assessment for user-centric grants may need to study program strategies to reach users as well as outcomes for those users. This, in turn, may mean that proper assessment should employ qualitative research approaches as well as quantitative ones.

APPENDIX B. COMMON ABBREVIATIONS

2G	Second-generation	CCHT	Care Coordination/Home Telehealth
3G	Third-generation	CDC	Centers for Disease Control and Prevention
4G	Fourth-generation	CEDS	Comprehensive Economic Development Strategy
AIP	Administrative Incentive Pricing	CFF	Computers for Families
ALI	Automated Location Information	CIO	Chief Information Officer
AMI	Advanced Metering	CIP	Critical Infrastructure Protection

			Infrastructure
AMT	Aeronautical Mobile Telemetry	CIRS	Cybersecurity Information Reporting System
AP	Advanced Placement	CITI	Columbia Institute for Tele-Information
APD	Advance Planning Document	CMS	Centers for Medicare and Medicaid Services
API	Application Programming Interface	CNCS	Corporation for National and Community Service
App	Application	CPE	Customer premises equipment
ATC	Ancillary Terrestrial Component	CSEA	Commercial Spectrum Enhancement Act
AWS	Advanced Wireless Services	CT scan	Computed tomography scan
BAS	Mobile Broadcast Auxiliary Service	CVD	Cardiovascular disease
BAWG	Broadband Accessibility Working Group	DARPA	Defense Advanced Research Projects Agency
BDIA	Broadband Data Improvement Act	DHS	Department of Homeland Security
BIP	Broadband Infrastructure Program	DIA	Dedicated Internet Access
BIS	Department for Business, Innovation and Skills	DOCSIS	Data Over Cable Service Interface Specification
BLS	Bureau of Labor Statistics	DoD	Department of Defense
BMAC	Broadband Measurement Advisory Council	DOE	Department of Energy
BRS	Broadband Radio Service	DOJ	Department of Justice
BSC	Broadband Strategy Council	DOL	Department of Labor
BTOP	Broadband Technology Opportunities Program	DOT	Department of Transportation
CAF	Connect America Fund	DS1	Digital Signal 1
Capex	Capital expenditures	DS3	Digital Signal 3
CARS	Mobile Cable TV Relay Service	DSL	Digital Subscriber Line
CBO	Community-based organization	DSLAM	Digital Subscriber Line Access Multiplexer
DSRC	Dedicated short-range communication	FTTN	Fiber-to-the-node
DTA	Digital Transport Adapter	FTTP	Fiber-to-the-premises
DTS	Distributed Transmission System	FY	Fiscal year
DTV	Digital television	GAO	Government Accountability Office
E911	Enhanced 911	Gbps	Gigabits per second
EAS	Emergency Alert System	GDP	Gross domestic product
EBS	Educational Broadband Service	GED	General Educational Development
EC	Enterprise Community	GPS	Global Positioning System
ECPA	Electronic Communications Privacy Act	GPT	General Purpose Technology
EDA	Economic Development Administration	GSA	General Services Administration
EHR	Electronic health record	GWU	George Washington University
EISA	Energy Independence and	HBCUs	Historically Black Colleges and

	Security Act of 2007		Universities
EMEA	Europe, the Middle East and Asia	HD	High definition
EO	Executive Order	HHS	Health and Human Services
EPSCoR	Experimental Program to Stimulate Competitive Research	HIPAA	Health Insurance Portability and Accountability Act
ERC	Engineering Research Center	HITECH Act	Health Information Technology for Economic and Clinical Health Act
ERIC	Emergency Response Interoperability	HL7 CDA	Health Level 7 Clinical Document Center Architecture
ET	Engineering and Technology	HPSA	Gealth professional shortage area
ETC	Eligible telecommunications carrier	HSIACs	Hispanic-Serving Institutions Assisting Communities
EZ	Empowerment Zone	HSPA	High Speed Packet Access
FCC	Federal Communications Commission	HUD	Department of Housing and Urban Development
FDA	Food and Drug Administration	IAS	Interstate Access Support
FDIC	Federal Deposit Insurance Corporation	IC3	Internet Crime Complaint Center
FERC	Federal Energy Regulatory Commission	ICAM	Identity, Credential, and Access Management
FHS	Framingham Heart Study	ICC	intercarrier compensation
FISMA	Federal Information Security Management Act	ICLS	Interstate Common Line Support
FLVS	Florida Virtual Schools	ICO	Implementation Coordination Office
FOIA	Freedom of Information Act	ICT	information and communications technology
FS-ISAC	Financial Services Information Sharing and Analysis Center	IHS	Indian Health Service
		ILEC	incumbent local exchange carrier
FTC	Federal Trade Commission	IMLS	Institute of Museum and Library Services
IP	Internet Protocol	MVPD	Multichannel video programming distributor
IPAWS	Integrated Public Alert and Warning System	NARUC	National Association of Regulatory Utility Commissioners
IPC	Informatization Promotion Committee	NASA	National Aeronautics and Space Administration
IPIA	Improper Payments Information Act	NATOA	National Association of Telecommunications Officers and Advisors
ISAC	Information Sharing and Analysis Center		
ISM	industrial, scientific and medical	NCS	National Communications System
ISO	Independent System Operator (ISO)	NECA	National Exchange Carrier Association
ISP	Internet service provider	NERC	North American Electric Reliability Corporation
IT	information technology	NG911	Next Generation 911

IT-ISAC	Information Technology Information Sharing and Analysis Center	NHTSA	National Highway Traffic Safety Administration
ITS	Intelligent Transportation System	NIA	National Institute on Aging
ITU	International Telecommunication Union	NIH	National Institutes of Health
		NIST	National Institute of Standards and Technology
JFO	Joint Field Office	NOFA	Notice of Funding Availability
K-12	Kindergarten through twelfth grade		
kbps	Kilobits per second	NPR	National Public Radio
kWh	Kilowatt-hour	NPRM	Notice of Proposed Rulemaking
LEA	Local educational agency	NS/EP	National Security/Emergency Preparedness
LEC	Local exchange carrier		
LEED	Leadership in Energy and Environmental Design	NSF	National Science Foundation
		NTIA	National Telecommunications and Information Administration
LMRS	Land mobile radio system	OATS	Older Adults Technology Services
LPTV	Low-power television	OEC	Office of Emergency Communications
LSTA	Library Services and Technology Act	OECD	Organisation for Economic Co-Operation and Development
LTE	Long Term Evolution		
M2M	Machine-to-machine	Ofcom	Office of Communications
Mbps	Megabits per second	OMB	Office of Management and Budget
MFN	Multi-Frequency Network	ONC	Office of the National Coordinator for Health
mpg	Miles per gallon	OOBE	Information Technology out-of-band emission
MRI	Magnetic resonance imaging	OSL	Online Skills Laboratory
MSA	Metropolitan service area	OSTP	Office of Science and Technology Policy
MS-ISAC	Multi-State Information Sharing and Analysis Center	PBS	Public Broadcasting Service
MSS	Mobile Satellite Services	PC	Personal computer
PCC	Public computing center	SME	Small and medium enterprise
PCS	Personal Communications Service	SMS	Short Message Service
		SOAR	Specialist Optimization Access and Results
PDF	Portable Document Format		
PET	Positron emission tomography	SSA	Social Security Administration
PHEV	Plug-in Hybrid Electric Vehicle	SSI	Supplemental Security Income
PISA	Programme for International Student Assessment	STEM	Science, technology, engineering and mathematics
POTS	Plain Old Telephone Service	TANF	Temporary Assistance for Needy Families
PSAP	Public safety answering point	TCUs	Tribal Colleges and Universities
PSBL	Public Safety Broadband Licensee	Telco	Telecommunications
PSTN	Public Switched Telephone Network	TLBC	Tribal Land Bidding Credit
PUC	Public utility commission	TOP	Technology Opportunity Program
R&D	Research and development	TRS	Telecommunications Relay

			Services
R&E	Research and Experimentation or	TSA	Transportation Security
RC	Renewal Community research		Administration
	and education	TV	Television
RFP	Request for Proposal	UCAN	Unified Community Anchor Network
RSA	Rural service area	UHF	ultra high frequency
RUS	Rural Utilities Service	USAC	Universal Service Administrative Company
SBA	Small Business Administration	USCIS	U.S. Citizenship and Immigration Services
SBDC	Small Business Development Center	USDA	U.S. Department of Agriculture
SBTDC	Small Business Technology Development Center	USF	Universal Service Fund
SCORE	Service Corps of Retired Executives	VHA	Veterans Health Administration
SCTCA	Southern California Tribal Chairmen's Association	VHF	Very high frequency
SD	Standard definition	VoIP	Voice over Internet Protocol
SDARS	Satellite Digital Audio Radio	WBC	Women's Business Center
SDB	Small disadvantaged business	WCS	Wireless Communications Service
SDV	Switched Digital Video	WiMAX	Worldwide Interoperability for Microwave Access
SFN	Single Frequency Network	WISP	wireless Internet service provider
SIM	Subscriber Identity Module	WPS	Wireless Priority Service
SLA	Service Level Agreement	WRC	World Radiocommunication
SLC	Subscriber line charge		Conference
SMB	Small or medium-sized business		

APPENDIX C. GLOSSARY[1612]

Accelerometer—An electromechanical device that measures acceleration forces or motion.

Advanced Metering Infrastructure (AMI) —Digital two-way communications hardware and software between smart meters and utility systems which can transmit energy usage, price, and control signals.

Air interface—The technical protocol that ensures compatibility between mobile radio service equipment, such as handsets, and the service provider's base stations.

Ancillary Terrestrial Component (ATC)—A ground-based infrastructure in a mobile satellite system to enhance the coverage of the satellite network.

Backhaul—The telecommunications link used to transport traffic from a geographically distant point, such as a wireless base station, to a significant aggregation point in the network, such as a mobile telephone switching office or Internet peering point.

Bluetooth—An industry standard using unlicensed radio frequency spectrum for wireless connectivity over short distances to link computers, wireless handsets, and other devices.

CableCARD—A credit card-sized device that contains the video provider's security information. When this card is plugged into a set-top box, it enables customers to access the video programming and services to which they have subscribed.

Carrier of last resort—The carrier that commits (or is required by law) to provide service to any customer in a service area that requests it, even if serving that customer would not be economically viable at prevailing rates.

Census block—The smallest geographic unit for which the Census Bureau collects and tabulates decennial census data.

Census tract—A small, relatively permanent statistical subdivision of a county, designed to contain roughly 1,000 to 8,000 people who are relatively homogeneous with respect to their demographics, economic status and living conditions.

Churn—The number of customers who leave a service provider over a given period of time, usually expressed as a percentage of total customers.

Commercial Mobile Alert System—A system established by the Commission that allows wireless service providers choosing to participate to send emergency alerts as text messages to their subscribers.

Commercial Mobile Radio Service—A mobile communications service that is provided for profit and makes interconnected service available to the public, usually in the form of mobile phone service.

Common carrier—A telecommunications provider, such as a telephone company, that offers its services for a fee to the public indiscriminately.

Competitive Local Exchange Carrier—A company that offers local telephone service in competition with the legacy telephone company.

Conditional access—Encrypting digital television services (e.g. premium channels) to limit access to authorized users.

Credentialing (or certification)—The process of establishing the qualifications of licensed professionals (e.g. physicians and teachers), organizational members, or organizations, and assessing their background and legitimacy.

Dark fiber—A fiber optic cable that is laid and ready for use, but for which the service provider has not provided modulating electronics; usually contrasted to lit fiber, which is fiber optic cable in use to provide wired communications.

Data Over Cable Service Interface Specification (DOCSIS)— A standard for the transmission of data over a cable network.

Emergency Alert System (EAS)—A national public warning system that requires broadcasters, cable television systems, wireless cable systems, satellite digital audio radio service (SDARS) providers, and direct broadcast satellite (DBS) providers to provide the communications capability to the President to address the American public during a national emergency. The system also may be used by state and local authorities to deliver important emergency information, such as AMBER alerts and weather information targeted to specific areas.

Encumbered—Spectrum that is burdened with occupancy, usage or congestion limitations or licenses that are subject to obligations or restrictions.

Ethernet—A type of digital transmission service. Traditionally, Ethernet operates at 10 megabits per second (Mbps) (also known as 10-Base-T), although 100-Base-T (100 Mbps) and Gigabit (1,000 Mbps) Ethernet are also available.

Extension arm—A support arm that extends from a telephone pole to hold communications lines at the same level as existing lines which are attached to the pole.

Gateway device—A network device that acts as an entrance to another network and often is used to connect two otherwise incompatible networks.

Grid computing—The linking of two or more computers in a way that allows efficient use of available resources. For example, grid computing could store a single database across multiple servers to allow efficient use of unused storage and parallel processing of database queries.

Independent System Operator (ISO)—An organization that coordinates, controls, and monitors the operation of the electrical power system, either within a single state or across multiple states.

Information service—The offering of a capability for generating, acquiring, storing, transforming, processing, retrieving, utilizing, or making available information via telecommunications.

Intelligent Transportation System (ITS)—A broad range of advanced communications technologies that, when integrated into transportation infrastructure and vehicles, relieves congestion, improves safety, and mitigates environmental impact.

Internet gateway—The closest peering point between a broadband provider and the public Internet for a given consumer connection. See diagram below.

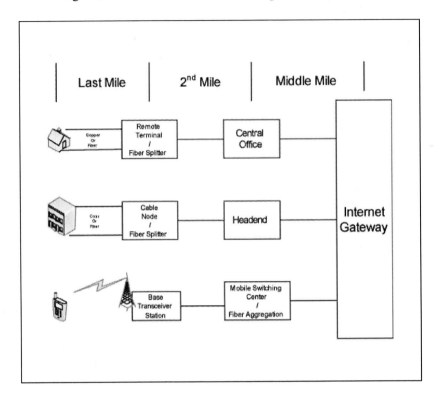

Linear channel—Video content that is delivered in a scheduled mode, such as through broadcast or cable network channels. Internet video (and other platforms such as Video On Demand, or VOD), on the other hand, delivers content upon request and often with pause/rewind/fast-forward capability.

Loop—The connection from the network central office to the customers' premises.

Microcell—Cell sites with extremely limited, but targeted, coverage. Microcells may provide indoor coverage in skyscrapers or may be placed in fire trucks, police cars and ambulances.

Mobile Earth Station—An earth station in the mobile-satellite service intended to be used while in motion or during halts at unspecified points.

Modem—A piece of customer premise equipment typically managed by a broadband provider as the last connection point to the managed network.

Multicast—Simultanous transmission of information/data to multiple receeipients.

Multichannel Video Programming Distributor (MVPD)—An entity that makes available for purchase, by subscribers or customers, multiple channels of video programming.

Multi-Frequency Network (MFN)—A network in which multiple stations consolidate their capacity and broadcast over different channels at different sites and times, similar to a frequency re-use pattern employed by mobile operators to avoid interference between cell sites.

Must-carry—A requirement that cable operators cablecast the broadcast signals of local commercial television stations that request carriage.

Near-Field communications device—A short-range high frequency wireless communication technology which enables simple two-way data interactions between devices.

Next Generation 911 (NG911)—An emergency response system that integrates the core functionalities of the E911 system and also supports multimedia communications (such as texting, e-mail, and video) to the PSAP and to emergency personnel on the ground.

Notice of Inquiry—A proceeding initiated by a federal agency to gather facts and public comment on an issue within the responsibility of the agency, which may lead to a Notice of Proposed Rulemaking.

Notice of Proposed Rulemaking (NPRM)—A notice containing a proposal for adoption of new rules. The Administrative Procedure Act (APA) requires that an agency, before promulgating a binding rule, must publish general notice of its proposal in the Federal Register.

Offload—Shifting telecommunications traffic from one network to another to relieve network congestion.

Open source—A software development model by which the source code to a computer program is made available publicly under a license that gives users the right to modify and redistribute the program.

Out-of-band emission (OOBE)—Any frequency outside of the frequency ranges covered by the adjacent channel power tables found in section 27.53 of the Commission's rules.

Over-builder—A facilities-based provider of cable service, telecommunications, or broadband that builds in an area already served by another facilities-based provider.

Overlay auction—An auction for licenses to unused portions of the spectrum already assigned to incumbent users.

Payload capacity—The amount of throughput possible using a given technology at certain specifications.

Penetration—The homes that are connected to a network, usually provided as a percentage of homes passed.

Point of Presence—A physical location where a communications carrier allows other carriers to access its network.

Pole attachment—Any attachment by a cable television system or provider of communications service to a pole, duct, conduit, or right-of-way owned or controlled by a utility.

Private Branch Exchange—Privately owned switch. A commercial building may have a PBX to route calls within the building.

Privileging—The process health care organizations (predominantly hospitals) employ to authorize practitioners to provide specific services and procedures for their patients.

Protocol stack—The ordered set of protocol types used in communications networks. At the lowest level, the protocol defines the physical interaction of the network components; at the highest level, the protocol defines the applications interacting with users. A protocol stack is designed so that protocols in each layer of the stack are substitutable for each other without affecting protocols higher up the stack.

Public Safety Answering Point (PSAP)—A call center responsible for answering emergency calls and dispatching emergency services.

Public Switched Telephone Network (PSTN)—The legacy circuit-switched telephone network.

Radiodetermination—The determination of the position, velocity or other characteristics of an object, or the obtaining of information relating to these parameters, by means of the propagation of radio waves.

Reband—To reconfigure the assignment of spectrum licenses regarding either who controls the license or how a licensee may use its spectrum.

Remote patient monitoring—Using devices and communications networks to remotely collect and send diagnostic data to a monitoring station for interpretation. For example, measuring blood pressure when a patient is at home.

Right-of-way—The right to pass over or occupy a particular piece of land. For example, utilities generally receive rightsof-way from municipalities to erect and wire poles to carry electricity, telecommunications services, and cable service.

Secondary market (for spectrum) —A mechanism for reapportioning allocated spectrum based on economic demand. The secondary market for spectrum enables licensees to lease their spectrum to third parties, which permits spectrum to flow more freely among users to the extent consistent with the Commission's public interest objectives.

Service Level Agreement (SLA)—An agreement between a user and a service provider defining the nature of the service provided and establishing metrics for that service, trouble reporting procedures and penalties if the service provider fails to perform.

Set-top box—A stand-alone device that receives and decodes programming so that it may be displayed on a television. Set- top boxes may be used to receive broadcast, cable, and satellite programming.

Side lobe—Distribution of microwave energy outside the main beam. Side lobes are measured in both the horizontal (E -plane) and the vertical (H-plane) directions. Normally, the E-plane has higher sidelobes, i.e., more energy distributed outside the main beam.

Single Frequency Network (SFN)—A network used in distributed transmission and differing from a cellular telephone system by using the same frequency in all adjacent cells.

Smart Grid—The electric delivery network, from electrical generation to end-use customer, integrated with sensors, software, and two-way communications technologies to improve grid reliability, security, and efficiency.

Smart meter—A digital meter (typically electric) located on the customer premises that records energy usage and has two-way communications capabilities with utility systems.

Spatial reuse—An efficiency measure that allows use of the same spectral link at the same time.

Subscriber Line Charge (SLC)—A federally regulated monthly service charge assessed by telephone companies to pay for a portion of the local telephone wires, poles and other facilities used to connect a local telephone exchange.

Substantially Underserved Trust Area—A community on land held in trust by the United States for Native Americans (or on certain other trust lands), which the Secretary of the Interior has determined has a high need for the benefits of certain federal programs.

Sufferance basis—The use of spectrum with no legal claim to tenancy. Using spectrum on a sufferance basis means that the use is subject to preemption at any time by the licensee.

Switched Digital Video (SDV)—A method of delivering video programming to subscribers in a given area only when at least one subscriber in that area actively requests that programming.

Switching—The process of connecting the transmission path that allows the calling party to connect to the called party.

Table of Allotments—A list of which television stations may broadcast a digital or analog signal over a given band of spectrum in a given community. The tables may be found in sections 73.606(b) and 73.622(b) of the Commission's rules.

Telecommunications Relay Service (TRS)— A telephone service that enables persons with TTYs, individuals who use sign language and people who have speech and hearing disabilities to use telephone services by having a third party transmit and translate a call. Consumers can access these services by using, for example, video phones, computers, web-enabled devices, captioned telephones, and TTYs.

Teletype or telephone typewriter—A type of machine that allows people with hearing or speech disabilities to communicate over the phone using a keyboard and a viewing screen.

Transcoding—The process of directly converting a digital media file or object from one format to another allowing one to view media that is otherwise not supported by his/her device.

Transport—The transmission facilities between the wire center or switch of an incumbent local exchange carrier and the wire center or switch of another carrier.

Use case—In software engineering and systems analysis, a methodology used to identify, clarify, and organize system requirements as it responds to a request that originates from outside of that system.

Video description—The insertion of audio-narrated descriptions of a television program's key visual elements into natural pauses between the program's dialogue so that the critical details of the information are accessible to persons with visual disabilities.

Video navigation device—A piece of equipment used by consumers within their premises to receive multichannel video programming and other services offered over multichannel video programming systems Converter boxes, interactive equipment, and other.

Wireless Priority Service (WPS)—A federal program that authorizes cellular communications service providers to prioritize calls over wireless networks. Participating service providers typically deploy WPS in stages until service is available in most coverage areas and functionality has reached full operating capability.

APPENDIX D. LIST OF WORKSHOPS AND FIELD HEARINGS

The FCC held 36 public workshops in Washington, D.C. and nine field hearings across the country as part of an extensive effort to engage the public in crafting the National Broadband Plan. These workshops and hearings attracted more than 10,000 in-person and online attendees. The panelists for the workshops and hearings included FCC staff and commissioners, other government officials and representatives from consumer groups, service providers, broadcasters, manufacturers, application providers and many other companies and organizations. The transcripts and videos for these events are all part of the National Broadband Plan record and are available at www.broadband.gov.

	Event	Date	Location
1	E-Gov/Civic Engagement Workshop	8/6/2009	Federal Communications Commission
2	Deployment: Wired-General Workshop	8/12/2009	Federal Communications Commission
3	Deployment: Wireless-General Workshop	8/12/2009	Federal Communications Commission
4	Deployment: Unserved-Underserved Workshop	8/12/2009	Federal Communications Commission
5	Technology/Fixed Broadband Workshop	8/13/2010	Federal Communications Commission
6	Technology/Wireless Workshop	8/13/2009	Federal Communications Commission
7	International Lessons Workshop	8/18/2009	Federal Communications Commission
8	Opportunities for Small and Disadvantaged Businesses Workshop	8/18/2009	Federal Communications Commission
9	Building the Fact Base: The State of Broadband Adoption and Utilization Workshop	8/19/2009	Federal Communications Commission
10	Low Adoption and Utilization: Importance of Broadband and Applications Workshop	8/19/2009	Federal Communications Commission
11	Programmatic Efforts to Increase Broadband Adoption and Usage: What Works and What Doesn't Workshop	8/19/2009	Federal Communications Commission
12	Broadband Opportunities for People with Disabilities Workshop	8/20/2009	Federal Communications Commission
13	Education Workshop	8/20/2009	Federal Communications Commission
14	Public Safety and Homeland Security Workshop	8/25/2009	Federal Communications Commission
15	Smart Grid, Broadband and Climate Change Workshop	8/25/2009	Federal Communications Commission
16	Economic Growth, Job Creation and Private Investment Workshop	8/26/2009	Federal Communications Commission
17	Job Training Workshop	8/26/2009	Federal Communications Commission
18	Technology/Applications and Devices Workshop	8/27/2009	Federal Communications Commission
19	State and Local Governments: Toolkits and Best Practices Workshop	9/1/2009	Federal Communications Commission
20	Benchmarks Workshop	9/2/2009	Federal Communications Commission
21	Big Ideas with Potential to Substantially Change the Internet Workshop	9/3/2009	Federal Communications Commission
22	Broadband Consumer Context Workshop	9/9/2009	Federal Communications Commission

(Continued)

	Event	Date	Location
23	Health Care Workshop	9/15/2009	Federal Communications Commission
24	The Role of Content in the Broadband Ecosystem	9/17/2009	Federal Communications Commission
25	Spectrum Workshop	9/17/2009	Federal Communications Commission
26	Public Field Hearing, National Broadband Plan, FCC Commissioner Meredith Atwell Baker	9/21/2009	The Thompson Conference Center, TCC 3.108 2405 Robert Dedman Drive Austin, Texas
27	Cybersecurity Workshop	9/30/2009	Federal Communications Commission
28	FCC Hearing on Capital Formation in the Broadband Sector	10/1/2009	Federal Communications Commission
29	Diversity and Civil Rights Issues In Broadband Deployment and Adoption Workshop	10/2/2009	Federal Communications Commission
30	FCC Hearing on Broadband Adoption, Commissioners Mignon Clyburn and Michael Copps	10/6/2009	Trident Technical College Palmer Campus 66 Columbus St. Charleston, S.C.
31	FCC Field Hearing: Mobile Applications and Spectrum	10/8/2009	Univ. of San Diego 5998 Alcala Park San Diego, Calif.
32	Economic Issues in Broadband Competition Workshop	10/9/2009	Federal Communications Commission
33	Broadband Accessibility for People with Disabilities II: Barriers, Opportunities and Policy Recommendations Workshop	10/20/2009	Federal Communications Commission
34	FCC Field Hearing on Broadband Access for People with Disabilities	11/6/2009	Gallaudet University Kellogg Conference Center 800 Florida Ave. N.E. Washington, D.C.
35	FCC Broadband Field Hearing on Improving Public Safety Communications and Emergency Response	11/12/2009	Georgetown University Leavey Center 3800 Reservoir Road N.W. Washington, D.C.
36	Capitalization Strategies for Small and Disadvantaged Businesses Workshop	11/12/2009	Federal Communications Commission
37	Future Fiber Architectures and Local Deployment Choices Workshop	11/19/2009	Federal Communications Commission
38	Research Recommendations for the Broadband Taskforce Workshop	11/23/2009	Federal Communications Commission
39	FCC Field Hearing on Energy and the Environment	11/30/2009	MIT Stratton Student Center Twenty Chimneys 84 Massachusetts Ave. Cambridge, Mass.

(Continued)

	Event	Date	Location
40	Lessons for the National Broadband Plan from Local Officials Representing Underserved Communities Workshop	12/9/2009	Federal Communications Commission
41	Global Broadband Connects America and the World: Infrastructure, Services and Applications Workshop	12/10/2009	Federal Communications Commission
42	Review and Discussion of Broadband Deployment Research Workshop	12/10/2009	Federal Communications Commission
43	FCC Field Hearing on Digital Inclusion	12/14/2009	National Civil Rights Museum Rose Room 450 Mulberry St. Memphis, Tenn.
44	FCC Broadband Field Hearing on Small Business	12/21/2009	Univ. of Chicago Gleacher Center 450 N. Cityfront Plaza Drive Chicago, Ill.
45	Broadband and New Media Strategies for Minority Radio Workshop	1/26/2010	Federal Communications Commission

APPENDIX E. LIST OF NATIONAL BROADBAND PLAN CONTRIBUTORS

The National Broadband Plan was created by the staff of the FCC

Omnibus Broadband Initiative

Rajeev Bajaj
Sharren Bates
Philip Bellaria
Kevin Bennett
Scott Berendt
Elana Berkowitz
Mialisa Bonta
Peter Bowen
Val Brock
Michael Broom
Thomas Brown
Paul Carroll
Mukul Chawla
Ronnie Cho
Robert Curtis
Brian David
Rohit Dixit
Vishal Doshi
Elizabeth Duncan

Pierce Graham-Jones
Rebecca Hanson
Joseph Heaps
Keyla Hernandez-Ulloa
John Horrigan
Shawn Hoy
Eugene Huang
Spencer Hutchins
Lyle Ishida
David Isenberg
Kristen Kane
Mohit Kaushal
Thor Kendall
Kevin King
Carlos Kirjner
Elise Kohn
Brian Korgaonkar
Thomas Koutsky
Anurag Lal

Janice Morrison
Byron Neal
Andrew Nesi
Stagg Newman
Karen Perry
Tom Peters
Marie Pharaoh
Sridhar Prasad
Steven Rosenberg
Ellen Satterwhite
Douglas Sicker
Michael Simkins
Nicholas Sinai
Joseph Soban
Jessica Strott
Elvis Stumbergs
Gayle Teicher
Jordan Usdan
Jing Vivatrat

Robert Eckert
Roger Fillion
Leo Fitzpatrick
Jennifer Flynn
John Erik Garr
Sheryl Gelfand
Adam Gerson
Roger Goldblatt
Rebekah Goodheart

Blair Levin
Elizabeth Lyle
Colleen Mallahan
Mark Maltais
Jennifer Manner
Carol Mattey
Nicholas Maynard
Kerry McDermott
Steve Midgley

Dave Vorhaus
Scott Wallsten
Christopher Walti
Stacey Weiss
Brian Weeks
Mark Wigfield
Charles Worthington
Phoebe Yang

Consumer & Governmental Affairs Bureau

Joel Gurin
Michael Jacobs
Karen Johnson
Donice Jones
Susan Kimmel

Cheryl King
Steve Klitzman
Lauren Kravetz
Yul Kwon
Celeste McCray

Mikelle Morra
Mark Stone
Gregory Vadas

Enforcement Bureau

Cynthia Bryant
P. Michele Ellison

Genaro Fullano
Nissa Laughner

Koyulyn Miller
Katherine Power

International Bureau

Donna Christianson
Mindel De La Torre
Anita Dey
Jerry Duvall
Kiran Duwadi
Gardner Foster
Pamela Gerr
Francis Gutierrez
Linda Haller Sloan

Narda Jones
Karl Kensinger
Carrie Lee Early
Hsin Mei Hsu
Robert Nelson
Kathryn O'Brien
Shelia S Crawley
Sean O'More
Roderick Porter

Caroline Schleh
Daniel Shiman
Marilyn Simon
Thomas Sullivan
Emily Talaga
Robert Tanner
Andrea Tutmarc
Irene Wu

Media Bureau

Simon Banyai
William Beckwith
Joyce Bernstein
Katie Costello
Heather Dixon
Marcia Glauberman
Roger Holberg

Alma Hughes
William Lake
Wayne McKee
Kris Monteith
Alison Neplokh
Michael Perko

Rodney Royse
Debra Sabourin
Dana Scherer
Krista Witanowski
John Wong

Office of Communications Business Opportunities

Gilberto DeJesus
Calvin Osborne

Belford Lawson
Thomas Reed

Carolyn Williams

Office of Engineering and Technology

Rashmi Doshi
WalterJohnston
Ira Keltz
Julius Knapp
Geraldine Matise

James Miller
Nicholas Oros
Nam Pham
Ron Repasi
Bruce Romano

Salomon Satche
Rodney Small
Alan Stillwell
Robert Weller
Anh Wride

Office of Legislative Affairs

Diane Atkinson
Jim Balaguer
Connie Chapman
Shomik Dutta
Terri Glaze

Solita Griffis
Christopher Lewis
Lori Maarbjerg
Joy Medley
Chris Moore

Aurelle Porter
Chelle Richmond
Timothy Strachan

Office of Managing Director

Kim Bassett
Walt Boswell
Gray Brooks
Toby Brown
Lavonia Connelly
Daniel Daly
Ruth Dancey
Arecio Dilone
Stephen Ebner
Bridget Gauer
Diane Graham
Noelle Green
Shoko Hair
Judith Herman

Joshua Wingard
Judy Herman
Diana Huynh
Eric Kanner
George Krebs
Vanessa Lamb
Andrew Martin
Lynn Moaney
Ann Pricci
Mercedes Ragland
Patricia Rinn
Richard Robinson
Juan Salazar
Erik Scheibert

Cynthia Schieber
Dana Shaffer
Larry Shields
Sheila Shipp
Wanda Sims
Mark Stephens
Geraldine Taylor
Jamie Thompson
Bonita Tingley
Haley Van Dyck
Steve VanRoekel
Carlyn Walker
Tenecia Williams
Darshan Williams

Office of Media Relations

Steve Balderson
Cozette Ballesteros
Charles Harrington
David Fiske

Jen Howard
Dann Oliver
Audrey Spivack

Jeffrey Riordan
David Kitzmiller
Meribeth McCarrick

Office of Strategic Planning and Policy Analysis

Adele Andrews	Sherille Ismail	Jon Peha
Jonathan Baker	Zachary Katz	William Sharkey
Robert Cannon	Evan Kwerel	Tamara Smith
Jared Cornfeld	Amaryllis Flores	John Williams
Paul de Sa		

Public Safety and Homeland Security Bureau

Pat Amodio	David Furth	William Lane
Kim Anderson	Aaron Garza	Richard Lee
Jamie Barnett	Behzad Ghaffari	Jennifer Manner
Tom Beers	Jeff Goldthorp	Tim May
Joe Casey	Brian Hurley	Susan McLean
Yoon Chang	Mike Iandolo	Ken Moran
Jeff Cohen	Greg Intoccia	Erika Olsen
Jean Ann Collins	Kurian Jacobs	Timothy Peterson
Eric Ehrenreich	Robert Kenny	Joy Ragsdale
Lisa Fowlkes	Deborah Klein	Deandrea Wilson

Wireless Telecommunications Bureau

Joan Andes	Beth Fishel	Paul Murray
Richard Arsenault	Benjamin Freeman	Roger Noel
Audrey Bashkin	Suzan B Friedman	Charles Oliver
Karen Black	Nese Guendelsberger	Michael Pollak
Cheryl Black	Mae Hall	Sayuri Rajapakse
Craig Bomberger	Kevin Holmes	Lynn Ratnavale
Ty Bream	William Huber	Annette Ritchie
Barret Brick	Jane Jackson	Mark Rossetti
James Brown	Elias Johnson	Erik Salovaara
Mary Bucher	Stephen Johnson	John Schauble
Steve Buenzow	Joyce Jones	Jim Schlichting
Saurbh Chhabra	Heidi Kroll	Blaise Scinto
Linda Chang	Yolanda Lee	Ziad Sleem
Michael Connelly	John Leibovitz	Michael Smith
Renee Crittendon	Joseph Levin	Martha Stancill
Lloyd Coward	Scott Mackoul	Jeff Steinberg
Howard Davenport	Eliot Maenner	Walt Strack
Peter Daronco	Paul Malmud	Joel Taubenblatt
Melvin Del Rosario	Charles Mathias	Ruth Taylor
Monica Delong	Nicole McGinnis	Jeffrey Tignor
Monica Desai	Gary Michaels	Peter Trachtenberg
Debra Dick	Chris Miller	Margaret Wiener
Sandra Eckenrode	Elizabeth Miller	Brian Wondrack

| Chelsea Haga Fallon | Ruth Milkman | Morasha Younger |
| Stacy Ferraro | Jackye Milne | Nancy Zaczek |

Wireline Competition Bureau

Claude Aiken	Zina Ellison	Jennifer Prime
Nicholas Alexander	Lynne Engledow	Jonathan Reel
James Bachtell	Irene Flannery	Vickie Robinson
Daniel Ball	Lisa Gelb	Catherine Seidel
Ernesto Beckford	Sharon Gillett	Cecilia Seppings
Amy Bender	Amy Goodman	Carol Simpson
Dana Bradford	Heather Hendrickson	Gina Spade
Val Brock	Terrance Judge	Cindy Spiers
Regina Brown	Katie King	Tim Stelzig
Thomas Buckley	Melissa Kirkel	Donald Stockdale
Kirk Burgee	Jim Lande	Craig Stroup
Ted Burmeister	Al Lewis	Jamie Susskind
Ellen Burton	Kenneth Lynch	Elizabeth Valinoti McCarthy
Thomas Butler	Marcus Maher	Cara Voth
Anita Cheng	Jennifer McKee	Geoff Waldau
Randy Clarke	Erica Meyers	Matthew Warner
Bryan Clopton	Jeremy Miller	Romanda Williams
Nicholas Degani	Alexander Minard	Rodger Woock
William Dever	Mark Nadel	Adrian Wright
Ian Dillner	Claudia Pabo	
James Eisner	Wesley Platt	

End Notes

[1] *See* The Pacific Railroad Act of July 1, 1862 § 5, 12 Stat. 489, 492–93, *available at* http://memory.loc.gov/ cgi-bin/ampage?collId=llsl&fileName=012/llsl012. db&recNum=524.

[2] *See* Federal Highway Administration, Eisenhower Interstate Highway System—Frequently Asked Questions, http://www.fhwa.dot.gov/interstate/faq. htm#question7 (last visited Feb. 12, 2010).

[3] Communications Act of 1934, Pub. L. No. 73-416, 48 Stat. 1064 (codified, as amended, at 47 U.S.C. § 151 et seq.).

[4] American Recovery and Reinvestment Act of 2009, Pub. L. No. 111-5, § 6001(k)(2)(D), 123 Stat. 115, 516 (2009) (Recovery Act).

[5] *See* John Horrigan, *Broadband Adoption and Use in America* 13 (OBI Working Paper No. 1, 2010) (Horrigan, *Broadband Adoption and Use in America*) (finding that only 67% of households have broadband).

[6] *See* Omnibus broadband initiative (OBI), the broadband availability Gap (forthcoming); In general, availability of access infrastructure capable of supporting a given download speed does not guarantee that service providers will offer service at those speeds. Note that these numbers do not take into account quality of service.

[7] Horrigan, *Broadband Adoption and Use in America* at 33. Since 75% of families have broadband at home, 25% of families do not. According to the U.S. Bureau of the Census, 17.4% of the U.S. population is between the ages of 5 and 17. 17.4% of 305 million (total U.S. population estimate) is 53 million people. 25% of 53 million is approximately 13 million. Population estimates come from U.S. Census Bureau, 2006–2008 American Community Survey 3-Year Estimates—Data Profile Highlights, http://factfinder.census.gov/servlet/ ACSSAFFFacts.

[8] Natalie Carlson, *National Survey Finds Kids Give High Marks to High Speed*, hispanic prWire (Apr. 2007), *available at* http://www.hispanicprwire.com/ generarnews.php?1=in&id=2774&cha=0.

[9] *See* Pew Research Center, Pew Internet & American Life Project, Data Sets, June 2003 and March 2007, http:// pewInternet.org/Data-Tools/Download-Data/ Data-Sets.aspx. (see variable BBW, which is the percentage of people saying they use broadband at work. The calculation compared the number of Americans saying they used broadband at work in 2003 to the number saying they used broadband at work in 2007.).

[10] *See* Bureau of Labor Stat., Occupational Projections and Training Data, 2008–2009 Edition (2008), *available at* http://www.bls.gov/emp/optd/optd_archive. htm (download from link).

[11] Horrigan, *Broadband Adoption and Use in America* at 5.

[12] Cathy Schoen et al., *Survey of Primary Care Physicians in Eleven Countries*, 28 Health Aff. w1171 (2009), *available at* http://content.healthaffairs.org/cgi/ reprint/28/6/w1171?ijkey=46Z9Be2ia7vm6&keytype=ref&siteid=healthaff (requires purchase). Count of 14 functions includes: (1) electronic medical record; (2, 3) electronic prescribing and ordering of tests; (4–6) electronic access to test results, Rx alerts, and clinical notes; (7–10) computerized system for tracking lab tests, guidelines, alerts to provide patients with test results, and preventive/follow-up care reminders; and (11–14) computerized list of patients by diagnosis, by medications, and due for tests or preventive care.

[13] Richard Hillestad et all., *Can Electronic Medical Record Systems Transform Healthcare? Potential Health Benefits, Savings, and Costs,* 24 Health Aff. 1103, 1103 (Sept./Oct. 2005), *available at* http://content. healthaffairs.org/cgi/reprint/24/5/1103.

[14] Pacific Northwest Nat'l Lab. (PNNL), DOE, Smart Grid: An Estimation of the Energy and CO_2 Benfits (2009), *available at* http://www.pnl.gov/main/ publications/external/technical_reports/PNNL-19112; pdf, Emission Facts: Greenhouse Gas Emissions from a Typical Passenger Vehicle (2005) (providing epa auto emission facts), *available at* http://www.epa.gov/OMS/ climate/420f05004.pdf.

[15] Marsha Lovett et al., *The Open Learning Initiative: Measuring the Effectiveness of the OLI Statistics Course in Accelerating Student Learning*, j. interact. media in educ., May 2008, *available at* http://jime.open. ac.uk/2008/14/jime-2008-14.pdf; Joel Smith, Vice Provost and CIO, Carnegie Mellon Univ., Remarks at FCC Education Workshop (Aug. 20, 2009), *available at* http://www.broadband.gov/docs/ws_education/ ws_education_smith.pdf.

[16] Richard Fry, Pew Research Center., College Enrollment Hits All-Time High, Fueled by Community College Surge (2009), *available at* http:// pewsocialtrends.org/assets/pdf/college-enrollment.pdf.

[17] Brian L. Hawkins & Julia A. Rudy, Educause , Fiscal Year, fiscal year 2007 Summary Report 35/29, *available at* http:// net.educause.edu/ir/library/pdf/PUB8005.pdf.

[18] Letter from Kathy Martinez, Exec. Director, World Inst. on Disability, to Michael J. Copps, FCC Acting Chairman, and Commission Members, GN Docket No. 09-51 (June 1, 2009) at 1–2.

[19] *See, e.g.*, Mitch Waldrop, *DARPA and the Internet Revolution*, in darpa: 50 years Of bridGinG the Gap 83 (2008), *available at* http://www.darpa.mil/Docs/ Internet_Development_200807180909255.pdf.

[20] *See, e.g.*, *Amendment of Section 64.702 of the Commission's Rules and Regulations (Second Computer Inquiry)*, Final Decision, 77 F.C.C.2d 384 (1980) (regulatorily separating "basic" from "enhanced" services to prevent owners of telecommunications infrastructure from impeding upon enhanced service growth); *MTS and WATS Market Structure*, Memorandum Opinion and Order, 97 F.C.C.2d 682, paras. 76–83 (1983) (allowing an exemption for access charges for enhanced service providers (ESP)); *Amendments of Part 69 of the Commission's Rules Relating to Enhanced Service Providers,* CC Docket No. 87-215, Order, 3 FCC Rcd 2631 (1988) (making the ESP access charge exemption permanent).

[21] Cable Television Consumer Protection and Competition Act of 1992, Pub. L. No. 102-385, 106 Stat. 1460 (1992) (codified at 47 U.S.C. § 533).

[22] *See* Omnibus Budget Reconciliation Act of 1993, Pub. L. No. 103-66, Title VI, § 6002(b), 107 Stat. 312 (1993) (amending the Communications Act of 1934 and codified at 47 U.S.C. §§ 153(n), 332(c)(1)).

[23] Letter from 21st Century Telecommunications et al., Members of the Consumer Electronic Association et al., to Chairman Julius Genachowski and Commissioners, FCC, GN Docket No. 09-51 (Dec. 2, 2009) at 1 (filed by Consumer Electronics Association on behalf of 115 parties).

[24] Omnibus Broadband Initiative, The Broadband Availability Gap (forthcoming).

[25] For the purposes of the plan, "Tribal lands" is defined as any federally recognized Tribe's reservation, pueblo, and colony, including former reservations in Oklahoma, Alaska Native regions established pursuant to the Alaska Native Claims Settlement Act, Pub. L. No. 92-203, 85 Stat. 688 (1971), and Indian allotments. The term "Tribe" means any American Indian or Alaska Native Tribe, Band, Nation, Pueblo, Village, or Community, which is acknowledged by the Federal government to have a government-to-government relationship with the United States and is eligible for the programs and services established by the United States. S*ee Statement of Policy on Establishing a Government-to-Government Relationship with Indian Tribes*, Policy Statement, 16 FCC Rcd 4078, 4080 (2000). Thus, "Tribal lands" includes American Indian Reservations and Trust Lands, Tribal Jurisdiction Statistical Areas, Tribal Designated Statistical Areas, and Alaska Native Village Statistical Areas, as well as the communities situated on such lands. This would also include the lands of Native entities receiving Federal acknowledgement or recognition in the future.

[26] 9/11 Comm'n, The 9/11 Commission Report 39 (2004), *available at* http://www.9-11commission.gov/report/911Report.pdf.

[27] Google Comments in re NBP PN #2 (*Comment Sought on the Implementation of Smart Grid Technology—NBP Public Notice #2*, GN Docket Nos. 09-47, 09-51, 09-137, Public Notice, 24 FCC Rcd 11747 (WCB 2009) (*NBP PN #2*)), filed Oct. 2, 2009, at 4.

[28] John Horrigan, *Broadband Adoption and Use in America* 16 (OBI, Working Paper No. 1, 2010) (Horrigan, *Broadband Adoption and Use in America*).

[29] comScore, Inc., Jan.–June 2009 Consumer Usage database (sampling 200,000 machines for user Web surfing habits) (on file with the Commission) (comScore database).

[30] Horrigan, *Broadband Adoption and Use in America* at 16.

[31] Nielsen Company, Viewership on the Rise as More Video Content Spans All Three Screens, A2/M2 Three Screen Report 2 (2Q 2009) (Nielsen, Viewership on the Rise), *available at* http://blog.nielsen.com/nielsenwire/ wp-content/uploads/2009/09/3ScreenQ209_US_ rpt_090209.pdf; Lee Rainie & Dan Packel, Pew Internet & Am. Life Project, More Online, Doing More3 (2001), *available at* http://www.pewinternet.org/~/media/ Files/Reports/2001/PIP_Changing_Population.pdf.pdf (last visited Feb. 19, 2009); *see also* Omnibus Broadband Initiative, Broadband Performance, (forthcoming) (OBI, broadband Performance).

[32] comScore database.

[33] comScore database; *see also* OBI, Broadband Performance; Cisco Sys., Cisco Visual Netwo rking Index: Forecast and Methodolog y, 2008–2013, at 4 (2009) (Cisco, Visual Networking Index), *available at* http://www. cisco.com/en/US/solutions/collateral/ns341/ns525/ ns537/ns705/ns827/white_paper_c11-481360.pdf; Letter from Craig Mundie, Chief Research & Strategy Officer, et al., Microsoft Corp., to Marlene H. Dortch, Secretary, FCC, GN Docket Nos. 09-47, 09-51, 09-137 (Sept. 22, 2009) at 3; University of Minnesota, Minnesota Internet Traffic Studies (MINTS), http://www.dtc.umn.edu/mints/home. php (last visited Feb. 19, 2009).

[34] comScore database.

[35] FCC, National Broadband Plan Survey of Businesses, Dec. 9, 2009–jan. 31, 2010 (FCC, NBP survey of businesses), http://fjallfoss.fcc.gov/ecfs/comment/ view?id=6015536973.

[36] Cisco Sys., Cisco IT Executive Presentation: TelePresence 6 (3Q 2009), *available at* http://www. cisco.com/web/about/ciscoitatwork/downloads/ ciscoitatwork/pdf/TelePresence_White.pdf.

[37] Dale Jorgenson et al., *Industry Origins of the American Productivity Resurgence*, 19 Econ. Sys. Res. 229–52 (2007).

[38] *See* Cisco, Visual Netwo rking Index 4; cisco Sys. Cisco Visual Netwo rking Index: Global Mobile Data Traffic Forecast Update, 2009–2014, at 1 (2009), *available at* http://www.cisco.com/en/US/solutions/collateral/ns341/ns525/ns537/ns705/ns827/white_paper_cll520862.pdf.

[39] Stevie Smith, *Skype 4.0 Looks to Expand Video Calling*, Tech Herald, June 18, 2008, http://www.thetechherald.com/article.php/200825/1273/Skype-4-0-looksto-expand-video-calling; Shamila Janakiraman, *Skype Supports Video Calls on PCs and Embeds Skype Software in HDTVs*, TMCnet.Com, Jan. 6, 2010, http:// voip-phone-systems.tmcnet.com/topics/voip-phone- systems/articles/72051-skype-supports-video-calls-pcs- embeds-skype-software.htm.

[40] Nielsen, Viewership on the Rise 2; Ctr. for Media Design, Nielsen, Video Consumer Mapping Study (2009) (Nielsen, Video Consumer Mapping Study).

[41] Letter from Susan L. Fox, Vice Pres. of Gov't Relations, Disney, to Marlene H. Dortch, Secretary, FCC, GN Docket No. 09-91, WC Docket No. 07-52 (Dec. 11, 2009) at 1.

[42] Horrigan, *Broadband Adoption and Use in America* at 17.

[43] RAND Corp., The Global Positioning System, App. B— GPS History, Chronolog y, and Budgets247–49 (1995), *available at* http://www.rand.org/pubs/monograph_ reports/MR614/MR614.appb.pdf.

[44] Sunlight Labs, Apps for America 2: The Data.gov Challenge, http://sunlightlabs.com/contests/appsforamerica2/ (last visited Feb. 19, 2010); FlyOnTime.us, http://flyontime.us (last visited Feb. 19, 2010).

[45] *See* Consumer Elec. Ass'n, US Consumer Electronics Sales & Forecasts 2005–2010, at 33 (2010) (CEA, Electronics Sales & Forecasts) (87 percent); Niki Scevak, Forrester Research, Inc., Forrester Research Online Population Access and Demog raphic Model (2010) (81 percent); Horrigan, Broadband Adoption and (2010) (81 percent); Horrigan, *Broadband Adoption and Use in America* at 13 (79 percent).

[46] CEA, Electronics Sales & Forecasts33.

[47] CEA, Electronics Sales & Forecasts 33 ("Netbooks will overtake all other notebooks by 2011"); GOldman sachs, adObe systems inc. (adbe) pc refresh beneficiary 15 (2009) (citing forecast of about 50 million units by 2013).

[48] Number calculated using Commission data. S*ee* Office of Engineering and Technology, FCC, Equipment Authorization Search, https://fjallfoss.fcc.gov/oetcf/ eas/reports/GenericSearch.cfm (last visited Feb. 22, 2010). The data represents applications for grants issued for new FCC IDs for equipment class parameters "PCE-PCS Licensed Transmitter held to ear" and "TNELicensed Non-Broadcast Transmitter Held to Ear." Data does not include applications for permissive changes and counts multiple entries for the same FCC ID only once.

[49] Carolina Milanesi et al., Gartner, Inc., Forecast: Mobile Devices, Worldwide,, 2003–2013, at tab 2 (Devices) (2009). We took the information from column L (2012 year), added rows 40 (Basic Phones) and 41 (Enhanced Phones) together (95 million) and compared the number with the number received when rows 43 (Smart Phones—Entry Level) and 44 (Smart Phone—Feature) are added together (109 million). This plan contains several references to Gartner. The Gartner Report(s) described herein, (the "Gartner Report(s)") represent(s) data, research opinion or viewpoints published, as part of a syndicated subscription service, by Gartner, Inc. ("Gartner"), and are not representations of fact. Each Gartner Report speaks as of its original publication date and the opinions expressed in the Gartner Report(s) are subject to change without notice.

[50] *See* OnStar Explained, http://www.onstar.com/us_ english/jsp/explore/index.jsp (last visited Mar. 1, 2010) (discussing OnStar).

[51] Section 629 covers equipment used to receive video programming—including cable set-top boxes, televisions, and DVRs—as well as equipment used to receive other services offered over MVPD systems, including cable modems. S*ee* 47 U.S.C. § 549 (codifying section 629 of the Telecommunications Act of 1996); *Implementation of Section 304 of the Telecommunications Act of 1996; Commercial Availability of Navigation Devices*, CS Docket No. 97-80, Report and Order, 13 FCC Rcd 14775 (1998).

[52] dell'OrO GrOup inc., set-tOp bOx repOrt 3Q09, at 89 (2009). Combined market shares for the two manufacturers (Motorola and Cisco) were 87% (2006), 86% (2007), and 92% (2008). S*ee id.*

[53] *Cf.* CableLabs, Certified, Verified and Self-Verified Cable Products, http://www.cablelabs.com/opencable/udcp/downloads/OC_PNP.pdf (Aug. 26, 2009) (reporting 11 certified set-top boxes), *with supra* note 22 (calculating 850 wireless devices).

[54] Letter from Neal M. Goldberg, Vice Pres. and Gen. Counsel, National Cable & Telecommunications Association, to Marlene H. Dortch, Secretary, FCC, CS Docket No. 97-80 (Dec. 22, 2009) at 1 (presenting report detailing CableCARD deployment and support).

[55] dell'OrO GrOup inc., set-tOp bOx repOrt 2Q09 at 89 (2009).

[56] These numbers include estimates for 4Q09. S*ee* GOldman sachs, telecOm/pay tv industry mOdel 23–25 (2009).

[57] Housing units are distinct from households. "A housing unit is a house, an apartment, a mobile home, a group of rooms, or a single room that is occupied (or if vacant, is intended for occupancy) as separate living quarters." U.S. Census Bureau, Households, Persons Per Household, and Households with Individuals Under 18 Years, 2000 http://quickfacts.census.gov/qfd/meta/long_71061.htm (last visited Feb. 28, 2010). In contrast, "A household includes all the persons who occupy a housing unit.... The occupants may be a single family, one person living alone, two or more families living together, or any other group of related or unrelated persons who share living arrangements." *Id.* There are 130.5 million housing units and 111.7 million households in the United States. U.S. Census Bureau, *Census Bureau Reports on Residential Vacancies and Homeownership* (press release), Feb. 2, 2010, at 3 tbl. 3, http://www.census.gov/hhes/www/ housing/hvs/qtr409/files/q 409press.pdf (Census Bureau, *Residential Vacancies and Homeownership*). Unoccupied housing units (the difference between the count of households and of housing units) include housing units vacant for sale or rent and those for occasional, temporary or seasonal use.

[58] *See* Obi, brOadband perfOrmance, the brOadband availability Gap (forthcoming) (Obi, the brOadband availability Gap).

[59] *See* Obi, the brOadband availability Gap. Note that this figure represents the capability of existing infrastructure, not current service offerings.

[60] *See* Obi, the brOadband availability Gap. Seven million housing units without access to 4 Mbps terrestrial service are outside the cable footprint and are more than approximately 11,000 feet from the nearest DSLAM location; 6 million housing units with 12 million people do not have access to any always-on service with actual download speeds of 768 Kbps or higher as they are more than approximately 16,000 feet from the nearest DSLAM. Note that the analysis excludes satellite broadband because satellite capacity is limited, as discussed in the working paper.

[61] *See* Obi, the brOadband availability Gap. In general, availability of access infrastructure capable of supporting a given download speed does not guarantee that service providers will offer service at those speeds. Note that these numbers do not take into account quality of service.

[62] *See* Obi, the brOadband availability Gap. Coverage reflects access at download speeds consistent with residential discussion; it does not necessarily reflect access to business-class broadband services.

[63] *See* Obi, the brOadband availability Gap; National Atlas of the United States, 2005-06, County Boundaries of the United States, 2001: National Atlas of the United States, Reston, VA (presenting map boundaries).

[64] natiOnal center fOr educatiOnal statistics, internet access in u.s. public schOOls and classrOOms: 1994–2005, at 4 (2006), *available at* http://nces.ed.gov/ pubs2007/2007020.pdf.

[65] dep't Of educ., evaluatiOn Of the enhancinG educatiOn thrOuGh technOlOGy prOGram: final repOrt 12 (2009), *available at* www.ed.gov/rschstat/eval/tech/netts/ finalreport.pdf.

[66] *See infra* Chapter 10; *see also* Letter from Theresa Cullen, Rear Admiral, U.S. Public Health Service, Chief Information Officer and Director, Indian Health Service, to Marlene H. Dortch, Secretary, FCC (Feb. 23, 2010) Attach. In this instance, "mass market" refers to non- dedicated line solutions for businesses, which are similar to residential broadband but called "small business" or "business packages" by carriers.

[67] Along with aggregate growth in broadband speeds, each technology has shown speed increases. For instance, cable typical advertised speeds have migrated from 1 Mbps in the late 1990s to roughly 10 Mbps today, a 20% annual growth rate. *See* Obi, brOadband perfOrmance.

[68] robert c. atkinsOn & ivy e. Schultz, cOlumbia institute fOr tele-infOrmatiOn, brOadband in america: Where it is and Where it is GOinG (accOrdinG tO brOadband service prOviders) at 8 (2009) (atkinsOn & schultz, brOadband repOrt), *available at* http:// www4.gsb.columbia.edu/citi/; *see also* Census Bureau, *Residential Vacancies and Homeownership* 3 tbl. 3.

[69] atkinson & schultz, broadband report at 8.

[70] atkinson & schultz, broadband report at 8 (top 5 cable company rankings based on subscribers).

[71] atkinson & schultz, broadband report 8.

[72] atkinson & schultz, broadband report 24.

[73] atkinson & schultz, broadband report 8.

[74] *See* Organisation for Economic Co-Operation and Development (OECD), Average advertised download speeds, by country (Sept. 2008) http://www.oecd. org/dataoecd/10/53/39575086.xls (last visited Dec. 22, 2009) (9.6 Mbps); FCC, 2008 Form 477 database (accessed Dec. 2009) (on file with the Commission) (6.7 Mbps). Note that 477 data is collected in speed "tiers" and reflects 2008 data. *See* Obi, the brOadband availability Gap.

[75] comScore database. The median speed is more representative of the speeds seen by the typical American consumer because the average speed is skewed upwards by a limited number of high-speed connections (>15 Mbps advertised). comScore monitored 200,000 computers for data usage and consumption, selected to represent American usage broadly (types of services, service providers, geographies, demographics, etc.). Speed testing was attempted every 36 hours at varying times of day and only done when a given computer was otherwise inactive. Speed tests were conducted using packets sent in ever-increasing size to measure average speeds experienced to end-users. Maximum speeds on each connection were determined based on maximum speeds achieved (+/- 10%) and with confirmation on a sample of bills in tandem with the FCC. Speed testing was conducted from the computer/device to the nearest Akamai server. This approach has been used for speed claims by 5 of the top 10 ISPs in America. *See* OBI, brOadband perfOrmance (discussing the methodology and data further).

[76] comScore database. *See* OBI, brOadband perfOrmance.

[77] Note that speeds experienced by the end-user can be impacted by many factors including the user's own equipment, the service provider network and the applications and sites being accessed online. In the first half of 2009, the median actual speed for those that subscribe to broadband in the United States was 3 Mbps download speed. comScore database. Given past annual growth rates in subscribed speed of approximately 20–25% per year, the median could exceed 4 Mbps by the end of 2010. *Cf.* akamai, the state Of the internet, 3rd Quarter, 2009, at 10 (Jan 2010) *available at* http:// www.akamai.com/dl/whitepapers/Akamai_State_Internet_Q3_2009.pdf?curl=/dl/whitepapers/Akamai_   State_Internet_Q3_2009.pdf&solcheck=1& (registration required) (finding average download speeds to be 3.9 Mbps in the third quarter of 2009); *see also* OBI, Broadband Performance (discussing past growth rates).

[78] comScore database. Note that fiber in the database refers to both fiber to the premises (FTTP) and short- loop fiber to the node (FTTN). According to the Form 477 database, FTTP advertised download speeds were 3-4 Mbps faster than comScore fiber average. For more data and detail on methodologies *see* OBI, brOadband perfOrmance.

[79] comScore database. Commission Form 477 data mirrors comScore advertised speed ranges of different technologies and relative advertised speeds, with important methodology differences for fiber. *See* OBI, brOadband perfOrmance.

[80] SamKnows Limited Comments in re NBP PN #24 (*Comment Sought on Broadband Measurement and Consumer Transparency of Fixed Residential and Small Business Services in the United States—NBP Public Notice #24*, GN Docket Nos. 09-47, 09-51, 09-137, Public Notice, DA 24 FCC Rcd 14120 (WCB, rel. Nov. 24, 2009) (*NBP PN #24*)), filed Dec. 16, 2009; OfcOm, uk brOadband speeds 2009, at 8 (2009), *available at* http://www.ofcom.org.uk/research/telecoms/   reports/broadband_speeds/broadband_speeds/   broadbandspeeds.pdf.

[81] Epitiro Comments in re NBP PN #24, filed Dec. 14, 2009, Apps. 1–4 (multiple appendices attached to comments detailing country results).

[82] *See* American Roamer Advanced Services database (accessed Aug. 2009) (aggregating service coverage boundaries provided by mobile network operators) (on file with the Commission) (American Roamer database); *see also* Geolytics Block Estimates and Block Estimates Professional databases (2009) (accessed Nov. 2009) (projecting census populations by year to 2014 by census block) (on file with the Commission) (Geolytics databases). The approximate of 60% is based on total landmass area. In 2008, this figure was 39.6%. *Implementation of Section 6002(b) of the Omnibus Budget Reconciliation Act of 1993; Annual Report and Analysis of Competitive Market Conditions With Respect to Commercial Mobile Services*, WT Docket No. 08-27, Thirteenth Report, 24 FCC Rcd 6185, 6257, tbl. 9 (WTB 2009).

[83] Data from American Roamer shows geographic coverage by technology. The actual service quality of data connections experienced by end-users will differ due to a large number of factors, such as location and

mobility. Further, the underlying coverage maps do not include information on the level of service (*i.e.*, signal quality and the speed of broadband service) provided; nor is coverage defined by providers in the same way. Thus, coverage as measured here does not correspond to a specific minimum signal quality or user experience. See American Roamer database; *see also infra* Chapter 4, Section 4.1 (Competition in Residential Broadband Networks) (discussing the American Roamer methodology). Population is based on projected census block figures from Geolytics. See Geolytics databases.

[84] *See infra* Chapter 4, Section 4.1 (Transparency in the retail broadband market) (discussing details on a possible new approach to measurement and disclosure of mobile services).

[85] *See* American Roamer database; Geolytics databases.

[86] Data from American Roamer applied to business locations will suffer from the same quality of service issues (in-building coverage, varying bit rates) as residential. See American Roamer database; *see also* GeoResults National Business and Telecom database (accessed Nov. 2009) (projecting business locations) (on file with the Commission) (GeoResults database).

[87] *See* American Roamer database; GEOResults database.

[88] *See* American Roamer database; GEOResults database.

[89] *See* American Roamer database at 8; *see also* Verizon Wireless, Network Facts, http://aboutus. vzw.com/bestnetwork/network_facts.html (last visited Feb. 28, 2010) (providing Verizon's 4G roll-out plan, and coverage of 285 million people by its 3G network).

[90] atkinsOn & schultz, brOadband repOrt at 27.

[91] *See* atkinsOn & schultz, brOadband repOrt at 27. The figures are in millions of people covered.

[92] *See* Letter from Dean R. Brenner, Vice Pres. Gov't Affairs, Qualcomm Inc., to Marlene H. Dortch, Secretary, FCC, GN Docket 09-51 (Dec. 9, 2009) Attach.

[93] *See* comScore database (discussing data on upload and download speeds); chetan sharma & sarla sharma, state Of the (mObile) brOadband natiOn: a benchmarkinG study (2009), *available at* http://www.chetansharma.com/ State%20of%20the%20Broadband%20Nation%20-%20 Chetan%20Sharma%20Consulting.pdf (Reprinted with permission. Copyright © 2009 Chetan Sharma Consulting. All rights reserved. Based on data compiled by Root Wireless, Inc.).

[94] Letter from Consumer Electronics Association et al., to Chairman Julius Genachowski and Commissioners, FCC, GN Docket No. 09-51 (Dec. 2, 2009) at 1.

[95] Horrigan, *Broadband Adoption and Use in America* at 13.

[96] Horrigan, *Broadband Adoption and Use in America* at 13.

[97] Horrigan, *Broadband Adoption and Use in America* at 13.

[98] For the purposes of the Plan, we define "Tribal lands" as any federally recognized Tribe's reservation, pueblo and colony, including former reservations in Oklahoma, Alaska Native regions established pursuant to the Alaska Native Claims Settlement Act (85 Stat. 688), and Indian allotments. The term "Tribe" means any American Indian or Alaska Native Tribe, Band, Nation, Pueblo, Village or Community which is acknowledged by the Federal government to have a government-togovernment relationship with the United States and is eligible for the programs and services established by the United States. See Statement of Policy on Establishing a Government-to-Government Relationship with Indian Tribes, 16 FCC Rcd 4078, 4080 (2000). Thus, "Tribal lands" includes American Indian Reservations and Trust Lands, Tribal Jurisdiction Statistical Areas, Tribal Designated Statistical Areas, and Alaska Native Village Statistical Areas, as well as the communities situated on such lands. This would also include the lands of Native entities receiving Federal acknowledgement or recognition in the future. While Native Hawaiians are not currently members of federally-recognized Tribes, they are intended to be covered by the recommendations of this Plan, as appropriate.

[99] Horrigan, *Broadband Adoption and Use in America* at 13. The survey offered a Spanish language option, so results for Hispanics include English- and Spanish-speaking Hispanics.

[100] FCC, NBP Survey of Businesses.

[101] *See, e.g.*, Paul Romer, *Endogenous Technological Change*, 98 J. Pol. Econ.. S71 (1990).

[102] Timothy Bresnahan & Manuel Trajtenberg, *General Purpose Technologies "Engines of Growth?"* 1 (Nat'l Bureau of Econ. Research, Working Paper No. W4148, 1995), *available at* http://www.nber.org/papers/w4148.pdf.

[103] Elhanan Helpman & Manuel Trajtenberg, *A Time to Sow and a Time to Reap: Growth Based on General Purpose Technologies*, *in* General Purpose Technolog ies and Economic Grow th 55–84 (1998).

[104] *See, e.g.*, Richard G. Harris, *The Internet as GPT: Factor Market Implications*, *in* General Purpose Technolog ies and Economic Grow th 145–66 (1998); Richard G. Lipsey et al., Economic Transformations: General Purpose Technolog ies and Long Term Economic Grow th 133 (2005).

[105] *See, e.g.*, *infra* Chapter 10.

[106] *See, e.g.*, *infra* Chapter 11.

[107] *See, e.g.*, *infra* Chapter 12.

[108] *See, e.g.*, *infra* Chapter 13.

[109] *See, e.g.*, *infra* Chapters 14–16.

[110] David B. Audretsch & Maryann Feldman, *R&D Spillovers and the Geography of Innovation and Production*, 86 Amer. Econ. Rev.. 630, 630 (1996).

[111] See, for example, Howard Shelanski, *Adjusting Regulation to Competition: Toward a New Model for U.S. Telecommunications Policy*, 24 Yale J. on ReG. 56 (2007), for a discussion. Even in the early days of high-speed access some recognized that the high-speed retail ISP market structure would differ from that of dialup ISP. Faulhaber and Hogendorn, for example, estimated that demand would support two or three wireline providers. See Gerald R. Faulhaber & Christiaan Hogendorn, *The Market Structure of Broadband Telecommunications*, 48 J. Indust. Econ.. 305, 321 (2000). Atkinson argues that the economics of "ultrabroadband" points to more concentration. Robert Atkinson, *Market Structure for Ultrabroadband*, Commcn's & Strategies, Special Issue 2008, at 35, 49 (2008).

[112] Bresnahan and Reiss's seminal article developed the model and tested its implications in a number of industries. See Timothy F. Bresnahan & Peter C. Reiss, *Entry and Competition in Concentrated Markets*, 99 j. pOl. ecON. 977 (1991). Similarly, Sutton introduced the concept of "endogenous sunk costs" (ESC) in which firms can choose how much to invest in sunk costs. John Sutton, Sunk Costs and Market Structure:Price Competition, Advertising, and the Evolution of Concentration (1991). The key insight is that in such industries the total number of firms is likely to be limited and may even shrink as the market grows. As Bresnahan and Greenstein state, "when ESC are important, demand growth does not lead to fragmentation; a larger market will have higher ESC, not more firms, in equilibrium." Timothy Bresnahan & Shane Greenstein, *Technological Competition and the Structure of the Computer Industry*, 47 j. indust. ecOn. 1, 6 (1999). Xiao and Orazem extend the Bresnahan-Reiss analysis to the broadband access market and find no additional competitive effects beyond a third competitor. See Mo Xiao & Peter F. Orazem, *Do Entry Conditions Vary over Time? Entry and Competition in the Broadband Market: 1999–2003* (Iowa State Univ., Working Paper No. 06004, 2006), *available at* http://www.econ.iastate.edu/research/webpapers/ paper_12500_06004.pdf. While suggestive, the research relies on the FCC's ZIP code counts from the old Form 477 data. Those data, discussed elsewhere, show that most ZIP codes have multiple high-speed providers, but those providers do not always serve the same area within the ZIP code.

[113] Imperfect competition occurs when goods or services are not perfect substitutes yet can impose some competitive discipline on each other due to the multidimensional nature of consumer preferences. For example, in this case, mobile broadband could provide some competitive pressure if enough people are willing to trade off speed for mobility.

[114] Department of Justice *Ex Parte* in re National Broadband Plan NOI, filed Jan. 4, 2010, at 11 ("We do not find it especially helpful to define some abstract notion of whether or not broadband markets are 'competitive.' Such a dichotomy makes little sense in the presence of large economies of scale, which preclude having many small suppliers and thus often lead to oligopolistic market structures. The operative question in competition policy is whether there are policy levers that can be used to produce superior outcomes, not whether the market resembles the textbook model of perfect competition. In highly concentrated markets, the policy levers often include: (a) merger control policies; (b) limits on business practices that thwart innovation (e.g., by blocking interconnection); and (c) public policies that affirmatively lower entry barriers facing new entrants and new technologies.").

[115] Department of Justice *Ex Parte* in re National Broadband Plan NOI, filed Jan. 4, 2010, at 7; Gregory L. Rosston, Deputy Director, Stanford Institute for Economic Policy Research, Remarks at FCC Benchmarks Workshop 5–17 (Sept. 2, 2009), *available at* http://www.broadband.gov/docs/ws_20_benchmarks. pdf; James Prieger, Professor of Pub. Policy, Pepperdine Univ., Remarks at FCC Economic Growth, Job Creation and Private Investment Workshop 4–15 (Aug. 26, 2009), *available at* http://broadband.gov/docs/ ws_16_economy.pdf; Ryan McDevitt, Lecturer, Dep't of Manag. & Strat., Northwestern Univ., Remarks at FCC Economic Growth, Job Creation and Private Investment Workshop 23–34 (Aug. 26, 2009), *available at* http:// broadband.gov/docs/ws_16_economy.pdf; Joseph Farrell, Director, Bureau of Econ., FTC, Remarks at FCC Economic Issues in Broadband Competition Workshop 55–66 (Oct. 9, 2009), *available at* http://broadband. gov/docs/ws_28_economic.pdf; Carl Shapiro, Deputy Ass't Attorney General for Economics, Antitrust Div., DOJ, Remarks at FCC Economic Issues in Broadband Competition Workshop 66–83 (Oct. 9, 2009), *available at* http://broadband.gov/docs/ws_28_economic.pdf.

[116] See FCC, 2008 Form 477 database (accessed Dec. 2009) (on file with the FCC) (Form 477 database). While much improved from past years, the new 477 data are not ideal for analyzing competition because the data identify providers that operate anywhere in a Census tract and not whether their service areas overlap geographically. We improve the 477 provider counts in two ways. First, we do not count providers with less than one percent of broadband subscriptions in a given Census tract under the assumption that a provider with such a small number of subscribers is probably not available to a large part of the tract. Second, we identify cable overbuilders (such as RCN) in the data, which allows us to make reasonable assumptions about where cable companies actually provide service to the same geographic areas. Specifically, we assume that any given area is served by a maximum of one facilities-based DSL provider and one cable provider unless a cable overbuilder is present, in which case we count both cable providers. We also count fiber-specific competitors, but do not double-count telco providers that offer both DSL and fiber in the same tract (i.e., Verizon DSL and

FIOs). Finally, we do not count CLECs providing service over another company's lines because we focus on facilities-based providers, and their inclusion would overstate the extent of competition.

[117] The limited useful data on availability make it difficult to estimate these figures with precision. The OBI team has used multiple inputs and analyses to better estimate the availability figures, as discussed *infra* Chapter 8. S*ee* Omnibus Braodband Initiative, OBI, The Broadband Availability Gap (forthcoming) (OBI, The Broadband Availability Gap).

[118] Robert C. Atkinson & Ivy E. Schultz, Columbia Inst. for Tele-Information, Broadband In America: Where It Is And Where It Is Going (According To Broadband Service Providers) 24 (2009) (Atkinson & Schultz, Broadband in America).

[119] According to Clearwire's November 10, 2009 earnings report, it expected to provide service in the following cities by the end of 2009: Atlanta, GA; Baltimore, MD; Boise, ID; Chicago, IL; Las Vegas, NV; Philadelphia, PA; Charlotte, Raleigh, and Greensboro, NC; Honolulu and Maui, HI; Seattle and Bellingham, WA; Portland and Salem, OR; and Dallas/Ft. Worth, San Antonio, Austin, Abilene, Amarillo, Corpus Christi, Killeen/Temple, Lubbock, Midland/Odessa, Waco and Wichita Falls, TX. Clearwire, *Clearwire Reports Third Quarter 2009 Results* (press release), Nov. 10, 2009, http://investors. clearwire.com/phoenix.zhtml? c=198722&p=irolnewsArticle&ID=1353840.

[120] Satellite-based broadband providers, because of limited satellite capacity, have Fair Access Policies (often termed usage caps) for their customers: the Hughes current limit is as low as 200 MB per day, while WildBlue's cap is as low as 7,500 MB per month. Next- generation satellites will have much higher capacities, in excess of 100 Gbps each, with download speeds per user of up to 25 Mbps. Larger capacities could allow for usage patterns that more-closely mirror terrestrial usage. However, the high fixed costs of designing, building and launching a satellite mean that satellite-based broadband is likely to be cheaper than terrestrial service only for the most expensive-to-serve areas. Atkinson &Schultz, Broadband in America in america at 57. As the report notes, however, actual speeds will depend on several factors, including intensity of use in any given area. For examples of commercial services with usage caps today, see HughesNet, Fair Access Policy, http://web.hughesnet. com/sites/legal/Pages/FairAccessPolicy.aspx (last visited Mar. 4, 2009) and WildBlue Communications, WildBlue Fair Access Policy, http://wildblue.com/legal/ fair.jsp (last visited Mar. 4, 2009).

[121] *See* Form 477 database. The figure is derived from econometric analysis of the FCC's December 2008 Form 477 data and controls for housing density, household income, and state fixed effects. Simple correlations between the number of providers and any particular outcome are not necessarily meaningful because some factors that affect the number of providers in an area may also affect outcomes. For example, providers may offer faster speeds in wealthier areas, and wealthier areas may tend to have more providers. A positive correlation between the two might therefore be an income, not a competition, effect. We handle this issue through econometric analyses, including modeling the number of firms in a market before estimating the effects of the number of firms on outcomes.

[122] *See* Form 477 database. This table is derived from FCC analysis of Form 477 data dated December 2008. Analysis controls for household income, housing density, and state-specific effects. The figure may understate the competitive effects due to the way Form 477 categorizes connection speeds our method of estimating speeds from those categories. In particular, rather than reporting actual advertised speeds, Form 477 identifies each connection as being in one of 8 groupings (200–768 Kbps, 768 Kbps–1.5 Mbps, 1.6–3 Mbps, 3.1–6 Mbps, 6.1–10 Mbps, 10.1–25 Mbps, 25–100 Mbps, and greater than 100 Mbps). We estimate speeds from these groupings by using the midpoint of each category as the advertised speed in our analyses. Therefore, increases in the figure may not appear to be especially large unless a large number of connections move from one category to another. For example, a connection that increases from 3.5 Mbps to 5.5 Mbps would not appear as an increase in our analysis. "Fiber" includes fiber-to-thehome connections (such as Verizon FiOS), but excludes fiber-to-the-node connections (such as AT&T U-verse). Furthermore, the analysis is based on advertised speeds, not actual delivered speeds. The highest available fiber speed in areas with three wireline providers is not statistically different from the speed in areas with two providers. This result is an artifact of the way Form 477 aggregates speed data. In particular, about two-thirds of all fiber connections in areas with two or three wireline competitors are grouped into the 10–25 Mbps tier. A 10 Mbps connection, therefore, would appear in the data identical to a 20 Mbps connection. As a result, we observe too little variation in the fiber speed data to identify differences in speeds between areas with two and three wireline providers

[123] Broadband providers can compete for customers in a number of ways. They can offer similar products and compete on price, they can improve their product so that people are willing to pay more for it, and they can offer products targeted to different groups. Chen and Savage find evidence that cable and DSL providers may compete by targeting different types of consumers rather than by lowering prices if preferences in the target population are sufficiently diverse. Yongmin Chen & Scott J. Savage, *The Effects of Competition on the Price for Cable Modem Internet Access* (NET Institute, Working Paper No. 07-13, 2007). Research on CLECs has found that they tend to target different types of consumers rather than lower prices. S*ee generally* Shane M. Greenstein & Michael J. Mazzeo, *The Role of Differentiation Strategy in Local Telecommunication Entry and Market Evolution: 1999–2002*, 54 j. indust. ecOn. 323 (2006); Nicholas Economides et al., *Quantifying the Benefits of Entry into Local Telephone Service*, 39 rand j. ecOn. 699 (2008).

[124] 2009 figures are estimates. *See* Atkinson & Schultz, Broadband in America at 66, tbl. 15.

[125] Atkinson & Schultz, Broadband in America at 4; *see also supra* Chapter 3.

[126] Atkinson & Schultz, Broadband in America at 24.

[127] Omnibus Braodband Initiative, Broadband Performance (forthcoming).

[128] Atkinson & Schultz, Broadband in America at 24.

[129] As noted, satellite-based broadband providers, because of limited satellite capacity, have Fair Access Policies (often termed usage caps) for their customers: the Hughes current limit is as low as 200 MB per day, while WildBlue's cap is as low as 7,500 MB per month. Next- generation satellites will have much higher capacities, in excess of 100 Gbps each, with download speeds per user of up to 25 Mbps. Larger capacities could allow for usage patterns that more-closely mirror terrestrial usage. However, the high fixed costs of designing, building and launching a satellite mean that satellite-based broadband is likely to be cheaper than terrestrial service only for the most expensive-to-serve areas. atkinsOn & schultz, brOadband in america at 57. As the report notes, however, actual speeds will depend on several factors, including intensity of use in any given area. For examples of commercial services with usage caps today, see HughesNet, Fair Access Policy, http://web.hughesnet. com/sites/legal/Pages/FairAccessPolicy.aspx (last visited Mar. 4, 2009) and WildBlue Communications, WildBlue Fair Access Policy, http://wildblue.com/legal/ fair.jsp (last visited Mar. 4, 2009).

[130] No definitive data source tracks whether consumers purchase broadband as a standalone product or as a bundle, but estimates of the share of subscribers with some type of bundle range from 65% (Yankee Group) to 90% (TNS). *See* TNS Bill Harvesting and other specific database (accessed Oct 2009) (on file with the FCC) (representing a custom, proprietary database of survey answers and corresponding household bills for a variety of products including voice, data and video services, including data from Q1 2002 to Q2 2009). *See,* Yankee Group, 2009 Consumer Survey Suite database (on file with the FCC). Both the Yankee Group and UBS estimate that about 21% of subscribers have a triple-play bundle. John Hodulik et al., UBS Securities, Q4 2009 Triple Play Consumer Model database (on file with the FCC).

[131] Berkman Center for Internet and Society, Harvard University, Next Generation Connectivity: A Review of Broadband Internet Transitions and Policy From Around the World (2010) (Berkman Broadband Report),, *available at* http://cyber.law.harvard.edu/sites/cyber. law.harvard.edu/files/Berkman_Center_Broadband_ Final_Report_15Feb2010.pdf.

[132] Gregory Rosston et al., Household Demand for Broadband Internet Service (2010), *available at* http:// siepr.stanford.edu/system/files/shared/Household_ demand_for_broadband.pdf; Int'l Telecomms. Union, Measuring the Information Society: The ICT Development Index 66 (2009), *available at* http://www. itu.int/ITU-D/ict/publications/idi/2009/material/ IDI2009_w5.pdf.

[133] Telogical High-Speed Internet Service Plans Offered database (Nov. 2009) (accessed Dec. 2009) (on file with the FCC) (representing data on high-speed Internet service plans offered in all select geographies covered by telogical clients).

[134] *See* Shane Greenstein & Ryan McDevitt, *Evidence of a Modest Price Decline in US Broadband Services* 1 (CSIO, Working Paper No. 0102, 2010) (Greenstein & McDevitt, *Evidence of a Modest Price Decline*), *available at* http:// www.wcas.northwestern.edu/csio/Papers/2010/CSIO-WP-0102.pdf.

[135] *See* Greenstein & McDevitt, *Evidence of a Modest Price Decline.*

[136] Specifically, Greenstein and McDevitt estimated a regression in which the dependent variable was the monthly price of the plan, and independent variables included upload speed, download speed, region dummy variables, and time dummy variables. Greenstein & McDevitt, *Evidence of a Modest Price Decline, passim.* The coefficients on the time dummies indicate the quality-adjusted change in price. The bundled price index cannot be calculated prior to 2006 due to the lack of available data on bundled plans. It is likely that some DSL plans that Point Topic did not identify as bundled prior to 2006 were, in fact, bundled with telephone service when the provider did not offer naked DSL service.

[137] Fisher price indices as calculated by Greenstein & McDevitt, *Evidence of a Modest Price Decline* tbls. 5a–b. The indices are based on all advertised plans recorded by Point Topic from 2004 through 2009 and calculated by regressing the advertised price on upload speed, download speed, and year dummy variables separately for DSL and cable plans and then using the number of subscriptions to each type of service as the weight for creating a single broadband index. The indices were set to 1 in 2006 to facilitate comparison.

[138] Bureau of Labor Statistics, Consumer Price Index: Internet Services and Electronic Information Providers (Series CUUR0000SEEE03), http://www.bls.gov/cpi/ (last visited Mar. 6, 2009). It is difficult to compare BLS Internet price indices before and after 2007 for at least two reasons. First, BLS's sampling method means that once included in the index a provider retains its weight for four years. Thus, AOL's decision to stop charging for its dialup service in 2006 caused the index to show a nearly 25% price decrease. Shane M. Greenstein & Ryan McDevitt, *The Broadband Bonus: Accounting for Broadband Internet's Impact on U.S. GDP* (Nat'l Bureau of Econ. Research, Working Paper No. 14758, 2009), *available at* http://www.nber.org/ papers/w14758.pdf. Second, as the previous point hints, the index includes dialup Internet service providers. The share of dialup ISPs presumably decreases steadily, but the further back in time one follows the index the more dialup ISPs were likely to be included.

[139] The forthcoming FCC Mobile Wireless Competition Report will provide a longer treatment of mobile broadband competition.

[140] *See* American Roamer Advanced Services database (accessed Aug. 2009) (aggregating service coverage boundaries provided by mobile network operators) (on file with the FCC) (American Roamer database); *see also* Geolytics Block Estimates and Block Estimates Professional databases (2009) (accessed Nov. 2009) (projecting Census populations by year to 2014 by Census block) (on file with the FCC) (Geolytics databases). The approximate of 60% is based on total landmass area. In 2008, this figure was 39.6%. *Implementation of Section 6002(b) of the Omnibus Budget Reconciliation Act of 1993; Annual Report and Analysis of Competitive Market Conditions With Respect to Commercial Mobile Services*, WT Docket No. 08-27, Thirteenth Report, 24 FCC Rcd 6185, 6257, tbl. 9 (WTB 2009).

[141] Data from American Roamer show geographic coverage by technology. The actual service quality of data connections experienced by end-users will differ due to a large number of factors, such as location and mobility. Further, the underlying coverage maps do not include information on the level of service (i.e., signal quality and the speed of broadband service) provided; nor is coverage defined by providers in the same way. Thus, coverage as measured here does not correspond to a specific minimum signal quality or user experience. See American Roamer database; *see also infra* Chapter 4, Section 4.1 (Competition in Residential Broadband Networks) (discussing the American Roamer methodology). Population is based on projected Census block figures from Geolytics. *See* Geolytics databases.

[142] *See infra* Chapter 4, Section 4.1 (Transparency in the retail broadband market) (discussing details on a possible new approach to measurement and disclosure of mobile services).

[143] *See* American Roamer database.

[144] comScore, Inc., Jan.–June 2009 Consumer Usage database (sampling 200,000 machines for user Web surfing habits) (on file with the FCC) (comScore database), *see* also chetan sharma & sarla sharma, state Of the (mObile) brOadband natiOn: a benchmarkinG study (2009), *available at* http:// www.chetanshar ma.com/State%20of%20the%20 Broadband%20Nation%20-%20Chetan%20 Sharma%20Consulting.pdf (Reprinted with permission. Copyright © 2009 Chetan Sharma Consulting. All rights reserved. Based on data compiled by Root Wireless, Inc.).

[145] Atkinson & Schultz, Broadband in America at 24. Note that some providers (such as AT&T) were not included in the report, although their networks have been upgraded. S*ee also supra* Chapter 3, Exhibit 3-H.

[146] Atkinson & Schultz, Broadband in Americaat 66.

[147] Atkinson & Schultz, Broadband in Americaat 66.

[148] Some of the largest providers of wireline broadband services have ownership stakes or commercial, go-tomarket relationships with wireless broadband service providers. For example, Verizon is the controlling shareholder of Verizon Wireless; AT&T owns AT&T Wireless; and several cable companies have ownership stakes or commercial relationships with Clearwire.

[149] As noted elsewhere in the plan, satellite coverage is available from two providers nearly everywhere. With prices exceeding $50 per month for 1 Mbps advertised download speeds usage caps as low as 200 MB per day, however, the current generation of satellite broadband is not ideal for consumers who live in areas with wireline access; for examples of usage caps see HughesNet, Fair Access Policy, http://web.hughesnet.com/ sites/legal/ Pages/FairAccessPolicy.aspx (last visited Mar. 4, 2009) and WildBlue Communications, WildBlue Fair Access Policy, http://wildblue.com/legal/fair.jsp (last visited Mar. 4, 2009).

[150] While technology will continue to improve, spectral efficiency of current OFDM-based 4G solutions is approaching the theoretical limit set by information theory.

[151] The chart only displays the GSM/3GPP family of technologies. Performance of EV-DO standards is comparable with HSPA. See Letter from Dean R. Brenner, Vice Pres., Gov't Aff., Qualcomm Inc., to Marlene H. Dortch, Secretary, FCC, GN Docket No. 09- 51 (Dec. 9, 2009) Attach. A at 2. Figure shows downlink capacities calculated for 2x10MHz spectrum availability. Estimates of spectral efficiency calculated for each technology with the following antenna configuration: WCDMA, 1x1 and 1x2; HSPDA, Rel.5, 1x1; HSPA Rel. 6, 1x2; HSPA, Rel. 7, 1x1 and 1x2; LTE, 1x1 and 1x2.

[152] Atkinson & Schultz, Broadband in America at 7 ("Wireless broadband service providers expect to offer wireless access at advertised speeds ranging up to 12 mbps downstream (but more likely 5 mbps or less due to capacity sharing) to about 94% of the population by 2013.").

[153] Atkinson & Schultz, Broadband in America at 7, 23–24.

[154] *See* OBI, The Broadband Availability Gap.. It is difficult to compare and categorize performance of different broadband access technologies. For example, in certain scenarios, some technologies may have better download performance than others but worse upload. In addition, the performance of different technologies will depend on different variables such as oversubscription levels at different aggregation points in the network such as number of users per node in the hybrid-fiber coax plant or oversubscription rates in the backhaul circuits of remote DSLAMs, loop lengths for FTTN, and specific technology choices. For example, there are material performance differences between G-PON, B-PON and other architectures, and FTTN networks performance will vary substantially depending on the specific type of DSL technology used, and whether or not copper pair bounding is used. For the purpose of these analyses, it is assumed that FTTP

deployments such as Verizon FiOS provide a "robust" competitor to DOCSIS 3.0, even though the performance of different technologies may not be the same.

[155] The disparity would likely appear even larger if the data did not exclude plans above the 95th percentile, which would show 50 Mbps and 100 Mbps plans offered by some cable providers.

[156] The figure is derived from data provided in Greenstein & McDevitt, *Evidence of a Modest Price Decline*, tbls. 3a–b, and shows the 5th percentile, mean, and 95th percentile of all prices advertised by cable and DSL providers and collected by the consultancy Point Topic from 2004–2009. The 95th percentile filter means that the figure does not show 50 Mbps and 100 Mbps plans offered by some cable providers.

[157] Grego ry Rosston et al., Household Demand for Broadband Internet Service (2010), *available at* http://siepr.stanford.edu/system/files/shared/Household_demand_for_broadband.pdf.

[158] The U.S. Department of Justice, in its filing to the FCC on the national broadband plan also recommends additional spectrum, better data collection, and more transparency of that data to help promote competition. Department of Justice *Ex Parte* in re National Broadband Plan NOI, filed Jan. 4, 2010, at 21–27.

[159] *See* 47 U.S.C. § 541 (a)(3).

[160] For example, certain U.S. Census data are made available to researchers in a controlled fashion at the U.S. Census Bureau's Center for Economic Studies and Research data center. S*ee* U.S. Census Bureau Ctr. for Econ. Studies, Research Program Overview, http://www.ces. census.gov/index.php/ces/researchprogram (last visited Feb. 14, 2010).

[161] Pew Campaign for Fuel Efficiency, History of Fuel Economy: One Decade of Innovation, Two Decades of Inaction 1 (2006), http://www.pewfuelefficiency. org/docs/cafe_history.pdf. For more detail on EPA's MPG disclosure actions, see Fueleconomy.gov, http:// www.fueleconomy.gov/ (last visited Feb. 12, 2010). *See also* U.S. Dep't of Energy & U.S. Env'tal Protection agency, 2010 MPG Fuel Economy Guide,, http://www. fueleconomy.gov/feg/FEG2010.pdf.

[162] American Heart Ass'n, A History of Trans Fat, http://www.americanheart.org/presenter. jhtml?identifier=3048193 (last visited Feb. 11, 2010); N.Y.C. Dep't of Health & Mental Hygiene, The Regulation to Phase Out Artificial Trans Fat (2007), http://www.nyc.gov/html/doh/downloads/pdf/cardio/cardio-transfat-bro.pdf

[163] New America Foundation Comments in re NBP PN #24 (*Comment Sought on Broadband Measurement and Consumer Transparency of Fixed Residential and Small Business Services in the United States—NBP Public Notice #24*, GN Docket Nos. 09-51, 09-47, 09-137, 24 FCC Rcd 14120 (2009) (*NBP PN #24*)), filed Dec. 14, 2009, at 2; Dharma Dailey et al., Soc. Sci. Research Council (SSRC), Broadband Adoption in Low –Income Communities at 25 (2010), ("No one seemed sure that they were getting what they are paying for (for example, if they were getting the speed that they should) or that charges were accurate."). The FCC has conducted some initial research regarding the information provided to consumers regarding—and consumers' understanding of—broadband speed, performance, pricing, and service terms and conditions. This research has implications for transparency issues as well as for the barriers consumers face to switching providers. To address gaps in the FCC's understanding of these issues, the FCC has prepared a consumer survey that will be launched later this spring (for a number of reasons, it was not possible to conduct the survey earlier). The results of this survey would ideally have been used as part of the formal report to Congress, as they are critical points in recommendations, but will now be concluded after the formal report is delivered. The FCC will obtain and analyze survey results and will present its analysis to Congress and the public during Fiscal Year 2010 as a supplement to the Plan.

[164] comScore database. The FCC, as part of the National Broadband Plan, will issue an RFP to potentially contract with a third party and conduct a six-month consumer panel to gather more detail on actual connection speeds and performance of U.S. broadband services. The results of this panel would ideally have been used as part of the formal report to Congress, as they are critical data points in recommendations, but will now be concluded after the formal report is delivered (for a number of reasons, it was not possible to conduct this panel earlier). Panel results will therefore be finalized after the formal report is delivered, and the FCC will submit results of this panel publicly and to Congress during Fiscal Year 2010 as a supplement to the Plan. Public comments on the record and data filed with the FCC, as noted, are sufficient for creating recommendations, but this panel will bolster and provide more detail necessary to complete the Plan's congressional charter.

[165] Speed (download and upload) is only one measure of performance—others include, but are not limited to, latency, jitter, availability, packet loss, etc.

[166] Verizon Comments in re NBP PN #24, filed Dec. 14, 2009, at 14–18; US Telecom Ass'n Comments in re NBP PN #24, filed Dec. 14, 2009, at 1–3; Intel Comments in re NBP PN #24, filed Dec. 14, 2009, at 2; New America Foundation Comments in re NBP PN #24, filed Dec. 14, 2009; Epitiro Comments in re NBP PN #24, GN Docket No. 09-137, filed Dec. 14, 2009; SamKnows Comments in re NBP PN #24, GN Docket No. 09-47, filed Dec. 16, 2009.

[167] Verizon Comments in re NBP PN #24, filed Dec. 14, 2009, at 14; SamKnows Comments in re NBP PN #24, GN Docket No. 09-47, filed Dec. 16, 2009, at 5; Epitiro Comments in re NBP PN #24, GN Docket No. 09-137, filed Dec. 14, 2009, at 7–14; NCTA Comments in re NBP PN #24, filed Dec. 14, 2009, at 9; Time Warner Cable Comments in re NBP PN #24, filed Dec. 14, 2009, at 5–6.

[168] Sandvine Comments in re NBP PN #24, filed Dec. 14, 2009, at 5–6.

[169] Epitiro Comments in re NBP PN #24, GN Docket No. 09-137, filed Dec. 14, 2009; SamKnows Comments in re NBP PN #24, GN Docket No. 09-47, filed Dec. 16, 2009; New America Foundation Comments in re NBP PN #24, filed Dec. 14, 2009.

[170] Verizon Comments in re NBP PN #24, filed Dec. 14, 2009, at 15 ("tests conducted using representative Internet file sizes").

[171] SamKnows Comments in re NBP PN #24, GN Docket No. 09-47, filed Dec. 16, 2009, at 4. As noted in many public notice comments, this measurement and reporting would focus on consumer fixed broadband connections by technology and provider, with geographic data provided at an aggregated level. As noted, this panel recruitment and measurement will be finalized during Fiscal Year 2010 but are critical to the recommendations of the plan and the completion of the plan's congressional charter.

[172] See, e.g., Epitiro Comments in re NBP PN #24, GNDocket No. 09-137, filed Dec. 14, 2009, Attachs.

[173] Gerald Faulhaber, Professor, Univ. of Penn. Wharton School, Presentation at the Open Internet Transparency Workshop (Jan. 19, 2010).

[174] In August 2009, the FCC issued a Notice of Inquiry on Consumer Information and Disclosure, which began a wide-ranging review of transparency in all communications services including broadband. See Consumer Information and Disclosure, CG Docket No. 09158, CC Docket No. 98-170, WC Docket No. 04-36, Notice of Inquiry, 24 FCC Rcd 11380 (2009).

[175] Letter from Thomas Cohen, Counsel, Fiber-to-theHome Council, to Marlene H. Dortch, Secretary, FCC, GN Docket Nos. 09-47, 09-51, 09-137 (Dec. 14, 2009) (FTTH Council GN Docket No. 09–137, filed Dec. 14, 2009 Ex Parte), Attach. at 24–25; Dr. Robert Pepper, Vice Pres. of Global Tech. Policy at Cisco, Presentation at FCC International Workshop (Aug. 18, 2009), available at http://www.broadband.gov/docs/ws_int_lessons/ws_int_lessons_pepper.pdf.

[176] Ron Dicklin, Root Wireless, Presentation at the Open Internet Transparency Workshop (Jan. 19, 2010), available at http://openinternet.gov/workshops/docs/ws-consumers-transparency- and-the-open-internet/FCC%20Round%20Table%20Root%20Wireless.pdf.

[177] Many respondents to Public Notice #24 on measurement of fixed broadband commented on the potential for measurement of wireless mobile broadband as well. See, for example, Epitiro Comments in re NBP PN #24, GN Docket No. 09–137, filed Dec. 14, 2009, Attachs., for examples of UK mobile broadband measurement.

[178] FTTH Council Dec. 14, 2009 Ex Parte at 55.

[179] The FCC continues to take action on retail entry and on competition. As a recent example of the FCC's actions to support competition, when Comcast proposed to acquire Cimco, a midwestern CLEC, for the purpose of entering SMB broadband markets, the FCC put forth an expedited process, consistent with the underlying provision of the Communications Act, for Comcast to obtain the required approvals from Local Franchising Authorities. See 47 U.S.C. § 572(d)(6)(B); Application Filed for the Acquisition of Certain Assets and Authorizations of CIMCO Communications, Inc. By Comcast Phone LLC, Comcast Phone of Michigan, LLC and Comcast Business Communications, LLC, WC Docket No. 09-183, Public Notice, 24 FCC Rcd 14815 (Dec. 1, 2009), clarified by Public Notice, DA 10-211 (WCB rel. Jan. 29, 2010).

[180] See, e.g., Review of the Section 251 Unbundling Obligations of Incumbent Local Exchange Carriers; Implementation of the Telecommunications Act of 1996; Deployment of Wireline Services Offering Advanced Telecommunications Capability, CC Docket Nos. 01-338, 96-98, 98-147, Report and Order and Order on Remand and Further Notice of Proposed Rulemaking, 18 FCC Rcd 16978, 17141–54, paras. 272–97 (2003) (subsequent history omitted); Petition of AT&T Inc. for Forbearance Under 47 U.S.C. § 160(c) from Title II and Computer Inquiry Rules with Respect to Its Broadband Services; Petition of AT&T Inc. for Forbearance Under 47 U.S.C. § 160(c) from Title II and Computer Inquiry Rules with Respect to Its Broadband Services, WC Docket No. 06-125, Memorandum Opinion and Order, 22 FCC Rcd 18705 (2007) (AT&T Fiber and Packet Services Forbearance Order). Lack of appropriate wholesale access to packet-based facilities in particular serves as a constraint on competition in broadband services, which can typically be provided more efficiently using packet- based inputs.

[181] See Reexamination of Roaming Obligations of Commercial Mobile Radio Service Providers, WT Docket No. 05-265, Report and Order and Further Notice of Proposed Rulemaking, 22 FCC Rcd 15817, 15836–39, paras. 52–60 (2007). Roaming is not available to mobile providers in markets in which they hold a spectrum license. Id. at 15835–36, paras. 48–51.

[182] See, e.g., Unbundled Access to Network Elements; Review of the Section 251 Unbundling Obligations of Incumbent Local Exchange Carriers, WC Docket No. 04-313, CC Docket No. 01-338, Order on Remand, 20 FCC Rcd 2533 (2005); Access Charge Reform; Price Cap Performance Review for Local Exchange Carriers; Interexchange Carrier Purchases of Switched Access Services Offered by Competitive Local Exchange Carriers; Petition of U.S. West Communications, Inc. for Forbearance from Regulation as a Dominant Carrier in the Phoenix, Arizona MSA, CC Docket Nos. 98-157, 96-262, 94-1, CCB/CPD File No. 98-63, Fifth Report and Order and Further Notice of Proposed Rulemaking, 14 FCC Rcd 14221 (1999).

[183] See, e.g., GAO, FCC Needs to Improve its Ability to Monitor and Determine the Extent of Competition in Dedicated Access Services,, GAO 07-80 (2006), available at http://www.gao.gov/new.items/d0780.pdf.

[184] *See Parties Asked to Comment on Analytical Framework Necessary to Resolve Issues in the Special Access NPRM*, WC Docket No. 05-25, Public Notice, 24 FCC Rcd 13638 (WCB 2009).

[185] *See Pleading Cycle Established for Comments on Petition for Expedited Rulemaking Filed by Cbeyond, Inc.*, WC Docket No. 09-223, Public Notice, 24 FCC Rcd 14517 (WCB 2009) (requesting a rulemaking to provide competitive carriers with access to packetized bandwidth of incumbent LEC hybrid fiber-copper loops, fiber-to-the-home (FTTH) loops and fiber-to-the-curb (FTTC) loops at the same rates that incumbent LECs charge their own retail customers).

[186] *Pleading Cycle Established For Comments On Petition For Expedited Rulemaking Regarding Section 271 Unbundling Obligations*, WC Docket No. 09-222, Public Notice, 24 FCC Rcd 14514 (WCB 2009); *Comment Sought On Maine Public Utilities Commission Petition For Declaratory Ruling Regarding Section 271 Access To Dark Fiber Facilities And Line Sharing*, WC Docket No. 10-14, Public Notice, 25 FCC Rcd 372 (WCB 2010).

[187] *See* 47 U.S.C. § 271

[188] A critical issue in establishing wholesale obligations is determining the appropriate price for wholesale access rights. Wholesale prices that are too high may deter efficient competitive entry, while prices that are too low may deter efficient investment by both incumbents and new entrants.

[189] A recent study by the National Regulatory Research Institute commissioned by NARUC provides a general discussion of special access services and a history of the FCC and state regulatory approach to these services. Peter Bluhm & Dr. Robert Loube, *Competitive Issues in Special Access Markets, Rev. Ed.* (Nat'l Reg. Research Institute, Working Paper No. 09-02, 2009). For a discussion of potential, non-incumbent alternatives, see generally Patrick Brogan & Evan Leo, *High-Capacity Services: Abundant, Affordable and Evolving* (2009), *attached to* Letter from Glenn T. Reynolds, Vice President, Policy, USTelecom, to Marlene H. Dortch, Secretary, FCC, WC Docket No. 05-25, GN Docket 09-51 (Jul. 16, 2009) at 8–41.

[190] For example, XO, a fiber-based competitive provider, reports that special access costs represent a "substantial portion" of their costs for serving customer that are not on their fiber network. XO Comments in re NBP PN #11 (*Comment Sought on Impact of Middle and Second Mile Access on Broadband Availability and Deployment—NBP Public Notice # 11*, GN Docket Nos. 09-47, 09-51, 09-137, Public Notice, 24 FCC Rcd 12470 (WCB 2009) (*NBP PN #11*)), filed Nov. 4, 2009, at 24; *see also* Letter from Thomas Jones, Counsel, tw telecom inc., to Marlene H. Dortch, Secretary, FCC, GN Docket Nos. 09-47, 09-51, 09-137 (Dec. 22, 2009) (tw telecom Dec. 22, 2009 *Ex Parte*).

[191] The Western Telecommunications Alliance estimates that these connections typically constitute 20 –40% of the cost of providing broadband service for its small, incumbent LEC members in the Western United States. Western Telecommunications Alliance Comments in re NBP PN #11, filed Nov. 4, 2009, at 6; *see also* Verizon Comments in re NBP PN #11, filed Nov. 4, 2009, at 4–5 (noting that "the cost and availability of middle- and second-mile facilities—generally together with other factors—have hindered the deployment of broadband in some instances" and that "high per-unit costs" for these connections "if passed on to consumers, would make broadband too expensive for most" consumers in low- density areas).

[192] *See, e.g.*, XO Comments in re NBP PN #11, filed Nov. 4, 2009, at 15–21 (arguing that restrictive terms and conditions on availability of certain pricing plans can effectively lock out customers from seeking competitive alternatives) ; tw telecom Dec. 22, 2009 *Ex Parte* at 9–11; GaO, fcc needs tO imprOve its ability tO mOnitOr and determine the extent Of cOmpetitiOn in dedicated access services, GaO 07-80 (2006), *available at* http://www.gao.gov/new.items/d0780.pdf.

[193] *See, e.g.*, *Qwest Petition for Forbearance Under 47 U.S.C. § 160(c) from Title II and Computer Inquiry Rules with Respect to Broadband Services*, WC Docket No. 06-125, Memorandum Opinion and Order, 23 FCC Rcd 12260 (2008); *Petition for of the Embarq Local Operating Companies for Forbearance Under 47 U.S.C. § 160(c) from Title II and Computer Inquiry Rules with Respect to Broadband Services; Petition of the Frontier and Citizens ILECs for Forbearance Under 47 U.S.C. § 160(c) from Title II and Computer Inquiry Rules with Respect to Broadband Services*, WC Docket No. 06-147, Memorandum Opinion and Order, 22 FCC Rcd 19478 (2007); *AT&T Fiber and Packet Services Forbearance Order*, 22 FCC Rcd 18705 (2007); *Petition of ACS of Anchorage, Inc. Pursuant to Section 10 of the Communications Act of 1934, as Amended (47 U.S.C. § 160(c)), for Forbearance from Certain Dominant Carrier Regulation of Its Interstate Access Services, and for Forbearance from Title II Regulation of Its Broadband Services, in the Anchorage, Alaska, Incumbent Local Exchange Carrier Study Area*, WC Docket No. 06-109, Memorandum Opinion and Order, 22 FCC Rcd 16304 (2007); *Verizon Telephone Companies' Petition for Forbearance from Title II and Computer Inquiry Rules with Respect to their Broadband Services Is Granted by Operation of Law*, WC Docket No 0440, News Release (rel. Mar. 20, 2006). Broadband providers have also asserted that as a result of these decisions, high-capacity Ethernet transport services have not been rolled out swiftly enough and at appropriate prices. *See, e.g.*, tw telecom Dec. 22, 2009 *Ex Parte* at 10–11 ("In the absence of effective regulation of incumbent LEC wholesale Ethernet prices, the incumbent LECs charge prices that are so high that they effectively preclude TWTC and other competitors from relying on these facilities to serve off-net locations.").

[194] CenturyLink notes that "Ethernet is rapidly replacing special access circuits, offering more capacity for less." Letter from Jeffrey S. Lanning, Director, Fed. Reg. Aff., CenturyLink, to Marlene H. Dortch, Secretary, FCC,

WC Docket No. 05-25 (Nov. 4, 2009) Attach.; *see also* Letter from Thomas Jones, Counsel, tw telecom inc., to Marlene H. Dortch, Secretary, FCC, GN Docket No. 09-51 (Oct. 14, 2009) Attach.; tw telecom Dec. 22, 2009 *Ex Parte* at 2 ("The remarkable efficiencies of Ethernet make high-bandwidth business applications as well as telemedicine and remote job training programs affordable").

[195] *See Parties Asked to Comment on Analytical Framework Necessary to Resolve Issues in the Special Access NPRM*, WC Docket No. 05-25, Public Notice, 24 FCC Rcd 13638 (WCB 2009).

[196] Estimates indicate that approximately 80% of business locations are served by copper because they are located in buildings that do not have fiber facilities. *See* Letter from Jerry Watts, Vice Pres., Gov't and Indus. Aff., DeltaCom, to Marlene H. Dortch, Secretary, FCC, GN Docket Nos. 09-29, 09-47, 09-51 (Oct. 20, 2009) Attach. 2 at 4 (citing Vertical Systems Group); XO Comments in re NBP PN #11, filed Nov. 4, 2009, at 10.

[197] *See* 47 C.F.R. §§ 51.325–51.335.

[198] *See, e.g.*, XO Comments in re NBP PN #11, filed Nov. 4, 2009, at 9; Letter from Karen Reidy, COMPTEL, to Marlene H. Dortch, Secretary, FCC, GN Docket Nos. 09-47, 09-51, 09-137, RM-11358 (Dec. 7, 2009). When a copper facility is retired, to continue providing service a competitor needs to redesign its network or purchase special access circuits from the incumbent LEC. These special access connections are typically more expensive, may have different service characteristics, and may limit the competitor's ability to differentiate its service.

[199] *See, e.g.*, Gerald W. Brock, The Telecommunications Industry, The Dynamics of Market Structure 148 (1981); 47 U.S.C. § 251(a), (c)(2).

[200] Letter from Matthew A. Brill, Counsel for Time Warner Cable, to Marlene H. Dortch, Secretary, FCC, GN Docket No. 09-51 (Nov. 12, 2009) (TWC Nov. 12, 2009 *Ex Parte*) (outlining examples where Time Warner Cable has had difficulty obtaining basic interconnection rights in rural areas); Letter from Jeremy M. Kissel, MetroCast Cablevision of New Hampshire, LLC, to Marlene H. Dortch, Secretary, FCC, GN Docket Nos. 09-51, 09-137 (Dec. 18, 2009); Time Warner Cable Comments in re NBP PN #256 (*Comment Sought on Transition from Circuit-Switched Network to All IP-Network—NBP Public Notice #5*, GN Docket Nos. 09-47, 09-51, 09-137, Public Notice, 24 FCC Rcd 14272 (WCB 2009) (*NBP PN #25*), filed Dec. 22, 2009, at 5–8; National Cable & Telecommunications Association Comments in re NBP PN #25, filed Dec. 22, 2009, at 5 n.12.

[201] TWC Nov. 12, 2009 *Ex Parte* at 2–3 (Nov. 12, 2009) (citing *Sprint Commc'ns Co. L.P. v. Pub. Util. Comm'n of Tex.*, No. A-06-CA-065-SS, 2006 U.S. Dist. LEXIS 96569 (W.D. Tex. Aug. 14, 2006)).

[202] *See, e.g.*, TWC Nov. 12, 2009 *Ex Parte* at 3 (citing a decision by the Maine Public Utilities Commission); Letter from Jeremy M. Kissel, MetroCast Cablevision of New Hampshire, LLC, to Marlene H. Dortch, Secretary, FCC, GN Docket Nos. 09-51, 09-137 (Dec. 18, 2009) Attach. 2 at 15 (citing decisions in Texas, Maine, and North Dakota); Time Warner Cable Comments in re NBP PN #25, filed Dec. 22, 2009, at 5–8 (describing difficulties Time Warner Cable has had obtaining basic interconnection rights in rural areas).

[203] *See Time Warner Cable Request for Declaratory Ruling that Competitive Local Exchange Carriers May Obtain Interconnection Under Section 251 of the Communications Act of 1934, As Amended, to Provide Wholesale Telecommunications Services to VoIP Providers*, WC Docket No. 06-55, Memorandum Opinion and Order, 22 FCC Rcd 3513 (2007). All telecommunications carriers have a basic duty to interconnect under Section 251(a). *See* 47 U.S.C. § 251(a). A rural carrier's rural exemption under Section 251(f) does not impact this obligation. *See* 47 U.S.C. § 251(f)(1).

[204] IP-to-IP interconnection is addressed in the inter- carrier compensation discussion in Chapter 8 *infra*.

[205] *See Reexamination of Roaming Obligations of Commercial Mobile Radio Service Providers*, WT Docket No. 05-265, Report and Order and Further Notice of Proposed Rulemaking, 22 FCC Rcd 15817 (2007).

[206] "Set-top box" is one example of video navigation devices, which are defined in 47 C.F.R. § 76.1200 as interactive communications equipment used by consumers to access multichannel video programming and other services offered over multichannel video programming systems. We use "set-top box" to broadly include set-top boxes, digital video recorders (DVRs), and home theater PCs (HTPCs).

[207] TiVo Comments in re NBP PN #27 (*Comment Sought on Video Device Innovation—NBP Public Notice #27*, GN Docket Nos. 09-47, 09-51, 09-137, Public Notice, 24 FCC Rcd 14280 (MB 2009) (*NBP PN #27*)), filed Dec. 22, 2009, at 12; Consumer Electronics Association Comments in re NBP PN #27, filed Dec. 21, 2009, at 15; Public Knowledge et al., Petition for Rulemaking, CS Docket No. 97-80, GN Docket Nos. 09-47, 09-51, 09-137, at 12–14 (Dec. 18, 2009) (Public Knowledge et al. Video Device Competition Petition) (asking "that the Commission initiate a rulemaking to address the lack of competition in the video device market"); Verizon Comments in re NBP PN #27, filed Dec. 22, 2009, at 6.

[208] Public Knowledge et al. Video Device Competition Petition at 11–12; Consumer Electronics Association Comments in re NBP PN #27, filed Dec. 21, 2009, at 15; Sony Electronics Inc. (Sony) Comments in re NBP PN #27, filed Dec. 21, 2009, at 3.

[209] TiVo Comments in re NBP PN #27, filed Dec. 22, 2009, at 4, 9; Public Knowledge et al. Video Device Competition Petition, filed Dec. 18, 2009, at 20–21; Consumer Electronics Association Comments in re NBP PN #27, filed Dec. 21, 2009, at 15.

[210] For example, innovation in computing devices, such as the creation of graphical user interfaces, contributed to the proliferation of software applications developed for the PC. Furthermore, innovation in mobile devices, such as the introduction of the iPhone and Android, has led to the development and launch of hundreds of thousands of new mobile applications.

[211] Consumer Electronic Retailers Coalition Comments in re NBP PN #27, filed Dec. 21, 2009, at 11–12; Sony Comments in re NBP PN #27, filed Dec. 21, 2009, at 3.

[212] Section 629 covers equipment used to receive video programming—including cable set-top boxes, televisions, and DVRs—as well as equipment used to receive other services offered over MVPD systems, including cable modems. See 47 U.S.C. § 549 (codifying section 629 of the Telecommunications Act of 1996).

[213] See 142 Cong. Rec. H1170 (daily ed. Feb. 1, 1996) (statement of Rep. Markey: "[The provision would] help to replicate for the interactive communications equipment market the success that manufacturers of customer premises equipment (CPE) have had in creating and selling all sorts of new phones, faxes, and other equipment subsequent to the implementation of rules unbundling CPE from common carrier networks.").

[214] Implementation of Section 304 of the Telecommunications Act of 1996; Commercial Availability of Navigation Devices, CS Docket No. 97-80, Report and Order, 13 FCC Rcd 14775 (1998).

[215] The FCC directly exempted satellite operators (e.g., DirecTV and DISH Network), since they operate throughout the United States and offer devices for retail sale through unaffiliated vendors, and certain Internet Protocol TV (IPTV) providers, primarily small telephone cooperatives. AT&T (an IPTV provider) has neither requested nor received a waiver for its U-Verse service. Verizon FiOS is considered a cable service for regulatory purposes and is not exempted from Section 629.

[216] Implementation of Section 304 of the Telecommunications Act of 1996; Commercial Availability of Navigation Devices, CS Docket No. 97-80, Second Report and Order, 20 FCC Rcd 6794, 6802–03, 6814, paras. 13, 31 (2005).

[217] Dell'Oro Group, Set-Top Box Report 89 (3Q 2009).

[218] The Hirfindahl-Hirschman Index (HHI) for cable set-top boxes in North America exceeds 5100, well above the threshold of 1800 for "concentrated" markets. See Dell'Oro Group, Set-Top Box Report 89 (3Q 2009). This is not typical for consumer electronics markets, which have relatively lower fixed costs when compared, for example, with network services markets. For example, the 2002 U.S. Census Economic Survey estimated that the four largest audio and video equipment manufacturers (NAICS 3343) held about 46% of the market and the HHI to be about 894, well below the DOJ's threshold of 1000 for "unconcentrated" markets Census Bureau, Concentration Ratios 2002, 2002 Economic Census: Manufacturing 51–52 (2006), available at http://www.census.gov/prod/ec02/ ec0231sr1.pdf.

[219] Dell'Oro Group, Set-Top Box Report 89 (3Q 2009). Annual figures from 2006 to 2009 (through Q3). The top two manufacturers in the European cable set-top box market during that time period were Thomson and Pace; three other manufacturers—Motorola, Cisco, and ADB Group—also each captured more than a 10% share.

[220] 456,000 CableCARDs have been deployed by the top 10 operators, who collectively have 90% share of overall cable customers. nat'l cable & telecOmm. ass'n, fcc cablecard Quarterly report, Sept.–Nov. 2009 (2009).

[221] 41.5 million digital cable subscribers, see SNL Kagan (a division of SNL Financial LC), Cable MSO Industry Benchmarks (June 2009), multiplied by a conservative range of 1.2–1.5 set-top boxes per household, totals 49.8–62.3 million set-top boxes.

[222] Examples include: gaming systems (e.g., Sony Playstation 3, Xbox 360), blu-ray DVD players, Internet video devices (e.g., AppleTV, Roku), Internet sites/applications (e.g., Google, Amazon, Netflix, Hulu), hybrid broadcast- broadband content providers (e.g., Sezmi).

[223] Estimated share of US households with Apple TV or Roku is 1%. Letter from Bruce Leichtman, President, Leichtman Research Group (LRG), to Marlene H. Dortch, Secretary, FCC, GN Docket Nos. 09-47, 09-51, 09-137 (Jan. 4, 2010) (LRG Jan. 4, 2010 Ex Parte).

[224] TiVo Comments in re NBP PN #27, filed Dec. 22, 2009, at 2–6; Public Knowledge et al. Video Device Competition Petition at 2–3, 6–10, 25–26; Consumer Electronics Association Comments in re NBP PN #27, filed Dec. 21, 2009, at 6–10, 13; Consumer Electronic Retailers Coalition Comments in re NBP PN #27, filed Dec. 21, 2009, at 4–9; Verizon Comments in re NBP PN #27, filed Dec. 22, 2009, at 10–11; Letter from Kyle McSlarrow, President and CEO, NCTA, to Carlos Kirjner, Senior Advisor to the Chairman on Broadband, and William Lake, Chief, Media Bureau, FCC, GN Docket Nos. 09-47, 09-51, 09-137, CS Docket No. 97-80 (Dec. 4, 2009) (NCTA Dec. 4, 2009 Ex Parte) at 3; Letter from Jeffrey Kardatzke, CTO & Founder, and Mike Machado, CEO, SageTV to Marlene H. Dortch, Secretary, FCC, GN Docket Nos. 09-47, 09-51, 09-137 (Jan. 29, 2010) (SageTV Jan. 29, 2010 Ex Parte).

[225] For example, Steve Jobs explains Apple's decision not to produce Apple TV as a set-top box with access to traditional TV content through MVPDs: "The minute you have an STB you have gnarly issues, CableCARD, OCAP.. . that just isn't something we would choose to do ourselves. We couldn't see a go-to-market strategy that makes sense." Ryan Block, Steve Jobs Live from D 2007, enGadget, May 30, 2007, http://www.engadget.com/2007/05/30/steve-jobs-live-from-d-2007; see also Auction Network Comments in re NBP PN #27, filed Dec. 18, 2009, at 1–3.

[226] All totals as of Sept. 30, 2009. *See* SNL Kagan (a division of SNL Financial LC), Cable MSO Industry Benchmarks (2009) (providing cable company totals), *available at* http:// www.snl.com/InteractiveX/TopCableMSOs.aspx (requires registration); DirecTV, *The DirecTV Group Announces Third Quarter 2009 Results* (press release), Nov. 5, 2009, http://investor.directv.com/releasedetail. cfm?ReleaseID=422185 (providing DirecTV totals); DISH Network, *DISH Network Corporation Reports Third Quarter 2009 Financial Results* (press release), Nov. 9, 2009, http://dish.client.shareholder.com/ releasedetail.cfm ?ReleaseID=422698 (providing DISH Network totals); verizOn, verizOn cOmmunicatiOns investOr Quarterly (3Q 2009) (providing Verizon FiOS totals), *available at* http://investor.verizon.com/ financial/quarterly/vz/3Q2009/3Q2009.pdf; AT&T, at&t financial and OperatiOnal results 13 (4Q 2009) (providing AT&T U-Verse totals), *available at* http:// www.att.com/Investor/Growth_Profile/download/master_Q4_09.pdf.

[227] TiVo Comments in re NBP PN #27, filed Dec. 22, 2009, at 9–10; Public Knowledge et al. Video Device Competition Petition at 36.

[228] Verizon Comments in re NBP PN #27, filed Dec. 22, 2009, at 10–12; Sony Comments in re NBP PN #27, filed Dec. 21, 2009, at 5; Netmagic Solutions Inc. (Netmagic) Comments in re NBP PN #27, filed Dec. 21, 2009, at 3; Nagravision Comments in re NBP PN #27, filed Dec. 21, 2009, at 2–3.

[229] The standards for the gateway device should be determined by industry standard-setting bodies, in consultation with the FCC. TiVo Comments in re NBP PN #27, filed Dec. 22, 2009, at 11, 13–15; Public Knowledge et al. Video Device Competition Petition at 31–33, 35; Verizon Comments in re NBP PN #27, filed Dec. 22, 2009, at 3, 5; Sony Comments in re NBP PN #27, filed Dec. 21, 2009, at 3.

[230] TiVo Comments in re NBP PN #27, filed Dec. 22, 2009, at 2, 5, 17; Consumer Electronics Association Comments in re NBP PN #27, filed Dec. 21, 2009, at 18; Public Knowledge et al. Video Device Competition Petition at 36. Retail devices may transcode or otherwise degrade the quality of the video signal as necessary to ensure compatibility with specific screen sizes, functionality, and form factors.

[231] TiVo Comments in re NBP PN #27, filed Dec. 22, 2009, at 2–3, 18–19; Public Knowledge et al. Video Device Competition Petition at 8–9, 34–35; Consumer Electronics Association Comments in re NBP PN #27, filed Dec. 21, 2009, at 16; SageTV *Ex Parte* in re NBP PN #27, filed Feb. 16, 2010, at 7, 12.

[232] Sony Comments in re NBP PN #27, filed Dec. 21, 2009, at 3.

[233] *See Implementation of Section 304 of the Telecommunications Act of 1996; Commercial Availability of Navigational Devices*, GN Docket No. 97-80, Notice of Proposed Rulemaking, 12 FCC Rcd 5639 (1997); *Implementation of Section 304 of the Telecommunications Act of 1996; Commercial Availability of Navigational Devices*, GN Docket No. 97-80, Further Notice of Proposed Rulemaking and Declaratory Ruling, 15 FCC Rcd 18199 (2000); *Implementation of Section 304 of the Telecommunications Act of 1996; Commercial Availability of Navigational Devices and Compatibility Between Cable Systems and Consumer Electronics Equipment*, GN Docket Nos. 97-80, 00-67, Further Notice of Proposed Rulemaking, 18 FCC Rcd 518 (2003); *Implementation of Section 304 of the Telecommunications Act of 1996; Commercial Availability of Navigational Devices and Compatibility Between Cable Systems and Consumer Electronics Equipment*, GN Docket Nos. 97-80, 00-67, Third Further Notice of Proposed Rulemaking, 22 FCC Rcd 12024 (2007); *A National Broadband Plan for Our Future*, GN Docket No. 09-51, Notice of Inquiry, 24 FCC Rcd 4342 (2009) (*National Broadband Plan NOI*).

[234] Public Knowledge Comments in re NBP PN #30 (*Reply Comments Sought in Support of National Broadband Plan —NBP Public Notice #30*, GN Docket Nos. 09-47,09-51, 09-137, Public Notice, 25 FCC Rcd 241 (WCB 2010) (*NBP PN #30*)), filed Jan. 27, 2010, at 11–13. We note that there are open questions to resolve as part of the rulemaking proceeding regarding the gateway architecture. *See, e.g.*, National Cable & Telecommunications Association Comments in re NBP PN #30, filed Jan. 27, 2010, at 11–15; Sage TV *Ex Parte* in re NBP PN #27, filed Feb. 16, 2010, at 1–11; TiVo Reply in re NBP PN #27, filed Feb. 17, 2010, at 9–15.

[235] TiVo Comments in re NBP PN #27, filed Dec. 22, 2009, at 4 (filed by Matthew Zinn); Public Knowledge et al. Video Device Competition Petition, filed Dec. 18, 2009, at 10.

[236] Public Knowledge et al. Video Device Competition Petition, filed Dec. 18, 2009, at 14, 26–27; Consumer Electronics Association Comments in re NBP PN #27, filed Dec. 22, 2009, at 14–15; Consumer Electronics Retailers Coalition Comments in re NBP PN #27, filed Dec. 22, 2009, at 9.

[237] Public Knowledge et al. Video Device Competition Petition, filed Dec. 18, 2009, at 6–7; Consumer Electronics Retailers Coalition Comments in re NBP PN #27, filed Dec. 22, 2009, at 7.

[238] TiVo Comments in re NBP PN #27, filed Dec. 22, 2009, at 3 (filed by Matthew Zinn); Public Knowledge et al. Video Device Competition Petition, filed Dec. 18, 2009, at 3, 9, 26; SageTV Jan. 29, 2010 *Ex Parte* at 1–2.

[239] TiVo Comments in re NBP PN #27, filed Dec. 22, 2009, at 4, 7 (filed by Matthew Zinn); Public Knowledge et al. Video Device Competition Petition, filed Dec. 18, 2009, at 10, 25–26; Consumer Electronics Association Comments in re NBP PN #27, filed Dec. 22, 2009, at 13; Letter from Matthew Zinn, Senior Vice President, General Counsel, Secretary & Chief Privacy Officer, TiVo, to Marlene H. Dortch, Secretary, FCC, GN Docket Nos. 09-47, 09-51, 09-137, CS Docket No. 97-80 (Feb. 17, 2010) (TiVo Feb. 17, 2010 *Ex Parte*), at 2–4. Cable headends with SDV would need to install a server that translates between standard IP signals from the

retail CableCARD device and the operator's proprietary network. The FCC may consider a two-step process in its rules: first, cable systems with SDV would need to deploy SDV tuning adapters immediately to support all retail CableCARD devices; second, within three to six months, those cable systems would need to install the servers required to allow IP communication without SDV tuners. Cable operators could voluntarily skip the first step if they are prepared to deploy servers in their headends immediately.

[240] Public Knowledge et al. Video Device Competition Petition, filed Dec. 18, 2009, at 14; Consumer Electronic Association Comments in re NBP PN #27, filed Dec. 21, 2009, at 3; Consumer Electronic Retailers Coalition Comments in re NBP PN #27, filed Dec. 21, 2009, at 9.

[241] Consumer Electronic Retailers Coalition Comments in re NBP PN #27, filed Dec. 21, 2009, at 12. For example, operators should make a self-install option available for retail CableCARD devices if such an option is available for leased set-top boxes.

[242] Public Knowledge et al. Video Device Competition Petition at 8–9; SageTV *Ex Parte* in re NBP PN #27, filed Feb. 16, 2010, at 9.

[243] TiVo Feb. 17, 2010 *Ex Parte* at 3.

[244] Network Advertising Initiative Comments in re NBP PN #29 (*Comments Sought on Privacy Issues Raised by the Center for Democracy and Technology—NBP Public Notice #29*, GN Docket Nos. 09-47, 09-51, 09-137, Public Notice, 25 FCC Rcd 244 (WCB 2010) (*NBP PN #29*)), filed Jan. 22, 2010, at 1–4.

[245] While online advertising rates are highly variable, this calculation is based on "generic" ad "cost per thousand impressions" (CPMs) being roughly $1–3, while targeted advertisements are estimated to command above $10 CPMs. Even more specialized types of advertising, such as targeted "cost-per-click" and search-based advertising, have been estimated at even higher rates. For example, according to eMarketer, Credit Suisse estimated average CPM at $2.39 in 2009 and $2.46 in 2008. *How Much Ads Cost*, emarketer. cOm, April 23, 2009, http://www.emarketer.com/ Article.aspx?R=1007053. JP Morgan forecasted aggregate (generic and targeted) CPM of $3.05 in 2009, down from a high of $3.50 in 2006. Erick Schonfeld, *JPMorgan Forecasts A 10.5 Percent Rebound In U.S. Display Advertising in 2010*, techcrunch, Jan. 4 2010, http://techcrunch.com/2010/01/04/jpmorganadvertising-2010/. The ad tracking firm Adify estimates CPM rates across several verticals from $3–12 CPM, although this is not split between behavioral and generic advertising. *The Average CPM Rates Across Different Verticals*, diGitalinspiratiOn, Nov. 25, 2009, http://www.labnol.org/internet/average-cpm-rates/11315/. However, more targeted advertising, such as video or search results, commanded far higher overall "per thousand impression" rates of $20–70 or more. *How Much Ads Cost*, emarketer.cOm, Apr. 23, 2009, http:// www.emarketer.com/Article.aspx?R=1007053. While dated, Advertising.com's study of non-targeted impressions versus targeted impressions demonstrated similar results of 3–6x benefits from targeting. *See* Robyn Greenspan, *Behavioral Targeting Study Reveals CPM Lift*, clickz, Aug. 17, 2004, http://www.clickz. com/3396431. For an easily readable overview of online advertising, see *Online Advertising: The Ultimate Marketing Machine*, ecOnOmist, July 6, 2006, *available at* http://www.economist.com/businessfinance/ displaystory.cfm?story_id=7138905.

[246] Charter Communications Comments in re NBP PN #29, filed Jan. 22, 2010, at 3; National Advertising Initiative Comments in re NBP PN #29, filed Jan. 22, 2009, at 6.

[247] Industry has realized the challenges of responsibly collecting this data and delivering targeted ads, and many groups have worked to create voluntary self- regulation standards, often alongside or spurred through FTC initiatives. *See, e.g.*, Am. Ass'n of Advertising Agencies et al., Self Regulatory Principles for Online Behavioral Advertising (2009), *available at* http://www. iab.net/media/file/ven-principles-07-01-09.pdf

[248] *See, e.g.*, Center for Democracy and Technology Comments in re NBP PN #29, filed Jan. 22, 2009, at 4, 19–26 (discussing "Trusted Identity Providers"); AT&T Comments in re NBP PN #29, filed Jan. 22, 2009, at 6 (discussing "OpenID" and Information Cards).

[249] *See* Network Advertising Initiative Comments in re NBP PN #29, filed Jan. 22, 2010, at 6 (providing more detail on "cookies"); CDT Comments in re NBP PN #29, filed Jan. 22, 2010; Data Foundry Comments in re NBP PN #29, filed Jan. 22, 2010, at 2–3.

[250] The latest version of the two most common browsers, Microsoft's Internet Explorer and Mozilla's Firefox, offer "secure" or "private" browsing that limits cookie activity. *See* Microsoft, Stay Safer Online, http://www. microsoft.com/windows/internet-explorer/features/ safer.aspx (last visited Mar. 6, 2010); Mozilla, Private Browsing, http://support.mozilla.com/en-US/kb/ Private+browsing (last visited Mar. 6, 2010). Other companies also help consumers track and understand data collection. *See* Ghostery, http://www.ghostery.com/ (last visited Mar. 6, 2010). But this is limited today. *See* AT&T Comments in re NBP PN #29, filed Jan. 22, 2010, at 6 (citing PrivacyChoice.org Comments, Analysis of Ad-Targeting Privacy Policies and Practices, Federal Trade Commission Exploring Privacy Roundtable Series, Dec. 4, 2009).

[251] David Vladeck, Director, Fed. Trade Comm'n Bureau of Consumer Protection, Privacy: Where Do We Go From Here?, Remarks at the International Conference of Data Protection and Privacy Commissioners 4 (Nov. 6, 2009) (Vladeck, Privacy: Where Do We Go From Here?) ("Disclosures are now as long as treatises, they are written by lawyers—trained in detail and precision, not clarity—so they even sound like treatises, and like some treatises, they are difficult to comprehend, if they are read at all. It is not clear that consent today actually reflects a conscious choice by consumers."), *available at* http://www.ftc.gov/ speeches/vladeck/0911

06dataprotection.pdf; Center for Democracy and Technology Comments in re NBP PN #29, filed Jan. 22, 2010, at 9–10.

[252] Center for Democracy and Technology Comments in re NBP PN #29, filed Jan. 22, 2010, at 8.

[253] John B. Horrigan, *Broadband Adoption and Use in America* 17 (OBI Working Paper No. 1, 2010) (Horrigan, *Broadband Adoption and Use in America*).

[254] *See* Fed. Trade Comm'n, Protecting Personal Information: a Guide for business, *available at* http://www.ftc.gov/bcp/edu/pubs/business/idtheft/bus69.pdf (2008). For example, the FTC has found violations of Section 5 of the Federal Trade Commission Act, 15 U.S.C. § 45, because a company's privacy practices were false and misleading, *see, e.g.*, In re Gateway Learning Corp, 2004 WL 1632833 (FTC July 7, 2004); In re GeoCities, 1998 WL 473217 (FTC Aug. 13, 1998), and for failure to implement reasonable and appropriate measures to protect personal information, *see, e.g.*, In re Life Is Good, Inc., 2008 WL 258309 (FTC Jan. 17, 2008); In re Petco Animal Supplies, Inc., 2004 WL 2682593 (FTC Nov. 8, 2004); In re MTS, Inc. d/b/a/ Tower Records/Books/ Video, 2004 WL 963226 (FTC Apr. 21, 2004); In re Guess?, Inc., 2003 WL 21406017 (FTC June 18, 2003); In re Eli Lilly, 133 F.T.C. 20 (2002). The FTC also has found violations of Section 5 and the Gramm-LeachBliley Act, 15 U.S.C §§ 6801–6809, for failure to provide reasonable and appropriate security for consumers' sensitive personal information, *see, e.g.*, In re Goal Financial, LLC, 2008 WL 625340 (FTC Mar. 4, 2008); In re Premier Capital Lending, Inc., 2008 WL 4892987 (FTC Nov. 6, 2008).

[255] 47 U.S.C §§ 222, 551.

[256] Center for Democracy and Technology Comments in re NBP PN #29, filed Jan. 22, 2010, at 4.

[257] 18 U.S.C. §§ 2510–2521 (protecting against acquisition of the content of communications without the consent of one of the parties to the communication).

[258] *See In re DoubleClick, Inc. Privacy Litigation*, 154 F. Supp. 2d 497 (S.D.N.Y. 2001), *available at* http://www.hbbllc.com/courses/infosec/ecpa/154_fsupp2d_497. pdf; *see also* Cybertelecom, Electronic Communications Privacy Act (ECPA), http://www.cybertelecom.org/ security/ecpaexception.htm (last visited Feb. 17, 2010) (explaining the ECPA).

[259] 15 U.S.C. § 6801 et seq.

[260] For example, a cable operator must inform its subscribers what personally identifiable information it collects, how it is used and for how long it is kept, and the cable operator may not disclose such information without the prior consent of the subscriber. See 47 U.S.C. § 551. Similarly, customers of telecommunications carriers have statutory protections against the non-consensual disclosure of information about the telecommunications service or habits of the customer, such as to or from whom the customer makes or receives calls, call location (if mobile), and the times that calls are made. See 47 U.S.C. § 222. Although privacy protections exist for traditional services and have even been applied to newer services like interconnected VoIP, *see* 47 C.F.R. § 64.2003(k), it is unclear whether, and to what extent, these protections apply to broadband ISPs. *See, e.g.*, *Klimas v. Comcast Cable, Inc.*, 465 F.3d 271, 276 (6th Cir. 2006) (finding that section 631 does not apply to the broadband ISP services offered by a cable operator).

[261] *See* 47 U.S.C. §§ 222, 531.

[262] *See generally* 45 C.F.R. Part 164, Subpart E (Privacy of Individually Identifiable Health Information).

[263] *See* 15 U.S.C. § 6809 (defining "nonpublic personal information").

[264] Wendy Davis, *Court: IP Addresses Are Not Personally Identifiable Information*, mediapOst, July, 6, 2009, http://www.mediapost.com/publications/?fa=Articles. ShowArticle&art_aid=109242.

[265] *See, e.g.*, 18 U.S.C. § 1514A (protecting employees who blow the whistle on publicly traded companies); 42 U.S.C. § 7622 (protecting employees who disclose possible violations of the Clean Air Act); 49 U.S.C. § 31105 (protecting employees who disclose possible violations of safety regulations for commercial motor vehicles); *see also* WhistleBlowerLaws, http://whistleblowerlaws.com/index.php?option=com_conte nt&task=view&id=141&Itemid=54 (last visited Feb. 17, 2010) (login is required).

[266] *See, e.g.*, *McIntyre v. Ohio Elections Comm'n*, 514 U.S. 334, 357 (1995) ("Anonymity is a shield from the tyranny of the majority.").

[267] *Reno v. ACLU*, 521 U.S. 844, 870 (1997).

[268] Vladeck, Privacy: Where Do We Go From Here? at 4.

[269] The FTC has begun a series of public roundtable discussions to explore the privacy challenges posed by the vast array of 21st century technology and business practices that collect and use consumer data. The first roundtable discussion took place on December 7, 2009. The second took place on January 29, 2010. The third is scheduled to take place on March 17, 2010. See Federal Trade Commission Comments in re NBP PN #21 (*Comment Sought on Data Portability and Its Relationship to Broadband—NBP Public Notice #21*, GN Docket Nos. 09-47, 09-51, 09-137, 24 FCC Rcd 13816 (WCB 2009) (*NBP PN #21*)), filed Dec. 9, 2009, at 2–3; FTC Comments in re NBP PN #29, filed Jan. 22, 2010; *see also* Fed. Trade Comm'n, Exploring Privacy: A Roundtable Series, http://www.ftc.gov/bcp/workshops/ privacyroundtables/index.shtml (last visited Mar. 5, 2010).

[270] *See* Fed. Trade Comm'n, Enforcing Privacy Promises: Section 5 of the FTC Act, http://www.ftc.gov/privacy/privacyinitiatives/promises.html (last visited Mar. 5, 2010).

271 *See* fed Fed. Trade Comm'n, FTC Staff Report: Self-Regulatory Principles for Online Behavioral Advertising 11–12, 46–47 (2009) (FTC Staff Report, *available at* http://www.ftc.gov/os/2009/02/ P085400b ehavadreport.pdf.

272 For example, a number of online search companies have developed policies and procedures to inform consumers about online tracking and provide additional protections and controls. *See* FTC Staff Report 2009 at 12 (noting that Yahoo! and Google allow consumers to opt out of targeted advertising). And industry coalitions and trade associations, including the largest online advertising networks, have developed self-regulating principles for online data management practices and begun cooperative efforts. *See* Netwo rk Advertising Initiative, 2008 NAI Principles, The Netwo rk Advertising Initiative's Self-Regulatory Code of Conduct (2008), *available at* http://www.networkadvertising.org/ networks/2008%20NAI%20Principles_final%20 for%20website.pdf; *see also* Best Practices and Guidelines for Location Based Services (2008), *available at* http://files.ctia.org/pdf/CTIA_LBS_ BestPracticesandGuidelines_04_08.pdf; ftc staff repOrt 2009 at 14; K.C. Jones, *Agencies to Self-Regulate Online Behavioral Ads*, info. Week, Jan. 13, 2009, http://www.informationweek.com/news/showArticle/ jhtml?articleID=212900156; Interactive Advertising Bureau, Privacy Principles, http://www.iab.net/iab_ products_and_industry_services/508676/508813/1464 (last visited Feb. 18, 2010).

273 *See* Federal Trade Commission Comments in re NBP PN #21, filed Dec. 9, 2009, at 2–3; Federal Trade Commission Comments in re NBP PN #29, filed Jan. 22, 2010; *see also* Fed. Trade Comm'n, Exploring Privacy: A Roundtable Series, http://www.ftc.gov/bcp/workshops/ privacyroundtables/index.shtml (last visited Mar. 5, 2010).

274 For details on how the Privacy Act and the collection of personal data impact other aspects of the broadband ecosystem, see *infra* Chapters 10 11, 12, and 14.

275 *See, e.g.*, Center for Democracy and Technology Comments in re NBP PN #29, filed Jan. 22, 2010, at 4, 19–26 (discussing "Trusted Identity Providers"); AT&T Inc. Comments in re NBP PN #29, filed Jan. 22, 2010, at 6 (discussing "OpenID" and Information Cards).

276 *See* Fed. Deposit Ins. Corp., Who is the FDIC?, http:// www.fdic.gov/about/learn/symbol/index.html (last visited Mar. 5, 2010) (providing the history of the FDIC).

277 As part of any rulemaking or processing, the FTC and FCC would also need to define "third parties" to account for the complex relationships of companies with affiliates, other subsidiaries and trusted intermediaries.

278 In fact, according to data from the Technology Policy Institute, only 11% of identity fraud cases involve the Internet. Thomas M. Lenard & Paul H. Rubin, Tech. Pol'y Inst., In Defense of Data: Information and the Costs of Privacy 7 (2009), *available at* http://www. techpolicyinstitute.org/files/in%20defense%20of%20 data.pdf.

279 Gartner, *Gartner Says Number of Identity Theft Victims Has Increased More Than 50 Percent Since 2003* (press release), Mar. 6, 2007, http://www.gartner.com/it/ page.jsp?id=501912; ConnectSafely, Online Safety 3.0: Empowering and Protecting Youth, Connect Safely, http://www.connectsafely.org/Commentaries-Staff/ online-safety-30-empowering-and-protecting-youth. html (last visited Feb. 18, 2010).

280 Fed. Trade Comm'n, Consumer Sentinel Netwo rk Data Book for January–December 2008 5 (2009) (FTC, Consumer Sentinel Netwo rk Data Book 2008); *see also*Internet Crime Complaint Center (IC3), 2008 Internet Crime Report 4 (2008) (IC3, 2008 Internet Crime Report) (showing that identity theft represented 2.5% of the total complaints received in 2008 by the Internet Crime Complaint Center), *available at* http://www.ic3.gov/media/annualreport/2008_ic3report.pdf.

281 FTC, Consumer Sentinel Netwo rk Data Book 2008 at 3.

282 A.F. Salam et al., *Consumer-Perceived Risk in E-Commerce Transactions*, 23 commc'ns of the acm 325 (2003), *available at* http://www.som.buffalo.edu/ isinterface/papers/Consumer-Perceived%20Risk%20 in%20E-Commerce.pdf.

283 IC3, 2008 Internet Crime Report at 1.

284 GAO, Identity Theft: Prevalence and Cost Appear to Be Grow ing 11,, GAO-02-363 (2002) ("Regarding state statutes, at the time of our 1998 report, very few states had specific laws to address identity theft. Now, less than 4 years later, a large majority of states have enacted identity theft statutes.").

285 *See* Fed. Trade Comm'n, ID Theft, Privacy, & Security: Identity Theft, http://www.ftc.gov/bcp/menus/ consumer/data/idt.shtm (last visited Mar. 5, 2010).

286 The Data Accountability and Trust Act, H.R. 2221, 111th Cong. (2009), would require entities that store personal information to protect the data through security policies and procedures and to provide nationwide notice in the event of a security breach while the Personal Data Privacy and Security Act of 2009, S. 1490, 111th Cong. (2009), would increase criminal penalties for identity theft involving electronic personal data and make it a crime to intentionally or willfully conceal a security breach involving personal data.

287 *U.S. v. Morris*, 928 F.2d 504 (2d Cir. 1991).

288 CERT, Meet CERT, http://www.cert.org/meet_cert/ (last visited Mar. 5, 2010).

289 APWG, Phishing Activity Trends Report: 1st Half 2009 (2009) (APWG, Phishing Activity Trends Report), *available at* http://www.antiphishing.org/reports/ apwg_report_h1_2009.pdf.

290 APWG, Phishing Activity Trends Report.

[291] White House, Cyberspace Policy Review: Assuring a Trusted and Resilient Information and Communications Infrastructure NSPD-54/HSPD23 (May 2009), *available at* http://www.whitehouse.gov/assets/documents/Cyberspace_Policy_Review_final.pdf.

[292] Horrigan, *Broadband Adoption and Use in America* at 17.

[293] Tanya Byron, Safer Children in a Digital World: The Report of the Byron Review 2 (2008), *available at* http://www.dcsf.gov.uk/byronreview/pdfs/Final%20 Report%20Bookmarked.pdf.

[294] Berkman Ctr. for Internet & Soc'y, Enhancing Child Safety & Online Technolog ies: Final Report of the Internet Safety Technical Task Force 5 (2008), *available at* http://cyber.law.harvard.edu/sites/cyber.law.harvard.edu/files/ISTTF_Final_Report.pdf.

[295] Anne Collier, *It's Time to Get Smart About Online Safety*, sch. libr. j., Nov. 1, 2009, *available at* http://www.schoollibraryjournal.com/article/CA6703696.html.

[296] Computer Sci. & Telecomm. Bd., Youth, Pornog raphy, and the Internet 9 (Dick Thornburgh & Herbert S. Lin, eds., 2002), *available at* http://www.nap.edu/openbook. php?isbn=0309082749 (requires purchase).

[297] Letter from Susan L. Fox, Vice Pres. of Gov't Relations, Disney, to Marlene H. Dortch, Secretary, FCC, GN Docket No. 09-191, WC Docket No. 07-52 (Dec. 11,2009) at 1; Letter from Harold Feld, Legal Dir., Public Knowledge, to Marlene H. Dortch, Secretary, FCC, CB Docket No. 97-80, MB Docket No. 08-82, GN Docket No. 09-51, MB Docket No. 09-168 (Oct. 28, 2009) at 1.

[298] Verizon, Reforming Federal and State Tax Policies Will Increase Investment in Broadband and Consumer Adoption 1–4, *attached to* Letter from Ann D. Berkowitz, Dir., Fed. Reg. Aff., Verizon, to Marlene H. Dortch, Secretary, FCC, GN Docket No. 09-51 (Feb. 12, 2010).

[299] *Preserving the Open Internet; Broadband Industry Practices*, GN Docket No. 09-191, WC Docket No. 07-52, Notice of Proposed Rulemaking, 24 FCC Rcd 13064, 13065, para. 3 (2009) (*Preserving the Open Internet NPRM*).

[300] *See Preserving the Open Internet NPRM*, 24 FCC Rcd at 13067, para. 4 ("As a platform for commerce, [the Internet] does not distinguish between a budding entrepreneur in a dorm room and a Fortune 500 company. As a platform for speech, it offers the same potential audience to a blogger on her couch and to a major newspaper columnist.").

[301] *Preserving the Open Internet NPRM*, 24 FCC Rcd at 13067, para. 95 ("The first principle in the Internet Policy Statement, and the first rule we propose to codify here, ensures that users are in control of the content that they send and receive.").

[302] *Preserving the Open Internet NPRM*, 24 FCC Rcd at 13067, para. 18.

[303] *See Preserving the Open Internet NPRM*, 24 FCC Rcd at 13067, para. 4 ("Because of the historically open architecture of the Internet, it has been equally accessible to anyone with a basic knowledge of its protocols. The Internet's accessibility has empowered individuals and companies at the edge of the network to develop and contribute an immense variety of content, applications, and services that have improved the lives of Americans. Such innovation has dramatically increased the value of the network, spurring—in a virtuous circle—investment by network operators, who have improved the Internet's reach and its performance in many areas."); *cf. id.* at 13067, para. 9 ("[B]roadband Internet access service providers may have both the incentive and the means to discriminate in favor of or against certain Internet traffic. . . in ways that negatively affect consumers, as well as innovators trying to develop Internet-based content, applications, and services. Such practices have the potential to change the Internet from an open platform that enables widespread innovation and entrepreneurship to an increasingly closed system with higher barriers to participation and reduced user choice and competition.").

[304] *See Preserving the Open Internet NPRM*, 24 FCC Rcd at 13067, para. 9.

[305] Chairman Genachowski and Commissioners Copps and Clyburn voted to adopt the NPRM. Commissioners McDowell and Baker concurred in part and dissented in part. *See Preserving the Open Internet NPRM*, 24 FCC Rcd at 13064.

[306] *See* NBP PN #25.

[307] AT&T Comments in re NBP PN #25, filed Dec. 22, 2009, at 11.

[308] AT&T Comments in re NBP PN #25, filed Dec. 22, 2009, at 11.

[309] AT&T Comments in re NBP PN #25, filed Dec. 22, 2009, at 12.

[310] AT&T Comments in re NBP PN #25, filed Dec. 22, 2009, at 14; OPASTCO Comments in re NBP PN #25, filed Dec. 17, 2009, at 3.

[311] Skype Comments in re NBP PN #25, filed Dec. 22, 2009, at 9; California Public Utilities Commission in re NBP PN #25, filed Dec. 18, 2009, at 10; OPASTCO Comments in re NBP PN #25, filed Dec. 17, 2009, at 3; Communications Workers of America Reply in re NBP PN #30, filed Jan. 27, 2010, at 3.

[312] Skype Comments in re NBP PN #25, filed Dec. 22, 2009, at 6; California Public Utilities Commission in re NBP PN #25, filed Dec. 18, 2009, at 7; Communications Workers of America Reply in re NBP PN #30, filed Jan. 27, 2010, at 3.

[313] Skype Comments in re NBP PN #25, filed Dec. 22, 2009, at 9, CTIA Comments in re NBP PN #25, Dec. 22, 2009, at 4; Massachusetts Department of Telecommunications and Cable Comments in re NBP PN #25, filed Dec. 22, 2009, at 1–6.

[314] AT&T Comments in re NBP PN #25, filed Dec. 22, 2009, at 17; California Public Utilities Commission in re NBP PN #25, filed Dec. 18, 2009, at 11; Massachusetts Department of Telecommunications and Cable Comments in re NBP PN #25, filed Dec. 22, 2009, at 1–6.

[315] AT&T Comments in re NBP PN #25, filed Dec. 22, 2009, at 19; Communications Workers of America Reply in re NBP PN #30, filed Jan. 27, 2010, at 3; CTIA Comments in re NBP PN #25, Dec. 22, 2009, at 4; Level 3 Comments in re NBP PN #25, filed Dec. 22, 2009, at 6; National Cable & Telecommunications Association Comments in re NBP PN #25, filed Dec. 22, 2009, at 3; OPASTCO Comments in re NBP PN #25, filed Dec. 17, 2009, at 3; Time Warner Cable Comments in re NBP PN # 25, filed Dec. 22, 2009, at 7.

[316] Communications Workers of America Reply in re NBP PN #30, filed Jan. 27, 2010, at 3.

[317] Org. for Econ. Co-operation & Dev., Information Technolog y Outlook 2008, fig. 2.1 (2008) (World trade in ICT goods, 1996–2007), *available at* http://dx.doi. org/10.1787/473254016535.

[318] Specifically, TeleGeography reports a 66% compound annual growth rate in average global traffic over Internet bandwidth connected across international borders over the past five years (2005–2009) and a 22% unweighted average compound annual price decrease for median IP transit prices per Mbps, Gigabit Ethernet, for select cities (Hong Kong, London, Los Angeles, New York, Sao Paulo, and Singapore) over the past five years (2005–2009). TeleGeog raphy Research, Global Internet Geog raphy (2010).

[319] The United States broke up AT&T in the 1980s into seven Regional Bell Operating Companies and a long distance provider. This liberalization and the subsequent passage of the Telecommunications Act of 1996 preceded similar reforms in other countries. CESifo Group, History of Telecommunication Liberalization,, *available   at*   http://www.cesifo-group.de/portal/page/portal/DICE_Content/INFRASTRUCTURE/ COMMUNICATION_NETWORKS/Liberalisation%20 Process/history-telecom-liber.pdf (last visited Mar. 5, 2010).

[320] *See generally* WTO, Post-Uruguay Round Negotiations on Basic Telecommunications, http://www.wto. org/english/tratop_e/serv_e/telecom_e/telecom_ posturuguay_neg_e.htm (last visited Feb. 26, 2010); WTO, Telecommunications Services: Reference Paper, Negotiating Group on Basic Telecommunications, Apr. 24, 1996, http://www.wto.org/english/tratop_e/serv_e/ telecom_e/tel23_e.htm; WTO, Services: Agreement, Annex on Telecommunications, http://www.wto.org/ english/tratop_e/serv_e/12-tel_e.htm (last visited Mar. 5, 2010).

[321] DataDyne has had success leveraging the mobile penetration rate in Africa, which approaches 50%, to create applications that allow real time data collection by health care workers and more effective mobilization of public health responses. Specifically, DataDyne designed an application that allowed public health workers in the rural areas of Kenya to collect patient health information using a form on their basic mobile phones and then sending it back to the main office in Nairobi where there is dependable broadband access to be included in a comprehensive database. S*ee generally* DataDyne, http://www.datadyne.org/ (last visited Feb. 26, 2010).

[322] There are over 1.7 billion Internet users worldwide. Miniwatts, Internet World Stats, Internet World Users by Language: Top Ten Languages (chart) (Sept. 30, 2009), http://www.internetworldstats.com/stats7. htm (Copyright © 2009, Miniwatts Marketing Group, all rights reserved worldwide). There are about 4.6 billion mobile phone subscriptions in the world. Int'l Telecomm. Union, The World in 2009: ICT Facts and Figures 1 (2009) (Int'l Telecomm. Union, ICT Facts), *available at* http://www.itu.int/ITU-D/ict/material/ Telecom09_flyer.pdf.

[323] In 2009, more than a quarter of the world's population is using the Internet. int'l telecOmm. uniOn, ict facts at 1. About 60% of the world's population has a mobile phone. Int'l Telecomm. Union, The World Telecommunication/ICT Indicators Database (13th ed. 2009), *available at* http://www.itu.int/ITU-D/ict/ publications/world/world.html (requires purchase).

[324] *See infra* Chapter 13, Section 13.4 (discussing broadband and local and regional economic development).

[325] *See* Suzanne Choney, *Mobile Banking On the Rise During Recession*, MSNBC, Oct. 5, 2009, http://www.msnbc. msn.com/id/33079970/ns/technology_and_sciencetech_and_gadgets//; Gautam Bandyopadhyay, *Banking the Unbanked: Going Mobile in Africa*, african exec., Sept. 17, 2008, *available at* http://www.africanexecutive. com/modules/magazine/articles.php?article=3541; *Mobile money in the poor world*, ecOnOmist, Sept. 24, 2009, *available at* http://www.economist.com/ printedition/displayStory.cfm?Story_ID=14505519.

[326] *See infra* Chapter 5.

[327] *See infra* Chapters 14 and 16.

[328] *See* Gerald R. Faulhaber & David Farber, *Spectrum Management: Property Rights, Markets, and the Commons* (AEI-Brookings Jt. Ctr., Working Paper No. 02-12, Dec. 2002).

[329] Harold Furchtgott-Roth, The Wireless Services Sector: A Key to Economic Growth in America 1 (Jan. 2009) (unpublished manuscript, on file with the Commission) (*Furchtgott-Roth Wireless Services Sector Report*).

[330] *See* Mary Meeker et al., Morgan Stanley Research, The Mobile Internet Report 1 (2009) (Meeker et al. The Mobile Internet Report) (Copyright 2009 Morgan Stanley. Courtesy of Morgan Stanley).

[331] *See* Meeker et al. The Mobile Internet Report 5.

[332] Kris Rinne, Sr. Vice Pres. of Architecture & Planning, AT&T, Remarks at the FCC Spectrum Workshop 11–12 (Sept. 17, 2009), *available at* http://www.broadband. gov/docs/ws_25_spectrum.pdf. Ms. Rinne added that in addition to increased data usage, voice usage continues to rise also. *Id.*

[333] Bill Stone, Executive Director of Network Strategy, Verizon Wireless, Remarks at the FCC Spectrum Workshop 14–15 (Sept. 17, 2009), *available at* http:// www.broadband.gov/docs/ws_25_spectrum.pdf. S*ee also* Verizon Wireless Comments in re NBP PN #6 (*Comment Sought on Spectrum for Broadband—NBP Public Notice #6*, GN Docket Nos. 09-51, 09-47, 09-137, 24 FCC Rcd 12032 (WTB 2009) (*NBP PN #6*)), filed Oct. 23, 2009, at 3.

[334] Cisco Sys., Cisco Visual Netwo rking Index Global Mobile Data Forecast 2009–2014 (2010) (Cisco, Global Mobile Data Forecast 2009–2014), *available at* http://www.cisco. com/en/US/solutions/collateral/ns341 /ns525/ns537/ ns705/ns827/white_paper_c11-520862.pdf .

[335] Philip Marshall, Yankee Group, Spectrum-Rich Players Are in the Driver's Seat for Mobile Broadband Economics (2009) (unpublished manuscript, on file with the FCC); Coda Res. Consultancy, US Mobile Traffic Forecasts: 2009–2015, at 25 (2009) (unpublished manuscript, on file with the FCC).

[336] Meeker et al., The Mobile Internet Report..

[337] For example, T-Mobile states that its G1 customers consume 300+ MB per month. Neville Ray, Sr. Vice Pres., Engineering Operations, T-Mobile USA, Presentation at the FCC Wireless Broadband Workshop (Aug. 12, 2009), *available at* http://www.broadband. gov/docs/ws_deployment_wireless/ws_deployment_ wireless_Ray.pdf. S*ee also* T-Mobile Comments in re NBP PN #6, filed Oct. 23, 2009, at 4–6. According to research conducted by Validas for Consumer Reports, iPhone users consume almost twice the data most other smart phones do, on average 273 MB per month. S*ee* Jeff Blyskal, *Exclusive: iPhones Hog Much More Data Than Other Smart Phones*, consumer reports electronics bloG, Feb. 10, 2010, http://blogs. consumerreports.org/ electronics/2010/02/iphone-data-usage-smart-phones- smartphones-blackberry-mb-network-att-carrier- istress.html.

[338] For example, Bill Stone stated that laptops consume "north of 1 GB per month." Bill Stone, Executive Director of Network Strategy, Verizon Wireless, Remarks at the FCC Spectrum Workshop 72 (Sept. 17, 2009), *available at* http://www.broadband.gov/docs/ ws_25_spectrum.pdf. According to research conducted by Validas for *Consumer Reports*, the average "aircard" user consumes 1.4 GB per month. S*ee* Jeff Blyskal, *Exclusive: iPhones Hog Much More Data Than Other Smart Phones*, cOnsumer repOrts electrOnics blOG, Feb. 10, 2010, http://blogs.consumerreports.org/ electronics/2010/02/iphone-data-usage-smart-phones- smartphones-blackberry-mb-network-att-carrier- istress.html.

[339] Cisco, Global Mobile Data Forecast 2009–2014.

[340] *See, e.g.*, Harbor Research, Announcing Harbor Research's 2009 Pervasive Internet/M2M Forecast Report, Feb. 24, 2009, http://www.harborresearch.com/ AnnouncementRetrieve.aspx?ID=17927 (last visited Feb. 18, 2010) ("[T]he number of intelligent device shipments will grow from 73 million units in 2008 to 430 million units in 2013").

[341] Colin Gibbs, *Multiple Mobile Devices = Increased Spending*, GiGaom, Oct. 8, 2009, http://gigaom. com/2009/10/08/multiple-mobile-devices-increasedspending/.

[342] Robert C. Atkinson & Ivy E. Schultz, Columbia Institute for Tele-Information, Broadband in America: Where It Is and Where It Is Going (According to Broadband Service Providers) 8 (2009) (Atkinson & Schultz, Broadband Report),, *available at* http://www4. gsb.columbia.edu/citi.

[343] *See* AdMob, AdMob Mobile Metrics Report 2 (2008), *available at* http://www.admob.com/marketing/pdf/ mobile_metrics_nov_08.pdf.

[344] Letter from Christine A. Varney, Ass't Atty. General, U.S. Dep't of Justice, to Marlene H. Dortch, Secretary, FCC, GN Docket No. 09-51 (Jan. 4, 2010) (DOJ Jan. 4, 2010 *Ex Parte*) at 21.

[345] *Implementation of Section 6002(b) of the Omnibus Budget Reconciliation Act of 1993; Annual Report and Analysis of Competitive Market Conditions with Respect to Commercial Mobile Services*, Fifth Report, 15 FCC Rcd 17660, 17677 (2000).

[346] *Implementation of Section 6002(b) of the Omnibus Budget Reconciliation Act of 1993, Annual Report and Analysis of Competitive Market Conditions with Respect to Commercial Mobile Services*, WT Docket No. 08-27, Thirteenth Report, 24 FCC Rcd 6185, 6276, tbl. 12 (2009).

[347] Robert F. Roche & Lesley O'Neill, CTIA 's Wireless Industry Indices—Semi-Annual Data Survey Results: A Comprehensive Report From CTIA Analyzing the U.S. Wireless Industry, Year End 2008 Results 27(2009) (Roche and O'Neill, CTIA Survey Results).

[348] Roche and O'Neill, CTIA Survey Results at 126.

[349] Roche and O'Neill, CTIA Survey Resultsat 150.

[350] Roche and O'Neill, CTIA Survey Resultsat 167.

[351] DOJ Jan. 4, 2010 *Ex Parte* at 17.

[352] NTIA allocates certain spectrum for restricted uses by federal entities for purposes such as defense, public safety, national security and scientific uses. Similarly, the FCC allocates spectrum for restricted uses by commercial entities and for use by state and local governments for purposes such as public safety and maintenance of critical infrastructure.

[353] This figure aggregates the values the four largest wireless providers placed on their holdings in the most recently filed 10-Q. S*ee* Verizon Communications Inc., Quarterly Report (Form 10-Q), at 10 (Oct. 29, 2009), *available at* http:// go.usa.gov/lEG (valuing wireless licenses at $71.9 billion); AT&T Inc., Quarterly Report (Form 10-

Q), at 3 (Nov. 5, 2009), *available at* http://go.usa.gov/lEo (valuing licenses at $47.9 billion); Sprint Nextel Corp., Quarterly Report (Form 10-Q), at 1 (Nov. 6, 2009), *available at* http://go.usa. gov/lEs (valuing FCC licenses and trademarks at $19.8 billion); T-Mobile, T-Mobile USA Reports Third Quarter 2009 Results, (press release Nov. 5, 2009 at 10), *available at* http://www.t-mobile.com/Cms/Files/Published/0000BD F20016F5DD010312E2BDE4AE9B/5657114502E70FF 30124C645BC1131D6/file/TMUS%20Q3%20Press%20 Release%20FINAL.pdf (valuing licenses at $15.2 billion).

[354] *See, e.g.*, Gregory L. Rosston, *The Long and Winding Road: The FCC Paves the Path with Good Intentions*, 27 telecOmms. pOl'y 501, 513 (2003); Coleman Bazelon, *The Need for Additional Spectrum for Wireless Broadband: The Economic Benefits and Costs of Reallocations*, *attached to* Consumer Electronics Association Comments in re NBP PN #6, filed Oct. 23, 2009, at 2.

[355] Legislation currently pending in Congress would require an inventory of radio spectrum bands managed by NTIA and the FCC. Radio Spectrum Inventory Act, H.R. 3125, 111th Cong. (2009); Radio Spectrum Inventory Act, S. 649, 111th Cong. (2009).

[356] Detailed information is available for: 700 MHz Band; Advanced Wireless Service (AWS); Broadband Personal Communications Service (PCS); Broadband Radio Service (BRS); Educational Broadband Service (EBS); Cellular; 2.3 GHz Wireless Communications Service (WCS); Full Power TV Broadcast; and Mobile Satellite Services (MSS). The FCC will also begin gathering data on state and local spectrum.

[357] Facilitating access to the FCC's spectrum dashboard will be a critical predicate for helping Tribal communities use spectrum or identify non-Tribal parties that hold licenses to serve Tribal lands. Letter from Loris Ann Taylor, Executive Director, Native Public Media, et al., to Marlene H. Dortch, Secretary, FCC, GN Docket Nos. 09-47, 09-51, 09-137 (Dec. 24, 2009) (Joint Native Filers Dec. 24, 2009 *Ex Parte*) at 7.

[358] NTIA has endorsed the idea of a spectrum inventory. *See* Letter from Lawrence E. Strickling, Ass't Sec'y for Commc'ns & Info., U.S. Dep't of Commerce, to Julius Genachowski, Chairman, FCC, GN Docket No. 09-51 (Jan. 4, 2010) (NTIA Jan. 4, 2010 *Ex Parte*) at 5.

[359] Congress is considering legislation that may specify a different frequency range for a spectrum inventory. *See* Radio Spectrum Inventory Act, H.R. 3125, 111th Cong. (2009) (requiring an inventory of spectrum between 225 MHz and 10 GHz as of February 18, 2010); Radio Spectrum Inventory Act, S. 649, 111th Cong. (2009) (requiring an inventory of spectrum between 300 MHz and 3.5 GHz as of February 18, 2010).

[360] New America Foundation Comments in re National Broadband Plan NOI, filed June 8, 2009, at 16. *But see* AT&T Comments in re NBP PN #6, filed Oct. 23, 2009, at 30.

[361] The FCC has developed Project Roll Call for the purpose of conducting spectrum usage analysis in areas affected by major emergencies such as hurricanes. With the acquisition of additional equipment, the capabilities of Project Roll Call could be expanded to provide more comprehensive data on spectrum usage nationwide. *See* FCC, Project Roll Call, http://go.usa.gov/lER (last visited Feb. 18, 2010).

[362] A fleet of vehicles was equipped to scan frequencies between 10 MHz and 5 GHz. Over a one-year period, the fleet drove 65,000 kilometers, measuring spectrum use 4.2 million times. *See* Ofcom, Capture of Spectrum Utilisation Information Using Moving Vehicles v (2009), *available at* http://www.ofcom.org.uk/research/ technology/research/state_use/vehicles/vehicles.pdf.

[363] Omnibus Budget Reconciliation Act of 1993, Pub. L. No. 103-66, § 6002, 107 Stat. 312, 387–92 (1993) (codified at 47 U.S.C. § 309(j)).

[364] Commercial Spectrum Enforcement Act, Pub. L. No. 108-494, 118 Stat. 3991 (2004).

[365] In addition, the FCC could grant incumbents more flexible rights to use the re-purposed spectrum as long as they agreed to participate in the auction. Requiring licensees to participate in the auction as a pre-condition for acquiring enhanced rights forces them to consider the opportunity cost of holding the repurposed licenses— since in the auction they will actually observe what other bidders are willing to pay for their licenses. *See* Evan Kwerel & John Williams, *A Proposal for a Rapid Transition to Market Allocation of Spectrum* 2 (Office of Strategic Planning & Policy Analysis, Working Paper No. 38, 2002), *available at* http://wireless.fcc. gov/auctions/conferences/combin2003/papers/ masterevanjohn.pdf.

[366] To provide further incentives for rapid aggregation of a significant spectrum block, a larger portion of proceeds could be offered to early participants.

[367] Commercial Spectrum Enhancement Act (CSEA), Pub. L. No. 108-494, 118 Stat. 3986, Title II (2004) (codified in different sections of Title 47 of the United States Code).

[368] CSEA §§ 201–209. Relocation costs are "costs incurred by a federal entity to achieve comparable capability of systems" and include "costs associated with the accelerated replacement of systems and equipment if such acceleration is necessary to ensure the timely relocation of systems to a new frequency assignment." *See* 47 U.S.C. § 923(g)(3).

[369] *See* Nat'l Telecomms. & Info. Agency, U.S. Dep't of Commerce, Relocation of Federal Radio Systems From the 1710–1755 MHz Spectrum Band, Second Annual Prog ress Report (2009), *available at* http://www. ntia.doc.gov/reports/2009/Final2ndAnnual RelocationReport20090416.pdf.

[370] Examples of flexible use bands include the Cellular, PCS, and AWS services.

371 For federal government users a similar effect could be achieved without any money changing hands. The relevant federal agency could simply include the value of its spectrum in its budget, and it could then decide whether to keep its spectrum allotment as is or use less spectrum and thus make money available in its budget for other priorities.

372 As the FCC has noted in other proceedings, it may lack the authority to impose certain user fees. *See Implementation of Sections 309(j) and 337 of the Communications Act of 1934 as Amended; Promotion of Spectrum Efficient Technologies on Certain Part 90 Frequencies; Establishment of Public Service Radio Pool in the Private Mobile Requencies Below 800 MHz*, Notice of Proposed Rulemaking, WT Docket No. 99-87, RM9332, RM-9405, 14 FCC Rcd 5206, 5244 (1999). The urgent need to make spectrum available for broadband heightens the importance of this authority at this time.

373 GAO, Options for and Barriers to Spectrum Reform, 11,GAO -06-526t (2006), *available at* http://www.gao.gov/ new.items/d06526t.pdf.

374 NTIA imposes fees to recover a portion of its spectrum management costs, but not fees that more closely resemble market prices and encourage greater spectrum efficiency among government users. Currently, NTIA does not have authority to impose fees that exceed its spectrum management costs.

375 *See* Ofcom, Ofcom Policy Evaluation Report: AIP (2009), *available at* http://www.ofcom.org.uk/research/ radiocomms/reports/policy_report/ (Ofcom AIP Report). Note that Australia and Canada also have adopted versions of spectrum incentive fees. *See* GaO, Comprehensive Review of U.S. Spectrum Management with Broad Stakeholder Involvement Is Needed 20–26 (2003).

376 Ofcom AIP Report at 7.

377 William Webb, Head of Research and Development, Ofcom, Remarks at FCC Spectrum Workshop (Sept. 17, 2009), *available at* http://www.broadband.gov/docs/ ws_25_spectrum.pdf.

378 See GPO Access, Budgets of the United States Government, Fiscal Years 2000 through 2011, http:// www.gpoaccess.gov/usbudget/browse.html (last visited Mar. 5, 2010). Every administration since 1999 has requested authority to impose user fees.

379 *See, e.g.,* Verizon and Verizon Wireless Comments, filed Sept. 30, 2009, at 110–17 (citing numbers of secondary market transactions providing spectrum access to non- nationwide providers); (*Comment Sought on Defining "Broadband"—NBP Public Notice #1*, GN Docket Nos. 09- 47, 09-51, 09-137, Public Notice, 24 FCC Rcd 10897 (WCB 2009) (*NBP PN #1*)); [[after National Broadband Plan NOI ((*A National Broadband Plan for Our Future*, GN Docket No. 09-51, Notice of Inquiry, 24 FCC Rcd 4342 (2009))

380 *See, e.g.,* National Telecommunications Cooperative Association Comments in re National Broadband Plan NOI, filed June 8, 2009, at 5 (would increase access to smaller providers in rural areas); MetroPCS Communications, Inc. Comments in re National Broadband Plan NOI, filed Sept. 30, 2009, at 14–15; United States Cellular Corporation Comments in re National Broadband Plan NOI, filed Sept. 30, 2009, at 24–26 (spectrum aggregation limits); *see also* Letter from Caressa D. Bennet, Counsel, NEP Cellcorp., Inc., to Ruth Milkman, Chief, Wireless Telecommunications Bureau, GN Docket No. 09-157 (Nov. 30, 2009) (asserting that reasonable efforts to obtain spectrum, either through a license transfer or a spectrum leasing arrangement, have been to no avail).

381 *See, e.g., Principles for Promoting Efficient Use of Spectrum By Encouraging the Development of Secondary Markets*, Policy Statement, 15 FCC Rcd 24178, 24178, para. 1 (2000) (*Secondary Markets Policy Statement*); *Promoting Efficient Use of Spectrum Through Elimination of Barriers to the Development of Secondary Markets*, WT Docket No. 00-230, Second Report and Order, Order on Reconsideration, and Second Further Notice of Proposed Rulemaking, 19 FCC Rcd 17503 (2004) (*Secondary Markets Second R&O*). The FCC's secondary market policies are not limited to wireless broadband services.

382 *See, e.g., Secondary Markets Policy Statement*, 15 FCC Rcd at 24183, para. 11.

383 These spectrum-leasing policies apply to spectrum license authorizations in which the licensee holds "exclusive use" rights. *Secondary Markets Second R&O*, 19 FCC Rcd 17503.

384 *See Secondary Markets Second R&O*, 19 FCC Rcd 17503.

385 *Secondary Markets Second R&O*, 19 FCC Rcd at 17547–49, paras. 88–90; *Service Rules for the 698–746, 747– 762 and 777–792 MHz Bands; Implementing a Nationwide, Broadband, Interoperable Public Safety Network in the 700 MHz Band*; WT Docket Nos. 06-150, 01-309, 03-264, 06-169, 06-229, 96-86, 07-166, CC Docket No. 94-102, PS Docket No. 06-229, Second Report and Order, 22 FCC Rcd 15289, 15374–80, paras. 231–48 (2007) (discussing the FCC's dynamic spectrum leasing policies).

386 The data shows, for instance, that the majority of cellular, broadband PCS, and AWS licenses has been assigned/transferred to different entities, including both the largest providers (who have consolidated their holdings into nationwide footprints), and regional and smaller providers. Similarly, many of these licenses have been partitioned or disaggregated, again transferring the spectrum to a wide range of entities of different sizes. There are many instances of spectrum leasing, although most of these are procedural in nature and none to date involve dynamic spectrum leasing arrangements.

387 *See, e.g., Secondary Markets Policy Statement*, 15 FCC Rcd at 24178, para. 1.

388 Timing and quantity depends on outcome of the investigation into possibility of reallocating federal spectrum in the 1755–1850 MHz band.

389 Timing and quantity depends on Congressional action to grant incentive auction authority as well as voluntary participation of broadcasters in an auction.

390 This does not include the 14 megahertz of licensed ESMR spectrum pending completion of the 800 MHz rebanding because broadband operations have not been shown to be viable under the interference protections provided to neighboring public safety operations per 47 CFR § 90.672.

391 Letter from 21st Century Telecommunications et al., Members of the Consumer Electronic Association et al., to Chairman Julius Genachowski and Commissioners, FCC, GN Docket No. 09-51 (Dec. 2, 2009) at 1 (filed by Consumer Electronics Association) (on behalf of 115 parties).

392 Clearwire states that 120 megahertz of contiguous spectrum is needed for true mobile broadband. John Saw, Senior Vice President and Chief Technology Officer, Clearwire, Remarks at FCC Spectrum Workshop (Sept. 17, 2009), *available at* http://www.broadband. gov/docs/ws_25_spectrum.pdf. Fibertower argues that 100 megahertz or more of spectrum will be needed for wireless backhaul in the next few years. Tarun Gupta, Vice President of Strategic Development, FiberTower, Remarks at FCC Spectrum Workshop (Sept. 17, 2009). T-Mobile's smartphone customers use 50 times more data than its average non-smartphone customers. T-Mobile Comments in re NBP PN #26, (*Data Sought on Users of Spectrum*—NBP Public Notice #26, GN Docket Nos. 09–47, 09–51, 09–137, Public Notice, 24 FCC Rcd 14275 (OBI 2009) (*NBP PN #26*)), filed Dec. 22, 2009, at 4. Verizon Wireless states that it might acquire more than 100 megahertz of spectrum within the next five years, if it were available. Bill Stone, Executive Director, National Strategy, Verizon Wireless, Remarks at FCC Spectrum Workshop (Sept. 17, 2009), *available at* http://www.broadband.gov/docs/ws_25_spectrum. pdf. WCAI states that 100 megahertz of new spectrum would be a substantial beginning for mobile broadband wireless providers to meet future needs. Wireless Communications Association International Reply in re NBP PN #6, filed Nov. 13, 2009, at 4.

393 CTIA Reply in re NBP PN #6, filed Nov. 13, 2009, at 2.

394 Int'l Telecomm. Union, Estimated Spectrum Bandwidth Requirements for the Future Development of IMT - 2000 and IMT -Advanced, Report ITU -R M.2078 (2006).

395 Ofcom, Predicting Areas of Spectrum Shortage (2009), *available at* http://www.ofcom.org.uk/research/ technology/research/spec_future/predicting/shortage.pdf.

396 The 2.3 GHz WCS spectrum includes two 15 megahertz bands (2305–2320 MHz, 2345–2360 MHz), which envelope the 25 megahertz SDARS band and is adjacent to the aeronautical telemetry band at 2360–2390 MHz. The WCS spectrum is licensed in two 10-megahertz blocks (each 5 megahertz paired) in 52 Major Economic Areas (MEAs), and in two 5 megahertz blocks in 12 Regional Economic Area Groupings (REAGs). The 52 MEA license areas encompass 172 Economic Areas (EAs). The FCC's 1997 auction of WCS spectrum netted $13.6 million.

397 *See* FCC, Amendment of Part 27 of the FCC's Rules to Govern the Operation of Wireless Communications Services in the 2.3 GHz Band, WT Docket No. 07-293, http://fjallfoss.fcc.gov/ecfs/proceeding/ view?name=07-293 (last visited Feb. 22, 2010). As of Feb. 22, 2010, the docket contained 282 filings, according to the Electronic Comments Filing System.

398 Time Division Duplex (TDD) is a technology where bi-directional communications occurs within the same frequency band as compared with Frequency Division Duplex technology where one band is used for transmission from base stations to mobile units and another band is used for transmission from mobile units to base stations. Orthogonal Frequency Division Multiplexing (OFDM) is a digital multi-carrier modulation scheme in which each signal is split into multiple smaller sub-signals that are then transmitted simultaneously at different frequencies to the receiver. WiMAX, for example, is being implemented today using TDD and OFDM technology.

399 *See, e.g.*, APCO Comments in re NBP PN #8, (*Additional Comments Sought on Public Safety, Homeland Security and Cybersecurity Elements of National Broadband Plan —NBP Public Notice #8*, GN Docket Nos. 09–47, 09–51, 09–137, PS Docket Nos. 06–229, 07–100, 07–114, WT Docket No. 06–150, CC Docket No. 94–102, WC Docket No. 05–196, Public Notice, 24 FCC Rcd 12136 (PSHSB 2009) (*NBP PN #8*). filed Nov. 12, 2009, at 11; AT&T Comments in re NBP PN #8, filed Nov. 12, 2009, at 2; Verizon Comments in re NBP PN #8, filed Nov. 12, 2009, at 6; Public Safety Spectrum Trust Comments in re *Public Safety and Homeland Security Bureau Seeks Comment on Petitions for Waiver to Deploy 700 MHz Public Safety Broadband Networks*, PS Docket No. 06- 229, Public Notice, 24 FCC Rcd 10814 (2009), filed Oct. 16, 2009, at 17.

400 Presently, the LTE specification designates "Band 14" as a single band class that incorporates both the Public Safety Broadband License (763–768 MHz and 793–798 MHz) and the Upper 700 MHz D Block (758–763 MHz and 788–793 MHz). *See* 3rd Generation Partnership Project, 3GPP TS 36.101 v8.8.0: 3rd Generation Partnership Project; Technical Specification Group Radio Access Netwo rk; Evolved Universal Terrestrial Radio Access (E-UTRA), User Equipment (UE) Radio Equipment and Reception, Release 8, at 14 & tbl. 5-5.1 (2009), available at http://www.quintillion. co.jp/3GPP/Specs/36101-880.pdf; 3rd Generation Partnership Project, 3GPP TS 36.104 v8.8.0 (2009-12) 3rd Generation Partnership Project; Technical Specification Group Radio Access Netwo rk; Evolved Universal Terrestrial Radio Access (E-UTRA), Base Station (BS) radio Equipment and Reception, Release 8, at 13 & tbl. 5-5.1 (2009).

[401] 3GPP band class 3 includes 1710–1785 MHz and is used in Europe, Asia, and Brazil. S*ee, e.g.*, Fred Christmas, on behalf of the GSM Association, Benefits of Frequency Harmonisation, Presentation at ITU Workshop on Market Mechanisms for Spectrum Management 8 (Jan. 2007), *available at* http://www.itu.int/osg/spu/stn/spectrum/ workshop_proceedings/Presentations_Abstracts_ Speeches_Day_1_ Final/ITU%20worshop%20jan%20 07%20v2%201+%20FAC%20comments%203.pdf.

[402] NTIA Jan. 4, 2010 *Ex Parte* at 5.

[403] *See* MetroPCS Comments in re NBP PN #6, filed Oct. 23, 2009, at 11–12; MetroPCS Reply in re NBP PN #6, filed Nov. 13, 2009, at 2–8; Sprint Comments in re NBP PN #6, filed Oct. 23, 2009, at 8–12; AT&T Reply in re NBP PN #6, filed Nov. 13, 2009 at 12–13 (filed as AT and T Inc.); CTIA Reply in re NBP PN #6, filed on Nov. 13, 2009, at 28–29; MSTV and NAB Comments in re NBP PN #6, filed Oct. 23, 2009, at 3–4; *but see* New DBSD Satellite Services Reply in re NBP PN #6, filed on Nov. 13, 2009, at 4–7; TerreStar *Ex Parte* Reply in re NBP PN #6, filed on Dec. 8, 2009, Attach. at 1–8; DISH Network and Echostar Corporation Reply in re NBP PN #6, filed on Nov. 13, 2009, at 7 (filed by Dish Network LLC); Satellite Industry Association Comments in re NBP PN #6, filed on Oct. 23, 2009, at 9.

[404] These numbers are current as of the end of third quarter, 2009. S*ee* SkyTerra Commc'ns, Inc., Quarterly Report (Form 10- Q), at 32 (Nov. 9, 2009) (number refers to telephony subscribers only); Inmarsat, Condensed Consolidated Financial Results 3 (Sept. 30, 2009), *available at* http://www.inmarsat.com/Downloads/ English/Investors/IHL_Q_3_2009.pdf (number refers to "active terminals," which Inmarsat describes as "the number of subscribers or terminals that have been used to access commercial services (except certain SPS [satellite phone service] terminals) at any time during the preceding twelve-month period and registered at 30 September [2009]. Active terminals also include the average number of certain SPS terminals . . . active on a daily basis during the period. Active terminals exclude our terminals (Inmarsat D+ and Isat M2M) used to access our Satellite Low Data Rate ("SLDR") or telemetry services."). As of 30 September 2009, Inmarsat had 231,486 SLDR terminals. Inmarsat, Condensed Consolidated Financial Results 3 (Sept. 30, 2009), *available at* http://www.inmarsat.com/ Downloads/English/Investors/IHL_Q_3_2009.pdf; Globalstar, Inc., Quarterly Report (Form 10-Q), at 27 (Nov. 16, 2009); Iridium Commc'ns Inc., Quarterly Report (Form 10-Q), at 37, 40, 43 (Nov. 16, 2009).

[405] In the bands 1544–1545 and 1645.5 –1646.5 MHz, the Mobile Satellite Service is limited to distress and safety communication and is not included in the 40 megahertz count.

[406] *Flexibility for Delivery of Communications by Mobile Satellite Service Providers in the 2 GHz Band, the L-band, and the 1.6/2.4 GHz Band; Review of the Spectrum Sharing Plan Among Non-Geostationary Satellite Orbit Mobile Satellite Service Systems in the 1.6/2.4 GHz Bands*, IB Docket No. 01-185, Report and Order and Notice of Proposed Rulemaking, 18 FCC Rcd 1962, 1964–65, para. 1 (2003).

[407] Globalstar Licensee LLC, filed December 14, 2009, IBFS File No. SAT-MOD-20091214-00152. SkyTerra Subsidiary LLC, filed April 29, 2009, IBFS File Nos. SATMOD-20090429-00046; SAT-MOD-20090429-00047; SES-MOD-20090429-00536.

[408] *See, e.g.*, Infineon, *Technology is Breakthrough for Mass-Market and Feature-Rich Multi-Mode Handsets* (press release), Apr. 1, 2009, http://www. infineon.com/cms/en/corporate/press/news/ releases/2009/INFWLS20090 3-047.html; Letter from Dean R. Brenner, Vice President, Government Affairs, Qualcomm, to Marlene H. Dortch, Secretary, FCC, GN Docket Nos. 09-47, 09-51, 09-137 (Oct. 23, 2009). For example, Globalstar has partnered with Open Range to lease spectrum for the deployment of wireless broadband service in underserved and rural areas using WiMAX technology; TerreStar has partnered with Nokia Siemens Networks to provide mobile broadband coverage in urban areas through a high-speed packet access (HSPA) network and recently announced roaming and distribution deals with AT&T. S*ee* Globalstar, Inc., *Globalstar Becomes The First Mobile Spectrum Satellite Services Authority to Utilize It's ATC Spectrum Authority* (press release), Jan. 12, 2009, http:// www.globalstar.com/en/news/pressreleases/press_ display.php?pressId=522; TerreStar Corp., *TerreStar Announces Nationwide Roaming Agreement with AT&T* (press release), Aug. 1, 2008, http://www.terrestar. com/press/archive/20080801.html; TerreStar Corp., *TerreStar Announces Distribution Agreement with AT&T* (press release), Sept. 30, 2009, http://www.terrestar. com/press/20090930.html.

[409] The 120 megahertz objective is based on the need for additional spectrum allocated to flexible, mobile broadband use outlined earlier in this chapter and on scenario modeling and analysis of the broadcast TV bands. For a more detailed analysis see Omnibus Broadband Initiative, Spectrum Reclamation: Options for Broadcast Spectrum (forthcoming) (OBI, Spectrum Reclamation).

[410] For example, Designated Market Areas (DMAs) with more than 1 million TV homes have a median of 16 full-power stations, while DMAs with fewer than 1 million TV homes have a median of 6. FCC, DTV Station Search, http://licensing.fcc.gov/cdbs/cdbs_docs/pa/ dtvsearch/dtv_search.cfm (last visited Jan. 21, 2010). The FCC is required to allocate channels among States and communities so as to provide a "fair, efficient, and equitable distribution" of service, 47 U.S.C. § 307(b), and should ensure minimum service levels in each market as determined by the rule-making proceeding and pursuant to its § 307(b) mandate.

[411] The 85–90% of U.S. households that subscribe to service through multichannel video programming distributors (MVPDs) pay for the programming that over-the-air television viewers receive for free. These households pay for broadcast network programming through retransmission fees that broadcast TV stations negotiate with

MVPDs—fees that MVPDs then pass on to their customers. SNL Kagan has forecasted total cash retransmission fees for 2009 at $738.7 million. *See* snl kaGan (a divisiOn Of snl financial lc), brOadcast investOr: deals & finance, brOadcast retrans fees On track tO break $1 bil. by 2011 (2009). Moreover, dedicating spectrum to broadcast use imposes on all consumers an implicit "opportunity cost" for that use of the spectrum over other potential uses.

[412] The following market value analysis does not take into account social value or other measures of consumer surplus associated with either over-the-air broadcast TV or mobile broadband use.

[413] *See generally* FCC, Summary for Auction 73 (700 MHz Band), http://wireless.fcc.gov/auctions/ default.htm?job=auction_summary&id=73 (last visited Feb. 20, 2010). Dollars per megahertz of spectrum, per person reached ($ per megahertz-pop) is the convention used to estimate the market value of spectrum. In the 700 MHz auction, $ per megahertz-pop values ranged from $0.03 in Paducah, Ken., Cape Girardeau, Mo., and Harrisburg- Mt. Vernon, Ill. to $3.86 in Philadelphia.

[414] This valuation assumes (1) that the total broadcast television industry enterprise value is $63.7B; (2) that the over-the-air audience is 14–19% of total TV viewership; (3) that the value of over-the-air broadcast television is $8.9–$12.2 billion; (4) that there is 294 megahertz of TV spectrum; and (5) that the United States has a population of 281.4 million people. These figures were calculated as follows. The total broadcast television industry's enterprise value equals industry revenue multiplied by average operating margin and by average EBITDA multiple. *See* BIA/Kelsey, *BIA/ Kelsey Expects TV Station Revenues to End Year Lower Than Anticipated; Levels Last Seen in 1990s Predicted Through 2013* (press release), Dec. 22, 2009, http:// www.bia.com/pr091222-IITV4.asp (BIA/Kelsey, *TV Station Revenues*) (estimating average broadcast television industry revenue to be $17.9 billion (2008 actual and 2009 estimate)). The average operating margin equals 35%, based on the average operating margin from company reports and the SEC filings of Belo Corp., Entravision Communications Corporation, Fischer Communications, Inc., Gannett Company, Gray Television, Hearst Corporation, LIN TV Corp., Nexstar Broadcasting Group, Sinclair Broadcast Group, Univision Communications, Inc., and Young Broadcasting, Inc. *See* U.S. Securities & Exchange Comm'n, EDGAR: Filings & Forms, http://www.sec.gov/ edgar.shtml (last visited Mar. 5, 2010) (U.S. Securities & Exchange Comm'n, EDGAR) (providing access to the filings of publicly held companies). The average EBITDA multiple equals 10.2, based on 2000–2009 monthly averages from the SEC filings of Gray Television, Inc., LIN TV Corp., Nexstar Broadcasting Group, and Sinclair Broadcast Group. *See* U.S. Securities & Exchange Comm'n, EDGAR; Yahoo! Finance, http:// finance.yahoo.com (last visited Mar. 5, 2010). Yahoo Finance was used to identify year-end stock share prices. The over-the-air TV audience is based on a range of estimates. *See* Nielsen Co., National Media Universe Estimate database (accessed Feb. 2010) (estimating 9.7% of viewers are over-the-air only); GAO, Digital Television Transition: Broadcasters' Transition Status, Low - Pow er Station Issues, and Information on Consumer Awareness of the DTV Transition 11, GAO- 08-881T (2008), (estimating 15% of viewers are overthe-air only and finding that ~21% of MVPD households have secondary TV sets that receive signals over-theair). Available at http://www.gao.gov/new.items/ d0888H.pdf. Assuming secondary TV sets are viewed 20% as often as primary sets, the overall over-the-air TV audience equals 9.7–15% plus 4.2%, or 14–19%. The value of over-the-air broadcast television equals the total enterprise value of the broadcast television industry times the over-the-air audience. The amount of TV spectrum equals 294 MHz, as allocated by the FCC. Off. Of enG. & tech. fcc Online table Of freQuency allOcatiOns 17–18, 22, 26 (rev. Jan. 25, 2010) (updating 47 C.F.R. § 2.106), *available at* http://www.fcc.gov/oet/ spectrum/table/fcctable.pdf.

[415] Economist Coleman Bazelon calculated value at $0.15 per megahertz-pop. *See* Consumer Electronics Association Comments in re NBP PN #6, filed Oct. 23, 2009, Attach. at 19.

[416] Nielsen Co., National Media Universe Estimates, Nov. 1998–Feb. 2010 (2010).

[417] BIA/Kelsey, *TV Station Revenues*.

[418] The latest employment figures from the U.S. Census Bureau for broadcast TV show a 0.3% decline in total from 2002 to 2007. *Compare* U.S. Census Bureau, 2002 Economic Census Television Broadcasting Industry Statistics, http://factfinder.census.gov/servlet/ IBQTable?-NAICS1997=513120&-ds_name=EC0251I2 (last visited Mar. 5, 2010), *with* U.S. Census Bureau, 2007 Economic Census Television Broadcasting Industry Statistics, http://factfinder.census.gov/servlet/ IBQTable?-NAICS2007=515120&-ds_name=EC0751I1 (last visited Jan. 21, 2010). Data are not yet available for 2008 or 2009, when the most meaningful declines are likely to have occurred. NAB data indicates a 4.5% decline in industry employment in 2008. *See* Nat'l Ass'n of Broad., NAB Television Financial Report 2 (2008); Nat'l Ass'n of Broad., NAB Television Financial Report 2 (2009).

[419] For example, full-power stations directly use a median of 120 megahertz (20 channels) out of 294 megahertz total in the top 10 DMAs; full-power stations in the most congested DMA, Los Angeles, directly use 156 megahertz (26 channels); across all 210 DMAs, full-power stations directly use a median of 42 megahertz (7 channels). FCC, DTV Station Search, http://licensing.fcc.gov/cdbs/ cdbs_docs/pa/dtvsearch/dtv_search.cfm (last visited Jan. 21, 2010).

[420] The DTV Table of Allotments is predicated on specific TV service areas established by FCC rules. *See* 47 C.F.R. § 73.623(b); *see also* Off. of Eng. & Tech., FCC, *Longley-Rice Methodology for Evaluating TV*

Coverage and Interference (OET Bulletin No. 69, 2004); 47 C.F.R. § 73.623(c)–(d) (establishing rules for required distance separations). TV service areas are defined by theoretical receiver antennas 10 meters off the ground that receive signals at given field strengths 90% of the time, in 50% of locations at the edge of a station's coverage (noise-limited) contour, where its signal is weakest. Stations wishing to establish broadcast operations that violate the allowable service areas or required distance separations must negotiate between themselves and obtain FCC approval.

[421] OBI, Spectrum Reclamation

[422] There are existing television broadcast agreements with Canada and Mexico. If the implementation of recommendations in the plan cause any broadcast TV station bordering on Canada or Mexico to alter its existing station structure (e.g., channel reassignment, relocation, change in transmission parameters), the FCC would need to coordinate these changes with Canada or Mexico.

[423] Data ranges represent upper and lower bounds from public filings and assume current technology; future technologies could reduce the bandwidth required. *See* Hampton Roads Educational Telecommunications Association, Inc. Comments in re NBP PN #26, filed Dec. 22, 2009, at 4; WITF, Inc. Comments in re NBP PN #26, filed Dec. 22, 2009, at 4; Iowa Public Broadcasting Board Comments in re NBP PN #26, filed Dec. 22, 2009, at 4.

[424] Each station may not have sufficient capacity to maintain current HD picture quality if both are transmitting highly complex HD programming simultaneously. Such incidences occur infrequently, however. Obi, spectrum reclamatiOn. Furthermore, any such infrequent incidences would not impact the quality of signals delivered to MVPDs that receive broadcast TV signals through direct fiber or microwave feeds— approximately 50% of cable headends and 27% of DirecTV local collection facilities. Letter from Jane E. Mago, Executive Vice President and General Counsel, Legal and Regulatory Affairs, National Association of Broadcasters, to Blair Levin, Executive Director, OBI, FCC, GN Docket Nos. 09-47, 09-51, 09-137 (Dec. 23, 2009) at 1. Stations have several options to mitigate the potential impact to over-the-air signal quality, including statistical multiplexing, bit grooming, and rate shaping. In addition, stations may be able to achieve at least a 15% improvement in MPEG-2 efficiency through more advanced encoding techniques. *See* mattheW s. GOldman, "it's nOt dead yet!"—mpeG-2 videO cOdinG efficiency imprOvements (2009), *attached to* Letter from Matthew Goldman, Vice President of Technology, TANDBERG Television, part of the Ericsson Group, to Marlene H. Dortch, Secretary, FCC (Jan. 22, 2010) (TANDBERG Jan. 22, 2010 *Ex Parte*); Matthew S. Goldman, "It's Not Dead Yet!"—MPEG-2 Video Coding Efficiency Improvements, Presentation at the Broadcast Engineering Conference (Apr. 22, 2009), *attached to* TANDBERG Jan. 22, 2010 *Ex Parte*.

[425] Letter from Craig Jahelka, Vice President and General Manager, WBOC 16, to Marlene H. Dortch, Secretary, FCC, GN Docket Nos. 09-47, 09-51, 09-137 (Jan. 15, 2010) at 1; *see also* Walt Disney Company Comments in re NBP PN #26, filed Dec. 22, 2009, at 1.

[426] For example, a station that broadcasts sports in HD and another that broadcasts talk shows during the same time period could agree on the best mechanisms to share their bandwidth dynamically to enable each to broadcast signals at certain quality levels, similar to how stations manage bandwidth allocations across multiple video streams today. These arrangements could further mitigate any risk to HD signal quality resulting from reduced bandwidth capacity per station.

[427] *See* 47 U.S.C. § 534.

[428] For example, stations could receive a portion of the proceeds from the megahertz-pops they contributed (megahertz-pops would equal the amount of megahertz contributed multiplied by the station's population coverage). The U.S. Treasury could receive proceeds from the adjacent channels recovered and auctioned as a result of stations clearing the band. In most markets, the number of adjacent channels recovered exceeds the bandwidth directly contributed by stations. *See* Recommendation 5.4, *supra*, for more details on incentive auctions.

[429] Petition for Writ of Certiorari, *Cablevision Sys. Corp. v. FCC*, No. 09-901 (Jan. 27, 2010).

[430] The FCC should continue to recognize that "Congress intended [47 U.S.C. § 307(b)] to check the inevitable economic pressure to concentrate broadcast service in urban areas at the expense of service to smaller communities and rural areas." *Educational Information Corporation For Modification of Noncommercial Educational Station WCPE (FM) Raleigh, North Carolina*, File No. BPED-930125IH, Memorandum Opinion and Order, 12 FCC Rcd 6917, 6920 (1997) (citing *Pasadena Broad. Co. v. FCC*, 555 F.2d 1046, 1049–50 (D.C. Cir. 1975)).

[431] 89.7% of revenue in 2010 for broadcast TV stations is forecast to come from advertising on the primary channel, 4.8% from retransmission consent, 4.4% from Internet, 0.9% from digital sub-channels, and 0.2% from Mobile. Television Bureau of Advertising, A Look at at 2010, at 34 (2009).

[432] 85–90% of the distribution reach of stations comes through MVPDs, and 10–15%comes through over-theair broadcasts. In general, stations with retransmission consent agreements with MVPDs earn more revenue from an MVPD viewer than from an over-the-air viewer—the same advertising revenue from each, but retransmission fee revenue only from the MVPD viewer.

[433] Repacking channels could result in declines in service areas for some stations, due to increased co-channel or adjacent channel interference, and in increases in service areas for others. Channel sharing would require

collocation of signal transmission, which would lead to coverage shifts for the station(s) moving to a new transmission location. In general, these shifts would expand the number of consumers who receive a given station's signal, as stations would choose to consolidate closer to population centers and at transmission facilities with the most favorable coverage attributes. Many broadcasters could also reduce transmission-related operating and capital expenses by sharing facilities. The FCC would have to ensure that shifts as a result of channel repacking or sharing comport with Section 307(b), and should work with affected stations on potential means to mitigate coverage losses, such as low power translators and boosters with off- and on-channel signal repeaters. In addition, the FCC would need to define "acceptable" thresholds for service loss as it did during and after the DTV Transition. In that situation, acceptable thresholds for service loss were 2.0% for evaluating channel and facilities changes during the DTV Transition, 0.1% during the process of stations electing their post-transition channel, and 0.5% for evaluating post-transition channel and facilities changes.

[434] There are several examples of stations multi-casting two HD streams in the broadcast TV market today. There is no universal technical standard for objectively measuring the quality of an HD picture, no HD reporting requirement, and thus no official database of HD streams. OBI, Spectrum Reclamation.. Section 2 (Viability of Channel Sharing for HD Programming).

[435] MSTV and NAB Comments in re NBP PN #26, filed Dec. 23, 2009, at 10. Some broadcasters are seeking to develop new nationwide audiences through airing or syndicating national programming over multicast channels (e.g., Live Well in HD, MHz Worldview, V-me, and ThisTV). Other stations are leasing capacity for ethnic programming or for hybrid broadcast-broadband competitive offerings to MVPD services, such as Sezmi Corporation. Sezmi Corporation Comments in re NBP PN #26, filed Dec. 23, 2009, at 1–2.

[436] Television Bureau of Advertising, A Look at 2010, at 34 (2009).

[437] Harris Corporation Comments in re NBP PN #26, filed Dec. 22, 2009, at 4.

[438] Japan and South Korea have 69 million mobile TV users, or 9 out of every 10 worldwide. Note that the largest subscription service in the world, run by South Korea Telecom's TU Media Corp., is a satellite-delivered service. Broadcasters in these countries, however, have yet to leverage this viewership into sustainable ad revenue to support free-to-air service. John Fletcher, SNL Kagan (a division of SNL Financial LC), Comparing Broadcast Mobile TV Services: Japan, South Korea, Italy, U.S.. (2009). The NAB issued base case projections, forecasting mobile DTV advertising would generate $2 billion in revenues in 2012, of which $1.1 billion would accrue to broadcasters, generating ~$9.1 billion in incremental market value. See Broadcast Engineering, OMVC Concurs with NAB Study; Mobile Digital TV Service Could Generate Billions (2008). A subscription-based domestic mobile broadcast TV service, MediaFlo, using spectrum bought at auction, also has generated varying opinions on the future of the format.

[439] These other mechanisms should also be implemented in a way that preserves minimum acceptable broadcast service levels and protects smaller and rural markets.

[440] For example, full-power stations directly use a median of 120 megahertz (20 channels) out of 294 megahertz total in the top 10 DMAs; full-power stations in the most congested DMA, Los Angeles, directly use 156 megahertz (26 channels); across all 210 DMAs, full-power stations directly use a median of 42 megahertz (7 channels). See FCC, DTV Station Search, http://licensing.fcc.gov/cdbs/ cdbs_docs/pa/dtvsearch/ dtv_search.cfm (last visited Jan. 21, 2010).

[441] Digital Television Distributed Transmission System Technologies, MB Docket No. 05-312, Report and Order, 23 FCC Rcd 16731, 16732, para. 1 (2008). For more information, see CTIA & CEA Comments in re NBP PN #26, filed Dec. 22, 2009, at 9–17.

[442] In an MFN, multiple stations consolidate their capacity and broadcast over different channels at different sites and times, similar to a frequency re-use pattern employed by mobile operators to avoid interference between cell sites. CTB Group, Inc. Comments in re NBP PN #26, filed Dec. 22, 2009, at 4. Letter from Peter Tannenwald, Counsel for CTB Group, Inc., to Marlene H. Dortch, Secretary, FCC, GN Docket No. 09-51, MB Docket No. 05-312, RM 11574 (Jan. 15, 2010) (CTB Group, Inc. Jan. 15, 2010 Ex Parte) at 10. An MFN would require the FCC to grant additional licenses and/or modify existing licenses.

[443] CTIA and CEA estimate the cost to implement this type of architecture at $1.4–$1.8 billion and the amount of spectrum that could be freed at 100–180 megahertz. CTIA & CEA Comments in re NBP PN #26, filed Dec. 22, 2009, at 3.

[444] See Amendment of Parts 21 and 74 of the Commission's Rules with regard to Filing Procedures in the Multipoint Distribution Service and in the Instructional Television Fixed Service; Implementation of Section 309(j) of the Communications Act—Competitive Bidding, MM Docket No. 94-131, PP Docket No. 93-253, Report and Order, 10 FCC Rcd 9589, 9612 (1995); Amendment of the Commission's Rules Regarding Multiple Address Systems, WT Docket No. 97-81, Report and Order, 15 FCC Rcd 11956, 11984 (2000); Amendment of the Commission's Rules Regarding the 37.0–38.6 GHz and 38.6–40.0 GHz Bands; Implementation of Section 309(j) of the Communications Act—Competitive Bidding, 37.0–38.6 GHz and 38.6–40.0 GHz, ET Docket No. 95-183, PP Docket 93-253, Report and Order and Second Notice of Proposed Rulemaking, 12 FCC Rcd 18600, 18637–38 (1997); Auction of Broadband Radio Service (BRS) Licenses Scheduled for October, AU Docket No. 09-56, Public Notice, 24 FCC Rcd 8277, 8288 (WTB 2009).

[445] Stations could clear the overlay license bands by ceasing to broadcast over-the-air or by relocating to another broadcast TV band with or without overlay licenses. As part of the agreement to cease over-the-air broadcasts, stations or overlay license winners could reach private contractual carriage agreements with MVPDs to reach the remaining 85–90% of households. Thomas Hazlett Comments in re NBP PN #26, filed Dec. 18, 2009, at 9. With FCC approval, relocating to another band could involve either occupying another available 6-megahertz channel or sharing a channel with another station.

[446] For example, Auctions 44, 49, and 60 of licenses in the 700 MHz band generated proceeds of $0.03–0.05 per megahertz-pop in 2002, 2003, and 2005, respectively, with these low valuations driven primarily by uncertainty over timing and cost to clear incumbent broadcast TV licensees in that band. Once the DTV Transition timeline was finalized, Auction 73 of similar licenses in the 700 MHz band generated proceeds of $1.28 per megahertz-pop. Auction data available on FCC auction website: FCC, Auctions Home, http://wireless. fcc.gov/auctions/default.htm?job=auctions_home (last visited Feb. 18, 2010). In addition, a holder of licenses from Auctions 44, 49, and 60, Aloha Partners, subsequently sold its licenses to AT&T for $1.06 per megahertz-pop. See Om Malik, *AT&T Buys 700 MHz Spectrum Licenses*, GiGaOm, Oct. 9, 2007, http:// gigaom.com/2007/10/09/att-buys-700-mhz-spectrumlicenses/.

[447] Subject to Congressional input and authorization, the FCC could consider loosening certain public interest obligations on commercial broadcasters as part of a broad review and potential rule-making involving spectrum fees. See Norman Ornstein Reply in re NBP PN #30 (*Reply Comments Sought in Support of National Broadband Plan—NBP Public Notice #30*, GN Docket Nos. 09-47, 09-51, 09-137, Public Notice, 25 FCC Rcd 241 (WCB, rel. Jan. 13, 2010) (*NBP PN #30*)), filed Jan. 20, 2010, at 10–13. The spectrum fees would be in addition to existing annual regulatory fees that broadcast TV stations pay. These regulatory fees vary depending on VHF/UHF placement and market location, ranging from $5,600 to $71,050 for VHF, and from $1,800 to $21,225 for UHF.

[448] Congress did not set a digital conversion date for low power stations when it established the date for full power stations. The FCC concluded that it has such authority in *Amendment of Parts 73 and 74 of the Commission's Rules to Establish Rules for Digital Low Power Television, Television Translator, and Television Booster Stations and to Amend the Rules for Digital Class A Television Stations*, Report & Order, 19 FCC Rcd 19331, 19336–39, paras. 11–19 (2004). Low power stations are licensed spectrum users, but most have secondary spectrum rights to full power stations; "Class A" stations operate at low power but have primary spectrum rights with interference protections.

[449] Since the transition to digital, many VHF stations have reported that some over-the-air viewers have experienced degraded reception due to the impact of environmental radio frequency noise on their digital signal.

[450] Currently, the following bands below 12 GHz are available for point-to-point microwave backhaul, either on a primary basis or secondary to other uses in the band: 3700–4200 MHz (Fixed Satellite—Space to Earth), 5925–6425 MHz (Fixed Satellite—Earth to Space), 6525–6875 MHz (Fixed Satellite—Earth to Space), 10550–10600 MHz (no other services sharing the band), 10600–10680 MHz (Earth Exploration Satellite, Space Research), and 10700–11700 MHz (Fixed Satellite).

[451] For frequencies below 15 GHz, National Spectrum Manager Association guidelines call for coordination within a 125-mile circle around a terrestrial microwave station and within 250 miles for the "keyhole" extending 5 degrees on either side of the main beam azimuth. See Nat'l Spectrum Managers Ass'n, Coordination Contours For Terrestrial Microwave Systems 2, Rec. WG 3.90.026 (2009), *available at* http://www.nsma.org/recommendation/WG3.90.026.pdf (last visited Feb. 18, 2010).

[452] Bands where sharing is currently and potentially viable include 6425–6525 MHz (Mobile Microwave, Broadcast Auxiliary Service (BAS), Cable Television Relay Service (CARS), Mobile Local Television Transmission Service (LTTS), Fixed Satellite Service (FSS)), 6875–7025 MHz (BAS, CARS, LTTS, FSS), 7025–7075 MHz (BAS, CARS, LTTS, FSS), and 7075–7125 MHz (BAS, CARS, LTTS).

[453] Letter from Michael Mulcay, Chairman, Wireless Strategies Inc., to Marlene H. Dortch, Secretary, FCC, GN Docket No. 09-51, WT Docket No. 07-121 (Nov. 4, 2009) at 1; Letter from Richard B. Engelman, Director, Spectrum Resources, Sprint Nextel Corp., to Marlene H. Dortch, Secretary, FCC, WT Docket No. 07-121 (Mar. 12, 2009) at 1–2.

[454] Letter from Mitchell Lazarus, Counsel, Alcatel-Lucent et al., to Marlene H. Dortch, Secretary, FCC, WT Docket No. 09-106 (May 8, 2009) at 3 (requesting interpretation of Section 101.141(a) (3) of the Commission's Rules to Permit the Use of Adaptive Modulation Systems); Fixed Wireless Communications Coalition Comments in re Adaptive Modulation PN (*Wireless Telecommunications Bureau Seeks Comment on Request of Alcatel-Lucent et al. For Interpretation of 47 C.F.R. §101.141 (a) (3) to Permit the Use of Adaptive Modulation Systems*, WT Docket No. 09-106, Public Notice, 24 FCC Rcd 8549 (WTB 2009) (*Adaptive Modulation PN*)), filed July 27, 2009, at 1–2; Fixed Wireless Communications Coalition Reply in re Adaptive Modulation PN, filed Aug. 11, 2009, at 2; Letter from Mitchell Lazarus, Counsel, Fixed Wireless Communications Coalition, to Marlene H. Dortch, Secretary, FCC, WT Docket Nos. 09-106, 09-114 (Oct. 30, 2009), Attach. at 7–9.

[455] Drago nWave Inc., Understanding the Total Cost of Ownership of Wireless Backhaul: Making the Right Choice at the Right Time 12, DWI-AP-190 (2010), *available at* http://www.wcai.com/images/pdf/wp_DragonWave_APP-190.pdf.

[456] *Amendment of Part 101 of the Commission's Rules to Modify Antenna Requirements for the 10.7–11.7 GHz Band*, WT Docket No. 07-54, Report and Order, 22 FCC Rcd 17153, 17161, para. 11 (2007).

[457] Opportunistic sharing techniques allow users to operate at low power simultaneously with incumbent users or during periods when incumbent users are not transmitting on their assigned frequencies.

[458] 47 C.F.R. Part 15.

[459] *See* 47 C.F.R. § 15.205 for a list of the restricted bands in which only spurious emissions are permitted. In many cases, these bands correspond to federal-only allocations that are used for passive spectrum sensing applications.

[460] 47 C.F.R. § 15.5(b).

[461] Public Interest Spectrum Coalition (PISC) Comments in re Wireless Innovation NOI (*Fostering Innovation and Investment in the Wireless Communications Market; A National Broadband Plan For Our Future*, GN Docket Nos. 09-157, 09-51, Notice of Inquiry, 24 FCC Rcd 11322 (2009) (*Wireless Innovation NOI*)), filed Nov. 5, 2009, at 20–25.

[462] NTIA has expressed the need to explore innovative spectrum access models, including opportunistic or dynamic use. *See* Letter from Kathy D. Smith, Chief Counsel, NTIA, to Marlene H. Dortch, Secretary, FCC, GN Docket No. 09-51 (Jan. 4, 2010) at 5.

[463] The ITU-R Study Group 1 has defined a cognitive radio system as a radio system employing technology that allows the system to obtain knowledge of its operational and geographical environment, established policies, and its internal state; to dynamically and autonomously adjust its operational parameters and protocols according to its obtained knowledge in order to achieve predefined objectives; and to learn from the results obtained. In layman's terms, this describes a radio and network that can react and self-adjust to local changes in spectrum use or environmental conditions. Cognitive radio is often confused with software defined radio (SDR). However, while often a cognitive radio will contain an SDR, an SDR does not necessarily imply a cognitive radio.

[464] A few of the more prominent projects are DARPA's neXt Generation Communications (XG) Program, the Federal Spectrum Sharing Innovation Test-Bed Pilot Program, and the European Commission's End-to-End Efficiency (E^3) Project. In April 2007, the IEEE created the IEEE Standards Coordinating Committee 41 (SCC41) on Dynamic Spectrum Access Networks. Finally, the IEEE 802.22 working group is developing a standard for wireless regional area networks for a cognitive radio- based air interface for use by unlicensed devices on a non-interfering basis in TV Broadcast spectrum.

[465] *See, e.g.*, Public Interest Spectrum Coalition Reply in re Wireless Innovation NOI, filed Nov. 5, 2009, at 20–30.

[466] *See* New America Foundation Comments in re National Broadband Plan NOI, filed Jun. 8, 2009, at 24. New America Foundation states that it believes, "the most promising mechanism for making substantial new allocations of spectrum available for wireless broadband deployments and other innovation is to leverage the TV Bands Database.. . ." *Id.*

[467] For the purposes of the Plan, we define "Tribal lands" as any federally recognized Tribe's reservation, pueblo and colony, including former reservations in Oklahoma, Alaska Native regions established pursuant to the Alaska Native Claims Settlement Act (85 Stat. 688), and Indian allotments. The term "Tribe" means any American Indian or Alaska Native Tribe, Band, Nation, Pueblo, Village or Community which is acknowledged by the Federal government to have a government-togovernment relationship with the United States and is eligible for the programs and services established by the United States. See Statement of Policy on Establishing a Government-to-Government Relationship with Indian Tribes, 16 FCC Rcd 4078, 4080 (2000). Thus, "Tribal lands" includes American Indian Reservations and Trust Lands, Tribal Jurisdiction Statistical Areas, Tribal Designated Statistical Areas, and Alaska Native Village Statistical Areas, as well as the communities situated on such lands. This would also include the lands of Native entities receiving Federal acknowledgement or recognition in the future. While Native Hawaiians are not currently members of federally-recognized Tribes, they are intended to be covered by the recommendations of this Plan, as appropriate.

[468] Letter from Native Public Media et al., to Marlene H. Dortch, Secretary, FCC, in re NBP PN #5, Docket Nos. 09-47, 09-51, 09-137 (Dec. 24, 2009) at 7.

[469] *See generally Extending Wireless Telecommunications Services to Tribal Lands*, WT Docket No. 99-266, Report and Order, 15 FCC Rcd. 11794 (2000).

[470] *See Policies to Promote Rural Radio Service and to Streamline Allotment and Assignment Procedures*, MB Docket No. 09-52, First Report and Order and Further Notice of Proposed Rulemaking, FCC 10-24 (rel. Feb. 3, 2010); 47 U.S.C. § 307(b).

[471] To the extent the FCC issues licenses or requires partitioning of licenses for very small tribal areas, however, consideration must be given to whether special technical or coordination rules are necessary in order to facilitate service to the tribal lands while minimizing the potential for interference among neighboring licensees.

[472] *Petition for Declaratory Ruling to Clarify Provisions of Section 332(c)(7)(B) to Ensure Timely Siting Review and to Preempt Under Section 253 State and Local Ordinances that Classify All Wireless Siting Proposals as Requiring a Variance*, WT Docket No. 08-165, Declaratory Ruling, 24 FCC Rcd 13994 (2009).

473 *See* Letter from Judith A. Dumont, Director, Massachusetts Broadband Initiative, to Marlene H. Dortch, Secretary, FCC, GN Docket Nos. 09-47, 09-51, 09-137 (Jan. 8, 2010) (Dumont Jan. 8, 2010 *Ex Parte*) at 2 (noting that permitting requirements and procedures for rights of way, poles, conduits and towers "are key to the efficient and streamlined deployment of broadband," and that difficulties in such access "often prove to be the greatest impediment to the efficient, cost-effective, and timely deployment of broadband.").

474 We derive this estimate from several sources. Omnibus Broadband Initiative, The Broadband Availability Gap.. (forthcoming) *See* Letter from Thomas Jones, Counsel to FiberNet, to Marlene H. Dortch, Secretary, FCC, GN Docket No. 09-51, WC Docket No. 07-245 (Sept. 16, 2009) (FiberNet Sept. 16, 2009 *Ex Parte*) at 20 (noting average cost for access to physical infrastructure of $4,611–$6,487 per mile); *Comment Sought on Cost Estimates for Connecting Anchor Institutions to Fiber—NBP Public Notice #12*, GN Docket Nos. 09-47, 09-51, 09-137, Public Notice, 24 FCC Rcd 12510 (2009) (*NBP PN #12*) App. A (Gates Foundation estimate of $10,500–$21,120 per mile for fiber optic deployment); *see also* Letter from Charles B. Stockdale, Fibertech, to Marlene H. Dortch, Secretary, FCC, GN Docket Nos. 09- 47, 09-51, 09-137 (Oct. 28, 2009) at 1–2 (estimating costs ranging from $3,000–$42,000 per mile).

475 One wireless carrier has cited instances in which it has been asked to pay a rental rate of $1,200–$3,000 per pole per year. S*ee, e.g.*, Letter from T. Scott Thompson, Counsel for NextG Networks, to Marlene H. Dortch, Secretary, FCC, WC Docket No. 07-245, RM-11293, RM11303 (June 27, 2008) Attach. at 11.

476 *See, e.g.*, Am. Cable Ass'n Comments in re National Broadband Plan NOI, filed June 8, 2009, at 8–9; *Amendment of the Commission's Rules and Policies Governing Pole Attachments*, WC Docket No. 07-245, Report and Order, 15 FCC Rcd 6453, 6507–08, para. 118 (2000) ("The Commission has recognized that small systems serve areas that are far less densely populated areas than the areas served by large operators. A small rural operator might serve half of the homes along a road with only 20 homes per mile, but might need 30 poles to reach those 10 subscribers.").

477 This analysis assumes that the customer purchases from an ILEC that rents all of its poles.

478 NCTA Comments in re American Electric Power Service Corp. et al., Petition for Declaratory Ruling that the Telecommunications Rate Applies to Cable System Pole Attachments Used to Provide Interconnected Voice over Internet Protocol Service, WC Docket No. 09-154 (filed Aug. 17, 2009) (Pole Attachments Petition), filed Sept. 24, 2009, App. B at 8–10; Letter from Thomas Jones, Counsel, Time Warner Telecom Inc., to Marlene H. Dortch, Secretary, FCC RM-11293, RM 11303 (Jan. 16, 2007) Attach., US Telecom Comments in re Pole Attachments Petition, filed Sept. 24, 2009, at 8; George S. Ford et al., Phoenix Ctr., The Pricing of Pole Amendment: Implications and Recommendations 7 (2008); Independent Telephone and Telecommunications Alliance (ITTA) Comments in re *implementation of Section 224 of the Act; Amendment of the Commission's Rules and Policies Governing Pole Attachments*, WC Docket No. 07–245, Notice of Proposed Rulemaking, 22 FCC Rcd 20195 (2007) (*Pole Attachments NPRM*), filed Mar. 7, 2008. As Pelcovits notes, monthly cost assumes 35 poles per mile and a 30% take rate. NCTA Comments in re Pole Attachments Petition, filed Sept. 24, 2009, App. B at 14. Additionally, this analysis assumes that all poles are rented by the broadband provider and not owned by it.

479 The variation in rates charged to incumbent LECs also can arise from the history of pole ownership by the incumbent LECs and certain "joint use" agreements that exist between some incumbent LECs and electric utilities.

480 *See, e.g.*, *Nat'l Cable & Telecom. Ass'n v. Gulf Power Co.*, 534 U.S. 327 (2002).

481 *See, e.g.*, *Alabama Power Co. v. FCC*, 311 F.3d 1357 (11th Cir. 2002); *FCC v. Florida Power Corp.*, 480 U.S. 245 (1987).

482 *See, e.g.*, Letter from Daniel L. Brenner, Counsel, Bright House Networks, to Marlene H. Dortch, Secretary, FCC, GN Docket Nos. 09-47, 09-51, 09-137 (Jan. 8, 2010) Attach. at 4; Letter from Daniel L. Brenner, Counsel, Bright House Networks, to Marlene H. Dortch, Secretary, FCC, GN Docket Nos. 09-47, 09-51, 09-137 (Feb. 16, 2010) Attach. (Affidavit of Nick Lenochi) (providing example of how application of higher telecommunications rate for poles would increase expense of deploying Fast Ethernet connections to a large school district by $220,000 annually); NCTA Comments in re Pole Attachments Petition, filed Sept. 24, 2009, at 15–17.

483 tw telecom et al. Comments in re NBP Staff Workshops PN (*The Commission Welcomes Responses to Staff Workshops*, GN Docket No. 09-51, Public Notice, 24 FCC Rcd 11592 (WCB 2009) (*NBP Staff Workshops PN*)), filed Sept. 15, 2009, at 14.

484 FiberNet Sept. 16, 2009 *Ex Parte* Attachs.; Letter from Thomas Jones, Counsel, FiberNet, LLC, to Marlene H. Dortch, Secretary, FCC, WC Docket No. 07-245, GN Docket No. 09-51 (Nov. 16, 2009) (filed by One Communications Corp.) (FiberNet Nov. 16, 2009 *Ex Parte*) at 3 (providing cost estimate breakdown). Similarly, Fibertech reports that it pays pole owners anywhere from $225–$780 to move a single cable on a pole, even though it estimates that it could do the work itself for $60. Fibertech Comments in re NBP PN #12, filed Oct. 26, 2009, at 2–3; *see also* Dumont Jan. 8, 2010 *Ex Parte* at 5–6 (proposing changes to pole attachment regulations so as to "facilitate easier access to existing infrastructure," including reform to the application and make-ready process).

485 FiberNet Nov. 16, 2009 *Ex Parte* Attach. C (providing cost estimate breakdown).

[486] Letter from Kelley A. Shields, Counsel, Fibertech and Kentucky Data Link, Inc. (KDL), to Marlene H. Dortch, Secretary, FCC, GN Docket Nos. 09-51, WC Docket No. 07-25, RM-11293, RM-11303 (Jan. 7, 2009) Attach. 2 at 2.

[487] Letter from Joseph R. Lawhon, Counsel, Georgia Power Co., to Marlene H. Dortch, Secretary, FCC, WC Docket No. 07-245, GN Docket Nos. 09-29, 09-51 (Nov. 17, 2009) Attach. B (noting one example covering 294 poles in Georgia in which the electric utility completed its work within 55 days but in which the process of coordinating with existing attachers took an additional 5 months).

[488] The FCC has already decided that utilities cannot discriminatorily prohibit such techniques when they use those techniques themselves. See *Salsgiver Commc'ns, Inc. v. North Pittsburgh Tel. Co.,* Memorandum Opinion and Order, 22 FCC Rcd 20536, 20543–44 (EB 2007); *Cavalier Tel. v. Virginia Elec. and Power Co.,* Order and Request for Information, 15 FCC Rcd. 9563, 9572 (EB 2000). One provider asserts that rules allowing these practices more generally in Connecticut has allowed it to deploy many more miles of fiber in its Connecticut markets. Fibertech & KDL Comments in re Pole Attachments NPRM, filed Mar. 25, 2009, at 7–8.

[489] Letter from John T. Nakahata, Counsel to Fibertech and KDL, to Marlene H. Dortch, Secretary, FCC, WC Docket No. 07-245, RM 11293, RM 11303, GN Docket Nos. 09- 29, 09-51 (July 29, 2009) at 7.

[490] *Implementation of Section 703(e) of the Telecommunications Act of 1996; Amendment of the Commission's Rules and Policies Governing Pole* Attachments, Report and Order, 13 FCC Rcd 6777, 6787–88, para. 17 (1998) (*1998 Pole Attachment Order*).

[491] *See, e.g.,* Crown Castle Comments in re Pole Attachments NPRM, filed Mar. 11, 2008, at 7 (12 month delay); Sunesys Comments in Petition for Rulemaking of Fibertech Networks, LLC, RM-11303 (Dec. 7, 2005) (Fibertech Petition), filed Jan. 30, 2006, at 11 (15 months); The DAS Forum Comments in re Pole Attachments NPRM, filed Mar. 7, 2008, at 11 (3 years); T-Mobile Comments in re Pole Attachments NPRM, filed Mar. 7, 2008, at 7 (4 years).

[492] *See, e.g.,* Fibertech & KDL Comments in re Pole Attachments NPRM, filed Mar. 25, 2009, at 4 (describing project to construct fiber to three rural school districts in Kentucky that KDL was unable to complete because of pole access delays); *1998 Pole Attachment Order*, 13 FCC Rcd. at 6788, para. 17 (delays in resolving access disputes can "delay a telecommunication's carrier's ability to provide service and unnecessar[ily] obstruct the process").

[493] *Order Adopting Policy Statement on Pole Attachments*, Case 03-M-0432 (New York Pub. Serv. Comm'n 2004) (*New York Timeline Order*) (requiring that all work be completed in 105 days), *available at* http://documents.dps.state.ny.us/public/Common/ViewDoc. aspx?DocRefId={C0C4902C-7B96-4E20-936B-2174CE0621A7}; *Review of the State's Public Service Company Utility Pole Make-Ready Procedures,* Decision, Docket No. 07-02-13 (Conn. Dep't of Pub. Util. Control, Apr. 30, 2008) (*Connecticut Timeline Order*) *available at* http://www.dpuc.state.ct.us/dockhist.nsf/8e6fc37a5411 0e3e85257619005 2b64d/69ccb9118f035bc38525755a 005df44a/$FILE/070213-043008.doc (90 days or 125 days when poles must be replaced).

[494] *See, e.g.,* Fibertech Comments in re NBP PN #12, filed July 21, 2009, Attach. (noting that since implementing timelines, in Connecticut it takes pole owners an average of 89 days to issue licenses and New York pole owners average 100 days for Fibertech's applications, compared to longer intervals elsewhere).

[495] *See, e.g.,* Connecticut *Timeline Order; New York Timeline Order;* Utah Admin. Code § R746-345-3; Vermont Public Service Board, Rules 3.708; *See also Utility Pole Make- Ready Procedures,* Docket No. 07-02-13 (Conn. Dep't of Pub. Util. Control, 2008), *available at* http://www.dpuc. State.ct.us/dockhist.nsf/8e 6fc37a54110e3e8525761900 52b64d/69ccb9118f035bc38525755a005df44a?OpenD ocument; Sunesys Comments in re National Broadband Plan NOI, filed June 8, 2009, at 6 ("By permitting pole owners to have an uncapped and unspecified period of time in which to issue a permit, many pole owners have caused tremendous delays in the process, thereby undermining broadband deployment."); Letter from Jacqueline McCarthy, Counsel, Broadband & Wireless Pole Attachment Coalition, to Marlene H. Dortch, Secretary, FCC, WC Docket No. 07-245 (Feb. 23, 2009) at 1–5.

[496] Wireless providers assert that negotiations with pole owners to attach wireless devices "often face a period of years in negotiating pole agreements." PTIA—The Wireless Infrastructure Association & The DAS Forum Comments in re National Broadband Plan NOI, filed June 8, 2009, at 7. As telecommunications providers, wireless providers have the right to attach to poles under Section 224 of the Act to provide service.

[497] Letter from Joshua Seidemann, Vice President, Regulatory Affairs, ITTA, to Marlene H. Dortch, Secretary, FCC, WC Docket No. 07-245, RM-11293, WC 09-154 (Dec. 22, 2009) (ITTA Dec. 22, 2009 *Ex Parte*) at 3 (noting a pole attachment dispute pending before a state for five years before the parties settled).

[498] *See* 47 C.F.R. §§ 1.1404–1.1410 (pole attachment complaint procedures).

[499] *See, e.g.,* ITTA Dec. 22, 2009 *Ex Parte* at 3 (noting that one provider alone deals with 600 separate entities and that the "lack of uniform rules, standards, and oversight makes negotiating reasonable attachment terms very difficult and extremely time consuming").

[500] Fed. Ministry of Econ. & Tech., Gov't of Germany, The Federal Government's Broadband Strategy 12 (2009), *available at* http://www.bmwi.de/English/Redaktion/ Pdf/broadband-strategy,property=pdf,bereich=bmwi,sp rache=en,rwb=true.pdf.

[501] For example, many pole owners utilize the National Joint Utilities Notification System (NJUNS) for maintaining and communicating data about their pole infrastructure. See generally National Joint Utilities Notification System— NJUNS, Inc., http://www.njuns.com/NJUNS_Home/ default.htm (last visited Mar. 2, 2010).

[502] NCTA Comments in re Pole Attachments Petition, filed Sept. 24, 2009, App. B (Declaration of Dr. Michael D. Pelcovits) Attach. 2 (Methodology and Sources) at 1–3.

[503] Nineteen states and the District of Columbia (representing approximately 45% of the U.S. population) have exercised this type of "reverse preemption" and have certified that they directly regulate utility-owned infrastructure in their regions. See Corrected List of States That Have Certified That They Regulate Pole Attachments, WC Docket No. 07-245, Public Notice, 23 FCC Rcd 4878 (WCB 2008). Section 224(a)(1) expressly excludes poles owned by cooperatives from regulation, an exemption that dates back to 1978. According to the National Rural Electric Cooperative Association, electric co-operatives own approximately 42 million poles. Letter from David Predmore, National Rural Electric Cooperative Association, to Marlene H. Dortch, Secretary, FCC, GN Docket Nos. 09-47, 09-51, 09-137, WC Docket No. 09-245 (Feb. 26, 2010). The exclusion of co-operatives from Section 224 regulation may impede broadband deployment in rural areas. For instance, one small broadband cable company claims that it ceased offering service in two rural communities in Arkansas because of an increase in pole attachment rates by unregulated electric cooperatives that owned the poles in those communities. Letter from Bennett W. Hooks, Jr., Buford Media Group, LLC, to Bernadette McGuire-Rivera, Assoc. Adm'r, Office of Telecom. & Info. Admin., Dep't of Comm. (Apr. 13, 2009) at n.2, 3, available at http://www.ntia.doc.gov/broadbandgrants/ comments/79C5.pdf.

[504] For a review of various approaches to state and local rights of way policies, see NTIA, state and lOcal riGhts Of Way success stOries, available at http://www.ntia. doc.gov/ntiahome/staterow/ROWstatestories.pdf .

[505] In 2003, the NTIA compiled a comprehensive survey of state rights-of-way approaches that may be found at NTIA, Rights-of-Way Laws by State, http://www.ntia. doc.gov/ntiahome/staterow/rowtableexcel.htm (last visited Feb. 18, 2010). In 2002, the National Association of Regulatory Utility Commissions undertook a similar project and issued a comprehensive report. See NARUC, Promoting Broadband Access Through Public Rightsof-Way and Public Lands (July 31, 2002).

[506] See, e.g., Level 3 Comments in re National Broadband Plan NOI, filed Jun. 8, 2009, at 19; Windstream Comments in re National Broadband Plan NOI, filed Jun. 8, 2009, at 2; Verizon Comments in re National Broadband Plan NOI, filed June 8, 2009, at 66; Qwest Comments in re National Broadband Plan NOI, filed June 8, 2009, at 27. Sunesys urges the FCC to "clarify the standards related to timely and reasonably priced access to necessary governmental rights of way." Sunesys Comments in re NBP PN #7 (Comment Sought on the Contribution of Federal, State, Tribal, and Local Government to Broadband—NBP Public Notice #7, GN Docket Nos. 09-47, 09-51, 09-137, Public Notice, 24 FCC Rcd 12110 (WCB 2009) (NBP PN #7)), filed Nov. 6, 2009, at 4.

[507] See, e.g., NATOA et al. Reply in re NBP PN #30, (Reply Comments Sought in Support of National Broadband Plan—NBP Public Notice #30, GN Docket Nos. 09–47, 09–51, 09–137, Public Notice 25 FCC Rcd 241 (2010) (NBP PN #30) filed Jan. 27, 2010, at 12–13; NATOA et al. Comments in re NBP PN #7, filed Nov. 7, 2009, at 46–47; City of New York Comments in re NBP PN #7, filed Nov. 6, 2009, at 8; City and County of San Francisco Comments in re NBP PN #7, filed Nov. 6, 2009, at 16–20. But cf. Dumont Jan. 8, 2010 Ex Parte at 2 (noting that "difficulties involved in negotiating and gaining access to the rights of way often prove to be the greatest impediment to the efficient, cost-effective, and timely deployment of broadband.").

[508] For example, the Broadband Principles adopted by the National Association of Telecommunications Officers and Advisors (NATOA), an organization for local government agencies, staff and public officials, states that "[t]he desired development of high capacity broadband networks and broadband services will require extensive collaboration among parties: local communities, regions, state governments, national government, the private sector, interest groups, and others." NATOA et al. Comments in re National Broadband Plan NOI, filed Jun. 8, 2009, at 3; see also Gary Gordier, CIO and IT Director, El Paso, Texas, Remarks at the FCC State and Local Government Workshop 161 (Sept. 1, 2009) ("There needs to be a lot better coordination across all jurisdictional levels to economize and share jointly in the infrastructure"), available at http://www.broadband.gov/docs/ ws_19_state_andjocal.pdf; Ray Baum, Comm'r, Oregon Pub. Util. Comm'n, Remarks at FCC State and Local Government Workshop 61 (Sept. 1, 2009) ("[W]e have a lot of infrastructure out there owned by utilities[,] both public and private[,] that sitting there that could be better utilized than it is today"); Lori Sherwood, Cable Adm'r, Howard County, Maryland, Remarks at the FCC State and Local Government Workshop 120 (Sept. 1, 2009) ("We have an opportunity to do this right and 25 years from now we don't want to say that we should have done a better job coordinating and talking to each other. For development of a national policy, the FCC should draw on its decade of government experiences including local governance.").

[509] See note 34, supra.

[510] See 47 U.S.C. § 253(c).

[511] A public record search by FCC Staff revealed that since passage of the 1996 Act, the FCC has taken an average of 661 days to resolve Section 253 disputes filed before it, and federal district court litigation of similar

disputes has taken an average of 580 days to conclude. Disputes often extend further through review by courts of appeal, as well.

[512] *See* NATOA et al. Reply in re NBP PN #30, filed Jan. 27, 2010, at 38 (recommending that the FCC "consider creating a special task force" of rights-of-way experts that would "catalog federal, state, and local right-of-way practices and fees in an effort to identify and articulate existing best practices being employed by federal, state, and local authorities for different categories of public rights of way and infrastructure."). As proposed by NATOA, the task force "could also examine and report to the Commission regarding the advantages and disadvantages of alternative forms of compensation for use of public rights of way, and other rights of way related infrastructure, such as poles and conduits." *Id.* at 39.

[513] *See* NATOA et al. Reply in re NBP PN #30, filed Jan. 27, 2010, at 38–39.

[514] Memorandum on Improving Rights-of-Way Management Across Federal Lands to Spur Greater Broadband Deployment, 40 Wkly. Comp. Pres. Doc.. 696 (May 3, 2004).

[515] Memorandum on Facilitating Access to Federal Property for the Siting of Mobile Services Antennas, 31 Wkly. Comp. Pres. Doc.. 1424 (Aug. 10, 1995).

[516] *See* Letter from Thomas Cohen, Counsel for the Fiber to the Home Council, to Marlene H. Dortch, Secretary, FCC, GN Docket No. 09-51 (Oct. 14, 2009).

[517] "Splicing" includes splice kit, installation of splicing enclosure, and splicing of fiber. Splice kit is excluded from "materials" cost. Cost of construction in joint deployment case refers to construction of a single 1-mile, 2" conduit containing 216-count fiber, when coordinated with a road construction project. Additional costs reflect the same project independent of road construction. Letter from Matthew R. Johnson, Legal Fellow, NATOA, to Marlene H. Dortch, Secretary, FCC, GN Docket No. 09-51 (Sept. 17, 2009) (attaching Columbia Telecomm. Corp. Brief Engineering Assessment: Efficiencies Available Through Simultaneous Construction And Co-Location Of Communications Conduit and Fiber tbls. 1, 2 (2009)).

[518] Moratoria on re-opening streets for further telecommunications facilities could impede broadband deployment in certain circumstances.

[519] Dep't of Public Works, City and County of San Francisco, Order No. 176,707 (rvsd): Regulations for Excavating and Restoring Streets in San Francisco § 5 (Mar. 26, 2007), *available at* http://www.sfgov.org/site/uploadedfiles/sfdpw/bsm/sccc/DPW_Order_176-707. pdf; *see also* City and County of San Francisco Department of Public Works, Coordinating Street Construction, http://www.sfgov.org/site/sfdpw_page. asp?id=32429 (last visited Jan. 4, 2010).

[520] Pub. Improvement Comm'n, City of Boston, Policy Relating to Grants of Location for New Conduit Network for the Provision of Commercial Telecommunications Services (Aug. 4, 1988), as amended.

[521] Hardik V. Bhatt, CIO, City of Chicago, Remarks at FCC State and Local Governments: Toolkits and Best Practices Workshop (Sept. 1, 2009), *available at* http:// www.broadband.gov/docs/ws_19_state_and_local.pdf; *see also id.* at 94 ("we have now started knowing every time a street gets dug up either for putting in a traffic signal interconnect, or putting some street light interconnects, or maybe a private utility has dug up the street, we have an opportunity to see if we could leverage that digging up of the street and maybe put conduit or if conduit is there to put fiber there").

[522] Gordon Cook, *Amsterdam's Huge FTTH Build*, Broadband Properties, Sept. 2006, at 68.

[523] NATOA et al. Comments in re NBP PN #7, filed Nov. 9, 2009, App. at 14.

[524] Dumont Jan. 8, 2010 *Ex Parte* at 3.

[525] Dumont Jan. 8, 2010 *Ex Parte* at 4 (recommending "a mechanism to ensure that all U.S. Department of Transportation projects are deploying conduit, and that space is created for four cables").

[526] Dumont Jan. 8, 2010 *Ex Parte*.

[527] United States Department of the Interior, National Atlas of the United States, http://www.nationalatlas.gov/printable/fedlands.html (last visited Jan. 7, 2010).

[528] General Services Administration, GSA Properties Overview, http://www.gsa.gov/Portal/gsa/ep/contentView.do?contentType =GSA_ OVERVIEW&contentId=8513 (last visited Jan. 7, 2010).

[529] Memorandum on Facilitating Access to Federal Property for the Siting of Mobile Services Antennas, 31 Weekly Comp. Pres. Doc. 1424 (Aug. 10, 1995).

[530] *See* Siting Antennas on Federal Property, 41 C.F.R. §§ 102-79.70–.100.

[531] GSA, *GSA's National Antenna Program Wins Vice President Al Gore's Hammer Award Agency's National Antenna Program Fosters Innovation and Saves Tax Dollars, Showing Government Can Work Better and Cost Less*, GSA #9552 (press release), Jan. 13, 1999 (GSA, *GSA's National Antenna Program*), http://www.gsa.gov/ Portal/gsa/ep/contentView.do?contentType=GSA_ BASIC&contentId=9125.

[532] GSA, *GSA's National Antenna Program.* These facts have been confirmed via follow-up e-mails and conversations with GSA.

[533] NTIA, Improving Rights-of-Way Management Across Federal Lands: A Roadmap for Greater Broadband Deployment 31–33, *available at* http://www.ntia.doc. gov/reports/fedrow/frowreport (discussing applicable statutes and agency procedures). For example, the Federal Land Policy Management Act of 1976, which applies to the Department of Interior Bureau of Land Management and National Forest Service, requires that "fair market value, as determined by the Secretary." 43 U.S.C. § 1764(g). In addition, OMB Circular A-25

(rvsd), § 6(a)(2)(b) requires that agencies assess "user charges based on market prices," although exceptions can be granted.

[534] Nat'l Research Council, Innovation in Information Technolog y 5–7 (2003).

[535] *See* Nat'l Research Council, Innovation in Information Technolog y 2–3 (2003).

[536] Case Western Reserve University, *A Smarter Region One Neighborhood at a Time: University Circle Innovation Zone* 2 (University Circle Innovation Zone), http://www.case.edu/its/publication/documents/BetaBlockPublic030210.pdf (last visited Mar. 4, 2010).

[537] University Circle Innovation Zone at 6.

[538] Minnie Ingersoll & James Kelly, *Think Big with a Gig: Our Experimental Fiber Network*, the Official GOOGle blOG, Feb. 10, 2010 (Ingersoll & Kelly, *Think Big with a Gig*), http://googleblog.blogspot.com/2010/02/think-big-with-gig-our-experimental.html.

[539] Cisco, *Cisco and Molina Healthcare Announce Transformative Telemedicine Pilot Program for Underserved and Underinsured Communities* (press release), Jan. 15, 2010, http://newsroom.cisco.com/dlls/2010/prod_011510b.html.

[540] Bronwyn H. Hall et al., *Measuring the Returns to R&D* (Nat'l Bur. of Econ. Res. Working Paper No. 16522, 2009), *available at* http://www.nber.org/papers/w15622 (requires purchase).

[541] David B. Audretsch & Maryann Feldman, *R&D Spillovers and the Geography of Innovation and Production*, 86 am. ecOn. rev. 630 (1996).

[542] A recent study prepared for the Technology Administration of the Department of Commerce noted the "persuasive research that shows that innovation drives economic growth and that the private sector will tend to underinvest in R&D, as the social value for innovation will outstrip private value." George S. Ford et al., Valley of Death in the Innovation Sequence: An Economic Investigation 2 (2007) (Ford et al., Valley of Death) 2 (2007) (fOrd et al., valley Of death), *available at* http://www.ntis.gov/ pdf/ValleyofDeathFinal.pdf. However, diffusion of basic research discoveries is not automatic—the study notes that government R&D efforts must be cognizant of and overcome "the roadblocks that may exist in the innovation process between basic research and commercialization." *Id.*

[543] Ford et al., Valley of Death at 11–14. This seminal insight was first provided by Nobel Laureate economist Kenneth J. Arrow. Kenneth J. Arrow, *Economic Welfare and the Allocation of Resources for Invention*, in the The Rate and Direction of Inventive Activity 609–25 (1962); *see also* Stephen Martin & John T. Scott, *The Nature of Innovation and Market Failure and the Design of Public Support for Private Innovation*, 29 res. pOl'y 437, 438 (2000); Scott Wallsten, *The Effects of Government- Industry R&D Programs on Private R&D: The Case of the Small Business Innovation Research Program*, 31 rand j. ecOn. 82 (2000).

[544] *See, e.g.*, Bronwyn Hall, *The Private and Social Returns to Research and Development: What Have We Learned?*, in technOlOGy, r&d, and the ecOnOmy 140 (L.R. Smith & Claude E. Barfield eds., 1996); Paul David et al., *Is Public R&D a Complement or Substitute for Private R&D? A Review of the Econometric Evidence*, 29 res. pOl'y 497 (2000).

[545] *See* Office of Sci. & Tech. Pol'y, Exec. Office of the Pres., A Strategy For American Innovation: Driving Tow ards Sustainable Grow th And Quality Jobs 1 (2009), *available at* http://www.whitehouse.go v/sites/default/files/ microsites/20090920-innovation-whitepaper.PDF.

[546] *See* Nat'l Res. Council, Renewing U.S. Telecommunications 23 (2006), *available at* http://books.nap.edu/openbook.php?record_ id=11711&page=23.

[547] *See, e.g.*, Adam Drobot, CTO & Pres., Advanced Tech. Solutions, Telcordia Techs., Remarks at FCC Research Recommendations for the Broadband Task Force Workshop (Nov. 23, 2009), *available at* http://broadband.gov/docs/ws_research_bb/ws_research_bb_ transcript.pdf.

[548] *See, e.g.*, David Clark, Senior Research Scientist, MIT, Remarks at FCC Research Recommendations for the Broadband Task Force Workshop (Nov. 23, 2009), *available at* http://broadband.gov/docs/ws_research_bb/ws_research_bb_transcript.pdf.

[549] Bronwyn Hall, *R&D Tax Policy During the Eighties: Success or Failure?* (NBER Working Paper No. 4240, 1993). Nat'l Bur. of Econ. Res.

[550] Kenneth J. Klassen et al., *A Cross-National Comparison of R&D Expenditure Decisions: Tax Incentives and Financial Constraints*. 21 cOntemp. acct. res. 639 (2003).

[551] Statement of Mr. Wayne Arny, Deputy Undersecretary of Defense (Installations and Environment) Before the Subcommittee on Military Construction, Veterans Affairs, and Related Agencies of the House Appropriations Committee (May 19, 2009), at 2.

[552] As noted in the 2007 Defense Installations Strategic Plan, this support is a "long-term, day-to-day commitment to deliver quality training, modern and well-maintained weapons and equipment, a safe, secure and productive workplace, a healthy environment, and good living conditions" for service personnel and their families. U.S. Department of Defense, *2007 Defense Installations Strategic Plan*, 10 (2007), available at: http://www.acq.osd.mil/ie/download/DISP2007_final. pdf.

[553] Department of Defense, Facilities and Vehicles Energy Use, Strategies, and Goals, May 11, 2009.

[554] Also known as "islanding," micro-grids are the concept of a base being able to disconnect from the grid and operate using only local renewable power and other on-base generation.

[555] *See* The National Academies, About The National Academies, http://www.nationalacademies.org/about (last visited Feb. 18, 2010).

[556] *See* The White House, Office of Science and Technology Policy: About OSTP, http://www.whitehouse.gov/administration/eop/ostp/about (last visited Feb. 18, 2010).

[557] *See* FCC Research Recommendations for the Broadband Taskforce Workshop (Nov. 23, 2009), *available at* http://www.broadband.gov/docs/ws_research_bb/ws_research_bb_transcript.pdf.

[558] *See* Charles Bostian, Alumni Distinguished Professor, Virginia Polytechnic Institute and State University, Remarks at FCC Research Recommendations for the Broadband Task Force Workshop (Nov. 23, 2009), *available at* http://broadband.gov/docs/ws_research_bb/ws_research_bb_transcript.pdf.

[559] Note that total funding for the individual ERCs from all sources in 2009 ranged from $4.1 to $8.8 million Nat'l Science Found., Engineering Research Centers: Linking Discovery to Innovation (2009), *available at* http://www.erc-assoc.org/factsheets/ERC%20Overview%20Fact%20Sheet_09-final.pdf.

[560] *See* Internet2 Home, http://www.internet2.edu/ (last visited Mar. 4, 2010); LambdaRail Home, http://www.nlr.net/ (last visited Mar. 4, 2010) ("National LambdaRail (NLR) is the innovation network for research and education. NLR's 12,000 mile, nationwide, advanced optical network infrastructure supports many of the world's most demanding scientific and network research projects."). For a description of a number of research and education networks in the United States, see U.S. R&E Networks Comments in re NBP PN # 22, (Comment Sought on research Necessary for Broadband Leadership—NBP PN #22, GN Docket Nos. 09–47, 09–51, 09–137, Public Notice, 24 FCC Rcd 13820 (2009) (NBP PN #22)) filed Dec. 8, 2009, at 2–10 (describing Internet2, NLR, CENIC, FLR, GPN, GlobalNOC, MAX, MCNC/NCREN, MCAN, NYSERNet, OARnet, OSHEAN, PNWGP, The Quilt, 3ROX, and UEN); SURFnet, About SURFnet: Mission, http://www.surfnet.nl/en/organisatie/Pages/Mission.aspx (last visited Mar. 4, 2010) ("It is SURFnet's mission to facilitate groundbreaking education and research through innovative network services. SURFnet combines the demand of the institutions connected to SURFnet. In doing so we create advantages of scale, innovation and collaboration from which they benefit. The SURFnet network services comprise five focus areas: Network infrastructure, Security, Authentication & authorisation, Group communication and Multimedia distribution.").

[561] *See* Digital Connections Council, Comm. for Econ. Dev., Harnessing Openness to Improve Research, Teaching, and Leaning in Higher Education (2009).

[562] TIA states that "[s]trengthening the robustness and resilience of our broadband networks is necessary not only to protect against attacks, but also to reduce the current drag on productivity caused by malware and attacks." Letter from Carolyn Holmes Lee, Dir., Legis. & Gov't Aff., TIA, to Marlene H. Dortch, Secretary, FCC, GN Docket Nos. 09-47, 09-51, 09-137 (Dec. 18, 2009), App. at 2; *see also* Subcomm. on Netwo rking & Info. Tech. Res. & Dev., Nat'l Sci. & Tech.Council, The Information Technolog y Research and Development Prog ram: Supplement to the President's Budget for Fiscal Year 2010, at 6–9 (2009).

[563] *See* 47 C.F.R. § 5.93 (2008). These limitations affect the size and scope of the marketing trial, as well as restrict ownership of equipment used in the trial to the licensee.

[564] *See Fostering Innovation and Investment in the Wireless Communications Market; A National Broadband Plan For Our Future*, GN Docket Nos. 09-51, 09-157, Notice of Inquiry, 24 FCC Rcd 11322 (2009).

[565] *See generally* Susan Mayer, What Money Can't Buy: Family Income and Children's Life Chances (1997).

[566] John Horrigan, *Broadband Adoption and Use in America* (OBI Working Paper No. 1, 2010); Omnibus Broadband Initiative, The Broadband Availability Gap (forthcoming). *See* U.S. Census Bureau, USA, http://quickfacts.census.gov/qfd/states/00000.html (last visited Feb. 26, 2010) (providing general population numbers).

[567] Toby Bell, Gartner Res., Success Factors Emerge from e-Forms Engagement for U.S. Army 3 (2008) ("The Army estimates that moving nearly 2,400 forms online will save $1.3 billion each year."). (The National Broadband Plan contains several references to Gartner. The Gartner Report(s) described herein, (the "Gartner Report(s)") represent(s) data, research opinion or viewpoints published, as part of a syndicated subscription service, by Gartner, Inc. ("Gartner"), and are not representations of fact. Each Gartner Report speaks as of its original publication date and the opinions expressed in the Gartner Report(s) are subject to change without notice.) irs, advancinG E-FilE study: phase 1 report—Executive Summary, v1.3, Case No. 08-1063, Doc. No. 0206.0209, at 13 (2008), *available at* http://www.irs.gov/pub/irs-utl/irs_advancing_e-file_study_phase_1_executive_summary_v1_3.pdf; Jill R. Aitoro, *IRS Continues to Pay Millions to Process Paper Tax Returns*, nextGOv, Sept. 23, 2009, http://www. nextgov.com/nextgov/ng_20090923_7490.php.

[568] Here, "access" refers only to the capability of the last- mile network. Service providers may, for any number of reasons, make only lower-speed services available to customers—in other words, the speeds or products to which consumers have access may not fully reflect network capabilities. Because access networks are the most capital-intensive elements of the broadband infrastructure, it is reasonable to expect that providers will meet demand for higher speeds once the access network is capable of supporting such speeds.

[569] For purposes of the plan, "actual speed" refers to the data throughput delivered between the network interface unit (NIU) located at the end-user's premises and the service provider Internet gateway that is the shortest administrative distance from that NIU. In the future, the technical definition of "actual speed" should be crafted by the FCC, with input from consumer groups, industry and other technical experts as is proposed in

Chapter 4. The technical definition should include precisely defined metrics to promote clarity and shared understanding among stakeholders. For example, "actual download speeds of at least 4 Mbps" may require certain achievable download speeds over a given time period. Acceptable quality of service should be defined by the FCC. See *supra* Chapter 4 (Transparency Section).

[570] In the first half of 2009, the median actual speed for those that subscribe to broadband in the United States was 3 Mbps download speed. comScore, Inc., Jan.–June 2009Consumer Usage database (sampling 200,000 machines for user Web surfing habits) (on file with the Commission) (comScore database). Given past annual growth rates in subscribed speed of approximately 20–25% per year, we expect the median to exceed 4 Mbps by the end of 2010. *Cf.* akamai, the state Of the internet, 3rd Quarter, 2009, at 10 (2010) (finding median download speeds to be 3.9 Mbps in the third quarter of 2009), *available at* http://www.akamai. com/dl/whitepapers/Akamai_State_Internet_Q3_2009. pdf?curl=/dl/whitepapers/Akamai_State_Internet_ Q3_2009.pdf&solcheck=1& (registration required); *see also* Omnibus Broadband Initiative, Broadband Performance (forthcoming) (discussing past growth rates).

[571] Countries use different incentive policies for "universalizing" speeds. For instance, Canada awards funding for rural build-out above 1.5 Mbps actual speeds, while Finland has mandated that incumbent providers deliver a minimum of 0.5–1.0 Mbps actual download speeds (varying by time of day) to all citizens. Gov't of Australia, Dep't of Broadband, Commc'ns & the Digital Econ., Australian Broadband Guarantee—Frequently Asked Questions, http://www.dbcde.gov.au/__data/ assets/pdf_file/0017/114281/ABG_FAQ-lowres.pdf (last visited Mar. 7, 2010) (particularly "speeds of at least 512 kbps download and 128 kbps upload, at least 3 GB monthly download limits, and a price of no more than $2500 (including GST) over a three year period, including all connection and equipment cost"); Gov't of Australia—Prime Minister of Australia, *New National Broadband Network* (press release), Apr. 7, 2009, http:// www.pm.gov.au/node/5233 (last visited Mar. 7, 2010) (specifically "[c]onnect 90 percent of all Australian homes, schools and workplaces with broadband services with speeds up to 100 megabits per second—100 times faster than those currently used by many households and businesses; Connect all other premises in Australia with next generation wireless and satellite technologies that will be deliver broadband speeds of 12 megabits per second"); Danish Gov't, Annual Broadband Mapping 2009, at 6 (2009) (English translation) ("The Government's target is for all Danes to have broadband access by the end of 2010 at the latest"), *available at* http://en.itst.dk/the-governments-it-and-telecommunications-policy/it-and-telecommunications-policy-reports/filarkiv/IT_and_Telecommunications_ Policy_Report_2009.pdf; danish Gov't, annual broadband mappinG 2009, at 6 (2009) (Danish) (referencing the measurement threshold for broadband as 512 kbit/s set in bilateral agreement between service providers and the government), *available at* http:// www.itst.dk/statistik/Telestatistik/Bredbandstatistik/ bredbandskortlegning-1/bredbandskortlegning-2009/ Bredbandskortlegning%202009.pdf; Ministry of Transp.& Commc'ns, Gov't of Finland, Making Broadband Available to Everyone 2–4 (2008) (English Version) particularly "[t]he report proposes that the public sector introduce business subsidies to enterprises that upgrade the public telecommunications network into a condition that makes available to most all citizens by 2015 an optical fiber or cable network supporting 100 Mbit connections. Prior to this goal, the speed of the broadband connection included in the universal service obligation must be raised to an average of 1 Mbit/s by the end of 2010 at the latest" with 100 Mbps target set to be delivered within 2 kilometers of all households), *available* at http://www.lvm.fi/c/document_library/get_file?folderId=57092&name=D LFE-4311.pdf; Éric bessOn, secRÉtariat d'État charGÉ de la prOspective, GOv't Of france, de Éric Besson, Secrétariat D'état Chargé De La Prospective, Gov't of France, De L'évaluation Des Politiques Publiques Et Du Développement De L'économie Numérique, Plan de développement de l'économie numérique 4 (2008) (French), *available at* http://lesrapports.ladocumentationfrancaise.fr/ BRP/084000664/0000.pdf; *see also* eurOpean cOmm'n, prOGress repOrt On the sinGle eurOpean electrOnic cOmmunicatiOns market 2008, 14th repOrt 4 (English) ("The Plan announced the launch of a call for tenders in the first half of 2009, for designating the provider that would ensure that service (a minimum of 512 kb/s) at an affordable price (35 euros/month) to all."), *available at* http://ec.europa.eu/information_society/ policy/ecomm/doc/implementation_enforcement/ annualreports/14threport/fr.pdf; Éric Besson, Secrétariat D'état Chargé De La Prospective, Gov't of France, De L'évaluation Des Politiques Publiques Et Du Développement De L'économie Numérique, Plan de développement de l'économie numérique 4 8 (2009) ("Gaps in broadband penetration are to be eliminated and capable broadband access made available nationwide by the end of 2010. . . . [Capable broadband connections] are currently defined as having transmission rates of at least 1MBit/s"; "A total of 75 percent of households are to have Internet access with transmission rates of at least 50MB/sec by 2014."), *available at* http://www.bmwi.de/English/Navigation/ Service/publications,did=294718.html; Gov't of Ireland, Dep't of Commc'ns, Energy, and Natural Resources, NBS Frequently Asked Questions, http://www.dcenr.gov.ie/ Communications/Communications+Development/ NBS+FAQs (last visited Mar. 7, 2010) (referencing Plan of December 23, 2008, particularly "3, the [National Broadband Scheme] Service Provider, will extend its network to provide mobile wireless broadband services into the NBS area. The mobile broadband service (I-HSPA) will have a minimum download speed of 1.2Mbps and a minimum upload speed of 200kpbs with a contention ratio of 36:1. In recognition of the fact that some areas will be very costly and difficult to reach, in a very limited number of cases, 3 will make available a satellite product of 1Mbps download and 128kbps

upload. This will cover up to a maximum of 8% of fixed residences and businesses in the NBS coverage area.... An uncharged monthly data cap of 15GB (12GB download and 3GB upload) will apply for the wireless product while 11GB (10GB download and 1GB upload) will be available for satellite users"); ministry Of internal aff. & cOmmc'ns, GOv't Of japan, diGital divide eliminatiOn strateGy 1 (2008) (Japanese, staff translation) (calling for elimination of all areas not served by broadband by 2010, and ultra high speed broadband coverage for 90% of households by 2010), *available at* http://www.soumu.go.jp/ menu_news/s-news/2008/pdf/080624_3_bt2.pdf. Also, note the inclusion of targets for fixed and mobile class infrastructures. S*ee* it strateGy headQuarters, GOv't Of japan, i-japan strateGy 2015, at 26 (2009) (English translation) ("The following measures will be carried out by 2015.. . further advances in ultra-high-speed broadband infrastructure will be made (in the Gbps class for fixed and in excess of 100 Mbps [class] for mobile) to allow everyone to easily obtain and exchange information safe[l]y and securely from anywhere at anytime."), *available at* http://www.kantei.go.jp/foreign/ policy/it/i-JapanStrategy2015_full.pdf; Letter from Young Kyu Noh, Minister Counselor of Broad. & ICT, Embassy of the Republic of Korea, to Marlene H. Dortch, Secretary, FCC, GN Docket Nos. 09-47, 09-51, 09-137 (Feb. 3, 2010) Attach. at 3, 6 (The 1.5–2M[bps] class high-speed network was completely established in 2008 with a goal of minimum 50Mbps to 95% of households by 2013; also shows that Korea served 99% of population with 1Mbps service by 2008.); Korean Commc'ns Comm'n, *Korean Internet Speeds to Be Ten Times Faster by 2012* (press release) (Mar. 28, 2009) (noting that 1Gbps is not an established download minimum for a percentage of the population at this time), *available at* http://eng.kcc.go.kr/user.do?mode=view&page=E040 10000&dc=E04010000 &boardId=1058&cp=1&search Key=ALL&searchVal=broadband+&boardSeq=15621; ministry Of enter., enerGy and cOmmc'ns, GOv't Of sWed., brOadband strateGy fOr sWeden 15 (2009) (particularly "In 2020... 90 per cent of all households and businesses have access to broadband at a minimum speed of 100 Mbps... . In 2015. . . 40 per cent of all households and businesses have access broadband at a minimum speed of 100 Mbps"), *available at* http:// www.sweden.gov.se/content/1/c6/13/49/80/112394be. pdf; dep't fOr culture, media and spOrts , GOv't Of the u.k., diGital britain 12 (2009) (particularly "[t]o ensure all can access and benefit from the network of today, we confirm our intention to deliver the Universal Service Broadband Commitment at 2Mbps by 2012"), *available at* http://www.culture.gov.uk/images/publications/ digitalbritain-finalreport-jun09.pdf.

[572] Section 254(c) (1)requires the FCC to establish periodically the definition of universal service that is supported by federal USF.

[573] Housing units are distinct from households. "A housing unit is a house, an apartment, a mobile home, a group of rooms, or a single room that is occupied (or if vacant, is intended for occupancy) as separate living quarters." In contrast, "A household includes all the persons who occupy a housing unit. . . . The occupants may be a single family, one person living alone, two or more families living together, or any other group of related or unrelated persons who share living arrangements." There are 130.1 million housing units and 118.0 million households in the United States. U.S. Census Bureau, Households, Persons Per Household, and Households with Individuals Under 18 Years, 2000, http:// quickfacts.census.gov/qfd/meta/long_71061.htm (last visited Mar. 7, 2010); Omnibus brOadband initiative, the brOadband availability Gap (forthcoming) (OBI, The Broadband Availability Gap).

[574] *See* OBI, The Broadband Availability Gap. Seven million housing units without access to 4 Mbps service are outside the cable footprint and are more than approximately 11,000–12,000 feet from the nearest DSLAM location. An FCC estimate shows that 12 million people in six million housing units do not have access to terrestrial broadband capable of 768 kbps actual download speeds; those 6 million housing units without access to any always-on service are more than approximately 16,000 feet from the nearest DSLAM.

[575] *See* OBI, The Broadband Availability Gap.

[576] The analysis depends on a variety of data sources. *See* OBI, The Broadband Availability Gap. Where the quality of data is limited, broadband-gap calculations will be affected. For example, there are 12 wire centers in Alaska that show no population within their boundaries, and an additional 18 wire centers that have no paved public-use roads (i.e., no roads other than 4WD or forest-service roads). All 30 of these wirecenters were excluded from wired broadband-gap calculations; however, all areas with population were covered by the wireless calculations. In addition, due to insufficient demographic and infrastructure data for Puerto Rico and the U.S. Virgin Islands in the Caribbean and Guam, American Samoa, and the Northern Marianas in the Pacific to calculate baseline availability, the broadband availability gap for these territories is not included.

[577] The estimate includes capital expenditure and 20 years of operating expenditure and revenue. All calculations use an annual discount rate of 11.25%. The calculation of the broadband availability gap does not include the cost of spectrum. Recent 700 MHz auctions in the A, B, C and E blocks had mean prices between $0.74 and $2.65 per MHz-POP, including a top price for a market of over $9.00 per MHz-POP; median prices for these same auctions were between $0.20 and $0.42 per MHz-POP. At $1.00 per MHz-POP, well above the median price of recent auctions, the cost of 40 megahertz of spectrum for serving 14 million unserved people would be $0.56 billion. See OBI, The Broadband Availability Gapfor more detail about the financial model and how it functions.

[578] Numbers may not add to 100% due to rounding.

[579] For more information about satellite broadband, see Obi, the brOadband availability Gap.

[580] Northern Sky Research, How Much HTS Capacity is Enough?, http://www.talksatellite.com/Americas-A781. htm (last visited Mar. 6, 2010).

[581] *See* OBI, The Broadband Availability Gap; American Roamer, Verizon Wireless 3g Coverage Area (2009); Robert C. Atkinson & Ivy E. Schultz, Columbia Inst. for Tele-Information, Broadband In America: Where It Is And Where It Is Going (According To Broadband Service Providers) 40 (2009) (Atkinson & Schultz, Broadband in America).

[582] *See* OBI, The Broadband Availability Gapp.

[583] *See* OBI, The Broadband Availability Gap.

[584] "Annual funding amount" refers to fiscal year 2008 funding for all programs except BTOP and BIP, which were one-time programs funded by the American Recovery and Reinvestment Act of 2009, and for the Universal Service Fund, which uses FY 2010 projected total outlays to ensure consistency with the rest of the document. The estimate of $2.5 billion under BTOP for infrastructure includes the $119 million in grants already awarded, plus the $2.35 billion announced in the January 2010 NOFA. GAO, Broadband Deployment Plan Should Include Performance Goals and Measures to Guide Federal Investment 13–14, GAO-09-494 (2009) (chart is modified from figure in this source), *available at* http://www.gao.gov/new.items/d09494.pdf; Broadband USA, The Portal To Apply for Broadband Funding under the American Recovery and Reinvestment Act of 2009, http://www.broadbandusa.gov (last visited Mar. 7, 2010); NTIA, *Commerce Department's NTIA and USDA's RUS Announce Availability of $4.8 Billion in Recovery Act Funding to Bring Broadband to More Americans* (press release), http://www.ntia.doc.gov/press/2010/ BTOP_BIP_NOFAII_100115.html (last visited Mar. 7, 2010); The White House, *Vice President Biden Kicks Off $7.2 Billion Recovery Act Broadband Program* (press release), http://www.whitehouse.gov/the-press-office/ vice-president-biden-kicks-72-billion-recovery-actbroadband-program (last visited Feb. 20, 2010); NTIA, Broadband Technology Opportunities Program Key Revisions in Second Notice of Funds Availability, http://www.ntia.doc.gov/press/2010/BTOP_NOFAII_ FACTSHEET_100115.pdf (last visited Mar. 7, 2010).

[585] "Other programs" include the Rural Utilities Service's Distance Learning and Telemedicine Loans and Grants Program and Community Connect Grant Program, the Appalachian Regional Commission's Telecommunications Initiative, the Economic Development Administration's program for Economic Development Facilities and Public Works, and the Delta Regional Authority's program for Delta Area Economic Development.

[586] Notice of Funds Availability for Broadband Initiatives Program and Broadband Technology Opportunities Program, 74 Fed. Reg. 33, 104 (July 9, 2009).

[587] Notice of Funds Availability for Broadband Initiatives Program and Broadband Technology Opportunities Program, 74 Fed. Reg. 33, 104 (July 9, 2009).

[588] NTIA, ION Upstate New York Rural Broadband Initiative Grant Award, http://www.ntia.doc.gov/broadbandgrants/ BTOPAward_IONHoldCoLLC_121709.pdf (last visited Feb. 20, 2010); NTIA, Project Connect South Dakota Grant Award, http://www.ntia.doc.gov/broadbandgrants/ BTOPAward_SDakotaNetwork_121709.pdf (last visited Feb. 20, 2010).

[589] 47 U.S.C. § 151.

[590] J.M. Bauer et al., *Whither Broadband Policy* (30th Annual Telecomms. Policy Research Conf. Paper, 2002), *available at* http://tprc.org/papers/2002/72/ Broadband_v1.pdf.

[591] The FCC has relied on the statutory language in section 254(h) to support internet access for schools, libraries and health care providers.

[592] Universal Serv. Admin. Co., Universal Service Fund, http://www.usac.org/about/universal-service/ (last visited Mar. 7, 2010). The estimated annual projected outlay for the federal USF can be found in the FY 2010 Federal budget. Office Of mGmt. & budGet, exec. Office Of the president, budGet Of the united states GOvernment, fiscal year 2010, at 1220 (2010), *available at* http://www.whitehouse.gov/omb/budget/fy2010/ assets/oia.pdf.

[593] While the E-rate program is capped by FCC regulation at $2.25 billion annually, unused funds from prior funding years may be rolled over to the future, enabling the FCC to disburse more than the annual cap in a given year. In addition, in a given year, the FCC may disburse more than the cap when invoices for funding commitments from prior years are presented for payment.

[594] Universal Serv. Admin. Co., Universal Service Fund, http://www.usac.org/about/universal-service/ (last visited Mar. 7, 2010). FCC total outlay estimates for FY 2010 submitted to OMB on December 15, 2009 based on Universal Service Administrative Company projections. S*ee* usac, federal universal service suppOrt mechanisms fund size prOjectiOns fOr secOnd Quarter 2010, at 2 (2010), *available at* http://www.universalservice.org/about/governance/   fcc-filings/2010/Q2/2Q2010%20Quarterly%20 Demand%20Filing.pdf.

[595] Peter Bluhm, et al. S*tate High Cost Funds: Purposes, Design, and Evaluation* 60 (Nat'l Regulatory Res. Inst. (NRRI), Working Paper No. 10-04, 2010), *available at* http://www.nrri.org/pubs/telecommunications/NRRI_ state_high_cost_funds_jan10-04.pdf. (Bluhm et al., *State High Cost Funds*); Public Utility Commission of Texas, Texas Universal Service Fund, http://puc.state.tx.us/ocp. telephone/choice/txunivserv.cfm (more recent data for Texas) (last visited Feb. 20, 2010).

[596] *See* Jing Liu & Edwin Rosenberg, *State Universal Service Funding Mechanisms: Results of the NRRI's 2005–2006 Survey* 43, 54 (NRRI, Working Paper No. 06-09, 2006), *available at* http://nrri.org/pubs/telecommunications/06-09.pdf (Liu & Rosenberg, *State Universal Service Funding Mechanisms*); Alliance for Pub. Tech. & Commc'ns Workers of Am., State Broadband Initiatives 3 (2009), *available at* http://www.apt.org/publications/reports-studies/state_ broadband_initiatives.pdf.

[597] Not all of these programs are administered by the state public utility commission. Bluhm et al. S*tate High Cost Funds* at 32. Examples of funding programs to support the build-out of advanced networks in unserved and underserved areas include the California Advanced Services Fund, ConnectME Authority, Illinois Technology Revolving Loan Program, Idaho Rural Broadband Investment Program (IRBIP), Louisiana Delta Development Initiative, and Massachusetts Broadband Initiative. S*ee* Alliance for Pub. Tech. & Commc'ns Workers of Am., State Broadband Initiatives 3, 47–49 (2009), *available at* http://www.apt.org/ publications/reports-studies/state_broadband_ initiatives.pdf.

[598] Alliance for Pub. Tech. & Commc'ns Workers of Am., State Broadband Initiatives 3, 44–56 (2009), *available at* http://www.apt.org/publications/reports-studies/ state_broadband_initiatives.pdf.

[599] Although several commenters submitted estimates into the record, not all commenters specified whether figures represented a percentage of total revenues or regulated revenues. S*ee* Western Telecommunications Alliance Comments in re NBP PN #19 (*Comment Sought on the Role of the Universal Service Fund and Intercarrier Compensation in the National Broadband Plan*, GN Docket No. 09-47, 09-51, 09-137, Public Notice, 24 FCC Rcd 13757 (WCB 2009) (*NBP PN #19*)), filed Dec. 7, 2009, at 25, 27 (stating that for small rural LECs, high cost represents 30–40% of regulated revenues, while intercarrier compensation represents 30–40% of regulated revenues); Organization for the Promotion and Advancement of Small Telecommunications Companies Comments in re NBP PN #19, filed Dec. 7, 2009, at 25 (stating that intercarrier compensation revenues together with high-cost USF support comprise approximately 60% of rate of return incumbent LECs' net telephone company operating revenue); Rural High Cost Carriers Comments in re NBP PN #19, filed Dec. 7, 2009, at 11 (noting that federal universal service support and intercarrier compensation account for between 40–62% of revenues for many rural carriers); Texas Statewide Telephone Company Comments in re NBP PN #19, filed Dec. 7, 2009, at 13 (intercarrier compensation revenues and high-cost support accounts for over 60% of rural LECs' revenue stream).

[600] Certain competitive aspects of special access will be addressed in Chapter 4 on a pro-competition framework for the high capacity circuit wholesale market.

[601] Figures based on USAC preliminary 2009 disbursement data.

[602] *See, e.g.*, National Exchange Carrier Association (NECA) Comments in re NBP PN #19, filed Dec. 7, 2009, at 5 (RLECs added gross investment of $1.2 billion in 2006–07, $1.6 billion in 2007–08, and $2.1 billion in 2008–09/10; "the vast majority of these investments in network upgrades are for fiber deployment and state-of- the-art softswitches"); Western Telecommunications Association Comments in re National Broadband Plan NOI, filed June 9, 2009, at 24–25 (USF support has permitted RLECs to install and operate digital switches and soft switches, and deploy and extend fiber optic and DSL facilities deeper into their networks).

[603] *See* OBI, The Broadband Availability Gap. Estimate does not take into account Frontier's proposed acquisition of Verizon lines.

[604] Funding levels for the larger carriers are based on a forward looking cost model that was designed to estimate the cost of providing circuit-switched voice service; it was never intended to address the investment necessary to extend broadband to unserved areas. In contrast, smaller carriers typically receive funding under formulas that allow them to recoup their actual costs of extending broadband to unserved areas, including the costs of deploying fiber and, for some companies, soft switches.

[605] *See, e.g.*, AT&T Inc. Comments in re NBP PN #19, filed Dec. 7, 2009, at 10 (describing Alabama and Mississippi requirements to report on use of high cost funds; AT&T reported its plans to spend funding on deployment of loop fiber and next generation digital loop carrier).

[606] Liu & Rosenberg, *State Universal Service Funding Mechanisms* at 43 & tbl. 26. For instance, in Maine, applicants seeking competitive ETC designation must file a plan describing with specificity, for the first two years, proposed improvements or upgrades to the applicant's network throughout the designated service area, projected start and completion date for each improvement, estimated amount of investment for each project that is funded by high cost support, specific geographic areas where improvements will be made, and the estimated population that will be served as a result of the improvements; only competitive ETCs are required to report annually on investments made with high cost support. Standards for Designating and Certifying Eligible Telecommunications Carriers Qualified to Receive Federal Universal Service Funding, 65-407-206 me. cOde r. § 3, § 6, *available at* www.maine.gov/sos/ cec/rules/65/407/407c206.doc.

[607] Jonathan E. Nuechterlein & Philip J. Weiser, Digital Crossroads: American Telecommunications Policy in the Internet Age 292 (2007). As noted above, ICC represents a significant revenue flow for many small carriers. S*ee* National Exchange Carrier Association Comments in re NBP PN #19, filed Dec. 7, 2009, at 27 (representing that, in 2005, an average 29% of its incumbent carriers' revenues came from intercarrier compensation, and some carriers received up to 49% of revenues from intercarrier compensation); Fred Williams and Associates Comments in re NBP PN #19, filed Dec. 7, 2009, at Attach. 1–2; Letter from

Genevieve Morelli, Counsel for XO et al., to Marlene H. Dortch, Secretary, FCC, GN Docket Nos. 09-47, 09-51, 09-137, WC Docket No. 05-337, CC Docket No. 01-92 (Dec. 9, 2009) Attach. at 1; Independent Telephone & Telecommunications Alliance Comments in re NBP PN #19, filed Dec. 7, 2009, at 6 ("A survey of ITTA members revealed that approximately 12% of member carrier revenues are obtained via ICC"); Alaska Telephone Association Comments in re NBP PN #19, filed Dec. 7, 2009, at 6.

[608] *See Economic Implications and Interrelationships Arising from Policies and Practices Relating to Customer Information, Jurisdictional Separations and Rate Structures*, Docket No. 20003, First Report, 61 FCC 2d 766, 796–97, paras. 81–82 (1976); Gerald W. brOck, the secOnd infOrmatiOn revOlutiOn 188 (2003).

[609] *See* Letter from Brian J. Benison, AT&T, to Marlene H. Dortch, Secretary, FCC, GN Docket No. 09-51, WC Docket Nos. 07-135, 05-337, 99-68, CC Docket Nos. 01-92, 96-45 (Jan. 6, 2010) Attach. at 2; *see also* fcc, universal service mOnitOrs repOrt 2009, at tbl. 7.10 (2009), *available at* http://hraunfoss.fcc.gov/edocs_public/attachmatch/DOC-295442A1.pdf (showing that interstate per-minute charges range up to 5.71 cents per minute).

[610] The FCC has set the rate for ISP-bound traffic at $0.0007 per minute. S*ee Implementation of the Local Competition Provisions in the Telecommunications Act of 1996, Intercarrier Compensation for ISP-Bound Traffic, Order on Remand and Report and Order*, CC Docket Nos. 99-68, 96-98, Order on Remand and Report and Order, 16 FCC Rcd 9151 (2001), *remanded* WorldCom Inc. v. FCC, 288 F.3d 429 (D.C. Cir. 2002); *High Cost Universal Service Reform; Federal-State Joint Board on Universal Service; Lifeline and Link Up; Universal Service Contribution Methodology; Numbering Resource Optimization; Implementation of the Local Competition Provisions in the Telecommunications Act of 1996; Developing a Unified Intercarrier Compensation Regime; Intercarrier Compensation for ISP-Bound Traffic; IP-Enabled Services*, CC Docket Nos. 96-45, 99-200, 96-98, 01-92, 99-68, WC Docket Nos. 05-337, 03-109, 06-122, 04-36, Order on Remand and Report and Order and Further Notice of Proposed Rulemaking, 24 FCC Rcd 6475 (2008), *aff'd*, Core Commc'ns Inc. v. FCC, No. 86-1365 (D.C. Cir. Slip op. Jan. 12, 2010). Other forms of ICC include LEC-CMRS traffic.

[611] Rates differ depending on if the terminating carrier is a rate-of-return carrier, price-cap carrier, competitive carrier or mobile wireless provider.

[612] PAETEC Communications et al. Comments in re NBP PN #19, filed Dec. 7, 2009, at 18 ("The Joint Commenters have invested substantial amounts to ensure proper billing.... These investments and the systems used to bill intercarrier compensation would be substantially simpler if Joint Commenters did not have to track and classify traffic based on artificial regulatory constructs"); US Telecom Comments in re NBP PN #19, filed Dec. 7, 2009, at 7; CenturyLink Comments in re NBP PN #19, filed Dec. 7, 2009, at 38 (citing Central Telephone Company of Virginia et al. v. Sprint Communications Company of Virginia, Inc. and Sprint Communications Company LP, Case No. 3:09-cv-00720 (E.D. Va.) (filed Nov. 16, 2009); CenturyTel of Chatham LLC et al. v. Sprint Communications Company LP, Case No. 3:09-cv-01951 (W.D. La.) (filed Nov. 23, 2009)).

[613] *See, e.g., Establishing Just and Reasonable Rates for Local Exchange Carriers*, WC Docket No. 07-135, Notice of Proposed Rulemaking, 22 FCC Rcd 17989 (2007) (*Access Stimulation NPRM*) (seeking comment on how to address access stimulation concerns); *Establishing Just and Reasonable Rates for Local Exchange Carriers; Call Blocking by Carriers*, WC Docket No. 07-135, Declaratory Ruling, 22 FCC Rcd 11629 (2007) (prohibiting self-help call blocking to address access stimulation concerns); *Qwest Commc'ns Corp. v. Farmers and Merchants Mut. Tel. Co.*, File No. EB-07-MD-001, Second Order on Reconsideration, 24 FCC Rcd 14801 (2009) Attach. (Second Petition for Reconsideration and Petition for Stay pending) (resolving dispute regarding payment of access charges in alleged access stimulation situation).

[614] AT&T Comments in re Access Stimulation NPRM, filed Dec. 17, 2007, Attach. (Decl. of Adam Panagia) at para. 11; *see also* Letter from Brian J. Benison, AT&T, to Marlene H. Dortch, Secretary, FCC, WC Docket No. 07- 135 (Nov. 20, 2009) Attach. at 4–6; Letter from Donna Epps, Verizon, to Marlene H. Dortch, Secretary, FCC, WC Docket No. 07-135 (June 4, 2008) at 2-3.

[615] *See* Cablevision Comments in re NBP PN #25 (*Comment Sought on Transition from Circuit-Switched Network to All-IP Network*, GN Docket No. 09-47, 09-51, 09-137, Public Notice, 24 FCC Rcd 14272 (WCB 2009) (*NBP PN #25*)), filed Dec. 22, 2009, at 2 ("[A]s incumbent local exchange carriers . . . upgrade their legacy networks to IP, they refuse to provide IP interconnection to their competitors on reasonable terms or at all. As a result, each IP voice call initiated on a competing carriers' network must be reduced to TDM, transmitted over an electrical DS-0 or similar connection, and routed to an ILEC customer over the legacy hierarchical circuit-switched network, with all of its associated costs, inefficiencies, and limitations."); Global Crossing Comments in re NBP PN #19, filed Dec. 7, 2009, at 6; Sprint Nextel Comments in re NBP PN #25, filed Dec. 22, 2009, at 10; PAETEC Comments in re NBP PN #25, filed Dec. 22, 2009, at 7-10.

[616] *See* Verizon Comments in re NBP PN #19, filed Dec. 7, 2009, at 18 ("Ongoing uncertainty regarding the compensation due to—and from—providers for IP traffic serves as a disincentive to further investment in the very next-generation services that consumers seek most").

[617] *See* Verizon Comments in re NBP PN #19, filed Dec. 7, 2009, at 17; AT&T Comments in re NBP PN #25, filed Dec. 22, 2009, at 12; Global Crossing Comments in re NBP PN #19, filed Dec. 7, 2009, at 5.

[618] *See* FCC , Universal Service Monitoring Report 2009, at tbl. 8.1 (2009), *available at* http://hraunfoss.fcc.gov/edocs_public/attachmatch/DOC-295442A1.pdf; *see also* AT&T Comments in re NBP PN #25, filed Dec. 21, 2009, at 10.

[619] Organization for the Promotion and Advancement of Small Telecommunications Companies Comments in re NBP PN #19, filed Dec. 7, 2009, at 23-24.

[620] Further, the FCC has not addressed whether VoIP traffic is subject to ICC charges, and, if so, what type of charges apply. Commenters in the record argue that the uncertainty regarding the treatment of VoIP traffic has resulted in significant disputes and costly litigation regarding the payment of intercarrier compensation for such traffic. CenturyLink Comments in re NBP PN #19, filed Dec. 7, 2009, at 38 (*citing Central Telephone Company of Virginia, et al v. Sprint Communications Company of Virginia, Inc and Sprint Communications Company LP*, Case No. 3:09-cv-00720 (E.D. Va.) (filed Nov. 16, 2009); *Century Tel of Chatham LLC, et al v. Sprint Communications Company LP*, Case No. 3:09-cv01951 (W.D. La.) (filed Nov. 23, 2009)).

[621] Wired special access circuits connect wireless towers to the rest of the network. Sprint estimates that one third of its total operating costs of a cell site are devoted to *second and middle*-mile connectivity. Sprint Comments in re NBP PN #11, (*Comments sought in Impact of Middle and Second Mile Access on Broadband Availability and Development-NBP Public Notice #11*, GN Docket Nos. 09-47, 09-51, 09-13, Public Notice 24 FCC Rcd 12470 (WCB 2009) NBP PN #11), filed Nov. 19, 2010, at 2.

[622] *See, e.g.*, Comments of Contact Communications, Inc. and wwyoming.com in re NBP PN #11, filed Nov. 3, 2009, at 4-6 (providing connectivity options and costs for served middle mile circuits in Wyoming of up to 231 miles); Letter from Thomas Jones, Counsel, tw telecom inc., to Marlene H. Dortch, Secretary, FCC, GN Docket Nos. 09-47, 09-51, 09-137 (Dec. 22, 2009).

[623] *See, e.g.*, Comments of Contact Communications, Inc. and Wyoming.com in re NBP PN #11, filed Nov. 3, 2009, at 4-6 (providing connectivity options and costs for several middle-mile circuits in Wyoming of up to 231 miles); National Exchange Carrier Association Comments in re NBP PN #11, filed Nov. 4, 2009; Wireless Internet Service Provider Association Comments in re NBP PN #11, filed Nov. 4, 2009.

[624] 47 U.S.C. § 201(b).

[625] National Telecommunications Cooperative Association Comments in re NBP PN #11, filed Nov. 20, 2009, at 5-13 (asserting that total middle-mile cost will rise as Internet demand increases, and small rural providers have per Mbps middle-mile costs higher than the larger providers).

[626] Per-megabit costs can vary significantly for small rural providers. The National Exchange Carrier Association reports that the price its members pay for a 45 Mbps DS3 connection ranges from $50–$375 per month. National Exchange Carrier Association Comments in re NBP PN# 11, filed Nov. 4, 2009, at 4.

[627] *See generally* Peter Bluhm & Robert Loube, *Competitive Issues in Special Access Markets* (NRRI, Working Paper No. 09-02, rev. ed. 2009), *available at* http://nrri. org/pubs/telecommunications/NRRI_spcl_access_mkts_jan09-02.pdf; XO Comments in re NBP PN #11, filed Nov. 4, 2009, at 15–27; Letter from Thomas Jones, Counsel, tw telecom inc., to Marlene H. Dortch, Secretary, FCC, GN Docket No. 09-51 (Oct. 14, 2009) Attach.; Letter from Thomas Jones, Counsel, tw telecom inc., to Marlene H. Dortch, Secretary, FCC, GN Docket No. 09-51 (Dec. 22, 2009) (regarding price, terms, and conditions of high-capacity Ethernet transport); Sprint Comments in re NBP PN # 11, filed Nov. 4, 2009, at 13–45; Wireless Internet Service Provider Association Comments in re NBP PN # 11, filed Nov. 4, 2009, at 25–28 (recommending fiber access policy), *But cf.* Verizon Comments in re NBP PN# 11, filed Nov. 4, 2009, at 4–5, 42 (noting that while "cost and availability of middle- and second-mile facilities—generally together with other factors—have hindered the deployment of broadband in some instances" to the point that broadband in those locations "would be too expensive for most," but asserting that "it is the distance such facilities must be deployed and the relatively small base of customers" that results in high costs); Letter from Jeffrey S. Lanning, Director, Federal Regulatory Affairs, CenturyLink, to Marlene H. Dortch, Secretary, FCC, GN Docket No. 09-51, WC Docket No. 05-25 (Nov. 4, 2009) at Attach. (noting that special access circuits "typically are sunk cost investments with considerable risk"); AT&T Comments in re NBP PN #11, filed Nov. 4, 2009, at 3–5, 9–13 (noting per-mile rates for special access second and middle-mile connections "typically vary little from urban to rural areas").

[628] *High-Cost Universal Service Support; Federal-State Joint Board on Universal Service*, WC Docket No. 05-337, CC Docket No. 96-45, Recommended Decision, 22 FCC Rcd 20477, 20490–92, paras. 55–62 (JB 2007).

[629] *See* National Exchange Carrier Association Comments in re NBP PN #19, filed Dec. 7, 2009, at 8 (as mechanisms are put in place to support broadband services, funding for existing voice-based programs can be phased down).

[630] *See, e.g.*, Letter from Mike Lovett, Executive Vice President and Chief Operating Officer, Charter, to Chairman Julius Genachoski, FCC, GN Docket Nos. 09-51, 09-47, 09-137, 09-919, 07-52, WC Docket No. 09-154, 05-337, RM-11584 (Feb. 24, 2010) at 8 (urging FCC to pinpoint support to unserved areas and prioritize applications that will deliver broadband to the greatest number of now unserved households per public dollar invested); Qwest Comments in re NBP PN #19, filed Dec. 7, 2009, at 4 (in early years of program, target unserved households where it is less costly to provide broadband service, in order to maximize the number of unserved households in every year).

[631] *See, e.g.*, Organization for the Promotion and Advancement of Small Telecommunications Companies Comments in re NBP PN #19, filed Dec. 7, 2009, at 10 (seven year transition period); TDS Telecommunications Corp. Comments in re NBP PN #19, filed Dec. 7, 2009, at 6–7 (supports OPASTCO proposal); Independent Telephone and Telecommunications Alliance Comments in re NBP PN #19, filed Dec. 7, 2009, at 16 (five to seven year transition); Free Press Comments in re National Broadband Plan NOI, filed June 8, 2009, at 29, 255 (ten years).

[632] Pennsylvania Public Utility Commission Comments in re NBP PN #25, filed Dec. 22, 2009, at 5–6 (stating that the FCC should seek input on how to reconcile national efforts with successful state programs).

[633] For the purposes of the Plan, we define "Tribal lands" as any federally recognized Tribe's reservation, pueblo and colony, including former reservations in Oklahoma, Alaska Native regions established pursuant to the Alaska Native Claims Settlement Act (85 Stat. 688), and Indian allotments. The term "Tribe" means any American Indian or Alaska Native Tribe, Band, Nation, Pueblo, Village or Community which is acknowledged by the Federal government to have a government-togovernment relationship with the United States and is eligible for the programs and services established by the United States. See Statement of Policy on Establishing a Government-to-Government Relationship with Indian Tribes, 16 FCC Rcd 4078, 4080 (2000). Thus, "Tribal lands" includes American Indian Reservations and Trust Lands, Tribal Jurisdiction Statistical Areas, Tribal Designated Statistical Areas, and Alaska Native Village Statistical Areas, as well as the communities situated on such lands. This would also include the lands of Native entities receiving Federal acknowledgement or recognition in the future. While Native Hawaiians are not currently members of federally-recognized Tribes, they are intended to be covered by the recommendations of this Plan, as appropriate.

[634] *See, e.g.* CenturyLink et al. Comments in re NBP PN #19, filed Dec. 7, 2009, at 3–4 (urging FCC to create expedited process to target additional support for broadband deployment in unserved areas pending resolution of longer term USF reform issues for areas that already have broadband and voice services).

[635] The Memorandum of Understanding is posted on the FCC's website. See Memorandum of Understanding between the Federal Communications Commission and the Universal Service Administrative Company (Sept. 9, 2008), http://www.fcc.gov/omd/usac-mou.pdf.

[636] *See, e.g.*, National Telecommunications Cooperative Association Reply in re NBP NOI, filed July 21, 2009, at 23–24 (target funding to "Market Failure Areas," defined as areas that lack the population base or economic foundation to justify build-out and ongoing maintenance without external monetary support); Nebraska Public Service Commission Comments in re NBP PN #19, filed Dec. 7, 2009, at 7–8 (need to target funding to "out of town" areas); National Cable & Telecommunications Association Comments in re NBP PN # 19, filed Dec. 7, 2009, at 2–3.

[637] *See, e.g.*, National Association of State Utility Consumers Advocates Comments in re NBP PN #19, filed Dec. 7, 2009, at 13; Rural Cellular Association Comments in re NBP PN #19, filed Dec. 7, 2009, at 14; Comcast Comments in re NBP PN #19, filed Dec. 7, 2009, at 2–3.

[638] *See, e.g.*, AT&T Comments in re NBP PN #19, filed Dec. 7, 2009, at 13; Organization for the Promotion and Advancement of Small Telecommunications Companies Comments in re NBP PN #19, filed Dec. 7, 2009, at 10, 16; TDS Telecommunications Corp. Comments in re NBP PN # 19, filed Dec. 7, 2009, at 6; Western Telecommunications Alliance Comments in re NBP PN #19, filed Dec. 7, 2009, at 20–21; Pioneer Comments in re NBP PN #19, filed Dec. 7, 2009, at 2.

[639] *See, e.g.*, Comcast Comments in re NBP PN #19, filed Dec. 7, 2009, at 3–4; New Jersey Rate Division of Counsel Comments in re NBP PN #19, filed Dec. 7, 2009, at 7–8; Letter from Ben Scott, Free Press to Marlene H. Dortch, Secretary, FCC, GN Docket No. 09-51 (Jan. 19, 2010) (need for high cost should be based on forward-looking infrastructure and total revenue earning potential); *see also* Sprint Comments in re *National Cable and Telecommunications Association Petition for Rulem akingTo Reduce Universal Service High-Cost Support Provided To Carriers In Areas Where There Is Extensive Unsubsidized Facilities-based Voice Competition*, WC Docket No. 05-337, GN Docket No. 09-51, RM-11584, filed Jan. 7, 2010, at 7 (FCC must recognize that USF recipients derive revenues from broadband and video services delivered over common network); National Cable & Telecommunications Association Petition for Rulemaking, Reducing Universal Service Support in Geographic Areas That Are Experiencing Unsupported Facilities-Based Competition (filed Nov. 5, 2009) (when considering need for ongoing support, FCC should consider whether ILEC costs, including costs attributable to provider of last resort obligations imposed under state law, cannot be recovered through the regulated and unregulated services provided over the network).

[640] *See, e.g.*, Florida Public Service Commission Comments in re NBP PN #19, filed Dec. 15, 2009, at 5 (carriers should not be able to double dip from different federal agencies for the same project); US Cellular Comments in re NBP PN #19, filed Dec. 7, 2009, at 15; Centurylink Comments in re NBP PN #19, filed Dec. 7, 2009, at 27.

[641] *See, e.g.*, US Cellular Comments in re NBP PN #19, filed Dec. 7, 2009, at 15–17; USA Coalition Comments in re NBP PN #19, filed Dec. 7, 2009, at 11.

[642] Liu & Rosenberg, *State Universal Service Funding Mechanisms* at 70, 76; alliance fOr pub. tech. & cOmmc'ns WOrkers Of am., state brOadband initiatives 3, 47–49 (2009), *available at* http://www.apt.org/publications/reports-studies/state_broadband_initiatives.pdf..

[643] *See* California Public Utility Commission Comments in re NBP PN #19, filed Dec. 7, 2009, at 4–5 (arguing that states that generate matching funding should get supplemental funding; states that do not should only get base level funding). In 2007, the Federal-State Joint Board on Universal Service recommended that the FCC adopt policies to encourage states to provide matching funds for the new Broadband Fund that it proposed be established. *Comprehensive Reform Recommended Decision*, 22 FCC Rcd at 20489, para. 50.

[644] *See, e.g.*, Letter from Ken Pfister, Vice Pres.–Strategic Pol'y, Great Plains Communications, Inc., to Marlene H. Dortch, Secretary, FCC, GN Docket No. 09-51 (Dec. 8, 2009) Attach. at 6 (arguing that the United States cannot afford to support more than one network; support should be targeted to where the market will not work); California Public Utility Commission Comments in re NBP PN #19, filed Dec. 7, 2009, at 6 (arguing that USF should provide support to only one provider in a given geographic area); Qwest Communications International Inc. Comments in re NBP PN #19, filed Dec. 7, 2009, at 3 (arguing that only a single provider of broadband, regardless of technology, should receive support); Maine Public Utility Commission and Vermont Public Service Board Comments in re NBP PN #19, filed Dec. 7, 2009, at 4; Charter Communications, Inc. Comments in re NBP PN #19, filed Dec. 7, 2009, at 5; *see also Comprehensive Reform Recommended Decision*, 22 FCC Rcd at 20481–82, para. 15 (recommending that Broadband Fund provide funding for "only one provider in any geographic area").

[645] *See, e.g.*, National Cable and Telecommunications Association Comments in re NBP PN #19, filed Dec. 7, 2009, at 7; American Cable Association Comments in re NBP PN #19, filed Dec. 7, 2009, at 32 (supporting the creation of a new support mechanism that is "competitively and technologically neutral").

[646] *See, e.g.*, Maine Public Utilities Commission and Vermont Public Service Board Comments in re NBP PN #19, filed Dec. 7, 2009, at 4 (arguing that the Commission should award support for broadband deployment through cost model, Request for Proposal or reverse auctions); Qwest Communications International Inc. Comments in re NBP PN #19, filed Dec. 7, 2009, at 3, 21 (recommending competitive bidding for onetime grant to cover cost of deploying and providing broadband in previously unserved area for finite time period, such as ten years); AT&T Inc. Comments in re NBP PN #19, filed Dec. 7, 2009, at 11 & App. A at 11 (suggesting competitive project-based funding approach); AdHoc Telecommunications Users Committee Comments in re NBP PN #19, filed Dec. 7, 2009, at 10.

[647] *See* National Association of State Utility Consumer Advocates Comments in re NBP PN #19, filed Dec. 7, 2009, at 22 (condition receipt of funding on acceptance of broadband carrier of last resort for the area where funding is accepted).

[648] *See, e.g.*, American Cable Association Comments in re NBP PN #19, filed Dec. 7, 2009, at 39 (arguing that funding should be provided with strict terms and conditions, including time limits for construction, reporting requirements, and annual audits); Kansas State Representative Tom Sloan and Mary Galligan Comments in re NBP PN #28 (*Comment Sought on Addressing Challenges to Broadband Deployment Financing—NBP Public Notice #28*, GN Docket No. 09-47, 09-51, 09-137, Public Notice, 24 FCC Rcd 14610 (WCB 2009) (*NBP PN #28*)), filed Jan. 8, 2010, at 1 (arguing that companies that receive financial support must be held accountable for actually signing up customers for broadband service, not simply installing infrastructure); Organization for the Promotion and Advancement of Small Telecommunications Companies Comments in re NBP PN #19, filed Dec. 7, 2009, at 10.

[649] *See, e.g.*, Cox Communications, Inc. Comments in re NBP PN #19, filed Dec. 7, 2009, at 10 ("[M]onopoly providers subject to COLR obligations should be required to meet service quality standards and reporting and oversight obligations to guarantee that they provide reasonable service in areas where customers have no competitive choice."); AT&T Inc. Comments in re NBP PN #19, filed Dec. 7, 2009, App. A at 19 (arguing that recipients should provide supported services at rates, terms and conditions reasonably comparable to those offered in urban areas); Qwest Communications International Inc. Comments in re NBP PN #19, filed Dec. 7, 2009, at 4 (arguing that winner bidders of subsidies to deploy broadband to unserved areas should be limited to charging no more than 125% of the state-wide average for comparable broadband service); OPASTCO Comments in re NBP PN #19, filed Dec. 7, 2009, at 21 (arguing that ETCs should be required to serve all customers at minimum broadband speeds and maximum rates).

[650] Data from American Roamer shows geographic coverage by technology. The actual service quality of data connections experienced by end-users will differ due to a large number of factors, such as location and mobility. Further, the underlying coverage maps do not include information on what level of service (*i.e.*, the quality of the signal and the speed of broadband service) provided; nor is coverage defined by providers in the same way. Thus, coverage as described here does not correspond to a specific minimum signal quality or user experience. *See* American Roamer Advanced Services database (accessed Aug. 2009) (aggregating service coverage boundaries provided by mobile network operators) (on file with the Commission) (American Roamer database). Population is based on projected census blocks figures from Geolytics. *See* Geolytics Block Estimates and BlockEstimates Professional databases (2009) (accessed Nov. 2009) (projecting census

populations by year to 2014 by census block) (on file with the Commission) (Geolytics databases). *See generally* OBI.

[651] *See* American Roamer database; Geolytics databases. *See generally* OBI, The Broadband Availability Gap..

[652] Atkinson & Schultz, Broadband in America at at 7.

[653] *See* Maine Public Utility Commission and Vermont Public Service Board Comments in re NBP PN #19, filed Dec. 7, 2009, at 3 (arguing the Commission should establish Mobility Fund to expand wireless mobile voice services in unserved areas). In 2007, the Federal-State Joint Board on Universal Service recommended that the FCC create a new Mobility Fund "tasked primarily with disseminating wireless voice services to unserved areas" with funding targeted to capital spending for new construction. *Comprehensive Reform Recommended Decision*, 22 FCC Rcd at 20482, 20486, paras. 16, 36.

[654] *See, e.g.*, United States Telecom Association Comments in re NBP PN #28, filed Jan. 8, 2010, at 9; Windstream Communications, Inc. Comments in re NBP PN #28, filed Jan. 8, 2010, at 5.

[655] *See* Letter from William J. Wilkins, Chief Counsel, U.S. Department of Treasury, to Cameron K. Kerry, General Counsel, U.S. Department of Commerce (Mar.4, 2010). The five factors outlined in the Treasury ruling are: (1) the contribution must become a permanent part of the transferee's working capital structure; (2) the contribution may not be compensation, such as a direct payment for a specific, quantifiable service; (3) the contribution must be bargained for; (4) the contributed asset foreseeably must result in a benefit to the transferee commensurate with its value; and (5) the contributed asset must ordinarily, if not always, be employed in or contribute to the production of additional income.

[656] For the definition of "Tribal lands" as used in this Plan, see Chapter 3 note 80, *supra*.

[657] *See* Traci L. Morris & Sacha D. Meinrath, Native Public Media & Open Technolog y Initiative, New America Foundation, New Media, Technolog y and Internet Use in Indian Country: Quantitative and Qualitative Analyses 36–37 (2009) (Morris & Meinrath, New Media, Technolog y and Internet Use in Indian Country); Letter from Mark Pruner, Pres., Native Am. Broadband Ass'n, to Marlene H. Dortch, Secretary, FCC, GN Docket Nos. 09- 47, 09-51, 09-137 (Dec. 17, 2009) Attach. at 2–3.

[658] Letter from Loris Ann Taylor, Exec. Dir., Native Public Media et al., to Marlene H. Dortch, Secretary, FCC, GN Docket Nos 09-47, 09-51, 09-137 (Dec. 24, 2009) (Joint Native Filers Dec. 24, 2009 *Ex Parte*) at 12.

[659] *Cf. Comprehensive Reform Recommended Decision*, 22 FCC Rcd at 20484, paras. 26–27 (recommending overall cap on the High Cost fund and a transition in which existing funding mechanisms would be reduced, and all, or a significant share, of savings transferred to proposed new funds for broadband and mobility).

[660] Verizon Wireless agreed to a five-year phase-out of its competitive ETC High-Cost support for any properties that it retained after mandated divestitures. *Applications of Cellco Partnership d/b/a Verizon Wireless and Atlantis Holdings LLC for Consent to Transfer Control of Licenses, Authorizations, and Spectrum Manager and De Facto Transfer Leasing Arrangements and Petition for Declaratory Ruling that the Transaction is Consistent with Section 310(b)(4) of the Communications Act*, WT Docket No. 08-95, File Nos. 0003463892 et al., ITC-T/C-20080613-00270 et al., ISP-PDR-20080613-00012, Memorandum Opinion and Order and Declaratory Ruling, 23 FCC Rcd 17444, 17529–17532, paras. 192–197 (2008). Similarly, Sprint agreed to a five-year phase-out of its competitive ETC high-cost support as part of its transaction with Clearwire. *Applications of Sprint Nextel Corporation and Clearwire Corporation For Consent to Transfer Control of Licenses, Leases and Authorizations*, WT Docket No. 08-94, File Nos. 0003462540 et al., Memorandum Opinion and Order and Declaratory Ruling, 23 FCC Rcd 17570, 17612, para. 108 (2008).

[661] Nat'l Telecomm. Coop. Ass'n, 2009 Broadband/Internet Availability Survey Report 3, 9 (2009) (89% of those surveyed face competition from at least one broadband provider in some portion of their service area; 47% face broadband competitors serving customers throughout their area).

[662] *Policy and Rules Concerning Rates for Dominant Carriers*, CC Docket No. 87-313, Second Report and Order, 5 FCC Rcd 6786, 6790, para. 32 (1990), *aff'd, Nat'l Rural Telecom Ass'n v. FCC*, 988 F.2d 174 (D.C. Cir. 1993).

[663] A number of the mid-sized telephone companies already have elected to convert to price cap regulation, with their ICLS support per line frozen; the proposed rule change would have no impact on those companies. *See Windstream Petition for Conversion to Price Cap Regulation and for Limited Waiver Relief*, WC Docket No. 07-171, Order, 23 FCC Rcd 5294 (2008); *Petition of Puerto Rico Telephone Company, In c. for Election of Price Cap Regulation and Limited Waiver of Pricing and Universal Service Rules; Consolidated Communications Petition for Conversion to Price Cap Regulation and for Limited Waiver Relief; Frontier Petition for Limited Waiver Relief upon Conversion of Global Valley Networks, Inc., to Price Cap Regulation*, WC Docket Nos. 07-292, 07-291, 08-18, Order, 23 FCC Rcd 7353 (2008); *Petition of Centurylink for Conversion to Price Cap Regulation and for Limited Waiver Relief*, WC Docket No. 08-191, Order, 24 FCC Rcd 4677 (2009).

[664] The Alliance of Rural CMRS Carriers has proposed that the FCC adopt an interim cap for incumbent telephone company support per line at either March 2010 or March 2008 levels, with estimated savings of $1.8 billion between 2010 and 2012 repurposed toward broadband programs, pending comprehensive USF reform. *See* Letter from David LaFuria, Counsel for Alliance for Rural CMRS Carriers, to Marlene H. Dortch, Secretary, FCC, CC Docket No. 96-45, WC Docket No. 05-337, GN Docket Nos. 09-47, 09-51, 09-137 (Mar. 3, 2010).

[665] *See* National Telecommunications Cooperative Association Comments in re NBP PN #19, filed Dec. 7, 2009, at 4 (suggesting that carriers receive future funding for broadband through Interstate Access Support and Interstate Common Line Support).

[666] Figures based on preliminary USAC 2009 disbursement data. *See* Universal Serv. Admin. Co., Disbursement Data (High Cost), http://www.usac.org/hc/tools/disbursements/ default.aspx (last visited Mar. 7, 2010) (providing disbursement data for all the high-cost programs).

[667] *Access Charge Reform; Price Cap Performance Review for Local Exchange Carriers; Low-Volume Long Distance Users; Federal-State Joint Board on Universal Service*, CC Docket Nos. 96-262, 94-1, 99-249, 96-45, Sixth Report and Order, Report and Order, and Eleventh Report and Order, 15 FCC Rcd 12962 (2000), *aff'd in part, rev'd in part, and remanded in part, Texas Office of Public Util. Counsel et al. v. FCC*, 265 F.3d 313 (5th Cir. 2001); *on remand, Access Charge Reform; Price Cap Performance Review for LECs; Low-Volume Long Distance Users; Federal-State Joint Board on Universal Service*, CC Docket Nos. 96-262, 94-1, 99-249, 96-45, Order on Remand, 18 FCC Rcd 14976 (2003). Interstate Access Support was created in 2000 as a first step in removing implicit support from the FCC's interstate access charge regime.

[668] Competitive ETC support per line is based on the incumbent telephone company's support per line. 47 C.F.R. § 54.307. As a consequence, the support a competitive ETC receives is not based on either its costs or the costs of the most efficient technology to support customers in a given area.

[669] *High-Cost Universal Service Support; Federal-State Joint Board on Universal Service*, WC Docket No. 05-337, CC Docket No. 96-45, Order, 23 FCC Rcd 8834 (2008), (2008), *aff'd, Rural Cellular Ass'n v. FCC*, 588 F.3d 1095 (D.C. Cir. 2009).

[670] *See* Letter from Michael J. Copps, Acting Chairman, FCC, to the Honorable Henry J. Waxman, Chairman, Committee on Energy and Commerce, U.S. House of Representatives, Part 4 (May 4, 2009), *available at* http://energycommerce.house.gov/index. php?option=com_content&view=article&id=1644.

[671] In 2007, the Federal-State Joint Board on Universal Service concluded, "We no longer believe it is in the public interest to use federal universal service support to subsidize competition and build duplicative networks in high cost areas." *Comprehensive Reform Recommended Decision*, 24 FCC Rcd at 6482, para. 12.

[672] *See* Letter from Melissa Newman, Vice Pres., Fed. Relations, Qwest Communications International, Inc., to Marlene H. Dortch, Secretary, FCC, CC Docket No. 96-45 (Feb. 4, 2010) (proposing that universal service support be limited to one handset per wireless family plan and suggesting that could yield savings of up to $463 million annually).

[673] *See, e.g.*, Broadview Networks et al. Comments in re NBP PN #19, filed Dec. 7, 2009, at 4–7; CenturyLink Comments in re NBP PN #19, filed Dec. 7, 2009, at 40; AT&T Inc. Comments in re NBP PN # 19, filed Dec. 7, 2009, App. A at 28–29; Verizon Comments in re NBP PN #19, filed Dec. 7, 2009 at 19–20.

[674] *See, e.g.*, AT&T Comments in re NBP PN #19, filed Dec. 7, 2009, App. A at 28–29.

[675] *See* AT&T Inc. Comments in re NBP PN #19, filed Dec. 7, 2009, App. A at 28–29; CenturyLink Comments in re NBP PN #19, Dec. 7, 2009, at 40.

[676] For example, 8 percent of local residential rates are $12 or less (excluding the SLC). *See* Letter from Mary L. Henze, Asst. Vice Pres., Fed. Reg., AT&T Inc., to Marlene H. Dortch, Secretary, FCC (Nov. 24, 2009) at 12.

[677] *See* Free Press Comments in re NBP PN#30 (*Reply Comments Sought in Support of National Broadband Plan — NBP Public Notice #30*, GN Docket No. 09-47, 09-51, 09-137, Public Notice, 25 FCC Rcd 241 (WCB 2010) (*NBP PN #30*)), filed Jan. 30, 2010, at 9–11 (suggesting the FCC could phase down over five years support for any lines receiving $20/month or less).

[678] *See* American Cable Association in re NBP PN #19, filed Dec. 7, 2009, at 40 (providers should not be able to draw from both broadband and high cost support mechanism for the same area).

[679] FCC , Universal Service Monitoring Report 2009, tbls. 1.1, 1.2 (2009), *available at* http://hraunfoss.fcc. gov/edocs_public/attachmatch/DOC-295442A1.pdf. According to USAC's most recent filing, the revenue base for second quarter 2010 is $16.6 billion, more than a $600 million drop from first quarter 2010. *See* usac, federal universal service cOntributiOn base prOjectiOns fOr secOnd Quarter 2010, at 7 (2010), *available at* http://www.universalservice.org/about/ governance/fcc-filings/2010/Q2/2Q2010%20 Quarterly%2 0Contribution%20Base%20Filing.pdf.

[680] According to the Telecommunications Industry Association (TIA), overall telecommunications revenues in the United States are projected to grow from $990 billion in 2010 to $1.133 billion in 2012, led by double digit growth in web conferencing, broadband, and support services. Broadband Internet access revenues alone are projected to grow from $39 billion in 2010 to $49 billion in 2012. TIA also projects that wireless data revenue will expand at a 24.6% compound annual growth rate between 2009 and 2012. Telecomm. Indus. Ass'n, TIA's 2009 ICT Market Review and Forecast 1–11 (2009).

[681] *See, e.g.*, National Telecommunications Cooperative Association Comments in re NBP PN #19, filed Dec. 7, 2009, at 2–3 (asking the Commission to expand contributions to all broadband providers and to assess contributions based on telecommunications and broadband revenues); TracFone Wireless Inc. Comments in re NBP PN #19, filed Dec. 7, 2009, at 4 (arguing that all providers of services which derive revenues from services which use telecommunications should be required to contribute to the support of universal service); Broadview Networks, Inc. et al. Comments in re NBP PN #19, filed Dec. 7, 2009, at 11 (arguing that all

information service revenues, including all broadband service provider revenues, should be included for contribution purposes); American Association of Paging Carriers Comments in re NBP PN #19, filed Dec. 7, 2009, at 5–6 (arguing that all revenues from high speed Internet access should be subject to direct assessment); National Association of State Utility Consumer Advocates Comments in re NBP PN #19, filed Dec. 7, 2009, at 7–8; Nebraska Public Service Commission Comments in re NBP PN #19, filed Dec. 7, 2009, at 6; Rural Cellular Comments in re NBP PN #19, filed Dec. 7, 2009, at 8–10; Vermont Public Service Board and Vermont Department of Public Service Comments in re National Broadband Plan NOI, filed June 8, 2009, at 7.

[682] *See, e.g.,* AT&T Inc. Comments in re NBP PN #19, filed Dec. 7, 2009, at 4, 6 (suggesting that if the Commission assessed $1.01 per month on both telephone numbers and residential broadband connections, the inclusion of mass market residential broadband connections in the contribution base would raise an additional $957 million per year); United States Telecom Association Comments in re NBP PN#19, filed Dec. 7, 2009, at 9–10; Western Telecommunications Alliance Comments in re NBP PN #19, filed Dec. 7, 2009, at 9.

[683] *See, e.g.,* AT&T Inc. Comments in re NBP PN #19, filed Dec. 7, 2009, at 4; Rural High Cost Carriers Comments in re NBP PN #19, filed Dec. 7, 2009, at 13–14; Rural Independent Competitive Alliance Comments in re NBP PN #19, filed Dec. 7, 2009, at 10; National Telecommunications Cooperative Association Comments in re NBP PN #19, filed Dec. 7, 2009, at 13.

[684] *See, e.g.,* National Cable & Telecommunications Association Comments in re NBP PN #19, filed Dec. 7, 2009, at 5; Access Humboldt et al. Comments in re NBP PN #19, filed Dec. 7, 2009, at 14–15 (filed as Rural/ Urban Commenters).

[685] *See Comprehensive Reform Recommended Decision,* 22 FCC Rcd at 20484, para. 25.

[686] USAC, Universal Service Fund, http://www.usac.org/about/universal-service/ (last visited Feb. 20, 2010). The estimated annual projected outlay for the federal USF can be found in the FY 2010 Federal budget. Office Of mGmt. & budGet, exec. Office Of the president, budGet Of the united states GOvernment, fiscal year 2010, at 1220 (2010), *available at* http://www. hitehouse.gov/omb/budget/fy2010/assets/oia.pdf. Total outlays for 2000 based the sum of disbursements for all four programs as reported in the FCC's Trends in Telephone Service report and USAC administrative expenses reported in USAC's 2000 annual report. FCC, trends in telephOne service, tbls 19.3, 19.10, 19.13, 19.15 (2008), *available at* http://hraunfoss.fcc.gov/edocs_ public/attachmatch/DOC-284932A1.pdf; universal service administrative cOmpany, 2000 annual repOrt 5 (2000), *available at* http://www.usac.org/about/ governance/annual-reports/2000/default.asp.

[687] Low Income Support total outlays were $930 million in FY 2009, and in annual projections to OMB submitted in December 2009, projected to be $1.2 billion in FY 2010. Based on USAC's most recent quarterly filing, total outlays for the Low Income programs are forecast to be approximately $1.4 billion in calendar year 2010. usac, federal universal service suppOrt mechanisms fund size prOjectiOns fOr secOnd Quarter 2010, at 15– 17 (2010), *available at* http://www.universalservice.org/ about/governance/fcc-filings/2010/Q2/2Q2010%20 Quarterly%20Demand%20Filing.pdf.

[688] A number of commenters suggest that the FCC should work within the existing budget of the current Universal Service Fund. S*ee, e.g.,* Florida Public Service Commission Comments in re NBP PN # 19, filed Dec. 15, 2009, at 3, 6–7 (opposing further growth in fund); Verizon and Verizon Wireless Comments in re NBP PN #19, filed Dec. 7, 2009, at 4–5 (noting that E-rate and rural health care programs are capped and urging FCC to set an overall cap for high cost funding); Benton Comments in re NBP PN #19, filed Dec. 7, 2009, at 6 (arguing that the size of the fund should remain the same but the FCC should redirect money to support broadband); New Jersey Division of Rate Counsel Comments in re NBP PN #19, filed Dec. 7, 2009, at 5, 7 (arguing to cap the high-cost fund and transition support to a Mobility Fund, a Broadband Fund and a Provider of Last Resort Fund, such that combined total of the three stays within cap).

[689] *See* Universal Service Reform Act of 2009, H.R. __, 111th Cong. (discussion draft), *available at* http://www. boucher.house.gov/images/usf%20discussion%20 draft.pdf. The discussion draft legislation to reform the universal services provisions of the Communications Act of 1934 would exempt a provider from having to serve all households in the service territory if the cost per line of deploying such service is at least three times the national average cost of providing broadband for all wire centers.

[690] *See* U.S. Department of Justice *Ex Parte* Submission, filed Jan. 4, 2010, at 28 (recommending that "the Commission monitor carefully those areas in which only a single provider offers—or even two providers offer— broadband service" and cautioning that price regulation may be appropriate only "to protect consumers from the exercise of monopoly power" and must be careful to avoid stifling "incentives to invest in infrastructure deployment").

[691] One factor that impacts the sensitivity of estimates is the amount of ICC revenue replacement that may ultimately be provided to carriers during implementation of long-term ICC reform. The need for explicit revenue replacement from the CAF depends in part on the rate benchmark ultimately selected and the extent to which costs are recovered from end-user customers. The estimates also do not include potential cost savings that would result from implementing other parts of the plan, such as lower pole and rights-of-way costs or spectrum reforms.

[692] Because the current USF contribution methodology assesses telecommunications revenues, the burden of funding universal service may ultimately fall more heavily on households that do not subscribe to broadband services, which on average may be lower income, older and more likely rural. In contrast, the federal income tax system is more progressive so that lower-income households pay a lower marginal rate than upper-income households.

[693] U.S. Metronets, LC Comments in re NBP PN #28, filed Mar. 3, 2010,at 4; Letter from Thomas Cohen, Kelley Drye & Warren LLP, counsel for Hiawatha Broadband Communications, Inc., to Marlene H. Dortch, Secretary, FCC (Nov. 25, 2009) at 1; *id.* Attach. at 1 (Hiawatha Broadband Communications Inc. White Paper); *see also* Letter from Thomas Cohen, Kelley Drye & Warren LLP, counsel for Hiawatha Broadband Communications, Inc., to Marlene H. Dortch, Secretary, FCC (Nov. 5, 2009) at 1; *id.* Attach. at 1–3 (Barriers to Broadband Rural Deployment—Challenges and Solutions).

[694] U.S. Dep't of Agric., *Agriculture Secretary Vilsack Seeks Applicants for Broadband Grants in Rural Areas* (press release), Apr. 28, 2009, *available at* http://www.usda.gov/wps/portal?contentidonly=true&contentid=2009/04/0135.xml; *see also* GaO, telecOmmunicatiOns: brOadband deplOyment plan shOuld include perfOrmance GOals and measures tO Guide federal investment 13, GaO-09-494 (2009).

[695] *See* Joint Native Filers Dec. 24, 2009 *Ex Parte* at 24–25. The Joint Native Filers estimate such a fund would require at least $310 million to effectively support deployment and adoption objectives. The New America Foundation's Open Technology Initiative (OTI) proposes an allocation of $1.2 billion to $4.6 billion for broadband deployment on Tribal lands. *See* Letter from Matthew F. Wood, Associate Director, Media Access Project, to Marlene H. Dortch, Secretary, FCC, GN Docket Nos. 09-47, 09-51, 09-137, CG Docket No. 09-158, CC Docket No. 98-170 (Jan. 20, 2010) at 2.

[696] *See* Joint Native Filers Dec. 24, 2009 *Ex Parte* at 25. The Joint Native Filers recommend an allocation of at least $30 million for funding small grants. *See id.* at 24.

[697] *See, e.g.*, Morris & Meinrath, New Media, Technolog y and Internet Use in Indian Country at 52, *available at* http://www.nativepublicmedia.org/images/stories/ documents/npm-nafnew-media-study-2009.pdf. Such entities include the Bureau of Indian Affairs, the Indian Health Service, and the Bureau of Indian Education. The connectivity at these locations is typically limited to T1 lines. *See generally id.* App. II.

[698] *See* California Association of Tribal Governments *Ex Parte* in re National Broadband Plan NOI, filed Dec. 18, 2009, at 12, 14 (filed by William Micklin); Joint Native Filers Dec. 24, 2009 *Ex Parte* at 9–10. This is consistent with the recommendation in Chapter 14 that Federal contracting power via Networx be made available to connect communities.

[699] *See* Native Public Media et al. Comments in re NBP PN #5 (*Comment Sought on Broadband Deployment and Adoption on Tribal Lands—NBP Public Notice #5*, GN Docket No. 09-47, 09-51,09-137, Public Notice, 24 FCC Rcd 12010 (CGB 2009) (*NBP PN #5*)), filed Nov. 9, 2009, at 2–3 (noting that "broadband penetration [on Tribal lands] may be as low as five percent (5%)"); Joint Native Filers Dec. 24, 2009 *Ex Parte* at 1 ("Broadband deployment on Tribal Lands is at less than a 10 percent penetration rate... ."); California Association of Tribal Governments *Ex Parte* in re National Broadband Plan NOI, filed Dec. 18, 2009, at 2 (filed by William Micklin) (stating that "broadband deployment in Indian Country is at less than a 10 percent penetration rate"); Native Public Media et al. Comments in re NBP PN #1 (*Comment Sought on Defining Broadband—NBP PN #1*, GN Docket Nos. 09-47, 09-51, 09-137, Public Notice, 24 FCC Rcd 10897 (WCB 2009) (*NBP PN #1*)), filed Aug. 31, 2009, at 3 ("There are no solid data on broadband deployment on Tribal lands, but it is estimated that Tribal penetration hovers somewhere around five percent (5%).").

[700] GAO, Challenges to Assessing and Improving Telecommunications for Native Americans on Tribal Lands 10,, GAO -06-189 (2006).

[701] *Extending Wireless Telecommunications Services to Tribal Lands*, WT Docket No. 99-266, Report and Order and Further Notice of Rule Making, 15 FCC Rcd 11794, 11798 (2000); *see also* General Communication, Inc. Comments in re NBP PN #5, filed Nov. 9, 2009, at 6; California Association of Tribal Governments Comments in re NBP PN #18 (*Comment Sought on Relationship Between Broadband and Economic Development—NBP PN #18*, GN Docket No. 09-47, 09-51, 09-137, Public Notice, 24 FCC Rcd 13736 (WCB 2009) (*NBP PN #18*)), filed Dec. 3, 2009, at 2; Letter from James M. Smith, Counsel, Davis Wright Tremaine LLP on behalf of Kodiak Kenai Cable Company, LLC, to Marlene Dortch, Secretary, FCC, GN Docket Nos. 09-47, 09-51, 09-137 (Dec. 18, 2009) at 9.

[702] *See* Joint Native Filers Dec. 24, 2009 *Ex Parte* at 25–26.

[703] *See* Joint Native Filers Dec. 24, 2009 *Ex Parte* at 25–26.

[704] *See, e.g.*, Alaska Telephone Association Reply in re NBP PN #5, filed Dec. 9, 2009, at 2; General Communication, Inc. Comments in re NBP PN #5, filed Nov. 9, 2009, at 16; Nemont Telephone Cooperative Comments in re NBP PN #5, filed Nov. 9, 2009, at 12–13.

[705] Fiber-to-the-Home Council, Municipal Fiber to the Home Deployments 2009 (2009), *available at* http://www.ftthcouncil.org/sites/default/files/Municipal%20 FTTH%20Systems%20October%202009%20 Final%20Oct09_1.pdf.

[706] *See, e.g.*, Commenters Supporting Anchor Institution Networks Reply in re NBP PN #30, filed Jan. 27, 2010, at 2–3; Comments of Nat'l Ass'n of Telecomm. Officers & Advisors, in re NBP PN #12, filed Oct. 28, 2009, at Ex. 2, 3 (providing case studies on many such local and regional networks).

[707] Baller Herbst Law Group, State Restrictions on Community Broadband Services or Other Public Communications Initiatives (2004), *available at* http://www.baller.com/pdfs/Barriers2004.pdf.

[708] *See* National Association of Telecommunications Officers and Advisors et al. Comments in re NBP PN #7 (*Comment Sought on the Contribution of Federal, State, Tribal, and Local Government to Broadband—NBP Public Notice #7*, GN Docket Nos. 09-47, 09-51, 09-137, Public Notice, 24 FCC Rcd 12110 (WCB 2009) (*NBP PN #7*)), filed Nov. 6, 2009, at 12, 17–20, 37–42.

[709] U.S. Energy Information Administration, Electric Power Industry Overview 2007, fig. 5, http://www.eia.doe.gov/ cneaf/electricity/page/prim2/toc2.html (Number of Ultimate Customers Served by U.S. Electric Utilities by Class of Ownership, 2007).

[710] Comments of Health Network Group Organized by Internet2 in re NBP PN #17, (*Comment Sought on Health Care Delivery Elements of National Broadband Plan —NBP Public Notice #17*, GN Docket Nos. 09-47, 09-51, 09-137, WC Docket No. 02-60, Public Notice, 24 FCC Rcd 13728 (WCB 2009) (*NBP PN #17*)), files Dec. 2, 2006, at 6; *see also* EDUCAUSE et al. Comments in re National Broadband Plan NOI, filed June 8, 2009, Attach. at 6 (recommending that state-led grants for institutional networks "should be done in a way that private sector companies can build upon them to extend connectivity to households in the future").

[711] For example, a group of participants in the FCC's Rural Health Care Pilot program have noted that the current program results in "the creation of independent special purpose networks that are usually not expected to interoperate." As a result, "[t]hese networks are often developed using the minimum bandwidth capabilities to meet the identified application" and in a manner that "does not encourage the aggregation of services" and "does not consider the community needs such as economic development." Health Network Group Organized by Internet2 Comments in re NBP PN #17 filed Dec. 2, 2009, at 4–5.

[712] U.S. R&E Networks and HIMSS Reply in re NBP PN #30, filed Jan. 27, 2010, at 43 (noting that state laws that limit networks to single purposes "thereby forc[e] the creation of more networks than are necessary," and noting that as a result, "[t]hese laws undermine the benefits from, and the efficiencies of, having one network, used by all community anchors who wish to be a part of it").

[713] *See* Liu & Rosenberg, *State Universal Service Funding Mechanisms* at 43, 54; Alliance for Public Technolog y and Communications Workers of America, State Broadband Initiatives 3 (2009), *available at* http:// www.apt.org/publications/reports-studies/state_ broadbandinitiatives.pdf.

[714] U.S. R&E Networks and HIMSS Reply in re NBP PN #30, filed Jan. 27, 2010, at 6, 43–45 ("With regard to community anchors, inclusion, rather than exclusion, should be the rule of the day.").

[715] American Library Association Comments in re NBP PN #15 (*Comment Sought on Broadband Needs in Education, Including Changes to E-Rate Program to Improve Broadband Deployment—NBP Public Notice #15*, GN Docket No. 09-47, 09-51, 09-137, Public Notice, 24 FCC Rcd 13560 (WCB 2009) (*NBP PN #15*)), filed Nov. 20, 2009, at 14 (noting that "current complexities associated with filing consortia applications can be overwhelming for even the most seasoned E-rate applicant" and that "[m]any consortia leaders are no longer willing to risk filing on behalf of member entities" as a result).

[716] Native Public Media et al. Comments in re: NBP PN #5, filed Nov. 9, 2009, at 18; mOrris & meinrath, new media, technOlOGy and internet use in indian cOuntry at 42; Joint Native Filers Dec. 24, 2009 *Ex Parte* at 13.

[717] U.S. R&E Networks and HIMSS Reply in re NBP PN #30, filed Jan. 27, 2010, at 16-18.

[718] Letter from John Windhausen, Jr., Coordinator, Schools, Health and Libraries Broadband Coalition, to Marlene H. Dortch, Secretary, FCC, GN Docket No. 09- 51, filed Mar. 5, 2010, at 2. Attach. (citing over 210,000 community anchor institutions).

[719] *See generally* Commenters Supporting Anchor Institution Networks Reply in re NBP PN #30, filed Jan. 27, 2010.

[720] U.S. R&E Networks and HIMSS Reply in re NBP PN #30, filed Jan. 27, 2010, at 35–36.

[721] Commenters Supporting Anchor Institution Networks Reply in re NBP PN #30, filed Jan. 27, 2010, at 3.

[722] John Horrigan, *Broadband Adoption and Use in America* 1 (OBI, Working Paper No. 1, 2010) (Horrigan, *Broadband Adoption and Use in America*); *see also* nat'l telecOmm. & infO. admin., diGital natiOn: 21st century america's prOGress tOWard universal brOadband internet access 4 (2010) (estimating that 64% of U.S. households used a broadband Internet access service), *available at* http://www.ntia.doc.gov/reports/2010/ NTIA_internet_use_report_Feb2010.pdf; lee rainie, peW internet & am. life, internet, brOadband and cell phOne statistics 1 (2010) (finding that 60–63% of American adults used broadband at home in 2009), *available at* http://www.pewinternet.org/~/media// Files/Reports/2010/PIP_December09_update.pdf.

[723] Horrigan, *Broadband Adoption and Use in America* at 1, 13–14. The table does not report results for Asian Americans or American Indians/Alaska natives because the survey did not have enough respondents in each of these groups to draw statistically reliable inferences.

[724] *See, e.g.*, Pew Research Ctr., Trend Data: Home Broadband Adoption Since 2000, http://www. pewinternet.org/Trend-Data/Home-BroadbandAdoption.aspx (last visited Mar. 4, 2010).

[725] John Horrigan, Pew Internet & Am. Life Project, Home Broadband Adoption 2009, at 8–11 (2009), *available at* http://www.pewinternet.org/~/media//Files/ Reports/2009/Home-Broadband-Adoption-2009.pdf.

[726] Indust. Analysis & Tech. Div., FC , High-Speed Service for Internet Access: Status as of December 31, 2008 (2010), *available at* http://hraunfoss.fcc.gov/edocs_ public/attachmatch/DOC-296239A1.pdf.

[727] In Fall 2009, the FCC fielded a national survey of Americans' technology use under authority granted by the Broadband Data Improvement Act (BDIA). This survey included an oversample of respondents who do not have broadband at home; of 5,005 survey respondents, 2,334 reported not having or using broadband at home. See Horrigan, *Broadband Adoption and Use in America* at 11.

[728] Horrigan, *Broadband Adoption and Use in America* at 5. The remaining 2% cite a combination of cost- related issues.

[729] Horrigan, *Broadband Adoption and Use in America* at 5.

[730] Horrigan, *Broadband Adoption and Use in America* at 5.

[731] Horrigan, *Broadband Adoption and Use in America* at 24, 7. The FCC Survey defined disability in accordance with OMB guidance. Disability status is related to respondents' answers to any of six questions and is aligned with the questions in upcoming American Community Surveys to be fielded by the Bureau of the Census.

[732] *See, e.g.*, Horrigan, *Broadband Adoption and Use in America* at 26 ("Some of the difference in adoption rates is due to individuals' disabilities and some is due to lower incomes, advanced age or other factors associated with low adoption."). FCC analysis of the data (a logit model) shows that having a disability is, independent of other factors, linked to lower broadband adoption. People with disabilities, for example, have employment rates that are less than half of those without disabilities (36.9% compared with 79.7%) and poverty rates that are nearly three times higher (24.7% compared with 9.0%). People with lower incomes are less likely to have broadband at home (35% compared with 65%). cOrnell university rehabilitatiOn research and traininG center On disability demOGraphics and statistics, 2007 disability status repOrt 24, 34 (2008), *available at* http://www.ilr.cornell.edu/edi/disabilitystatistics/ StatusReports/2007-PDF.

[733] *See, e.g.*, Eric Bridges, American Council of the Blind, Statement, Remarks at the FCC Broadband Accessibility for People with Disabilities Workshop II, at 81–84 (Oct. 20, 2009) (noting the first Smartphone that had features built in allowing it to be used by a person who was blind was introduced in July 2009), *available at* http://www. broadband.gov/docs/ws_accessibility_disabilities/ ws_accessibility_disabilities_transcript.pdf.

[734] *See* American Foundation for the Blind, Technology, Assistive Technology, Braille Technology, http://www. afb.org/Section.asp?SectionID=4&TopicID=31&Docum entID=1282 (last visited Jan. 9, 2010).

[735] For example, people with hearing and speech disabilities who have transitioned from using TTYs to text and video communications cannot call 911 directly. See Telecommunications for the Deaf and Hard of Hearing, Inc. Comments in re NBP PN #14 (*Comment Sought on Public Safety Issues Related to Broadband Deployment in Rural and Tribal Areas and Communications to and from Persons with Disabilities —NBP PN #14*, GN Docket Nos. 09-47, 09-51, 09-137, Public Notice, 24 FCC Rcd 13512 (WCB 2009) (*NBP PN #14*)), filed Dec. 1, 2009, at 2.

[736] *See, e.g.*, Webaim, screen reader user survey results 23 (2009) (finding that only about 8% of the 665 screen reader users surveyed found that social media sites were "very accessible"), *available at* http://www.webaim.org/ projects/screenreadersurvey2/.

[737] *See* Rehabilitation Engineering Research Center on Telecommunications Access Comments in re NBP PN #4 (*Comment Sought on Broadband Accessibility for People with Disabilities Workshop II: Barriers, Opportunities, and Policy Recommendations—NBP PN #4*, GN Docket Nos. 09-47, 09-51, 09-137, Public Notice, 24 FCC Rcd 11968 (CGB 2009) (*NBP PN #4*)), filed Oct. 6, 2009, at 3. Video description is "the insertion of verbal descriptions of on-screen visual elements during natural pauses in a program's audio content." Karen Peltz Strauss, *Past and Present: Making the Case for a Regulatory Approach to Addressing Disability Discrimination in the Provision of Emerging Broadband and Cable Technologies, Broadband and Cable Television Law 2010 Developments in Cable Technolog y 6* n.17 (2010).

[738] Horrigan, *Broadband Adoption and Use in America* at 6. The FCC survey found 86% of Americans have premium television, 86% have a cell phone and 80% have a working computer at home.

[739] Horrigan, *Broadband Adoption and Use in America* at 6.

[740] *See* Letter from William J. Cirone, Superintendent, Santa Barbara County Education Office, to Marlene H. Dortch, Secretary, FCC, GN Docket No. 09-51 (June 30, 2009).

[741] *See* Computers for Families, http://www.sbceo. org/~sbceocff/ (last visited Feb. 22, 2010); Cox Comments in re National Broadband Plan NOI, filed June 8, 2009, Attach. at 5–6.

[742] Carmen DeNavas-Walt et al., U.S. Census Bureau, Current Population Reports, Income, Poverty, and Health Insurance Coverage in the United States:: 2008, at 4 (2009), *available at* http://www.census.gov/ prod/2009pubs/p60-236.pdf.

[743] United States Department of Labor, WB-Previous Projects, Strengthening the Family Initiatives 2008, http://www.dol.gov/wb/programs/family1.htm(last visited Mar. 4, 2010).

[744] *See* Native Public Media (NPM) and the National Congress of American Indians (NCAI) Comments in re NBP PN #5 (*Comment Sought on Broadband Deployment and Adoption on Tribal Lands—NBP Public Notice #5*, GN Docket Nos. 09-47, 09-51, 09-137, Public Notice, FCC Rcd 12010 (CGB 2009) (*NBP PN #5*)) filed Dec.

9, 2009 (NCAI-NPM Dec. 9, 2009, Comments), Attach. at 4–5; *Tribes Take to Wireless Web*, bbc neWs, Mar. 3, 2004, *available at* http://news.bbc.co.uk/2/hi/ technology/3489932.stm.

[745] *See, e.g.*, Paul DiMaggio et al., From Unequal Access to Differentiated Use: A Literature Review and Agenda for Research on Digital Inequality (2001) (recommending research agendas focused on the extent and causes of different returns to Internet use for different kinds of uses), *available at* http://citeseerx.ist. psu.edu/viewdoc/download?doi=10.1.1.85.6001&rep= rep1&type=pdf; Eszter Hargittai & Amanda Hinnant, *Digital Inequality: Differences in Young Adults' Use of the Internet*, 35 cOmm. res. 602 (2008) (discussing the impact of differentiated Internet use and capital- enhancing activities by young people).

[746] Horrigan, *Broadband Adoption and Use in America* at 33.

[747] Horrigan, *Broadband Adoption and Use in America* at 31–33.

[748] *See, e.g.*, Digital Impact Group Comment in re NBP PN #16, (*Comment Sought on Broadband Adoption—NBP Public Notice #16*, GN Docket Nos. 09-47, 09-51, 09-137, Public Notice, 24 FCC Rcd 13692 (WCB 2009) (*NBP PN #1 6*)) filed Dec. 2, 2009, at 4–5 (noting that relevant uses of broadband technology provide both the initial motivation for broadband adoption and sustained use thereafter); Windstream Communications, Inc. Comments in re NBP PN #16, filed Dec. 2, 2009, at ii ("[W]hether and what amount a consumer is willing to pay for broadband service is largely a function of the value a consumer places on the service . . ."); National Black Caucus of State Legislators Comments in re National Broadband Plan NOI, filed Jan. 8, 2010, Attach. at 23.

[749] Horrigan, *Broadband Adoption and Use in America* at 19.

[750] *See generally* Everett Rog ers, Diffusion of Innovations (Free Press 4th ed. 1995) (rOGers, diffusiOn Of innOvatiOns).

[751] *See, e.g.*, National Black Caucus of State Legislators Comments in re National Broadband Plan NOI, filed Jan. 8, 2010, Attach. at 13.

[752] Awardees must contribute support equal to 15% of the requested grant amount. Further information, including application materials and guidelines, is available at USDA, Rural Development, www.usda.gov/rus/telecom/ commconnect.htm (last visited Mar. 9, 2010).

[753] *See* U.S. Dep't Agric., Community Connect Broadband Prog ram, Grant Application Guide, Fiscal Year 2009, at 22 (2009), *available at* http://www.usda.gov/rus/ telecom/commconnect/2009/2009CommConnectApp Guideb.pdf.

[754] National Telecommunications and Information Administration, Technology Opportunities Program About TOP, http://www.ntia.doc.gov/top/ about.html (last visited Feb. 22, 2010); National Telecommunications and Information Administration, Technology Opportunities Program, Grants, http:// www.ntia.doc.gov/top/ grants/grants.htm (last visited Feb. 22,2010).

[755] *See* Austen Free-Net, About AFN, http://www. austinfree.net/about/index.html (last visited Feb. 22, 2010); Mountain Area Information Network, About Main, http://www.main.nc.us/about/. (last visited Feb. 22, 2010).

[756] American Recovery and Reinvestment Act of 2009, Pub. L. No. 111-5, div. A, tit. II, 123 Stat. 115, 128 (2009) (Recovery Act).

[757] *See* Broadband Technology Opportunities Program, 75 Fed. Reg. 3,792 (Jan. 22, 2010); NTIA, Broadband Technology Opportunities Program, BTOP Project Information, http://www.ntia.doc. gov/broadbandgrants/projects.html (last visited Feb. 20, 2010). *See* BroadbandUSA, Fast-Forward New Mexico—Project Description, http://www. ntia.doc.gov/broadbandgrants/BTOPAward_ NewMexicoSt ateLibrary_121709.pdf.

[758] *See* National Telecommunications and Information Administration, BroadbandUSA, Fast-Forward New Mexico, http://www.ntia.doc.gov/broadbandgrants/ BTOPAward_NewMexicoStateLibrary_121709.pdf (last visited Feb. 23, 2010).

[759] *See* National Telecommunications and Information Administration, Broadband USA, Spokane Broadband Technology Alliance, http://www.ntia.doc.gov/ broadbandgrants/BTOPAward_TINCANWA_121709. pdf (last visited Feb. 23, 2010).

[760] *See* National Telecommunications Information Administration, BroadbandUSA, Los Angeles Computer Access Network, http://www.ntia.doc.gov/ broadbandgrants/LA_BTOP_Factsheet_FINAL.pdf (last visited Feb. 23, 2010).

[761] *See* Advanced Communications Law & Policy Institute Comments in re NBP PN #16, filed Dec. 2, 2009, at 6.

[762] Ultra High-Speed Broadband Task Force, Minnesota Ultra High-Speed Broadband Report 66 (2009) (Minnesota Ultra High-Speed Broadband Report), *available at* http://www.ultra-high-speed-mn.org/CM/ Custom/UHS%20Broadband%20Report_Full.pdf.

[763] Minnesota Ultra High-Speed Broadband Report at 71.

[764] *See, e.g.*, Advanced Communications Law & Policy Institute Comments in re NBP PN #16, filed Dec. 2, 2009, at 6–7.

[765] City of Seattle, *Community Technology Overview* available at http://seattle.gov/tech/overview/

[766] Cmty. Tech. Prog ram, Dep't of Info. Tech., City of Seattle, Information Technolog y Access and Adoption in Seattle (2009), *available at* http://www.cityofseattle.net/tech/indicators/docs/2009_ TechAccessA ndAdoptionInSeattleReport.pdf.

[767] Puget Sound Off, *Empower, Encourage SOUNDING OFF in your community*, http://pugetsoundoff.org/

[768] City of Seattle, *Community Technology Overview* available at http://seattle.gov/tech/overview/ of Chicago Comments in re NBP PN #16, filed Dec. 3, 2009, at 15–16.

[769] *See, e.g.*, City of Chicago Comments in re NBP PN #16, filed Dec. 3, 2009, at 2; Connected Nation Comments in re NBP PN #16, filed Dec. 2, 2009, at 7.

[770] *See, e.g.*, City of Chicago Comments in re NBP PN #16, filed Dec. 3, 2009, at 4–5; Connected Nation Comments in re NBP PN #16, filed Dec. 2, 2009, at 7.

[771] Letter from Rep. Calvin Smyre, George House of Representatives and President of the National Black Caucus of State Legislators (NBCSL), to Hon. Julius Genachowski Chairman, FCC, GN Docket No. 09–51 (filed Jan. 8, 2010) (NBCSL Jan. 8, 2010 Letter) Attach. at 10 ("Broadband in the home can help minimize the socio-economic disparities that persist among low income, minority or socially disadvantaged populations, which tend to be disparately impacted by a lack of access to quality information or essential services.").

[772] Horrigan, *Broadband Adoption and Use in America* at 19.

[773] *See generally* rOGers, diffusiOn Of innOvatiOns.

[774] NBCSL Jan. 8, 2010, Attach. at 7.

[775] *See* Greenlining Institute Comments in re NBP PN #13 (*Comment Sought on Broadband Study Conducted by the Berkman Center for Internet and Society—NBP Public Notice #13*, GN Docket Nos. 09-47, 09-51, 09-137, Public Notice, 24 FCC Rcd 12609 (WCB 2009) (*NBP PN #13*)), filed Nov. 16, 2009, Attach. at 3, 6–12; Advanced Communications Law & Policy Institute Comments in re NBP PN #16, filed Dec. 2, 2009, at 6; Broadband Diversity Supporters Comments in re National Broadband Plan NOI (*A National Broadband Plan for Our Future*, GN Docket No. 09-51, Notice of Inquiry, 24 FCC Rcd 4342 (2009)), filed Jun. 8, 2009, at 23.

[776] *See* Janice Hauge & James Prieger, Demand-Side Prog rams to Stimulate Adoption of Broadband: What Works? 59 (2009) (Hauge & Prieger, Prog rams to Stimulate Adoption of Broadband).

[777] Horrigan, *Broadband Adoption and Use in America* at 13.

[778] *See* Federal-State Joint Board on Universal Service Staff, 2009 Universal Service Monitoring Report, CC Docket Nos. 96-45, 98-62, at 2-2 (2009 universal service mOnitOrinG repOrt), *available at* http://hraunfoss.fcc.gov/edocs_public/attachmatch/DOC295442A1.pdf.

[779] *See* USAC, Federal Universal Service Support Mechanisms Fund Size Projections for Second Quarter 2010, at 2 (2010) (USAC, 2Q 2010 fund size prOjectiOns), *available at* http://www.universalservice.org/about/governance/fcc-filings/2010/ Q2/2Q2010%20Quarterly%20Demand%20Filing.pdf.

[780] 2009 Universal Service Monitoring Report at tbl. 2.1; *see also* USAC, 2008 Lifeline Participation Rate Data, http://www.usac.org/li/about/participation-rateinformation.aspx (last visited Feb. 19, 2010).

[781] In 2008, five states—Alaska, California, Colorado, Montana and Oklahoma—has an estimated Lifeline participation rate in excess of 50%. See USAC, 2008 Lifeline Participation Rates by State Map, http://www.usac.org/li/about/participation-rate-information.aspx (last visited Feb. 19, 2010).

[782] *See, e.g.*, Mark Burton et al., *Understanding Participation in Social Programs: Why Don't Households Pick up the Lifeline?*, 7 b.e. j. ecOn. anal & pOl'y, Art. 57 (2007), *available at* http:www.bepress.com/bejeap/vol7/iss1/art57 (purchase required); Janice A. Hague et al., *Whose Call Is It? Targeting Universal Service Programs to Low-Income Households' Telecommunications Preferences*, 33 telecOmm. pOl'y 129, 136–38 (2009), *available at* http://warrington.ufl.edu/purc/purcdocs/papers/0805_ Hauge_Whose_Call_is.pdf (pages 8–10 in this version).

[783] The FCC's rules impose one limitation on eligibility criteria for states that have their own programs: the criteria must be linked to income. See 47 C.F.R. § 54.409(a).

[784] *See, e.g.*, Cox Comments in re NBP PN #19, (*Comment Sought on the Role of the Universal Service Fund and Intercarrier Compensation in the National Broadband Plan—NBP Public Notice #19*, GN Docket Nos. 09-47, 09-51, 09-137, Public Notice, 24 FCC Rcd 13757 (OSP 2009) (*NBP PN #19*)) filed Dec. 7, 2009, at 12 (Lifeline customer should be able to use broadband virtual vouchers for fixed dollar amount of subsidy for any service tier that meets customer needs).

[785] *See, e.g.*, AT&T Comments in re NBP PN #19, filed Dec. 7, 2009, at 31.

[786] *See, e.g.*, AT&T Comments in re NBP PN #19, filed Dec. 7, 2009, at 31.

[787] *See, e.g.*, Letter from Jaime M. Tan, Director, Federal Regulatory, AT&T, to Marlene H. Dortch, Secretary, FCC, WC Docket No. 03-109, GN Docket Nos. 09-47, 09- 51, 09-137 (Dec. 22, 2009); State of New York Comments in re NBP PN #19, filed Dec. 7, 2009, at 2 (filed by David B. Salway on behalf of Melodie Mayberry-Stewart).

[788] Fl. Pub. Serv. Comm'n, Florida Lifeline & Link-Up Assistance: Number of Customers Subscribing to Lifeline Service and the Effectiveness of Procedures to Promote Participation 1 (2009), *available at* http://www.psc.state.fl.us/publications/pdf/telecomm/telelifelinereport2009.pdf.

[789] *See, e.g.*, Time Warner Cable Comments in re NBP PN #23 (*Comments Sought on Network Deployment Study Conducted by The Columbia Institute for Tele- Inform ation—NBP Public Notice #23*, GN Docket Nos. 09-47. 09-51, 09-137, Public Notice, 24 FCC Rcd 13890 (WCB 2009) (*NBP PN #23*)), filed Dec. 4, 2009, at 3; Free Press Reply in re NBP PN #30 (*Reply Comments Sought in Support of National Broadband Plan —NBP*

Public Notice #30, GN Docket Nos. 09-47, 09-51, 09-137, Public Notice, DA 10-61 (WCB, rel. Jan. 13, 2010) (*NBP PN #30*), filed Jan. 27, 2010, at 12.

[790] *See, e.g.*, Tracfone Comments in re NBP PN #19, filed Dec. 7, 2009, at 7.

[791] USAC, 2Q 2010 fund size prOjectiOns at 3, 15–17.

[792] *See, e.g.*, Cox Comments in re NBP PN #19, filed Dec. 7, 2009 (urging the FCC to promote digital literacy in other ways, such as partnerships between service providers and community organizations, schools and community colleges).

[793] *See, e.g.*, Robert D. Atkinson, Info.Tech. & Innovation Found., Policies to Increase Broadband Adoption At Home 3–4 (2009) (suggesting a market-based competition that would spur innovative adoption strategies by rewarding ISPs for attracting new subscribers in low-income communities), *available at* http://www.itif.org/files/2009- demand-side-policies.pdf.

[794] *Annual Assessment of Status of Competition in the Market for the Delivery of Video Programming*, MB Docket No. 06-189, Thirteenth Annual Report, 24 FCC Rcd 542, 546, para. 8 (2009).

[795] The cost of providing this wireless broadband service is not reflected in the broadband availability gap discussed in Chapter 8.

[796] Horrigan, *Broadband Adoption and Use in America* at 5.

[797] Immigrant and minority communities were heavily represented in the study, as the sizes of these populations tend to be too small to survey accurate. In this study, for example, 5% of the sample were Hmong—a population of relatively recent immigrants from Laos and Cambodia. Researchers attempted to explore possible regional differences, conducting interviews across the country, in urban and rural areas.

[798] Horrigan, *Broadband Adoption and Use in America* at 24, 26 ("Among current 'not-at-home' Internet users, 22% live with someone who uses the Internet at home. These nonusers often ask their online housemates to carry out tasks online for them"); Jon P. Gant et al., National Minority Broadband Adoption: Comparative Trends in Adoption, Acceptance and Use, Jt. Ctr. For Pol. & Econ. Stud. 3 (2010) (Gant et al., National Minority Broadband Adoption)), *available at* http:// www.jointcenter.org/publications1/publication-PDFs/MTI_BROADBAND_REPORT_2.pdf.

[799] Dharma Dailey et al., Broadband Adoption in Low -Income Communities 27 (2010), (Dharma Dailey et al., Broadband Adoption) *available at* http://www.ssrc. org/programs/broadband-adoption-in-low-income-communities/.

[800] Letter from Rey Ramsey, Chief Executive Officer, One Economy Corporation, to Julius Genachowski, Chairman, FCC, GN Docket No. 09-51 (Nov. 3, 2009).

[801] For a sampling of digital literacy programs in the European Union that are offered in a variety of formats, including face-to-face training, see knud erik hildinGhamann et al., danish tech. inst., suppOrtinG diGital literacy: analysis Of GOOd practice initiatives, tOpic 1 repOrt annexes (April 2008), *available at* http:// ec.europa.eu/information_society/eeurope/i2010/docs/ benchmarking/dl_topic_report_1.pdf.

[802] *See, e.g.*, Letter from Rey Ramsey, CEO, One Economy Corp., to Chmn. Julius Genachowski, FCC, GN Docket No. 09-51 (Nov. 3, 2009), Attach. (Comments of One Economy Corporation [on] National Digital Literacy Initiative) at 17; Senior Connects Corporation, Senior Connects, http://www.seniorconnects.org/index.html (last visited Mar. 3, 2010), *as referenced in* Net Literacy Corporation Reply in re NBP PN #16, filed Dec. 6, 2009, at 25–29 (filed by Daniel Kent).

[803] *See, e.g.*, Senior Connects Corporation, Senior Connects, http://www.seniorconnects.org/index. html (last visited March 3, 2010), *as referenced in* Net Literacy Corporation Reply in re NBP PN #16, filed Dec. 6, 2009, at 25– 29 (filed by Daniel Kent); Net Literacy, Community Connects Program, http:// www.communityconnects.org/netliteracy.html (last visited March 3, 2010) ("Net Literacy's programs are independently beginning to be developed by students from New York to California. The European Union's Commission on Digital Inclusion has nominated Net Literacy to be one of their 85 "Best of Class" digital inclusion models, based upon the Senior Connects programs established in Germany.").

[804] *See, e.g.*, Nat'l Telecomms. & Info. Admin., Tech. Opportunities Program, Grambling State University (Award Number 22-60-01064), *available at* http:// ntiaotiant2.ntia.doc.gov/top/details.cfm?oeam= 226001064 (last visited March 4, 2010); Nat'l Telecomms. & Info. Admin., Broadband Tech. Opportunities Program, Lowell Internet, Networking and Knowledge: Sustaining Broadband Access Across the Generations, http://www.ntia.doc.gov/broadbandgrants/factsheets/ UMassLowell_BTOP_Factsheet_LES_011910.pdf (last visited March 4, 2010).

[805] NBCSL Jan. 8, 2010, Attach. at 12.

[806] OBI, 2009 Broadband Adoption and Use Survey database (providing data of 5,005 respondents). 16% of Hispanic non-adopters who took the survey in Spanish cite relevance and 19% cite digital literacy as the main barriers to broadband adoption. Number of cases of Hispanic non- adopters who answered the survey in Spanish is 126.

[807] Corp. for Nat'l and Cmty. Serv., AmeriCorps: Changing Lives, Changing America 8 (2007), *available at* www. Serve.illinois.gov/national_service/pdfs/AmeriCorps_ Lives_America.pdf.

[808] For more about the CyberNavigators, see Chicago Pub. Library Found., Programs, http://www. chicagopubliclibraryfoundation.org/programs/ (last visited Mar. 4, 2010).

[809] Gant et al., natiOnal minOrity brOadband adOptiOn at 3.

[810] *See, e.g.*, American Library Association Comments in re NBP PN #16, filed Dec. 2, 2009, at 3.

[811] Dharma Dailey et al., broadband adoption at 27–28.

[812] *See* Dharma Dailey et al., broadband adoption at 4.

[813] *See generally* Dharma Dailey et al., broadband adoption.

[814] American Library Association Comments in re NBP PN #16, filed Dec. 2, 2009, at 9.

[815] am. library ass'n, libraries cOnnect cOmmunities 3: public library fundinG & technOlOGy access study 45–46 (2009), *available at* http://ala.org/ ala/research/initiatives/plftas/2008_2009/ librariesconnectc ommunities3.pdf.

[816] Letter from Elvis Stumbergs, National Broadband Taskforce, FCC, on behalf of Learning Express: Top 25 Products Usage, to Marlene H. Dortch, Secretary, FCC, GN Docket Nos. 09-47, 09-51, 09-137 (Jan. 14, 2010) Attach. at 1.

[817] Dharma Dailey et al., broadband adoptiOn at 28.

[818] Letter from Elvis Stumbergs, National Broadband Taskforce, FCC, on behalf of Inst. for Museum & Libr. Serv., to Marlene H. Dortch, Secretary, FCC, GN Docket Nos. 09-47, 09-51, 09-137 (Jan. 13, 2010) (IMLS Nov. 3, 2009 *Ex Parte*) Attach. at 1.

[819] Inst. of Museum and Library Serv., A Catalyst for Change: LSTA Grants to States Prog ram Activities and the Transformation of Library Services to the Public (2009), *available at* http://www.imls.gov/pdf/ CatalystForChange.pdf.

[820] Dharma Dailey et al., broadband adoption at 31.

[821] IMLS Nov. 3, 2009 *Ex Parte*, Attach. at 132.

[822] *See* OnGuard Online, About Us, http://www. onguardonline.gov/about-us/overview.aspx (last visited Feb. 22, 2010).

[823] U.S. Dep't of Hous. & Urban Dev., Office of Policy Dev. and Research, Minority-Serving Institutions of Higher Education: Developing Partnerships to Revitalize Communities 7–9 (2003), *available at* http://www.oup. org/files/pubs/minority-report.pdf.

[824] Horrigan, *Broadband Adoption and Use in America* at 30.

[825] American Library Association Comments in re NBP PN #16, filed Dec. 2, 2009, at 7.

[826] Horrigan, *Broadband Adoption and Use in America* at 30.

[827] *See* Letter from David E. Chase, Dir., Program Monitoring and Res. Div., Off. of Pol'y Dev. & Res., U.S. Dep't of Housing & Urban Dev., to Marlene H. Dortch, Secretary, FCC, GN Docket No. 09-51 (Feb. 25, 2010).

[828] *See, e.g.*, Video: U.S. Dep't of Hous. & Urban Dev., Choice Neighborhoods Stakeholder Meeting Presentation (Nov. 10, 2009), *available at* http://link.onlinevideoservice. com/hud/2009/1110/Archive_20091110_edited-1.wmv.

[829] OBI, 2009 Broadband Adoption and Use Survey database (providing data of 5,005 respondents). Survey respondents who reported having a child (under 18) living at home and who make less than $20,000 per year have a broadband adoption rate of 50% compared to families earning between $50,000 and $75,000 per year (85%). A crosstab of variable RECINC7 and ADOPTERS yields these results.

[830] AT&T, *National Survey Finds Kids Give High Marks to High Speed* (press release), Aug. 4, 2004, http://www.att. com/gen/press-room?pid=4800&cdvn=news&newsarti cleid=21284.

[831] Social Sec. Admin., Annual Report of the Supplemental Security Income Prog ram Prog ram 106–07 (2009), *available at* http://www.ssa.gov/OACT/ssir/SSI09/ ssi2009.pdf.

[832] Horrigan, *Broadband Adoption and Use in America* at 43.

[833] Horrigan, *Broadband Adoption and Use in America* at 35, 37.

[834] Older Adults Technology Services Comments in re NBP PN #16, filed Dec. 2, 2009, at 4.

[835] For example, multiple projects have been proposed that would use remote monitoring to assess and assist Alzheimer's patients and low-income, underserved elderly populations. *See, e.g.*, Letter from Alice Borelli, Dir., Global Healthcare & Workforce Pol'y, Intel Corp., to Marlene H. Dortch, Secretary, FCC, GN Docket Nos. 09-47, 09-51, 09-137, WC Docket No. 02-60 (Jan. 15, 2010) Attachs.; Letter from Alice Borelli, Dir., Global Healthcare & Workforce Pol'y, Intel Corp., to Marlene H. Dortch, Secretary, FCC, GN Docket Nos. 09-47, 09-51, 09-137, WC Docket No. 02-60 (Dec. 16, 2009) Attachs.; *see also* Oregon Health & Science University, Orcatech Research Studies, http://www.orcatech.org/research/ studies (last visited Jan. 19, 2010).

[836] *See, e.g.*, Advanced Communications Law & Policy Institute Reply in re National Broadband Plan NOI, filed July 21, 2009, at 4–5; Consumer Policy Solutions Comments in re National Broadband Plan NOI, filed June 8, 2009, at 3–4 (filed by Debra Berlyn).

[837] BBC, *Internet use 'Good for The Brain'*, bbc neWs, Oct. 14, 2008, *available at* http://news.bbc. co.uk/2/hi/health/7667610.stm; *see also* Advanced Communications Law & Policy Institute Reply in re National Broadband Plan NOI, filed July 21, 2009, at 4–5; Benedict Carey, *At the Bridge Table, Clues to a Lucid Old Age*, n.y. times, May 22, 2009 (mental engagement may delay the onset of symptoms of dementia), *available at* http://www.nytimes.com/2009/05/22/health/ research/22brain.html.

[838] *See* Joseph C. Kvedar, M.D., *Is Facebook the Up and Coming Health IT Application?*, health it neWs, Feb. 3, 2009, http://www.healthcareitnews.com/blog/ facebook-and-coming-health-it-application; *see also* Shereene

Z. Idress et al., *The Role of Online Support Communities, Benefits of Expanded Social Networks to Patients with Psoriasis*, 145 arch. Of dermatOl. 46 (2009), *available at* http://archderm.ama-assn.org/cgi/content/full/145/1/46.

[839] Amanda Lenhart, *Senior Citizens Not Flocking to Social Networking Sites: Just 7% Have Posted Profile*, seniOrjOurnal.cOm, Jan. 22, 2009, http://seniorjournal. com/NEWS/WebsWeLike/2009/20090122-SenCitNotFlocking.htm.

[840] Horrigan, *Broadband Adoption and Use in America* at 35–36. Some 39% of African Americans have gone online with their cell or Smartphone (defined as e-mailing, accessing the web for information or downloading an application), 39% of Hispanics have done this, and 27% of whites have done this.

[841] Horrigan, *Broadband Adoption and Use in America* at 35–36. The 20% of African Americans without broadband at home have used the Internet on their handheld devices, and 25% of Hispanics without broadband at home have done this.

[842] *See, e.g.*, Robert C. Atkinson & Ivy E. Schultz, Columbia Inst. for Tele-Information, Broadband In America: Where It Is And Where It Is Going (According To Broadband Service Providers) 10 (2009).

[843] Horrigan, *Broadband Adoption and Use in America* at 17.

[844] NBCSL Jan. 8, 2010, Attach. at 13.

[845] NBCSL Jan. 8, 2010, Attach. at 13.

[846] Common Sense Media Nov. 23, 2009 *Ex Parte* at 2–3.

[847] Workforce Investment Act of 1998, § 508, Pub. L. No. 105-220, 112 Stat. 936 (1998) (codified as § 504 of the Rehabilitation Act, 29 U.S.C. § 794d) (Workforce Investment Act).

[848] Workforce Investment Act § 508 (a) (1)(A).

[849] *See, e.g.*, Eric Bridges, American Council of the Blind, Remarks at FCC Broadband Accessibility for People with Disabilities II: Barriers, Opportunities, and Policy Recommendations Workshop (Oct. 20, 2009), *available   at*   http://broadband.gov/docs/ws_accessibility_ disabilities/ws_accessibility_disabilities_transcript.pdf; Karen Peltz Strauss, Co-Chair, Coalition of Organizations for Accessible Technologies, Remarks at FCC Broadband Accessibility for People with Disabilities II: Barriers, Opportunities, and Policy Recommendations Workshop (Oct. 20, 2009), *available at* http://broadband.gov/docs/ws_accessibility_ disabilities/ws_accessibility_disabilities_transcript.pdf.

[850] Workforce Investment Act § 508(d)(2).

[851] *See* Dep't of Justice, Civil Rights Div., Section 508 Homepage, http://www.justice.gov/crt/508/508home. php (last visited Feb. 20, 2010).

[852] Workforce Investment Act § 508(d)(2).

[853] For example, under Medicare's regulations, coverage of assistive technologies is limited to "durable medical equipment" that is "primarily and customarily used to serve a medical purpose" and "generally is not useful to a person in the absence of an illness or injury." 42 C.F.R. § 414.202.

[854] Karen Peltz Strauss, Co-Chair, Coalition of Organizations for Accessible Technologies, Remarks at FCC Broadband Accessibility for People with Disabilities II: Barriers, Opportunities, and Policy Recommendations Workshop   (Oct.   20,   2009),   *available   at*   http://broadband.gov/docs/ws_accessibility_ disabilities/ws_accessibility_disabilities_transcript. pdf; *see also* Letter from Gregg Vanderheiden, Dir., Rehabilitation Eng. Res. Ctr. on Universal Interface & Info. Tech. Access, Trace R&D Ctr., Univ. of Wisc. et al., to Marlene H. Dortch, Secretary, FCC, GN Docket Nos. 09-47, 09-51, 09-137 (Jan. 6, 2010) at 1.

[855] Broadband Data Improvement Act of 2008, Pub. L. No. 110-385, 122 Stat. 4097 (2008) (codified at 47 U.S.C. §§ 1301–1304) (BDIA).

[856] *See, e.g.*, Twenty-First Century Communications and Video Accessibility Act of 2009, H.R. 3101, 111th Cong. §2 (2009).

[857] 47 C.F.R. § 6.1 et seq. The rules implementing Section 255 require telecommunications and interconnected VoIP service providers and manufacturers to consider accessibility issues in the design and development phase and to include accessibility features in their products when it is readily achievable to do so.

[858] Advanced services as defined in H.R. 3101 include non- interconnected VoIP, electronic messaging, and video conferencing (as well as interconnected VoIP, which is covered by Section 255). The FCC should assure itself of its jurisdiction to extend Section 255 to all advanced services or, if it cannot do so, seek authorization from Congress.

[859] H.R. 3101 requires advanced services providers and equipment manufacturers to make their products accessible unless doing so would cause an undue burden. H.R. 3101 should be a starting point for discussion of both the scope of coverage and the legal standard of the accessibility obligation applied to service providers and manufacturers. We encourage stakeholders to work toward a long-term goal of having as much inclusion as possible for people with disabilities.

[860] *See, e.g.*, Twenty-first Century Communications and Video Accessibility Act of 2009, H.R. 3101, 111th Cong. § 102 (2009).

[861] This proceeding should be coordinated with the FCC proceeding, which addresses the future roles of 911 and NG911 as communications technologies, networks and architectures expand beyond traditional voice-centric devices. As part of the proceeding, the FCC should assess its jurisdiction to adopt rules with respect to (i)

captioning and emergency information of video programming on the Internet and devices which display such programming; and (ii) related user interfaces, video programming guides and menus.

[862] The Americans with Disabilities Act of 1990, Pub. L. No. 101-336, 104 Stat. 327 (1990) (codified at 42 U.S.C. §12101) (ADA).

[863] This recommendation is similar to a provision in H.R. 3101, § 201.

[864] In *Motion Picture Ass'n of America, Inc. v. FCC*, 309 F.3d 796 (D.C. Cir. 2002), the D.C. Circuit vacated the FCC's video description rules, finding that the FCC lacked the authority to adopt such rules. H.R. 3121 should be a starting point for discussion with respect to the scope of the FCC's authority to adopt video description rules.

[865] *See, e.g.*, Twenty-first Century Communications and Video Accessibility Act of 2009, H.R. 3101, 111th Cong. § 105 (2009).

[866] *See* FCC, *FCC Announces Agenda and Panelists for Workshop on VRS Reform To Be Held on December 17, 2009* (press release), Dec. 15, 2009, http://hraunfoss.fcc. gov/edocs_public/attachmatch/DOC-295208A1.doc.

[867] *See* FCC, FCC Telecommunications Relay Services, Consumer Facts, http://www.fcc.gov/cgb/consumerfacts/trs.html (last visited Jan. 6, 2010).

[868] *See* Rebecca Ladew, East Coast Representative, Speech Communications Assistance by Telephone, Inc., Remarks at FCC Broadband Accessibility for People with Disabilities II: Barriers, Opportunities, and Policy Recommendations Workshop (Nov. 6, 2009), *available at* http://broadband.gov/docs/ws_accessibility_disabilities/ws_accessibility_disabilities_transcript.pdf; Letter from Monica Martinez, Commissioner, Mich. Pub. Serv. Comm'n, to Julius Genachowski, Chairman, FCC, GN Docket Nos. 09-47, 09-51, 09-137, CS Docket No. 97-80 (Dec. 23, 2009) at 1.

[869] BDIA § 106(i)(2) (codified at 47 U.S.C. § 1304(i)(2)); *see also* California Public Utilities Commission Comments (filed July 30, 2009) at 4.

[870] California Emerging Technology Fund, History, http:// cetfund.org/aboutus/history (last visited Mar. 4, 2010).

[871] BDIA § 102(4) ("The Federal Government should also recognize and encourage complementary State efforts to improve the quality and usefulness of broadband data and should encourage and support the partnership of the public and private sectors in the continued growth of broadband services and information technology for the residents and businesses of the Nation.")

[872] BDIA § 106(a)(1)–(2).

[873] Esme Vos, *Ten States Receive Broadband Mapping and Planning Grants from the NTIA*, muniWireless, Jan, 12, 2010, http://www.muniwireless.com/2010/01/12/ ten-states-receive-broadband-mapping-and-planninggrants-from-the-ntia/.

[874] *See* BDIA § 106(e)(5)(B)(iii) (codified at 47 U.S.C. § 1304(e) (5) (B)(iii)).

[875] BDIA § 106(e)(7) (codified at 47 U.S.C. §1304(e)(7)).

[876] *See* BDIA § 106(e)(6)–(7) (codified at 47 U.S.C. § 1304(e)(6)–(7)). S*ee also* Sen. Kay Bailey Hutchinson, *Broadband Plan Must be Daring, Comprehensive*, hill, Jan. 5, 2010, *available at* http://thehill.com/special-reports/technology-january-2010/74481-broadband- plan-must-be-daring-comprehensive.

[877] *See* BDIA § 106(e)(5)(B)(ii), (e)(7).

[878] *See* BDIA § 106(e)(5)(B)(i), (e)(6)–(7).

[879] *See* BDIA § 106(e)(5)–(7) (codified at 47 U.S.C. §1304(e)–(7)).

[880] *See generally* Westat, Collected Case Study Evaluations: Summary of Findings 20 (1999), *available at* http://www.ntia.doc.gov/top/research/EvaluationReport/case_studies/casestudysummary.pdf.

[881] Hauge & Prieger, Prog rams to Stimulate Adoption of Broadband at 59.

[882] Hauge & Prieger, Prog rams to Stimulate Adoption of Broadband at 62.

[883] *See Statement of Policy on Establishing a Governmentto-Government Relationship with Indian Tribes*, Policy Statement, 16 FCC Rcd 4078 (2000).

[884] *See* Food, Conservation and Energy Act of 2008, Pub. L. No. 110-246, § 6105, 122 Stat. 1651, 1957–58 (2008) (codified at 7 U.S.C. § 936f).

[885] *See* California Association of Tribal Governments Ex Parte in re NBP PN #5, filed Dec. 17, 2009, at 7; Letter from Loris Ann Taylor, Executive Director, Native Public Media et al., to Marlene H. Dortch, Secretary, FCC, GN Docket Nos 09-47, 09-51, 09-137 (Dec. 24, 2009) (Native Public Media et al. Dec. 24, 2009 *Ex Parte*) at 24.

[886] *See* California Association of Tribal Governments *Ex Parte* in re NBP PN #5, filed Dec. 17, 2009, at 12; Native Public Media et al. Dec. 24, 2009 *Ex Parte* at 5–6; Native Public Media & the National Congress of American Indians Comments in re NBP PN #5, filed Dec. 9, 2009, Attach. 1 at 4, 39, 44.

[887] Erik Brynjolfsson & Adam Saunders, Wired for Innovation: How Information Technolog y is Reshaping the Economy (2010) (Brynjolfsson & Saunders, Wired for Innovation)..

[888] Brynjolfsson & Saunders, Wired for Innovation at xii–xiii.

[889] Paul A. David, *The Dynamo and the Computer: An Historical Perspective on the Modern Productivity Paradox*, 80 AEA Paper & Proceeding 355 (1990) (David, The Dynamo and the Computer).

[890] David, *The Dynamo and the Computer* at 358–59.

[891] David, *The Dynamo and the Computer* at 357.

[892] David, *The Dynamo and the Computer* at 356–57.

[893] David, *The Dynamo and the Computer* at 357.

[894] American Recovery and Reinvestment Act of 2009, Pub. L. No. 111-5, § 6001(k)(2)(D), 123 Stat. 115, 516 (2009).

[895] *See generally* U.S. R&E Networks and HIMSS Reply in re NBP PN #30 (*Reply Comments Sought in Support of National Broadband Plan—NBP Public Notice #30*, GN Docket Nos. 09-47, 09-51, 09-137, Public Notice, 25 FCC 241 (WCB 2010) (*NBP PN #30*)), filed Jan. 27, 2010; Commenters Supporting Anchor Institution Networks Reply in re NBP PN #30, filed Jan. 27, 2010.

[896] Ctr. for Medicare & Medicaid Serv., National Health Expenditure Projections 2008–2018, http://www. cms.hhs.gov/NationalHealthExpendData/downloads/ proj2008.pdf (last visited Jan. 21, 2010).

[897] Office of the Surgeon General, Overweight and Obesity: At a Glance, http://www.surgeongeneral.gov/topics/ obesity/calltoaction/fact_glance.htm (last visited Jan. 21, 2010). Obesity is a particular challenge for African Americans, as approximately four out of five African American women are overweight or obese. *See* Department of Health and Human Services, Office of Minority Health, http://minorityhealth.hhs.gov/ templates/content.aspx?ID=6456 (last visited Feb. 28, 2010). In addition, African American adults are twice as likely as non-Hispanic white adults to have been diagnosed with diabetes by a physician, and in 2005, African American men were 30% more likely to die from heart disease, as compared to non-Hispanic white men. *See* Department of Health and Human Services, Office of Minority Health, http://minorityhealth.hhs.gov/ templates/browse.aspx?lvl=2&lvlID=51 (last visited Feb. 28, 2010).

[898] Susan Dentzer, *Reform Chronic Illness Care? Yes, We Can*, 28 health aff. 12, 12 (Jan./Feb. 2009), *available at* http://content.healthaffairs.org/cgi/reprint/28/1/12.

[899] Shin-Yi Wu & Anthony Green, Rand Corp., Projection of Chronic Illness Prevalence and Cost Inflation (2000).

[900] Am. Heart Ass'n, Heart Failure, http://www. americanheart.org/chf (last visited Jan. 21, 2010)

[901] Inst. of Medicine, Preventing Medical Errors 5 (Philip Aspden et al. eds., 2007), *available at* http://books.nap. edu/openbook.php?record_id=11623&page=5.

[902] Robert A. Weinstein, *Nosocomial Infection Update*, 4 emerGinG infectiOus disease 416 (July–Sept. 1998), *available at* ftp://ftp.cdc.gov/pub/EID/vol4no3/adobe/ weinstein.pdf.

[903] *See* Health Res. & Serv. Admin., U.S. Dep't of Health & Human Serv., The Physician Workforce: Projections and Research into Current Issues Affecting Supply and Demand (2008), ftp://ftp.hrsa.gov/bhpr/workforce/ physicianworkforce.pdf (HRSA, Physician Workforce); Michael J. Dill & Edward S. Salsberg, Ass'n of Am. Med. Coll., The Complexities of Physician Supply and Demand: Projections Through 2025, at 6 (2008) (estimating a shortage of 124,000 physicians by 2025), https://services.aamc.org/publications/ index. cfm?fuseaction=Product.displayForm&prd_id=244 (download report from this page).

[904] HRSA, Physician Workforce at iv.

[905] Institute of Medicine, Unequal Treatment: Confronting Racial and Ethnic Disparities in Health Care 29 (Brian Smedley et al. eds., 2003.)

[906] *Id.*

[907] Medicare and Medicaid Programs; Electronic Health Record Incentive Program; Proposed Rule, 75 Fed. Reg. 1851 (Jan. 13, 2010).

[908] This is not a definitive taxonomy. Electronic health record definition used by Health Information and Management Systems Society (HIMSS). *See* HIMSS, Electronic Health Record, http://www.himss.org/ASP/ topics_ehr.asp (last visited Feb. 27, 2010).

[909] Todd Park & Peter Basch, Ctr. for Am. Prog ress, A Historic Opportunity: Wedding Health Information Technolog y to Care Delivery Innovation and Provider Payment Reform 6 (2009), *available at* http://www. americanprogress.org/issues/2009/05/pdf/ health_it.pdf (Park & Basch, Wedding Health Information Technology).

[910] *See* Catherine DesRoches et al., *Electronic Health Records in Ambulatory Care—A National Survey of Physicians*, 359 New Eng. J. Med. 50 (2008), *available at* http://content.nejm.org/cgi/reprint/359/1/50.pdf.

[911] James Bigelow et al., Rand Corp., Analysis of Healthcare Interventions That Change Patient Trajectories, at xxiv (2005), http://www.rand.org/ pubs/monographs/2005/RAND_MG408.pdf.

[912] Richard Hillestad et al., *Can Electronic Medical Record Systems Transform Healthcare? Potential Health Benefits, Savings, and Costs*, 24 health aff. 1103, 1103 (Sept./Oct. 2005), *available at* http://content. healthaffairs.org/cgi/reprint/24/5/1103.

[913] Hillestad et al., *Can Electronic Medical Record Systems Transform Healthcare?* at 1114.

[914] American Academy of Pediatrics Comments in re NBP PN #17 (*Comment Sought on the Health Care Delivery Elements of National Broadband Plan—NBP Public Notice #17*, GN Docket Nos. 09-47, 09-51, 09-137, Public Notice, 24 FCC Red 137 28 (WCB 2009) (*NBP PN #17*)), filed Dec. 4, 2009, at 1; Rural Wisconsin Health Cooperative Information Technology Network (RWHC ITN) Comments in re NBP PN #17, filed Dec. 4, 2009, at 3.

[915] Sixty-five million people reside in Primary Care Health Professional Shortage Areas (HPSAs). HHS designates HPSAs as having a shortage of primary medical care, dental, or mental health providers. They may be urban or

rural areas, population groups, or medical or other public facilities. See HRSA, Shortage Designation: HPSAs, MUAs & MUPs, http://bhpr.hrsa.gov/shortage/ (last visited Jan. 29, 2010).

[916] Lee Schwamm et al., *Recommendations for the Implementation of Telemedicine within Stroke Systems of Care: A Policy Statement from the American Heart Association*, 40 strOke 2635 (2009), *available at* http://stroke.ahajournals.org/cgi/reprint/40/7/2635. The United States has approximately 4.0 neurologists per 100,000 persons, who ideally need to be caring for over 780,000 acute strokes per year, and many parts of the country lack access to acute stroke services entirely. *Id.* at 2638. Tissue plasminogen activator (tPA) is a clot-busting drug that must be administered within three hours of ischemic stroke onset to be effective. *Id.* at 2641.

[917] Richard Knox, *Drug Can Stop Strokes, But Most Patients Don't Get It*, NPR, Dec. 14, 2009, http://www.npr.org/templates/story/story. php?storyId=121032269&sc=emaf.

[918] Store-and-forward technologies represent the collection and storage of clinical data or images that are forwarded for interpretation at a time distant from an in-person clinical encounter. Ctr. for Info. Tech. Leadership, Health Care Information and Management Systems Society (Himss), the Value of Provider-to-Provider Telehealth Technologies (2007) (Citl, the Value of Provider-to-Provider Telehealth Technologies), *available at* http://www.citl.org/_pdf/CITL_Telehealth_ Report.pdf.

[919] CITL, The Value of Provider-to-Provider Telehealth Technologies. CITL modeled pre- and post-telehealth costs based on national baseline number of transports, transport cost, and number of avoided transports. Annual savings were calculated by subtracting posttelehealth costs from pre-telehealth costs for each provider-to-provider setting. These savings sum to $1.2 billion.

[920] Juan M. Aranda, Jr. et al., *Current Trends in Heart Failure Readmission Rates: Analysis of Medicare Data*, 32 Clinical Cardiology 47, 47 (2009), *available at* http://www3.interscience.wiley.com/cgi-bin/fulltext/121637973/ PDFSTART.

[921] Robert Litan, Better Health Care Together Coalition, Vital Signs via Broadband: Remote Health Monitoring Transmits Savings, Enhances Lives (2008), *available at* http://www.betterhealthcaretogether.org/Library/Documents/VITAL%20SIGNS%20via%20BROADBAND%20FINAL%20with%20FOREWORD%20and%20 TITLE%20pp%2010%2022.pdf

[922] Adam Darkins et al., *Care Coordination/Home Telehealth: The Systematic Implementation of Health Informatics, Home Telehealth, and Disease Management to Support the Care of Veteran Patients with Chronic Conditions*, 10 telemed. & e-health 1118, 1118 (2008), *available at* http://www.liebertonline.com/doi/pdf/10.1089/tmj.2008.0021?cookieSet=1.

[923] Qualcomm Inc. Comments in re NBP PN #17, filed Dec. 4, 2009, at 20.

[924] Body sensor networks are very short-range networks consisting of multiple body-worn sensors and/or nodes and a nearby hub station. The sensors and/or nodes take readings of key patient-specific information, such as temperature, pulse, blood glucose level, the heart's electrical activity, blood pressure, and respiratory function. Antenna components embedded in the sensors and/or nodes make it possible to wirelessly transmit data to body- worn or closely-located hub devices. Hub devices may process the data locally and/or transmit it wirelessly for centralized processing, display, and storage.

[925] West Wireless Health Institute Comments in re NBP PN #17, filed Dec. 4, 2009, at 2.

[926] West Wireless Health Institute Comments in re NBP PN #17, filed Dec. 4, 2009, at 5.

[927] Wireless cardiovascular solutions are designed to enable early detection, prevention, and treatment of cardiovascular conditions. These solutions consist of wearable sensors, global wireless capabilities, and comprehensive web-based infrastructures. *See, e.g.*, Corventis Home Page, http://www.corventis.com/US/ (last visited Jan. 30, 2009).

[928] Medtronic, REAL-Time Continuous Glucose Monitoring, http://www.minimed.com/products/insulinpumps/components/cgm.html (last visited Feb. 1, 2010).

[929] *Amendment of Parts 2 and 95 of the Commission's Rules to Establish the Medical Micropower Network Service in the 413–457 MHz Band*, RM-11404, Petition for Rulemaking (filed Sept. 5, 2007).

[930] Letter from Cheryl A. Tritt, Counsel to the Alfred Mann Foundation, Morrison & Foerster LLP, to Marlene H. Dortch, Secretary, FCC, ET Docket No. 09-36 (Feb. 2, 2010) (Alfred Mann Feb. 2, 2010 *Ex Parte*) at 12.

[931] Cathy Schoen & Robin Osborn, the Commonwealth Fund, the Commonwealth Fund 2009 internatiOnal health pOlicy survey Of primary care physicians in eleven cOuntries 10 (2009), http://www. commonwealthfund.org/~/media/Files/Publications/ In%20the%20Literature/2009/Nov/PDF_ Schoen_2009_ Commonwealth_Fund_11country_intl_ survey_chartpack_white_bkgd_PF.pdf. Count of 14 functions includes: (1) electronic medical record; (2, 3) electronic prescribing and ordering of tests; (4–6) electronic access to test results, Rx alerts, and clinical notes; (7–10) computerized system for tracking lab tests, guidelines, alerts to provide patients with test results, and preventive/follow-up care reminders; and (11–14) computerized list of patients by diagnosis, by medications, and due for tests or preventive care.

[932] Joint Advisory Committee on Communications Capabilities of Emergency Medical and Public Health Care Facilities, Report To Congress (2008), *available at* http://energycommerce.house.gov/Press_110/JAC. Report_FINAL%20Jan.3.2008.pdf.

[933] Hospitals & Health Networks, Figure 10: Home Telemonitoring, http://www.hhnmag.com/hhnmag_ app/jsp/ articledisplay.jsp?dcrpath=HHNMAG/Article/data/07JUL2008/0807HHN_MW_MainArticle_Fig10& domain=HHNMAG (last visited Jan. 30, 2010).

[934] Phillips, National Study on the Future of Technology and Telehealth in Home Care 105 (2008).

[935] David Blumenthal, *Launching HITECH*, New Eng. J. Med., Dec. 30, 2009, at 3, (Blumenthal, *Launching HITECH*) *available at* http://content.nejm.org/cgi/ reprint/NEJMp0912825v1.pdf.

[936] Hit Policy Committee, Meaningful Use Objectives and Measures: 2011-2013-2015 (early version) (2009), http://healthit.hhs.gov/portal/server.pt/gateway/ PTARGS_0_10741_887553_0_0_18/Proposed_ Revisions_ to_Meaningful_Use_post_7_16_2009_ FINAL_PT1_508.pdf.

[937] American Recovery and Reinvestment Act of 2009, Pub. L. No. 111-5, § 13101, 123 Stat. 115, 231 (2009) (Recovery Act).

[938] Blumenthal, *Launching HITECH* at 3.

[939] Park & Basch, Wedding Health Information Technology at 2, 5–7, 14; Medicare Payment Advisory Comm'n (Medpac), Report to the Congress: Promoting Greater Efficiency in Medicare 17 (2007), *available at* http://www.medpac.gov/documents/jun07_ EntireReport.pdf; Great Call, Inc. Comments in re NBP PN #17, filed Dec. 4, 2009, at 7; Palmetto State Providers Network Comments in re PN #17, filed Nov. 24, 2009, at 12 (filed by W. Roger Poston, II)

[940] Cal. Telemed. and eHealth Ctr., Optimizing Telehealth in California: An Agenda for Today and Tomorrow 4–7 (2009) (CTEC, Optimizing Telehealth in California), *available at* http://www.cteconline.org/_pdf/Findings-Report.pdf.

[941] Ashish K. Jha et al., *Use of Electronic Health Records in U.S. Hospitals*, 360 neW enG. j. med. 1628 (2009), *available at* http://content.nejm.org/cgi/ reprint/360/16/1628.pdf.

[942] CTEC, Optimizing Telehealth in California at 4.

[943] *See* Box 10.3, *supra.*

[944] Park & Basch, Wedding Health Information Technology at 2, 5–7, 14.

[945] MedPAC, Report to the Congress: Medicare in Rural America 36 (2001), *available at* http://www.medpac. gov/documents/Jun01%20Entire%20report.pdf.

[946] Park & Basch, Wedding Health Information Technology at 2, 5–7, 14.

[947] CMS uses the term telehealth, which is similar to the term e-care. See Box 10.1 for further explanation.

[948] Payment Policies Under the Physician Fee Schedule and Other Revisions to Part B for CY 2010, 74 Fed. Reg. 33519, 33663 (July 13, 2009) (codified at 42 C.F.R. pts. 410, 411, 414, etc.), *available at* http://edocket.access. gpo.gov/2009/E9-15835.htm; Office of the Nat'l Coordinator for Health Info. Tech., HHS, Justification of Estimates for Appropriations Committees: Fiscal Year 2011 (2010), *available at* http://healthit.hhs.gov/portal/ server.pt/gateway/PTARGS_0_11673_910512_0_0_18/ ONC_FY2011_CJ.pdf.

[949] For instance, the Health IT Policy Council cites "Incorporate data from home monitoring device" as one of the objectives for defining Meaningful Use in 2013. HIT Policy Committee, Meaningful Use Objectives and Measures: 2011-2013-2015 (final version) 6 (2009), available at http://healthit.hhs.gov/portal/server.pt/ gateway/PTARGS_0_10741_888532_0_0_18/FINAL%20MU%20RECOMMENDATIONS%20TABLE.pdf.

[950] CMS, *CMS Announces Demonstration to Encourage Greater Collaboration and Improve Quality Using Bundled Hospital Payments* (press release), May 16, 2008, *available at* http://www.cms.hhs.gov/apps/media/ press/release.asp?Counter=3109&intNumPerPage=10&checkDate=&checkKey=&srchType=1&numDays=35 00 &srchOpt=0&srchData=&keywordType=All&chkNewsType=1%2C+2%2C+3%2C+4%2C+5&intPage=& showA ll=&pYear=&year=&desc=&cboOrder=date.

[951] CMS, Medical Home Demonstration Fact Sheet (2009), http://www.cms.hhs.gov/demoprojectsevalrpts/ downloads/medhome_factsheet.pdf.

[952] *See* Patient Protection and Affordable Care Act, H.R. 3590, 111th Cong. §§ 2703, 3022–24 (2009), *available at* http://frwebgate.access.gpo.gov/cgi-bin/getdoc. cgi?dbname=111_cong_bills&docid=f:h3590eas.txt.pdf; Affordable Health Care for America Act, H.R. 3962, §§ 1152, 1301–02, 1312 (2009), *available at* http://docs. house.gov/rules/health/111_ahcaa.pdf.

[953] IHS is an agency within the Department of Health and Human Services responsible for providing federal health services to 1.9 million American Indians and Alaska Natives in predominantly rural and remote areas. Indian Health Service, IHS Fact Sheets Year 2009 Profile, http://info.ihs. gov/Profile09.asp (last visited Dec. 29, 2009).

[954] The Connected Care Telehealth Program is a pilot of video consultation services—sponsored by UnitedHealth Group, Cisco, Centura Health, the Colorado Rural Health Center and the Colorado Community Health Network—that will make available 4,800 remote specialist visits per year in Colorado. UnitedHealthcare, *UnitedHealthcare to Launch Connected Care Telehealth Program in Rural Colorado* (press release), (Apr. 17, 2009), http://www.connectedcareamerica.com/news-and-resources.php (link available to the article from this page).

[955] The Community Partnerships and Mobile Telehealth to Transform Research in Elder Care is a National Institute of Health Challenge Grant awarded to the University of Virginia to test connecting seniors to health IT applications.

[956] American Telemedicine Association Comments in re NBP PN #17, filed Dec. 3, 2009, at 14.

[957] Letter from Alice Borrelli, Director, Global Healthcare and Workforce Policy, Intel Corporation, to Marlene H. Dortch, Secretary, FCC, GN Docket Nos. 09-57, 09-51, 09-137 (Dec. 16, 2009) (Intel Dec. 16, 2009 *Ex Parte*) at 32.

[958] The Joint Commission (TJC), which accredits roughly 82% of the nation's hospitals, is permitted by CMS to implement a proxy process for credentialing between TJC-accredited hospitals. This leaves the new privileging requirement for TJC-accredited hospitals, and both the credentialing and privileging requirements for non-TJC accredited hospitals. Telehealth Leadership Initiative, Question and Answer: Credentialing and Privileging of Telehealth Providers, http://www.telehealthleadership. org/Credentialing%20Privileging%20Q&A%20.DOC (last visited Jan. 30, 2010).

[959] Nat'l Ass'n of Children's Hospitals & Related Inst., *National Shortages of Pediatric Subspecialists Impede Children's Access to Care* (press release), Jan. 13, 2010, *available at* http://www.childrenshospitals.net/AM/Template.cfm?Section=Home3&CONTENTID=49773 &TEMPLATE=/CM/ContentDisplay.cfm.

[960] American Telemedicine Association Comments in re NBP PN #17, filed Dec. 3, 2009, at 16.

[961] Ctr. for Info. Tech. Leadership, The Value of Computerized Provider Order Entry in Ambulatory Settings: Executive Preview 6–7 (2009), http://www. citl.org/research/ACPOE_Executive_Preview.pdf.

[962] John Moore, *Doctors and the DEA*, GOv't health it, Oct. 1, 2008, http://www.govhealthit.com/newsitem. aspx?tid=77&nid=69394.

[963] *See* FDA, Medical Devices, Is the Product a Medical Device?, http://www.fda.gov/MedicalDevices/DeviceRegulationandGuidance/Overview/Classify YourDevice/ucm051512.htm (last visited Jan. 30, 2010) (providing the FDA definition of a medical device).

[964] HHS, Comparative Effectiveness Research Funding, http://www.hhs.gov/recovery/programs/cer/index. html (last visited Feb. 20, 2010).

[965] *See* Framingham Heart Study, About the Framingham Heart Study, http://www.framinghamheartstudy.org/about/index.html (last visited Feb. 19, 2010) (providing an overview of the Framingham Heart Study).

[966] Am. Heart Ass'n, Framingham Heart Study, http://www. americanheart.org/presenter.jhtml?identifier=4666 (last visited Feb. 20, 2010).

[967] Joy Pritts & Nina L. Kudszus, Health Pol'y. Ins., Your Medical Record Rights in Alabama (A Guide To Consumer Rights Under HIPAA), *available at* http:// medicalrecordrights.georgetown.edu/stateguides/al/al.pdf.

[968] Lawrence P. Casalino et al., *Frequency of Failure to Inform Patients of Clinically Significant Outpatient Test Results*, 169 arch intern med. 1123, 1123 (2009), *available at* http://archinte.ama-assn.org/cgi/reprint/169/12/1123.

[969] *See* Sharon B. Arnold, Acad. Health & Robert Wood Johnson Found., Improving Quality Health Care: The Role of Consumer Engagement (2007).

[970] *See supra* Chapter 4.

[971] There are sensible exemptions, which can be found at 45 C.F.R. § 164.524(a)(3). See citation for more circumstances whereby a covered entity can deny access but individuals also have a right to request a review of the denial.

[972] *See* Michael Arrington, *AsthmaMD Helps Asthma Sufferers, Gathers Aggregate Research Data*, TechCrunch, Jan. 10, 2010, http://www.techcrunch. com/2010/01/10/asthmamd-helps-asthma-sufferersgathers-aggregate-research-data/.

[973] Letter from Chuck Parker, Executive Director, Continua Health Alliance, to Marlene H. Dortch, Secretary, FCC, GN Docket Nos. 09-57, 09-51 (Nov. 16, 2009) (Continua Nov. 16, 2009 *Ex Parte*) Attach. at 13. Bandwidth thresholds are actual (i.e., not advertised) speeds.

[974] GE Healthcare Comments in re NBP PN #17, filed Dec. 4, 2009, at 8; Euclid Seeram, *Digital Image Compression*, radiOlOGic tech., July–Aug. 2005, http://www. entrepreneur.com/tradejournals/article/134676840. html; Human Genome Project Information, Frequently Asked Questions, http://www.ornl.gov/sci/techresources/ Human_Genome/faq/faqs1.shtml (last visited Jan. 31, 2010); Ichiro Mori et al., *Issues for Application of Virtual Microscopy to Cytoscreening, Perspectives Based on Questionnaire to Japanese Cytotechnologists*, diaGnOstic pathOlOGy, July 15, 2008, http://www. diagnosticpathology.org/content/pdf/1746-1596-3-S1-S15. pdf. *See, e.g.*, DICOM sample image sets, http://pubimage. hcuge.ch:8080/ (last visited Jan. 31, 2010).

[975] Mbps recommendations reflect compilation of the record. Numbers are guidelines, not precise measures. *See, e.g.*, Letter from Alice Borelli, Director, Global Healthcare and Workforce Policy Intel, to Marlene H. Dortch, Secretary, FCC, GN Docket Nos. 09–47, 09–51, 09–137, WC Docket No 07–10 (Dec. 16, 2009). S*ee also, e.g.*, Fiberutilities Group, A Practical Review of Broadband Requirements for Healthcare Clinical Applications 6–7 (2009), *available at* http://www. fiberutilities.com/documents/FG_Press_Release_FCC_ Briefing_Healthcare_Application_Requirements_for_ Broadband_110609.pdf.

[976] GE Healthcare Comments in re NBP PN #17, filed Dec. 4, 2009, at 2.

[977] Ascension Health Comments in re NBP PN #17, filed Dec. 4, 2009, at 5; RWHC ITN Comments in re NBP PN #17, filed Dec. 4, 2009, at 3.

[978] AT&T and Verizon both call these offerings "Small Business" offerings. Comcast calls it their "Business Internet Service." TimeWarner Cable calls it their "Business Class Wideband Internet" service.

[979] DIA is used as an umbrella term for all components of monthly broadband service for DSL lines. This includes both circuit and access services.

[980] Telegeography, Enterprise Network Pricing Service (Q3 2009). (Regarding methodology, "the Enterprise Network Pricing Service presents and analyzes responses to hypothetical bid scenarios for MPLS IP VPN WAN services. Carriers complete surveys on a bi-annual basis supplying prices for each component of a ten-city wide area network, breaking out core network, equipment and management fees from local access."). Price cited here is an average of the five lowest bids for Los Angeles.

[981] TimeWarner Business Class Professional 10 x 2, listed at $399.95 / month on TimeWarner Los Angeles website. Taxes and other surcharges not factored in. *See* Time Warner Cable Business Class, https://www.twcbc.com/ LA/buyflow/buyflow.ashx (last visited Feb. 27, 2010) (requires providing additional information to access).

[982] Statement reflects compilation of the record and is a guideline, not a comprehesive requirement. Record in response to NBP PN #17 and *ex parte* filings (*see e.g.*, Letter from Winfred Y. Wu, Director, Public Health Informatics, New York City Department of Health and Mental Hygiene, to Marlene H. Dortch, Secretary, FCC, GN Docket Nos. 09-47, 09-51, 09-137 (Mar. 1, 2010).

[983] Access to mass-market broadband is used here to mean passed by terrestrial broadband access facilities such as those used to deliver DSL or cable modem service. This analysis does not predict how many of the 307,000 small providers purchase the appropriate level of broadband; only the mass-market broadband available to them. The analysis is a predictive estimate combining the FCC's statistical network model and provider databases, as shown below. Gap is calculated based on connectivity requirement threshold of 4 Mbps for Single Physician Practices and 10 Mbps for all other practices. AMA small provider locations (four physicians or less) were assigned to an appropriate census block, based on their street address, and then reconciled with the model showing connectivity availability for that census block. About 24,000 (or 7%) of the health care locations in the AMA database had addresses that were impossible to convert accurately to census blocks; results for these locations were modeled to complete the analysis. A small percentage of the records (less than 1.5%) were geographically located outside of the Master Broadband Availability data (e.g., Puerto Rico), and therefore were dropped from consideration in the connectivity analysis. The analysis does not take into account other network quality requirements. Some of these locations may have alternative networks or commercial services, where residential broadband is unavailable.

- Omnibus Broadband Initiative (OBI) The Broadband Gap (forthcoming) (OBI, The Broadband Availability Gap).

- Database of all locations in the United States with practicing physicians: Am. Med. Ass'n, AMA Physician Masterfile Database (2009) on file with the FCC, "The Physician Masterfile includes current and historical data for more than 940,000 residents and physicians and approximately 77,000 students in the United States." Includes all active practicing physicians in the US (655,630) and the addresses where they practice. Sorting by address sorts 655,630 physicians into 351,172 locations, with a size metric for each one based on how many physician entries are associated with each location entry. Removed 5,077 locations in Puerto Rico and other locations that were not included in the Statistical Model, leaving 346,095 locations for our analysis. Detailed information on this database is available from the AMA. AMA Physician Masterfile, http://www.ama-assn.org/ama/pub/about-ama/ physician-data-resources/ physician-masterfile. Shtml (last visited Feb. 27, 2010). FCC's Rural definition, 47 CFR § 54.5: "For purposes of the rural health care universal service support mechanism, a 'rural area' is an area that is entirely outside of a Core Based Statistical Area; is within a Core Based Statistical Area that does not have any Urban Area with a population of 25,000 or greater; or is in a Core Based Statistical Area that contains an Urban Area with a population of 25,000 or greater, but is within a specific census tract that itself does not contain any part of a Place or Urban Area with a population of greater than 25,000. 'Core Based Statistical Area' and 'Urban Area' are as defined by the Census Bureau and 'Place' is as identified by the Census Bureau."

[984] *See supra* note 88.

[985] Difference between DS3 purchased in Wyoming versus Vermont for one year of service, according to rates listed in Exhibit 10-E, *infra*, is $27,384.

[986] Wyoming, Mississippi, New York, and Vermont prices: USAC, Urban Rate Search Tool, http://www.usac.org/ rhc/tools/rhcdb/UrbanRates/search.asp (last visited Feb. 8, 2010) (use 2009 data).

[987] Letter from William England, Vice President, Rural Health Care Division, Universal Service Administrative Company, to Marlene H. Dortch, Secretary, FCC, GN Docket Nos. 09-47, 09-51, 09-137 (Feb. 23, 2010) (USAC Feb. 23, 2010 *Ex Parte*) at 1. If locations in Alaska are excluded, the participants' broadband price still averages 3x the price of their urban benchmarks.

[988] Federally funded providers include provider networks that are directly administered by the federal government (e.g., Veterans Health Administration, NASA, Bureau of Prisons, Indian Health Service), as well as recipients of federal subsidies.

[989] *See* Letter from Theresa Cullen, RADM, U.S. Public Health Service, Chief Information Officer and Director, Indian Health Service, to Marlene H. Dortch, Secretary, FCC, GN Docket Nos. 09-47, 09-51, 09-137 (Feb. 23, 2010) (IHS Feb. 23, 2010 *Ex Parte*), Attach.

[990] "FQHCs are 'safety net' providers such as community health centers, public housing centers, and programs serving migrants and the homeless. The main purpose of the FQHC Program is to enhance the provision of primary care services in underserved urban and rural communities." CMS, federally Qualified health center fact sheet 1 (2009), http://www.cms.hhs.gov/ MLNProducts/downloads/fqhcfactsheet.pdf. FQHCs qualify for cost-based CMS reimbursement and other benefits.

[991] "The Rural Health Clinic Program was established in 1977 to address an inadequate supply of physicians who serve Medicare and Medicaid beneficiaries in rural areas." CMS, rural health clinic fact sheet 1 (2007), http://www.cms.hhs.gov/MLNProducts/Downloads/ rhcfactsheet.pdf. Clinics must meet criteria established by HHS, including being located in rural area and in a Health Provider Shortage Area or a Medically Underserved Area. RHC institutions qualify for cost- based CMS reimbursement and other benefits.

[992] Critical Access Hospitals are hospitals qualified to receive cost-based reimbursement from Medicare and are important components of states' rural health networks. *See generally* CMS, critical access hOspitals fact sheet (2009) (discussing what qualifies as a Critical Access Hospital), *available at* http://www.cms.hhs.gov/MLN Products/downloads/ CritAccessHospfctsht.pdf.

[993] Access to mass-market broadband is used here to mean passed by terrestrial broadband access facilities such as those used to deliver DSL or cable modem service; business-class service, including business-grade service level agreements, is likely available currently but at much higher prices (potentially including large one time special-construction costs). This analysis does not predict how many of the providers purchase the appropriate level of broadband; only if mass-market broadband is available to them. The analysis is a predictive estimate combining the FCC's statistical network model and provider databases as shown below. Gap is calculated based on connectivity requirement threshold of 4 Mbps for Single Physician Practices (from either DSL/FTTN or Cable) and 10 Mbps for all other practices (from cable service only). Health care locations were assigned to an appropriate census block, based on their street address, and then reconciled with the model showing connectivity availability for that census block. For each database, a percentage of the health care locations had addresses that were impossible to convert accurately to census blocks; results for these locations were modeled to complete the analysis. For the AMA, this accounted for ~24,000 (or 7%) of total entries. For IHS, this accounted for ~350 (or 52%) of entries. Additionally, the FQHC database contained duplicate location records, which were excluded from the connectivity analysis. A small percentage of the records (less than 1.5%) were geographically located outside of the Master Broadband Availability data (e.g., Puerto Rico), and therefore were dropped from consideration in the connectivity analysis.The analysis does not take into account other network quality requirements. Some of these locations may have alternative networks or commercial services, where residential broadband is unavailable.

- OBI, The Broadband Availability Gap. The OBI deployment team created a nationwide model for broadband availability from wired and wireless technologies. Database of all locations in the United States with practicing physicians: AMA, AMA Physician Masterfile Database (2009) on file with the FCC, "The Physician Masterfile includes current and historical data for more than 940,000 residents and physicians and approximately 77,000 students in the United States." Includes all active practicing physicians in the US and the addresses where they practice. Sorting by address sorts 655,630 physicians into 346,095 locations, with a size metric for each one based on how many physician entries are associated with each location entry. Removed 5,077 locations in Puerto Rico and other locations that were not included in the Statistical Model, leaving 346,095 locations for our analysis. Detailed information on this database is available from the AMA. AMA Physician Masterfile, http://www. ama-assn.org/ama/pub/about-ama/physician-dataresources/physician-masterfile.shtml (last visited Feb. 27, 2010).
- Federally Qualified Health Center Database: HRSA Electronic Handbooks, Bureau of Primary Health Care Management Information System, Scope Repository retrieved via the HRSA Geospatial Data Warehouse's Health Care Service Delivery Sites report at http://datawarehouse.hrsa.gov/ HGDWReports/RT_App.aspx?rpt=HS, retrieved on Oct. 24, 2009.
- Rural Health Clinic Database: CMS, Name and Address Listing For Rural Health Clinic Database (accessed Oct. 6, 2009). Updated versions are available at http://www.cms.hhs.gov/ MLNProducts/ downloads/rhclistbyprovidername. pdf.
- Critical Access Hospitals Database: HHS, Health Resources and Services Administration, HRSA Geospatial Data Warehouse—Report Tool, http:// datawarehouse.hrsa.gov/HGDWReports/RT_App. aspx?rpt=P2 (providing data snapshot from Sept. 30, 2009).
- IHS Database: IHS Feb.23, 2010*Ex Parte*, Attach.

[994] *See* 47 U.S.C. §§ 254(h)(1)(A), 254(h)(2)(A); 47 C.F.R. Part 54, Sbpt. G—Universal Service Support for Health Care Providers.

[995] 47 C.F.R. §§ 54.605–613.

[996] 47 C.F.R. § 54.621.

[997] *See Rural Health Care Support Mechanism*, WC Docket No. 02-60, Order, 21 FCC Rcd 11111 (2006) (*2006 Pilot Program Order*); *Rural Heath Care Support Mechanism*, WC Docket No. 02-60, Order, 22 FCC Rcd 20360 (2007) (*2007 RHC PP Selection Order*).

[998] There were 2,570 locations that participated in the FCC's Rural Health Care Program, excluding the Pilot Program, in 2009. Eligibility was determined by matching the locations of non-public and public institutions with the FCC's geographic definition of rural. *See* 47 C.F.R. § 54.5. Estimate of 10,660 unique locations include 1,851 nonprofit hospitals, 2,612 Federally Qualified Health Centers (FQHCs), 3,349 Rural Health Clinics (as defined by HHS), 358 Indian Health Service (by HHS rules, all IHS sites are also FQHCs), 607 Veterans Health Affairs, 106 Federal Prisons (BOP), and 3,219 public health departments. At the time of publication, we did not have addresses for individual BOP and VHA sites, so we assumed a rural/urban split in the same proportions to IHS and hospitals, respectively. Public Health Departments were estimated as one location per county that was deemed totally rural by the FCC. All other locations were geo-coded by census block to determine eligibility. These categories may be inconsistent with FCC terminology, since it has traditionally used its own definition of "hospital" and "rural health clinic." Also, 10,660 is likely an underestimate of eligible institutions because it does not count community mental health centers, postsecondary medical education, or state prisons.

[999] *See* RWHC ITN Comments in re NBP PN #17, filed Dec. 4, 2009, at 7; USF Consultants Comments in re NBP PN #17, filed Dec. 7, 2009, at 5.

[1000] USAC Feb. 23, 2010 *Ex Parte* at 1.

[1001] *See* PSPN Comments in re NBP PN #17, filed Dec. 4, 2009, at 13; HNG Comments in re NBP PN #17, filed Dec. 4, 2009, at 5–6; MDH Comments in re NBP PN #17, filed Dec. 4, 2009, at 4; RWHC ITN Comments in re NBP PN #17, filed Dec. 4, 2009, at 7–8.

[1002] *See* IHS Comments in re NBP PN #17, filed Dec. 4, 2009, at 13; PSPN Comments in re NBP PN #17, filed Dec. 4, 2009, at 15; PMHA et al. Comments in re NBP PN #17, filed Dec. 4, 2009, at 6; State of New York Comments in re NBP PN #17, filed Dec. 4, 2009, at 12.

[1003] Total rural health care providers determined by geocoding of the American Medical Association's physician master-file (38,403), which includes every location where a licensed physician practices. Am. Med. Ass'n, AMA Physician Masterfile Database (2009). The 10,660 locations that are eligible under the FCC's Rural Health Program (see endnote 103, *supra*) only represent 28% of the total locations.

[1004] FCC, *Rural Telemedicine Program Funds 16 More Broadband Telehealth Networks* (press release), Feb. 18, 2010, *available at* http://hraunfoss.fcc.gov/edocs_public/attachmatch/DOC-296348A1.pdf.

[1005] Safety net institutions are defined by the Health Resource and Services Administration (HRSA). HRSA, HRSA and the Safety-Net, http://answers.hrsa. gov/cgi-bin/hrsa.cfg/php/enduser/std_adp.php?p_faqid=1702&p_created=1243947992&p_topview=1 (last visited Jan. 31, 2010).

[1006] *See 2007 RHC PP Selection Order*, 22 FCC Rcd at 20381–82, para. 47.

[1007] The Rural Health Care Program uses the statutory definition of "health care provider" established in section 254(h)(7)(b) of the 1996 Act, which defines health care providers as: (i) post-secondary educational institutions offering health care instruction, teaching hospitals, and medical schools; (ii) community health centers or health centers providing health care to migrants; (iii) local health departments or agencies; (iv) community mental health centers; (v) not-for-profit hospitals; (vi) rural health clinics; and (vii) consortia of health care providers consisting of one or more entities described in clauses (i) through (vi).

[1008] See for example, 47 U.S.C. § 254(h)(4), which allows E-rate support to private schools that have an annual endowment of less than $50,000,000.

[1009] *See* 47 U.S.C. §§ 254(h)(1)(A), 254(h)(2)(A) (limiting support to public and nonprofit health care providers).

[1010] See also HHS stipulation that any physician (including private practice physicians) can qualify for meaningful use incentives, provided such physicians accepts Medicare or derives more than 20% of their billing from Medicaid patients.

[1011] The Recovery Act provides Medicare and Medicaid incentive payments to eligible providers, such as physicians and hospitals, in order to increase the adoption of electronic health records (EHRs). To receive the incentive payments, providers must demonstrate "meaningful use" of a certified EHR. Building upon the work done by the HIT Policy Committee, the Centers for Medicare & Medicaid Services (CMS), along with the Office of the National Coordinator for Health Information Technology (ONC), are developing a proposed rule that provides greater detail on the incentive program and proposes a definition of meaningful use. *See* HHS, *Important First Step to Expand the Use of Information Technology to Improve the Health and Care of Every American* (press release), June 16, 2009, http://www.hhs.gov/news/ press/2009pres/06/20090616a.html.

[1012] As opposed to, for instance, a private hospital network, where the hospital shareholders directly realize financial gains from using such technologies. In such an example, the government only indirectly realizes the gains, where they result in reductions to overall CMS reimbursements.

[1013] *See* Box 10-3, "How Health IT Saves Veterans Affairs Billions Each Year," *supra*.

[1014] IHS Feb. 23, 2010 *Ex Parte*, Attach. Indian Health Service calculated the annual cost to upgrade its broadband networks to the minimum requirements in Exhibit 10.3, *supra*. Estimates were made using median prices paid across its 600+ location system. Competitive bidding and selective network deployment similar to the FCC's universal service programs will likely reduce prices. Also, as ARRA funding through BIP and BTOP is spent on Tribal lands, the prices for service may decline.

[1015] Am. Society for Training & Development, Bridging the Skills Gap 5 (2006), http://www.astd.org/ NR/rdonlyres/FB4AF179-B0C4-4764-9271-17FAF86A8E23/0/BridgingtheSkillsGap.pdf.

[1016] Bureau of Labor Stat., *Employment Projections—2008–18* (press release), Dec. 10, 2009, at 3, http://www.bls. gov/news.release/pdf/ecopro.pdf; Inst. for a Competitive Workforce, U.S. Chamber of Com., The Skills Imperative 4 (2008), *available at* http://www. uschamber.com/NR/rdonlyres/eciaj45n6o5jxdngkik p6zg phwy4gqbkt3vyv7q4eu5xlcpms7escmdu5koxwfyvrgdpxukqamx35ljclqfydbuob2g/CTEPaperFINAL.pdf.

[1017] Nat'l Ctr. on Educ. And The Econ., Tough Choices or Tough Times 7–9 (2007), available at http://www. skillscommission.org/pdf/exec_sum/ToughChoices_ EXECSUM.pdf; Partnership for 21st Century Skills, Results that Matter 2–6 (2006), available at http:// www.21stcenturyskills.org/documents/RTM2006.pdf.

[1018] Gary Orfield et al., Civil Rights Project at Harvard Univ. et al., Losing Our Future: How Minority Youth are Being Left Behind by the Graduation Rate Crisis 2 (2004), *available at* http://www.urban.org/ UploadedPDF/410936_LosingOurFuture.pdf.

[1019] Mckinsey & Co., The Economic Impact of the Achievement Gap in America's Schools 9 (2009) (Mckinsey & Co., The Economic Impact of the Achievement Gap), *available at* http://www.mckinsey.com/App_Media/ Images/Page_Images/Offices/ SocialSector/PDF/achievement_gap_report.pdf.

[1020] Mckinsey & Co., The Economic Impact of the Achievement Gap at 7.

[1021] Johnny J. Moye, *Technology Education Teacher Supply and Dem and—A Critical Situation*, 69 Tech. Tchr. 30 (2009); Business-Higher Education Forum (Bhef), The American Competitiveness Initiative: Addressing the Stem Teacher Shortage and Improving Student Academic Readiness 1 (2006), *available at* http://www. eric.ed.gov/ERICDocs/data/ericdocs2sql/content_ storage_01/0000019b/80/42/e8/38.pdf.

[1022] Anthony Picciano & Jeff Seaman, Sloan Consortium, K–12 Online Learning: A 2008 Follow -up of the Survey of U.S. School District Administrators 5 (2009) (Picciano & Seaman, K–12 Online Learning),), *available at* http://www.sloan-c.org/publications/survey/k12online2008.

[1023] Rebeca Gajda & Matthew Militello, *Recruiting and Retaining School Principals: What We Can Learn from Practicing Administrators,* 5 AASA J. Scholarship & Prac. 14 (2008); Rebecca H. Goodwin et al., *The Changing Role of the Secondary Principal*, 87 nassp bulletin 26 (2003), *available at* http://bul.sagepub. com/cgi/content/abstract/87/634/26.

[1024] Stephanie Moller & Elizabeth Stearns, Retention and School Dropout: Examining Connectivity Between Children and Schools 2 (Aug. 14, 2004) (paper presented at Am. Sociological Association Meeting), *available at* http://www.allacademic.com//meta/p_mla_apa_ research_citation/1/0/8/7/6/pages108764/p108764-1. php.

[1025] Regional Education Laboratory for the Central Region, Research in Brief: High School Standards & Expectations for College & the High-Skills Workplace 1–3 (2009), *available at* http://www.mcrel.org/topics/ Standards/products/321/; ACT, dO current state standards and assessments reflect cOlleGe readiness?: a case study 5–6 (2005) (ACT, state standards case study), *available at* www.act.org/research/policymakers/pdf/ current_standards.pdf.

[1026] Wee Chuen Tan et al., *GLOOTT Model: A Pedagogically- Enriched Design Framework of Learning Environment to Improve Higher Order Thinking Skills*, 14 AACE J. 139, 141, 143 (2006), *available at* http://www.editlib. org/?fuseaction=Reader.ViewFullText&paper_id=6198.

[1027] Annenberg Inst. Sch. Reform, Professional Development Strategies that Improve Instruction: Professional Learning Communities 4 (2004), *available at* http://www.annenberginstitute.org/Products/ PDStrategies.php; Kenneth Tye & Barbara Benham Tye, *Teacher Isolation and School Reform,* 65 phi beta kappan 319 (1984), *available at* http://www.jstor.org/ pss/20387022 (requires purchase).

[1028] Mckinsey & Co., The Economic Impact of the Achievement Gap at 9.

[1029] U.S. Dep't of Educ., The American Recovery and Reinvestment Act of 2009: Saving and Creating Jobs and Reforming Education (2009), *available at* http:// www2.ed.gov/policy/gen/leg/recovery/presentation/ arra.pdf.

[1030] Natl'l CTR. on Educ & the Econ, New Commission on the Skills of the American Workforce, Tough Choice or Tough Times 6–9 (2006).

[1031] Catherine A. Little et al., *Constructing Complexity for Differentiated Learning*, 15 Mathematics Teaching in the Middle Sch. 34, 34–42 (2009).

[1032] Nancy Protheroe, *Technology and Student Achievement*, principal, Nov.–Dec. 2005, at 46, *available at* http:// www.naesp.org/resources/2/Principal/2005/N-Dp46. pdf; Clayton Christensen et al., Disrupting Class: How Disruptive Innovation Will Change the Way the World Learns (2008).

[1033] Laura D'Amico, Ctr. for Learning Tech. in Urban Schools, A Case of Design-Based Research in Education 32 (2005), http://www.sfu.ca/~ldamico/LeTUS_ FullCase_Final.pdf; Robert Geier et al., Standardized Test Outcomes of Urban Students Participating in Standards and Project Based Science Curricula 206–13 (June 25, 2004) (paper presented at the 6th Int'l Conf. on Learning Sciences); Ctr. for ICT , Pedagogy, and Learning, Manchester Metropolitan Univ., Evaluation of the ECT Test Bed Project Final Report 19–20 (2007); Byron Review, Safer Children in a Digital World 8–9, 126–32 (2008), *available* at http://www. dcsf.gov.uk/byronreview/pdfs/Final%20Report%20 Bookmarked.pdf.

464   Federal Communications Commission

[1034] Marsha Lovett et al., *The Open Learning Initiative: Measuring the Effectiveness of the OLI Statistics Course in Accelerating Student Learning*, j. interact. Media in educ., May 2008 (Lovett et al., *The Open Learning Initiative*), *available at* http://jime.open.ac.uk/2008/14/ jime-2008-14.pdf; Joel Smith, Vice Provost and CIO, Carnegie Mellon Univ., Remarks at FCC Education Workshop (Aug. 20, 2009), *available at* http://www.broadband.gov/docs/ws_education/ws_education_smith.pdf.

[1035] Ctr. for Educ. Performance & Accountability, Florida TaxWatch, Final Report: A Comprehensive Assessment of Florida Virtual School 17 (2007), *available at* http://www.floridataxwatch.org/resources/pdf/110507FinalReportFLVS.pdf.

[1036] Oregon Connections Academy, *Oregon Connections Academy Earns "Outstanding" Grade* (press release), PR Newswire, Nov. 13, 2009, http://www. connectionsacademy.com/news/orca-state-report-card. aspx.

[1037] Reuters, *K12's Florida Virtual Academy Posts High Scores on 2009 State Tests* (press release), June 1, 2009, http://www.reuters.com/article/pressRelease/ idUS117026+01-Jun-2009+PRN20090601

[1038] Missouri Virtual Instruction Program (MOVIP), Annual Evaluation Report 2007–2008, at 39 (2008), http://www.movip.org/about/evalreport2007-2008.pdf.

[1039] Michigan Department of Education Comments in re NBP PN #15 (*Comment Sought on Broadband Needs in Education, Including Changes to E-Rate Program to Improve Broadband Deployment—NBP Public Notice #15*, GN Docket Nos. 09-47, 09-51, 09-137, Public Notice 24 FCC Rcd 13560 (WCB 2009) (*NBP PN #15*)), filed Nov. 20, 2009, at 4 (filed by Jeannene Hurley).

[1040] John Watson & Butch Gemin, N. American Council for Online Learning, Promising Practices in Online Learning: Using Online Learning for At-Risk Students and Credit Recovery 8–9 (2008), http://www. inacol.org/research/promisingpractices/NACOL_CreditRecovery_PromisingPractices.pdf.

[1041] *See* Lower Kuskokwin School District Comments in re NBP PN #15, filed Nov. 20, 2009, at 2. Or in some models of learning, the teacher and student need to spend much less time face-to-face, and states such New Mexico are also experimenting with remote high quality real-time video conferencing as a solution.

[1042] Pamela E. Harrell & Mary Harris, *Teacher Preparation Without Boundaries: A Two-Year Study of an Online Teacher Certification Program*, 14 J. Technology & Teacher Education. 755 (2006), *available at* http://www.thefreelibrary.com/Teacher+preparation+without+boundaries:+a+two-year+study+of+an+online...   -a0151387501.

[1043] Gordon Freedman, The Blackboard Inst., Is the Tipping Point for Education in Sight?, at 4 (2009) (Freedman, Is the Tipping Point for Education in Sight?), http:// www.inacol.org/research/docs/Blackboard%20 K20%20CouncilSummaryReport.pdf; Patricia Deubel, *K–12 Online Teaching Endorsements: Are They Needed?*, The Journal, Jan. 10, 2008 (Deubel, *K–12 Online Teaching Endorsements*), http://thejournal. com/Articles/2008/01/10/K12-Online-Teaching- Endorsements-Are-They-Needed.aspx?; Lorraine Sherry, *Issues in Distance Learning*, 1 Int'l J. Educ. Telecom. 337 (1996) (Sherry, *Issues in Distance Learning*), *available at* http://carbon.cudenver. edu/~lsherry/pubs/issues.html.

[1044] Verizon and Verizon Wireless Comments in re NBP PN #15, filed Nov. 20, 2009, at 5; American Association of School Administrators and the Association of Educational Service Agencies Comments in re NBP PN #15, filed Nov. 20, 2009, at 3 (filed by Noelle Ellerson).

[1045] *See, e.g.*, Apple, iTunes, http://www.apple.com/itunes (last visited Dec. 22, 2009).

[1046] *See, e.g.*, Netflix Home Page, http://www.netflix.com (last visited Dec. 22, 2009); YouTube Home Page, http:// www.youtube.com (last visited Dec. 22, 2009); Hulu Home Page, http://www.hulu.com (last visited Dec. 22, 2009).

[1047] *See, e.g.*, Wikipedia, Amazon Kindle, http://en.wikipedia. org/wiki/Amazon_Kindle (last visited Dec. 22, 2009); Wikipedia, Sony Reader, http://en.wikipedia.org/wiki/ Sony_Reader (last visited Dec. 22, 2009); Wikipedia, Barnes & Noble nook, http://en.wikipedia.org/wiki/ Barnes_and_Noble_nook (last visited Dec. 22, 2009).

[1048] Doug McKessock et al., *Dynamic Online Homework System: An Enabler of Learning* 399 (Dec. 4–7, 2005) (paper presented at the Ascilite Conference), *available at* http://www.ascilite.org.au/conferences/brisbane05/ blogs/proceedings/47_McKessock.pdf; Deborah Hellman, Implementing Differentiated Instruction in Urban, Title I Schools (2007) (unpublished Ph.D. dissertation, U. So. Fla.), http://kong.lib.usf.edu:8881/ usfldc/71/170176.html.

[1049] MIT, Tufts, Yale, Utah State, Stanford, UC Berkeley and Carnegie Mellon are all current examples of open course publishers.

[1050] H. Jerome Keisler, Elementary Calculus (2d ed. 2000), http://www.math.wisc.edu/~keisler/calc.html.

[1051] Katie Dean, *Bleary Days for Eyes on the Prize*, Wired, Dec. 22, 2004, *available at* http://www.wired.com/ culture/lifestyle/news/2004/12/66106.

[1052] Renee Hobbs et al., CTR. for Soc. Media, The Cost of Copyright Confusion for Media Literacy 16–17 (2007), *available at* http://www.centerforsocialmedia.org/files/ pdf/Final_CSM_copyright_report.pdf.

[1053] Geoffrey Fowler, *New Kindle Audio Feature Causes a Stir*, Wall st. j., Feb. 10, 2009, *available at* http://online. wsj.com/article/SB123419309890963869.html; Don Reisinger, *Universities Reject Kindle Over Inaccessibility for the Blind*, CNET, Nov. 12, 2009, http://news.cnet. com/8301-13506_3-10396177-17.html.

[1054] 17 U.S.C. § 504.

[1055] 17 U.S.C. §§ 110(2), 112(f).

[1056] Cathy Cavanaugh, CTR. for Am. Progress, Getting Students More Learning Time Online: Distance Education In Support Of Expanded Learning Time in K–12 Schools 4 (2009), *available at* http:// www.americanprogress. org/issues/2009/05/pdf/ distancelearning.pdf.

[1057] Long distance learning is a form of online learning where teachers and students frequently interact live over an audio/video link.

[1058] Picciano & Seaman, k–12 Online Learning at 5.

[1059] U.S. Dep't of Educ., Fact Sheet: New No Child Left Behind Flexibility: Highly Qualified Teachers (Mar. 2004), *available at* http://www2.ed.gov/nclb/methods/ teachers/hqtflexibility.pdf; u.s. dep't Of educ., a summary Of hiGhly Qualified teacher data 3–5 (2009),

[1060] Picciano & Seaman, K–12 Online learning at 5.

[1061] U.S. Dep't of Educ., Connecting Students To Advanced Courses Online 4–8 (2007) (Dep't Of Educ., Connecting students), *available at* http://www.ed.gov/admins/lead/ academic/advanced/coursesonline.pdf.

[1062] Freedman, is the Tipping Point for Education in Sight? at 4; Deubel, *K–12 Online Teaching Endorsements*; Sherry, *Issues in Distance Learning*; jOhn WatsOn et al., everGreen educatiOnGrOup, keepinG pace With k–12 Online learninG 11 (Nov. 2008) (Watson et Al., Keeping] Pace with K–12 Online Learning), *available at* http://www.kpk12.com/downloads/KeepingPace_2008.pdf.

[1063] Watson et al., keepinG pace With k–12 Online learninG, at 49.

[1064] Joel Smith, Vice Provost and CIO, Carnegie Mellon Univ., Remarks at FCC Education Workshop (Aug. 20, 2009), *available at* http://www.broadband.gov/docs/ ws_education/ws_education_smith.pdf.

[1065] Department Of Education., Connecting Students at 4–8.

[1066] Lovett et al., *The Open Learning Initiative* at 2.

[1067] Utah Education Network Comments in re NBP PN #15, filed Nov. 20, 2009, at 4.

[1068] Rich Kaestner, Consortium on School Networking, The Real Cost of Open Source Software 1 (2006), *available at* http://www.cosn.org/Portals/7/docs/The%20 Real%20Cost%20of%20Open%20Source%20Software. pdf.

[1069] Ryan Paul, *Department of Defense Study Urges Open Source Adoption*, ars technica, Aug. 20, 2006, http:// arstechnica.com/old/content/2006/08/7545.ars (citing j.c. herz et al., department Of defense, Open technOlOGy develOpment, rOadmap plan (2006), http:// www.acq.osd.mil/actd/articles/OTDRoadmapFinal. pdf).

[1070] Federation American Scientists, The FAS Learning Technologies Program Policy Initiative, http://www.fas. org/programs/ltp/policy_and_publications/index.html (last visited Feb. 15, 2010).

[1071] Henry Kelly, *Games, Cookies and the Future of Education*, Issues in Sci. and Tech., Summer 2005, at 33, *available at* http://www.fas.org/programs/ltp/policy_ and_publications/publications/games_cookies1.pdf.

[1072] Antonio Cordella & Kai A. Simon, The Impact of Information Technology on Transaction and Coordination Cost (Aug. 9–12, 1997) (paper presented at the Conference on Information Systems Research in Scandinavia, Oslo), *available at* http://www.instantscience.net/pub/tracost.pdf.

[1073] Sean M. Kerner, *IDC: Linux-Related Spending Could Top $49B by 2011*, internetneWs.cOm, Apr. 8, 2008, http:// www.internetnews.com/software/article.php/3739491.

[1074] Similar in purpose to Advanced Research Projects Agency-Energy (ARPA-E) and DARPA.

[1075] Cornell Univ., Digital Literacy Resource, http:// digitalliteracy.cornell.edu/ (last visited Feb. 15, 2009).

[1076] Barbara R. Jones-Kavalier & Suzanne L. Flannigan, *Connecting the Digital Dots: Literacy of the 21st Century*, 29 educause Q. 8 (2006), *available at* http://www.educause.edu/EDUCAUSE+Quarterly/ EDUCAUSEQuarterlyMagazineVolum/ ConnectingtheDigitalDotsLitera/157395.

[1077] David Buckingham, *Digital Media Literacies: Rethinking Media Education in the Age of the Internet*, 2 res. in cOmp. & int'l educ. 43–44 (2007), *available at* http:// www.wwwords.co.uk/pdf/validate.asp?j=rcie&vol =2&is sue=1&year=2007&article=4_Buckingham_RCIE_2_1_ web (requires entering text); City of Chicago Comments in re NBP PN #15, filed Nov. 20, 2009, at 7; Albuquerque Public Schools Comments in re NBP PN #15, filed Nov. 20, 2009, at 3.

[1078] Verizon and Verizon Wireless Comments in re NBP PN #15, filed Nov. 20, 2009, at 5.

[1079] Verizon and Verizon Wireless Comments in re NBP PN #15, filed Nov. 20, 2009, at 5; American Association of School Administrators and the Association of Educational Service Agencies Comments in re NBP PN #15, filed Nov. 20, 2009, at 3 (filed by Noelle Ellerson).

[1080] Univ. College London, Information Behaviour of the Researcher of the Future 20 (2008), *available at* http://www.jisc.ac.uk/media/documents/programmes/ reppres/gg_final_keynote_11012008.pdf.

[1081] European Comm'n, Digital Literacy Report: A Review for the I2010 Einclusion Initiative 3 (2008), http:// www.digital-literacy.eu/_root/media/36395_digital_ literacy_review.pdf.

[1082] Rodney K. Marshall, *Review*, 8 J. Literacy & Tech. 49 (2007) (reviewing Mark Warschauer, Laptops And Literacy: Learning In The Wireless Classroom (2006)), *available at* http://www.literacyandtechnology.org/ volume8/no1/JLTv8bookrev.pdf.

[1083] The Commission has an open proceeding wherein it is considering the issue of media literacy for both parents and children and what actions it should take concerning this issue. See *Empowering Parents and Protecting Children in an Evolving Media Landscape*, MB Docket No.09-194, Notice of Inquiry, 24 FCC Rcd 13171 (2009).

[1084] Partnership for 21st century skills, results that matter 2–6 (2006), http://www.21stcenturyskills.org/documents/RTM2006.pdf.

[1085] BHEF, An American Imperative: Transforming the Recruitment, Retention and enewal of Our Nation's Mathematics and Science Teaching Workforce 2 (2007), *available at* http://www.bhef.com/solutions/stem/americanimperative.asp.

[1086] BHEF, The American Competitiveness Initiative: Addressing the Stem Teacher Shortage and Improving Student Academic Readiness 1–2 (2006), *available at* http://www.bhef.com/publications/documents/brief3_s06.pdf.

[1087] Anthony G. Picciano & Jeff Seaman, Sloan Consortium, K–12 Online Learning: A 2008 Follow-Up of the Survey of U.S. School District Administrators 5–6 (2009), http://www.sloan-c.org/publications/survey/k-12online2008.

[1088] Nick Anderson, *White House Announces $250M Effort for Science and Math Teachers*, the WashinGtOn pOst, Jan. 6, 2010, http://www.washingtonpost.com/wpdyn/content/article/2010/01/06/AR2010010602063.html?hpid=moreheadlines.

[1089] U.S. Dep't of Educ., Evaluation of the Enhancing Education Through Tech. Program: Final Report 33 (2009), www.ed.gov/rschstat/eval/tech/netts/ finalreport.pdf.

[1090] Kathleen Kennedy Manzo, *Whiteboards Impact on Teaching Seen as Uneven*, diGital directiOns, Jan. 8, 2010, http://www.edweek.org/dd/articles/2010/01/08/0 2whiteboards.h03.html.

[1091] John Watson & Butch Gemin, Inacol, Promising Practices in Online Learning: Funding and Policy Frameworks for Online Learning 14 (2009), http:// www.inacol.org/research/bookstore/detail.php?id=13.

[1092] Meris Stansbury, *Panelists: Online Learning Can Help Minority Students*, eschOOlneWs, Apr. 11, 2008, at 1, http://www.eschoolnews.com/2008/04/11/panelistsonline-learning-can-help-minority-students/.

[1093] Cathy Cavanaugh et al., *Effectiveness of Online Algebra Learning: Implications for Teacher Preparation*, 38 J. Educ. Computing Research 70, 70–71 (2008), *available at* http://www.flvs.net/areas/aboutus/Documents/Research/OnlineAlgebraTeacherPrep05.pdf.

[1094] MarGaret Hilton, Nat'l Academics Es Press, Protecting Student Records and Facilitating Education Research: A Workshop Summary 75 (2008), http://www.nap. edu/catalog.php?record_id=12514; Charles A. Walls, *Providing Highly Mobile Students with an Effective Education*, ERIC Clearinghouse On Urban Educ., Nov. 2003, *available at* http://www.ericdigests.org/2004-3/ mobile.html.

[1095] Lawrence Gallagher et al., Teachers' Use of Student Data Systems to Improve Instruction 2005–2007, at 26 (2008), http://www.ed.gov/rschstat/eval/tech/teachersdata-use-2005-2007/teachers-data-use-2005-2007.pdf.

[1096] Lawrence Gallagher et al., Teachers' use of Student Data Systems To Improve Instruction 2005–2007, at 26 (2008), http://www.ed.gov/rschstat/eval/tech/teachersdata-use-2005-2007/teachers-data-use-2005-2007.pdf.

[1097] Econorthwest, Issue Paper: Improving K–12 Business Practices and Maximizing Available Revenues 3 (2005), *available at* http://www.chalkboardproject.org/images/ PDF/K12BusinessPractices.pdf; Elizabeth Millard, *E-Procurement,* district administratOr, Feb. 2008, http://www.districtadministration.com/viewarticle.aspx?articleid=1470; Mike Kennedy, *Getting More for Less*, am. Sch. & univ., Jan. 1, 2004, http://asumag.com/ mag/university_gettingjess/.

[1098] Debra Sherman, *U.S. grants $1.2 billion for electronic health records,* reuters, Aug. 20, 2009, http://www.reuters.com/article/topNews/ idUSTRE57J21J20090820.

[1099] Integrity Tech. Solutions, Mclean County Community Unit School District No. 5 Uses SIF to Streamline District Information Exchange 1 http://www.sifinfo.org/upload/story/76CF27_Unit5SIFCaseStudy.pdf.

[1100] *See, e.g.*, Schools Interoperability Framework, SIF Association, http://www.sifinfo.org (last visited Feb. 15, 2010).

[1101] *See* Nat'l Ctr. for Educ. Stat., Digest of Education Statistics, http://nces.ed.gov/programs/digest/d08/tables/dt08_363.asp (last visited Feb. 15, 2010).

[1102] McKinsey & Co., The Economic Impact of the Achievement Gap in America's Schools 9 (2009), http:// www.mckinsey.com/App_Media/Images/Page_Images/Offices/SocialSector/PDF/achievement_gap_report.pdf.

[1103] Svend Albaek et al., *Government–Assisted Oligopoly Coordination? A Concrete Case*, 45 j. indus. ecOn. 429 (1997), *available at* http://ideas.repec.org/p/kud/ kuieci/1997-03.html.

[1104] Specifically, the Department of Education should ensure that it is not making it easier for its suppliers to artificially adjust prices using the collection, aggregation and analysis of transaction specific information that includes pricing information.

[1105] There might be circumstances where local bidding only is important for any number of reasons.

[1106] 47 U.S.C. § 254(h)(1)(B).

[1107] 47 U.S.C. § 254(c)(3), 254(h)(2)(A).

[1108] Nat'l Ctr. for Educ. Stat., Internet Access in U.S. Public Schools And Classrooms: 1994–2005, at 4 (2006), *available at* http://nces.ed.gov/pubs2007/2007020.pdf.

[1109] Nat'l Ctr. for Educ. Stat., Internet Access in U.S. Public Schools And Classrooms: 1994–2005, at 4 (2006), *available at* http://nces.ed.gov/pubs2007/2007020.pdf.

[1110] Amanda Lenhart et al., The Internet and Education: Findings of the Pew Internet & American Life Project 3 (2001), *available at* http://wwww.pewinternet.org/~/ media//Files/Reports/2001/PIP_Schools_Report.pdf. pdf;

Marianne Bakia et al., Evaluation of the Enhancing Education Through Technology Program: Final Report 33, exh. 18 (2009), *available at* http://www.ed.gov/ rschstat/eval/tech/netts/finalreport.html.

[1111] Lucinda Gray & Laurie Lewis, Educational Technology in Public School Districts: Fall 2008, at 3 (2009) (Gray & Lewis, Educational Technology), http://nces.ed.gov/ pubs2010/2010003.pdf. Schools may have more than one type of connection.

[1112] Gray & Lewis, Educational Technology, at 3 (2009), http://nces.ed.gov/pubs2010/2010003.pdf.

[1113] *See* Alaska Department of Education Comments in re NBP PN #15, filed Nov. 20, 2009, at 6–7; American Association of School Administrators and the Association of Educational Service Agencies Comments in re NBP PN #15, filed Nov. 20, 2009, at 2 (filed by Noelle Ellerson); Iowa Department of Education Comments in re NBP PN #15, filed Nov. 20, 2009, at 2–3; Oregon Department of Education Comments in re NBP PN #15, filed Nov. 20, 2009, at 2–3.

[1114] Tom Greaves, Chairman, The Greaves Group, Remarks at FCC Education Workshop 2 (Aug. 20, 2009) (Greaves Aug. 20, 2009 Remarks), *available at* http://www. broadband.gov/docs/ws_education/ws_education_greaves.pdf.

[1115] Greaves Aug. 20, 2009 Remarks at 2, *available at* http:// www.broadband.gov/docs/ws_education/ws_education_ greaves.pdf.

[1116] Greaves Aug. 20, 2009 Remarks at 2, *available at* http:// www.broadband.gov/docs/ws_education/ws_education_ greaves.pdf.

[1117] 47 C.F.R. § 54.504(b)(2)(v), (c)(1)(vii); *see also* 47 C.F.R. § 54.500(b) (defining "educational purposes").

[1118] Alaska Department of Education and Early Development Comments in re NBP PN #15, filed Nov. 20, 2009, at 72; American Association of School Administrators and the Association of Educational Service Agencies Comments in re NBP PN #15, filed Nov. 20, 2009 (filed by Noelle Ellerson), at 5; Anchorage School District Comments in re NBP PN #15, filed Nov. 20, 2009, at 18; AT&T Comments in re NBP PN #15, filed Nov. 20, 2009, at 5; California K–12 High Speed Network Comments in re NBP PN #15, filed Nov. 20, 2009, at 9 (filed by Imperial County Office of Education); CenturyLink Reply in re NBP PN #15, filed Dec. 11, 2009, at 6; City of Chicago Comments in re NBP PN #15, filed Nov. 20, 2009, at 24; Council of the Great City Schools Comments in re NBP PN #15, filed Nov. 20, 2009, at 3; Dell, Inc. Comments in re NBP PN #15, filed Nov. 20, 2009, at 4; Education and Libraries Networks Coalition Comments in re NBP PN #15, filed Nov. 20, 2009, at 5; ENA Comments in re NBP PN #15, Nov. 20, 2009, at 6; Funds for Learning Comments in re NBP PN #15, filed Nov. 20, 2009, at 3; International Association for K–12 Online Learning Reply in re NBP PN #15, filed Dec. 11, 2009, at 16; Iowa Department of Education Comments in re NBP PN #15, filed Nov. 20, 2009, at 4; Kellogg & Sovereign Consulting Comments in re NBP PN #15, filed Nov. 20, 2009, at 8; Miami-Dade County Public School Comments in re NBP PN #15, filed Nov. 17, 2009, at 1; Microsoft Corp. Reply in re NBP PN #15, filed Dec. 11, 2009, at 7–8; The National Internet2 K–20 Initiative Comments in re NBP PN #15, filed Nov. 20, 2009, at 1 (filed by Louis Fox); Software & Information Industry Association Reply in re NBP PN #15, filed Dec. 11, 2009, at 12; State E–rate Coordinators Alliance Comments in re NBP PN #15, filed Nov. 20, 2009, at 11; University of Alaska Comments in re NBP PN #15, filed Nov. 19, 2009, at 2.

[1119] AT&T Comments in re NBP PN #15, filed Nov. 20, 2009, at 9; International Association for K–12 Online Learning Reply in re NBP PN #15, filed Dec. 11, 2009, at 18; Kellogg & Sovereign Consulting Comments in re NBP PN #15, filed Nov. 20, 2009, at 11; Northeastern Regional Information Center Comments in re NBP PN #15, filed Dec. 10, 2009, at 12; State E–rate Coordinators Alliance Comments in re NBP PN #15, filed Nov. 20, 2009, at 19; West Virginia Department of Education Comments in re NBP PN #15, filed Nov. 20, 2009, at 15 (filed by Julia Benincosa); Wisconsin Department of Public Instruction Comments in re NBP PN #15, filed Nov. 20, 2009, at 5.

[1120] *See* Letter from Jeff Donley, Director of Information Systems, Mukilteo School District, Washington, GN Docket No. 09-51 (filed Jan. 29, 2010) (comparing 100 Mbps service connecting 6 schools for $180,000 per year plus T-1 service connecting 14 additional schools for $114,000 per year (all eligible for a 65% E-rate discount) with a 1 Gbps fiber offering connecting all 20 schools for $65,000 (and yet not eligible for the E-rate discount)).

[1121] Council of the Great City Schools Comments in re NBP PN #15, filed Nov. 20, 2009, at 5.

[1122] Wisconsin Department of Public Instruction Comments in re NBP PN #15, filed Nov. 20, 2009, at 3.

[1123] American Association of School Administrators and the Association of Educational Service Agencies Comments in re NBP PN #15, filed Nov. 20, 2009, at 7, 9 (filed by Noelle Ellerson); City of Chicago Comments in re NBP PN #15, filed Nov. 20, 2009, at 27–28; South Kitsap School District Comments in re NBP PN #15, filed Nov. 19, 2009, at 1; Oregon Department of Education Comments in re NBP PN #15, filed Nov. 20, 2009, at 3; National Internet2 K–20 Initiative Comments in re NBP PN #15, filed Nov. 20, 2009, at 1; Texas Education Telecommunications Network Ex Parte in re NBP PN #15, filed Feb. 19, 2010, at 1; School District of Palm Beach County Comments in re NBP PN #15, filed Nov. 20, 2009, at 4, 7–8.

[1124] This figure was calculated using publicly available GDP deflators from 1997 to 2009, yielding total monetary deflation of $676 million.

[1125] State E-rate Coordinators Alliance Comments in re NBP PN #15, filed Nov. 20, 2009, at 29.

[1126] American Association of School Administrators and the Association of Educational Service Agencies Comments in re NBP PN #15, filed Nov. 20, 2009, at 8 (filed by Noelle Ellerson); American Library Association Comments in re NBP PN #15, filed Nov. 20, 2009, at 4; Berkeley County School District Comments in re NBP PN #15, filed Nov. 19, 2009, at 1; Bill and Melinda Gates Foundation Reply in re NBP PN #15, filed Dec. 9, 2009, at 4; California K–12 High Speed Network Comments in re NBP PN #15, filed Nov. 20, 2009, at 13 (filed by Imperial County Office of Education); Dell Comments in re NBP PN #15, filed Nov. 20, 2009, at 4; EdLinc Comments in re NBP PN #15, filed Nov. 20, 2009, at 4; Education Networks of America Comments in re NBP PN #15, filed Nov. 20, 2009, at 10 (filed as ENA); Funds for Learning Comments in re NBP PN #15, filed Nov. 20, 2009, at 10; International Association for K12 Online Learning Reply in re NBP PN #15, filed Dec 11., 2009, at 20; Iowa Department of Education Comments in re NBP PN #15, filed Nov. 20, 2009, at 10; Miami- Dade County Public Schools Comments in re NBP PN #15, filed Nov. 17, 2009, at 1; National Association of Telecommunications Officers and Advisors Comments in re NBP PN #15, filed Nov. 20, 2009, at 10; New York Office of Children and Family Services Reply in re NBP PN #15, Dec. 10, 2009, at 4; Northeastern Regional Information Center Reply in re NBP PN #15, filed Dec. 10, 2009, at 13; Oneida-Herkimer-Madison Board of Cooperative Educ. Services Comments in re NBP PN #15, filed Nov. 19, 2009, at 2; Oregon Department of Education Comments in re NBP PN #15, filed Nov. 20, 2009, at 10; School District of Palm Beach County Comments in re NBP PN #15, filed Nov. 20, 2009, at 3; Schools, Health and Libraries Broadband Coalition Comments in re NBP PN #15, filed Nov. 20, 2009, at 4; Software & Information Industry Association Reply in re NBP PN #15, filed Dec. 11, 2009, at 13; State E-rate Coordinators Alliance Comments in re NBP PN #15, filed Nov. 20, 2009, at 29; Quilt and StateNets Reply in re NBP PN #15, filed Dec. 11, 2009, at 2; Washington State Office of Superintendent of Public Instruction Reply in re NBP PN #15, filed Dec. 10, 2009, at 1 (filed by Dennis Small).

[1127] American Association of School Administrators and the Association of Educational Service Agencies Comments in re NBP PN #15, filed Nov. 20, 2009, at 6 (filed by Noelle Ellerson); American Library Association Comments in re NBP PN #15, filed Nov. 20, 2009, at 16; Bill and Melinda Gates Foundation Reply in re NBP PN #15, filed Dec. 9, 2009, at 4; California K–12 High Speed Network Comments in re NBP PN #15, filed Nov. 20, 2009, at 11 (filed by Imperial County Office of Education); CenturyLink Comments in re NBP PN #15, filed Nov. 20, 2009, at 11; City of Chicago Comments in re NBP PN #15, filed Nov. 20, 2009, at 22; Dell Comments in re NBP PN #15, filed Nov. 20, 2009, at 5; Iowa Department of Education Comments in re NBP PN #15, filed Nov. 20, 2009, at 8; Michigan Department of Education Comments in re NBP PN #15, filed Nov. 20, 2009, at 7 (filed by Jeannene Hurley); Microsoft Reply in re NBP PN #15, filed Dec. 4, 2009, at 9; Oregon Department of Education Comments in re NBP PN #15, filed Nov. 20, 2009, at 9; Pelican City School District Comments in re NBP PN #15, filed Nov. 20, 2009; Schools, Health and Libraries Broadband Coalition Comments in re NBP PN #15, filed Nov. 20, 2009, at 5; State Educational Technology Directors Association Reply in re NBP PN #15, filed Dec. 11, 2009, at 2; State E-rate Coordinators Alliance Comments in re NBP PN #15, filed Nov. 20, 2009, at 26; Washington State Office of Superintendent of Public Instruction Comments in re NBP PN #15, filed Nov. 20, 2009, at 1 (filed by Dennis Small).

[1128] *See* Schools and Libraries Universal Service Description of Services Ordered and Certification Form 471, OMB 3060–0806 (Nov. 2004) (FCC Form 471) at 2, blocks 2, 3 (November 2004), *available at* http://www.usac.org/_ res/documents/sl/pdf/471_fy05.pdf (requesting filers to explain the impact of E-rate funds on the number of buildings connected to the Internet at up to 10 Mbps, up to 200 Mbps, and over 200 Mbps).

[1129] Oregon Department of Education Comments in re NBP PN #15, filed Nov. 20, 2009, at 8; State E-rate Coordinators Association Comments in re NBP PN #15, filed Nov. 20, 2009, at 33; South Kitsap School District Comments in re NBP PN #15, filed Nov. 20, 2009, at 2–3.

[1130] Oregon Department of Education Comments in re NBP PN #15, Nov. 20, 2009, at 11; State E-rate Coordinators Association Comments in re NBP PN #15, Nov. 20, 2009, at 19–20, 22; Iowa Department of Education Comments in re NBP PN #15, filed Nov. 20, 2009, at 11–12.

[1131] Pelican City School District Comments in re NBP PN #15, filed Nov. 20 2009, at 9.

[1132] *See* Letter from Loris Ann Taylor, Executive Director, Native Public Media et al., to Marlene H. Dortch, Secretary, FCC, GN Docket Nos. 09-47, 09-51, 09-137 (Dec. 24, 2009) (Joint Native Filers Dec. 24, 2009 Ex Parte) at 13–16.

[1133] *See* GAO, Challenges to Assessing and Improving Telecommunications for Native Americans on Tribal Lands 30–32, GAO-06-189 (Jan. 2006) (January 2006 GAO Report).

[1134] Texas Center for Education Research, Evaluation of the Texas Technology Immersion Pilot: Final Outcomes for a Four–Year Study (2004–05 to 2007–08), at vi–vii (2009), http://www.etxtip.info/y4_etxtip_final.pdf.

[1135] Gill Valentine et al., Children and Young People's Home Use of ICT for Educ. Purposes: the Impact on Attainment at Key Stages 1–4, at 8–9 (2005), *available at* http://www.dcsf.gov.uk/research/data/uploadfiles/RR672.pdf; mizukO itO et al., livinG and learninG With neW media summary Of findinGs frOm the diGital yOuth prOject 1–3 (2008), *available at* http://digitalyouth. ischool.berkeley.edu/files/report/digitalyouth-WhitePaper.pdf; dOn passey et al., the mOtivatiOnal effect Of ict On pupils 3 (2004), *available at* http://www.dcsf.gov.uk/research/data/uploadfiles/RR523new. pdf; becta, minister's taskfOrce On hOme access tO tech., extendinG OppOrtunity 4 (2008), *available at* http://partners.becta.org.uk/upload-dir/downloads/page_documents/partners/home_access_report.pdf.

[1136] E-rate currently supports wireless data services to mobile devices for educators. That support should be harmonized with this support for student devices during any rulemaking.

[1137] Albuquerque Public Schools Comments in re NBP PN #15, filed Nov. 20, 2009, at 6; City of Chicago Comments in re NBP PN #15, filed Nov. 20, 2009, at 28; Michigan Department of Education Comments in re NBP PN #15, filed Nov. 20, 2009, at 4; National Internet2 K–20 Alliance Comments in re NBP PN #15, filed Nov. 20, 2009, at 1; Oregon Department of Education Comments in re NBP PN #15, filed Nov. 20, 2009, at 3; San Diego Unified School District Comments in re NBP PN #15, filed Nov. 20, 2009, at 2; Sprint Nextel Comments in re NBP PN #15, filed Nov. 20, 2009, at 5.

[1138] *See, e.g.,* Open Learning Initiative, Open Courses Backed by Learning Research, http://oli.web.cmu.edu/openlearning (last visited Feb. 28, 2010).

[1139] Steve Lohr, *Study finds that online education beats the classroom*, N.Y. Times, Aug. 19, 2009, http://bits.blogs.nytimes.com/2009/08/19/study-finds-that-onlineeducation-beats-the-classroom.

[1140] American Association of Community Colleges and Educause Comments in re NBP PN #15, filed Nov. 20, 2009, at 4; California Public Utilities Commission Comments in re NBP PN #15, filed Nov. 20, 2009, at 3; City of Chicago Comments in re NBP PN #15, filed Nov. 20, 2009, at 25; National Internet2 K–20 Initiative Comments in re NBP PN #15, filed Nov. 20, 2009, at 1; New York Education Department Comments in re NBP PN #15, filed Nov. 20, 2009, at 2; Texas Education Telecommunications Network Ex Parte in re NBP PN #15, filed Feb. 19, 2010, at 1; Texas State Library and Archives Comments in re NBP PN #15, filed Nov. 20, 2009, at 1; Quilt and StateNets Reply in re NBP PN #15, filed Dec. 11, 2009, at 2.

[1141] *See* National Center for Education Statistics, Integrated Postsecondary Data System, http://nces.ed.gov/IPEDS/ (last visited Feb. 28, 2010).

[1142] Educause Core Data Service, Fiscal Year 2007 Summary Report 35 (2007) (reporting that only 16.1% of colleges offering an associate's degree have more than 45 Mbps in bandwidth, whereas 90.4% of institutions offering a doctorate have that level of connectivity), *available at* http://net.educause.edu/ir/library/pdf/PUB8005.pdf.

[1143] Madeline Patton, Community Colleges Impact k–12 STEM TEACHING 4 (2008), *available at* http://www.aacc.nche.edu/Resources/aaccprograms/Documents/ impactk12_2008.pdf.

[1144] *See* U.S. Energy Info. Admin. (Eia), Doe, Annual Energy Review 2008, at 13 (2009) (providing energy expenditures and consumption), *available at* http:// www.eia.doe.gov/emeu/aer/pdf/aer.pdf.

[1145] Steven Chu, Secretary, Doe, Presentation at Copenhagen: Meeting the Energy and Climate Challenge (Dec. 14, 2009), *available at* http://www.energy.gov/news/documents/ Chu_Climate_Challenge_12-14-09.pdf.

[1146] Boston Consulting Group (Bcg), Global E-Sustainability Initiative, Smart 2020: Enabling the Low Carbon Economy in the Information Age, United States Report Addendum (2008) (BCG, smart 2020), *available at* http://www.smart2020.org/_assets/files/Smart2020UnitedStatesReportAddendum.pdf. For comparison, note that all the coal fired power plants in the United States generate about 2 billion metric tons of greenhouse gas emissions per year. epa, inventOry Of u.s. GreenhOuse Gas emissiOns and sinks: 1990–2007 (2009) (EPA, Inventory of Emissions and Sinks), *available at* http://www.epa.gov/climatechange/emissions/downloads09/GHG2007entire_report-508.pdf.

[1147] Elec. Power Res. Inst. (EPRI), Report to NIST on the smart Grid Interoperability Standards Roadmap (2009), *available at* http://www.nist.gov/smartgrid/ InterimSmartGridRoadmapNISTRestructure.pdf.

[1148] EPRI, Power Delivery System of the Future: a preliminary estimate Of cOsts and benefits (2004), *available at* http://mydocs.epri.com/docs/ public/000000000001011001.pdf.

[1149] Consortium for Elec. Infrastructure to Support A Digital Soc'y, The Cost of Power Disturbances to Industrial & Digital Economy Companies (2001), *available at* http://www.epri-intelligrid.com/intelligrid/docs/Cost_of_Power_Disturbances_to_Industrial_and_ Digital_Technology_Companies.pdf.

[1150] Edison Electric Institute Comments in re NBP PN #2 (*Comment Sought on the Implementation of Smart Grid Technology—NBP Public Notice #2*, GN Docket Nos. 09-47, 09-51, 09-137, Public Notice, 24 FCC Red 11747 (WCB 2009) (*NBP PN #2*)), filed Oct. 2, 2009.

[1151] Pacific Northwest Nat'l Lab. (PNNL), Doe, Smart Grid: An Estimation of the Energy and CO_2 Benefits (2009), *available at* http://www.pnl.gov/main/ publications/external/technical_reports/PNNL-19112. pdf; epa, emissiOn facts: GreenhOuse Gas emissiOns frOm a typical passenGer vehicle (2005) (providing EPA auto emission facts), *available at* http://www.epa. gov/OMS/climate/420f05004.pdf.

[1152] Lawrence Livermore Nat'l Lab., United States Energy Flow in 2007 (chart) (2009), *available at* https://publicaffairs.llnl.gov/news/energy/content/ international/United_States_Energy_2007.png.

[1153] EPA, Inventory of Emissions and Sinks.

[1154] Electric power generation is responsible for 34 percent of U.S. greenhouse gas emissions. epa, inventOry Of emissiOns and sinks at ES-16, tbl. ES-7.

[1155] Philip Giudice, Commissioner, Mass. Dep't of Energy Res., Presentation at FCC Energy Field Hearing: Our Energy Future and Smart Grid Communication (Nov. 30, 2009), *available at* http://www.broadband.gov/fieldevents/fh_energy_environment/giudice.pdf.

[1156] FERC, National Assessment Of Demand Response Potential (2009) (FERC, demand respOnse pOtential), *available at* http://www.ferc.gov/legal/staff-reports/06- 09-demand-response.pdf.

[1157] PNNL, Impacts and Assessment of Plug-in Hybrid Vehicles on Electric Utilities and Regional U.S. Power Grids: Part 1: Technical Analysis 12 (2006) (PNNL, assessment Of pluG-in hybrid vehicles On electric utilities), *available at* http://www.ferc.gov/about/commem/wellinghoff/5-24-07-technical-analy-wellinghoff. pdf.

[1158] PNNL, Assessment of Plug-In Hybrid Vehicles on Electric Utilities at 13.

[1159] D. M. Lemoine et al., *An Innovation and Policy Agenda for Commercially Competitive Plug-in Hybrid Electric Vehicles*, envtl. res. letters, Jan.–Mar. 2008, at 6 ("No additional capacity needed to charge 10 million vehicles from 11p.m.–8a.m." and "Charging 10 million vehicles from 6p.m.–12a.m. requires approximately 30 percent more capacity" superimposed on the graphs by the Commission), *available at* http://www.iop.org/ EJ/article/1748-9326/3/1/014003/erl8_1_014003. pdf?request-id=ebf87cfb-96ec-4f5b-bccbea307197f80d.

[1160] FERC, Demand Response Potential.

[1161] Steven Chu, Secretary, DOE, Presentation to the GW Solar Institute: Investing in Our Energy Future (Sept. 21, 2009), *available at* http://solar.gwu.edu/index_files/ Variability/chu%20presentation%20at%20gridweek. pdf.

[1162] Memorandum from Vice Pres. Joseph Biden, Jr., to Pres. Barack Obama on the Transformation to a Clean Energy Economy 5 (Dec. 15, 2009) (stating that there will be 877 sensors installed by 2013), *available at* http://www. whitehouse.gov/sites/default/files/administrationofficial/vice_president_memo_on_clean_energy_ economy.pdf.

[1163] Tropos Networks Comments in re NBP PN #2, filed Oct. 2, 2009, at 2.

[1164] Sempra Energy Utilities (Sempra) Comments in re PN #2, filed Oct. 2, 2009, at 15; Southern Company Comments in re NBP PN #2, filed Oct. 2, 2009, at 13.

[1165] DTE Energy (DTE) Comments in re NBP PN #2, filed Oct. 2, 2009, at 7; American Electric Power (AEP) Comments in re NBP PN #2, filed Oct. 2, 2009, at 25.

[1166] *See* AEP Comments in re NBP PN #2, filed Oct. 2, 2009; Centerpoint Comments in re NBP PN #2, filed Oct. 2, 2009; Cleco Comments in re NBP PN #2, filed Oct. 5, 2009; DTE Comments in re NBP PN #2, filed Oct. 2, 2009; Florida Power and Light Comments in re NBP PN #2, filed Oct. 2, 2009; Sempra Comments in re NBP PN #2, filed Oct. 2, 2009; Southern Company Comments in re NBP PN #2, filed Oct. 2, 2009; *See also* Letter from Andres E. Carvallo, CIO, Austin Energy et al., to Pres. Barack Obama (June 29, 2009), *available at* http:// fjallfoss.fcc.gov/ecfs/document/view?id=7020356770; Alcatel Lucent Comments in re NBP PN #2, filed Oct. 2, 2009; Aclara Comments in re NBP PN #2, filed Oct. 2, 2009; GE Energy Comments in re NBP PN #2, filed Oct. 2, 2009; Gridnet Comments in re NBP PN #2, filed Oct. 2, 2009; Hewlett-Packard Comments in re NBP PN #2, filed Oct. 2, 2009, *attached to* Letter from Tony Erickson, Industry Leader, Utilities, to Chmn. Julius Genachowski, FCC, GN Docket Nos. 09-47, 09-51, 09-137 (Oct. 2, 2009); Motorola Comments in re NBP PN #2, filed Oct. 2, 2009; On-Ramp Wireless Comments in re NBP PN #2, filed Oct. 2, 2009; Tropos Networks Comments in re NBP PN #2, filed Oct. 2, 2009.

[1167] Sempra Comments in re NBP PN #2, filed Oct. 2, 2009, at 11.

[1168] DTE Comments in re NBP PN #2, filed Oct. 2, 2009, at 14.

[1169] Southern California Edison (SCE) Comments in re NBP PN #2, filed Oct. 2, 2009, at 14.

[1170] AEP—a major investor-owned utility—estimates that 59 percent of its substations do not have 3G wireless access. AEP Comments in re NBP PN #2, filed Oct. 2, 2009, at 16.

[1171] FCC fact findings and the record provide examples of incidents when commercial networks were unable to provide communications during and immediately following emergencies. When remnants of Hurricane Ike struck Ohio, congestion rendered commercial networks "nearly useless" in large parts of Columbus. AEP Comments in re NBP PN #2, filed Oct. 2, 2009, at 14. During Hurricane Katrina, the communications systems operated by utility subsidiaries were for a time the sole source of wireless communications in Gulfport, Miss. Southern Company Comments in re PN #2, filed Oct. 2, 2009, at 10. Cleco has experienced similar problems with commercial network connectivity during hurricanes. Cleco Comments in re NBP PN #2, filed Oct. 2, 2009, at 2. The FCC has previously found that commercial networks are often disrupted in emergencies and that hardened networks (including many utility networks) are less susceptible to failure due to their site hardening, onsite backup power, redundant backhaul, and staff dedicated to maintenance of backup capabilities. independent panel revieWinG the impact Of hurricane katrina On cOmmunicatiOns netWOrks, repOrt and recOmmendatiOns tO the federal cOmmunicatiOns cOmmissiOn (2006), *attached to* Letter from Nancy J. Victory, Chair, Indep. Panel Reviewing the Impact of Hurricane Katrina on Communications Networks, to Chmn. Kevin J. Martin, FCC, EB Docket No. 06-119 (June 12, 2006), *available at* www.fcc.gov/ pshs/docs/advisory/hkip/karrp.pdf.

[1172] *See* Alcatel Lucent Comments in re NBP PN #2, filed Oct. 2, 2009; Sempra Comments in re NBP PN #2, filed Oct. 2, 2009; Utilities Telecom Council (UTC) Comments in re NBP PN #2, filed Oct. 2, 2009.

[1173] *See* Rob Curtis et al., Omnibus brOadband initiative, (OBI) the brOadband availability Gap 3G coverage figures rely on American Roamer data for HSPA and EV-DO coverage and Geolytics data for population; covered population for partially covered census blocks is calculated based on the fraction of area covered by the American Roamer 3G coverage shapefiles American Roamer Advanced Services database (accessed Aug.

2009) (aggregating service coverage boundaries provided by mobile network operators) (on file with the Commission) (American Roamer database). For detail on American Roamer methodology, see Chapter 4.

[1174] Stephen J. Blumberg & Julian V. Luke, Nat'l CTR For Health Stat., Wireless Substitution: Early Release Of Estimates from the National Health Interview Survey, January–June 2009 (2009), http://www.cdc.gov/nchs/data/nhis/earlyrelease/wireless200912.pdf.

[1175] T-Mobile Comments in re NBP PN #2, filed Oct. 2, 2009, at 5–6.

[1176] EISA requires that each state consider authorizing rate recovery for "capital, operating expenditure, or other costs . . . for the deployment of the qualified smart grid system." 16 U.S.C. § 2621(d)(18)(B).

[1177] California Public Utility Commission Comments in re NBP PN #2, filed Oct. 2, 2009, at 9.

[1178] There are notable exceptions to this, principally in states where the regulators have implemented decoupling. Publicly and cooperatively owned utilities, which deliver roughly 30 percent of the nation's electricity, are another exception.

[1179] Kate Galbraith, *Why is a Utility Paying Customers?*, n.y. times, Jan. 23, 2010, *available at* http://www.nytimes.com/2010/01/24/business/energy-environment/24idaho.html?pagewanted=2&emc=eta1.

[1180] For example, Centerpoint cites NERC CIP as a reason it did not select commercial networks to provide wide area connectivity. Centerpoint Comments in re NBP PN #2, filed Oct. 2, 2009, at 10. On the other hand, AT&T touts its NERC CIP-compliant network security capabilities as a benefit for utilities that choose its networking services. Letter from Joseph P. Marx, Assistant to the President, Federal Regulatory, AT&T Inc., to Marlene H. Dortch, Secretary, FCC, GM Dockets Nos. 09–47, 09–51, 09–137 (Dec. 18, 2009) AT&T Dec. 18, 2009 *Ex Parte* at 3. Sempra believes commercial networks can meet NERC CIP requirements if the operators have "the ability to prove that each communications device and link in the path is properly managed, configured and secure under the terms of national standards and regulations related to critical infrastructure" (*e.g.*, FERC and NERC standards). Sempra Comments in re NBP PN #2, filed Oct. 2, 2009, at 9. DTE Comments in re NBP PN #2, filed Oct. 2, 2009, (expresses a similar sentiment). Alcatel Lucent believes it may not be possible to use Internet Protocol for some applications due to NERC CIP requirements. Alcatel Lucent Comments in re NBP PN #2, filed Oct. 2, 2009, at 12. CTIA points out that irrespective of NERC CIP requirements, commercial networks can be made very secure and are commonly used for very sensitive communications, including the communications of the U.S. Department of Treasury, the U.S. Secret Service, and the U.S. Department of Homeland Security. CTIA Comments in re NBP PN#2, filed Oct. 2, 2009, at 9.

[1181] Marc Pallans, *Public Safety and Private Utility ... a Unique Partnership*, laW & Order, July 2009, *available at* http://www.pspc.harris.com/news/published_articles/ Law%20Jul09%20pg42%20w%20ad.pdf.

[1182] *See* Centerpoint Comments in re NBP PN #2, filed Oct. 2, 2009; Sempra Comments in re NBP PN #2, filed Oct. 2, 2009.

[1183] AEP Comments in re NBP PN #2, filed Oct. 2, 2009, at 23; UTC Comments in re NBP PN #2, filed Oct. 2, 2009, at 21.

[1184] *See* AEP Comments in re NBP PN #2, filed Oct. 2, 2009; Centerpoint Comments in re NBP PN #2, filed Oct. 2, 2009; UTC Comments in re NBP PN #2, filed Oct. 2, 2009; Edison Electric Institute in re NBP PN #2, filed Oct. 2, 2009.

[1185] Sempra Comments in re NBP PN #2, filed Oct. 2, 2009, at 15.

[1186] Examples include Arcadian Networks, Gridnet, and Sensus.

[1187] Sempra Comments in re NBP PN #2, filed Oct. 2, 2009, at 15.

[1188] *See* AEP Comments in re NBP PN #2, filed Oct. 2, 2009; Centerpoint Comments in re NBP PN #2, filed Oct. 2, 2009; UTC Comments in re NBP PN #2, filed Oct. 2, 2009.

[1189] Ahmad Faruqui & Sanem Sergici, Household Response of Dynamic Pricing To Electricity—A Survey of the Experimental Evidence (2009), *available at* http:// www.loadeconomics.com/files/The_Power_of_Experimentation.pdf.

[1190] Google Comments in re NBP PN #2, filed Oct. 2, 2009, at 4.

[1191] Table 5. Average Monthly Bill Data by Census Division, and State 2008, *attached to* eia, electric sales, revenue, and averaGe price 2008 (2010), http://eia.doe. gov/cneaf/electricity/esr/table5.html.

[1192] *See* GE Energy Comments in re NBP PN #2, filed Oct. 2, 2009, at 23 (discussing the appliances they are developing that will "change their operating model" based on real-time price and consumption information).

[1193] Candace Lombardi, *Whirlpool Wants to Pull the Plug on 'Dumb' Appliances*, cnet, Oct. 29, 2009, http://news.cnet.com/8301-11128_3-10386123-54.html.

[1194] Tendril Comments in re NBP PN #2, filed Oct. 2, 2009.

[1195] *See* Google Comments in re NBP PN #2, filed Oct. 2, 2009; Letter from Paula Boyd, Regulatory Counsel for Microsoft, to Marlene H. Dortch, Secretary, FCC, GN Docket Nos. 09-47, 09-51, 09-137 (Nov. 9, 2009).

[1196] Dan Johnson, Founder & CEO, Verisae, Remarks at FCC Energy Field Hearing (Nov. 30, 2009), *available at* http:// www.fcc.gov/live/archive/2009_11_30-workshop.html.

[1197] Rick Counihan, Vice President Regulatory Affairs, EnerNOC, Remarks at the FCC Energy Field Hearing (Nov. 30, 2009) (Counihan Energy Hearing Remarks), *available at* http://www.fcc.gov/live/ archive/2009_11_30-workshop.html.

[1198] Counihan Energy Hearing Remarks; Adrian Tuck, CEO, Tendril Networks, Remarks at FCC Energy Field Hearing (Nov. 30, 2009), *available at* http://www.fcc. gov/live/archive/2009_11_30-workshop.html.

[1199] eMeter survey of 25 utilities with plans to deploy 16.7 million AMI meters in the next four years (does not include California utilities) eMeter Comments in re NBP PN #2, filed Oct. 2, 2009, at 3. This finding is supported by GE Energy. GE Energy Comments in re NBP PN #2, filed Oct. 2, 2009, at 22.

[1200] *See* AT&T Comments in re NBP PN #2, filed Oct. 2, 2009; Letter from David M. Don, Senior Director Public Policy, Comcast, to Marlene H. Dortch, Secretary, FCC, GN Docket No. 09-51 (Oct. 19, 2009); Google Comments in re NBP PN #2, filed Oct. 2, 2009; Honeywell Comments in re NBP PN #2, filed Oct. 2, 2009; Qwest Comments in re NBP PN #2, filed Oct. 2, 2009; Tendril Networks Comments in re NBP PN #2, filed Oct. 2, 2009; Verizon & Verizon Wireless Comments in re NBP PN #2, filed Oct. 2, 2009.

[1201] Images from left to right: Visible Energy, Inc., Control4, Tendril, ecobee.

[1202] EISA requires consumers be given "direct access" to their energy price, consumption, and generation mix data "through the Internet." 16 U.S.C. § 2621(d)(17).

[1203] There are a number of possible applications from this data, including empowering customers to use energy when renewable power sources are plentiful and helping businesses track their own greenhouse gas emissions impact.

[1204] *Decision Adopting Policies and Findings Pursuant to the Smart Grid Policies Established by the Energy Information and Security Act of 2007*, Rulemaking 08-12-009, Decision 09-12-046 (Cal. PUC Dec. 17, 2009), *available at* http://docs.cpuc.ca.gov/word_pdf/ FINAL_DECISION/111856.pdf.

[1205] As directed by Congress in the Energy Security and Independence Act of 2007, DOE must submit a report to Congress "concerning the status of Smart Grid deployments nationwide and any regulatory or government barriers to continued deployment." Pub. L. No. 110-140, § 1302, 121 Stat. 1492, 1784 (2007).

[1206] National Rural Electric Cooperative Association Comments in re NBP PN #2, filed Oct. 2, 2009, at 2.

[1207] Joseph Badin, Representative of RUS for the Federal Smart Grid Task Force, Remarks at Smart Grid Task Force Meeting (Dec. 16, 2009).

[1208] DOE, *Secretary Chu Announces $47 Million to Improve Efficiency in Information Technology and Communications Sectors* (press release), Jan. 6, 2010, http://www.energy.gov/news2009/8491.htm.

[1209] BCG, smart 2020.

[1210] EIA, 2005 Energy Data (unpublished, on file at EIA). For each year, the "Residential Electricity Use: PCs, Laptops and Peripherals" plus "Commercial Electricity Use: Office Equipment (PC)" the divided by the sum of "Residential: Grand Total" and "Commercial: Delivered Energy" equals the approximate amount of residential and commercial electricity use consumed by PCs and Peripherals.

[1211] EIA, Annual Energy Outlook 2010 early release, at tbls. 4 (Residential Sector), 5 (Commercial Sector) (Dec. 14, 2009), *available at* http://www.eia.doe.gov/oiaf/ aeo/aeoref_tab.html. For each year, the "Residential Electricity Use: PCs, Laptops and Peripherals" plus the "Commercial Electricity Use: Office Equipment (PC)" divided by the sum of "Residential: Grand Total" and "Commercial: Delivered Energy" equals the approximate amount of residential and commercial electricity use consumed by PCs and Peripherals.

[1212] Judy Roberson et al., After-hours Power Status of Office Equipment and Inventory Of Miscellaneous Plug-Load Equipment (2004), *available at* http://dx.doi. org/10.2172/821675.

[1213] GSM World, *Mobile Industry Unites to Drive Universal Charging Solution for Mobile Phones* (press release), Feb. 17, 2009, http://www.gsmworld.com/newsroom/pressreleases/2009/2548.htm.

[1214] H. Scott Matthews et al., *Electricity Use of Wired and Wireless Telecommunications Networks in the United States*, ieee sympOsium On elec. & env't 131 (2003), *available at* http://ieeexplore.ieee.org/stamp/stamp.jsp? tp=&isnumber=27162&arnumber=1208061.

[1215] Chris Carruth & Clint Wheelock, Green Telecom Networks: Energy Efficiency, Renewable Power, and Carbon Emissions Reductions For Fixed And Mobile Telecommunication Networks (2009).

[1216] Sprint, Company Info, Corporate Responsibility, Sustainable Operations, Energy, http://www.sprint. com/responsibility/environment/energy.html/ (last visited Feb. 2, 2010).

[1217] Sprint, Company Info, Corporate Responsibility, Sustainable Operations, Energy, http://www.sprint. com/responsibility/environment/energy.html/ (last visited Feb. 2, 2010).

[1218] EPA, Report To Congress on Server and Data Center Energy Efficiency, Public Law 109-431 (2007) (EPA, Data Center Energy Efficiency), *available at* http://www.energystar.gov/ia/partners/prod_development/ downloads/EPA _Datacenter_Report_Congress_Final1.pdf.

[1219] EPA, Data Center Energy Efficiency; John Laitner & Karen Ehrhardt-Martinez, Information and Communication Technologies: The Power of Productivity (2008), http://www.aceee.org/pubs/e081.htm (requires purchase).

[1220] Epa, Data Center Energy Effiency Report to Congress On Server and Data Center Energy Efficiency (2007), *available at* http://www.energystar.gov/index. cfm?c=prod_development.server_efficiency_study.

[1221] BCG, Smart 2020. As a case in point, the U.S. Postal Service eliminated 791 of its 895 physical servers by using virtualization, reducing its electricity consumption by 3.5 GWh per year. Tim Kauffman, *Obama Targets Data Centers for Energy Cuts*, fed. times, Nov. 15, 2009, http://www.federaltimes.com/article/ 20091115/ FACILITIES04/911150311/1031/FACILITIES04 (Kauffman, *Data Center Energy Cuts*).

[1222] BCG, Smart 2020. Microsoft's Dublin data center uses many other best practices, including 24/7 temperature monitoring and outside air for cooling, which has led the facility to use 50 percent less energy than traditional data centers. Microsoft, *Greening the Dublin Data Center*, 2009, http://www.microsoft.eu/Stories/Viewer/tabid/77/articleType/ArticleView/articleId/329/ Menu/8/Greening-the-Dublin-data-center.aspx.

[1223] Yahoo!, *Serving Up Greener Data Centers* (press release), June 30, 2009, http://ycorpblog.com/2009/06/30/serving-up-greener-data-centers/2009/.

[1224] Kauffman, *Data Center Energy Cuts*.

[1225] EPA, Data Center Energy Efficiency.

[1226] Lawrence Livermore Nat'l Lab., Energy, Carbon Emissions, And Water Flow Charts (2008), https://publicaffairs. llnl.gov/news/energy/energy.html; epa, inventOry Of emissiOns and sinks.

[1227] BCG, smart 2020.

[1228] Joseph Schwieterman et al., Chaddick Inst. Pol. Study, 2008 Update on Intercity Bus Service: Summary of Annual Change (2009), *available at* http://las.depaul.edu/chaddick/docs/Docs/2008_Update_on_Intercity_Bus_Service.pdf.

[1229] Katie Johnston-Chase, *All's Fare in Travel by Bus*, bOstOn GlObe, Nov. 17, 2009, *available at* http://www.boston.com/business/articles/2009/11/17/cheaper_fares_web_access_draw_many_to_bus_travel/.

[1230] More than 37,000 Americans were killed in traffic accidents in 2008. Fatality Analysis Reporting System, National Statistics, http://www-fars.nhtsa.dot.gov/ Main/index.aspx (last visited Feb. 2, 2010).

[1231] Mary Madden & Sidney Jones, Pew Internet & Am. Life Project, Networked Workers 3 (2008).

[1232] Bureau of Labor Stat., Occupational Projections and Training Data, 2009–2010 editiOn (2009), *available at* http://www.bls.gov/emp/optd/ (available for download in various parts). Based on these data, jobs that were broadband related were identified, and a growth rate was calculated for that subset of jobs compared to national projected employment growth.

[1233] U.S. Census Bureau, American Fact Finder (enter "Diller," "Nebraska" in "Fast Access to Information"), http://factfinder.census.gov/home/saff/main.html?_ lang=en (last visited Feb. 13, 2010)

[1234] Letter from Dave Vorhaus, National Broadband Taskforce, FCC, on behalf of Blue Valley Brand Meats, to Marlene H. Dortch, Secretary, FCC (Jan. 13, 2010) at 1 (filed as Federal Communications Commission).

[1235] WCPN.org, Upside/Downside: Youngstown Business Incubator a Bright Spot in Region, http://www.wcpn.org/index.php/WCPN/news/24955/ (last visited Jan. 12, 2010).

[1236] John Tozzi, *New Orleans: A Startup Laboratory*, bus. Wk., Aug. 27, 2007, http://www.businessweek.com/print/smallbiz/content/aug2007/sb20070823_490984. htm; Abby Ellin, *Entrepreneurs Leverage New Orleans's Charm to Lure Small Businesses*, n.y. times, Jul. 29, 2009, http://www.nytimes.com/2009/07/30/business/ smallbusiness/30sbiz.html?pagewanted=all.

[1237] Office of Advocacy, SBA, Advocacy: The Voice of Small Business in Government, Frequently Asked Questions 1 (2009), (SBA, Small Business Economy) *available at* http://www.sba.gov/advo/stats/sbfaq.pdf.

[1238] SBA, Small Business Economy at 99.

[1239] Applied percentages of minority- and women-owned businesses from the small business administratiOn for 2002 Census Bureau data to totals of non-employer and employer businesses from 2006 to create estimates for 2006 totals of minority- and women-owned businesses. *See* minOrities in business at 5, 28. *See also* SBA, small business ecOnOmy at 99.

[1240] U.S. Senate Comm. on Small Bus. & Entrepreneurship, Democratic Page, Minority Entrepreneurs, http://sbc.Senate.gov/public/index.cfm?p=MinorityEntrepreneurs (last visited Mar. 3, 2010).

[1241] John Tozzi, *The Rise of the 'Homepreneur,'* bus. Wk., Oct. 23, 2009, http://www.businessweek.com/smallbiz/content/oct2009/sb20091023_263258.htm.

[1242] Verizon and Verizon Wireless Comments in re NBP PN #18 (*Comment Sought on Relationship Between Broadband and Economic Opportunity—NBP Public Notice #18*, GN Docket Nos. 09-47, 09-51, 09-137, Public Notice, 24 FCC Rcd 13736 (WCB 2009) (*NBP PN #18*)), filed Dec. 14, 2009, at 95.

[1243] Maija Renko & Paul Reynolds, Profiling The Growth Oriented Nascent Entrepreneur in the US—Evidence from Representative Samples 12 (2006) (renkO & reynOlds, prOfilinG the GrOWth Oriented).

[1244] U.S. Census Bureau, Dep't of Com., Women- Owned Firms: 2002, at 1–2 (2006), *available at* http://www2.census.gov/econ/sbo/02/sb0200cswmn.pdf.

[1245] *How Companies Are Benefiting From Web 2.0: McKinsey Global Survey Results*, mckinsey Q., Sept. 2009.

[1246] Org. for Econ. Co-Operation and Dev. [OECD], *Broadband and the Economy: Ministerial Background Report*, at 15, DSTI/ICCP/IE(2007)3/FINAL (2008) (OECD, *Broadband and the Economy*), *available at* http://www.oecd.org/dataoecd/62/7/40781696.pdf.

[1247] *See* Margot Dorfman, CEO, US Women's Chamber of Commerce, Remarks at FCC Opportunities for Small and Disadvantaged Business Workshop 15 (Aug. 18, 2009) ("On any given day, 20 percent of all Americans go online to look for a service or product they are thinking of buying."), *available at* http://www.broadband.gov/ docs/ws_08_op_small_dis_biz.pdf.

[1248] FCC, natiOnal brOadband plan survey Of businesses, dec. 9, 2009–jan. 31, 2010 (2010) (fcc, NBP survey Of businesses) (on file with the Commission).

[1249] To illustrate this point, ThomasNet conducted a case study of Orr & Orr, Inc., a 14-person business that distributes hardware and accessories to the automotive industry. Due to the introduction of an online product catalog, the company can now serve much larger businesses and generate additional revenue. President of Orr & Orr, Hank Hines, describes the benefits as follows: "The online catalog levels the playing field for a small company like ours. The customer on the 'other end' of the Internet doesn't know how big we are—just the products that we have to offer. We can now attract them to our business and let our expertise take over." Orr & Orr, Inc., A ThomasNet Case Study, http://promoteyourbusiness.thomasnet.com/case_studies/ orr-and-orr.html (last visited Feb. 14, 2010).

[1250] OECD, *Broadband and the Economy* at 24.

[1251] Justin Jaffe, Int'l Data Corp. (IDC), Smb Cluster Analysis: SMB 2.0s Lead The Way Toward Nextgeneration Technology, Doc # 219830 (2009) (jaffe, smb cluster analysis).

[1252] OECD, *Broadband and the Economy* at 47.

[1253] Jaffe, SMB cluster analysis.

[1254] FCC, NBP survey Of businesses.

[1255] E.J. Ourso College of Business, LBTC Mobile Classroom, http://www.bus.lsu.edu/centers/lbtc/mobile classroom.asp (last visited Feb. 14, 2010).

[1256] Dep't of Bus. Innovation & Skills, Digital Britain 185–86 (2009), *available at* http://www.culture.gov.uk/ images/publications/digitalbritain-finalreport-jun09.pdf.

[1257] The SBA partners with states and educational institutions to operate nearly 1,000 SBDCs across the country. These SBDCs offer counseling, mentoring, support, and training for small business owners and entrepreneurs.

[1258] The SBA's Office of Women's Business Ownership (OWBO) exists to establish and oversee a network of WBCs throughout the United States and its territories, which provide comprehensive training and counseling on a vast array of topics in many languages to help entrepreneurs, especially women, start and grow their own businesses.

[1259] SBA, FY 2011 Congressional Budget Justification and FY 2009 Annual Performance Report 53 (2010), *available at* http://www.sba.gov/idc/groups/public/ documents/sba_homepage/fy_2011_cbj_09_apr.pdf.

[1260] U.S. Census Bureau, 2006–2008 American Community Survey, S1603. Characteristics of People by Language Spoken at Home, http://factfinder.census.gov/servlet/ STTable?_bm=y&-geo_id=01000US&-qr_name=ACS_ 2008_3YR_G00_S1603&-ds_name=ACS_2008_3YR_ G00_&-_lang=en&-redoLog=false&-format=&- CONTEXT=st (last visited Feb. 9, 2009). According to these data, 55 million people over the age of 5 speak a language other than English at home. The total population estimate of those 5 years and over is 280.5 million.

[1261] ASBDC, America's Small Business Development Center Network, About Us, http://www.asbdc-us.org/About_ Us/aboutus.html (last visited Feb. 14, 2010).

[1262] SCORE: Counselors to America's Small Businesses, Ask SCORE, http://www.score.org/ask_score1.html (last visited Feb. 14, 2010).

[1263] SBA, Counseling & Assistance, http://www.sba.gov/ services/counseling/index.html (last visited Feb. 14, 2010).

[1264] ASBDC, About Us, http://www.asbdc-us.org/About_Us/ aboutus.html (last visited Feb. 14, 2010)

[1265] Letter from Sridhar Prasad, National Broadband Taskforce, FCC, on behalf of SBA, to Marlene H. Dortch, Secretary, FCC GN Docket 09–47, 09–51, 09–137, (Jan. 14, 2010) at 1 (filed as Federal Communications Commission).

[1266] Cisco, Virtual Sales Expertise Case Study: How Cisco Supports Virtual Access to Technical Experts (2009).

[1267] Initiative For a Competitive Inner City, State of the Inner City Economies: Small Businesses in the Inner City 1 (2005), *available at* http://www.sba.gov/advo/ research/rs260tot.pdf.

[1268] Stephan J. Goetz, *Self-Employment in Rural America: The New Economic Reality*, rural realities, 2008, iss. 3 at 1, *available at* http://ruralsociology.org/ StaticContent/Publications/Ruralrealities/pubs/ RuralRealities2-3.pdf.

[1269] Rural Broadband Policy Group Comments in re NBP PN #18, filed Dec. 4, 2009, at 11.

[1270] Asian American Justice Center et al. Comments in re NBP PN #18, filed Dec. 4, 2009, at 5.

[1271] The SBA defines a small disadvantaged business as a business that is at least 51% owned by one or more individuals from groups that have been in a socially disadvantaged position, including the following: African Americans, Asian Pacific Americans, Hispanic Americans, Native Americans, and Subcontinent Asian Americans.

[1272] David Ferreira, Vice Pres. of Gov't Aff., US Hispanic Chamber of Commerce, Remarks at FCC Opportunities for Small and Disadvantaged Businesses Workshop 37 (Aug. 18, 2009), *available at* http://www.broadband.gov/ docs/ws_08_op_small_dis_biz.pdf.

[1273] Letter from Dave Vorhaus, National Broadband Taskforce, FCC, on behalf of Service Corps of Retired Executives (SCORE), to Marlene H. Dortch, Secretary, FCC GN Docket 09–47, 09–51, 09–137, (Jan. 25, 2010) at 1 (filed as Federal Communications Commission).

[1274] Renko & Reynolds, Profiling the Growth Oriented at 6.

[1275] Jason Henderson, *Building the Rural Economy with High-Growth Entrepreneurs*, Econ. Rev.—Fed. Reserve Bank of K.C., July 1, 2002, *available at* http://www. kc.frb.org/PUBLICAT/ECONREV/PDF/3q02hend.pdf.

[1276] Dep't of Labor, *Number of Jobs Held, Labor Market Activity, and Earnings Growth Among the Youngest Baby Boomers: Results From a Longitudinal Survey* (press release), June 27, 2008, *available at* http://www.bls.gov/news.release/nlsoy.nr0.htm.

[1277] Nat'l Skills Coal (Formerly Workforce Alliance), Toward Ensuring America's Workers and Industries the Skills to Compete 6 (2007), *available at* http://www.nationalskillscoalition.org/assets/reports-/towardensuring-americas.pdf.

[1278] J.D. Fletcher, Why Technology? Why ADL? Report frOm a 30-year (so far) Campaign 16 (2009) (Fletcher, Why Technology?).

[1279] Council Of Econ. Advisors, Executive Office of the President, Preparing the Workers of Today for the Jobs of Tomorrow 19 (July 2009) (Council of Econ. Advisors, Preparing the Workers) (describing the workforce development system as an "often conflicting and confusing, maze of job training programs spread across several Federal agencies."), *available at* http://www.whitehouse.gov/assets/documents/Jobs_of_the_Future.pdf.

[1280] Participants in most government-funded employment assistance services are serviced through the One-Stop Delivery System, a set of 2,995 physical centers across the country operated by DOL. CareerOneStop, America's Service Locator, http://www.servicelocator. org/ (last visited Feb. 14, 2010).

[1281] Analysis using U.S. Bureau of Labor Statistics data and annual reports from states submitted to the U.S. Department of Labor Employment Training Administration show that One-Stop centers in cities that suffered job losses greater than 100,000 between July 2008 and July 2009 served an average of only 3,379 people in each One-Stop in 2008. Based on OBI team analysis of locations of One-Stop centers from the Department of Labor, compared to geographic areas of job losses exceeding 100,000 between July 2008 and July 2009. *See* U.S. Bureau of Labor Stat., Metropolitan Area Employment and Unemployment, tbl. 1 (July 2009). *See also* U.S. Dep't of Labor, PY 2008 WIA Annual Reports, Employment & Training Administration, http://www.doleta.gov/performance/results/ AnnualReports/annual-report-08.cfm(last visited Feb. 20, 2010) (click on map for respective state report).

[1282] Community Service Society, the unheard third 2009: jOb lOss, ecOnOmic insecurity, and a decline in jOb Quality 41 (2009), *available at* http://www.cssny. org/userimages/downloads/Unheard%20Third%20 2009%20Report%2010-7-09.pdf.

[1283] American Library Association, Public Library Use, http://www.ala.org/ala/professionalresources/ libfactsheets/alalibraryfactsheet06.cfm (last visited Mar. 3, 2010).

[1284] John Horrigan, *Broadband Adoption and Use in America* 36-37 (OBI Working Paper No. 1, 2010)

[1285] Joseph Cohen's Education Dominance program at the Defense Advanced Research Projects Agency (DARPA), where new naval recruits were inducted into an online IT training program, found that online trained cadets who completed a 14 week program tested at a 7-year IT Navy technician level. The Institute of Defense Analysis has found that using technology-based instruction reduces cost of instruction by about a third, and either reduces time of instruction by about a third or increases effectiveness of instruction by about a third. Fletcher, Why Technology? at 16.

[1286] Skills2Compete, Middle-Skill Jobs Demand 7 (2009).

[1287] U. S. Dep't of Labor, Tools for America's Job Seekers Challenge, http://dolchallenge.ideascale.com/a/panel.do?id=5847 (last visited Dec. 14, 2009).

[1288] Council of Econ. Advisors, Preparing The Workers at 14.

[1289] White House, *Fact Sheet on American Graduation Initiative* (press release), July 14, 2009 (discussing Online Skills Laboratory, consisting of new open online courses to be developed by the U.S. Departments of Defense, Education, and Labor), *available at* http:// www.whitehouse.gov/the_press_office/Excerpts-of- the-Presidents-remarks-in-Warren-Michigan-and-fact- sheet-on-the-American-Graduation-Initiative/.

[1290] Stephen A. Wandner, *Employment Programs for Recipients of Unemployment Insurance*, mOnthly labOr rev., Oct. 2008, at 17, 18, http://www.bls.gov/opub/ mlr/2008/10/art2full.pdf.

[1291] GAO, Human Capital: Opportunities To Improve Federal Continuity Planning Guidance 12–13, GaO-04-384 (2004), *available at* http://www.gao.gov/new.items/ d04384.pdf.

[1292] Reid Forgrave, *Living on the Edge: Disabled Become Able to Work*, desmOinesreGister.cOm, Mar. 20, 2008, http://www.desmoinesregister.com/apps/pbcs.dll/ article?AID=/20080320/NEWS/803200376/-1/ SPORTS09.

[1293] Ryan Wallace Comments in re NBP PN #3 (*Comment Sought on Telework—NBP Public Notice #3*, GN Docket Nos. 09-47, 09-51, 09-137, Public Notice, 24 FCC Rcd 11752 (WCB 2009) (*NBP PN #3*)), on behalf of Citrix Online (Citrix Comments in re NBP PN #3), filed Sept. 30, 2009, Attach. at 4.

[1294] *See* Connected Nation Comments in re NBP PN #3, filed Sept. 22, 2009, at 16–17.

[1295] Sue Shellenbarger, *The Five Second Commute*, Wsj. cOm, Nov. 25, 2009, http://online.wsj.com/article/SB 10001424052748703819904574555710881471416. html?mod=WSJ_hpp_sections_careerjournal.

[1296] Washington State University, Rural Telework Project, http://cbdd.wsu.edu/telework/overview.html (last visited Mar. 3, 2010).

[1297] Global e-Sustainability Initiative, SMART 2020, united states repOrt addendum 49 (2008), *available at* http://www.gesi.org/LinkClick.aspx?fileticket=cOArprY nXWY%3D&tabid=60.

[1298] AT&T Comments in re NBP PN #3, filed Sept. 22, 2009, at 25.

[1299] Toni Kistner, *Fighting for Fair Telework Tax*, netWOrk WOrld, June 7, 2004, http://www.networkworld.com/net.worker/news/2004/0607netlead.html.

[1300] Telecommuter Tax Fairness Act, H.R. 2600, 111th Cong. 2009.

[1301] Office of Personnel Management (OPM), Federal Employment Statistics: Total Government Employment Since 1962, http://www.opm.gov/feddata/ HistoricalTables/TotalGovernmentSince1962.asp (last visited Feb. 14, 2010).

[1302] OPM, status Of teleWOrk in the federal GOvernment 3 (2009), *available at* http://www.telework.gov/Reports_and_Studies/Annual_Reports/2009teleworkreport.pdf.

[1303] OPM, *OPM Director Berry Drives Plan to Increase Telework among Federal Employees* (press release), Apr. 29, 2009, *available at* http://www1.opm.gov/news/ opm-director-berry-drives-plan-to-increase-telework-among-federal-employees,1460.aspx.

[1304] Am. Elec. Ass'n, Telework in the Information Age (2008), *available at* http://www.aeanet.org/Publications/AeA_CS_Telework.asp.

[1305] Timothy McNeil, Director of Development, National Conference of Black Mayors, Remarks at FCC Opportunities for Small and Disadvantaged Businesses Workshop 26–27 (Sept. 2, 2009), *available at* http://www.broadband.gov/docs/ws_08_op_small_dis_biz.pdf.

[1306] Karen Mills et al., Brookings Inst., Clusters and Competitiveness: A New Federal Role for Stimulating Regional Economies 24 (2008) (mills et al., clusters and cOmpetitiveness).

[1307] BroadbandUSA, http://www.broadbandusa.gov/ (last visited Feb. 15, 2010).

[1308] 13 C.F.R. § 303.6(a).

[1309] Econ. Dev. Admin., U.S. Dep't of Com., Comprehensive Economic Development Strategies: Summary of Requirements, http://www.eda.gov/PDF/ CEDSFlyer081706.pdf (last visited Nov. 24, 2009).

[1310] The U.S. Department of Agriculture designates and oversees rural EZs and ECs while U.S. Department of Housing and Urban Development designates and oversees RCs and urban EZs. To qualify, communities must demonstrate economic distress, including poverty rates and unemployment rates higher than the national average. Designation as an EZ/RC/EC confers a range of tax incentives and block grants over an initial 10-year period. See irs, tax incentives fOr distressed cOmmunities, Pub. 954, Cat. No. 20086A (2004), *available at* http:// www.irs.treas.gov/pub/irs-pdf/p954.pdf.

[1311] FCC, 2008 Form 477 database (accessed Nov. 2009) (on file with the Commission). The Commission used the Form 477 data to estimate, for individual census tracts, the share of households with highspeed connections over fixed-location technologies. Combining reported numbers of total lines in a Census Tract with GeoLytics, Inc census block-level estimates of households in 2009, the Commission determined the number of lines per 1,000 households—or, broadband penetration rates. We filtered the data by census tract, and we flagged census tracts for Empowerment Zones, Enterprise Communities, Renewal Communities, and Hope VI Communities. Filtering by EZ/EC/RC's and Hope IV census tracts, the Commission was able to determine the average broadband penetration rate for each classification. Housing and Urban Development provided the appropriate census tracts. For more detail on the Form 477 results and Commission analysis, please see Indus. Analysis & Tech. Div., FCC, High- Speed Services for Internet Access: Status as of December 31, 2008, at 1 (2010), *available at* http://hraunfoss. fcc.gov/edocs_public/attachmatch/DOC-296239A1. pdf. For more information on Empowerment Zones, Enterprise Communities, Renewal Communities, including maps and locations, please see HUD, Tour EZ/RC/ECs by State, http://www.hud.gov/offices/cpd/ economic development/programs/rc/tour/index.cfm (last visited Feb. 20, 2010).

[1312] *See* note 81, *supra*.

[1313] *See* note 81, *supra*.

[1314] *See* note 81, *supra*.

[1315] Empowerment Zones: Performance Standards for Utilization of Grant Funds, 72 Fed. Reg. 71 008–018 (Dec. 13, 2007).

[1316] Mills et al., Clusters and Competitiveness at 33.

[1317] Data would come from multiple agency databases, including the Bureau of Labor Statistics, the Census Bureau, the Bureau of Economic Analysis, the US Department of Education, the Employment and Training Administration, and the U.S. Patent Office, among others.

[1318] Mills et al., Clusters and Competitiveness at 9.

[1319] Grants.gov, Agencies that Provide Grants, http://grants. gov/aboutgrants/agencies_that_provide_grants.jsp (last visited Feb. 15, 2010).

[1320] Krisztina Holly, *IMPACT: Innovation Model Program for Accelerating the Commercialization of Technologies—A Proposal for Realizing the Economic Potential of University Research*, ssrn, Aug. 3, 2009, http://ssrn. com/abstract=1480449.

[1321] NSF, Academic Research and Development Expenditures: Fiscal Year 2007, NSF 09-303 (Mar. 2009) (NSF, Academic Research and Development Expenditures: Fiscal Year 2007), *available at* http:// www.nsf.gov/statistics/nsf09303/pdf/nsf09303.pdf.

[1322] NSF, Academic Research and Development Expenditures: Fiscal Year 2007.

[1323] Internet2, Research and Commercial Network: capacity at u.s. research universities (2009), *available at* http://www.internet2.edu/government/files/200911- IS-NSF-survey3.pdf.

[1324] Div. of Sci. Resources Stat., NSF, survey Of science and enGineerinG research facilities, fiscal year 2006, tbl. 78, *available at* http://www.nsf.gov/statistics/nsf07325/ pdf/tab78.pdf.

[1325] Jason Baumgarten & Michael Chui, *E-Government 2.0*, mckinsey On GOv't, Summer 2009, at 26–27, *available at* http://www.mckinsey.com/clientservice/publicsector/pdf/TG_MoG_Issue4_egov.pdf.

[1326] Shelley Waters-Boots, Ford Found. Et al., Improving Access to Public Benefits: Helping Individuals and Families Get the Income Supports They Need (2010), *available at* http://www.opportunityatwork.org/pdf/Improving_Access_To_Public_Benefits_1_12_10.pdf.

[1327] *See* Jane Patterson, Executive Director, e-NC Authority, State of North Carolina, Remarks at the FCC State and Local Government Workshop (Sept. 1, 2009), *available at* http://www.broadband.gov/docs/ws_19_state_and_local.pdf; John Conley, Deputy State Chief Information Officer, State of Colorado, Remarks at FCC State and Local Government Workshop (Sept. 1, 2009), *available at* http://www.broadband.gov/docs/ws_19_state_and_local.pdf; Gary Gordier, Chief Information Officer and IT Director, El Paso, TX, Remarks at the FCC State and Local Government Workshop (Sept. 1, 2009), *available at* http://www.broadband.gov/docs/ws_19_state_and_local.pdf; FiberTower Comments in re National Broadband Plan NOI, filed June 8, 2009, at 2, 6.

[1328] Transportation, Treasury, Independent Agencies, and General Government Appropriations Act of 2005, Pub. L. No. 108-447, Div. H, 118 Stat. 2809 (2004).

[1329] Memorandum from Joshua B. Bolten, Director, Office of Mgmt. & Budget (OMB), to Heads of Departments and Agencies, Regulation on Maintaining Telecommunication Services During a Crisis or Emergency in Federally-owned Buildings, M-05-16 (June 30, 2005), *available at* http://www.whitehouse. gov/omb/assets/omb/memoranda/fy2005/m05-16.pdf; Memorandum from Clay Johnson III, Deputy Director for Management, OMB, to Heads of Departments and Agencies, Implementation of Trusted Internet Connections (TIC), M-08-05 (Nov. 20, 2007), *available at* http://www.whitehouse.gov/omb/assets/omb/memoranda/fy2008/m08-05.pdf; Nat'l Commc'ns Sys., Dep't of Homeland Security, the National Communications System Directive (Ncsd) 3-10, Minimum Requirements For Continuity Communications Capabilities (July 25, 2007).

[1330] E-Government Act of 2002, Pub. L. No. 107-347, 116 Stat. 2899 (2002).

[1331] *See* Alaska Dep't of Educ. Comments in re NBP PN #15 (*Comment Sought on Broadband Needs in Education, Including Change to E-rate Program to Improve Broadband Deployment—NBP Public Notice #15*, GN Docket Nos. 09-47, 09-51, 09-137, CC Docket No. 02-6, WC Docket No. 05-195, Public Notice, 24 FCC Rcd 13560 (WCB 2009) (*NBP PN #15*)), filed Nov. 20, 2009, at 79; State E-Rate Coordinators Alliance Comments in re NBP PN #15, filed Nov. 20, 2009, at 19–20; Am. Ass'n of Sch. Adm'rs & Ass'n of Educ. Serv. Agencies Comments in re NBP PN #15, filed Nov. 20, 2009, at 5–6; Nat'l Ass'n of Telecomm. Officers & Advisors (NATOA) Comments in re NBP PN #15, filed Nov. 20, 2009, at 11–12; AT&T Comments in re NBP PN #15, filed Nov. 20, 2009, at 8–9; City of Chicago Comments in re NBP PN #15, filed Nov. 20, 2009, at 26; Dell Comments in re NBP PN #15, filed Nov. 20, 2009, at 4; Mich. Dep't of Educ. Comments in re NBP PN #15, filed Nov. 20, 2009, at 7; Tex. Educ. Telecomm. Network Comments in re NBP PN #15, filed Nov. 20, 2009, at 3–4; Ohio Pub. Library Info. Network Comments in re NBP PN #15, filed Nov. 17, 2009, at 1–2; Alaska E-Rate Coordinator Comments in re National Broadband Plan NOI, filed June 8, 2009, at 10; U.S. R&E Networks and HIMSS Reply in re NBP PN #30 (*Reply Comments Sought in Support of National Broadband Plan*, GN Docket Nos. 09-47, 09-51, 09-137, Public Notice, 25 FCC Rcd 2417 (WCB 2010) (*NBP PN #30*)), filed Jan. 28, 2010, at 43–44.

[1332] *See* Alaska Dep't of Educ. Comments in re NBP PN #15, filed Nov. 20, 2009, at 7.

[1333] IBM, Smarter Cities, http://www.ibm.com/ smarterplanet/us/en/sustainable_cities/ideas/ *(last visited Feb. 17, 2010)*; Steve Lohr, *To Do More With Less, Governments Go Digital*, N.Y. Times, Oct. 10, 2009, http://www.nytimes.com/2009/10/11/ business/11unboxed.html.

[1334] Cisco, Literature: Cisco Connected Communities for State and Local Governments, http://www.cisco.com/web/strategy/government/local_connected_ communities.html (last visited Feb. 17, 2010).

[1335] Richard Whitt, *Experimenting with New Ways to Make Broadband Better, Faster, and More Available*, Google Pub. Pol'y Blog, Feb. 10, 2010, http://googlepublicpolicy.blogspot.com/2010/02/experimenting-with-new-ways-to-make.html.

[1336] Benton Found. Comments in re NBP PN#22 (*Comment Sought on Research Necessary for Broadband Leadership—NBP Public Notice #22*, GN Docket Nos. 09-47, 09-51, 09-137, Public Notice, 24 FCC Rcd 138207 (WCB 2009) (*NBP PN #22*)), filed Dec. 8, 2009, at 9–11; Free Press Reply in re NBP PN #30, filed Jan. 27, 2010, at 13.

[1337] Pew Res. Ctr. for the People and the Press, Trends in Political Values and Core Attitudes: 1987–2007, at 49 (2007), http://people-press.org/reports/pdf/312.pdf.

[1338] Ctr. for Digital Gov't, Renovation Nation: Improving Government Service Delivery in Smart and Sustainable Ways 10 (2009), *available at* http://www.govtechblogs. com/fastgov/CDG09RenovationNation.pdf.

[1339] The National Institute for Standards and Technology (NIST) defines cloud computing as "a model for enabling convenient, on-demand network access to a shared pool of configurable computing resources (e.g., networks, servers, storage, applications, and services) that can be rapidly provisioned and released with minimal management effort or service provider interaction." National Institute of Standards and Technology, Cloud Computing, the NIST Definition of Cloud Computing (2009), http://csrc.nist.gov/groups/SNS/ cloud-computing/ (last visited Feb. 17, 2010). While there is not universal agreement on the definition, this plan will use the NIST definition. For a full discussion of the definition of cloud computing, see AT&T Comments in re NBP PN #21 (*Comment Sought on Data Portability and Its Relationship to Broadband—NBP Public Notice #21*, GN Docket Nos. 09-47, 09-51, 09-137, Public Notice, 24 FCC Rcd 13816 (WCB 2009) (*NBP PN #21*)), filed Dec. 9, 2009, at 3–5; DataPortability Project Comments in re NBP PN #21, filed Dec. 9, 2009, at 5; FTC Comments in re NBP PN #21, filed Dec. 9, 2009, at 1–2; InCommon Steering Committee Comments in re NBP PN #21, filed Dec. 9, 2009, at 4; Qwest Comments in re NBP PN #21, filed Dec. 9, 2009, at 2–4; Letter from Paula Boyd, Regulatory Counsel, Microsoft, to Marlene H. Dortch, Secretary, FCC, GN Docket No. 09–51 (Jan. 20, 2010) (Microsoft Jan. 20, 2010 *Ex Parte*), Attach. 2 (B. Smith) at 1.

[1340] Gwen Morton & Ted Alford, Booz Allen Hamilton, the Economics of Cloud Computing 1 (2009) (Morton & Alford, The Economics of Cloud Computing), *available at* http://www.boozallen.com/media/file/Economics-of-Cloud-Computing.pdf.

[1341] Jason Miller, *Data Center Proliferation Must End, Kundra Says*, Fed. News Radio, Oct. 28, 2009, http://www.federalnewsradio.com/?sid=1796664&nid=263.

[1342] Morton & Alford, the Economics of Cloud Computing at 5, 9.

[1343] C.G. Lynch, *How Vivek Kundra Fought Government Waste One Google App at a Time*, CIO.com, Sept. 22, 2008, http://www.cio.com/article/450636/How_Vivek_Kundra_Fought_Government_Waste_One_Google_App_At_a_Time_.

[1344] Gautham Nagesh, *OPM Claims Victory in Huge e-Payroll System Consolidation*, nextGOv, Oct. 21, 2009, http://www.nextgov.com/nextgov/ng_20091021_4165.php.

[1345] *See* GSA, IT Schedule 70: Maximizing the Speed and Value of it Acquisition Solutions (2007), *available at* http://www.gsaadvantage.gov/images/products/elib/ pdf_files/70.pdf.

[1346] Patrick Thibodeau, *CIA Endorses Cloud Computing, But Only Internally*, Computerworld, Oct. 7, 2009, http://www.computerworld.com/s/article/9139016/ CIA_endorses_cloud_computing_but_only_internally.

[1347] Elise Castelli, *DISA Expands Cloud Computing Services*, fed. times, Oct. 5, 2009, http://www.federaltimes.com/article/20091005/IT03/910050304/1036/IT.

[1348] MeriTalk & Merlin Federal Cloud Initiative, the 2009 Cloud Consensus Report 10 (2009), *available at* http://www.meritalk.com/2009-cloud-consensus.php (must register to download); David Talbot, *Security in the Ether*, mit tech. rev., Jan./Feb. 2010, *available at* http://www.technologyreview.com/web/24166/page1/; Letter from Paula Boyd, Regulatory Counsel, Microsoft Corp., to Marlene H. Dortch, Secretary, FCC, GN Docket Nos. 09-47, 09-51, 09-137 (Nov. 12, 2009) (Microsoft Nov. 12, 2009 *Ex Parte*) at 8; InCommon Steering Committee Comments in re NBP PN #21, filed Dec. 9, 2009, at 5; FTC Staff Comments in re NBP PN #21, filed Dec. 9, 2009, at 2; DataPortability Project in re NBP PN #21, filed Dec. 9, 2009 at 6 (filed as Elias Bizannes); Miguel Helft, *Now, Even the Government Has an App Store*, n.y. times, Sept. 15, 2009, http://bits.blogs.nytimes.com/2009/09/15/noweven-the-government-has-an-app-store/; OnLive Reply in re National Broadband Plan NOI, filed July 21, 2009, at 4; Yaana Reply in re National Broadband Plan NOI, filed July 21, 2009, at 4.

[1349] The Federal CIO Council was created by Executive Order 13011 on July 16, 1996. *See* Exec. Order No. 13011, 61 Fed. Reg. 37657 (July 16, 1996). This order was subsequently revoked. *See* Exec. Order No. 13403, 71 Fed. Reg. 28543 (May 12, 2006). The Council's existence was codified by the E-Government Act of 2002, Pub. L. No. 107-347, 116 Stat. 2899 (2002) (codified at 44 U.S.C. § 101).

[1350] Aliya Sternstein, *Feds Offer 38,484 Budget Cuts*, nextGOv, Oct. 19, 2009, http://techinsider.nextgov.com/2009/10/feds_offer_38484_budget_cuts.php; OMB, SAVE Award, http://www.saveaward.gov (last visited Feb. 20. 2010).

[1351] Jason Miller, *Idea to Reuse Medication at VA Hospitals Wins SAVE Award*, fed. neWs radiO, Dec. 11, 2009, http://www.federalnewsradio.com/index. php?nid=110&sid=1837851.

[1352] Memorandum from Peter Orszag, Director, OMB, to the Heads of Departments and Agencies Responding to General Government Proposals from the President's SAVE Award, M-10-09 (Dec. 21, 2009), *available at* http://www.whitehouse.gov/omb/assets/ memoranda_2010/m10-09.pdf.

[1353] OMB, USASpending.gov., http://www.usaspending.gov/ faads/tables.php?tabtype=t1&subtype=atf&rowtype=a (last visited Feb. 20, 2010).

[1354] Jason Miller, *OMB Calls for a Review of Grant Application Systems* (federal neWs radiO broadcast March 11, 2009), *available at* http://www. federalnewsradio.com/index.php?nid=35&sid=1621782.

[1355] Larry Freed, Foresee Results, E- Government Satisfaction Index 2, 6, 18 (2009), *available at* http://www.foreseeresults.com/_downloads/acsicommentary/ ACSI_EGov_Report_Q1_2009.pdf.

[1356] GAO, Grants.gov Has Systemic Weaknesses That Require Attention 5, 24, GaO-09-589 (2009), *available at* http://www.gao.gov/new.items/d09589.pdf.

[1357] As used here, social media refers to the use of applications within government to facilitate collaboration and information sharing within the federal workforce. See Chapter 15: Civic Engagement for further discussion of the use of social media in government.

[1358] Jennifer L. Dorn, *Rebooting the Public Square*, fed. cOmputer Wk., Dec. 3, 2007, at 30, *available at* http://fcw.com/articles/2007/11/30/web-20-rebooting-thepublic-square.aspx?scjang=en.

[1359] Nora Ganim Barnes & Eric Mattson, Ctr. For Marketing Res., Social Media in the 2009 Inc. 500: New Tools & New Trends (2009), *available at* http://www. umassd.edu/cmr/studiesresearch/socialmedia2009.pdf.

[1360] Ben Bain, *4 Studies in Collaboration—Case 2: TSA's IdeaFactory*, fed. cOmputer Wk., Feb. 29, 2008, *available at* http://fcw.com/articles/2008/02/29/4-studies-incollaboration-151-case-2-tsa146s-ideafactory.aspx; The White House, IdeaFactory, http://www.whitehouse.gov/ open/innovations/IdeaFactory/ (last visited Feb. 20, 2010).

[1361] The White House, IdeaFactory, http://www. whitehouse.gov/open/innovations/IdeaFactory/ (last visited Feb. 20, 2010).

[1362] Jill R. Aitoro, *Defense More Likely Than Civilian Agencies To Use Social Networking Tools*, nextGOv, Jan. 15, 2010, http://www.nextgov.com/nextgov/ ng_20100115_4048.php?oref=mostread.

[1363] *See* Andrea Di Maio, Gartner, Inc., Citizen-Driven Government Must Be Employee-Centric, Too (2009), *available at* http://www.gartner.com/ DisplayDocument?doc_cd=168334 (purchase required); Fed. Web Managers Council, Social Media And The Federal Government: Perceived And Real Barriers And Potential Solutions 2 (2008), *available at* http://www. usa.gov/webcontent/documents/SocialMediaFed%20Govt_BarriersPotentialSolutions.pdf.

[1364] Fed. Chief Info. Officers Council, Guidelines for Secure Use of Social Media by Federal Departments and Agencies 9 (2009), *available at* http://www.cio.gov/ Documents/Guidelines_for_Secure_Use_Social_Media_v01-0.pdf.

[1365] Massimo Calabresi, *Wikipedia for Spies: The CIA Discovers Web 2.0*, Time, Apr. 8, 2009 (Calabresi, *Wikipedia for Spies*), http://www.time.com/time/ nation/article/0,8599,1890084,00.html.

[1366] Calabresi, *Wikipedia for Spies*.

[1367] Cyberspace Policy Review: Assuring a Trusted and Resilient Information and Communications Infrastructure Review, iii (2009), *available at* http:// www.whitehouse.gov/assets/documents/Cyberspace_ Policy_Review_final.pdf.

[1368] President's National Security Telecommunications Advisory Committee, Cybersecurity Collaboration Report: Strengthening Government and Private Sector Collaboration Through a Cyber Incident Detection, Prevention, Mitigation, and Response Capability 4 (2009) (Advisory Committee Cybersecurity Collaboration Report), *available* at http://www.ncs.gov/ nstac/reports/2009/NSTAC%20CCTF%20Report.pdf..

[1369] Ellen Nakashima, *More Than 75,000 Computer Systems Hacked in One of Largest Cyber Attacks, Security Firm Says*, Wash. Post, Feb. 18, 2010 (Nakashima, *More Than 75,000 Computer Systems Hacked*), http://www.washingtonpost.com/wp-dyn/content/ article/2010/02/17/AR2010021705816.html.

[1370] Ellen Nakashima, *War Game Reveals U.S. Lacks Cyber-Crisis Skills*, Wash. pOst, Feb. 17, 2010, http:// www.washingtonpost.com/wp-dyn/content/ article/2010/02/16/AR2010021605762.html.

[1371] Nakashima, *More Than 75,000 Computer Systems Hacked*.

[1372] David Drummond, *A New Approach to China*, Official GOOGle blOG, Jan. 12, 2010, http://googleblog.blogspot. com/2010/01/new-approach-to-china.html.

[1373] Mark Clayton, *US Oil Industry Hit by Cyberattacks: Was China Involved?*, christian sci. Monitor, Jan. 25, 2010, *available at* http://www.csmonitor.com/USA/2010/0125/US-oil-industry-hit-by-cyberattacksWas-China-involved.

[1374] Statement of Liesyl I. Franz, Vice President, TechAmerica, before *the Subcommittee on Research and Science Education, House Committee on Science and Technology*, 111th Cong. (June 10, 2009), *available at* http://democrats.science.house.gov/Media/file/Commdocs/hearings/2009/Research/10jun/Franz_Testimony.pdf.

[1375] Statement of Dr. Fred B. Schneider, Samuel B. Eckert Professor of Computer Science, Cornell University), before *the Subcommittee on Research and Science Education, House Committee on Science and Technology*, 111th Cong. (June 10, 2009) *available at* http://democrats.science. house.gov/Media/file/Commdocs/hearings/2009/Research/10jun/Scheider_Testimony.pdf.

[1376] Ellen Nakashima & John Pomfret, *China Proves to be an Aggressive Foe in Cyberspace*, Wash. pOst, Nov. 11, 2009, http://www.washingtonpost.com/wp-dyn/content/ article/2009/11/10/AR2009111017588_pf.html (last visited Feb. 19, 2010).

[1377] Advisory Committee Cybersecurity Collaborative Report at 5.

[1378] *See* DOJ, International Criminal Investigative Training Assistance Program, http://www.justice.gov/criminal/icitap/ (last visited Feb. 21, 2010).

[1379] This should include, at a minimum, representatives from the intelligence community, Department of Defense, Department of Justice, Department of Homeland Security, Department of Energy, Department of State, Department of Treasury, Department of Education, Department of Commerce, the Federal Communications Commission, and the Federal Trade Commission.

[1380] FCC, Connecting the Globe: A Regulator's Guide to Building a Global Information Community, http://www. fcc.gov/connectglobe/ (last visited Feb. 21, 2010).

[1381] Comcast, *Comcast Launches Comprehensive Internet Security Solution to Help Keep Customers Safe Online* (press release), Aug. 16, 2005, http://www.comcast. com/About/PressRelease/PressReleaseDetail. ashx?PRID=132.

[1382] Comcast, *Com cast Unveils Comprehensive "Constant Guard" Internet Security Program* (press release), Oct. 8, 2009, http://www.comcast.com/About/PressRelease/ PressReleaseDetail.ashx?PRID=926.

[1383] Dan Goodin, *Anti-virus Protection Gets Worse: What Is This Thing You Call Heuristics?*, channel reG., Dec. 21, 2007, http://www.channelregister.co.uk/2007/12/21/ dwindling_antivirus_protection/ (last visited Feb. 18, 2010).

[1384] Alex Goldman, *Top 23 U.S. ISPs by Subscriber: Q3 2008*, ISP Planet, Dec. 2, 2008, http://www.isp-planet.com/ research/rankings/usa.html.

[1385] GAO, Information Security: Progress Reported, but Weaknesses at Federal Agencies Persist, GAO-08-571T (Mar. 12, 2008), *available at* http://www.gao.gov/new. items/d08571t.pdf; *see* Carolyn Duffy Marsan, *GAO: Common Desktop Configuration Holds Promise for Better Security*, fed. cOmputer Wk., Mar. 13, 2008 (Duffy, *GAO: Common Desktop Configuration Holds Promise*), *available at* http://fcw.com/Articles/2008/ 03/13/GAO-Common-desktop-configuration-holds-promise-for- better-security.aspx.

[1386] Carolyn Duffy Marsan, *U.S. Internet Security Plan Revamped*, netWOrk WOrld, Feb. 11, 2010, http://www. networkworld.com/news/2010/021110-cybersecuritydefense-revamped.html, *see* Duffy, *GAO: Common Desktop Configuration Holds Promise*.

[1387] Judi Hasson, *Agencies Must Submit FISMA Reports Online*, fierce GOv't it, Aug. 25, 2009, http://www. fiercegovernmentit.com/story/agencies-must-submitfisma-reports-online/2009-08-25; Vivek Kundra et al., *Moving Beyond Compliance: The Status Quo Is No Longer Acceptable*, it dashbOard blOG, Sept. 28, 2009, http://it.usaspending.gov/?q=content/blog&pageno=2.

[1388] Connected Nation Reply in re NBP PN #30, filed Jan. 27, 2010, at 16–17.

[1389] GAO, Means-Tested Programs: Determining Financial Eligibility is Cumbersome and Can Be Simplified 3 (2001), *available at* http://www.gao.gov/new.items/ d0258.pdf.

[1390] Fed. Chief Info. Officers Council, Federal Identity, Credential, and Access Management (Ficam), Roadmap and Implementation Guidance (2009), *available at* http://www.idmanagement.gov/documents/FICAM_ Roadmap_Implementation_Guidance.pdf.

[1391] IDManagement.gov, Open ID solutions for Open Government, http://www.idmanagement.gov/ drilldown.cfm? action=openID_openGOV (last visited Feb. 20, 2010).

[1392] Assurance levels indicate the level of confidence in a user's identity. Low assurance level applications might include a customized "My Page" on federal Web sites. Higher assurance level applications might include filing taxes online. For more information, see Memorandum from Joshua B. Bolton, Director, OMB, to the Heads of All Departments and Agencies, E-Authentication Guidance for Federal Agencies, Memo M-04-04, Attach. A (Dec. 16, 2003), *available at* http://www.whitehouse. gov/OMB/memoranda/fy04/m04-04.pdf.

[1393] This functionality would allow users to save content relevant to them on one page that would be available every time a user signed on.

[1394] Nat'l Inst. of Health, Open Identity for Open Government at NIH, http://datacenter.cit.nih.gov/ interface/interface245/open_gov.html (last visited Feb. 20, 2010).

[1395] Center for Democracy and Technology Comments in re NBP PN #21, filed on Dec. 9, 2009, at 3 (filed as Heather West); OpenID Foundation Comments in re NBP PN #21, filed Dec. 9, 2009, at 8; AT&T Comments in re NBP PN #29, (*Comment Sought on Privacy Issues Raised by the Center for Democracy and Technology—NBP PN #29*, GN Docket Nos. 09-47, 09-51, 09-137, Public Notice, 25 FCC Rcd 244 (2010) (NBP PN #29), filed Jan. 22, 2010, at 5–8; Microsoft Jan. 21, 2010 *Ex Parte* at 1–13.

[1396] Center for Democracy and Technology Comments in re PN #21, filed on Dec. 9, 2009, at 6 (filed by Heather West).

[1397] Privacy Act of 1974, Pub. L. No. 93-579, 88 Stat. 1896 (1974) (codified at 5 U.S.C. § 552a).

[1398] *See* Andrea Di Maio, Gartner, Inc., The Case for Citizen Data Vaults 3, 4 (2009), *available at* http://www. gartner.com/DisplayDocument?id=1031315 (purchase required); DataPortability Project Comments in re NBP PN #21, filed Dec. 9, 2009, at 7.

[1399] Randall Stross, *Why Can't the IRS Help Fill in the Blanks?*, n.y. times, Jan. 23, 2010, http://www.nytimes. com/2010/01/24/business/24digi.htm.

[1400] *See generally* info. Sec. and Privacy Advisory Bd., Toward a 21st Century Framework for Federal Government Privacy Policy (2009), *available at* http://csrc.nist.gov/groups/SMA/ispab/ documents/correspondence/ispab-report-may2009. pdf; Ctr. for Democracy and Tech., E-Privacy Act Amendments Wiki, http://eprivacyact.org/index. php?title=Welcome (last visited Feb. 20, 2010); Center for Democracy and Technology Comments filed in re NBP PN #29, Jan. 22, 2010, at 12; Microsoft Jan. 21, 2010 *Ex Parte* at 1–13.

[1401] InCommon Steering Committee Comments in re NBP PN #21, filed Dec. 9, 2009, at 2–3.

1402 John Horrigan, *Broadband Adoption and Use in America* 16 exh. 3 (OBI Working Paper No. 1, 2010), *Horrigan, Broadband Adoption and Use in America.*

1403 Jason Baumgarten & Michael Chui, *How We Get to E-Government 2.0*, mckinsey Q., July 28, 2009, *available at* http://www.ciozone.com/index.php/GovernmentIT/How-We-Get-to-E-government-2.0.html; larry freed, e-GOverment satisfactiOn index 6 (2009), *available at* http://fcg.nbc.gov/documents/ACSI-EGov-commentary_Q2-2008.pdf.

1404 U.S. Customs and Immigration Services, http://www. uscis.gov (last visited Nov. 27, 2009).

1405 U.S. Office of Science and Technology Policy, http:// www.whitehouse.gov/open (last visited Nov. 27, 2009).

1406 Massimiliano Claps, Case Study: The Ecitygov Alliance Provides Cross-County Online Services Portals (2009).

1407 Paperwork Reduction Act of 1980, Pub. L. No. 96-511, 94 Stat. 2812 (1980), *codified at* 44 U.S.C. §§ 3501–21.

1408 Vivek Kundra & Michael Fitzpatrick, *Enhancing Online Citizen Participation through Policy*, Open GOv't blOG, June 16, 2009, http://www.whitehouse.gov/blog/ Enhancing-Online-Citizen-Participation-Through-Policy.

1409 Aliya Sternstein, *Government Seeks to Update Paperwork Rule*, nextGOv, Oct. 26, 2009 (Sternstein, *Government Seeks to Update Paperwork Rule*), http://www.nextgov. com/nextgov/ng_20091026_1611.php.

1410 *See* Sternstein, *Government Seeks to Update Paperwork Rule; see also* Improving Implementation of the Paperwork Reduction Act, 74 Fed. Reg. 55269 (proposed Oct. 27, 2009), *available at* http://www.whitehouse.gov/ omb/assets/fedreg_2010/10272009_pra.pdf.

1411 *See* OMB, The President's Management Agenda 24 (2002), *available at* http://www.whitehouse.gov/omb/ budget/fy2002/mgmt.pdf.

1412 E- Government Act of 2002, Pub. L. No 107-347 § 3606, 116 Stat. 2899 44 U.S.C. § 3606 (2002).

1413 John Horrigan, *Broadband Adoption and Use in America* (OBI, Working Paper No. 1, 2010).

1414 *See, e.g.*, Randy Albelda & Heather Boushey, Ctr. for Econ. & Pol'y Res., Bridging the Gaps: a Picture of How Work Supports *Work* in Ten States 29 (2007), *available at* http://www.bridgingthegaps.org/publications/ nationalreport.pdf.

1415 GAO, Food Stamp Program: Use of Alternative Methods to Apply for and Maintain Benefits Could Be Enhanced by Additional Evaluation and Information on Promising Practices 27, GaO-07-573 (2007), *available* at http:// www.gao.gov/cgi-bin/getrpt?GAO-07-573.

1416 Sean Coffey et al., Nat'l League of Cities, Screening Tools to Help Families Access Public Benefits 6 (2005), *available at* http://www.nlc.org/ASSETS/E2DF31BA4AFF4ADEB19BA434142B0545/iyefscreening tools. pdf.

1417 comScore, Inc., *Google Sites Surpass 10 Billion Video Views in August* (press release), Sept. 28, 2009, http:// comscore.com/Press_Events/Press_Releases/2009/9/ Google_Sites_Surpasses_10_Billion_Video_Views_in_ August.

1418 Sean Corcoran, Forrester Research, Inc., the Broad Reach of Social Technologies 1 (2009).

1419 Aaron Smith et al., Pew Internet & Am. Life Project, The Internet and Civic Engagement 9 (2009), *available at* http://www.pewinternet.org/~/media//Files/Reports/2009/The%20Internet%20and%20Civic%20Engagement.p df.

1420 Letter from Thomas Jefferson to Edward Carrington (Jan. 16, 1787), *available at* http://en.wikisource. org/w/index.php?title=Special:Book&bookcmd=download&collection_id=5abd8f3c1473eb5f&writer=rl&retu rn_to=Letter+to+Edward+Carrington+- +January+16%2C+1787.

1421 *See* Knight Comm'n, Informing Communities: Sustaining Democracy in the Digital Age 38 (2009) (Knight Comm'n, Informing Communities), *available at* https:// secure.nmmstream.net/anon.newmediamill/aspen/ kcfinalenglishbookweb.pdf.

1422 Society of Professional Journalists Comments in re NBP PN #20 (*Comment Sought on Moving Toward a Digital Democracy—NBP Public Notice #20*, GN Docket Nos. 09-47, 09-51, 09-137, Public Notice, 24 FCC Rcd 13810 (WCB 2009) (*NBP PN #20*)), filed Dec. 9, 2009, at 1 (filed by Kevin Z. Smith).

1423 *See* Public Access To Court Electronic Records— Overview, http://pacer.psc.uscourts.gov/pacerdesc.html (last visited Jan. 7, 2010).

1424 Carl Malmud, President and CEO, Public.Resource. Org., By the People, Address at the Gov 2.0 Summit, Washington, D.C. 25 (Sept. 10, 2009), *available at* http:// resource.org/people/3waves_cover.pdf.

1425 *See* Letter from Sen. Joseph I. Lieberman to Carl Malamud, President and CEO, Public.Resources.Org (Oct. 13, 2009), *available at* http://bulk.resource.org/ courts.gov/foia/gov.senate.lieberman_20091013_from. pdf.

1426 Knight Comm'n, Informing Communities at 38.

1427 Ellen S. Miller Comments in re National Broadband Plan NOI, filed June 8, 2009, at 5.

1428 *See* Knight Comm'n, Informing Communities at 38.

1429 *See* Broadband for the Deaf and Hard of Hearing Corporation Comments in re NBP PN #20, filed Dec. 9, 2009, at 2.

1430 Society of Professional Journalists Comments in re NBP PN #20, filed Dec. 9, 2009, at 1 (filed by Kevin Z. Smith).

1431 *See* New York State Senate, NYSenate Markup, http:// www.nysenate.gov/markup (last visited Feb. 21, 2010).

1432 *See* PublicMarkup.Org, Welcome to Public Markup, www.publicmarkup.org (last visited Nov. 28, 2009).

[1433] Jonathan Rintels, Benton Found., An Action Plan for America: Using Technology and Innovation to Address Our Nation's Critical Challenges 34 (2008) (rintels, an actiOn plan fOr america), *available at* http://www.benton.org/sites/benton.org/files/Benton_Foundation_Action_Plan.pdf.

[1434] *See* Christopher Eliott, *Is the Transportation Department Really 'Open'? No, But It's Getting There*, cOnsumer traveler, Feb. 11, 2010, http://www.consumertraveler. com/today/is-the-transportation-department- really-%e2%80%9copen%e2%80%9d-no-but- it%e2%80%99s-getting-there/.

[1435] *See* Rintels, An Action Plan for America at 35.

[1436] Vivek Kundra, *Changing the Way Washington Works*, Omb blOG, Dec. 16, 2009, http://www.whitehouse.gov/omb/blog/09/12/16/Changing-the-Way-WashingtonWorks/.

[1437] Joab Jackson, *CIO Council Set up Data.gov in Two Months, and Third Parties are Putting Data to Use*, Governing Comp. News, Oct. 12, 2009, http://www.gcn.com/Articles/2009/10/12/GCN-Awards-DataGov.aspx.

[1438] City of San Francisco, DataSF, http://www.datasf. org/ (last visited Feb. 21, 2010); Fritz Nelson, *Open Government: A San Francisco Treat*, infO. Week, Nov. 19, 2009, http://www.informationweek.com/blog/main/archives/2009/11/open_government_1.html.

[1439] Clay Johnson, *Get Your Act Together, Data.gov*, sunliGht labs blOG, Nov. 13, 2009, http://sunlightlabs.com/blog/2009/get-your-act-together-datagov/.

[1440] sunshineweek.org, Sunshine Week 2009 survey Of state GOvernment infOrmatiOn Online (2009), *available at* http://www.spj.org/pdf/sw09-surveyreport.pdf.

[1441] Freedom of Information Act of 1966, Pub. L. No. 89-554, 80 Stat. 383 (1966) (codified at 5 U.S.C. § 552) (FOIA).

[1442] DHS, 2008 Annual Freedom of Information Act Report to the Attorney General of the United States 6 (2008), *available at* http://www.dhs.gov/xlibrary/assets/foia/ privacy_rpt_foia_2008.pdf; GAO, DHS Has Taken Steps to Enhance its Program, but Opportunities Exist to Improve Efficiency and Cost-Effectiveness 22 (2009), *available at* http://www.gao.gov/new.items/d09260.pdf.

[1443] Dep't of Justice Office of Info. Policy, FOIA Post: Summary of Annual FOIA Reports for Fiscal Year 2008, http://www.justice.gov/oip/foiapost/2009foiapost16. htm (last visited Feb. 21, 2010).

[1444] Data Quality Act of 2001, Pub. L. No. 106-554, § 515, 114 Stat. 2763A-153 (2001) (codified at 44 U.S.C. § 3516) (Data Quality Act).

[1445] GAO, Expanded Oversight and Clearer Guidance by the Office of Management and Budget Could Improve Agencies' Implementation of the act 4–5, GAO-06-765 (2006), *available at* http://www.gao.gov/new.items/d06765.pdf.

[1446] Aliya Sternstein, *White House Bars Agencies From Posting Some Statistics*, nextGOv, Jan. 27, 2010, http://www.nextgov.com/nextgov/ng_20100127_9912.php.

[1447] Pew Project for Excellence in Journalism, Pew Res. Ctr., the State of the News Media 2009: An Annual Report on American Journalism 3 (2009), *available at* http://www.stateofthemedia.org/2009/chapter%20pdfs/COMPLETE%20EXEC%20SUMMARY%20PDF. pdf.

[1448] Leonard Downie, Jr., & Michael Schudson, *The Reconstruction of American Journalism*, Colum. Journalism Rev., Oct. 19, 2009, at 3, *available at* http:// www.cjr.org/reconstruction/the_reconstruction_ of_american.php; *see also id.* at 2 (Accountability journalism "holds government officials accountable to the legal and moral standards of public service and keeps business and professional leaders accountable to society's expectations of integrity and fairness.").

[1449] *See* voiceofsandiego.org, About Us, http://www. voiceofsandiego.org/support_us/about_us/ (last visited Mar. 3, 2009); MinnPost.com, About Us, http://www. minnpost.com/about/ (last visited Mar. 3, 2009).

[1450] *See* The American Independent, About, http://tainews. org/about/ (last visited Mar. 3, 2009); ProPublica, About Us, http://www.propublica.org/about/ (last visited Mar. 3, 2009).

[1451] *See, e.g.*, Dan Gillmor, We the Media: Grassroots Journalism, by the People, for the People 18–19 (2006) (Discussing the Transformative Power of Open Source Journalism); Henry Jenkins et al., Confronting the Challenges of Participatory Culture: Media Education for the 21st century 8–9 (2009) (developing a toolkit for future participatory media creation and consumption). *See generally* Henry Jenkins, Convergence Culture: Where Old and New Media Collide (2006); Yochai Benkler, The Wealth of Networks: How Social Production Transforms Markets And Freedom (2006).

[1452] *See, e.g.*, c. Edwin Baker, Media, Markets, and Democracy 285–307 (2005) (discussing why the emergence of digital technologies does not eliminate the problems with media markets); Jon M. Garon, *Media & Monopoly in the Information Age: Slowing the Convergence at the Marketplace of Ideas*, 17 Cardozo Arts & Ent. L.J. 491, 591–92 (1999) (discussing the problem of convergence in traditional media); Richard T. Karcher, *Tort Law and Journalism Ethics*, 40 lOy. u. chi. l.j. 781, 797–801 (2009) (discussing the problem of infotainment and sensationalism in journalism); Christa Corrine McLintock, Comment, *The Destruction of Media Diversity, or: How the FCC Learned to Stop Regulating and Love Corporate Dominated Media*, 22 j. marshall j. cOmp. & infO. l. 569, 602 (2004) ("The premise that the Internet alone will solve all media consolidation and diversity problems is fundamentally flawed because in order for this proposed solution to work, we must assume the same market pressures and problems that exist in the traditional media market will not infiltrate the Internet.").

1453 *See* Eagle Creek Broadcasting et al. Reply in re NBP PN #30 (*Reply Comments Sought in Support of National Broadband Plan—NBP Public Notice #30*, GN Dockets No. 09-47, 09-51, 09-137, 23 FCC Rcd 241 (WCB, 2010) (*NBP PN #30*)), filed Jan. 27, 2010, at 6 (filed as Joint Broadcast Parties); Pew Project for Excellence In Journalism, Pew Res. Ctr., How News Happens: A Study of the News Ecosystem of One American City (2010) (finding that 95% of news stories that contained new information came from traditional media), *available at* http://www.journalism.org/analysis_report/how_news_happens; Gary Kamiya, *The Death of the News*, salOn, Feb. 17, 2009, http://www.salon.com/opinion/ kamiya/2009/02/17/newspapers/ (estimating that "80% of all online news originates in print").

1454 *See, e.g.*, Graham Murdock, *Redrawing the Map of the Communication Industries: Concentration and Ownership in the Era of Privatization*, *in* Public Communication: The New Imperatives (Marjorie Ferguson ed., 1990); Peter Dahlgren, *Introduction*, cOmmunicatiOn and citizenship 10 (Peter Dahlgren & Colin Sparks eds., 1991); Dan Gillmor, We the Media: Grassroots Journalism, by the People, for the People, at xxvii (2006).

1455 *See* FCC, Reboot.FCC.gov, Future of Media, http:// reboot.fcc.gov/futureofmedia/ (last visited Feb. 2, 2010); FCC, *FCC Launches Initiative to Examine Future of Media: Issues Public Notice and Launches FCC.Gov/FutureofMedia* (press release), GN Docket No. 10-25 (Jan. 21, 2010).

1456 Knight Comm'n, Informing Communities at 3.

1457 Knight Comm'n, Informing Communities at 35.

1458 Knight Comm'n, Informing Communities at 35.

1459 Elysa Gardner, *At 40, 'Sesame Street' Is in A Constant State of Renewal*, usa tOday, Nov. 10, 2009, *available at* http://www.usatoday.com/life/television/news/2009- 11-06-sesame06_CV_N.htm.

1460 Pub. Broad. Serv., *PBS KIDS Web Sites Break Video View Records* (press release), Jan. 13, 2009, *available at* http://www.pbs.org/aboutpbs/news/20100113_ pbskidssitesbreakvideorecords.html.

1461 Ellen P. Goodman Comments in re National Broadband Plan NOI, filed Nov. 7, 2009, at 9.

1462 Ellen P. Goodman Comments in re National Broadband Plan NOI, filed Nov. 7, 2009, at 13.

1463 Ellen P. Goodman Comments in re National Broadband Plan NOI, filed Nov. 7, 2009, at 4.

1464 Knight Comm'n, Informing Communities at 38.

1465 Ellen P. Goodman Comments in re National Broadband Plan NOI, filed Nov. 7, 2009, at 4; Ellen P. Goodman, *Public Service Media 2.0*, *in* And Communications for All: A Public Policy Agenda for a New Administration 263 (Amit M. Schejter ed., Lexington Books 2009) (Goodman, *Public Service Media 2.0*).

1466 Letter from Robert M. Winteringham, Deputy General Counsel, Corp. for Public Broad., to Marlene H. Dortch, Secretary, FCC, GN Docket Nos. 09-47, 09-51, 09-137 (Dec. 30, 2009) at 2; Letter from Kinsey Wilson, Sr. Vice President, Digital Media, NPR & Michael Riksen, Vice President, Policy and Representation, NPR, to Marlene H. Dortch, Secretary, FCC, GN Docket Nos. 09-47, 09- 51, 09-137 (Dec. 28, 2009) at 4–5.

1467 Corp. for Pub. Broad, Public Broadcasting Revenue, Fiscal Year 2008 (2009), *available at* http://www.cpb.org/stations/reports/revenue/2008PublicBroadcasting Revenue.pdf.

1468 Am. Univ. Ctr. for Social Media, Report On Public Media 2.0: Dynamic, Engaged Publics 21 (2009).

1469 William W. Fisher & William Mcgeveran, Berkman Ctr. for Internet & Soc'y, The Digital Learning Challenge: Obstacles to Educational Uses of Copyrighted Materials in the Digital Age, 6, 50 (2006) (Fisher & Mcgevern, Digital Learning Challenge); Goodman, *Public Service Media 2.0* at 263, 270–71, 276; Letter from Susan L. Kantrowitz, Vice President and General Council, WGBH Educ. Found. et al., to Marlene H. Dortch, Secretary, FCC, GN Docket No. 09-51 (Feb. 22, 2010) at 15 (filed on behalf of the Association of Public Television Stations, Corporation for Public Broadcasting, Public Broadcasting Service, and National Public Radio); *see also* Kim Hart, *Public Knowledge Proposes Copyright Reform Bill*, hill's tech. blOG, Feb. 16, 2010, http://thehill.com/blogs/hillicon-valley/ technology/81117-public-knowledge-proposescopyright-reform-bill.

1470 At least one Congressional committee has already recognized the tremendous value of this opportunity. On June 22, 2007, the Senate Appropriations Committee stated that "[t]he Committee has strongly supported the conversion of public broadcasting stations to digital formats and continues to do so." It also "recognize[d] that this conversion to digital transmission leaves a great number of stations with limited programming and makes a substantial proportion of the public broadcasting library unusable." It concluded that "this archive of material is a valuable asset to the public and to historians." Ass'n. of Pub. Television Stations, Senate Committee Endorses American Archive Project (press release), June 22, 2007, http://www.apts.org/news/ senateendorses Americanarchive1.cfm.

1471 William W. Fisher & William Mcgeveran, the Berkman Cetenr for Internet & Society, the Digital Learning Challenge: Obstacles to Educational Uses of Copyrighted Materials in the Digital Age 94–95 (2006); Ellen P. Goodman, *Public Service Media 2.0* at 263, 277; Ellen P. Goodman National Broadband Plan NOI Comments, filed Nov. 7, 2009, at 29.

1472 amanda lenhard, peW internet & am. life prOject, adults and sOcial netWOrkinG Websites 1 (2009).

1473 CDC, CDC.gov, Social Media & CDC-INFO Metrics, http://www.cdc.gov/metrics/campaigns/reports/Biweekly_SocialMediaWebandCDC-INFO_ Metrics_12-07-09.pdf (last visited Dec. 9, 2009).

[1474] Posting of Kip Hawley to The TSA Blog, http://www.tsa. gov/blog/2008/01/welcome.html (Jan. 30, 2008 10:00 AM).

[1475] Posting of Craig Newmark to The Hill's Pundit's Blog, http://thehill.com/blogs/pundits-blog/technology/31046-tsa-blog-a-good-example-of- providing-improved-citizencustomer-service (Mar. 30, 2009, 4:27 AM EDT); Posting of Blogger Bob to The TSA Blog, http://www.tsa.gov/blog/2009/09/tsa-blog1000000-hits.html (Sept. 10, 2009 9:33 AM).

[1476] Twitter, FCC, http://twitter.com/fcc (last visited Feb. 21, 2010); Twitter, The White House, http://twitter.com/whitehouse (last visited Feb. 21, 2010); Twitter, CDC Emergency, http://twitter.com/cdcemergency (last visited Feb. 21, 2010).

[1477] FCC, Broadband.gov IdeaScale, http://broadband. ideascale.com/ (last visited Feb. 19, 2010); FCC, OpenInternet.gov Ideascale, http://openinternet. ideascale.com/ (last visited Feb. 19, 2010).

[1478] FCC, Reboot Blog, http://reboot.fcc.gov/blog/ (last visited Feb. 19, 2010); FCC, Broadband Blog, http://blog.broadband.gov/ (last visited Feb. 19, 2010); FCC, Future of Media Blog http://reboot.fcc.gov/futureofmedia/blog (last visited Feb. 19, 2010); FCC, Open Internet Blog, http://Blog.openinternet.gov (last visited Feb. 19, 2010).

[1479] Athena Kwey, *U.S. Embassy Uses New Media to Reach Chinese Netizens*, dipnOte, Nov. 16, 2009, http://blogs.State.gov/index.php/entries/new_media_netizens/.

[1480] U.S. State Dep't, Virtual Student Foreign Service, http:// www.state.gov/vsfs (last visited Feb. 21, 2010); Olivia Jung, *Dorm Room Diplomacy Group Pairs Penn, Middle East College Students*, the daily pennsylvanian, Sept. 20, 2009, http://thedp.com/article/dorm-room-diplomacy-grouppairs-penn-middle-east-college-students.

[1481] Anna P. Mussman, *Online Conversation Connects Students in Afghanistan and Massachusetts*, dipnOte, Nov. 19, 2009, http://blogs.state.gov/index.php/entries/ students_boston_jalalabad/.

[1482] Ai-Mei Chang and P.K. Kannan, Leveraging Web 2.0 in Government 22 (2008).

[1483] Paterson Announces Web Site, Town Hall Meeting to Address New York State Budget Problems, Gov't Tech., Nov. 10, 2008, http://www.govtech.com/gt/429195 (Nov. 10, 2008).

[1484] State of Maine, Office of the Governor, The Budget, http://www.maine.gov/governor/baldacci/policy/ budget/index07-07.html (last visited Mar. 4, 2009).

[1485] Beth Noveck, U.S. Deputy Chief Tech. Officer for Open Gov't, Presentation at FCC Open Government and Civic Engagement Workshop 10 (Aug. 6, 2009), http:// www.broadband.gov/docs/Fcc_OpenGov_Noveck.pdf; Beth Noveck, Open Gov't Directive, Phase III: Drafting, http://www.whitehouse.gov/blog/Open-Government-Directive-Phase-III-Drafting (June 22, 2009).

[1486] This group should include all relevant senior stakeholders at the federal level and could consist of the same positions represented by the working group on transportation, accountability, participation and collaboration that was created by the Open Government Directive. Memorandum from Peter R. Orszag, Director, Open Gov't Directive to the Heads of Executive Departments and Agencies (Dec. 8, 2009), *available at* http://www.whitehouse.gov/omb/assets/ memoranda_2010/m10-06.pdf.

[1487] New York Law School, *Peer-to-Patent Pilot Releases Report Demonstrating Success of Public Participation in Patent Process* (press release), June 18, 2008, *available at* http://www.nyls.edu/news_and_events/peer_to_patent.

[1488] White House, White House Fellows Program, http:// www.whitehouse.gov/about/fellows (last visited Feb. 23, 2010).

[1489] Pew Ctr.on the States, Bringing Elections into the 21st Century: Voter Registration Modernization 1–2 (2009) (Pew, Voter Registration Modernization), *available at* http://www.pewtrusts.org/uploadedFiles/ wwwpewtrustsorg/Reports/Election_reform/Voter_ Registration_Modernization_Brief_web.pdf.

[1490] Overseas Vote Found., 2008 OVF Post Election Uocava Survey Report and Analysis 5 (2009), *available at* https://www.overseasvotefoundation.org/ files/OVF_2009_PostElectionSurvey_Report.pdf.

[1491] Pew, Voter Registration Modernization at 1–2.

[1492] Open Source Digital Voting Foundation Comments in re NBP PN #20, filed Dec. 10, 2009, at 6–7; Broadband for the Deaf and Hard of Hearing Corp. Comments in re NBP PN #20, filed Dec. 9, 2009, at 1–2.

[1493] National Association of Counties Comments in re NBP PN #20, filed Dec. 10, 2009, at 4; Open Source Digital Voting Foundation Comments in re NBP PN #20, filed Dec. 10, 2009, at 5.

[1494] Governor's Comm'n on Strengthening Democracy, Final Report 22 (2009), *available at* http://www.strengthendemocracy.org/uploads/2009/12/governorscommission-final-report.pdf.

[1495] Pew, Voter Registration Modernization at 3–4.

[1496] Pew, Voter Registration Modernization at 3.

[1497] IBM, Travis County Tax Office Develops a New Way to Vote and Access Other Key Government Services (2007), *available at* http://www-01.ibm.com/software/ success/cssdb.nsf/CS/JSTS-78RR3D.

[1498] Military and Overseas Voter Empowerment Act, Subtitle H of the National Defense Authorization Act for Fiscal Year 2010, Pub. L. No. 111-84, §§ 575–589, 123 Stat. 2190, 2318–35 (2009).

[1499] Democrats Abroad Comments in re NBP PN #20, filed Dec. 10, 2009, at 1 (filed by Christine Marques).

[1500] Overseas Vote Found., *2008 OVF Post Election UOCAVA Survey Report And Analysis: A Detailed Look at How Overseas and Military Voters And Election Officials Fared in the 2008 General Election and What to Do About It* 5 (2009).

[1501] Pew Ctr. on the States, *Bringing Elections into the 21st Century: Voter Registration Modernization* 2 (2009).

[1502] William Jackson, *Offshore Voting Gets Assist from the Web*, Gov't Computer News, Nov. 11, 2008, http://gcn.com/Articles/2008/11/11/Offshore-voting-gets-assistfrom-the-Web.aspx.

[1503] William Jackson, *States Launch Online Voter Registration Sites*, Gov't Computer News, Jan. 16, 2008, http://gcn.com/Articles/2008/01/16/States-launchonline-voter-registration-sites.aspx.

[1504] Nat'l Conf. of State Legislatures, *Arizona Leads the Way As Military and Overseas e-Voting Gains Momentum*, the canvass, Mar. 2009, http://www.ncsl.org/Default. aspx?TabId=16466.

[1505] Under this approach, for example, the public safety licensee(s) is afforded the flexibility to enter into agreements with commercial partners for construction and operation of their 700 MHz network.

[1506] Based on the results of the 2006 National Interoperability Baseline Survey, the 2007 UASI Tactical Interoperability scorecards, and 2008/2009 information provided by each state regarding its Statewide Communications Interoperability Plans, it is possible to estimate that a majority of the UASIs and states are at approximately an intermediate level of interoperability. *See generally* Dep't of Homeland Sec., 2006 National Interoperability Baseline Survey (2006), *available at* http://www.safecomprogram. gov/NR/rdonlyres/ 40E2381C-5D30-4C9C-AB81- 9CBC2A478028/0/2006NationalInteroperability BaselineSurvey.pdf; dep't hOmeland sec., tactical interOperable cOmmunicatiOns scOrecards summary repOrt and findinGs (2007), *available at* http:// www.dhs.gov/xlibrary/assets/grants-scorecardreport-010207.pdf; Dep't Homeland Sec., Tactical Interoperable Communications Scorecards Summary Report and Findings (SCIPS) (2009), *available at* http://www.safecomprogram.gov/NR/rdonlyres/ C6C0CD6A-0A15-4110-8BD4-B1D8545F0425/0/ NationalSummaryofSCIPs_February2009.pdf. As set forth in the Goals of the National Emergency Communications Plan, DHS plans to assess each of the nation's 60 largest urban areas' ability to clearly achieve response-level communications by September 30, 2010, and will evaluate each of the more than 3,000 counties in the United States by September 30, 2011. *See* Dep't of Homeland Sec., National Emergency Communications Plan 6–7 (2008), *available at* http://www.dhs.gov/ xlibrary/assets/ national_emergency_communications_ plan.pdf.

[1507] Eur. Telecomm. Standards Inst. [ETSI], Project MESA; *Technical Specification Group—System; System and Network Architecture*, at 20, ETSI TR 102 653 V3.1.1 (2007–2008), *available at* http://www.etsi.org/ deliver/etsi_tr/102600_102699/102653/03.01.01_60/ tr_102653v030101p.pdf.

[1508] *See Implementing a Nationwide, Broadband, Interoperable Public Safety Network in the 700 MHz Band*, PS Docket No. 06-229, Second Report and Order, 22 FCC Rcd 15289 (2007).

[1509] Comments submitted in the Commission's 700 MHz D block proceeding suggest a number of possible explanations. *See, e.g.*, Association of Public Safety Communications Officials-International, Inc. (APCO) Comments in re 700 MHz Third Further Notice (*Service Rules for the 698–746, 747–762 and 777–792 Bands; Implementing a Nationwide, Broadband, Interoperable Public Safety Network in the 700 MHz Band*, WT Docket No. 06-150, PS Docket No. 06-229, Third Further Notice of Proposed Rulemaking, 23 FCC Rcd 16661 (2008) (*700 MHz Third Further Notice*)), filed June 20, 2008, at 3; Verizon Wireless Comments in re 700 MHz Third Further Notice, filed June 20, 2008, at 2 .

[1510] The record includes proposals, for example, for public safety agencies to use existing core infrastructure while individuating end-user devices and other aspects of the edge network to meet public safety requirements, and also to employ satellite, aircraft or other technologies to extend coverage to rural areas. *See, e.g.*, Letter from Lucian Randolph, CEO, Planet TV Air-Tower Systems, to Marlene H. Dortch, Secretary, FCC GN Docket No. 09-51, (Nov. 12, 2009) (Planet TV Nov. 12, 2009 *Ex Parte*) at 9; Space Data Reply in re National Broadband Plan NOI, filed July 21, 2009, at 3; Iridium Satellite Comments in re National Broadband Plan NOI, filed June 8, 2009, at 4–5; MSS/ATC Coalition Comments in re National Broadband Plan NOI, filed June 8, 2009, at 5–6; Spacenet Inc. Comments in re National Broadband Plan NOI, filed June 8, 2009, at 9. The Commission should also explore how to best meet public safety requirements through a variety of means, including the use of commercial infrastructure to be procured by the public safety broadband licensee.

[1511] This serves the added purpose of allowing the public safety licensee(s) to leverage infrastructures that utilities might currently have. Therefore, access to utilities' towers and other structures may be part of any secondary usage program.

[1512] *See, e.g.*, APCO Comments in re NBP PN #8, (*Additional Comment Sought on Public Safety, Homeland Security, and Cybersecurity Elements of National Broadband Plan —NBP Public Notice #8*, GN Docket Nos. 09-47, 09-51, 09-137, PS Docket Nos. 06-229, 07-100, 07-114, WT Docket No. 06-150, CC Docket No. 94-102, WC Docket No. 05-196, Public Notice, 24 FCC Rcd 12136 (PSHSB 2009) (*NBP PN #8*)) filed Nov. 12, 2009, at 11; AT&T Comments in re NBP PN #8, filed Nov. 12, 2009, at 2; Verizon and Verizon Wireless Comments in re NBP PN #8, filed Nov. 12, 2009, at 6; Public Safety Spectrum Trust Comments in re *700 MHz Public Safety Broadband Networks Waiver PN (Public Safety and Homeland Security Bureau Seeks Comment on Petitions for Waiver to Deploy 700 MHz Public Safety Broadband Networks*, PS Docket No. 06-

229, Public Notice, DA 09-1819 (rel. Aug. 4, 2009) (*700 MHz Public Safety Broadband Networks Waiver PN*) at 11.

[1513] *See* New and Emerging Technologies 911 Improvement Act of 2008, Pub. L. No. 110-283, 122 Stat. 2620 (2008) (NET 911 Act) amending Wireless Communications and Public Safety Act of 1999, Pub.L. No. 106-81, 113 Stat. 1286 (1999) (Wireless 911 Act).

[1514] To the extent that other users are permitted on a public safety network, ERIC will also be responsible for working on establishing common priorities.

[1515] ERIC's mission can also be extended over time to serve other functions, such as coordinating PSAP access to the network and improving interoperability for mission critical voice.

[1516] The FCC should consider a membership comprised of representatives of state and local public safety agencies, public safety trade associations, the Public Safety Spectrum Trust, federal user groups, and SAFECOM. The FCC should also consider appropriate representation from industry representatives and representatives of equipment vendors and service providers. The FCC should also establish a federal partners coordinating committee that includes DHS, the Department of Justice, NIST and the National Telecommunications and Information Administration (NTIA) and that leverages the Emergency Communications Preparedness Center (ECPC).

[1517] This includes 20 new engineering and technical personnel, travel and office expenses, computing and simulation equipment and contracting with NIST for standards development and testing Omnibus Broadband Initiative, The Public Safety Broadband Wireless Netwo rk (forthcoming) (OBI, The Public Safety Broadband Wireless Netwo rk).

[1518] This advisory committee should be made exempt from the Federal Advisory Committee Act. Secondly, Congress should ensure appropriate funding for ERIC to enable the FCC to pay for reasonable travel expenses of the public safety advisory committee members.

[1519] Under this model, public safety entities, as authorized by the FCC, should be allowed to select entities they want to partner with to construct and operate their networks, consistent with FCC, including ERIC, requirements.

[1520] Many state and local jurisdictions have enacted regulations requiring the installation of transmitters or other equipment within buildings to improve in- building coverage for public safety narrowband voice networks. State and local governments should consider implementing similar in-building coverage requirements for public safety broadband communications.

[1521] To achieve the 99% population coverage, externally mounted antennas are assumed for use in highly rural areas of the country.

[1522] The cost basis for this funding request will be released subsequently in an OBI , The Public Safety Broadband Wireless Network . These capital costs include leveraging approximately 41,600 commercially deployed sites, 3,200 rural sites (a blend of new and upgraded sites, with vehicles being mounted with externally deployed antennas), hardening of all sites, and providing deployable caches of equipment at the state and local level.

[1523] This figure is based on an annual RAN fee for managed services, additional costs for rural services and an annual OA&M including transport managed services fee. OBI, The Public Safety Broadband Wireless Netwo rk.

[1524] Most of these jobs will be in services and operations, while a smaller percentage will be in product development and manufacturing. OBI, The Public Safety Broadband Wireless Netwo rk..

[1525] Such a fee should be modest. Operating expenses for the first 2 years of network operation are estimated at $500 million.

[1526] *See* 6 U.S.C. § 575. This statute mandates the formation of RECC working groups, *id.* at § 575(a), and charges them with, among other duties, "assessing the survivability, sustainability and interoperability of local emergency communications systems." *Id.* at § 575(d)(1). This section does not direct the working groups to focus on broadband infrastructure.

[1527] These surveys should include information to be provided to ERIC on the current status of interoperability for the public safety broadband network.

[1528] FCC , FCC Preparedness for Major Public Emergencies, Chairman's 30 Day Review (2009), *available at* http://hraunfoss.fcc.gov/edocs_public/attachmatch/DOC293332A1.pdf.

[1529] *See* Letter from Diane Cornell, Vice President of Government Affairs, Inmarsat, to Marlene H. Dortch, Secretary, FCC, GN Docket Nos. 09-47, 09-51, 09-137, WC Docket No. 02-60 (Dec. 4, 2009) at 7.

[1530] *See* 47 U.S.C. § 5172(a)(1)(B); Office of the President, The Federal Response to Hurricane Katrina: Lessons Learned 58–59 (2006), *available at* http:// georgewbush-whitehouse.archives.gov/reports/katrinalessons-learned.pdf.

[1531] Ann Arnold, President, Tex. Ass'n of Broadcasters, Statement at FCC Summit: Lessons Learned: Hurricane Seasons 2008 (Dec. 11, 2008) *available at* http://www. fcc.gov/realaudio/mt121108.ram (1:00:35).

[1532] For-profit entities should be deemed eligible for assistance only when the need for their services exceeds the capabilities of the private sector and any relevant state, Tribal and local governments, or relates to an immediate threat to life and property, is critical to disaster response or community safety, or relates to essential federal recovery measures.

[1533] *See* Mike McConnell, Op.-Ed., *Mike McConnell on How to Win the Cyber- War We're Losing,* Wash. pOst, Feb. 28, 2010, http://www.washingtonpost.com/wp-dyn/ content/article/2010/02/25/AR2010022502493.html. (McConnell, How to Win the Cyber-War).

[1534] McConnell, *How to Win the Cyber- War.*

[1535] Steven Chabinsky, Deputy Ass't Director-Cyber Division, Fed. Bureau of Investigation (FBI), Testimony before the U.S. Senate Judiciary Committee, Subcommittee on Terrorism and Homeland Security (Nov. 17, 2009). The FBI considers the cyber threat against the nation to be "one of the greatest concerns of the 21st century." *Id.*

[1536] Verizon Business, 2008 Data Breach Investigations Report 2–3 (2008), *available at* http://www. verizonbusiness.com/resources/security/ databreachreport.pdf.

[1537] The Commission will have to allocate funding to obtain a vendor to develop audit criteria and to accredit third-party certification bodies. Congress should consider public funding for the FCC in its next budget and on an ongoing basis as required.

[1538] In fact, estimates of residential-access network capacity suggest that current networks can carry between 1/100 and 1/10 of their advertised per-user capacity. *See also* AT&T Comments in re National Broadband Plan NOI, filed June 8, 2009, at 67–69; Telcordia Comments in re National Broadband Plan NOI, filed June 8, 2009, at 19.

[1539] Research and Innovative Tech. Admin., Next Generation 911 Concept of Operations, Fig. 4-2, http://www.its.dot. gov/ng911/pubs/concept_operations.htm (last visited Feb. 15, 2010).

[1540] *See generally* NENA Comments in re NBP PN #8, filed Nov. 12, 2009.

[1541] PSST Comments in re NBP PN #8, filed Nov. 12, 2009, at 2.

[1542] U.S. Dep't of Transp., Next Generation 911 (NG9-1-1) System Initiative, Final Analysis of Cost, Value, and Risk (Mar. 5, 2009) (DOT NG911 Cost study).

[1543] Intrado Comments in re NBP PN #8, filed Nov. 12, 2009, at 11.

[1544] For instance, through the 911 Access Program, the Rural Utilities Service provides low-interest loans to state and local governments, Indian tribes and other entities for facilities and equipment to improve 911 access in rural areas. Food, Conservation, and Energy Act of 2008, Pub. L. No. 110-246, §6107, 122 Stat. 1651, 1959 (2008); *see* E911 Grant Program, 74 Fed. Reg 29,967 (June 5, 2009).

[1545] U.S. Dep't of Transp., Next Generation 911 (NG9-1-1) System Initiative, Final Analysis of Cost, Value, and Risk (Mar. 5, 2009) (DOT NG911 Cost Study).

[1546] *See* Peter Svensson, *Iowa 911 Call Center Becomes First to Accept Texts*, abc neWs, Aug. 5, 2009, http:// abcnews.go.com/Technology/wireStory?id=8259735 .

[1547] Real-Time Text is a feature that allows users to see text as it is typed into a text interface.

[1548] NENA Comments in re NBP PN #8, filed Nov. 12, 2009, at 18–20.

[1549] *See* NENA Comments in re NBP PN #8, filed Nov. 12, 2009, at 18 –20.

[1550] *See Wireless E911 Location Accuracy Requirements; Revision of the Commission's Rules to Ensure Compatibility with Enhanced 911 Emergency Calling Systems; 911 Requirements for IP-Enabled Service Providers*, PS Docket No. 07-114, CC Docket No. 94-102, WC Docket No. 05-196, Notice of Proposed Rulemaking, 22 FCC Rcd 10609 (2007).

[1551] *See* Federal Emergency Management Agency, Integrated Public Alert and Warning System (IPAWS): Success Stories, http://www.fema.gov/emergency/ ipaws/successstories.shtm (last visited Mar. 5, 2010) (IPAWS Success Stories).

[1552] *See* Federal Emergency Management Agency, Integrated Public Alert and Warning System (IPAWS), http://www.fema.gov/emergency/ipaws/ (last visited Feb. 15, 2010).

[1553] GAO, Emergency Preparedness: Improved Planning and Coordination Necessary for Modernization and Integration of Public Alert and Warning System 14 (2009) (GAO Emergency Preparedness Report), *available at* http://www.gao.gov/new.items/d09834.pdf (noting that capabilities to distribute emergency alerts and warnings through e-mails, telephones, text message devices, cell phones, pagers and Internet-connected desktops have not been implemented).

[1554] GAO Emergency Preparedness Reportat 20–24.

[1555] GAO Emergency Preparedness Reportat 24–26. Challenges identified by GAO included lack of redundancy, gaps in coverage, systems integration, standards development, development of geo-targeted alerting and alerts for people with disabilities and those who do not speak English. In response to the report, the DHS agreed with GAO's recommendations for addressing these concerns and has begun to address many of these challenges. *See* Written Statement of Damon Penn, Assistant Administrator, FEMA before the Committee on Transportation and Infrastructure, Subcommittee on Economic Development, Public Buildings and Emergency Management, U.S. House of Representatives (Sept. 30, 2009), http:// republicans.transportation.house.gov/Media/file/ TestimonyEDPB/2009-09-30-Penn.pdf.

[1556] *See* Radio World, EAS Trigger Saved Lives in Samoa Tsunami (Sept. 20, 2009), http://www.radioworld. com/article/87954; Bill Hoffman, *Lucky To Be Alive After Tsunami Destroys Dream Resort*, neW zealand herald, Oct. 1, 2009, *available at* http://www. nzherald.co.nz/american-samoa/news/article.cfm?l_ id=500605&objectid=10600668.

[1557] *See* Plan | Define Plan at Dictionary.com, http:// dictionary.reference.com/browse/plan (last visited Mar. 1, 2010).

[1558] *See generally* Letter from Andrew Blau et al., Monitor Group, to Marlene H. Dortch, Secretary, FCC, GN Docket No. 09-51 (Feb. 19, 2010) Attach.

[1559] Baljit Singh Grewal, *Neoliberalism and Discourse: Case Studies of Knowledge Policies in the Asia-Pacific* 130–33 (2008) (unpublished Ph.D. thesis, Auckland University of Technology), *available at* http://aut.researchgateway. ac.nz/bitstream/10292/407/4/GrewalB.pdf.

[1560] Republic of Korea Nat'l Info. Soc'y Agency, Korea Informatized 1–3 (2007), *available at* http://old.nia. or.kr/open_content/board/fileDownload.jsp?tn=PU_00 00100&id=53923&seq=1&fl=7.

[1561] Jong Sung Hwang & Sang-Hyun Park, Digital Review of Asia Pacific 2009–2010: .kr Korea, Republic of, http:// www.idrc.ca/pan/ev-140957-201-1-DO_TOPIC.html (last visited Feb. 19, 2010).

[1562] Jong Sung Hwang & Sang-Hyun Park, Digital Review of Asia Pacific 2009–2010: .kr Korea, Republic of, http:// www.idrc.ca/pan/ev-140957-201-1-DO_TOPIC.html (last visited Feb. 19, 2010).

[1563] Letter from Young Kyu Noh, Minister Counselor of Broadcasting and ICT, Embassy of the Republic of Korea, to Marlene H. Dortch, Secretary, FCC, GN Docket Nos. 09-47, 09-51, 09-137 (filed Dec. 15, 2009) at 1.

[1564] The full name of the IT Strategy Headquarters is the Strategic Headquarters for the Promotion of an Advanced Information and Telecommunications Network Society.

[1565] Prime Minister of Japan, Basic Law on the Formation of an Advanced Information and Telecommunications Network Society, http://www.kantei.go.jp/foreign/it/ it_basiclaw/it_basiclaw.html (last visited Mar. 1, 2010).

[1566] Letter from Masaru Fujino, Counselor for Communications Policy, Embassy of Japan, to Marlene H. Dortch, Secretary, FCC, GN Docket Nos. 09-47, 09-51, 09-137 (filed Dec. 15, 2009) App. A (National Broadband Strategies in Japan (2001–09)).

[1567] Department for Culture, Media and Sport and Department for Business, Innovation and Skills, Digital Britain: Final Report (2009), *available at* http://www. culture.gov.uk/images/publications/digitalbritainfinalreport-jun09.pdf.

[1568] Department for Culture, Media and Sport and Department for Business, Innovation and Skills, Digital Britain: Implementation Plan 1–4 (2009) (Digital Britain: Implementation Plan), *available at* http://www.culture.gov.uk/images/publications/ DB_ImplementationPlanv6_Aug09.pdf.

[1569] Digital Britain: Implementation Plan 1–4.

[1570] Department for Culture, Media and Sport and Department for Business, Innovation and Skills, Digital Britain: Implementation Update—December 2009 (2009), *available at* http://www.culture.gov.uk/images/ publications/DB_Implementationplan_Dec09.pdf.

[1571] At a minimum, membership should include representatives from the Federal Communications Commission, the Department of Commerce and the National Telecommunications and Information Administration, the Department of Homeland Security, the Department of Education, the Department of Energy, the Department of Health and Human Services, the Department of Housing and Urban Development, the Department of Agriculture, the Small Business Administration, and other agencies who may be responsible for implementing recommendations set forth in the plan.

[1572] *See also* Ch. 14, Recommendation 14.3, *supra* (recommending that federal agencies coordinate grants that have a broadband connectivity requirement).

[1573] For example, certain U.S. Census data are made available to researchers in a controlled fashion at the U.S. Census Bureau's Center for Economic Studies and Research data center. *See* U.S. Census Bureau Center for Economic Studies, Research Program Overview, http:// www.ces.census.gov/index.php/ces/researchprogram (last visited Feb. 14, 2009).

[1574] *See* Ch. 4, *supra* (recommending more detailed and accurate broadband data collection).

[1575] *See, e.g.*, Gov't of Germany, Bundesministerium für Wirtschaft und Technolog ie, Breitbandatlas 2009_02 (2009) (detailing Germany's broadband mapping efforts), *available at* http://www.zukunft-breitband.de/ Dateien/BBA/PDF/breitbandatlas-bericht-2009-02,pr operty=pdf,bereich=bba,sprache=de,rwb=true.pdf; Nat'l IT and Telecom Agency, Gov't of Denmark, Mapping of Broadband Access Services in Denmark (2004), *available at* http://en.itst.dk/the-governments-it-and- telecommunications-policy/publications/mapping- of-broadband-access-services-in-denmark-status-by-mid-2004/Mapping%20of%20Broadband%20 Access%20Services%20in%20Denmark%20-%20 Status%20by%20mid-2004.pdf; The National Broadband Map: New Zealand's Broadband Landscape, Home, www.broadbandmap.govt.nz/map (last visited Feb. 19, 2010).

[1576] *See, e.g.*, Swedish Post and Telecom Agency, PTS Statistics Portal, http://www.statistics.pts.se/start_en/ (last visited Feb. 19, 2010); Finnish Commc'ns Reg. Auth., Gov't of Finland, Use of Telecommunications Services (2009),http://www.ficora.fi/attachments/suomiry/5n2kRC9zk/Tutkimusraportti_2009_Telepalveluiden_ kayttotutkimus.pdf (in Finnish).

[1577] *See, e.g.*, Startsida, Telepriskollen, http://www. telepriskollen.se/ (last visited Feb. 19, 2010) (Swedish site for comparing tariffed services); Finnish Commc'ns Reg. Auth., Gov't of Finland, Use of Telecommunications Services, http://www.ficora.fi/ en/index/tutkimukset/puhelinjalaajakaistapalvelut/ markkinatieto.html (last visited Mar. 7, 2010).

[1578] *See, e.g.*, IT-Borger, Bredbåndsmåleren, http://www. it-borger.dk/verktojer/bredbaandsmaaleren (last visited Feb. 19, 2010) (providing a broadband speed test in Danish); Nettimitari–etusivu, http://nettimittari.ficora. fi/nettimittari/mainPage.aspx (last visited Feb. 19, 2010) (Finland's broadband speed test).

[1579] *See* Ch. 4, *supra* (recommending more detailed and accurate broadband data collection).

[1580] Paul A. David, *The Dynamo and the Computer: An Historical Perspective on the Modern Productivity Paradox*, 80 aea papers & prOceedinG, 355, 356–60 (1990), *available at* http://elsa.berkeley.edu/~bhhall/ e124/David90_dynamo.pdf.

[1581] Robert Solow, *We'd Better Watch Out*, N.Y. Rev. of Books, July 12, 1987, at 36.

[1582] The Broadband Data Improvement Act (BDIA) mandates that the FCC conduct periodic surveys of national characteristics of the use of broadband services. *See* BDIA, Pub. L. No. 110-385, 122 Stat. 4096 (2008).

[1583] Fed. Energy Reg. Comm'n, Assessment of Demand Response & Advanced Metering (2009), *available at* http://www.ferc.gov/legal/staff-reports/ sep-09-demand-response.pdf; Dep't of Energy, Smart Grid Systems Report (2009), *available at* http://www.oe.energy.gov/DocumentsandMedia/ SGSRMain_090707_lowres.pdf.

[1584] *See* 47 U.S.C. §§ 201–76.

[1585] *Amendment of Section 64.702 of the Comm'n's Rules & Regs, Second Computer Inquiry*, Docket No. 20828, Final Decision, 77 F.C.C. 2d 384, 417–35, 86–132, 461–75, paras. 201–31 (1980) (*Computer II Final Decision*), *aff'd sub nom. Computer & Commc'ns Indus. Ass'n v. FCC*, 693 F.2d 198 (D.C. Cir. 1982), *cert. denied*, 461 U.S. 938 (1983); *Amendment of Section 64.702 of the Comm'n's Rules & Regs.* (*Third Computer Inquiry*), CC Docket No. 85-229, Phase I, Report and Order, 104 F.C.C. 2d 958, para. 4 (1986) (*Computer III Phase I Order*) (subsequent history omitted); *see also* 47 C.F.R. § 64.702 (defining "enhanced service").

[1586] 47 U.S.C. §§ 151–61.

[1587] *United States v. Southwestern Cable Co.*, 392 U.S. 157 (1968); *United States v. Midwest Video Corp.*, 406 U.S. 649 (1972). The Commission has ancillary authority when a matter falls within its jurisdiction over interstate communications by wire or radio and the proposed action is "reasonably ancillary to the effective performance of the Commission's various responsibilities." *See* 47 U.S.C. § 152(a); *Southwestern Cable*, 392 U.S. at 172–73; *Midwest Video*, 406 U.S. at 667–68 ("[T]he critical question . . . is whether the Commission has reasonably determined that its [regulatory action] will further the achievement of long established regulatory goals." (quotation marks omitted)); *United Video Inc. v. FCC*, 890 F.2d 1173 (D.C. Cir. 1989) (upholding ancillary authority where action was taken to further "a valid communications policy goal").

[1588] *Nat'l Cable & Telecomms. Ass'n v. Brand X Internet Servs.*, 545 U.S. 967, 986–87 (2005).

[1589] *See Framework for Broadband Access to the Internet over Wireline Facilities; Universal Service Obligations of Broadband Providers*, CC Docket Nos. 02-33, 01-337, 95-20, 98-10, WC Docket Nos. 04-242, 05-271, Report and Order and Notice of Proposed Rulemaking, 20 FCC Rcd 14853, 14864, 14909–11, para. 15, paras. 103–04 (2005) (*Wireline Broadband Report and Order* and *Broadband Consumer Protection Notice*), *aff'd sub nom. Time Warner Telecom, Inc. v. FCC*, 507 F.3d 205 (3d Cir. 2007).

[1590] *United Power Line Council's Petition for Declaratory Ruling Regarding the Classification of Broadband over Power Line Internet Access Service as an Information Service*, WC Docket No. 06-10, Memorandum Opinion and Order, 21 FCC Rcd 13281, 13286, 13288, paras. 9, 12 (2006) (*BPL-Enabled Broadband Order*).

[1591] *Appropriate Regulatory Treatment for Broadband Access to the Internet Over Wireless Networks*, WT Docket No. 07–53, Declaratory Ruling, 22 FCC Rcd 5901, 5909–10, 5912, paras. 22, 29 (2007) (*Wireless Broadband Order*).

[1592] *Wireline Broadband Report and Order*, 20 FCC Rcd at 14858, para. 5, 14900–03, paras. 89–95, 14909–10, para. 103; *BPL-Enabled Broadband Order*, 21 FCC Rcd at 13289, para. 15; *Wireless Broadband Order*, 22 FCC Rcd at 5913–14, para. 33. The Commission, however, has interpreted the *Wireline Broadband Order* to mean that rate-of-return ILECs are not entitled to offer broadband Internet access transmission on a non-common carrier basis until the Commission has addressed associated cost allocation issues. S*ee Petition of ACS of Anchorage, Inc. Pursuant to Section 10 of the Communications Act of 1934, as Amended (47 USC Section 160(c)), for Forbearance from Certain Dominant Carrier Regulation of Its Interstate Access Services, and for Forbearance from Title II Regulation of its Broadband Services, in the Anchorage, Alaska, Incumbent Local Exchange Carrier Study Area*, WC Docket No. 06–109, Memorandum Opinion and Order, 22 FCC Rcd 16304, 16339–40, paras. 75, 80 (2007).

[1593] *See, e.g.*, Letter from Gary L. Phillips, General Attorney & Associate General Counsel, AT&T Services, Inc., to Marlene H. Dortch, Secretary, FCC (Jan. 29, 2010), GN Docket Nos. 09–51, 09–47, 09–137, WC Docket Nos. 05–337, 03–109, attach. at 6 ("[F]unding of universal service lies at the heart of the Commission's core statutory mission, and widespread deployment and subscribership of broadband Internet access is without question a high national priority. The Commission's jurisdictional analysis [for use of Title I authority] could stop here.").

[1594] *See, e.g.*, Electronic Privacy Information Center Comments in re National Broadband Plan NOI (*A National Broadband Plan for Our Future*, GN Docket No. 09-41, Notice of Inquire, 24 FCC Rcd 4342 (2009) (*National Broadband Plan NOI*), filed June 8, 2009, at 3 (filed by Mark Rotenberg) ("[T]he Commission should exercise its ancillary jurisdiction to ensure that the national broadband plan includes robust privacy safeguards, lest consumers' critical broadband privacy interests go unaddressed.").

1595 *See, e.g.*, RERC on Telecom Access Comments in re NBP PN#4 (*Comment Sought on Broadband Accessibility for People with Disabilities Workshop II: Barriers, Opportunities, and Policy Recommendations—NBP Public Notice #4*, GN Docket Nos. 09-47, 09-51, 09-137, Public Notice, 24 FCC Rcd 11968 (GCB 2009) (*NBP PN#4*)), filed Oct. 6, 2009, at 11 ("In order to ensure that individuals who use hearing aids and cochlear implants are not left out again, it is critical for the FCC to use its ancillary jurisdiction to carry over the protections now afforded under existing [Hearing Aid Compatibility] laws to handsets used with broadband communication technologies.").

1596 *See, e.g.*, National Telecommunications Cooperative Association Reply in re National Broadband Plan NOI, filed July 21, 2009, at 5 ("[S]ome argue that [universal service funding to support broadband] should be allowed without a finding that broadband is a supported service. However, there is no rational, policy or legal arguments [sic] to support that position." (citation omitted)), 28 ("Defined under Title II, the FCC can look at broadband and determine what access regulations are appropriate and necessary, and refrain from regulations that are inappropriate and unnecessary. The FCC does not have sthis is harming consumers, education, public health and safety, and national security."); Public Knowledge Reply in re NBP PN#30 (*Reply Comments Sought in Support of the National Broadband Plan—NBP Public Notice #30*, GN Docket Nos. 09-47, 09-51, 09-137, Public Notice, 25 FCC Rcd 241 (WCB 2010)), filed Jan. 26, 2010, at 5 ("[R]eliance on Title I creates uncertainty for the Commission and imperils the goals of the National Broadband Plan." (citation omitted)).

1597 *See, e.g.*, The Nebraska Rural Independent Companies Comments in re NBP PN #19 (*Comment Sought on the Role of the Universal Service Fund and Intercarrier Compensation in the National Broadband Plan —NBP Public Notice #19*, GN Docket Nos. 09-47, 09-51, 09-137, Public Notice, 24 FCC Rcd 13757 (OSP 2009) (*NBP PN #19*)), filed Dec. 7, ("[N]ow is the time to fulfill the mandate of Section 254 and to expand USF support to broadband Internet access services by finding they are properly considered telecommunications services, or telecommunications at a minimum." (citation omitted)); National Telecommunications Cooperative Association Reply in re National Broadband Plan NOI, filed July 21, 2009, at 5, 26 ("By classifying all broadband Internet access service as a telecommunications service regulated under Title II, the base of USF contributors will expand to include all broadband service providers as required under section 254."); Public Knowledge Reply in re NBP PN #30, filed Jan. 26 2010. *See generally* 47 U.S.C. §§ 254(b) (3), (j); *Federal-State Joint Board on Universal Service*, Report and Order, 12 FCC Rcd 8776, 8956–57, paras. 332–40 (1997).

1598 *See, e.g.*, Metro PCS Communications Comments in re Consumer Information and Disclosure NOI (*Consumer Information and Disclosure; Truth-in-Billing and Billing Format; IP-Enabled Services*, CG Docket No. 09-158, CC Docket No. 98-170, WC Docket No. 04-36, Notice of Inquiry, 24 FCC Rcd 11380 (2009) (*Consumer Information and Disclosure NOI*)), filed Oct. 13, 2009, at 19 (noting that if the Commission proceeds with disclosure rules for communications services, "the better alternative, in the view of MetroPCS, is to find other statutory bases in addition to Title I to justify its actions or not impose regulation. One possible approach, which may go beyond this particular proceeding, is for the Commission to revisit its earlier determinations that high speed broadband Internet access services are not common carrier services."). *But see* Letter from Mark A. Stachiw, Executive Vice President, General Counsel & Secretary, MetroPCS Communications Inc., to Marlene H. Dortch, Secretary, FCC, GN Docket Nos. 09-47, 09- 51, 09-137, 09-191, WC Docket No. 07-52 (Feb. 3, 2010) (opposing reclassification of broadband Internet access services on policy grounds). *See generally* 47 U.S.C. §§ 201(b), 258.

1599 *See, e.g.*, Letter from Michael Weinberg, Staff Attorney, Public Knowledge, to Marlene H. Dortch, Secretary, FCC GN Docket Nos. 09-51, 09-137, 09-157, 09-191, WC Docket Nos. 04-36, 07-52, CC Docket No. 98-170, CG Docket No. 09-158, WT Docket No. 09-66 (Feb. 19, 2010) at 1.

1600 *See, e.g.*, National Telecommunication Cooperative Association Reply in re National Broadband Plan NOI, filed July 21, 2009, at 28 ("Defined under Title II, the FCC can look at broadband and determine what access regulations are appropriate and necessary, and refrain from regulations that are inappropriate and unnecessary."); Public Knowledge Reply in re NBP PN #30, filed Jan. 26, 2010, at 6 (citing *Wireline Broadband Order*). *See generally* 47 U.S.C. § 160(a).

1601 47 U.S.C. § 332(c)(1)(A). That provision does not grant the Commission the authority to forbear from sections 201, 202 or 208.

1602 *See, e.g.*, Verizon and Verizon Wireless Comments in re National Broadband Plan NOI, filed June 8, 2009, at 87–91 ("[T]he Commission's recommendations . . . Should not move backward by supporting . . . ill-fitting common carrier or Title II regulation."); Alcatel-Lucent Reply in re National Broadband Plan NOI, filed July 21, 2009, at 2 ("The Commission must reject calls to repudiate the entire framework under which broadband services are regulated by re-classifying those offerings . . . ."), 4–9.

1603 Pres. Franklin Delano Roosevelt, Address at Barnesville, Georgia (Aug. 11, 1938) (Pres. F. D. Roosevelt 1938 Address at Barnesville, Georgia), *available at* http://georgiainfo.galileo.usg.edu/FDRspeeches/ FDRspeech38–6.htm.

1604 Pres. F. D. Roosevelt 1938 Address at Barnesville, Georgia.

1605 Nat'l Telecomm. & Information Admin., Office of Telecommunications and Information Applications, http://www.ntia.doc.gov/otiahome/otiahome.html (last visited Mar. 1, 2010).

[1606] Nat'l Telecomm. & Information Admin. BTOP Project Information, http://www.ntia.doc.gov/broadbandgrants/projects.html (last visited Feb. 20, 2010).

[1607] Exec. Off. of the Pres., Nat'l Econ. Council, Recovery Act Investments in Broadband: Leveraging Federal Dollars to Create Jobs and Connect America (2009), *available at* http://www.whitehouse.gov/sites/default/files/20091217- recovery-act-investments-broadband.pdf.

[1608] Nat'l Telecomm. & Information Admin., *Secretary Locke Announces Recovery Act Grants to Expand Broadband Internet Access and Spur Economic Growth* (press release), Feb. 18, 2010, *available at* http://www.ntia.doc.gov/ press/2010/02182010_Locke_BTOP_Announcement.pdf.

[1609] Nat'l Telecomm. & Information Admin., *Secretary Locke Announces Recovery Act Grants to Expand Broadband Internet Access and Spur Economic Growth* (press release), Feb. 18, 2010.

[1610] Scott J. Wallsten, *Measuring the Effectiveness of the Broadband Stimulus Plan,* The Economists' Voice 6:6, art. 3 (2009).

[1611] Janice Hauge & James Prieger, Demand-side Programs to Stimulate Adoption: What Works? (Oct. 22, 2009) (unpublished working paper), *available at* http://papers. Ssrn.com/sol3/papers.cfm?abstract_id=1492342.

[1612] The National Broadband Plan provides this glossary solely as a reader aid. These definitions do not necessarily represent the views of the FCC or the United States Government on past, present, or future technology, policy, or law and thus have no interpretive or precedential value.

In: The National Broadband Plan: Analysis and Strategy for...   ISBN: 978-1-61122-024-7
Editor: Daniel M. Morales   © 2011 Nova Science Publishers, Inc.

Chapter 3

SPECTRUM POLICY IN THE AGE OF BROADBAND: ISSUES FOR CONGRESS

Linda K. Moore

SUMMARY

The convergence of wireless telecommunications technology with the Internet Protocol (IP) is fostering new generations of mobile technologies. This transformation has created new demands for advanced communications infrastructure and radio frequency spectrum capacity that can support high-speed, content-rich uses. Furthermore, a number of services, in addition to consumer and business communications, rely at least in part on wireless links to broadband backbones. Wireless technologies support public safety communications, sensors, smart grids, medicine and public health, intelligent transportation systems, and many other vital communications.

Existing policies for allocating and assigning spectrum rights may not be sufficient to meet the future needs of wireless broadband. A challenge for Congress is to provide decisive policies in an environment where there are many choices but little consensus. In formulating spectrum policy, mainstream viewpoints generally diverge on whether to give priority to market economics or social goals. Regarding access to spectrum, economic policy looks to harness market forces to allocate spectrum efficiently, with spectrum license auctions as the driver. Social policy favors ensuring wireless access to support a variety of social objectives where economic return is not easily quantified, such as improving education, health services, and public safety. Both approaches can stimulate economic growth and job creation.

Deciding what weight to give to specific goals and setting priorities to meet those goals pose difficult tasks for federal administrators and regulators and for Congress. Meaningful oversight or legislation may require making choices about what goals will best serve the public interest. Relying on market forces to make those decisions may be the most efficient and effective way to serve the public but, to achieve this, policy makers may need to broaden the concept of what constitutes competition in wireless markets.

The National Broadband Plan (NBP), a report on broadband policy mandated by Congress, has provided descriptions of perceived issues to be addressed by a combination of regulatory changes and the development of new policies at the Federal Communications Commission, with recommendations for legislative actions that Congress might take.

Among the spectrum policy initiatives that have been proposed in Congress in recent years are: allocating more spectrum for unlicensed use; auctioning airwaves currently allocated for federal use; and devising new fees on spectrum use, notably those collected by the FCC's statutory authority to implement these measures is limited. The NBP reiterates these proposals and adds several more.

Substantive modifications in spectrum policy would almost surely require congressional action. The Radio Spectrum Inventory Act introduced in the Senate (S. 649, Kerry) and the similar House-introduced Radio Spectrum Inventory Act (H.R. 3125, Waxman) would require an inventory of existing users on prime radio frequencies, a preliminary step in evaluating policy changes. The Spectrum Relocation and Improvement Act of 2009 (H.R. 3019, Inslee) and the Spectrum Relocation Act of 2010 (S. 3490, Warner) would amend the Commercial Spectrum Enhancement Act of 2004 (P.L. 108-494, Title II). The Broadband for First Responders Act (H.R. 5081, King) would allocate additional radio frequencies for public safety use.

THE ROLE OF SPECTRUM POLICY

Wireless broadband[1] can play a key role in the deployment of broadband services. Because of the importance of wireless connectivity, radio frequency spectrum policy is deemed by the National Broadband Plan[2] (NBP) to be a critical factor in national broadband policy and planning. Wireless broadband, with its rich array of services and content, requires new spectrum capacity to accommodate growth. Spectrum capacity is necessary to deliver mobile broadband to consumers and businesses and also to support the communications needs of industries that use fixed wireless broadband to transmit large quantities of information quickly and reliably.

The purpose of spectrum policy, law, and regulation is to manage a natural resource[3] for the maximum possible benefit of the public. Radio frequency spectrum is managed by the Federal Communications Commission (FCC) for commercial and other non-federal uses and by the National Telecommunications and Information Administration (NTIA) for federal government use. International use is facilitated by numerous bilateral and multilateral agreements covering many aspects of usage, including mobile telephony.[4]

Although radio frequency spectrum is abundant, usable spectrum is currently limited by the constraints of applied technology. Spectrum policy therefore requires making decisions about how radio frequencies will be allocated and who will have access to them.[5] Spectrum policy also entails encouraging innovation in wireless technologies and their applications. Arguably, the role of technology policy in crafting spectrum policy has increased with the need to reduce or eliminate capacity constraints that may deter the expansion of broadband mobile services. The adoption of spectrum-efficient technologies is likely to require a rethinking of spectrum management policies and tools. Policies for channel management to control interference might be superseded by managing interference through guidelines for networks and devices. The assignment and supervision of licenses might give way to policies and procedures for managing pooled resources. Auctioning licenses might be replaced by auctioning access; the static event of selling a license replaced by the dynamic auctioning of spectrum access on a moment-by-moment basis.

Current spectrum policy relies heavily on auctions to assign spectrum rights through licensing. Economy of scale in wireless communications has become an important determinant in the outcome of these auctions. Companies that have already made substantial investments in infrastructure have been well placed to maximize the value of new spectrum acquisitions. Corporate mergers and acquisitions represent another way to improve scale economies. Efficiencies through economy of scale have contributed to creating a market for wireless services where four companies—Verizon Wireless LLC, AT&T Inc., Sprint Nextel Corporation, and T-Mobile USA Inc.—had approximately 90% of the customer base of subscribers at the end of 2009.[6] These companies also own significant numbers of spectrum licenses covering major markets nationwide.

The leading position of these few companies in providing a critical distribution channel—wireless—for information and services may need to be considered in plans for national broadband deployment. One approach to ensuring wireless access to meet national broadband goals might be to tighten the regulatory structure under which wireless communications are managed. Other approaches might seek ways to modify spectrum policies to increase market competition and to accommodate the age of broadband. In the NBP, the FCC has emphasized the latter course and committed to a number of actions intended to increase opportunities for competition and innovation in mobile broadband.

Competition

With the introduction of auctions for spectrum licenses in 1994, the United States began to shift away from assigning spectrum licenses based on regulatory decisions and toward competitive market mechanisms. One objective of the Telecommunications Act of 1996 was to open up the communications industry to greater competition among different sectors. One outcome of the growth of competition was the establishment of different regulatory regimes for information networks and for telecommunications.[7] As a consequence of these and other legislative and regulatory changes, the wireless industry has areas of competition, e.g. for spectrum licenses, within a regulatory shell, such as the rules governing the Public Switched Telephone Network (PSTN).[8] As the bulk of wireless communications traffic moves from voice to data, companies will likely modify their business plans in order to remain competitive in the new environment. A shift in infrastructure technology and regulatory environment[9] might open wireless competition to companies with business plans that are not modeled on pre-existing telecommunications industry formulae. Future providers of wireless broadband might include any company with a robust network for carrying data and a business case for serving broadband consumers. Potential new entrants, however, may lack access to radio frequency spectrum, the essential resource for wireless broadband.

The FCC, in the NBP, has concluded that an effective way to improve competition among wireless broadband providers is to increase the amount of spectrum available. This approach was validated by a number of filings with the FCC; for example, the Department of

Justice provided arguments as to why the "primary tool for promoting broadband competition should be freeing up spectrum."[10] Policy tools that might be used to increase the availability of radio frequency spectrum for wireless broadband include allocating additional spectrum, reassigning spectrum to new users, requiring that wireless network infrastructure be shared, pooling radio frequency channels, moving to more spectrum-efficient technologies, and changing the cost structure of spectrum access.

Innovation

From a policy perspective, actions to speed the arrival of new, spectrally efficient technologies might have significant impact on achieving broadband policy goals over the long term. In particular, support for technologies that enable sharing could pave the way for dramatically different ways of managing the nation's spectrum resources. The NBP has laid out several opportunities for the FCC, the NTIA, and other government agencies to contribute to and encourage the development of new technologies for more efficient spectrum access.[11] Among the technologies that facilitate spectrum sharing are cognitive radio and Dynamic Spectrum Access (DSA).[12] Enabling technologies such as these allow communications to switch instantly among network frequencies that are not in use and therefore available to any radio device equipped with cognitive technology. Among the steps that might be taken to encourage spectrum-efficient technologies, the NBP has recommended that the FCC identify and free up a "new, contiguous nationwide band for unlicensed use"[13] and provide spectrum and take other steps to "further development and deployment"of new technologies that facilitate sharing.[14]

The NTIA has recommended exploring "ways to create incentives for more efficient use of limited spectrum resources, such as dynamic or opportunistic frequency sharing arrangements in both licensed and unlicensed uses."[15] This suggestion was incorporated into the 2011 Budget prepared by the Office of Management and Budget. The budget document directed the NTIA to collaborate with the FCC "to develop a plan to make available significant spectrum suitable for both mobile and fixed wireless broadband use over the next ten years. The plan is to focus on making spectrum available for exclusive use by commercial broadband providers or technologies, or for dynamic, shared access by commercial and government users."[16]

The NTIA's Commercial Spectrum Management Advisory Committee is actively looking at policy and technology issues in a series of subcommittee reports. The reports are addressing spectrum inventory, transparency, dynamic spectrum access, incentives, unlicensed spectrum, and sharing.[17]

THE NATIONAL BROADBAND PLAN AND SPECTRUM POLICY

In the American Recovery and Reinvestment Act of 2009 (ARRA), Congress required the FCC to prepare a national broadband plan, to be delivered not later than February 17, 2010 (later extended to mid-March). The primary objective of the plan is "to ensure that all people of the United States have access to broadband capability...." The plan is to include "an analysis of the most effective and efficient mechanisms for ensuring broadband access...." and "a plan for use of broadband infrastructure and services in advancing consumer welfare...."[18]

On March 16, 2010, the FCC publically released its report, *Connecting America: The National Broadband Plan.*[19] The National Broadband Plan is presented as three major policy areas.

- *Innovation and Investment* "discusses recommendations to maximize innovation, investment and consumer welfare, primarily through competition. It then recommends more efficient allocation and management of assets government controls or influences." The recommendations address a number of issues, including spectrum policy.
- *Inclusion* "makes recommendations to promote inclusion—to ensure that all Americans have access to the opportunities broadband can provide."
- *National Purposes* "makes recommendations to maximize the use of broadband to address national priorities. This includes reforming laws, policies and incentives to maximize the benefits of broadband in areas where government plays a significant role." National purposes include health care, education, energy and the environment, government performance, civic engagement, and public safety.

Spectrum Policy Recommendations

The section in the NBP on spectrum policy (Chapter 5) has taken particular note of the convergence of the Internet with mobile devices and the resulting increased demand for spectrum capacity to support mobile broadband services.

The NBP has proposed to increase spectrum capacity by

- Making more spectrum licenses available for mobile broadband.
- Increasing the amount of spectrum available for shared use.
- Encouraging and supporting the development of spectrum-efficient technologies, particularly those that facilitate sharing spectrum bands.
- Instituting new policies for spectrum management, such as assessing fees on some spectrum licenses, to encourage more efficient use.

To facilitate the deployment of broadband in rural areas, the NBP also has proposed

- Improving the environment for providing wireless components to build out infrastructure.

Many of the NBP proposals for wireless broadband may be achieved through changes in FCC regulations governing spectrum allocation and assignment. Other actions may require changes by federal agencies, state authorities, and commercial owners of spectrum licenses. To assist the implementation of the NBP there are also a number of areas where congressional action might be required to change existing statutes or to give the FCC new powers. Legislation has been proposed that would create an inventory of existing users on prime radio frequencies, a preliminary step in evaluating policy changes.[20] The NBP included the announcement of plans for the FCC to create what it refers to as a Spectrum Dashboard.[21] The initial release of the FCC's Spectrum Dashboard provided an interactive tool to search for information about how some nonfederal frequency assignments are being used.[22] The dashboard could be expanded to meet requirements set by Congress for a spectrum inventory. In addition to the dashboard, the NBP has proposed that the FCC and the NTIA should create methods for recovering spectrum[23] and that the FCC maintain an ongoing spectrum strategy plan.[24] All of these steps will facilitate decisions about spectrum management by providing detailed information about the current and potential use of spectrum resources.

Spectrum Licenses

One of the management tools available to the FCC is its power to assign spectrum licenses through auctions. Auctions are regarded as a market-based mechanism for rationing spectrum rights. Before auctions became the primary method for distributing spectrum licenses the FCC used a number of different approaches, primarily based on perceived merit, to select license-holders. The FCC was authorized to organize auctions to award spectrum licenses for certain wireless communications services in the Omnibus Budget Reconciliation Act of 1993 (P.L. 103-66). Following passage of the act, subsequent laws that dealt with spectrum policy and auctions included the Balanced Budget Act of 1997 (P.L. 105-33), the Auction Reform Act of 2002 (P.L. 107-195), the Commercial Spectrum Enhancement Act of 2004 (P.L. 108-494, Title II), and the Deficit Reduction Act of 2005 (P.L. 109-171). The Balanced Budget Act of 1997 gave the FCC auction authority until September 30, 2007. This authority was extended to September 30, 2011, by the Deficit Reduction Act of 2005 and to 2012 by the DTV Delay Act (P.L. 111-4).

In the NBP, the FCC has proposed taking steps to add 300 MHz of licensed spectrum for broadband within five years and a total of 500 MHz of new frequencies in ten years.[25] Approximately 50 MHz would be released in the immediate future by the completion of existing auction plans. An additional 20 MHz might be reassigned from federal to commercial use and made available for auction. Reallocating some spectrum from over-the-air broadcasting to commercial spectrum might provide an additional 120 MHz of spectrum. Final rulings on existing proceedings would release 110 MHz, of which 90 MHz would be for Mobile Satellite Services (MSS); resolution of interference issues between Wireless Communications Services (WCS) and satellite radio would free up 20 MHz of new capacity.

The spectrum assignment proposals put forth in the NBP are contentious in that the various parties affected by the decisions have diverging views on how technology should be used to provide access to these frequencies. Although Congress has shown interest in all of these debates, three proposals that are the most likely to generate pressure for congressional action are the plans for: repurposing and auctioning an estimated 120 MHz of airwaves assigned to over-the-air digital television broadcasting; auctioning the D Block (10 MHz in the 700 MHz band); and auctioning up to 60 MHz of spectrum for Advanced Wireless Services.

Television Broadcast Spectrum

The Balanced Budget Act of 1997, which mandated the eventual transition to digital television, represented the legislative culmination of over a decade of policy debates and negotiations between the FCC and the television broadcast industry on how to move the industry from analog to digital broadcasting technologies. To facilitate the transition, the FCC provided each qualified broadcaster with 6 MHz of spectrum for digital broadcasting to replace licenses of 6 MHz that were used for analog broadcasting. The analog licenses would be yielded back when the transition to digital television was concluded. The completed transition freed up the 700 MHz band for commercial and public safety communications in 2009.

The FCC has revisited the assumptions reflected in the 1997 act and has made new proposals, and decisions based on, among other factors, changes in technology and consumer habits. The NBP announced that a new proceeding would be initiated to recapture up to 120 MHz of spectrum from broadcast TV allocations for reassignment to broadband communications. This proceeding would propose four sets of actions to achieve the goal; a fifth set of actions to increase efficiency would be pursued separately.[26] The FCC stipulated in the NBP that its recommendations "seek to preserve [over-the-air television] as a healthy, viable medium going forward, in a way that would not harm consumers overall, while establishing mechanisms to make available additional spectrum for flexible broadband uses."[27]

Many of the proposals for redirecting TV broadcast capacity are based on refinements in the way frequencies are managed and are procedural in nature. Because over-the-air digital broadcasting does not necessarily require 6 MHz of spectrum, the NBP has proposed that some stations could share a single 6 MHz band without significantly reducing service to over-the-air TV viewers. Among the proposals for how broadcasters might make better use of their TV licenses, the NBP has raised the possibility of auctioning unneeded spectrum and sharing the proceeds between the TV license-holder and the U.S. Treasury. The FCC has called on Congress to provide new legislation that would allow these "incentive auctions."

D Block

The D Block refers to a set of frequencies within the 700 MHz band that were among the frequencies made available after the transition from analog to digital television in 2009. In compliance with instructions from Congress to auction all unallocated spectrum in this band, the FCC conducted an auction, which concluded on March 18, 2008. As part of its preparation for the auction (Auction 73), the FCC sought to increase the amount of spectrum available to public safety users in the 700 MHz band. Congress had previously designated 24 MHz of radio frequencies in the 700 MHz band for public safety channels. In 2007, the FCC proposed to designate 10 MHz – part of the original 24 MHz designated for public safety use – specifically for public safety broadband communications. Of the balance, 12 MHz were designated for mission critical voice communications on narrowband networks and 2 MHz were set aside as a guard band to protect against interference. In the FCC plan for Auction 73, the Public Safety Broadband License would be matched with a commercial license for 10 MHz, known as the D Block. The D Block was to be auctioned under rules that would require the creation of a public-private partnership to develop the two 10-MHz assignments as a single broadband network, available to both public safety users and commercial customers.

The D Block license was offered for sale in 2008 but did not find a buyer. The FCC then set about the task of writing new service rules for a reauction of the D Block.[28]

In the NBP, the FCC announced its decision to auction the D Block under rules that would not require a partnership with public safety but would establish a framework for priority access to the D Block network by public safety users.[29] Based mainly on FCC efforts to create a public-private partnership, public safety officials have, by and large, anticipated that the D Block would be an integral part of a public safety broadband network. Since the failed D Block auction of 2008, there has also been growing pressure on the FCC and on Congress to take the steps necessary to reallocate the D Block from commercial to public safety use.[30] The NBP announcement regarding the D Block is considered by many to be a reversal of announced policy, creating controversy and renewed calls for Congress to take action to release the D Block to public safety. Although funding and control are critical elements of the debate, the controversy is rooted in contradictory assumptions about the level of service and reliability that new, largely untried, and in some cases undeveloped technology will be able to deliver for public safety broadband communications.

The FCC would address public safety needs such as developing standards and establishing procedures through the newly established Emergency Response Interoperability Center (ERIC).[31] ERIC would work closely with the Public Safety Communications Research program, jointly managed by the National Institute of Standards and Technology (NIST) and the NTIA, to develop and test the technological solutions needed for public safety broadband communications.[32] The Department of Homeland Security will participate in the areas of public safety outreach and technical assistance, as well as best practices development.[33]

ERIC would take on the role of creating and implementing a federal plan to assist in building a nationwide, interoperable network for public safety. As the lead agency, the FCC would rely on its authority to require the D Block and other commercial license-holders in the 700 MHz band to accommodate public safety needs. Although public safety users would be charged for access to commercial networks, proponents of the plan have argued that overall costs would be less than if a network were built primarily or exclusively for public safety use, because of greater economies of scale. One of the expectations is that ERIC will be able to guide the development of standards for crucial radio components, with the participation of commercial providers and public safety representatives. The participation of commercial carriers in developing and deploying, for example, a common radio interface, is expected to put the cost of public safety radios in the same price range as commercial high-end mobile devices ($500). By contrast, interoperable radios for the narrowband networks at 700 MHz cost $3,000 and up, each.

Advanced Wireless Service Auctions

During 2007, the FCC was petitioned by several companies, led by M2Z Networks Inc., to release 20 MHz of spectrum licenses at 2155-2175 MHz for a national broadband network. M2Z offered to provide free basic service to consumers and public safety and to offer content filtering for family-friendly access. In return for the grant of the license, which would be assigned without auction, M2Z offered to pay a percentage of gross revenues to the U.S. Treasury. In September 2007, the FCC issued a Notice of Proposed Rulemaking to establish service rules for the auction of a license or licenses at 2155-2175 MHz, designated as Auction AWS-3.[34] Proposed provisions for the auction included obligations to offer free broadband service similar to that proposed by M2Z and family-friendly access. The proposed spectrum band is adjacent to bands previously auctioned in the Advanced Wireless Service (AWS-1) auction that concluded in 2006. T-Mobile, a major winner in the AWS-1 auction, has stated to the FCC that the network proposed by M2Z would cause "pervasive harmful interference" to licensees of the AWS-1 frequencies.[35]

The FCC did not act on the AWS-3 auction proposal but announced new plans in the NBP that included the 2155-2175 MHz frequencies.[36] As outlined in the NBP, the FCC would seek to pair the AWS-3 frequencies with an additional 20 MHz of frequencies reassigned from federal use. The plan has recommended that the NTIA, in consultation with the FCC, assess the possibility of such a reallocation and, if the reallocation appears feasible, that they move ahead with plans to organize an auction. If reallocation and auction is not deemed feasible, the FCC would proceed "promptly" to auction the AWS-3 frequencies, according to the plan. The plan had proposed using frequencies in the 1755-1780 MHz range, but the NTIA has instead offered to assess the feasibility of using frequencies in the 1675-1710 MHz band.[37] The FCC subsequently requested comments to evaluate approaches to making this band available for shared use that would include wireless broadband services.[38]

In addition to the AWS-3 frequencies, there are two blocks of spectrum under the designation of AWS-2 "H" and "J" that have been under consideration for auction since 2004. The AWS-2 "J" band, with paired frequency assignment at 2020-2025 MHZ and 2175-2180 MHz, might be paired with AWS-3 or with an adjacent Mobile Satellite Service band.

The process of finalizing auction plans for licenses to use the AWS-2 and AWS-3 frequencies might renew the debate over interference. FCC proceedings also might provide an opportunity to revisit the possibility of including a requirement for auction winners to offer basic broadband service at no cost to the consumer. The concept of a lifeline broadband service has received support from many policy makers in Congress.

Shared Resources

The FCC has stated in the NBP that it would facilitate sharing resources through a number of regulatory means.[39] Among the methods of sharing wireless connectivity currently practiced in the United States are sharing network facilities, sharing network operations, and sharing spectrum. Examples of sharing include nationwide roaming,[40] selling packages of minutes purchased from a facilities-based network, leasing network capacity and spectrum access from a facilities-based network to create a new service provider—known as a Virtual Mobile Network Operator—and spectrum sharing. In general, access is leased from an owner—of a tower, a network, or a spectrum license. Another option is to allocate spectrum for unlicensed use; any device authorized by the FCC may operate on the designated frequencies.

The primary difficulty for regulators in overseeing the sharing of spectrum is to minimize interference among devices operating on the same or nearby frequencies. It was primarily to prevent interference to wireless messages that spectrum licensing was first instituted. Today, a number of administrative and technological methods are available to minimize interference of wireless transmissions. In theory, all spectrum bands can be shared if interference can be managed.

Open Access

In the 2008 auction of spectrum licenses at 700 MHz,[41] several companies associated with Silicon Valley and Internet ventures petitioned the FCC to set aside a block of spectrum as a national license with a requirement that the network be available—open—to all. Open access was defined as open devices, open applications, open services, and open networks.[42] The position put forward by these companies was that access of unlicensed airwaves was not enough to stimulate innovation and competition for new devices, services, and applications. They argued that innovators, especially start-up companies, were often closed out of markets unless they could convince a wireless network operator to accept and market their inventions.[43] The FCC subsequently ruled to auction licenses for 22 MHz of spectrum (designated as the C Block) with service rules requiring the first two criteria: open devices and open applications. The winning bidders, most notably Verizon Wireless,[44] are required to allow their customers to choose their own handsets and download programs of their choice, subject to reasonable conditions needed to protect the network from harm.

Wholesale Networks

The FCC was also petitioned to designate spectrum licenses at 700 MHz for networks that would operate on a wholesale business model. It was argued that the wholesale business model would be the most viable for a small business new entrant and that the auction rules and conditions adopted by the FCC were prejudicial to small business.[45] Wireless incumbents, in particular, have challenged the concepts of open access and wholesaling. They have claimed that the unproven nature of a wholesale business model makes it risky and that therefore the auction value of licenses with a wholesaling requirement would be diminished. They have argued that imposing requirements that would create a wholesale network introduces an extra level of regulatory oversight, covering such areas as handset compatibility, applications standards, market access regulation, and interconnection rules.[46]

Proponents of open access argue that only an open network that anyone can use—not just subscribers of one wireless company—can provide consumer choice. From this perspective, a wholesale network could provide more market opportunities for new wireless devices, especially wireless devices that could provide unrestricted access to the Internet. A wholesale network would allow customers to choose their own wireless devices without necessarily committing to a service plan from a single provider. The network owner would operate along the same principles used for shopping malls, providing the infrastructure for others to retail their own products and services.

Unlicensed Use

Unlicensed spectrum is not sold to the highest bidder and used for the services chosen by the license-holder but is instead accessible to anyone using wireless equipment certified by the FCC for those frequencies. Both commercial and non-commercial entities use unlicensed spectrum to meet a wide variety of monitoring and communications needs. Suppliers of wireless devices must meet requirements for certification to operate on frequency bands designated for unlicensed use. Examples of unlicensed use include garage door openers and Wi-Fi communications.

New technologies that can use unlicensed spectrum without causing interference are being developed for vacant spectrum designated to provide space between the broadcasting signals of digital television, known as white spaces. On September 11, 2006, the FCC announced a timetable for allowing access to the spectrum so that devices could be developed.[47] Devices using the white-space frequencies would be required to incorporate geolocation technology to signal when and where potential interference was detected.[48] A geolocation database would be created and maintained to facilitate sharing of the white space by authorized devices. The design and operation of this database is being studied by the FCC.[49] The National Association of Broadcasters (NAB), and others, have protested the use of white space for consumer devices on the grounds that they could interfere with digital broadcasting and with microphones used for a variety of purposes.[50] Companies such as Microsoft, Dell, and Motorola, however, have stated the belief that solutions can be found to prevent interference. In November 2008, the FCC established rules that permit the unlicensed use of the white spaces, with special provisions to protect microphone use.[51] One of the recommendations of the NBP is that the FCC complete the proceeding that would allow use of the white spaces for unlicensed devices.[52]

Spectrum-Efficient Technology

Mobile communications became generally available to businesses and consumers in the 1980s. The pioneering cell phone technologies were analog.[53] Second-generation (2G) wireless devices were characterized by digitized delivery systems. Third-generation (3G) wireless technology represents significant advances in the ability to deliver data and images. The first commercial release of 3G was in Japan in 2001; the technology successfully debuted in the United States in 2003. 3G technologies can support multi-function devices, such as the BlackBerry and the iPhone. Successor technologies, often referred to as 4G, are expected to support broadband speeds that will rival wireline connections such as fiber optic cable, with the advantage of complete mobility. 4G wireless broadband technologies include WiMAX[54] and Long Term Evolution (LTE) networks. Both are based on TCP/IP, the core protocol of the Internet.[55]

Wireless technologies to facilitate broadband deployment for which spectrum may need to be allocated that were identified by the NBP include 4G networks; fixed wireless as an alternative to fiber optic cable; and broadband on unlicensed frequencies.

The NBP spectrum assignment proposals are based on managing radio channels as the way to maximize spectral efficiency while meeting common goals such as minimizing interference among devices operating on the same or nearby frequencies. Today, channel management is a significant part of spectrum management; many of the FCC dockets deal with assigning channels and resolving the issues raised by these decisions. In the future, channel management is likely to be replaced by technologies that operate without the need for designated channels. In the NBP, the FCC refers to these spectrum-seeking technologies as opportunistic. Identifying an opportunity to move to an open radio frequency is more flexible—and therefore more productive—than operating on a set of pre-determined frequencies. The primary benefit from these new technologies will be the significant increase in available spectrum but new efficiencies in operational and regulatory costs will also be realized.

The concept of channel management dates to the development of the radio telegraph by Guglielmo Marconi and his contemporaries. In the age of the Internet, however, channel management is an inefficient way to provide spectrum capacity for mobile broadband. Innovation points to network-centric spectrum management as an effective way to provide spectrum capacity to meet the bandwidth needs of fourth-generation wireless devices. [56] Network-centric technologies organize the transmission of radio signals along the same principle as the Internet. A transmission moves from origination to destination not along a fixed path but by passing from one available node to the next. Pooling resources, one of the concepts that powers the Internet now, is likely to become the dominant principle for spectrum management in the future.

New Technologies

The iPhone 3G and 3GS provide early examples of how the Internet is likely to change wireless communications as more and more of the underlying network infrastructure is converted to IP-based standards. The iPhone uses the Internet Protocol to perform many of its functions; these require time and space—spectrum capacity—to operate. The next generation of wireless networks, 4G, for Fourth Generation, will be supported by technologies structured and managed to emulate the Internet. The wireless devices that operate on these new, IP-powered networks will be able to share spectrum capacity in ways not currently used on commercial networks, greatly increasing network availability on licensed bandwidths. Another technological boost will come from improved ways to use unlicensed spectrum. Unlicensed spectrum refers to bands of spectrum designated for multiple providers, multiple uses, and multiple types of devices that have met operational requirements set by the FCC. Wi-Fi is an example of a current use of unlicensed spectrum.

More efficient spectrum use can be realized by integrating adaptive networking technologies, such as DSA, with IP-based, 4G commercial network technologies such as LTE. Adaptive networking has the potential to organize wireless communications to achieve the same kinds of benefits that have been seen to accrue with the transition from proprietary data networks to the Internet. These enabling technologies allow communications to switch instantly among network frequencies that are not in use and therefore available to any wireless device equipped with cognitive technology. Adaptive technologies are designed to use pooled spectrum resources. Pooling spectrum licenses goes beyond sharing. Licenses are aggregated and specific ownership of channels becomes secondary to the common goal of maximizing network performance.

New Policies

Among the steps that might be taken to encourage "opportunistic" technologies, the NBP recommends that the FCC identify and free up a "new, contiguous nationwide band for unlicensed use" by 2020;[57] and provide spectrum and take other steps to "further development and deployment" of new technologies that facilitate sharing.[58] Unlike its recommendations for auctioning spectrum licenses in the near future, the FCC's plans for bringing new technologies into play provide few details. The NBP provides a glimpse through the keyhole of the horizons beyond, but not the key that might open the door.

The testing of new technologies that increase spectrum capacity, and the policy changes they are likely to bring, has been designated by the NBP as a future event. Its immediate plans for spectrum policy are to fine-tune existing spectrum assignments to increase the availability of licensed capacity. The level of opposition to most of these spectrum assignment plans might suggest that current spectrum management practices have reached the point of diminishing returns. The FCC might consider first identifying the new technologies mobile broadband will require before it begins the hunt for more spectrum.

Management Tools

In the NBP, the FCC has asked Congress to consider granting it authority to impose spectrum fees on license holders as a means of addressing inefficient use.[59] The report has presented the hypothesis that "Fees may help to free spectrum for new uses such as broadband, since licensees who use spectrum inefficiently may reduce their holdings once they bear the opportunity cost" of holding the spectrum.[60]

The Obama Administration also has proposed that the FCC be given the authority to levy fees, and to use other economic mechanisms, as a spectrum management tool.[61] The 2011 fiscal year budget prepared by the Office of Management and Budget projects new revenue from spectrum license user fees of $4.775 billion for fiscal years 2011 through 2020.[62] Similar projections were made in the 2010 budget[63] and in budget proposals during the administration of President George W. Bush.[64]

Although Congress never took up legislation in response to the Bush Administration proposals, the 108th Congress instructed the GAO to take note of the possible impact of changing the spectrum license fee structure. In the Commercial Spectrum Enhancement Act, the GAO was instructed to examine "national commercial spectrum policy as implemented by the Federal Communications Commission" and report on its findings in 2005.[65] The GAO was to examine the impact of auctioning licenses on the economic climate for broadcast and wireless technologies and to assess whether the holders of spectrum licenses received before the auction process was instituted (i.e., largely for free) have an economic advantage over license holders that purchased spectrum through the auction process. The GAO was also to evaluate whether the disparate methods of allocating spectrum had an adverse impact on the introduction of new services. The conclusions of the study were to be reviewed in the context of an Administration proposal to introduce license user fees on licenses that had not been auctioned. The GAO was also to provide an evaluation for Congress regarding the impact of assessing license fees on the competitive climate in the wireless and broadcast industries.

After consultation with the committees of jurisdiction, the GAO did not include an analysis of license fees in its report. Instead it focused on the impact of auctions on factors such as end-user prices, investment in infrastructure, and competition. One of the report's conclusions was that the cost of purchasing licenses did not affect price and competition in the long run because the cost was a one-time, sunk cost.[66] New licensing regimes were mentioned in the report as a possible means of increasing spectral efficiency but the suggestion received no discussion in the report.[67]

The FCC's statutory authority to impose new spectrum user fees is limited. The FCC was authorized by Congress to set license application fees[68] and regulatory fees to recover costs.[69] A new fee structure seeking recovery beyond costs would require Congressional authorization, either through an appropriations bill or new legislation. New fees could be difficult to devise as many of the licenses originally assigned at little cost to the acquirer were subsequently sold to other carriers.

Wireless Backhaul

Most mobile communications depend on fixed infrastructure to relay calls to and from wireless networks. The infrastructure that links wireless communications to the wired world is commonly referred to as backhaul. In situations where installing communications cables is impractical, fixed wireless infrastructure may be used to provide the needed backhaul. Microwave technologies, for example, are used in a number of applications to extend coverage to areas not served by fiberoptic or other wire links.

The NBP has predicted that the importance of backhaul will increase with the implementation of 4G technologies, as mobile access to the Internet and other wired networks becomes increasingly prevalent.[70] The FCC therefore has proposed to take a number of procedural steps to increase the flexibility and capacity for point-to-point wireless backhaul technologies.[71] On June 7, 2010, the FCC adopted an order intended to "enhance the flexibility and speed" of acquiring spectrum for wireless backhaul.[72]

NATIONAL PURPOSES

Among the requirements for the National Broadband Plan, Congress specified that it should include

> a plan for use of broadband infrastructure and services in advancing consumer welfare, civic participation, public safety and homeland security, community development, health care delivery, energy independence and efficiency, education, worker training, private sector investment, entrepreneurial activity, job creation and economic growth, and other national purposes.[73]

In the plan, the Federal Communications Commission (FCC) has made recommendations that might fulfill both social and economic goals. In the section of the plan titled "National Purposes," it has focused on social goals with an agenda of actions for federal, state, and local agencies. The areas covered in this section are

- Health care. The NBP has identified stated goals of the Department of Health and Human Services that might be effectively supported with technologies that are enhanced by access to broadband communications.
- Education. The NBP has proposed that broadband can provide an effective tool for meeting the educational needs and ambitions of educators, students, and parents of young children as well as support the Department of Education's strategies to improve educational achievement.

- Energy and the Environment. Broadband has multiple applications in the field of energy, conservation, and environmental protection. For example, SmartGrid goals set by Congress[74] might not be achievable without broadband communications.
- Economic Opportunity. Actions proposed in the NBP to further economic opportunity are centered on increasing access to Information Technology for small and medium-sized businesses. The role of broadband in providing job training and employment services and supporting telework are also addressed in recommendations.
- Government Performance. The recommendations for federal government actions encompassed both ways that broadband might improve the effectiveness of government and also steps the federal government might take to increase the availability of broadband networks. The latter included federal actions to improve cybersecurity and ways that federal agencies might assist communities and state and local governments in building broadband infrastructure.
- Civic Engagement. The NBP has described concepts such as government transparency that can lead to greater participation by all in the democratic process. Broadband access has been described in the plan as a useful tool for encouraging civic engagement because of the part it plays in interactive communication and providing information.
- Public Safety. The NBP recommendations dealt primarily with delivering wireless broadband to the radios of first responder. It also considered the role of broadband in upgrading the nation's 911 services and emergency alert systems.

Meeting Policy Goals

Each of the sections on national purposes has mentioned the existing legislative and regulatory framework and trends in the field that might benefit from better broadband access and services. Although each sector serves different needs and goals, the NBP recommendations are fairly similar for each. In general, stakeholders have been encouraged to

- Create incentives to achieve broadband goals.
- Leverage broadband technology, including wireless broadband.
- Encourage innovation and improved productivity.
- Provide or increase funding for programs that support broadband policy goals.
- Modify regulations.

The NBP has recommended that the Executive Branch create a Broadband Strategy Council.[75] This council would coordinate efforts by the many agencies that the FCC has identified as having a role in the plan's implementation. The NBP has suggested that the President could require that federal departments and agencies submit broadband implementation plans to the council. The council could also act as an intermediary between the agencies and Congress regarding legislation that might facilitate meeting the NBP's goals. Another recommendation of the NBP would require the FCC to track progress in meeting the plan's goals.[76]

Community Broadband

Rural communities have on occasion used their resources to install fiber-optic networks in part because they were too small a market to interest for-profit companies. Networks that depend on a fiber-optic cable backbone are capital-intensive and usually more profitable in high-density urban areas. Increasingly, communities of all sizes are looking at wireless technologies to support their networks. Municipalities, for example, are installing free Wi-Fi zones. Among the reasons often cited for installing wireless facilities are that generally available access to the Internet through wireless connections has become an urban amenity, a necessity in sustaining and developing the local economy, and a part of essential infrastructure with many public benefits.[77]

Opponents to community-owned networks contend that they provide unfair competition, distorting the marketplace and discouraging commercial companies from investing in broadband technologies. In particular, the fact that urban areas are creating Wi-Fi networks and providing, among other services, free wireless links to the Internet is viewed as a threat to commercial companies.

Several states have passed laws prohibiting or limiting local governments' ability to provide telecommunications services. An effort to challenge such a law in Missouri by municipalities offering local communications services in the state was heard before the U.S. Supreme Court in 2004.[78] In the Telecommunications Act of 1996, Congress barred states from "prohibiting the ability of any entity to provide any interstate or intrastate telecommunications service."[79] The Court ruled that "entity" was not specific enough to include state political divisions; if Congress wished specifically to protect both public and private entities, they could do so by amending the language of the law. This Court decision and the steady improvement in broadband communications technologies that municipalities wish to have available in their communities have provided fuel for a policy debate about access to broadband services.

Because community broadband networks can help with NBP goals for inclusion and national purposes, the NBP has considered the possibility that federal investment in broadband infrastructure might be leveraged for community and state broadband services. For example, the NBP has recommended that federal agencies could open their broadband networks to state and local agencies and to unserved and underserved areas.[80] It has also made recommendations to help reduce costs and improve funding to community broadband programs.[81]

ISSUES FOR THE 111TH CONGRESS

The NBP has made a number of recommendations for Congressional action to grant new authorities to the FCC, including several related to spectrum policy and auctions. The Radio Spectrum Inventory bills being considered by the 111th Congress parallel NBP recommendations. The NBP has recommended changes to the Commercial Spectrum Enhancement Act; not all of these recommendations are included in the bill that would amend that act, H.R. 3019, the Spectrum Relocation and Improvement Act of 2009. Congress may decide to consider NBP recommendations for legislation that would permit incentive auctions, particularly as they apply to recommended reassignments of digital TV broadcast frequencies. H.R. 5081, the Broadband for First Responders Act of 2010, would require the reallocation of 700 MHz D Block frequencies, contrary to announced plans to auction the frequencies.

Spectrum Inventory

Similar versions of a Radio Spectrum Inventory Act (S. 649, Senator Kerry and H.R. 3125, Representative Waxman) would require the FCC and NTIA to prepare an inventory of spectrum allocations and assignments in prime radio frequency bands. The information from the detailed report on users and uses would help policy makers evaluate whether spectrum is being allocated and used effectively. The bills would require an accounting of spectrum allocation in the designated bands that would identify commercial license-holders, government agency spectrum allocations, and the number of devices deployed in those bands. If available, information would be provided on the types of wireless devices used on licensed frequencies and on unlicensed frequencies within each band. Contour maps and information on the location of base stations and other fixed transmitters might also be included in the inventory. The inventory results would be available over the Internet. Exemptions from public access to some information may be granted for reasons of national security, although the two bills vary on terms for these exemptions. The inventory is to be completed and submitted in reports to Congress within 180 days of passage into law.[82]

The Commercial Spectrum Enhancement Act and Federal Relocation

The Spectrum Relocation and Improvement Act of 2009 (H.R. 3019, Representative Inslee) and the similarly worded Spectrum Relocation and Improvement Act of 2010 (S. 3490, Senator Warner) would address issues arising from current experiences in relocating federal users to clear space for commercial license-holders. The bills would define the rights and responsibilities of federal entities in the spectrum relocation process, especially obligations for sharing, and their eligibility for payments from the Spectrum Relocation Fund. The NBP has recommended that Congress consider improvements to the Commercial Spectrum Enhancement Act (CSEA) to provide "adequate incentives and assistance" to support relocation. The NBP has further recommended that the compensation to federal agencies be structured to provide additional incentives "for using commercial services and non-spectrum based operations." In particular, the NBP has recommended that "Congress revise the CSEA to provide for payments of relocation funds to federal users that vacate spectrum and make use of commercial networks instead...."[83]

The Spectrum Relocation Fund was created by the Commercial Spectrum Enhancement Act in 2004, to provide a mechanism whereby federal agencies could recover the costs of moving from one spectrum band to another. The fund is administered by the Office of Management and Budget. Following procedures required by the act, the FCC scheduled an auction of designated federal frequencies for commercial use as Advanced Wireless Services (AWS). The AWS auction was completed on September 18, 2006, attracting nearly $13.9

billion in completed bids.[84] The FCC ruled that auction winners wishing to put acquired licenses to immediate use would in most cases be able to share with current federal users under guidance from the FCC.[85] The act was written to be applied to most transfers of spectrum from federal to commercial use.

Incentive Auctions

The NBP has asked Congress to consider expanding the FCC's auction authority to permit incentive auctions.[86] As outlined in the NBP, such authority would enable spectrum license-holders to return spectrum for auction, with the expectation of receiving part of the proceeds as compensation for the costs associated with relinquishing the asset. Although most spectrum license auction revenues are deposited as general funds,[87] Congress has passed laws, such as the Commercial Spectrum Enhancement Act, that permit the proceeds to be used for other purposes.

Broadband for First Responders Act

The Broadband for First Responders Act of 2010 (H.R. 5081, Representative King) would amend the Communications Act of 1934 by requiring the FCC to allocate an additional 10 MHz of spectrum, known at the D Block, at 758-763 MHz and 788-793 MHz, for public safety services and assign these paired bands for public safety broadband use. The bill would require the FCC to establish rules to encourage the rapid deployment of an interoperable national wireless broadband network.

Service rules would provide priority access to mission critical public safety applications; provide for roaming on public safety spectrum by authorized users; encourage public-private partnerships by requiring consideration of the use of existing or planned commercial infrastructure; meet technical and operational standards for interoperability and roaming; require that networks are built to survive most large scale disasters; ensure networks have the appropriate level of cyber security; and facilitate the shared use of the public safety broadband spectrum and infrastructure with commercial and other entities.

Standards would be set in consultation with the National Institute of Standards and Technology (NIST) and others to enable nationwide interoperability and roaming across any communications system using public safety broadband spectrum.

CONCLUSION

Telephone service was once considered a natural monopoly, and regulated accordingly. The presumption was that redundant telephone infrastructure was inefficient and not in the public interest. State and federal regulators favored granting operating rights to a single company, within a specific facilities territory, to benefit from economies of scale, facilitate interoperability, and maximize other benefits. In return for the monopoly position, the selected provider was expected to fulfill a number of requirements intended to benefit society.

Thus, for decades, the regulated monopoly was seen by most policy-makers as (1) ensuring that costly infrastructure was put in place and (2) meeting society's needs, as interpreted by regulations and the law.[88] Past policies to regulate a monopolistic market may have influenced current policies for promoting competition. The FCC's emphasis on efficiency for delivering services to a pre-determined market could be leading wireless competition toward monopoly; new regulatory regimes might be a consequence of this trend, if it continues.

Current spectrum policy seeks to maximize the value of spectrum by encouraging economies of scale and appears to treat spectrum assets as an extension of existing infrastructure (spectrum license ownership and network management, for example) instead of an alternative infrastructure (Wi-Fi and wireless backhaul are examples). This policy course has provided a form of workable competition that has brought wireless services (until 2006, almost exclusively voice) at affordable prices to most of the country. However, wireless technology has reached an inflection point and is shifting from voice to data. Some argue that wireless policy should also shift, placing a greater value on innovation to achieve goals deemed to be in the public interest. A policy that prioritizes providing spectrum to spur innovation, for example, could create new markets, new models for competition, and new competitors. If spectrum policy serves broadband policy and broadband policy serves multiple sectors of the economy, then perhaps spectrum should be more readily available for a wider pool of economic participants.

The amount of spectrum needed for fully realized wireless access to broadband is such that meeting the needs of broadband policy goals could be difficult to achieve through the market-driven auction process unless large amounts of new radio frequencies can be identified and released for that purpose.[89] Without abandoning competitive auctions, spectrum policy could benefit from including additional ways to assign or manage spectrum that might better serve the deployment of wireless broadband and the implementation of a national broadband policy.

Legislation geared to improve auction mechanisms might benefit from the consideration of measures that would use technology to increase the amount of spectrum available, thereby opening the field to new players, fostering competition, and spurring innovation.

To further the transition to new technologies, Congress might choose to require performance goals for improved spectrum efficiency, not unlike the way federal goals have been set for energy conservation or transportation safety.

APPENDIX A. SPECTRUM-HUNGRY TECHNOLOGIES

Enabling technologies that are fueling both the demand for mobile broadband services and the need for radio frequency spectrum include Long Term Evolution, WiMAX; fixed wireless; Wi-Fi; high performance mobile devices such as smartphones and netbooks; and cloud computing. Fixed wireless and Wi-Fi are not new technologies but mobile broadband has given them new roles in meeting consumer demand. Future technologies include network-centric technologies, which include opportunistic solutions such as Dynamic Spectrum Access (DSA).

Long Term Evolution (LTE)

LTE is the projected development of existing 3G networks built on Universal Mobile Telephone System (UMTS) standards.[90] Like all fourth-generation wireless technologies, LTE's core network uses Internet protocols. The network architecture is intended to facilitate mobile broadband deployment with capabilities that can deliver large amounts of data, quickly and efficiently, to large numbers of simultaneous users. LTE will likely be implemented in stages through modifications to networks using frequencies in bands already allocated for commercial wireless networks.[91] LTE might operate on spectrum bands at 700 MHz, 1.7 GHz, 2.3 GHz, 2.5 GHz, and 3.4 GHz.[92]

WiMAX

WiMAX provides mobile broadband but its earliest applications were for fixed wireless services. WiMAX (Worldwide Interoperability for Microwave Access) refers to both a technology and an industry standard, the work of an industry coalition of network and equipment suppliers.[93] WiMAX uses multiple frequencies around the world in ranges from 700 MHz to 66 GHz. In the United States, available frequencies include 700 MHz, 1.9 GHz, 2.3 GHz, 2.5 GHz and 2.7 GHz. The introduction of WiMAX in the United States is being jointly led by Sprint Nextel Corporation and Clearwire Corporation under the name Clearwire. Clearwire Wi-MAX, branded CLEAR, plans to serve 80 markets by the end of 2010.[94]

Fixed Wireless Services

Fixed wireless services have taken on new importance as a "backhaul" link for 4G. Backhaul is the telecommunications industry term that refers to connections between a core system and a subsidiary node. An example of backhaul is the link between a network—which could be the Internet or an internetwork that can connect to the Internet—and the cell tower base stations that route traffic from wireless to wired systems. Two backhaul technologies well-suited for mobile Internet access are fiber optic cable and point-to-point microwave radio relay transmissions.[95] Network expansion plans for WiMAX and LTE include microwave links as a cost-effective substitute for fiber optic wire under certain conditions. Radio frequencies available in the United States for microwave technologies of different types start in the 930 MHz band and range as high as the 90 GHz band.

Wi-Fi

The popularity of Wi-Fi is often cited as a successful innovation that was implemented using unlicensed frequencies.[96] Wi-Fi provides wireless Internet access for personal computers and handheld devices and is also used by businesses to link computer-based communications within a local area. Links are connected to a high-speed landline either at a business location or through hotspots. Hotspots are typically located in homes or convenient public locations, including airports and café environments such as Starbucks. Wi-Fi uses radio frequencies in the free 2.4 GHz and 5.4/5.7GHz spectrum bands. Many 3G and 4G wireless devices that operate on licensed frequencies can also use the unlicensed frequencies set aside for Wi-Fi.[97]

Smartphones and Netbooks

Two of the fastest growing segments in the category of mobile Internet devices are smartphones and netbooks. The introduction of Apple Inc.'s iPhone, in 2007, is widely viewed as heralding a new era in wireless smartphones. The smartphone market is predicted to thrive on growing demand for downloadable applications,[98] interactive websites, and imaginative videos—all delivered wirelessly. A parallel development has been the accelerating use of netbooks. These book-sized laptop computers are designed to provide broadband wireless access to the Internet. The line between smartphone and netbook technologies is fading as the newer generations of these devices provide many of the same features. The iPad, introduced by Apple Inc. in 2010, provides an example of the interchangeability of features across different platforms. The majority of these devices can operate on Wi-Fi as well as over 3G and 4G networks using licensed frequencies.

Cloud Computing

Cloud computing is a catch-all term that is popularly used to describe a range of information technology resources that are separately stored for access through a network, including the Internet. An Internet search on Google, for example, is using cloud computing to access a rich resource of data and information processing. Network connectivity to services is another resource provided by cloud computing. Google Inc. also offers word processing, e-mail and other services through Google Docs. Although off-site data processing and information storage are not new concepts, cloud computing benefits from the significant advances in network technology and capacity that are hallmarks of the broadband era. Cloud computing can provide economies of scale to businesses of all sizes. Small businesses in particular can benefit from forgoing the costs of installing and managing hardware and software by buying what they need from the cloud. Consumers also can benefit because they no longer need to buy personal computers in order to run complex programs or store large amounts of data. The convergence of 4G wireless technology—with its smartphones and netbooks—and the growing accessibility of cloud computing to businesses and consumers alike will contribute to the predicted explosive growth in demand for wireless bandwidth.[99]

Network-Centric Technologies

More efficient spectrum use can be realized by integrating adaptive networking technologies, such as dynamic spectrum access, with IP-based commercial network technologies such as LTE. Radios using DSA chipsets are more effective at managing interference and congestion than the channel management techniques currently in use. If a channel's link fails, the radio is cut off. When radios are networked using DSA, individual communications nodes continue to operate and can compensate for failed links. The effects of interference are manageable rather than catastrophic. The network is used to overcome radio limitations.

Adaptive networking has the potential to organize radio communications to achieve the same kinds of benefits that have been seen to accrue with the transition from proprietary data networks to the Internet. Adaptive technologies are designed to use pooled spectrum resources. Pooling spectrum licenses goes beyond sharing. Licenses are aggregated and specific ownership of channels becomes secondary to the common goal of maximizing network performance.

The Department of Defense (DOD) is working to implement network-centric operations (NCO) through a number of initiatives.[100] Leadership and support to achieve DOD goals in the crucial area of spectrum management is provided by the Defense Spectrum Organization (DSO) created in 2006 within the Defense Information Systems Agency (DISA). The DSO is leading DOD efforts to transform spectrum management in support of future net-centric operations and warfare, and to meet military needs for dynamic, agile, and adaptive access to spectrum. The DSO is guiding DOD spectrum management along a path that envisions moving away from stove-piped systems to network-centric spectrum management and, ideally, to bandwidth on demand and cognitive self-synching spectrum use.

Among the steps to advance toward the goal of spectrum access that is fully adaptable to any situation is the testing of network-centric technologies developed by the Defense Advanced Research Projects Agency (DARPA) within the Wireless Network After Next (WNaN) program.

WNaN is evaluating DSA, Disruptive Tolerant Networking, and other tools, possibly to replace the existing Joint Tactical Radio System (JTRS) now in use. JTRS uses software-programmable radios to provide interoperability, among other features.[101] WNaN's testing and evaluation of network-centric technologies is expected to lead to a decisison in late 2010. WNAN technology is planned for transition to the Army in 2010.[102]

APPENDIX B. COMPETITION

A combination of policy and market forces has divided the commercial wireless market into sharply different tiers. Policies that have encouraged economies of scale have favored mergers and acquisitions of wireless companies. There are now four facilities-based[103] wireless companies in the United States that the FCC describes as nationwide: AT&T, Verizon Wireless, Sprint Nextel, and T-Mobile, [104] which had approximately 90% of the subscriber market at the beginning of 2010. Another four providers had subscriber bases of between 1 million and 15 million Over one hundred smaller carriers serve niche markets.[105]

Barriers to Competition

In evaluating competition within an industry, economists and policy makers examine barriers to entry, among other factors.[106] Barriers might come from high costs for market entry such as investment in infrastructure or there might be legal and regulatory barriers to entry. As part of its evaluation of competition for mobile services, the FCC has identified three factors that could constitute barriers to entry to the commercial mobile communications industry. These barriers affect not only competitiveness but also access to networks and investment in new technology. The factors are: "first-mover advantages, large sunk costs, and access to spectrum."[107] All three of these factors are subject to regulations that have been influenced by past or existing policies regarding spectrum allocation and assignment.

First-mover advantages[108] have accrued primarily to the early entrants in the wireless industry. Early in the development of the cell phone industry, the FCC created cellular markets and assigned two spectrum licenses to each market; one license went automatically to the incumbent provider in that market. The second license was made available to a competing service provider (not the market incumbent); the difficulties in choosing the competitors that would receive licenses contributed to the subsequent move to auctions as a means for assigning spectrum rights.[109] These early entrants, and the successor companies that acquired them and their licenses, have maintained their core customer base and benefit from early investments in infrastructure. Many first movers into the wireless market, therefore, acquired their market-leader status through regulatory decisions that provided them with spectrum licenses, not through market competition.

Large sunk costs refer to the high levels of investment needed to enter the wireless market. Not including the price of purchasing spectrum, billions of dollars are required to build new infrastructure. The sunk costs of incumbent wireless service providers set a high bar for new entrants to match if they are to compete effectively in major markets. In the mobile telephone industry, the FCC has observed that most capital expenditures are spent on existing networks: to expand and improve geographic coverage; to increase capacity of existing networks; and to improve network capabilities. Performance requirements for spectrum license-holders, such as the size of a market that must be served or deadlines for completing infrastructure build-outs, are some of the policy decisions that can add to the cost of entry.

Spectrum Auctions and Competition

The FCC, acting on the statutory authority given to it by Congress, has broad regulatory powers for spectrum management. The FCC was created as part of the Communications Act of 1934[110] as the successor to the Federal Radio Commission, which was formed under the Radio Act of 1927.[111] The first statute covering the regulation of airwaves in the United States was the Radio Act of 1912, which gave the authority to assign usage rights (licenses) to the Secretary of the Department of Commerce and Labor.[112] Licensing was necessary in part because, as radio communications grew, it became crucial that frequencies be reserved for specific uses or users, to minimize interference among wireless transmissions.[113]

A key component of spectrum policy is the allocation of bands of frequencies for specific uses and the assignment of licenses within those bands. Allocation refers to the decisions, sometimes reached at the international level, that set aside bands of frequencies for categories of uses or users; assignment refers to the transfer of spectrum rights to specific license-holders. Radio frequency spectrum is treated as a natural resource that belongs to the American people. The FCC, therefore, licenses spectrum but does not convey ownership. Before auctions became the primary method for assigning spectrum licenses the FCC used a number of different approaches, primarily based on perceived merit, to select license-holders.

Auctions are regarded as a market-based mechanism for assigning spectrum. The FCC was authorized to organize auctions to award spectrum licenses for certain wireless communications services in the Omnibus Budget Reconciliation Act of 1993 (P.L. 103-66). The act amended the Communications Act of 1934 with a number of important provisions affecting the availability of spectrum. The Licensing Improvement section[114] of the act laid out the general requirements for the FCC to establish a competitive bidding methodology and consider, in the process, objectives such as the development and rapid deployment of new technologies.[115] The law prohibited the FCC from making spectrum allocation decisions based "solely or predominately on the expectation of Federal revenues...."[116] The Emerging Telecommunications Technologies section[117] directed the NTIA to identify not less than 200 MHz of radio frequencies used by the federal government that could be transferred to the commercial sector through auctions.[118] The FCC was directed to allocate and assign these released frequencies over a period of at least ten years, and to reserve a significant portion of the frequencies for allocation after the ten-year time span.[119] Similar to the requirements for competitive bidding, the FCC was instructed to ensure the availability of frequencies for new technologies and services, and also the availability of frequencies to stimulate the development of wireless technologies.[120] The FCC was further required to address "the feasibility of reallocating portions of the spectrum from current commercial and other non-federal uses to provide for more efficient use of spectrum" and for "innovation and marketplace developments that may affect the relative efficiencies of different spectrum allocations."[121] Over time, auction rules have been modified in accordance with the changing policy goals of the FCC and Congress but subsequent amendments to the Communications Act of 1934 have not substantively changed the above-noted provisions regarding spectrum allocation.[122]

The rules set by the FCC for using spectrum licenses (service rules) may have been oriented toward the concepts of building and managing networks that were formed in the days of the telephone, favoring traditional telecommunications business plans over those of companies with different business models. Some companies that might be well suited to meet social goals, such as access in rural areas, might have been precluded from bidding at all because of constraints not considered relevant to market-driven allocations. For example, public utilities, municipal co-operatives, commuter railroads, and other public or quasi-public entities face a variety of legal, regulatory, and structural constraints that limit or prohibit their ability to participate in an auction or buy spectrum licenses. Many of these constraints exist at the state level but federal spectrum policy plays a role in perpetuating the status quo.

Auction winners are deemed to be the companies that can maximize the value of the spectrum to society by maximizing its value as a corporate asset. However, auction-centric spectrum policies appear to have generally focused on assigning licenses to commercial competitors in traditional markets that serve consumers and businesses. Auctioning spectrum licenses may direct assets to end-use customers instead of providing wireless services where the consumer may be the beneficiary but not the customer. Wireless networks are an important component of smart grid communications. Spectrum resources are also needed for railroad safety,[123] for water conservation,[124] for the safe maintenance of critical infrastructure industries,[125] and for many other applications that may not have an immediate commercial value but can provide long-lasting value to society as a whole.

Spectrum Caps

As part of its preparations for the first spectrum license auctions, the FCC decided to set caps on the amount of spectrum any one company could control in any geographically designated market.[126] The theory behind spectrum capping is that each license has an economic value and a foreclosure value. The economic value is derived from the return on investment in spectrum licenses and network infrastructure. The foreclosure value is the value to a wireless company that already has substantial market share and wants to keep its dominant position by precluding competition. Spectrum caps were chosen as the method to prevent foreclosure bidding. The intent was to ensure multiple competitors in each market and to restrict bidding to only the licenses that could be used in the near term.

Beginning in 2001, spectrum policy placed increased emphasis on promoting spectrum and market efficiency through consolidation. The FCC ruled to end spectrum caps, citing greater spectral efficiency from larger networks as one benefit of the ruling. Spectrum caps were seen as barriers to mergers within the wireless industry, to the growth of existing wireless companies, and to the benefits of scale economies. The spectrum caps were eliminated on January 1, 2003.[127] Auction rules requiring the timely build-out of networks became a key policy tool to deter hoarding. The FCC instituted a policy for evaluating spectrum holdings on a market-by-market, case-by-case basis—a practice referred to as spectrum screening—as a measure of competitiveness.

In 2008, the Rural Telecommunications Group, Inc. (RTG) petitioned the FCC to impose a spectrum cap of 110 MHz for holdings below 2.3 GHz. In October 2008, the FCC sought comments on the RTG petition for rulemaking.[128] RTG argued that competition in the industry was declining as it became more concentrated. It claimed that the larger carriers were warehousing their spectrum holdings in rural areas while rural carriers were struggling to acquire spectrum capacity for mobile broadband and expansion. Rural carriers, RTG reported, were being shut out of opportunities to acquire new spectrum holdings and were being outbid in spectrum auctions.[129] Opponents to the spectrum cap cited data to support their claims that the wireless communications market is competitive. They argued that additional amounts of spectrum are needed to support the growth in mobile broadband and that a spectrum cap could cut off growth and innovation.[130] Implementing spectrum caps as a tool for regulating competition would represent a significant shift in policy for the FCC, were it to take that course.

In comments filed regarding the National Broadband Plan, the Department of Justice considered the possibility that "the foreclosure value for incumbents in a given locale could be very high."[131] Although it recognized some form of spectrum caps as an option for assuring new market entrants, it observed that "there are substantial advantages to deploying newly available spectrum in order to enable additional providers to mount stronger challenges to broadband incumbents."[132]

Market Competition

There are many ways to view competition. Although competitiveness may be evaluated by factors such as barriers to entry or number of market participants, a key measure of whether market competition is working is an assessment of the dynamic of a specific market: its prices, variety, level of service, and other indicators that are considered hallmarks of competitive behavior. The Federal Trade Commission, for example, promotes competition as "the best way to reduce costs, encourage innovation, and expand choices for consumers."[133] Viewpoints about the level of competitiveness in providing wireless services to the U.S. market differ.[134] However, telecommunications business analysts generally describe the U.S. market for wireless services as competitive because consumers benefit in many ways from competition on price, service, coverage, and the availability of new devices.

Both the wireless industry and its regulator have focused on "wireless consumer welfare"[135] in evaluating competition and the effectiveness of spectrum policies for assigning spectrum licenses. Auctions are judged to be an efficient way of assigning spectrum for commercial uses that adhere to traditional business plans.[136]

Competition in Rural Markets

Over the years, various legislative and policy initiatives have created a number of requirements to help small and rural carriers acquire spectrum licenses.[137] Some of the FCC's efforts to encourage spectrum license ownership for small, rural, or entrepreneurial businesses are in response to Congressional mandates.[138] These and other statutory and regulatory programs may have allowed many small carriers to remain in business even though many others have been absorbed by larger carriers.[139] As wireless traffic, revenue, and profits migrate to broadband, business models that were effective for voice traffic may no longer be viable, especially for companies that have relied on the regulatory environment to protect their markets. This change in operating environment may have disproportionately affected the ability of rural wireless carriers, in particular, to compete effectively.[140] A study of how new technologies might be affecting the competitiveness of small and rural carriers might be useful in reviewing the effectiveness of policies intended to aid them.[141]

APPENDIX C. INTERNATIONAL POLICIES FOR SPECTRUM MANAGEMENT

Wireless companies also compete as providers in global markets. Although international traffic may be a small part of wireless voice communications, competition in providing services is global.[142] AT&T, Verizon, and T-Mobile are major players internationally as well as in the United States.[143] Corporate decisions such as the introduction of new technologies and services are made for both the United States and international markets. Actions taken for domestic markets may influence decisions made to enhance global competition and vice versa. Therefore, policies for assigning spectrum assets might incorporate U.S. goals for global competiveness.

Spectrum allocation is not a uniquely domestic process. Some spectrum allocations are governed by international treaty. Additionally, there is a trend to harmonize spectrum allocations for commercial use across countries through international agreements. Harmonization of radio frequencies is achieved by designating specific bands for the same category of use worldwide. With harmonization, consumers and businesses are able to benefit from the convenience and efficiency of having common frequencies for similar uses, thus promoting development of a seamless, global communications market. Spectrum allocation at the national level, therefore, is sometimes coordinated with international spectrum allocation agreements. The Advanced Wireless Services (AWS) auction in the United States, completed in 2006,[144] was the conclusion of a process initiated by an agreement for international harmonization of spectrum bands.[145] At this auction, T-Mobile was able to acquire new spectrum licenses that improved its competitiveness in the United States[146] and, consequently, the worldwide competiveness of its owner, Deutsche Telekom.[147]

The International Telecommunications Union (ITU), the lead United Nations agency for information and communication technologies, has been vested with responsibility to ensure interference-free operations of wireless communication through implementation of international agreements.[148] The ITU adopts a Table of Frequency Allocations in conjunction with International Radio Regulations. This International Table allocates spectrum for various radio services and includes, directly or indirectly, conditions for the use of the allocated spectrum.[149] There is also a domestic table for each country. The United States Table of Allocations is maintained by the National Telecommunications and Information Administration (NTIA). The U.S. Table of Allocations is modified to correspond with changes in international spectrum allocations agreed to under the auspices of the ITU. These agreements are reached through processes such as the World Radiocommunications Conferences (WRC). Each WRC provides an opportunity to revise the International Radio Regulations and International Table of Frequency Allocations in response to changes in technology and other factors. Modifications to rules from one WRC to the next are part of an ongoing process of technical review and negotiations. WRC meetings are held approximately every two years. Provisions that require changes in frequency allocation to accommodate new technology will typically take effect 10 to 15 years after agreement is reached. These delays give time to phase out older technologies and to formulate new investment strategies.

The possibility of allocating additional spectrum for mobile broadband was among the deliberations of WRC-07 (October 22-November 16, 2007) and may be considered at the next WRC, scheduled to be held in January 2012.[150] Future decisions about spectrum allocation for broadband in the United States might be influenced by international agreements. Worldwide harmonization of frequencies for mobile broadband may be sought in bands at 3 GHz and higher.

In the NBP, the FCC has briefly discussed its participation in world forums and the role of the ITU in the development of new wireless technologies and services.[151] The NBP has recommended that the FCC should work within the ITU to promote innovative and flexible approaches to global spectrum allocation.[152]

End Notes

[1] Broadband refers here to the capacity of the radio frequency channel. A broadband channel can quickly transmit live video, complex graphics, and other data-rich information as well as voice and text messages, whereas a narrowband channel might be limited to handling voice, text, and some graphics.

[2] Federal Communications Commission, *Connecting America: The National Broadband Plan*, March 17, 2010, at http://download.broadband.gov/plan/national-broadband-plan.pdf.

[3] The Code of Federal Regulations defines natural resources as "land, fish, wildlife, biota, air, water, ground water, drinking water supplies and other such resources belonging to, managed by, held in trust by, appertaining to, or otherwise controlled by the United States.... " (15 CFR 990, Section 990.30).

[4] The International Telecommunication Union (ITU), an agency of the United Nations, is the primary organization for coordinating global telecommunications and spectrum management.

[5] Spectrum allocation and assignment is addressed in **Appendix B**, Competition.

[6] Subscribers are customers who have signed up for a plan, including those with more than one plan subscription; prepaid and pay-as-you go customers may not be included in reported totals. FCC, *Fourteenth Report; annual report and analysis of competitive market conditions with respect to commercial mobile services*, FCC, WT Docket No. 09-66, released May 20, 2010, Table 3, p. 31, reported for second Quarter 2009.

[7] For a discussion of policy issues, see CRS Report R40234, *The FCC's Authority to Regulate Net Neutrality after Comcast v. FCC* , by Kathleen Ann Ruane, and CRS Report R40616, *Access to Broadband Networks: The Net Neutrality Debate* , by Angele A. Gilroy.

[8] PSTN is a global system; rights of access and usage in the United States are regulated by the FCC.

[9] On December 1, 2009, the FCC published a public notice seeking comments on the "appropriate policy framework to facilitate and respond to the market-led transition in technology and services, from the circuit-switched PSTN system to an IP-based communications world." "Comment Sought on Transition from Circuit-Switched Network to All-IP Network," NBP Public Notice #25, DA 09-2517 at http://hraunfoss.fcc.gov/edocs_public/attachmatch/DA-09-2517A1.pdf.

[10] *Ex Parte* Submission of the United States Department of Justice, In the matter of Economic Issues in Broadband Competition: A National Broadband Plan for Our Future, GN Docket 09-51, January 4, 2010, p. 21 at http://fjallfoss.fcc.gov/ecfs/document/view?id=7020355122.

[11] *Connecting America*, Recommendations 5.13 and 5.14. The NBP proposed that the National Science Foundation "should fund wireless research and development that will advance the science of spectrum access." p. 96.

[12] Dynamic Spectrum Access, Content-Based Networking, and Delay and Disruption Technology Networking, along with cognitive radio, and decision-making software, are examples of technologies that can enable Internet-like management of spectrum resources. DSA is part of the neXt Generation program, or XG, a technology development project sponsored by the Strategic Technology Office of the Defense Advanced Research Projects Agency (DARPA). The main goals of the program include developing both the enabling technologies and system concepts that dynamically redistribute allocated spectrum.

[13] *Connecting America,* Recommendation 5.11.

[14] *Connecting America*, Recommendation 5.13.

[15] Letter to the FCC, Re: National Broadband Plan, GN Doc. No. 09-51, January 4, 2010 at http://www.ntia.doc.gov/ filings/2009/FCCLetter_Docket09-51_20100104.pdf.

[16] Office of Management and Budget, *Budget of the U.S. Government, Fiscal Year 2011, Appendix,* "Other Independent Agencies," p. 1263. See also, FCC, *Fiscal Year 2011 Budget Estimates Submitted to Congress,* February 2010 at http://hraunfoss.fcc.gov/edocs_public/attachmatch/DOC-296111A1.pdf.

[17] See Spectrum Management Advisory Committee website at http://www.ntia.doc.gov/advisory/spectrum/.

[18] P.L. 111-5, Division B, Title VI, Sec. 6001 (k); 123 STAT. 515.

[19] Available at http://www.broadband.gov/plan/.

[20] Radio Spectrum Inventory Act introduced in the Senate (S. 649, Kerry) and the similar House-introduced Radio Spectrum Inventory Act (H.R. 3125, Waxman).

[21] *Connecting America*, Recommendation 5.1.

[22] For more information on the Spectrum Dashboard, go to http://reboot.fcc.gov/reform/systems/spectrum-dashboard/ about.

[23] *Connecting America*, Recommendation 5.2.

[24] *Connecting America*, Recommendation 5.3.

[25] *Connecting America*, Recommendation 5.8.

[26] *Connecting America*, Recommendation 5.8.5.

[27] *Connecting America*, p. 89.

[28] Background information regarding the D Block is provided in CRS Report R40859, *Public Safety Communications and Spectrum Resources: Policy Issues for Congress*, by Linda K. Moore

[29] *Connecting America*, Recommendation 5.8.2.

[30] A bill that would assign the D Block for public safety communications has been introduced (Broadband for First Responders Act of 2010, H.R. 5081, King).

[31] FCC News, "The Federal Communications Commission Establishes New Emergency Response Interoperability Center," April 23, 2010 at http://hraunfoss.fcc.gov/edocs_public/attachmatch/DOC-297707A1.pdf.

[32] NIST, "Demonstration Network Planned for Public Safety 700 MHz Broadband," December 15, 2009 at http://www.nist.gov/eeel/oles/network_121509.cfm.

[33] FCC News, "The Federal Communications Commission Establishes New Emergency Response Interoperability Center," April 23, 2010.

[34] FCC, *Notice of Proposed Rulemaking*, WT Docket No. 07-195, released September 19, 2007.

[35] See for example, comments by T-Mobile USA, Inc. filed July 25, 2008, FCC, Docket No. 07-195.

[36] *Connecting America*, Recommendation 5.8.3.

[37] Remarks of Lawrence E. Strickling, June 3, 2010 at http://www.ntia.doc.gov/presentations/2010/PublicKnowledge_Spectrum_06032010.html.

[38] FCC, Public Notice, "Office of Engineering and Technology Requests Information on Use of 1675-1710 MHz Band," DA 10-1035, released June 4, 2010, Docket No. 10-123 at http://fjallfoss.fcc.gov/edocs_public/attachmatch/ DA-10-1035A1.pdf.

[39] Connecting America, p. 79 and Recommendation 5.7.

[40] The practice of transferring a wireless call from one network to another—or roaming—is described in *Understanding Wireless Telephone Coverage Areas,* FCC Consumer Facts at http://www.ifap.ru/library/book385.pdf.

[41] For information, see Auction 73 at http://wireless.fcc.gov/auctions/default.htm?job=auction_summary&id=73.

[42] FCC filings, WT Docket No. 96-86, by Frontline Wireless, LCC, Google, Inc., the 4G Coalition, and the Public Interest Spectrum Coalition.

[43] Comments, for example, made by Ram Shriram and Vanu Bose at the Frontline Town Hall, July 12, 2007, Washington, DC, and by Jason Devitt at a panel discussion during the State of the Net conference, January 30, 2008, Washington, DC.

[44] Of the 10 licenses of the C Block, seven were auctioned to Verizon Wireless: all six licenses covering the continental United States and a seventh license for Hawaii. Licenses providing coverage for Alaska, Puerto Rico, and the Gulf of Mexico were won by other bidders. See "FCC 700 MHz Band Auction, Auction ID:73, Winning Bids," attachment A, p. 63, at http://hraunfoss.fcc.gov/edocs_public/attachmatch/DA-08-595A2.pdf.

[45] Petition for Reconsideration of Frontline Wireless, LLC, WT Docket No. 96-86.

[46] FCC filings, WT Docket No. 96-86, by CTIA-The Wireless Association, AT&T, and others.

[47] FCC, *First Report and Order and Further Notice of Proposed Rule Making*, ET Docket No. 04-186, released October 18, 2006, at http://fjallfoss.fcc.gov/edocs_public/attachmatch/FCC-06-156A1.pdf.

[48] Geolocation associates a geographic location with a device using embedded information such as an IP address, Wi-Fi address, GPS coordinates, or other, perhaps self-disclosed information. Geolocation usually works by automatically looking up an IP address.

[49] FCC, Public Notice, "Office of Engineering and Technology Invites Proposals from Entities Seeking to be Designated TV Band Device Database Managers," ET Docket No. 04-186, released November 25, 2009 at http://fjallfoss.fcc.gov/edocs_public/attachmatch/DA-09-2479A1.pdf.

[50] In addition to filed comments with the FCC. NAB, the Association for Maximum Service Television, and a coalition of theater groups, sports leagues, and TV networks have challenged the FCC white spaces order in the U.S. Court of Appeals for the District of Columbia. Requirements intended to protect microphone use in the white spaces are proposed in the Wireless Microphone Users Interference Protection Act (H.R. 4353, Representative Rush).

[51] FCC, *Second Report and Order and Memorandum Opinion and Order*, ET Docket No. 04-185, released November 14, 2008 at http://fjallfoss.fcc.gov/edocs_public/attachmatch/FCC-08-260A1.pdf.

[52] *Connecting America*, Recommendation 5.12.

[53] A wireless analog signal uses a continuous transmission form. Digital signals are discontinuous (discrete) transmissions.

[54] WiMAX stands for Worldwide Interoperability for Microwave Access.

[55] Key technologies for mobile broadband are summarized in **Appendix A,** Spectrum-Hungry Technologies.

[56] A leading advocate for replacing channel management of radio frequency with network-centric management is Preston Marshall, the source for much of the information about network-centric technologies in this chapter. Mr. Marshall is Director, Information Sciences Institute, University of Southern California, Viterbi School of Engineering, Arlington, Virginia.

[57] *Connecting America,* Recommendation 5.11.

[58] *Connecting America*, Recommendation 5.13.

[59] *Connecting America*, Recommendation 5.6.

[60] *Connecting America*, p. 82.

[61] Office of Management and Budget, *Budget of the U.S. Government, Fiscal Year 2011, Appendix,* "Other Independent Agencies," p. 1263. See also, FCC, *Fiscal Year 2011 Budget Estimates Submitted to Congress,* February 2010 at http://hraunfoss.fcc.gov/edocs_public/attachmatch/DOC-296111A1.pdf.

[62] Office of Management and Budget, *Budget of the U.S. Government, Fiscal Year 2011,* Summary Tables, Table S-8, p. 169.

[63] Office of Management and Budget, *A New Era of Responsibility: Renewing America's Promise,* Table S-6, p. 126.

[64] For example, the President's budget for FY2004 and again for 2006 proposed that (1) the FCC's authority to conduct auctions be extended indefinitely; (2) user fees be levied on unauctioned licensed spectrum; and (3) broadcasters pay an annual lease fee on analog TV spectrum that they are holding as part of the Congressionally-mandated transition to digital television. In his budget for 2005, the President supported proposals for indefinitely extending the FCC's auction authority and giving the FCC the authority to set user fees on unauctioned spectrum.

[65] P.L. 108-494, Title II, Sec. 209 (a).

[66] GAO, *Telecommunications: Strong Support for Extending FCC's Auction Authority Exists, but Little Agreement on Other Options to Improve Efficient Use of Spectrum,*" December 20, 2005, GAO-06-236, p. 2.

[67] *Ibid.,* p. 10, footnote 15.

[68] 47 USC § 158 (a).

[69] 47 USC § 159 (a).

[70] *Connecting America,* p. 93.

[71] *Connecting America,* Recommendations 5.9 and 5.10.

[72] FCC, *Report and Order,* released June 8, 2010, Docket No. 09-114 at http://fjallfoss.fcc.gov/edocs_public/attachmatch/FCC-10-109A1.pdf.

[73] P.L. 111-5, § 6001 (k) (2) (D); 123 Stat. 516.

[74] P.L. 110-140, Sec. 1301; 123 Stat. 1783.

[75] *Connecting America,* Recommendation 17.1.

[76] *Connecting America,* Recommendation 17.2.

[77] The Federal Trade Commissions' Internet Access Task Force has published a report discussing many aspects of municipal broadband implementation and related issues, at http://www.ftc.gov/opa/2006/10/muniwireless.htm.

[78] U.S. Supreme Court, Docket Number 02-1238.

[79] 47 U.S.C. 253 (a).

[80] *Connecting America,* Recommendation 14.1.

[81] *Connecting America,* Recommendation 14.2, 14.3, and 14.4.

[82] The status of the bills can be monitored through the Legislative Information System (LIS) at http://www.congress.gov.

[83] *Connecting America,* Recommendation 5.5.

[84] FCC News, "FCC's Advanced Wireless Services (AWS) Spectrum Auction Concludes," September 18, 2006.

[85] FCC Public Notice "Coordination Procedures in the 1710-1755 MHz Band," FCC 06-50, April 20, 2006 (WTB Docket No. 02-353).

[86] *Connecting America,* Recommendation 5.4.

[87] As required by the Communications Act of 1934, 47 U.S.C. 309 j (8).

[88] The original Communications Act of 1934 codified many regulations for monopolies as practiced at the time.

[89] International Telecommunications Union projects an estimated need for additional spectrum capacity that could reach nearly 1,000 MHz in the United States, as reported in "Summary of Results of ITU-R Report M. 2079," p. 13, presented by Cengiz Evci, Chief Frequency Officer, Wireless Business Group, Alcatel-Lucent, August 28, 2007. Available at http://standards.nortel.com/spectrum4IMT/Geneva/R03-WRCAFR07-C-0024.pdf. See also CTIA-The Wireless Association, *Written Ex Parte Communication,* FCC, GN Docket No. 09-51, September 29, 2009, which suggests a goal of at least 800 MHz, based on extrapolations from the ITU research.

[90] See, for example, "Mobile Broadband Evolution: the roadmap from HSPA to LTE," UMTS Forum, February 2009, Universal Mobile Telephone System Forum at http://www.umts-forum.org/.

[91] Implementation summarized in *Connecting America,* Exhibit 5-B, p. 77.

[92] Spectrum is segmented into bands of radio frequencies and typically measured in cycles per second, or hertz. Standard abbreviations for measuring frequencies include kHz—kilohertz or thousands of hertz; MHz—megahertz, or millions of hertz; and GHz—gigahertz, or billions of hertz.

[93] Founding members of the WiMAX Forum include Airspan, Alvarion, Analog Devices, Aperto Networks, Ensemble Communications, Fujitsu, Intel, Nokia, Proxim, and Wi-LAN. For additional information, see http://www.wimaxforum.org/.

[94] Implementation summarized in *Connecting America,* Exhibit 5-B, p. 77.

[95] A discussion of backhaul technology is part of the testimony of Ravi Potharlanka, Chief Operating Officer, Fiber Tower Corp., at House of Representatives, Committee on Energy and Commerce, Subcommittee on

Communications, Technology, and the Internet, "An Examination of Competition in the Wireless Industry," May 7, 2009.

[96] Unlicensed frequencies are bands set aside for devices approved by the FCC. The frequencies are effectively managed by the FCC instead of by a license-holder.

[97] "Wi-Fi Popular Now in Smartphones, Set to Boom," by Matt Hamblen, Computerworld, April 1, 2009.

[98] See, for example, "Smart Phones are Edging Out Other Gadgets," by Christopher Lawton and Sara Silver, The Wall Street Journal, March 25, 2009, for a discussion of how "beefed up cellphones" are replacing some electronic devices as their functions are incorporated into smart phones.

[99] The many factors driving demand for mobile broadband and the impact of growth in data and video services on demand for spectrum are reviewed in Mobile Spectrum Broadband Demand, Rysavy Research, December 2008.

[100] A discussion of the goals of NCO is included in CRS Report RL32411, Network Centric Operations: Background and Oversight Issues for Congress, by Clay Wilson.

[101] Information at http://jpeojtrs.mil/.

[102] Information about the WNaN program is based on comments by Bob Wilson, Deputy Program Manager for Army WNaN program, Communications-Electronics Command at DoD Spectrum Symposium, Arlington, VA, October 14-15, 2009.

[103] Facilities-based mobile telephone operators own and operate their network facilities.

[104] Fourteenth Report, paragraph 27.

[105] Fourteenth Report, paragraph 29.

[106] For example, U.S. Department of Justice and the Federal Trade Commission, "Horizontal Merger Guidelines," Jointly issued April 2, 1992, revised April 8, 1997.

[107] FCC, "Wireless Telecommunications Bureau Seeks Comment on Commercial Mobile Radio Services Market Competition," Public Notice, February 25, 2008, DA 08-453, WT Docket No. 08-27 at http://hraunfoss.fcc.gov/ edocs_public/attachmatch/DA-08-453A1.pdf. Earlier annual reports have also cited these barriers.

[108] The initial occupant of a market segment may benefit from a number of advantages such as preemption of resources, advantageous relationships with customers and suppliers, and early profits for reinvestment in infrastructure.

[109] The distribution of licenses for cell phone networks from the early days of the technology until the introduction of auctions is described in Wireless Nation: The Frenzied Launch of the Cellular Revolution in America, by James B. Murray, Jr., Perseus Press, 2001, 2002.

[110] 47 U.S.C. § 151.

[111] P.L. 632, Sec. 3.

[112] P.L. 264, "License."

[113] An "Act to regulate radio communications," usually referred to as the Radio Act of 1912, was passed partly in response to radio problems—including interference—associated with the sinking of the Titanic. Hearings Before a Subcommittee of the Committee on Commerce, 62nd Congress, 2nd Session, pursuant to S. Res. 283, "Directing the Committee on Commerce to Investigate the Cause Leading to the Wreck of the White Star Liner 'Titanic,'" testimony of Guglielmo Marconi, et al.

[114] P.L. 103-66 Title III, Subtitle C, Chapter 1.

[115] 47 U.S.C. § 309 (j), especially (1), (3), and (4).

[116] 47 U.S.C. § 309 (j) (7) (A).

[117] P.L. 103-66 Title III, Subtitle C, Chapter 2.

[118] 47 U.S.C. § 923 (b) (1).

[119] 47 U.S.C. § 925 (b) (1).

[120] 47 U.S.C. § 925 (b) (2).

[121] 47 U.S.C. § 925 (b) (3).

[122] See United States Code Annotated, Title 47, sections as footnoted, WEST Group, 2001 and the 2007 Cumulative Annual Pocket Part.

[123] The railroad industry uses wireless communications as part of their information networks to monitor activity.

[124] For example, sensors buried at the level of plant roots recognize when watering is needed and communicate this information over wireless networks.

[125] In general, critical infrastructure industries facilitate the production of critical goods and services such as safe drinking water, fuel, telecommunications, financial services, and emergency response. A discussion of key issues appears in CRS Report RL30153, Critical Infrastructures: Background, Policy, and Implementation, by John D. Moteff.

[126] Licenses are designated for a specific geographic area, such as rural areas, metropolitan areas, regions, or the entire nation.

[127] FCC News, "FCC Announces Wireless Spectrum Cap to Sunset Effective January 1, 2003," November 8, 2001. Report and Order FCC-01-328. See Docket No. 01-14, Notice of Proposed Rulemaking, released January 23, 2001 at http://hraunfoss.fcc.gov/edocs_public/attachmatch/FCC-01-28A1.pdf.

[128] FCC RM No. 11498, October 10, 2008. Comments supporting and opposing the petition are published in this proceeding.

[129] Those supporting the RTG petition included the Organization for the Promotion and Advancement of Small Telecommunications Companies (OPASTCO), the National Telecommunications Cooperative Association, the Public Interest Spectrum Coalition, and a number of smaller (non-dominant) wireless carriers.

[130] Opponents to spectrum caps that filed comments were AT&T Inc., Verizon Wireless, CTIA – The Wireless Association, the Telecommunications Industry Association, and the Wireless Communications Association International.

[131] *Ex Parte* Submission of the United States Department of Justice, In the matter of Economic Issues in Broadband Competition: A National Broadband Plan for Our Future, GN Docket 09-51, January 4, 2010, p. 23 at http://fjallfoss.fcc.gov/ecfs/document/view?id=7020355122.

[132] Ibid., p. 24.

[133] "Competition in the Technology Marketplace" at http://www.ftc.gov/bc/tech/index.htm.

[134] Different assessments of competition in the wireless market have been filed as comments in FCC Docket No. 09-66, part of the process for the preparation of the FCC's *Fourteenth Report; annual report and analysis of competitive market conditions with respect to commercial mobile services.*

[135] This phrase is used in the written statement of AT&T Inc. submitted for a hearing before the House of Representatives, Committee on Energy and Commerce, Subcommittee on Communications, Technology, and the Internet, "An Examination of Competition in the Wireless Industry," May 7, 2009. In written testimony submitted by Verizon Wireless for the same hearing, comments stated that wireless providers need suitable and sufficient spectrum because of "consumers' reliance on broadband services."

[136] The GAO has reported this viewpoint in several reports, including *Telecommunications: Strong Support for Extending FCC's Auction Authority Exists, but Little Agreement on Other Options to Improve Efficient Use of Spectrum,*" December 20, 2005, GAO-06-236 and *Telecommunications: Options for and Barriers to Spectrum Reform,* March 14, 2006, GAO-06-526T.

[137] For example, most auctions have provided bidding credits for small businesses.

[138] In 47 USC § 309 (j) (3) (B), the FCC is instructed to promote "economic opportunity and competition and ensuring that new and innovative technologies are readily available to the American people by avoiding excessive concentration of licenses and by disseminating licenses among a wide variety of applicants...."

[139] The Congressional Budget Office (CBO) reported in a 2005 study that a significant number of small companies that acquired spectrum licenses through preferential programs later transferred the licenses to larger companies: *Small Businesses in License Auctions for Wireless Personal Communications Services,* A CBO Paper, October 2005, at http://www.cbo.gov/ftpdocs/68xx/doc6808/10-24-FCC.pdf.

[140] A number of rural wireless carriers and their associations have filed comments on the increasing difficulties they face in competing for wireless customers. Comments are in a number of FCC dockets, such as RM11498, regarding spectrum caps, and WT Docket No. 09-66, on the state of wireless competition.

[141] The CBO study cited above was prepared at the request of the Senate Budget Committee to examine the impact of small-bidder preferences on federal revenue and was completed before data traffic became a significant factor in providing wireless services.

[142] The international framework for spectrum management and wireless competition is summarized in Appendix C, International Policies for Spectrum Management.

[143] Verizon Wireless is 45% owned by the British telecommunications giant Vodafone, PLC. T-Mobile is 100% owned by Deutsche Telecom.

[144] FCC News, "FCC's Advanced Wireless Services (AWS) Spectrum Auction Concludes," September 18, 2006.

[145] The WRC-2000 agreed on spectrum bands to be harmonized for advanced wireless services, referred to as IMT 2000. See FCC News, "International Bureau Reports on Success of the 2000 World Radio Communications Conference," June 8, 2000, http://www.fcc.gov/Bureaus/International/News_Releases/2000/nrin0009.html.

[146] FCC, *Twelfth Report; annual report and analysis of competitive market conditions with respect to commercial mobile services,* Docket No. 07-71, released February 4, 2008, p. 9 and paragraph 75, at http://hraunfoss.fcc.gov/ edocs_public/attachmatch/DA-08-453A1.pdf.

[147] Deutsche Telekom owns 100% of T-Mobile International, which includes T-Mobile USA. For information see "Global Player on the Mobile Communications Market" at http://www.telekom .com/dtag/cms/content/dt/en/530494.

[148] The GAO notes that "The federal government considers ITU the principal, competent, and appropriate international organization for the purpose of formulating international treaties and understandings regarding certain telecommunications matters*." Better Coordination and Enhanced Accountability Needed to Improve Spectrum Management,* GAO-02-906, September 2003, p. 19, fn. 26.

[149] There are 39 internationally defined wireless services, including broadcasting, meteorological satellite, and mobile services. Description of ITU-R functions are at http://www.itu.int/ITU-/index.asp?category=information&rlink= rhome&lang=en.

[150] The NTIA and FCC websites carry information about planning for WRC 2012. For FCC, see IB Docket No. 04-286, Public Notice at http://hraunfoss.fcc.gov/edocs_public/attachmatch/DA-09-763A1.pdf. An NTIA

overview is at http://www.ntia.doc.gov/osmhome/wrc/ntia.html. The ITU site is at http://www.itu.int/ITU-R/index.asp?category= conferences&rlink=wrc-11&lang=en.

[151] The International Telecommunications Union (ITU) is considering how policies and regulations may need to be changed in response to new technologies. A World Telecommunication Policy Forum in April 2009, organized by the ITU, addressed these and other topics. See http://www.itu.int/osg/csd/wtpf/wtpf2009/about.html.

[152] *Connecting America,* Recommendation 5.16.

In: The National Broadband Plan: Analysis and Strategy for...   ISBN: 978-1-61122-024-7
Editor: Daniel M. Morales   © 2011 Nova Science Publishers, Inc.

Chapter 4

THE EVOLVING BROADBAND INFRASTRUCTURE: EXPANSION, APPLICATIONS, AND REGULATION

*Patricia Moloney Figliola, Angele A. Gilroy and
Lennard G. Kruger*

SUMMARY

Over the past decade, the telecommunications sector has undergone a vast transformation fueled by rapid technological growth and subsequent evolution of the marketplace. Much of the U.S. policy debate over the evolving telecommunications infrastructure is framed within the context of a "national broadband policy." The way a national broadband policy is defined, and the particular elements that might constitute that policy, determine how and whether various stakeholders might support or oppose a national broadband initiative. The issue for policymakers is how to craft a comprehensive broadband strategy that not only addresses broadband availability and adoption problems, but also addresses the long term implications of next-generation networks on consumer use of the Internet and the implications for a regulatory framework that must keep pace with evolving telecommunications technology.

Consumers have been integrating communications technologies into their lives at unprecedented rates. Trends include increased use of smartphones, increased subscribership on social networking sites such as Facebook and MySpace, increased expectations of cross-platform accessibility, and development of "cloud computing" applications. Each of these trends taken alone likely would have had a significant impact on consumer behavior, but taken together they create a heretofore unseen demand for real-time access to information and an ability to share that information from wherever the consumer happens to be. Policy choices related to consumer use of the Internet, such as user authentication, privacy, digital rights management, filtering of unwanted information, wireless Internet standards, instant messaging, the deployment of IPv6 ("Internet protocol version 6"), and how to link the telephone network to the Internet will all have a profound impact on how broadband and next generation networks evolve.

The challenge facing today's policymakers is to develop a regulatory environment that not only addresses these more recent trends, but that also contains the flexibility to accommodate future and possibly unanticipated changes in technology, applications, and consumer demands. The growth of broadband networks and the proliferation of applications and devices has placed increasing pressure on policy makers to formulate a framework to address a broadband-based world. Many of these developments were not anticipated when the 1996 Telecommunications Act (P.L. 104-104) was passed and have led to the need to update the regulatory assumptions and subsequent regulatory framework upon which the act was based.

Technological changes such as the advancement of Internet technology and the melding of data, voice, and video have resulted in additional trends which must be considered. These trends include the transition from a circuit switched to a packet switched network, thereby enabling the integration of voice, video, and data; the transition from fixed to mobile service; and the transition from one-way to interactive service. Additionally, as broadband becomes an integral component of society, regulators have been called upon to consider how these trends may affect social goals that may or may not have been associated with traditional telephony. Social objectives such as the advancement of universal service goals, timely and accurate emergency services, disability access, and consumer protection that are part of traditional telephony regulatory policies are migrating to the broadband policy environment.

INTRODUCTION

Over the past decade, the telecommunications sector has undergone a vast transformation fueled by rapid technological growth and subsequent evolution of the marketplace. A wide range of new services have become available, offered by a growing list of traditional as well as nontraditional providers. One of the results of this transformation is that the nation's expectations for communications services have also grown.

For nearly a century, access to the public switched network through a single wireline connection, enabling voice service, was the standard of communications. Today the desire for simple voice connectivity has been replaced by the demand, on the part of consumers, business, and government, for access to a vast array of multifaceted fixed and mobile services. Consumers are also demanding greater flexibility and may choose to gain access to identical content over a variety of technologies, whether it be a computer, a television, or a mobile telephone. The trend towards sharing information, such as music, movies, or photographs, is also growing, making it necessary to ensure that network upload speeds match download capabilities. These advances require that networks transition into converged next-generation wireline and wireless broadband networks capable of meeting these demands.

Much of the policy debate over the evolving telecommunications infrastructure is framed within the context of a "national broadband policy." The issue for policymakers is how to craft a comprehensive broadband strategy that not only addresses broadband availability and adoption problems, but also addresses the long term impacts of next-generation networks on consumer use of the Internet and a regulatory framework that must keep pace with evolving telecommunications technology.

EVOLUTION OF BROADBAND IN THE UNITED STATES

Prior to the late 1990s, American homes accessed the Internet at maximum speeds of 56 kilobits per second by dialing up an Internet Service Provider (such as AOL) over the same copper telephone line used for traditional voice service. A relatively small number of businesses and institutions used broadband or high speed connections through the installation of special "dedicated lines" typically provided by their local telephone company. Starting in the late 1990s, cable television companies began offering cable modem broadband service to homes and businesses. This was accompanied by telephone companies beginning to offer DSL (digital subscriber line) service (broadband over existing copper telephone wireline). Figure 1 shows the Federal Communication Commission's (FCC's) tracking of high-speed lines[1] in the United States between December 1999 (the initial broadband deployment data point reported) and December 2007 (the most recent data available). Growth has been steep, rising from 2.8 million high speed lines reported as of December 1999 to 121.2 million lines as of December 31, 2007. Of the 121.2 million high speed lines reported by the FCC, 74.0 million serve residential users.[2] Since the initial deployment of residential broadband in the United States, the primary residential broadband technologies deployed continue to be cable modem and DSL.

December 2008 survey data from the Pew Internet and American Life Project found that 57% of Americans have broadband at home.[3] It is estimated that less than 10% of U.S. households have no access to any broadband provider whatsoever (not including satellite).[4] While the broadband *adoption* or *penetration* rate stands at close to 60% of U.S. households, broadband *availability* is much higher, at more than 90% of households. Thus, approximately 30% of households have access to some type of terrestrial (non-satellite) broadband service, but do not choose to subscribe. According to the FCC, possible reasons for the gap between broadband availability and subscribership include the lack of computers in some homes, price of broadband service, lack of content, and the availability of broadband at work.[5] According to Pew, non-broadband users tend to be older, have lower incomes, have trouble using technology, and may not see the relevance of using the Internet to their lives. Between 2007 and 2008, low income Americans (under $20,000 annual income) and African Americans showed no significant growth in home broadband adoption after strong growth in previous years.[6] Pew also found that about one-third of adults without broadband cite price and availability as the reasons why they don't have broadband in their homes, while two-thirds cite reasons such as usability and relevance.[7]

Broadband speeds (and prices) are important factors that can determine which technologies are deployed, which applications will be enabled, and how widespread deployment will be. The FCC's fifth and latest "706 report," which is prepared pursuant to section 706 of the Telecommunications Act of 1996 to periodically determine whether broadband is being deployed in a reasonable and timely fashion, found that, "In the future, we anticipate ever-greater demand for services and applications requiring greater bandwidth over an ever-expanding area."[8] Table 1 shows a compilation by the California Broadband Task Force showing different broadband speed ranges and the applications they make possible. Table 2 shows advertised speed ranges offered by different broadband technologies that are currently commercially available.

As part of any discussion over national broadband policy, a distinction is often made by industry and policymakers between "current generation" and "next generation" broadband (commonly referred to as next generation networks or NGN). "Current generation" typically refers to currently deployed cable, DSL, and many wireless systems, while "next generation" refers to dramatically faster download and upload speeds offered by fiber technologies and also potentially by future generations of cable, DSL, and wireless technologies.

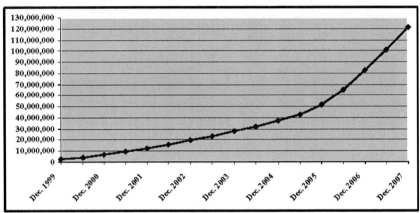

Source: FCC.

Figure 1. Total High-Speed Lines in the United States

Table 1. Broadband Speeds and Applications

Upstream and Downstream Speeds	Applications
500 kbps - 1 Mbps	voice over IP, SMS, basic email, web browsing (simple sites), streaming music (caching), low quality video (highly compressed)
1 Mbps - 5 Mbps	web browsing (complex sites), email (larger size attachments), remote surveillance, IPTV-SD (1-3 channels), file sharing (small/medium), telecommuting (ordinary), digital broadcast video (1 channel), streaming music
5 Mbps - 10 Mbps	telecommuting (converged services), file sharing (large), IPTV-SD (mul-tiple channels), switched digital video, video on demand SD, broadcast SD video, video streaming (2-3 channels), HD video downloading, low definition telepresence, gaming, medical file sharing (basic), remote diagnosis, (basic), remote education, building control & management
10 Mbps - 100 Mbps	telemedicine, educational services, broadcast video SD and some HD, IPTV-HD, gaming (complex), telecommuting (high quality video), high quality telepresence, HD surveillance, smart/intelligent building control
100 Mbps - 1 Gbps	HD telemedicine, multiple educational services, broadcast video full HD, full IPTV channel support, video on demand HD, gaming (immersion), remote server services for telecommuting
1 Gbps - 10 Gbps	research applications, telepresence using uncompressed HD video streams, live event digital cinema streaming, telemedicine remote control of scientific/medical instruments, interactive remote visualization and virtual reality, movement of terabyte datasets, remote supercomputing

Source: California Broadband Task Force, January 2008. Available at http://www.calink.ca.gov /pdf/CBTF_FINAL_Report.pdf.

Table 2. Advertised Broadband Speeds by Technology

Technology	Advertised Broadband Product Speed Ranges (downstream rate)
Mobile Wireless	200 kbps - 1.4 Mbps
DSL	384 kbps - 6 Mbps
Satellite	512 kbps - 2 Mbps
Fixed Wireless	768 kbps - 3 Mbps
Cable	768 kbps - 15 Mbps
Fiber-to-the-home	1 Mbps - 50 Mbps

Source: California Broadband Task Force, January 2008.

In general, more sophisticated (and potentially valuable) applications are available with faster download and upload connection speeds. The most recent FCC broadband status report to Congress characterized future advances in broadband networks as follows:

At the same time that broadband demand increases, network technology continues to evolve and improve. Previously distinct networks are now converging and overlapping to form competing broadband networks that perform all of the network applications once only possible by purchasing services from multiple service providers. Competition between broadband platform providers attempting to keep up with their competitors will drive higher speed technologies and service offerings to the marketplace. Coverage too will continue to become more ubiquitous as a diversity of technologies mature.[9]

Subsequently, as increasingly sophisticated and innovative applications become enabled, the impacts on consumers, the economy, and society become potentially more profound and far- reaching.

THE CONSUMER-ORIENTED INTERNET

Businesses and government have always had a stake in ensuring they have reliable communications available for their employees and the Internet has helped extend the reach of those networks. However, having been given a glimpse at the services available to them at work, consumers began to demand similar services for personal use. Just how are consumers using the Internet today? How did the Internet evolve from a government experiment to share computer resources to a consumer "destination"?[10]

Access to e-mail, text messages, and social networks have become at least as imperative for a significant number of users as the original intent of the cell phone – voice calls. In the days leading up to and following the 2008 presidential election, the press made much of then-Senator Barack Obama's reliance on his Blackberry. Commentators wondered aloud whether he would become the first "connected President." President Obama initially said that he would reluctantly relinquish his Blackberry upon assuming office due to security concerns, but eventually reached a compromise and will have a new secure Blackberry for his correspondence. President Obama's decision illustrates the extent to which Americans have come to rely on "24-7" access to their information.

A Day in the Life[11]

When her Blackberry[12] alarm goes off, "Tina" – a typical "innovative adopter"[13] of technology products – grabs her Blackberry off the nightstand and, before she's even out of bed, glances at the screen to see what types of messages may have come in during the night: e-mail, "text,"[14] or Facebook. After scrolling through her e-mails to see if any urgent work messages have come in (her work voicemail system also sends her an e-mail with the content of any messages), she checks her calendar for the day, which was e-mailed to her overnight, and her friends' most recent Facebook "status updates." Just to make sure she's ready for the day, she checks the weather forecast before heading to the kitchen to get her coffee.

Once at work, Tina logs onto her computer and brings up her work e-mail and her Web browser, opening up three windows automatically – her workplace homepage, the *Washington Post*, and her Gmail. On some days, she may also bring up a Web-based unified "chat" program, such as IMO,[15] and her Facebook page.

Flexible work arrangements at her office – which became possible because of greater technology adoption by employees – allow Tina to do some of her work from home. This day, Tina needs to be home for a technician to install a new speaker system throughout her house. Before leaving the office for the day, therefore, Tina needs to decide which project or projects she will "take home" with her. Previously, Tina may have taken home her files on a floppy disk or a flash drive, or perhaps e-mailed her files to herself. Today, however, Tina can work across platforms and locations using "cloud computing" applications, such as Google Docs. After she uploads her documents, she heads home and works for about another two hours on her desktop computer in her home office. When her work is done, she simply saves her work to her Google Docs account and can be assured she will be able to access it again in the morning from the office.

Once the work day is over, Tina settles onto her couch for a full evening of mostly online activities: "socializing" with her friends via her ultra-compact "Netbook."[16] Most of Tina's online social communication takes place via her Facebook account, which allows her to chat online with her friends, play Scrabble and other games, and comment on her friends' activities and postings—in fact, communicating with her friends has become a new and richer form of entertainment.

Tina will likely also watch TV (or other video entertainment) and perform other online activities, such as

- checking her stock portfolio and making online trades;[17]
- checking her bank or credit card account to pay bills;[18]
- shopping for clothes, books, or other items;[19]
- ordering her groceries; and
- searching for health information regarding a medical procedure she has been discussing with her doctor.[20]

This particular night, Tina also accesses two e-government sites, the first to pay a parking ticket and another to find information on a city board vacancy she is interesting in filling. Additionally, she will spend some time doing research for a distance learning class she is taking from a university in another state.

THE INTERNET EVERYWHERE: WHO

62% of all Americans are part of a wireless, mobile network

58% of Americans adults have used a cell phone or smart phone to do at least one of ten mobile non-voice activities, such as texting, e-mailing, taking a picture, or looking for a map or directions.

41% of American adults have logged onto the Internet "on the go," that is, away from home or work either with a wireless laptop connection or handheld device.

15% of American teens own a smartphone with Internet access.

Source: Pew Internet & American Life, "Mobile Access to Data and Information," March 2008 (Data collected December 2007).

This particular night, Tina also accesses two e-government sites, the first to pay a parking ticket and another to find information on a city board vacancy she is interesting in filling. Additionally, she will spend some time doing research for a distance learning class she is taking from a university in another state.

In deciding what to watch on TV, Tina can choose from her regular cable line-up, her recorded programs on her TiVo, or an instantly downloadable movie from her Netflix account to her television over her Roku box. While Tina watches her program and "chats" online, she receives an incoming video call from her mother, who is coming to visit. She pauses her program and takes the call. Afterwards, she logs onto two different travel sites to search for the best flight deal for her mother's visit and purchases a ticket.[21]

THE INTERNET EVERYWHERE: HOW

Computers: Desktop, Laptop/Notebook,Subnotebook/"Netbook"

Smartphones: Blackberry (Various devices), Palm (Various devices), Apple iPhone, Samsung (Various devices)

Discussion

As illustrated in the above examples, always-on[22] broadband enables disintermediation between the consumer and the product sought, eliminating the "middle man" that was previously required to obtain a good or service: a stock broker, a bank, a medical professional (for information), a travel agent, a store, or a university, to name a few. Instead, the consumer can search for and obtain information, services, and products (and, when appropriate, compare them) for oneself.

For example, with respect to "E-Health," the Pew Internet Project estimated in the Fall of 2008 that between 75% and 80% of Internet users have looked online for health information. In fact, home broadband users are twice as likely as home dial-up users to do health research on a typical day – 12% vs. 6%.[23] Further, 75% of consumers with a chronic condition say their last health search affected a decision about how to treat an illness or condition, compared with 55% of patients without a chronic condition. Newly diagnosed individuals and those who have experienced a health crisis in the past year are also particularly likely to use information found online: 59% say the information they found online led them to ask a doctor new questions or get a second opinion, compared with 48% of those who had not had a recent diagnosis or health crisis. Some 57% of recently diagnosed e-patients say they felt eager to share their new health or medical knowledge with others, compared with 45% of other patients.[24]

In addition to always-on broadband at home, many consumers also rely on their mobile devices to maintain contact with friends and colleagues and to search for information while away from their homes or offices. For example, if it had been a Friday night instead of a weeknight, Tina may have headed straight out from work to meet friends. In that case, she might have posted her plans on Facebook via her status message so that her friends would know where to find her. As the night wore on and the places she went became too loud for a telephone conversation, she would be able to stay in touch with people via e-mail, text or Facebook messaging, or the unified chat program on her Blackberry. Additionally, the global positioning capabilities of her phone would ensure she was able to find the places she wanted to go.

For early adopters of technology, as well as an increasing number of Internet users overall, these scenarios are not out of the ordinary. The Internet is no longer used only to find information and communicate with coworkers, it is now also a social, educational, and entertainment medium: a means to socialize with friends (some likely "known" only online), research issues, watch television programming and movies, play games, and maybe even get a little work done on off hours that would not have been done before.

THE EVOLUTION OF THE CONSUMER-ORIENTED INTERNET

When early-adopting consumers downloaded Mosaic, the first widely-available web browser in December 1993,[25] it gave them access to "the Internet:" the World Wide Web (Web). Previously, the Internet had been the domain of a small cadre of defense and university researchers. The Web, however, provided a graphical, "what-you-see-is-what-you-get" (WYSIWYG) experience to information previously available only as plain text, as well as the ability to refer a reader to another page via a "hyperlink" that the reader would "click" on with her computer mouse. Browsers also provided a more user-friendly interface to File Transfer Protocol ("FTP"), Gopher, and similar information sources. These early-adopters could not have foreseen how much the Web would transform the way they worked, shopped, searched for information on current events, and conducted other day-to-day activities – in short, how the Web would change their lives.

In the early days of the Web, most Web and e-mail use took place at work, where Internet connections were provided via dedicated high-speed circuits. Most people did not have any Internet access at home; those who did had dial-in access to a corporate network, usually to access e-mail. Accessing the Web was excruciatingly slow via dial-up and there wasn't that much content of interest on the Web. Further, there was not much information on the Web that consumers would be interested in accessing at home: the earliest information available on the Web was not aimed to appeal to a broad audience and tended towards national, local, and technology news. In fact, the Web address "http://www.news.com" is not for a general news website, but for CNET News, a still-popular technology news site.

During its early years, the Web was without exception a "one-way street" as far as information was concerned: A user would access the Web and download the information he was seeking. Later, many Web sites began requiring users to set up accounts to identify themselves and log in when using the site, but no one would classify such a feature as "interactivity" as we now know it.

In 2002, the FCC removed a regulatory barrier that had limited modem speeds to 53.3 kilobits per second, which prompted equipment manufacturers and Internet service providers to begin offering higher speed dial-up access. Soon after, equipment manufacturers began selling to the growing number of home users of the Internet and Web. Also during this time, telephone and cable companies were increasing the number of subscribers to their digital subscriber line (DSL) and cable modem services, respectively.

The Web slowly grew more interactive as more consumers gained access to the Internet at work as well as at home:

- personal Web sites grew in popularity;
- news sources added interactive components to their sites, such as chats with news makers and journalists;
- online shopping options increased;
- individuals began writing and keeping their own Web logs – or "blogs"; and
- photo-sharing sites increased their subscribers.

One driver of home adoption of high-speed Internet service was that consumers became accustomed to faster speeds while at work and were less willing to tolerate slower dial-up speeds at home; the number of home broadband subscribers surpassed the number of dial-up subscribers in early 2005.

At the same time that home-based Internet access was becoming faster and more a part of consumers' daily lives, wireless service providers were also beginning to deploy increased speeds for Internet access, which also increased consumer demand for Internet-ready wireless devices. Between December 2004 and June 2008, for example, use of "smartphones" grew from 2.9 million users to 20.3 million users.[27] Also, many non-smartphones also offer some degree of access to the Internet, whether it be e-mail or limited Web access. These devices allow users to access much of the same information they access on their desktop or laptop computers. This cross-platform accessibility has led to growing expectations by consumers that they should be able to seamlessly access information through multiple devices.

THE INTERNET EVERYWHERE: WHAT, WHY, AND WHERE[26]

What and Why: **News** sites to stay abreast of current news while on the go.
Where: All major news organizations.
What and Why: **Social networking** to connect with friends, family, and colleagues.
Where: MySpace, Facebook, Bebo, LinkedIn.
What and Why: **Chat** applications and online services to communicate with others in real time.

Where: AOL's AIM, MSN's WindowsLive, IMO, Google Chat, and Yahoo's Y!.

What and Why: **Music** services to download music and, in some cases, share it with others.

Where: iTunes, Rhapsody, Pandora.

What and Why: **Video** websites to share videos and/or legally access copyrighted programming.

Where: YouTube, Hulu.

What and Why: **Photo sharing** sites to share photos and order prints.

Where: Kodakgallery (previously Ofoto), Snapfish, Photobucket.

What and Why: **Blog** sites to share one's thoughts on matters big and small.

Where: Blogger (now owned by Google), Blogspot.

What and Why: **Personal websites** to share personal, professional and educational information with the public.

Where: Geocities, Tripod.

What and Why: **Shopping** sites to compare prices and make purchases without having to visit a "brick and mortar" store.

Where: eBay, Amazon, most major retailers

What and Why: **Cloud computing**, which allows users to store, and in some cases edit, documents and other files regardless of location (e.g., save a document online at work, then open it up and edit it at home).

Where: Google Apps, Drop.io, Symantec's goEverywhere, Amazon's Simple Storage Service and CloudFront.

THE CONSUMER-ORIENTED INTERNET: THE PAST AS PRELUDE

Between 2005 and 2007, the synergies between advanced services and technologies drove increased demand for and availability of advanced services and technologies. By 2008, consumers had begun integrating communications technologies into their lives at unprecedented rates. Several trends gained momentum nearly simultaneously during 2008:

- increased use of smartphones by consumers (the domain previously used primarily by business users);
- increased membership on social networking sites such as Facebook and MySpace;
- increased expectations of cross-platform accessibility;[28] and
- development of "cloud computing" applications, in which computing and file storage functions are moved off the user's computer and instead provided over the Web as a service.

Separately, each of these trends may have had a significant impact on consumer behavior, but taken together they created a previously unseen expectation for real-time access to information and an ability to share that information from virtually anywhere.

In the past, Web use was primarily one way, with users limited to accessing information (i.e., "downstream" to the user). Today, the consumer has become, in many cases, a producer as well as a consumer of content and can operate as his or her own information hub. This emerging "state" of the Internet is often referred to as "Web 2.0."

The Internet is on the cusp of a new stage, much like it was in 1993 when the Mosaic browser became popular. Now, however, there are many more decisions to be made by industry and policymakers about technology and, to some extent, service development. Choices made today and in the near future regarding user authentication, privacy, digital rights management, filtering of unwanted information, wireless Internet standards, instant messaging, the deployment of IPv6 ("Internet Protocol version 6"), and how to link the telephone network to the Internet will all have a significant impact on the "future Internet" we see in coming years.[29]

FAST FACTS

Every day people post more than 65,000 videos on YouTube.

In 2006, MySpace surpassed 100 million profiles.

Since 1999, the number of blogs has grown from 50 to 50 million.

More than 50 percent of blogs are authored by children younger than 19.

Source: Daniel J. Solove, "The End of Privacy?," *Scientific American*, September 2008, pp. 101-106.

To expand on just one of those elements, for example, once it is fully implemented, IPv6 will allow a virtually unlimited number of devices to be connected to the Internet, each with a unique and permanent IP address.[30] Currently, many devices share what amounts to a pool of addresses, making true end-to-end connectivity impossible.[31] Today, consumers can remotely do such tasks as control their TiVo through their Blackberry or their computer at any time; with IPv6, larger concepts of the "smart home" can be realized. For example, consumers will be able to better manage their energy consumption by having remote access to their heating and air conditioning systems, their light fixtures, and other electric appliances.

As with technological leaps that have come before, the evolution into the next stage of connectedness may necessitate that policymakers assess new directions for regulation that will encourage innovation while still protecting consumers. An appropriate regulatory environment will provide the crucial third element in the "deployment-applications-regulation" triad that will ensure American consumers will have access to the technologies and applications they desire.

NATIONAL BROADBAND POLICY AND THE EVOLVING TELECOMMUNICATIONS INFRASTRUCTURE

As broadband technologies and applications evolve, the policy debate in Congress is likely to hinge on how a national broadband policy may be characterized and structured. Particularly, Congress is likely to address the problems of how access to fast and affordable broadband across all sectors of society may be encouraged, how new telecommunications infrastructures should (or shouldn't) be regulated, and how certain societal impacts of new applications that are enabled and increasingly pervasive may have to be regulated or managed.

What is a "National Broadband Policy"?

A variety of stakeholders have called for a "national broadband policy" to help provide ubiquitous broadband coverage throughout the United States. Many argue that a national broadband policy is necessary to ensure the future prosperity of the United States, and in particular, that economically disadvantaged areas of the United States can maintain or recapture economic viability. Although most agree that a national broadband policy or strategy may be necessary and that a goal of "universal broadband" is worthy, stakeholders diverge when the debate focuses on specific policies and measures the federal government should take to reach those goals. Because broadband in the United States is largely deployed by the private sector,[32] any discussion of a governmental broadband policy will by definition lead to issues of how government intervention in the marketplace may affect that private sector deployment. Support for a "national broadband policy" depends largely on how the phrase is defined and characterized, and the public policies that are adopted to shape and support a national policy.

A working definition of "national broadband policy" or "national broadband strategy" is inherently imprecise.[33] "Broadband" can encompass a wide variety of industries, technologies, applications, and individual telecommunications policy debates. **Table 3** shows various broadband technologies (both deployed and potential) and general types of applications, as well as the many specific and discrete policy issues that could arguably be categorized under a "national broadband policy." Specific broadband technologies and applications (which in turn can be tied to specific industries and interest groups) can lead to specific policy issues, sometimes unique to that particular technology or application. For example, deployment of wireless broadband technologies can be dependent on how spectrum policies and issues are resolved. Deployment of fiber (for example Verizon's FIOS and AT&T's U-verse) are affected by regulatory issues such as cable franchising and unbundling. Entertainment applications can be affected by how intellectual property issues are managed. E-commerce applications can thrive or be impeded depending on how privacy and security concerns are addressed.

On the other hand, there are issues and policies that are more "technology neutral" and can affect all broadband technologies, industries, and applications. These could include financial assistance policies (such as tax incentives, loans and grants, expansion of universal service), data mapping (determining where broadband is deployed at a granular level), and community broadband – all of which are intended to enhance broadband deployment generally.

Essentially, one can frame a national broadband policy in response to two separate public policy challenges. First, what are the policies necessary to ensure that broadband is deployed to all Americans in a reasonable and timely fashion, as is called for in section 706 of the Telecommunications Act of 1996? A second – and much broader question – is how the future of broadband will transform the economy and society, and whether and to what extent those transformations should be managed by policymakers.

The Policy Debate in Context

As discussed above, the way a "national broadband policy" is defined, and the particular elements that might constitute that policy, determine how and whether various stakeholders might support or oppose a national broadband initiative. However, in the ongoing broadband debate, there are general areas of agreement that are usually cited, as well as areas of controversy where policymakers and stakeholders diverge.

Areas of General Agreement

There are three basic areas on which most observers seem to agree in any discussion of a national broadband policy. First, broadband is generally viewed as vital public infrastructure, increasingly significant to the nation's (as well as regional, state, and local) economic growth and vitality. The most recent FCC "706 report" acknowledges the link between broadband and economic development:

> local communities report that a key to their future is broadband. In order to attract business and residents, they must be able to provide the necessities, and this increasingly includes broadband. The future of a community's economy, employment opportunities, telecommuting, and opportunities for individuals with disabilities are related directly to the future of broadband in that community.[34]

With broadband initially being deployed in the United States about ten years ago, quantitative data on its impact has just recently begun to be collected and evaluated. A February 2006 study by the Massachusetts Institute of Technology for the Department of Commerce's Economic Development Administration marked the first attempt to measure the impact of broadband on economic growth. The study found that "between 1998 and 2002, communities in which mass-market broadband was available by December 1999 experienced more rapid growth in employment, total number of businesses, and businesses in IT-intensive sectors, relative to comparable communities without broadband at that time."[35] Subsequently, a June 2007 report from the Brookings Institution found that for every one percentage point increase in broadband penetration in a state, employment is projected to increase by 0.2 to 0.3% per year. For the entire U.S. private non-farm economy, the study projected an increase of about 300,000 jobs, assuming the economy is not already at full employment.[36]

A second area of agreement is that there exist some areas and populations of the U.S. which are unserved or markedly underserved by broadband providers. A particularly pronounced disparity in broadband service persists between urban/suburban areas and rural areas. Although there are many examples of rural communities with state of the art telecommunications facilities,[37] recent surveys and studies have indicated that, in general, rural areas tend to lag behind urban and suburban areas in broadband deployment.[38] The comparatively lower population density of rural areas is likely the major reason why broadband is less deployed than in more highly populated suburban and urban areas. Particularly for wireline broadband technologies—such as cable modem and DSL—the greater the geographical distances among customers, the larger the cost to serve those customers. Thus, there is often less incentive for companies to invest in broadband in rural areas than, for example, in an urban area where there is more demand and less cost to wire the market area.[39]

Table 3. Selected Ingredients of a National Broadband Policy

Technologies	
• cable modem	• fixed wireless
• next gen cable (DOCSIS 3.0)	• mobile wireless
• DSL (copper telephone line)	• wifi
• fiber to the home	• satellite
• fiber to the curb	• broadband over powerline (BPL)
Applications	
• voice over the Internet protocol (voIP)	• e-commerce
• telehealth	• social networking
• distance learning	• teleconferencing
• e-government	• telework
• entertainment (video, music, gaming)	• surveillance
• smart electric grids	• public safety communications
Policies/Issues	
deregulation; cable franchising; federal broadband coordination/broadband "czar"; spectrum and wireless policy; Universal Service Fund reform; financial assistance (grants, loans, tax incentives); data mapping and collection; rights of way;	community broadband; intercarrier compensation reform; net neutrality/network management; content issues (privacy, copyright, decency); demandside issues (training and education, computers for low income families); R&D; Internet2; others

Source: Compiled by CRS.

Access to affordable broadband service is viewed as particularly important for the economic development of rural areas because it enables individuals and businesses to participate fully in the economy regardless of geographical location. For example, aside from enabling existing businesses to remain in their rural locations, broadband access could attract and grow new businesses drawn by lower costs and what some may consider a more desirable lifestyle. Essentially, broadband potentially allows businesses and individuals in rural America to live locally while competing globally.

Finally, there is agreement that data regarding broadband deployment in the United States is inadequate, and that policymakers have an incomplete picture of where broadband service is available (and at what speeds and prices). States have begun to address this with a number of mapping and data collection efforts. On the federal level, the FCC, in March 2008, adopted a significantly more detailed data collection protocol.[40] Similarly, the 110th Congress enacted S. 1492, the Broadband Data Improvement Act (P.L. 110-385), which requires the FCC to collect demographic information on unserved areas, data comparing broadband service with 75 communities in at least 25 nations abroad, and data on consumer use of broadband. The act also directs the Census Bureau to collect broadband data, the Government Accountability Office to study broadband data metrics and standards, and the Department of Commerce to provide grants supporting state broadband initiatives. Looking forward, as broadband data improves, it is hoped that a more detailed and granular picture will emerge of where and to what extent broadband deployment shortfalls exist and how they might be addressed.

Areas of Controversy

Although most agree that some form of government intervention may be necessary to help provide broadband in chronically unserved areas, stakeholders disagree over the appropriate level and nature of government intervention in the broadband marketplace. The overarching issue is how to strike a balance between providing federal assistance for unserved and underserved areas where the private sector may not be providing acceptable levels of broadband service, while at the same time minimizing any deleterious effects that government intervention in the marketplace may have on competition and private sector investment. Those who favor increased government intervention argue that measures such as setting a formal national goal (with respect to penetration or speed, for example), expanding universal service, or mandating an "open" Internet (net neutrality) are necessary to ensure a competitive broadband economy. Those who favor a more limited government role argue that markets and the private sector can best deploy broadband with a minimum of government intervention, that deregulatory policies will unleash private sector investment in the broadband infrastructure, and that excessive or inappropriate government intervention in the marketplace is likely to be inconsequential if not deleterious.[41]

The question of current versus next generation broadband also raises important issues for policymakers when formulating broadband policies.[42] For example, as broadband technologies develop, and as speeds increase and applications become more sophisticated, what exactly constitutes "underserved" when assessing areas of the nation that might need some type of government intervention? While most agree that any broadband policy should be "technology neutral" (i.e., not favoring any particular technology or industry), should minimum speed thresholds (and/or upgradeability) be encouraged, and how far-reaching should those thresholds be – to meet the needs of consumers today, or what may be anticipated for the future? As always, the countervailing question is: to the degree that government policies encourage or prescribe specific broadband capacities, to what extent does this disrupt the marketplace and impede private sector broadband deployment efforts?

Meanwhile, the debate over government intervention in broadband markets is accompanied by disagreement over how broadband deployment in the United States compares with broadband deployment in other nations. Many supporters of a national broadband policy featuring an increased level of government intervention argue that statistics ("broadband rankings") compiled by the Organisation for Economic Co-operation and Development (OECD)[43] and the International Telecommunications Union (ITU)[44] show that the United States is progressively falling behind other nations in broadband penetration, speeds, and pricing, and that this comparatively low ranking has ominous implications for U.S. economic competitiveness.[45] Those supporting less government intervention assert that the OECD and ITU data is flawed and undercounts U.S. broadband deployment,[46] and that cross-country broadband deployment comparisons involving penetration, speeds, and prices are not necessarily meaningful and inherently problematic.[47]

Looking Ahead: The Future of Broadband

Much of the discussion above concerns a national broadband policy in response to the challenge of providing broadband to unserved and underserved regions and populations of the

United States. A national broadband policy can also be viewed from a broader perspective by considering how the future of broadband, accompanied by increasingly sophisticated and pervasive applications, might transform the economy and society and how those transformations might be managed by policymakers. In other words, as broadband technologies, speeds, and applications advance, what are the regulatory issues that may confront policymakers not only with respect to deployment, but also with respect to consumer applications and societal impacts?

REGULATORY FRAMEWORK

As our telecommunications environment continues to evolve, regulators are forced to accommodate the realities of a changing infrastructure as well as the changing expectations of both suppliers and consumers. One of the challenges facing this transition is how to establish a regulatory framework to address this increasingly interrelated and complex environment. Over the past few decades, laws and regulations have been formulated in an attempt to address the growth of competition in what were previously considered to be monopolistic markets. Much attention has been given to attempts to formulate a regulatory environment to incorporate and encourage competition based on the implementation of the 1996 Telecommunications Act (P.L. 104-104).[48] This act and its subsequent implementation has largely focused on the development of a regulatory structure to accommodate the growth of intramodal and intermodal competition in the provision of services.[49]

However, the telecommunications sector is dynamic and technological changes such as the advancement of Internet technology and the melding of data, voice, and video have resulted in additional trends which must be addressed. These trends include:

- the transition from a circuit switched to a packet switched network, thereby enabling the integration of voice, video, and data;
- the transition from fixed to mobile service;
- the transition from narrowband to broadband, thereby enabling greater interactivity.

The challenge facing today's policymakers is to develop a regulatory environment that not only addresses current trends, but contains the flexibility to accommodate future and as yet unanticipated changes in technology, applications, consumer expectations, and policy objectives. The growth of broadband networks and the proliferation of applications and devices has placed increasing pressure on policymakers to formulate a framework to address a broadband-based world. Many of these developments were not anticipated when the 1996 Telecommunications Act was passed and have led to the need to update the regulatory assumptions and subsequent regulatory framework the act was based on. A further challenge results from the Internet's lack of national boundaries. Regulations established in one country may be circumvented, since the World Wide Web is global. Activities that may be declared illegal in one country may be undertaken with relative ease by accessing foreign web sites.[50] However, as broadband access continues to become more vital to both the economic and social well-being of the nation, increased attention will be placed on the degree to which regulators should help to shape this constantly evolving environment.

From Monopoly to Competition

As the sector continues its transition from monopoly to competition, regulatory bodies are confronted with the task of establishing a regulatory environment that does not favor one player over another, nor establish regulatory obstacles to deployment and access. The regulatory treatment of broadband technologies, whether offered by traditional or emerging providers, or incumbents or new entrants, has become a major focal point in this transition. Whether present laws and regulatory policies are necessary to ensure the development of competition and its subsequent consumer benefits, or are overly burdensome and only discourage needed investment in and deployment of broadband services, continues to be an issue. The policy debate focuses on issues such as the extent to which legacy regulations should be applied to traditional providers as they enter new markets; the extent to which legacy regulations should be imposed on new entrants as they compete with traditional providers in their markets; and the appropriate treatment of new and converging technologies. Additional concerns over how the role of local, state, and federal regulators should be determined, and under what circumstances federal preemption may be evoked, also arise.

Barriers to Competition

In an attempt to level the playing field and encourage the benefits of marketplace competition, regulators are called upon by policy makers and stakeholders to develop a range of policies that promote competition. Such regulations can take many forms, including subjecting providers of like or competing services to similar regulations; establishing new regulations to protect or nurture new competitors; developing new regulations, if deemed necessary, to address the entrance of new services; or when certain market conditions are met, removing legacy regulations from incumbents.

Technological advances and the growth of competition have had a profound impact on market structure and subsequently their established regulatory framework, requiring regulators to address a wide range of issues. Some of the issues that regulators are grappling with include modifying universal service goals and obligations to address the growth of new and/or competing services; ensuring access to existing infrastructure such as ducts, poles, and rights-of-way; developing portability requirements for subscriber numbers to ease the ability to switch among competitors; implementing technology-neutral regulations; and establishing guidelines to remove, or forbear, regulations in competitive markets.

Integration and Interactivity

Legacy policies and regulatory frameworks have come under increasing strain as technological advances have led to the ability to provide new and integrated services. Historically, a provider was identified by the service it offered and its regulatory destiny was determined by its service classification. For example, a provider of voice telephone service classified as a telecommunications common carrier is regulated under Title II (Common Carriers) of the Communications Act of 1934.[51] Similarly, a provider of video service classified as a cable television system operator is regulated under Title VI (Cable Communications) of the 1934 Act. However, the world of distinct services and applications is disappearing as networks transition from a circuit switched to a packet switched network, enabling the integration of voice, video, and data. As providers move to Internet protocols and advanced broadband options continue to grow, the lines among distinct industry sectors continue to blur. Providers seek to integrate or bundle their services into triple play and in some cases quadruple play offerings.[52] New, previously undefined services such as Voice over Internet Protocol (VoIP) further strain traditional regulatory perimeters. The challenge facing regulators is to adapt the regulatory framework to this new environment to ensure that legacy regulations do not inhibit the development of and subsequent benefits derived from next-generation broadband networks but still balance economic and social policy objectives.

The growth in interactive, or two-way applications, has expanded significantly, placing increased demands on the broadband infrastructure. Services and applications are moving from static one- way uses, such as e-mail or web surfing, to interactive two-way applications such as video and voice services that are more dependent on uninterrupted streams of data. Additionally the growth in peer-to-peer activity has placed increasing demands on the existing broadband infrastructure as upload speeds now need to match download speeds and peak usage may cause congestion. As the popularity of such services expand, leading to an increase in the demand for bandwidth, the need to address issues relating to capacity and the subsequent need to manage network traffic have come to the attention of regulators. Policies to balance the needs of subscribers and suppliers, as well as network operators, are among the issues being debated. How to establish a policy framework that ensures effective management of networks facing capacity shortages, protects users of the network, and encourages both innovation and future investment for expansion are among the issues under consideration.

Mobility

Wireless broadband networks offer the ability to access broadband anytime and anywhere and may also offer a solution to providing broadband access to underserved and unserved areas. As access to broadband networks becomes increasingly mobile, regulators may be called upon to address policies to facilitate this connectivity. Wireless providers have expanded and continue to expand their service offerings beyond the traditional voice and ringtones to include text messaging, mobile search, e-mail, games, photo messaging, music, and video. Users, whether they be individual consumers, businesses, or government, expect to receive reliable high-quality service to meet these growing mobile expectations. As wireless broadband applications and expectations increase, the need to develop policies to ensure both sufficient radio frequency spectrum capacity and the ability to offer a seamless mobile experience become paramount. Key to these objectives is the development of policies that ensure effective spectrum management and connectivity among networks.

Spectrum Management

The demand for spectrum is intense and the FCC has responsibility for allocating spectrum among the various users. As wireless broadband networks continue to shift from traditional voice networks to those that incorporate a wide range of advanced broadband services, they will need more spectrum in wider contiguous bandwidths. How this spectrum should be allocated among the myriad users, the size of the blocks allocated, and whether any special considerations should be given to specific entities (e.g., rural or small providers) that bid in spectrum auctions are among the issues confronting policymakers. The need to harmonize spectrum allocations worldwide is also a key policy issue. Harmonization enables network operators and equipment manufacturers to realize significant economies of scope and scale and facilitates global interoperability for consumers.[53]

Roaming

Directly related to the issue of spectrum management is the need to ensure that users have the ability to connect seamlessly among providers. Mobility depends on the ability of users to roam, or move from location to location outside the provider's service area, without their signal dropping or degrading. In the mobile world, there is a greater tendency to need to share networks, making the ability to interconnect a vital component of the mobile experience. The capability to provide nationwide coverage, absent owning such a network, is dependent on the provider's ability to negotiate roaming commitments with other carriers. Subscribers served by these carriers need access to other networks for voice, data, and broadband traffic when roaming outside of their carrier's home market. The absence of, or potentially exorbitant costs associated with, such commitments can place a carrier in a negative competitive position. Furthermore, subscribers in rural areas, who are often served by small and regional carriers, can be particularly vulnerable in the absence of favorable roaming agreements, subjecting them to more limited and/or more costly service. Although these commitments are largely negotiated without regulatory intervention, some concern has been expressed, particularly on the part of rural, small, and regional carriers, that regulators should intervene to protect their ability to negotiate "reasonable" roaming commitments. Regulators have been called upon to address roaming issues, particularly in the context of the recent trend towards industry consolidations. Concerns that existing roaming commitments may be terminated or degraded during a change in ownership, the lack of roaming obligations among in-market carriers, and the application of roaming commitments to data are among the issues that are under examination.

Open Access

As the number of new providers, products, and services proliferate, the ability to gain access to the marketplace becomes paramount. Considerable debate has focused on what has been termed "open access," a term generally defined to mean the ability of suppliers and users to gain unfettered access to networks, content, applications, devices, and ultimately consumers.

Networks

Much of the recent debate over open access has focused on the ability to gain access to content, applications, and services on the Internet. The ability of subscribers to gain access to and use the Internet, in any legal manner, and the ability of applications and service providers to gain access to those consumers has become a focal point for the open access debate. Today's residential market for broadband delivery is dominated by two platform providers: cable television companies that provide cable modem service and landline telephone companies that provide Internet access service (i.e., wireline broadband Internet access, or Digital Subscriber Line Service [DSL]).[54]

The movement to place restrictions on the owners of the networks that comprise and provide access to the Internet, to ensure equal access and non-discriminatory treatment is referred to as "net neutrality." There is no single accepted definition of net neutrality. However, most agree that any such definition should include the general principles that owners of the networks that comprise and provide access to the Internet should not control how consumers lawfully use that network; and should not be able to discriminate against content provider access to that network. Most people acknowledge that networks have always been managed and that a certain degree of management is necessary and may even be desirable. The challenge, however, is to distinguish between what is needed or appropriate management versus discrimination. A balance must be struck between the ability of network operators to manage and maintain their infrastructure responsibly and the ability of suppliers and users to access the network in a nondiscriminatory manner.

In an attempt to strike such a balance the FCC adopted a policy statement outlining four principles to "encourage broadband deployment and preserve and promote the open and interconnected nature of [the] public Internet": (1) consumers are entitled to access the lawful Internet content of their choice; (2) consumers are entitled to run applications and services of their choice (subject to the needs of law enforcement); (3) consumers are entitled to connect their choice of legal devices that do not harm the network; and (4) consumers are entitled to competition among network providers, application and service providers, and content providers. Former FCC Chairman Martin did not called for their codification. However, they have been incorporated into the policymaking and oversight activities of the Commission.[55]

The question of what, if any, action should be taken to ensure "net neutrality" has become a major focal point in the debate over broadband regulation. As the marketplace for broadband continues to evolve, some contend that no new regulations are needed and, if enacted, will slow deployment of and access to the Internet, as well as limit innovation. Others, however, contend that the consolidation and diversification of broadband providers into content providers has the potential to lead to discriminatory behaviors that conflict with net neutrality principles.[56]

Historically, however, regulatory policies regarding access to broadband service have focused on wired networks. However, with the onset and growth of wireless broadband capabilities as a third broadband network option,[57] and the potential of wireless networks to provide broadband access to unserved and underserved areas, pressure to apply some type of open access principles to the wireless network has increased. It appears that under the present regulatory environment wireless carriers, when providing broadband access, are not subject to broadband access policies. Whether the move to an open wireless network will become widespread through voluntary industry efforts, due to the development and adoption of open architecture technologies, or perhaps due to regulatory pressure or mandate, is yet to be determined. As broadband access continues to migrate to the wireless world, increased attention will be focused on what role regulators may have in helping to ensure that wireless networks, like their wired counterparts, are adequately open.

Devices and Applications

In addition to concerns over access to networks, open access principles have also been applied to devices. In the public switched wireline world, the debate over opening the network to devices, or terminal equipment, has a long and complex history. The long-debated issue of whether or not consumers would be permitted to attach their own equipment to the telephone network was largely resolved by the 1968 Carterfone Decision.[58] This decision was issued by the FCC in response to a petition filed by Thomas Carter, an inventor, who developed a device, known as the Carterfone. The Carterfone enabled consumers to connect a two-way mobile radio system to the telephone network. In accordance with its long held policy, AT&T, the parent company of the Bell System, denied the attachment of the Carterfone to its telecommunications network citing concerns about the potential of foreign devices, i.e., non-Bell System devices, to harm the network. Given the Bell System's status as the monopoly provider of telecommunications, the refusal to allow non-Bell System equipment to attach to the network, in effect, resulted in a de facto monopoly over devices as well as the transmission platform. The FCC, however, determined that the Carterfone device and other customer-supplied equipment could be attached to the public telephone network as long as the devices were "privately beneficial, but not publically harmful." These rules on connection were later codified as Part 68 of the FCC's rules.[59] As a result, consumers are free to attach any equipment they desire to the public switched network, as long as it meets Part 68 standards.

Some are pressing the FCC to apply Carterfone-type rules to the wireless network as well. However, the application of such rules may prove to be more difficult. Unlike the public switched telephone network, which is basically supported by a common technology, wireless networks are supported by a variety of different technologies. As the technologies that support wireless networks converge, the application of a Carterfone-type solution may become more feasible.

The FCC has taken some steps to encourage the opening up of the wireless broadband network to devices and applications. When the FCC auctioned spectrum licenses in the 700 MHz band it required the winner of 22 MHz for advanced wireless services, known as the C Block,[60] to adopt an open concept with respect to devices and applications.[61] Consumers are permitted to use the device of their choice, as long as it is not harmful to the network, as well as any legal software or applications on these networks. The FCC reaffirmed and clarified this two-pronged open access decision in an order adopted November 4, 2008.[62] The FCC stated that its open access requirements, which were codified in Title 47 Section 27.16 of the Code of Federal Regulations, apply to all auctioned licenses in the 700 MHz C Block. Section 27.16 states that a C Block licensee "shall not deny, limit, or restrict the ability of their customers to use the devices and applications of their choice on the licensee's C Block Network," unless reasonably necessary for network management or protection, or to comply with applicable law and also prohibits the above licensees from "disabl[ing] features on handsets" that they provide to their customers.[63]

There are some signs that industry players have begun to voluntarily embrace, to a limited degree, openness principles for wireless devices and applications. For instance, the new Clearwire Corporation,[64] a competitor in the broadband wireless market, has chosen to operate a nationwide network based on WiMax technology. WiMax is an open source technology in the sense that it is designed to permit both applications and devices of the consumer's choice.[65] At the time of filing for FCC approval, the petitioners (Sprint Nextel and Clearwire) proposed a number of voluntary commitments that the new entity, the new Clearwire, would adhere to that were open access based. The FCC, in a November 4, 2008 action, approved the petition, but did not require any open access requirements as a condition of that approval.[66] However, the new Clearwire by choosing to deploy WiMax is committed to pursuing a network which is open access for both devices and applications.

Despite this movement however, the wireless industry continues to be restrictive, to varying degrees, with regard to devices. For example, issues such as handset locking[67] and exclusivity arrangements[68] between commercial wireless carriers and handset manufacturers continue to remain contentious.

Social versus Economic Regulation

The regulatory issues discussed above are largely classified as those that deal with economic regulation; that is, regulations that address such issues as competition, innovation, and investment. As broadband becomes an integral component of society, however, regulators have been called upon to consider the application of social goals, that may or may not have been identified with traditional telephony, to the broadband world. Social objectives such as the advancement of universal service goals, timely and accurate emergency services, access for those with disabilities, and consumer protection that are part of traditional telephony regulatory policies, are migrating to the broadband policy environment.

Whether universal service objectives, which have been a basic tenet of wireline telephony, should include universal access to broadband service is one of the more significant social issues under debate. The concept of universal service, when applied to telecommunications, refers to the ability to make available a basket of telecommunications goods to the public, across the nation, at a reasonable price. As the importance and growing acceptance of broadband services permeates our lives, a consensus is forming that the definition of what should be included in the package of services should be expanded to include access to advanced (i.e., broadband) services. Others however, have expressed concern over the uncertainty and costs associated with mandating nationwide deployment of such services as a universal service policy goal.[69]

Assurance that 911 emergency access is of comparable quality whether a consumer is using a wireline, wireless, or broadband connection is also under scrutiny. Providing effective 911 service as we migrate from analog to digital technology has proved challenging. For example, considerable effort is being invested on the part of policymakers to ensure that effective enhanced 911 (E-911) capabilities are available to all users so that their location can be determined in an accurate and timely manner regardless of the technology used. There is a growing consensus that a modernized emergency system should incorporate IP networks and standards.[70]

One outcome of the net neutrality debate has been a focus on consumer protection. A growth in information transparency and assurance that consumers are aware of the impact that network management may have on their broadband usage, full disclosure of broadband speeds and usage caps, the ability of consumers to monitor and track their personal usage, as well as the protection of consumer usage information from tracking by other parties, have been significant outgrowths of the debate. Additional social goals being addressed by policymakers include the degree to which broadband devices and services should be fully accessible to those with disabilities (e.g., hearing, speech, or sight deficits); protection from identity theft; and the protection of minors from inappropriate material. Although there is considerable agreement that all of these social goals are worthwhile and must be addressed, the policies needed to achieve them may prove to be controversial and increasingly complex when adapted to a broadband environment.

End Notes

[1] Defined as a line providing a customer over 200 kbps in at least one direction.

[2] FCC, *High-Speed Services for Internet Access: Status as of December 31, 2007,* January 2009. Available at http://hraunfoss.fcc.gov/edocs_public/attachmatch/DOC-287962A1.pdf.

[3] Horrigan, John, Pew Internet & American Life Project, "Barriers to Broadband Adoption – The User Perspective," December 19, 2008, available at http://otrans.3cdn.net/fe2b6b302960dbe0d7_bqm6ib242.pdf.

[4] S. Derek Turner, Free Press, *Down Payment on Our Digital Future*, December 2008, p. 8.

[5] Federal Communications Commission, *Fourth Report to Congress*, "Availability of Advanced Telecommunications Capability in the United States," GN Docket No. 04-54, FCC 04-208, September 9, 2004, p. 38. Available at http://hraunfoss.fcc.gov/edocs_public/attachmatch/FCC-04-208A1.pdf.

[6] "Barriers to Broadband Adoption – The User Perspective," p. 1.

[7] Horrigan, John, Pew Internet & American Life Project, "Obama's Online Opportunities II: If You Build It Will They Log On?" January 21, 2009, available at
http://www.pewinternet.org/pdfs/PIP_Broadband%20Barriers.pdf.

[8] FCC, Fifth Report, p. 36.

[9] FCC, Fifth Report, p. 36.

[10] The Internet Society provides "A Brief History of the Internet," online at http://www.isoc.org/internet/history/brief. shtml.

[11] Examples are illustrative and are not intended as an endorsement of any particular technology, service, or device.

[12] A Blackberry or other "smartphone," such as an iPhone.

[13] Everett Rogers, "Innovativeness and Adopter Categories," in *Diffusion of Innovations*, 3rd ed. (New York: The Free Press, 1962), pp. 247-25 1.

[14] Text messages are technically called "Short Message Service," or SMS, messages.

[15] See http://imo.im.

[16] A "netbook" is a very small, light-weight, low-cost, energy-efficient laptop, primarily used for Internet-based services such as web browsing, e-mailing, and instant messaging.

[17] A survey by the Pew Internet and American Life Project found that 11% of online users have bought or sold stocks online, with that figure jumping to 21% who earn over $100,000 a year. Online Shopping, Pew Internet and American Life Project, February 2008, p. 2, online at http://www.pewinternet.org/pdfs/PIP_Online%20Shopping.pdf.

[18] A survey by the Pew Internet and American Life Project found that 53% of online users have done banking online, or 39% of all adult Americans. When the Pew Internet Project first asked about banking online in 2000, 18% of internet users (or 9% of all Americans) had at some point done banking online. By 2002, that number had risen to 30% of online users (or 18% of Americans) and when asked again in February 2005, 41% of Internet users had done some banking online (or 27% of all Americans). Online Shopping, Pew Internet and American Life Project, February 2008, p. 6, online at http://www.pewinternet.org/pdfs/PIP_Online%20Shopping.pdf.

[19] A survey by the Pew Internet and American Life Project reported that 66% of online users said they had bought something online. Online Shopping, Pew Internet and American Life Project, February 2008, p. 2, online at http://www.pewinternet.org/pdfs/PIP_Online%20Shopping.pdf.

[20] A survey by the Pew Internet and American Life Project found that between 75% and 80% of Internet users have looked online for health information. The Engaged E-patient Population, The Pew Internet and American Life Project, p. 4, online at http://www.pewinternet.org/pdfs/PIP_Health_Aug08.pdf.

[21] In Fall 2007, about half of all Americans had purchased airline or other travel tickets online. Online Shopping, Pew Internet and American Life Project, February 2008, p. 6, at http://www.pewinternet.org/pdfs/PIP_Online%20Shopping.pdf.

[22] In other words, a non-dial-up connection that is never turned off.

[23] The Engaged E-patient Population, The Pew Internet and American Life Project, p. 4, at http://www.pewinternet.org/ pdfs/PIP_Health_Aug08.pdf.

[24] The Engaged E-patient Population, The Pew Internet and American Life Project, p. 4, at http://www.pewinternet.org/ pdfs/PIP_Health_Aug08.pdf.

[25] Depending on the operating system (OS) used, earlier releases were compatible with OSs not commonly used by consumers.

[26] These examples are illustrative and not intended to be all encompassing.

[27] Proprietary data provided by CTIA – The Wireless Association.

[28] Meaning, consumers increasingly expect to be able to access their data from any device they use, whether it be a mobile device, their desktop home or work computer, or perhaps a laptop or netbook.

[29] Michael R. Nelson, Ph.D., "The Grid, the Cloud, and the Next Phase of the Internet," Presentation at Google Cloud Computing Seminar, Washington, DC, September 12, 2008.

[30] IPv6 supports 2^{128} (about 3.4×10^{38}) addresses. Additional information about IPv6 is available online at http://www.isoc.org/briefings/001.

[31] For a more technical explanation of this process, see "How Stuff Works: Network Address Translation," online at http://computer.howstuffworks.com/nat.htm.

[32] According to the Federal Communications Commission (FCC), telecommunications companies expect to make $50 billion in capital expenditures in 2008 and 2009. See Federal Communications Commission, Fifth Report, "In the Matter of Inquiry Concerning the Deployment of Advanced Telecommunications Capability to All Americans in a Reasonable and Timely Fashion, and Possible Steps to Accelerate Such Deployment Pursuant to Section 706 of the Telecommunications Act of 1996," GN Docket No. 07-45, FCC 08-88, adopted March 19, 2008, released June 12, 2008, p. 37. Available at http://hraunfoss.fcc.gov/edocs_public/attachmatch/FCC-08-88A1.pdf.

[33] In fact, "broadband" itself has a definition that has evolved as technologies and applications evolved. For example, in earlier days of broadband deployment, "broadband" was seen as synonymous with "high-speed Internet access," which implied access by a computer via a web browser. However today, broadband also includes voice and video directly delivered to telephones and televisions by providers through Internet protocol.

[34] FCC, Fifth Report, p. 74.

[35] Gillett, Sharon E., Massachusetts Institute of Technology, Measuring Broadband's Economic Impact, report prepared for the Economic Development Administration, U.S. Department of Commerce, February 28, 2006, p. 4. Available at http://www.eda.gov/ImageCache/EDAPublic/documents/pdfdocs2006/mitcmubbimpactreport_2epdf/v1/mitc mubbimpa ctreport.pdf.

[36] Crandall, Robert, William Lehr, and Robert Litan, The Effects of Broadband Deployment on Output and Employment: A Cross-sectional Analysis of U.S. Data, June 2007, 20 pp. Available at http://www3.brookings.edu/views/papers/crandall/200706litan.pdf.

[37] See for example: National Exchange Carrier Association (NECA), Trends 2006: Making Progress With Broadband, 2006, 26 p. Available at http://www.neca.org/media/trends_brochure_website.pdf.

[38] For example, 2008 data from the Pew Internet & American Life Project indicate that while broadband adoption is growing in urban, suburban, and rural areas, broadband users make up larger percentages of urban and suburban users than rural users. Pew found that the percentage of all U.S. adults with broadband at home is 60% for suburban areas, 57% for urban areas, and 38% for rural areas. See Horrigan, John B., Pew Internet & American Life Project, Home Broadband Adoption 2008, July 2008, p. 3, available at http://www.pewinternet.org/pdfs/PIP_Broadband_2008.pdf.

[39] The terrain of rural areas can also be a hindrance to broadband deployment because it is more expensive to deploy broadband technologies in a mountainous or heavily forested area. An additional added cost factor for remote areas can be the expense of "backhaul" (e.g., the "middle mile") which refers to the installation of a dedicated line which transmits a signal to and from an Internet backbone which is typically located in or near an urban area.

[40] FCC, News Release, "FCC Expands, Improves Broadband Data Collection," March 19, 2008. Available at http://hraunfoss.fcc.gov/edocs_public/attachmatch/DOC-280909A1.pdf.

[41] Some argue that government policies have demonstrated a minimal impact on broadband penetration rates and that variables such as household income, education, and general economic factors are much more determinative. See Ford, George, Phoenix Center, The Broadband Performance Index: What Really Drives Broadband Adoption Across the OECD?, Phoenix Center Policy Paper Number 33, May 2008, 27 pp; available at http://www.phoenixcenter.org/pcpp/PCPP33Final.pdf.

[42] To some extent, the federal government has already had an impact on which speeds are considered "broadband." Starting in 1999, and for many years following, the FCC defined broadband (or more specifically "high-speed lines") as over 200 kilobits per second (kbps) in at least one direction, which was roughly four times the speed of conventional dial-up Internet access. In recent years, the 200 kbps threshold was considered too low, and on March 19, 2008, the FCC adopted a report and order (FCC 08-89) establishing new categories of broadband speed tiers for data collection purposes. Specifically, 200 kbps to 768 kbps will be considered "first generation," 768 kbps to 1.5 Mbps as "basic broadband tier 1," and increasingly higher speed tiers as broadband tiers 2 through 7 (tier seven is greater than or equal to 100 Mbps in any one direction). Tiers can change as technology advances.

[43] Data from the OECD ranks the United States 15th among OECD nations in broadband access per 100 inhabitants as of June 2008. OECD, OECD Broadband Statistics, June 2008. Available at http://www.oecd.org /sti/ict/broadband.

[44] According to the ITU, the United States ranks 24th worldwide in broadband penetration (subscriptions per 100 inhabitants in 2007). International Telecommunications Union, Economies by Broadband Penetration, 2007. Available at http://www.itu.int/ITU-D/ict/statistics/at_glance/top20_broad_2007.html.

[45] See Benton Foundation, Using Technology and Innovation to Address Our Nation's Critical Challenges: A Report for the Next Administration, November 2008, pp. 5-6. Available at http://www.benton.org/ sites/benton.org/files/ Benton_Foundation_Action_Plan.pdf.

[46] National Telecommunications and Information Administration, Fact Sheet: United States Maintains Information and Communication Technology (ICT) Leadership and Economic Strength. Available at http://www.ntia.doc.gov/ntiahome/ press/2007/ICTleader_042407.html.

[47] See Wallsten, Scott, Progress and Freedom Foundation, Towards Effective U.S. Broadband Policies, May 2007, 19 pp. Available at http://www.pff.org/issues-pubs/pops/pop14.7usbroadbandpolicy.pdf.

[48] Provisions in the 1996 Telecommunications Act required the Federal Communications Commission (FCC) to initiate more than 80 rulemakings to address the changing telecommunications landscape.

[49] Intramodal competition refers to competition among identical technologies in the provision of the same service (e.g., a cable television company competing with another cable television company in the offering of video services) where intermodal competition refers to provision of the same service by different technologies (i.e., a cable television company competing with a telephone company in the provision of video services).

[50] For example, Internet gambling is generally prohibited in the United States, but such activities are easily accessed through offshore websites. For further information on unlawful Internet gambling see CRS Report RS22749, *Unlawful Internet Gambling Enforcement Act and Regulations Proposed for Its Implementation*, by Charles Doyle.

[51] Communications Act of 1934, as amended, 47 U.S.C. 151 et seq.

[52] A typical triple play offering would include video, data, and fixed voice services. A typical quadruple play offering would include video, data, and fixed and mobile voice services.

[53] For a discussion of spectrum policy and other issues relating to wireless broadband policies see A National Wireless Broadband Strategy, issued by the Wireless Communications Association, available at http://www.wcai.com/images/pdf/2008_wcai_wb_strategy.pdf.

[54] For FCC market share data for high-speed connections see High-Speed Services for Internet Access: Status as of June 30, 2007, Federal Communications Commission, released March 2008. View report at http://hraunfoss.fcc.gov/edocs_public/attachmatch/DOC-280906A1.pdf. For the most recent data see Local Telephone Competition and Broadband Deployment, High-Speed Services for Internet Access available at http://www.fcc.gov/wcb/iatd/comp.html.

[55] See http://www.fcc.gov/headlines2005.html. August 5, 2005. FCC Adopts Policy Statement on Broadband Internet Access.

[56] For a more detailed discussion of the net neutrality concept and issues see CRS Report RS22444, *Net Neutrality: Background and Issues*, by Angele A. Gilroy.

[57] According to FCC data, mobile wireless is the provider of almost 35 percent of high speed lines (over 200 kbps in at least one direction), as of June 30, 2007. See High-Speed Services for Internet Access: Status as of June 30, 2007, Federal Communications Commission, Table 6, released March 2008. View report at http://hraunfoss.fcc.gov/edocs_public/attachmatch/DOC-280906A1.pdf. For the most recent data, see Local Telephone Competition and Broadband Deployment, High-Speed Services for Internet Access, available at http://www.fcc.gov/wcb/iatd/comp.html.

[58] See In the Matter of Use of the Carterfone Device in Message Toll Service, 13 FCC 2d 420 (1968).

[59] See Part 68 of Title 47 of the Code of Federal Regulations. Additional information on Part 68 Regulations can be found at http://www.fcc.gov/wcb/iatd/part-68.html.

[60] The C block is a nationwide block of spectrum of which Verizon Wireless was the winner of the majority.

[61] For a copy of former Chairman Martin's March 18, 2008 statement on the 700 MHz auction and its open access provisions see http://hraunfoss.fcc.gov/edocs_public/attachmatch/DOC-280887A1.pdf.

[62] See, In the Matter of Union Telephone Company, Cellco Partnership d/b/a/ Verizon Wireless, Applications for 700 MHZ Band Licenses, Auction No. 73 available at http://hraunfoss.fcc.gov/edocs_public/attachmatch/FCC-08-257A1 .pdf.

[63] See paragraph 20 of In the Matter of Union Telephone Company, Cellco Partnership d/b/a/ Verizon Wireless, Applications for 700 MHZ Band Licenses, Auction No. 73. For a more detailed discussion of the 700MHz auction and the open access debate see CRS Report RS22218, *Spectrum Use and the Transition to Digital TV*, by Linda K. Moore.

[64] The new Clearwire is composed of merged assets from Sprint Nextel Corporation and Clearwire Corporation with the majority ownership (51 percent) resulting company owned by Sprint Nextel.

[65] WiMax (Worldwide Interoperability for Microwave Access) is an IP-based wireless technology based on an open non-proprietary standard for the delivery of non-line-of-sight wireless broadband services based upon the IEEE802.16 standard. For detailed information on WiMax see WiMax Questions & Answers available at http://www.wimaxforum.org/news/wimax_faq_10-2007.pdf.

[66] See, In the Matter of Sprint Nextel Corporation and Clearwire Corporation, Applications For Consent to Transfer Control of Licenses, Leases, and Authorizations (WT Docket No. 08-94.) Available at http://hraunfoss.fcc.gov/edocs_public/attachmatch/FCC-08-259A1.pdf.

[67] Handset locking refers to the practice of limiting the handset to the applications and features sold by that service provider thereby making them not readily portable to other carriers.

[68] For example, AT&T successfully negotiated a five-year exclusivity deal with Apple Inc. to market, in the United States, the popular iPhone.

[69] For a detailed analysis of the issues surrounding the expansion and reform of the Federal Universal Service Fund see CRS Report RL33979, *Universal Service Fund: Background and Options for Reform*, by Angele A. Gilroy.

[70] For a detailed analysis of the policy issues facing emergency communications see CRS Report RL34755, *Emergency Communications: The Future of 911*, by Linda K. Moore.

CHAPTER SOURCES

The following chapters have been previously published:

Chapter 1 – This is an edited, excerpted and augmented edition of a United States Congressional Research Service publication, Report Order Code R41324, dated July 9, 2010.

Chapter 2 – This is an edited, excerpted and augmented edition of a Federal Communications Commission publication.

Chapter 3 – This is an edited, excerpted and augmented edition of a United States Congressional Research Service publication, Report Order Code R40674, dated June 21, 2010.

Chapter 4 – This is an edited, excerpted and augmented edition of a United States Congressional Research Service publication, Report Order Code R40230, dated February 19, 2009.

INDEX

C

E

F

M

P

T

U

V

W

Y

Z